The Chemistry of the

p-Block Elements

Syntheses, Reactions and Applications

Anil J Elias

Professor, Department of Chemistry
Indian Institute of Technology Delhi

Universities Press

THE CHEMISTRY OF THE *p*-BLOCK ELEMENTS: SYNTHESES, REACTIONS AND APPLICATIONS

UNIVERSITIES PRESS (INDIA) PRIVATE LIMITED

Registered Office
3-6-747/1/A & 3-6-754/1. Himayatnagar, Hyderabad 500 029, Telangana, India
info@universitiespress.com; www.universitiespress.com

Distributed by
Orient Blackswan Private Limited

Registered Office
3-6-752, Himayatnagar, Hyderabad 500 029, Telangana, India

Other Offices
Bengaluru / Chennai / Guwahati / Hyderabad / Kolkata / Mumbai / New Delhi / Noida / Patna

First published 2019
Reprinted 2024, 2025, 2026

ISBN 978-93-86235-71-8

Cover and book design
© Universities Press (India) Private Limited 2019

Typeset in Minion Pro 10/12 by
MacroTex Solutions, Chennai 600 091

Printed at
Kensington Printing & Packaging, Noida 201 301

Published by
Universities Press (India) Private Limited
3-6-747/1/A & 3-6-754/1, Himayatnagar, Hyderabad 500 029, Telangana, India

"Yeah, we're always talking about following your passion, but we're all part of the flow of history … you've got to put something back into the flow of history that's going to help your community, help other people … so that 20, 30, 40 years from now … people will say, this person didn't just have a passion, he cared about making something that other people could benefit from." —Steve Jobs

Foreword

This book of Anil Elias covers the elements and their reactions from group 13 to the noble gases of group 18 with the title, 'The chemistry of the *p*-block elements'. The content of the book exhibits the largest part of the main group elements of the periodic system. Three of these elements oxygen (50%), silicon (27%), and aluminum (7%) belong to the most available elements in the earth's crust and are ubiquitous. For the sustainable development of chemistry they play a very important role.

The first chapter gives an overview of the main group elements used in a broad range of applications and their historic landmarks. The second chapter is a very well written demonstration of the concepts of structure and bonding of the *p*-block elements.

The chapters from the boron group to the noble gases cover the traditional chemistry of these elements. Furthermore, very recent new research results with unforeseen properties are included. This demonstrates the deep and competent knowledge of the author. Included are also historic stories, they give the reader an additional background of the development of a certain field. In summary, I recommend this book for the undergraduates as well as for graduate students. The text is well written and easily understandable. Moreover, at the end of each chapter are questions for self-study, which are very useful for the reader.

Finally, I wish that this book will have a broad resonance in the chemical community.

Prof. Dr Herbert W Roesky
Georg-August-Universität, Göttingen

Preface

p-Block elements, also known as main group elements belong to groups 13 to 18 of the periodic table. Chemistry of the *p*-block elements is an essential and integral part of the chemistry curriculum all over the world. The wide variety of proven applications of the compounds of these elements makes them standout from the rest of the elements. In addition, these elements show physical and chemical properties and reactivity which are significantly different from the rest of the elements of the periodic table. In this context, it is to be noted that most of the currently available textbooks that discuss *p*-block chemistry are large compilations of reactions with no specific focus and many of these are also outdated with their last edition belonging to the 1980–1995 period. The large volume of reactions presented in these books confuse the readers regarding their significance or lack thereof, especially students for whom learning all these reactions without a specific focus may seem an insurmountable task. Most of these books have also been written from the perspective of an inorganic chemist, thus giving little importance to the organic chemistry of the *p*-block elements, which has grown exponentially. In addition, the *d*-orbital participation in the bonding of *p*-block compounds has been proved theoretically to be non-existent or negligible. This is yet to enter common textbooks. Topics such as low-valent *p*-block compounds, stable radicals, frustrated Lewis acid–base pairs, catalysis involving nonmetals and chemistry of multiple-bonded main group compounds, have become highly sought after research areas in the last two decades, thus rekindling new research interest in *p*-block chemistry. New applications of *p*-block-based materials such as GaN light emitting diodes, GaAs solar cells, ITO transparent conducting oxides and silicone polymers have in the recent past entered our daily life in a big way. Based on these reasons, the author felt that there is an urgent requirement for an up-to-date textbook dealing with the chemistry of the *p*-block elements especially with a focus on synthesis and reactions, new developments and industrial applications

This book is the outcome of more than 30 years of learning, research and teaching of the chemistry of the *p*-block elements by the author, which began at IIT Madras with a doctoral thesis work on sulphur–nitrogen chemistry. Experience gathered as a post-doctoral researcher at IISc Bangalore, University of Göttingen and University of Idaho in the areas of silicon, phosphorus and fluorine chemistry and research initiated as a faculty at IIT Kanpur and IIT Delhi has enriched and helped immensely in the shaping of this book. The writing of this book began in 2013 with lecture notes being made for two courses which were taught by the author at IIT Delhi. Efforts have been made to lay emphasis on fundamentals in each chapter so that it is useful to a wider student community.

While learning *p*-block chemistry, a fresh student is inundated with a deluge of reactions, making her wonder as to what is important, why it is important and what to study. Therefore, for the reactions given in this book, a very careful selection has been made by the author keeping in mind proven applications, potential for future applications, unprecedented and outstanding chemistry, reactions representing fundamental concepts as well as path-breaking research. In addition, historically important work and how the chemistry developed from such work has been included. Chemistry which seemed important in the 70's and 80's but did not show much future promise or development has been kept to the bare minimum.

The author takes this opportunity to thank teachers, colleagues, present and former students who helped him make this book a reality. Foremost among them is Dr Sushabhan Sadhukan, faculty at IIT Indore who, while at Cornell University, helped in collecting and making available a good amount of chemical literature and information not accessible from India which was essential for formulating many chapters. Prof. Jeanne M Shreeve, University of Idaho, was kind enough to spare time to read through chapters and suggest modifications. Prof Herbert W Roesky, University of Göttingen, went through the final form of the book, made valuable suggestions for modifications and was gracious enough to write the foreword for this book. Thanks are due to my former and present students, Prof. N D Reddy, Prof. K Muralidharan, Dr Nem Singh, Dr Dheeraj Kumar, Dr Jatinder Singh, Mayukh Deb, Susanta Hazra and Pritam Duloi for helping with collecting literature and adding information to the chapters. Special thanks to Ms Priti Pilania, M.Sc. student of IIT Delhi, who read through the chapters and suggested modifications from a student's perspective. Thanks are also due to Prof. J D Singh, Prof. N D Kurur, Prof. A Ramanan and Prof. N G Ramesh of IIT Delhi, Dr P A Joy, NCL Pune, Prof. Sangit Kumar, IISER Bhopal and Prof. Sanjay Singh, IISER Mohali, for valuable suggestions and help in authenticating parts of the material given. Special thanks to Prof. S Nagendran of IIT Delhi and Dr Sakya Sen of NCL Pune who reviewed the book and provided valuable inputs for modification of the manuscript.

Finally a word of thanks to Mr Madhu Reddy, Dr Gita S Dattatri and her excellent editorial team at Universities Press who made this book a reality. Thanks are due to my family—Latha, Aleesha and Ankitha who always gave encouragement and strength to complete this assignment with patience and love.

Anil J Elias
IIT Delhi
2018

Contents

INTRODUCTION

CHAPTER 1

1.1 IMPORTANCE OF COMPOUNDS BASED ON *p*-BLOCK ELEMENTS

1.1.1 Industrial Importance as Bulk Chemicals

The elements of the *p*-block (groups 13 to 18 of the periodic table) and compounds based on them exhibit physical and chemical properties, reactivity and applications which are significantly different from the rest of the periodic table. Unlike *s*, *d* and *f* block elements, the *p*-block is constituted not only by metals but also by non-metals and metalloids, some of which are conductors, non-conductors and semiconductors. Most of the compounds based on the elements of this block possess empty or completely filled *d* and *f* orbitals. Thus, this results in such compounds exhibiting no magnetic properties or colour resulting from electronic transitions involving *d* and *f* orbitals. Unlike metals which are electropositive, *p*-block elements are relatively more electronegative and also have the elements with the highest electronegativities. The rare gas elements also show the highest first ionisation enthalpies among all elements of the periodic table.

The presence of both electron rich and electron deficient elements in this block leads to the most well known Lewis acids and Lewis bases. The chemistry of the first row of the *p*-block elements, shows unique differences from the elements present in the rest of the block primarily due to inaccessibility of *d* orbitals and hardness. An additional unique and significant attribute is the inert pair effect shown by the heavier *p*-block elements which reduces their dominant valence by two units compared to the lighter *p*-block elements.

What makes the *p*-block elements different from transition and inner transition elements is the wide range of applications of some of its elements and compounds as bulk chemicals of industrial importance. World chemicals turnover in 2010 was valued at 2353 billion Euros with China being the major producer of most of the bulk chemicals.[1] Among industrially important inorganic chemicals and chemical compositions, those based on the *p*-block elements constitute the major component (other than organic carbon-based compounds). Many of the *p*-block elements and compounds/compositions based on them are produced in large quantities and used in agriculture, industry and

Name **Symbol** **Elec. config.** **At. No. At. Wt.**	**The *p*-block elements** **(groups 13 to 18)** **Electronic configuration, atomic number and atomic weights**				Helium He $1s^2$ 2 4.00
Boron B [He]$2s^22p^1$ 5 10.81	Carbon C [He]$2s^22p^2$ 6 12.01	Nitrogen N [He]$2s^22p^3$ 7 14.01	Oxygen O [He]$2s^22p^4$ 8 16.00	Fluorine F [He]$2s^22p^5$ 9 19.00	Neon Ne [He]$2s^22p^6$ 10 20.18
Aluminium Al [Ne]$3s^23p^1$ 13 26.98	Silicon Si [Ne]$3s^23p^2$ 14 28.09	Phosphorus P [Ne]$3s^23p^3$ 15 30.97	Sulfur S [Ne]$3s^23p^4$ 16 32.06	Chlorine Cl [Ne]$3s^23p^5$ 17 35.45	Argon Ar [Ne]$3s^23p^6$ 18 39.95
Gallium Ga [Ar]$4s^24p^1$ 31 69.72	Germanium Ge [Ar]$4s^24p^2$ 32 72.64	Arsenic As [Ar]$4s^24p^3$ 33 74.92	Selenium Se [Ar]$4s^24p^4$ 34 78.96	Bromine Br [Ar]$4s^24p^5$ 35 79.90	Krypton Kr [Ar]$4s^24p^6$ 36 83.80
Indium In [Kr]$5s^25p^1$ 49 114.82	Tin Sn [Kr]$5s^25p^2$ 50 118.71	Antimony Sb [Kr]$5s^25p^3$ 51 121.76	Tellurium Te [Kr]$5s^25p^4$ 52 127.60	Iodine I [Kr]$5s^25p^5$ 53 126.90	Xenon Xe [Kr]$5s^25p^6$ 54 131.29
Thallium Tl [Xe]$6s^26p^1$ 81 204.38	Lead Pb [Xe]$6s^26p^2$ 82 207.20	Bismuth Bi [Xe]$6s^26p^3$ 83 208.98	Polonium *Po* [Xe]$6s^26p^4$ 84 209	Astatine *At* [Xe]$6s^26p^5$ 85 210	Radon *Rn* [Xe]$6s^26p^6$ 86 222

daily life. As Table 1.1 indicates, among the top fifteen bulk inorganic chemicals/elements produced by chemical industries across the world, nine of them belong to the *p*-block. Among materials for building infrastructure, aluminosilicates dominate with materials such as clay bricks (200 billion bricks per year), cement (3600 mmt) and glass (56 mmt) representing the *p*-block (mmt = million metric tons). Natural zeolites, which are environmentally friendly aluminosilicates with a range of applications in ion-exchange, catalysis and water softening are produced in much lesser quantities (3 mmt, 2010). On a comparable scale, the three major organic chemicals produced in the word in 2013 were ethylene (156 mmt), propylene (80 mmt) and methanol (65 mmt), the first two mostly as monomers for the polymer industry.[2,3]

A major component of the bulk industrial chemicals of *p*-block elements produced, goes as plant nutrients and fertilisers (ammonia, phosphoric acid and phosphates). To a much lesser extent, organophosphates and chlorinated organic compounds find use in crop protection as pesticides and fungicides. The major industrial use of sulfuric acid is to generate phosphoric

Table 1.1 The top 15 bulk inorganic chemicals produced by the world and their major uses

S. No.	Bulk inorganic chemical/ element/composition	Year of reference	Quantity (world production in million metric tons)	Major uses and applications
1	Cement	2016	4100	Construction
2	Steel	2017	1675	Construction, ships, containers…
3	Calcium carbonate/ lime stone	2016	350	Metallurgy, building materials, cement
4	Salt	2016	255	Highway de-icing, Cl_2 and NaOH
5	Sulfuric acid	2012	231	Production of phosphate fertilisers (56%)
6	Ammonia	2016	146	Fertilisers (85%), nitric acid synthesis, polyamides
7	Sulfur	2016	70	Sulfuric acid manufacture
8	Sodium hydroxide	2016	70	Paper, detergents, Bayer process for aluminium
9	Chlorine	2016	65	PVC (30%), polyurethanes, epoxy resins, chloromethanes, disinfectants
10	Glass	2010	56	Construction, buildings, automobiles
11	Nitric acid	2016	55	Nitrates for use in fertilisers (80%) and explosives
12	Sodium carbonate	2016	52	Glass, detergents and cleaners, as source of sodium ion
13	Hydrogen	2016	50	Ammonia (53%) and methanol synthesis, petroleum refining
14	Phosphoric acid	2014	43	Phosphate fertilisers (67%), trisodium polyphosphate
15	Aluminium	2016	33	Construction, airplanes, trains…

1 million metric tons (mmt) = 1,000,000,000 kilograms; Source: Davis, S (April 2011). Chemical Economics Handbook Product Review: Petrochemical Industry Overview. SRI Consulting. Global Chemicals Outlook - Towards Sound Management of Chemicals ISBN: 978-92-807-3320-4 © United Nations Environment Programme, 2013 http://www.essentialchemicalindustry.org/chemicals. *US Geological Survey, Mineral Commodity Summaries, 2016.*

acid by the wet process for further use in the fertiliser industry. Nitric acid – itself prepared industrially by the catalytic oxidation of ammonia – is used for making ammonium nitrate (a fertiliser as well as explosive) and many explosives containing nitro groups. Chlorine finds use in the production of PVC, polyurethanes, epoxy resins and disinfectants. Fluorine is used in producing nuclear fuels, polymers, refrigerants and pharmaceuticals. In recent times, the importance of elements such as silicon and gallium has grown exponentially with applications in areas such as solar cells, LED lights and silicone polymers.

1.1.2 Fertilisers and Crop Protection Agents[4–6]

Table 1.2 Common fertilisers with their NPK values

Chemical name of fertiliser	Industrial name	Formula	NPK value
Ammonia	Ammonia	NH_3	82–00–00
Urea	Urea	$(H_2N)_2CO$	46–00–00
Ammonium sulfate	Ammonium sulfate	$(NH_4)_2SO_4$	21–00–00
Calcium nitrate	Norwegian saltpetre	$Ca(NO_3)_2$	15–00–00
Ammonium nitrate	Ammonium nitrate	NH_4NO_3	33–00–00
Mono ammonium phosphate	MAP	$(NH_4)H_2PO_4$	11–48–00
Diammonium phosphate	DAP	$(NH_4)_2HPO_4$	18–46–00
Monocalcium phosphate	Triple superphosphate	$Ca(H_2PO_4)_2$	00–44–00
Monocalcium phosphate + gypsum	Single superphosphate	$Ca(H_2PO_4)_2 + 2\ CaSO_4$	00–17–00

Selected *p*-block and *s*-block elements constitute both the primary macronutrients required by plants (nitrogen, phosphorus and potassium) as well as secondary macronutrients (calcium, sulfur and magnesium). The agricultural revolution of the developed and some of the developing countries has been made possible by the large scale production of fertilisers supplying these macronutrients. The nitrogen required by plants has been supplemented by ammonia, diammonium phosphate, monoammonium phosphate, ammonium sulfate, ammonium nitrate, calcium cyanamide, calcium nitrate and urea. Phosphorus is generally supplemented by diammonium phosphate, calcium dihydrogen phosphate or superphosphate of lime (a mixture of calcium dihydrogen phosphate and gypsum). Potassium has been supplemented from potassium sulfate or potassium chloride.

In 2012, the world used approximately 5.8 billion pounds of pesticides, with herbicides constituting the biggest part of the world pesticide use at 49%, followed by insecticides

(18%) and fungicides (14%). Organochloro pesticides dominated the industry in the early decades—these have now been mostly banned and are being replaced by organophosphate pesticides. First introduced in 1944, organophosphorus compounds (OPs) are economically still the most successful and diverse chemical class of insecticides. Currently one of the most successful and maximum produced organophosphorus herbicides (chemicals which selectively prevent growth of weeds) is glyphosate. Organophosphate pesticides such as chlorpyriphos irreversibly inactivate acetylcholinesterase, a key enzyme which is essential to nerve function in insects, humans and many other animals. This prevents hydrolysis of the neurotransmitter acetylcholine in the central nervous system (CNS) and leads to prolonged periods of nerve excitation. This then results in paralysis and subsequently death of the insects and their predators.[7,8]

Organophosphate pesticides affect this enzyme in varied ways, and have the potential to poison. Parathion was one of the first commercialised organophosphorus pesticides. Unlike organochlorine pesticides such as DDT, endosulfan and aldrin which are not easily degraded and often enter the ground water systems, organophosphate pesticides to a large extent degrade rapidly by hydrolysis on exposure to sunlight (UV), air and soil. Although organophosphates degrade faster than the organochlorides, they have greater acute toxicity, posing serious health risks to people who may be exposed to large amounts. Many of the notorious nerve gases, which have been suspected to be used as chemical warfare agents (such as Tabun and Sarin), are also organophosphates.[9]

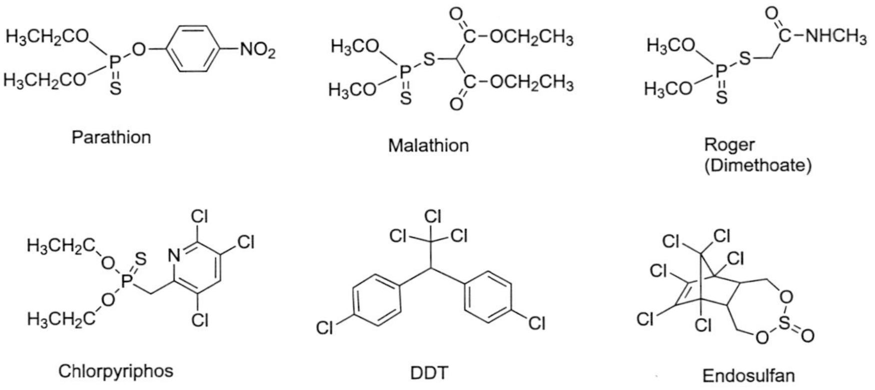

Parathion

Malathion

Roger (Dimethoate)

Chlorpyriphos

DDT

Endosulfan

Fig. 1.1 Examples of common phosphorus and chlorine-based pesticides

Commonly used organophosphate pesticides include parathion, malathion, methyl parathion, chlorpyrifos, diazinon, dichlorvos, phosmet, fenitrothion, tetrachlorvinphos and azamethiphos. Unlike parathion which is highly toxic, malathion – which is relatively less toxic to humans – is widely used in agriculture, residential landscaping and in public mosquito eradication programmes.

1.1.3 Semiconductors, Solar Cells and Light Emitting Diodes

Table 1.3 *p*-Block based solar cell materials, their efficiencies, advantages and disadvantages

Material	Bandgap (300K) eV	Conversion efficiency (%)	Advantages	Disadvantages
c-Si Monocrystalline silicon Wafers (cut from cylindrical ingots)	1.1	19–22.5	High conversion efficiency, long life, performs in low light	Very expensive, loss of material during cutting of ingots
p-Si Polycrystalline silicon Made by casting molten silicon	1.1	13.0–16.0	Process simple and less costly than c-Si, lower heat tolerance than c-Si	Less efficient due to lower purity, lower space efficiency
a-Si Amorphous silicon Thin films	1.7–1.8	11.1	Mass production easy, flexible, cheap	Mostly small scale applications (such as calculators)
CdTe Cadmium telluride Thin films	1.4	16.5–17.0	Cost efficient. large scale installation possible	Cadmium is highly toxic; degrades
CIGS Copper indium gallium selenide Thin films	1.0–1.7	19.5–20.8	Flexible substrate, more efficient than CdTe	Less toxic than CdTe; degrades
GaAs Gallium arsenide	1.43	28.8	Insensitive to overheating	Expensive
Multi Junction (MJ) Indium gallium phosphide; Indium gallium arsenide; Germanium Thin films	 1.9 1.4 0.7	40.7	Highest conversion efficiency, application preferred in space	Most expensive; degrades fast

Semiconductors are the basic materials of modern electronics, including transistors, solar cells, LEDs (light-emitting diodes), quantum dots (nanocrystals whose electronic properties are intermediate between bulk semiconductors and discrete molecules and also exhibit quantum mechanical properties) as well as digital and analog integrated circuits. It will not be an exaggeration to state that more than 95% of the semiconductors are based on *p*-block elements or are a combination of *p*-block elements with or without late transition metals such as cadmium and copper. Industrial grade photovoltaic solar cells are nowadays made of monocrystalline silicon wafers, polycrystalline silicon, and thin films made of materials such as amorphous silicon, cadmium telluride, copper indium gallium selenide (CIGS) as well as multi-junction materials (based mostly on indium gallium arsenide, indium gallium phosphide and germanium) (see Table 1.3).

These materials differ primarily in their bandgap energies. Bandgap energy is the amount of energy required to displace an electron from its covalent bond and allow it to become part of an electrical circuit. To free an electron, the energy of a photon must be at least as high as the bandgap energy. Photons with more energy will squander the extra energy as heat while freeing the electrons. The complete spectrum of sunlight, from infrared to ultraviolet, covers a bandgap range from approximately 0.5 eV to about 2.9 eV (red light 1.7 eV and blue light 2.7 eV). Most solar cells do not use more than half of the energy of sunlight because this energy is either below the bandgap of the material used or it has excess energy. Almost 90% of the world's photovoltaics (as of 2017) are based on some type of silicon. Crystalline silicon has a bandgap energy of 1.1 eV. Crystalline silicon forms the basis of mono- and polycrystalline silicon solar cells. Amorphous silicon layers can be made thinner than crystalline silicon and is prepared by plasma enhanced chemical vapour deposition using silane, SiH_4, as the source gas.

Cadmium telluride (CdTe) is the only thin film photovoltaic material to surpass crystalline silicon in cost in multi-kilowatt photovoltaic systems. It is also considered to be the ecologically most efficient photovoltaic technology in commercial rooftop applications or large scale ground mount applications. On a life cycle basis, CdTe PV has the smallest carbon footprint, lowest water use and fastest energy payback time of all solar technologies.[10]

CIGS is a solid solution of copper indium selenide and copper gallium selenide. It has a chemical formula of $CuIn_xGa_{(1-x)}Se_2$ where the value of x can vary from 1 to 0. CIGS has the advantages of being able to be deposited on flexible substrate materials, producing highly flexible, lightweight solar panels. Improvements in efficiency have made CIGS a leader among alternative cell materials. CIGS has a bandgap varying continuously, with x from about 1.0 eV (for copper indium selenide) to about 1.7 eV (for copper gallium selenide). It has an exceptionally high absorption coefficient of more than 10^5/cm for 1.5 eV and higher energy photons.

Gallium arsenide (GaAs) has a band gap of 1.43 eV. GaAs naturally performs better than silicon in converting solar energy into electricity under normal conditions but is more expensive. Further, GaAs solar cells deliver more energy in high heat or low light, operate near their full efficiency at levels of illumination that are only one tenth of the sun. At high temperature, GaAs continues to deliver energy at near its rated output, while the energy output of a silicon cell declines approximately by 30%. A multi-junction solar cell will produce electric current at multiple wavelengths of light, increasing the conversion efficiency. Indium gallium phosphide (1.8 eV) and indium gallium arsenide (1.4 eV) along with germanium (0.7 eV) forms the basic units of multi-junction solar cells which have a conversion efficiency in the range of 34%–41%.[11]

Among LEDs, GaN and InGaN are the most sought after materials as they are used for making white LEDs. Gallium nitride, due to its wide band gap emits brilliant light when a small electric current is passed through it. By varying the band gap by adding more and more indium to gallium nitride, light of any colour can be produced. Gallium nitride based light emitting diodes are already in widespread use replacing regular incandescent and CFL light bulbs.

On a related topic, transparent conducting oxides (TCOs) are electrical conductive materials with comparably low absorption of electromagnetic radiation within the visible region of the spectrum. Indium tin oxide is one of the most widely used transparent conducting oxides because of its unique properties such as good electrical conductivity, very high optical transparency, as well as ease with which it can be deposited as a thin film. ITO which is stable up to 150°C is omnipresent in modern touch screen technology starting from automatic teller machines, touch screen mobile phones and tablet PCs. It is also used in flat panel LCD, OLED plasma and electro-chromatic displays and solar panels.[12]

1.1.4 In the Design and Functioning of Nuclear Reactors

A few of the *p*-block elements find significant use in the functioning of nuclear reactors. The most well documented among these is lead, which is useful as a material capable of blocking γ and X-radiations, and boron whose ^{10}B isotope has been found to be useful as a material for absorbing thermal neutrons. ^{10}B also is useful in the now well known non-invasive therapy to treat cancer, namely, boron neutron capture therapy (BNCT).

Lead can effectively attenuate γ and X-radiations because of its relatively high density (11.3) and high atomic number. The high density of lead is caused by the combination of its high atomic mass and the relatively small size of its bond lengths and atomic radius. The measure of the effectiveness of the material as a gamma shield is called the *half-thickness*—the thickness of a material that will reduce its gamma ray exposure by half. The half-thickness values for gamma radiation emitted by U-235 for a variety of materials are listed in Table 1.4.[13,14]

Table 1.4 Half-thickness for γ radiation for selected materials

Lead	0.06 cm	Tungsten	0.04 cm
Concrete	4.24 cm	Lead glass	0.08 cm
Water	12.44 cm	Aluminium	2.10 cm
Iron	0.80 cm	Depleted uranium	0.04 cm

Many other metals which have greater density than lead such as tungsten (19.3), uranium (19.1) and gold (19.3) have also the potential to act as γ radiation shields. However, compared to lead, these elements are much more expensive and this deters their bulk usage.

While three specific isotopes of *p*-block elements (along with Cd-113) have been found to show appreciable cross section for thermal neutrons, the one which has been found to have practical use is boron-10 (Table 1.5). It is one of the non-radioactive isotopes of boron with a natural abundance close to 20%. While xenon-135 has the maximum ability to absorb thermal neutrons, it has a short half life of 9.1 hours and does not occur naturally, being a fission product of nuclear reaction. Indium-115, although having a high natural abundance, is scarcely available and is nowadays widely used in the preparation of indium tin oxide (ITO), a transparent conducting thin film. The toxicity of cadmium-113 makes it less acceptable although its half life is quite high.[15]

The ease of availability of boron-10 and its non-radioactivity makes it an ideal choice for neutron capture. Control rods of nuclear reactors are made of boron-10. It is reported that 40 tons of boron compounds, mostly as boron carbide (along with 2400 tons of lead) were dumped using helicopters on the burning Chernobyl nuclear reactor (in April 1986) to absorb neutrons. Boron neutron capture therapy (a non-invasive method to specifically target cancer cells) involves concentrating cancer cells with boron-10 and bombarding the affected area by thermal neutrons from outside to generate α particles and lithium nuclei which has a range of approximately the diameter of a cell.

Table 1.5 Isotopes of elements with good cross section for thermal neutrons

Isotope	Cross section for thermal neutron capture (barn)	Half life	Natural abundance
Indium-115	100	4.41×10^{14} years	95.71%
Boron-10	200	Non radioactive	19.9%
Cadmium-113	30,000	7.7×10^{15} years	12.22%
Xenon-135	2,600,000	9.14 hours	0 (Xe-135 is a decay product of I-135)

1.1.5 Inorganic Polymers

A major section of inorganic polymers with diverse properties and application potential is based on the *p*-block elements, especially silicon and phosphorus. In addition, the properties of perfluorinated carbon-based polymers such as Teflon can be credited to the presence of the highly electronegative and relatively small sized fluorine atoms present in it. The three major inorganic polymers with significant application potential are the silicone polymers, polyphosphazenes and polysilanes.

$$-[SiR_2-O]_n- \qquad -[PR_2=N]_n- \qquad -[SiR_2]_n-$$

Silicones Polyphosphazenes Polysilanes

Fig. 1.2 Common examples of inorganic polymers based on silicon and phosphorus

Among inorganic polymers, silicones, especially polydimethylsilicone (PDMS), stands out with a very diverse application potential due to its properties such as hydrophobicity, oxygen permeability, non-toxicity, wide temperature range and low glass transition temperature. One of the reasons for the development of PDMS as an industry-friendly polymer is that its precursor, dimethyldichlorosilane can be prepared inexpensively by the Rochow–Müller process, a reaction involving MeCl and a Si/Cu alloy. Polyphosphazenes, although more expensive compared to silicones, have the advantage of having different properties based on the type of substituents present on the polymer back bone. For example, polyphosphazenes with polyfluoroalkoxy substituents are highly stable against oxidation and hydrolysis and are water repellent, while polyphosphazenes with primary amino substituents are hydrolytically degradable. The phosphazene backbone upon hydrolysis gives NH_4OH and H_3PO_4 which are non-toxic in contrast to the carbon-based decomposition products of many organic polymers. Relatively less known polysilanes show unique electronic properties such as σ-electron delocalisation are precursors for polycarbosilanes and are used in silicon carbide fabrication. This property stems from the low ionisation energy for electrons in Si–Si σ bonds relative to that of C–C σ bonds. Accordingly, they absorb strongly in the UV-region (300–400 nm) leading to intense σ–σ* electronic transitions.[16,17]

Among the three, the one that has the highest market value in terms of quantity and diversity is the silicones. The global market for silicones in terms of revenues was estimated to be worth $14.2 billion in 2017 and is expected to reach $18.9 billion by 2022. The silicone market is dominated by elastomers and fluids, together accounting for over 80%. Silicones possess superior properties and characteristics which are extremely useful in automotive, construction, medical and personal care and, electrical and electronics end-user industries.[18]

1.1.6 Explosives Based on Nitrogen

More than 90% of all explosives are based on compounds having nitrogen and oxygen often having $C–NO_2$, $O–NO_2$, $N–NO_2$ units or simply $M–N_3$ groups. Such compounds have very weak N–O (55 kcal/mol), and N–N (38 kcal/mol) bonds which, during an explosion, readily break releasing energy and highly stable gases in high volume. Especially N_2 and to some extent CO_2 and CO are released which have the strongest multiple bonds among simple gases [N≡N (226 kcal/mol) C≡O (258 kcal/mol), C=O 192 kcal/mol]. Measuring velocity of detonation (VoD) which is the velocity at which a shock wave front travels through a detonated explosive, gives a good indication of the strength and performance of an explosive (Fig. 1.3).[19] There is a growing worldwide market for novel explosives, not only for defence but also for many infrastructure projects.

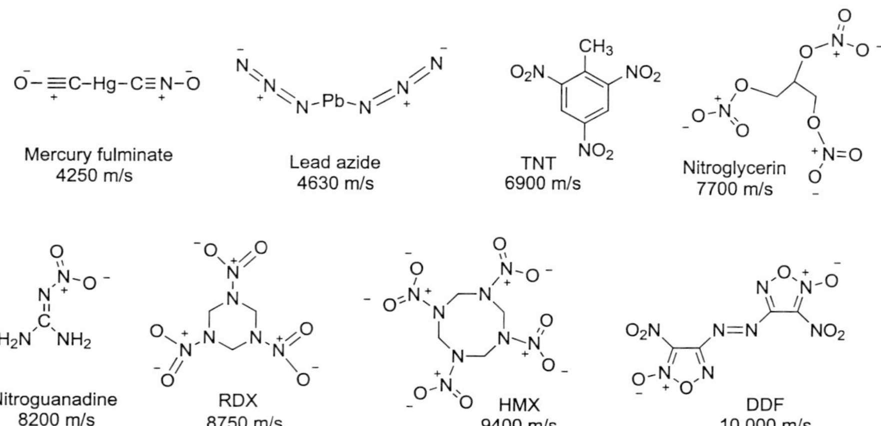

Fig. 1.3 Examples of some common explosives with their velocities of detonation

1.1.7 Materials Based on Carbon

After the discovery of fullerenes, a new area has come into prominence which centres around nanomaterials based on carbon. In addition to fullerenes, the major types of materials in this category are carbon nanotubes and graphene (Fig. 1.4). These materials have been found to possess unusual physical properties (not found in any other materials), such as high electrical and thermal conductivity of carbon nanotubes or transparency, and two dimensionality of graphene. However, extending their use to practical applications is still in their infancy.[20] Another material which is effectively replacing aluminium in applications requiring a high strength to weight ratio such as aircraft manufacture is carbon fibre reinforced plastics (CFRP).

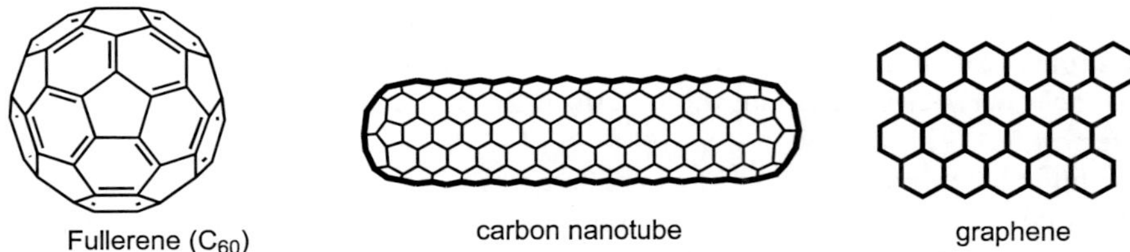

Fig. 1.4 Carbon based nanomaterials

1.1.8 Disinfectants and Germicides Based on Chlorine

The utility of hypochlorous acid (HClO, obtained when Cl_2 is dissolved in water) as a germicide, has been known from 1847. Ignaz Semmelweis, who is now known as the pioneer of antiseptic procedures, while working in the maternity department of the Vienna lying-in hospital was able to stop the death of women who had been through child birth and were afflicted with an unknown disease called child bed fever. He instituted a policy for doctors and medical students of using a solution of chlorinated lime (bleaching powder) for washing hands between autopsy work and the examination of patients. Since then, chlorine gas is extensively used for water purification and killing of bacteria. Another chlorine-based oxidiser, chlorine dioxide (ClO_2, possessing an unpaired electron) is fast replacing chlorine as a mild, harmless and selective oxidising agent for many industrial and household uses. The active ingredients of many disinfectants are chlorinated phenols. For example, para-chloro-meta-xylenol (PCMX) is the active ingredient of the most common household disinfectant 'Dettol'.[21]

HClO ClO_2 PCMX

1.1.9 Fluorine-Based Volatile Compounds and Drugs with Enhanced Activity

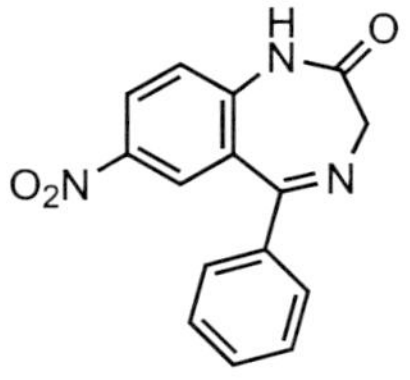

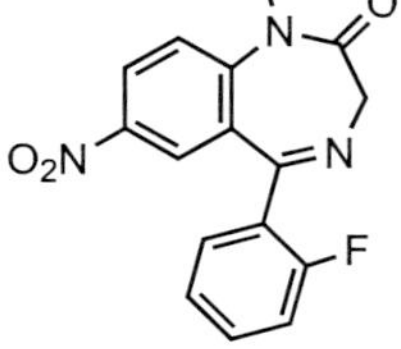

About 20% of all FDA approved drugs have fluorine present in them as a CF, CF_2 or CF_3 unit. Introducing fluorine atoms or CF_3 groups into pharmaceuticals is often accompanied by higher binding affinity, enhanced metabolic stability, improved bioavailability and in many cases enhanced selective activity of the drug. C–F bonds have high oxidation and thermal stability. In the body, unlike C–H bonds, C–F bonds have been found to be unsusceptible to oxidation by the enzyme cytochrome P-450, thereby improving the bioavailability of fluorinated drugs. For example, between the two variations of the nitrazepam drugs shown, the fluorinated one is more potent.

The high electronegativity and low polarisability of the C–F bond translates into very less secondary interactions resulting in high volatility for small molecules having covalently bound fluorine substituents. This volatility along with stability are the basic reasons why CFCs, HFCs and HFOs find application as refrigerants (Table 1.6). The non-stick nature of fluoropolymers such as Teflon is also a consequence of very less secondary interactions.[22]

Table 1.6 Evolution of fluorine based refrigerants

Refrigerants: past, present and future	Chlorofluorocarbons (Banned from1987)	Hydrofluorocarbons (Being phased out 2011–2017)	Hydrofluoro-olefins (Proposed refrigerant after hydrofluorocarbons)
Code	CFC-12	HFC-134a	HFO-1234yf
Name	Dichlorodifluoro methane	1,1,1,2-Tetrafluoroethane	2,3,3,3-Tetrafluoropropene
Formula	CCl_2F_2	CF_3–CFH_2	CH_2=CF–CF_3
Boiling point	–29.8°C	–26.3°C	–30°C

1.1.10 As Reagents and Catalysts

The most extensive and diverse applications of compounds based on *p*-block elements are as reagents and catalysts in synthesis. Lewis acid catalysts based on *p*-block elements such as aluminium, boron, silicon, tin and antimony are used extensively in reactions such as Friedel–Crafts and aldol reactions, in various pericyclic processes that proceed slowly at room temperature, such as the Diels–Alder reaction, Claisen rearrangement and acetal formation. In addition to accelerating the reactions, Lewis acid catalysts are able to impose regioselectivity and stereoselectivity in many cases. Early developments in Lewis acid reagents focussed on easily available compounds such as BF_3, $SnCl_4$, SiF_4, SbF_5 and $AlCl_3$. Over the years, versatile catalysts bearing ligands designed for specific applications have facilitated improvement in both reactivity and selectivity of Lewis acid-catalysed reactions. More recently, Lewis acid catalysts with chiral ligands have become an important class of tools for asymmetric catalysis. Lewis acid catalysis, especially in reactions having polar mechanisms, often involve binding of the catalyst to the basic heteroatoms of the substrate. They withdraw electron density, thus facilitating heterolytic bond cleavage (as in Friedel–Crafts reaction). In other cases, it can directly activate the substrate towards nucleophilic attack (as in carbonyl addition reactions).[23,24]

Activation by a Lewis base can enhance the chemical reactivity by increasing the nucleophilicity or electrophilicity, or modulate the electrochemical properties. Amines, phosphines, arsines, N-heterocyclic carbenes and some hypervalent compounds of silicon contribute significantly as Lewis base catalysts and reagents. Many Lewis bases which are non-nucleophilic act as proton abstractors. Acyclic phosphazene bases represent a family of extremely strong, non-ionic, non-charged nitrogen-bases with an impressive list of advantages. Highly reactive 'naked' enolates produced by acyclic phosphazene bases proved to be more stable towards side reactions than lithium enolates gained from conventional lithium dialkylamides.

Preventing the formation of conventional Lewis acid–base pairs by increasing steric bulkiness has resulted in a new class of compounds called frustrated Lewis pairs (FLPs) with interesting uses as metal free reagents and catalysts.

The unusual catalytic activity shown by aluminosilicates, such as zeolites, is unique to the *p*-block elements. The open framework stable structures such as zeolites act as dehydration agents and shape-selective catalysts. Methylating agents such as $AlMe_3$ and its partially hydrolysed product methyl alumoxane (MAO) find application as cocatalysts in olefin polymerisation reactions.

1.2 Major Landmark Discoveries in the Chemistry of *p*-Block Elements

~3000 bc Bronze, an alloy of tin and copper was the first alloy to be used around 3000 BC. After 600 BC, pure metallic tin was produced.

~1570 bc Tyrian purple, a reddish purple organobromine natural dye was used by ancient Phoenicians. The dye was produced from the secretions of a species of sea snails from the family, Muricidae. The chemical component (main ingredient: 6,6′-dibromoindigo) was discovered in 1909 by Paul Friedlaender.[25,26]

~800 BC Fumigating power of sulfur was mentioned by Homer. Sulfur is also mentioned in the Bible as brimstone.[27]

940 AD Sulfuric acid was mentioned by the Persian writer Abu Bekr al Rases.

1250 Albertus Magnus, a catholic saint (Albert the Great, 1193–1280), is believed to have been the first to isolate arsenic by heating soap together with arsenic trisulfide.

1669 Hennig Brandt discovered phosphorus from reductive distillation of urine concentrate by evaporating urine to dryness and distilling the residue with sand.[28]

1694 Robert Boyle made P_4O_{10} by oxidation of yellow phosphorus and phosphoric acid H_3PO_4, by the hydrolysis of P_4O_{10}.[29]

1746 John Roebuck introduced the *lead chamber process* for sulfuric acid manufacture where lead-lined chambers that were stronger and less expensive replaced soda glass reaction vessels which were prone to break with temperature variations during large scale synthesis.[30]

1772 Daniel Rutherford isolated N_2 gas after he removed oxygen and carbon dioxide from the air. He named the isolated gas *noxious air* or *phlogisticated air* and believed it was exhaled by the mouse which he used in the experiment to remove oxygen by breathing.[31]

1772 Scheele isolated oxygen gas by heating manganese dioxide (and later potassium nitrate and mercuric oxide) but did not publish his findings immediately. It was also independently discovered by Joseph Priestly by heating and decomposing mercuric oxide in 1774.[32]

1783 PH_3 was discovered by the French chemist P Gengembre by the action of potash on bones.[33]

1785 Claude Louis Berthollet determined the composition of NH_3 gas. J Priestly had isolated it by using mercury in a pneumatic trough (1774) and termed it alkaline air. Berthollet also discovered the bleaching ability of chlorine.

1808 Boron trifluoride was discovered by Gay-Lussac and Thénard, who were trying to isolate hydrofluoric acid by combining calcium fluoride with vitrified boric acid. The resulting vapours failed to etch glass, so they named it fluoroboric gas.[34]

1810 Peter Durand, a British merchant received the first patent for preserving food using tin cans. Durand made containers of tinplate (out of iron sheets coated with tin) to prevent rusting and corrosion. Tinplate could be sealed and made airtight and was unbreakable unlike glass.

1812 Pierre Louis Dulong prepared nitrogen trichloride NCl_3, and lost two fingers and an eye in two explosions. An NCl_3 explosion also blinded Humphry Davy (famous for Davies safety lamp) triggering him to hire Michael Faraday as a coworker. [35]

1818 Thénard described hydrogen peroxide. He produced it by treating barium peroxide with nitric acid.[36]

1824 Silicon was discovered by Berzelius from the reduction of potassium fluorosilicate with metallic potassium.[37]

1831 Contact process for sulfuric acid was patented by the British vinegar merchant Peregrine Phillips. It was a better process than the previous lead chamber process and made concentrated sulfuric acid.[38]

1834 J Liebig and F Wohler prepared the inorganic heterocycle $N_3P_3Cl_6$ from the reaction of PCl_5 and NH_4Cl.[39]

1835 S_4N_4, the first example of a sulfur nitrogen heterocycle, was prepared by W Gregory by the reaction of S_2Cl_2 and NH_3.[40]

1837 John Bennet Lawes, an English entrepreneur, began to experiment on the effects of various manures on plants growing in pots in 1837, and a year or two later the experiments were extended to crops in the field. One immediate consequence was that in 1842 he patented a manure formed by treating phosphates with sulfuric acid, and thus was the first to create the first chemical fertiliser.[41]

1838 The French chemist Henri Victor Regnault prepared sulfuryl chloride (SO_2Cl_2), the well known chlorinating agent.[42]

1840 C F Schoenbein detected and named ozone from its characteristic odour after a lightning storm.[43]

1840 Justus von Liebig identified that the chemical elements, nitrogen (N), phosphorus (P) and potassium (K), were essential to plant growth. Liebig suggested that nitrogen could be supplied in the form of ammonia and predicted the possibility of substituting chemical fertilisers for natural ones.[44,45]

1859 The French physicist Gaston Planté discovered the lead–acid battery, the oldest type of rechargeable battery. Its ability to supply high surge currents along with low cost made it attractive for use in motor vehicles.

1863 Friedel and Craft synthesised the first organosilicon compound tetraethylsilane by the reaction of diethylzinc and silicon tetrachloride.[46]

1865 G D Buckton and W Odling reported the synthesis of trimethylaluminium from the reaction of Me_2Hg and aluminium.[47]

1886 Henri Moissan prepared fluorine gas F_2, by the electrolysis of a solution of potassium hydrogen difluoride in liquid hydrogen fluoride. For this discovery he received the Nobel prize in 1906.[48]

1890 The Belgian scientist Frédéric Swarts pioneered the synthesis of CFCs in the 1890s. He developed an effective exchange agent to replace chloride in carbon tetrachloride with fluoride to synthesise CFC-11 (CCl_3F) and CFC-12 (CCl_2F_2).[49]

1895 Ramsay and Lord Rayleigh discovered argon as a new constituent of the atmosphere.[50]

1900 A New York physician developed a liquid preparation having bismuth subsalicylate, zinc salts and phenyl salicylate (with a red dye) as a remedy named 'mixture cholera infantum' to treat small children suffering and often dying with a fatal form of gastroentitis.[51,52]

1901 F S Kipping coined the word 'silicone' to describe polydiphenylsiloxane by analogy of its formula, with the formula of benzophenone.[53] However, the silicone did not have the expected Si=O bond.

1901 Henri Moissan and Paul Lebeau prepared a very unreactive and dense gas SF_6 by the fluorination of S_8 with F_2. The compound was later found to be useful as a convenient dielectric material.[54]

1904 Sir William Ramsay discovered neon, krypton and xenon and received the Nobel Prize in 1904. He also isolated helium which had been observed in the spectrum of the sun but had not been found on earth till then. In 1910 he also made and characterised radon.[55]

1908 Ostwald's process for the synthesis of HNO_3 by the catalytic oxidation of ammonia on an industrial scale was discovered by W Ostwald.[56]

1909 Fritz Haber reported the Haber process for the catalytic synthesis of NH_3 which was made into a large scale industrial process along with C Bosch in 1913. He received the Nobel Prize in Chemistry in 1918.[57]

1912 Alfred Stock prepared a series of boron hydrides including B_2H_6 and separated them using the first sophisticated vacuum manifold. Due to their very high air and moisture sensitivity and flammability, they were not separable till then. He published his work in 1933.[58–59]

1920 Thomas Midgley Jr, replaced the commonly used refrigerants of that time such as ammonia, chloromethane and sulfur dioxide with chlorofluorocarbons.[60]

1926 Alfred Stock and Pohland discovered borazine $B_3N_3H_6$, also known as inorganic benzene by the reaction of diborane and ammonia.[61]

1927 The term *inert pair effect* was first proposed by Nevil Sidgwick.[62]

1929 C H Finske and Y Subbarow discovered adenosine triphosphate (ATP) in muscle tissue. This was synthesised 20 years later by A Todd (Nobel Prize 1957).[63]

1935 H J Taylor showed that boron-10 nuclei could capture thermal neutrons. In 1936, G L Locher, of the Franklin Institute in Pennsylvania, realised the therapeutic potential of this discovery and suggested that neutron capture could be used to treat cancer—thus beginning the boron neutron capture therapy (BNCT).[64]

1938 Polytetrafluoroethylene (Teflon) was accidentally discovered by Roy J Plunkett of kinetic chemicals (a subsidiary of Dupont) while attempting to make a new CFC using tetrafluoroethylene.[65]

1943 H Christopher Longuet Higgins proposed the actual bridging structure of B_2H_6 after due theoretical considerations (an undergraduate student at that time). This ended the controversy on the structure of B_2H_6.[66]

1943 E G Rochow discovered the direct synthesis of organochlorosilanes from a silicon–copper alloy and chloromethane which initiated easy large scale synthesis of Me_2SiCl_2 monomer needed for silicone polymers.[67]

1949 The first polysilane, poly(dimethylsilylene) $[(CH_3)_2Si]_x$, was reported by Charles A Burkhard of the General Electric Company. It was prepared by heating sodium metal with dimethyldichloro silane.

1954 G Rupprecht reported the semiconducting properties of indium oxide In_2O_3. This was followed by the report by V A Williams in 1966 on the potential of tin doped indium oxide (indium tin oxide, ITO) as a transparent conducting material.[68]

1957 Ronald Gillespie and Ronald Sydney Nyholm proposed the VSEPR theory for predicting the shape of *p*-block element based compounds.[69]

1960 P B Weisz and V J Frilette observed the first 'shape selective catalysis' by zeolites (Faujasite 13X) in the selective dehydrogenation of *n*-butanol in the presence of isobutanol.[70]

1960 M F Hawthorne discovered the dodecaborate anion $[B_{12}H_{12}]^{2-}$ the first icosahedral boron cluster.[71]

1962 In June 1962, Niel Bartlett reacted Xe with PtF_6 and obtained a new compound which he proposed as xenon hexafluoroplatinate $Xe^+PtF_6^-$. The identity of this compound was later proved as $XeF^+Pt_2F_{11}^-$. This was the first real compound of any noble gas. In September and November 1962, Howard Claasen and Rudolf Hoppe prepared XeF_4 and XeF_2 respectively.[72]

1963 Carboranes (specifically 1,2-*closo*-dicarbadodecaboranes), were reported simultaneously by groups at Olin Corporation and the Thiokol Chemical Corporation working under the US Air Force. These groups demonstrated the unprecedented stability of the 1,2-*closo*-dicarbadodecaborane unit.[73]

1965 The first metal complexes of the dicarbollide ion $[B_9C_2H_{11}]^{2-}$ (which has a close similarity to the cyclopentadienyl ligand) was discovered by M Frederick Hawthorne and coworkers. This anion forms sandwich compounds, referred to as bis(dicarbollides).[74]

1967 Allen K Breed used sodium azide NaN_3, along with a collision sensor in automobile airbags (instead of compressed air) and achieved inflation of the airbag in under 30 milliseconds (required for collision safety).[75]

1968 George A Olah prepared the first examples of superacids called magic acid, so-named for its ability to attack saturated hydrocarbons. The magic acid was obtained by mixing the Lewis acid antimony pentafluoride (SbF_5) with the Brønsted acid fluorosulphonic acid (FSO_3H).[76]

1971 Kenneth Wade proposed the Wades rules which provide a straightforward rationalisation of the shapes of electron-deficient cluster compounds in terms of the number of skeletal electron pairs (SEPs) these molecules possess.[77]

1976 Discovery of selenocysteine, the selenium containing amino acid was made by Thressa Stadtman.[78]

1980 Robert F Furchgott discovered nitric oxide's signaling role in the cardiovascular nervous systems which later fetched him the Nobel Prize in 1998.[79]

1981 Robert West made the first stable silicon–silicon double bonded compound.[80]

1986 Karl O Christe discovered a chemical synthesis of elemental fluorine which was earlier accessible only by electrolysis.[81]

$$K_2MnF_6 + 2\,SbF_5 \longrightarrow 2KSbF_6 + MnF_3 + 1/2\,F_2$$

1991 Anthony J Arduengo III prepared the first stable carbene. It is now well known as the NHC or N-heterocyclic carbene.[82]

1994 Herbert W Roesky discovered the fluorinating ability and recyclability of Me_3SnF for a wide range of substrates especially organometallic compounds.[83]

1997 G M Robinson proposed the first gallium–gallium triple bonded compound which evoked the famous comment from F A Cotton, 'That's no more a triple bond than I'm the Dalai Lama!'[84]

1999 The first N_5^+ (pentazenium) compound $N_5^+AsF_6^-$ was discovered by Karl O Christe. The N_5^+ cation is a highly energetic ion with a calculated endothermicity of 1471.8 kJ/mol.[85]

$$cis\text{-}N_2F_2 + SbF_5 \longrightarrow [N_2F]^+[SbF_6]^-$$

$$[N_2F]^+[SbF_6]^- + HN_3 \xrightarrow[-78^\circ C]{\text{dry HF}} [N_5]^+[SbF_6]^- + HF$$

2000 Herbert W Roesky and co-workers prepared the first stable monomeric and dicoordinate Al(I) compound, which is equivalent to a carbene. Schnöckel, in 1996, had prepared the first tetrameric Al(I) compound.[86]

2000 Markku Räsänen and coworkers prepared the first argon compound HArF by mixing argon and HF on a CsF surface at –265°C and exposing to UV. The compound is stable below –256°C.[87]

2000 Konrad Seppelt discovered the first metal–xenon compound where xenon acts as a ligand. This was achieved by reduction of AuF_3 with xenon. The $AuXe_4^{2+}(Sb_2F_{11}^-)_2$ complex had a square planar gold with a Au–Xe bond distance of 274 pm.[88]

2001 E D Jemmis proposed the Jemmis *mno* rule, a simple and effective rule for predicting the stability of boranes, carboranes, metallaboranes and even macropolyhedral clusters.[89]

2004 Akiro Sekiguchi prepared the first silicon–silicon triple bonded compound.[90]

2006 Douglas W Stephan discovered the first example of a frustrated Lewis acid base pair based on boron and phosphorus. Due to their unquenched reactivity, such molecules are able to heterolytically cleave dihydrogen making them potential metal-free hydrogenation catalysts.[91]

Mes = 2, 4, 6 $(CH_3)_3C_6H_2$

2007 G H Robinson prepared the first neutral diborene containing a B=B bond.[92]

Dip = 2,6 diisopropylphenyl

2008 G H Robinson prepared the first stable silicon(0) compound.[93]

2008 G Bertrand and coworkers prepared an extremely bent acyclic allene (carbodicarbene) which passes the test for zerovalent carbon compounds of the type ($L \rightarrow C^0 \leftarrow L$) predicted earlier by G Frenking as 'carbone' compounds.[94]

2010 Andre Geim received the 2010 Nobel Prize in physics jointly with Novoselov for groundbreaking experiments discovering the two dimensional material graphene.[95]

2012 T Matsuo and K Tamao prepared a stable germanone molecule having a Ge=O bond. This is the first example of a heavier element ketone analogue.[96]

[Eind.Li] $\xrightarrow[\text{THF}]{GeCl_2\text{.Dioxane}}$ Eind₂Ge: $\xrightarrow{Me_3NO}$ Eind₂Ge=O

2012 Florian Kraus identified the first occurrence of free F_2 (difluorine) in nature, in samples of the mineral Antozonite, a variation of CaF_2 which gives off a foul smell when crushed.[97]

2012 The first example of a stable diboryne having a B≡B and stabilised by N-heterocyclic carbenes was prepared by Holger Braunschweig.[98]

Sod. Naphthalenide, –78°C

Dip = 2,6 diisopropylphenyl

2013 Herbert W Roesky and co-workers prepared the first examples of silylones (L→Si⁰←L) and germylones (L→Ge⁰←L) using cyclic alkyl (amino) carbene (cAAC) ligands.[99, 100]

2014 Shuji Nakamura along with Isamu Akasaki and Hiroshi Amano received the Nobel Prize in physics for the blue LED based on gallium indium nitride.[101]

2014 Christopher A Reed and coworkers prepared the carborane-based acid $H(CHB_{11}F_{11})$ which is the strongest isolable Brønsted acid (Lewis acid free) presently known. This has the capability to protonate simple alkanes at room temperature.[102]

2014 The first stable silanone having a tri-coordinate silicon atom was prepared by A C Filippou.[103]

$[BAr^F{}_4]$ $\xrightarrow{N_2O}$ $[BAr^F{}_4]$

2015 Herbert W Roesky and coworkers discovered that aluminium hydrides can function as efficient catalysts for reactions such as hydroboration similar to transition metal catalysts.[104, 105]

2015 Drozodov, Eremets and coworkers discovered that compressed hydrogen sulfide (H_2S) shows superconductivity with the highest T_c observed for any material so far, 203 K (–70°C) at 153 GPa pressure.[106]

2016 F H Arnold demonstrated the first natural C–Si bond formation catalysed using a biological metalloenzyme, cytochrome C.[107]

2017 The first example of a stable pentazolate (cyclo-N_5^-) anion as $(N_5)_6(H_3O)_3(NH_4)_4Cl$ was prepared and characterised by Chong Zhang and coworkers. The methodology was modified by others to make $[Na(H_2O)(N_5)]\cdot 2H_2O$ and a host of transition metal complexes of the cyclic pentazolate.[108–110]

Supplementary reading

1. McCoy M, Facts and figures of the chemical industry, *Chemical & Engineering News,* 2006, Vol. 84, 35.
2. Chandler A D, *Shaping the industrial century: The remarkable story of the evolution of the modern chemical and pharmaceutical industries*, Harvard University Press, 2005, 366.
3. Higgins S, European chemicals industry: A review, *Chemical News*, 2013, 18.
4. Dittmar H, Drach M, Vosskamp R, Trenkel M E, Gutser R and Steffens G, Fertilisers, 2 Types, *Ullmann's Encyclopedia of Industrial Chemistry*, Wiley VCH, 2009.
5. Jones J B Jr, *Inorganic chemical fertilisers and their properties in plant nutrition and soil fertility manual*, 2nd Ed., CRC Press, 2012.
6. Gowariker V, Krishnamurthy V N, Gowariker S, Dhanorkar M and Paranjape K, *The fertiliser encyclopedia*, John Wiley & Sons, 2009.
7. Wyant K A, Corman J E, Corman J R and Elser J J, *Phosphorus, food, and our future*, Oxford University Press, 2013.
8. Wolfgang Kramer W, Schirmer U, Jeschke P and Witschel M, *Modern crop protection compounds*, 2nd Ed., Wiley VCH, 2011.
9. Bohmart B L, *The standard pesticide user's guide*, 4th Ed., London: Prentice-Hall International, 1997.
10. Markvart T and Castaner L, *Solar cells: Materials, manufacture and operation*, Elsevier, 2010.
11. Mickey C D, Solar photovoltaic cells, *J. Chem. Educ.*, 1981, Vol. 58, 418.
12. Du J, Chen X, Liu C, Ni J, Hou G, Zhao Y and Zhang X, Highly transparent and conductive indium tin oxide thin films for solar cells grown by reactive thermal evaporation at low temperature, *Appl. Phys. A*, 2014, Vol. 117, 815.
13. Najam M R, Khan Z U and Ahmed N M, Selection of lead for γ-ray shielding and in-pile collimation, *U.S. At. Energy Comm.*, 1967, 12.
14. Maurin A, Tiles for shielding against atomic radiations, *Fr. Demande, FR 2482761 A1 19811120*, 1981.
15. Shaeffer N M, Reactor shielding for nuclear engineers, *US dept. of energy*, 1973.
16. Mark J E, Allcock H R and West R, *Inorganic polymers*, Prentice Hall, Englewood, NJ, 1992.
17. R D and Michl J, Polysilane high polymers, *Chem. Rev.* 1989, Vol. 89, 1359.
18. Chandrasekhar V, *Inorganic and organometallic polymers*, Springer, 2005.
19. Akhavan J, The chemistry of explosives, *The royal society of chemistry*, 2004.
20. Dresselhaus M S, Dresselhaus G and Eklund P C, *Science of fullerenes and carbon nanotubes*, Academic Press, 1996.
21. Darnall C R, Purification of water by anhydrous chlorine, *Am. J. Public Health*, 1911, Vol. 1, 783.
22. O' Hagen D, Understanding organofluorine chemistry: An introduction to the C–F bond, *Chem. Soc. Rev.*, 2008, Vol. 37, 308.

23. Yamamoto H, *Lewis acids in organic synthesis*, Wiley-VCH, 2000.
24. Olah G A, *Friedel–Crafts and related reactions*, Interscience, New York, 1964.
25. McGovern P E and Michel R H, Royal purple dye: tracing the chemical origins of the industry, *Anal. Chem*., 1985, Vol. 57, 1514A.
26. Cooksey C J, Tyrian purple: 6,6'-Dibromoindigo and related compounds, *Molecules*, 2001, Vol. 6, 736.
27. Genesis 19:24, Then the LORD rained on Sodom and Gomorrah brimstone and fire from the LORD out of heaven, *The Holy Bible*, King James Version.
28. Emsley J, *The shocking history of phosphorus: A biography of the devil's element*, Pan books, 2001.
29. Crass M F Jr, A history of the match industry, Part 9, *J. Chem. Educ*., 1941, Vol. 18, 428.
30. Kiefer D M, Sulfuric acid: Pumping up the volume, *Todays chemist, American chemical society*, 2001, Vol. 10, 57.
31. Conant J B, *The overthrow of Phlogiston theory: The chemical revolution of 1775–1789*, Cambridge: Harvard University Press, 1950.
32. Priestley J, An account of further discoveries in air, *Philosophical Transactions*, 1775, Vol. 65, 384.
33. Gengembre, Mémoiresurun nouveau gas obtenu, par l'action des substances alkalines, sur le phosphore de Kunckel" (Memoir on a new gas obtained by the action of alkaline substances on Kunckel's phosphorus), *Mémoires de mathématiqueet de physique*, 1783, Vol. 10, 651.
34. Gay-Lussac J L and Thénard L J, Sur l'acidefluorique, *Annales de Chimie*, 1809, Vol. 69, 204.
35. Thénard J L and Berthollet C L, Report on the work of Pierre Louis Dulong, *Annales de Chimie et de Physique*, 1813, Vol. 86, 37.
36. Thénard L J, Observations sur des nouvelles combinaisons entre l'oxigèneet divers acides, *Annales de chimie et de physique, 2nd series*, 1818, Vol. 8, 306.
37. Berzelius, On the mode of obtaining silicium, and on the characters and properties of that substance, *Philosophical Magazine*, 1825, Vol. 65, 254.
38. Cook E and Peregrine Phillips, the inventor of the contact process for sulfuric acid, *Nature*, 1926, Vol. 117, 419.
39. Allcock H R, *Phosphorus–Nitrogen compounds*, Academic Press, New York, 1972, 5.
40. Valkenburgh H B V and Bailar J C Jr, Nitrogen tetrasulfide and nitrogen tetraselenide, *J. Am. Chem. Soc*., 1925, Vol. 47, 2134.
41. Justus von Liebig, *An address to the agriculturalists of Great Britain* (Liverpool, 1845) Lawes patent was BP 9353/42 (1842).
42. Regnault V, Sur l'acide chlorosulfuriqueet la sulfamide, (On sulfuryl chloride and sulfamide), *Annales de chimieet de physique*, Series 2, 1838, Vol. 69, 170.
43. Schönbein C F, On the odour accompanying electricity and on the probability of its dependence on the presence of a new substance, *Philosophical Magazine*, 1840, Vol. 17, 293.
44. Die organische Chemie in ihrer Anwendung auf Agricultur und Physiologie (Organic Chemistry in its Application to Agriculture and Physiology) (1840), Justus von Liebig (Braunschweig, 1860).
45. Brock W H, *Justus von Liebig: the chemical gatekeeper*, 1st Ed., Cambridge, U.K.: Cambridge University Press, 1997.
46. Pawlenko S, *Organosilicon Chemistry*, Walter de Gruyter New York, 1986.
47. Mole T and Jeffery E A, *Organoaluminium compounds*, Elsevier, Amsterdam, 1972.
48. Tressaud A, Henri Moissan: winner of the Nobel Prize for Chemistry 1906, *Angew. Chem. Int. Ed*., 2006, Vol. 45, 6792.
49. Swarts, *Acad. Roy. Belg*., 1892, Vol. 3, 474.
50. Rayleigh L and Ramsay W, Argon, a new constituent of the atmosphere, *Proceedings of the Royal Society of London*, 1894–1895, Vol. 57, 265.

51. Bierer D W, Bismuth subsalicylate: history, chemistry, and safety, *Reviews of Infectious Diseases*, Oxford University Press, 1990, Vol. 12, S3.
52. Gorbach S L, Bismuth therapy in gastrointestinal diseases, *Gastroenterology*, 1990, Vol. 99, 863.
53. Thomas N R, Frederic Stanley Kipping-Pioneer in silicon chemistry: His life &legacy, *Silicon*, 2010, Vol. 2, 187.
54. Moissan H and Lebeau P, Sulfur hexafluoride, *C.R. Acad. Sci.*, Paris, 1900, Vol. 130, 869.
55. Ramsay W and Travers M W, On the companions of argon, *Proceedings of the Royal Society of London*, 1898, Vol. 63, 437.
56. Ostwald W, Improvements in the manufacture of nitric acid and nitrogen oxides, *GB 190200698*, 1902.
57. Smil V, *Enriching the earth: Fritz Haber, Carl Bosch, and the transformation of world food production*, 1st Ed., Cambridge, MA: MIT, 2004.
58. Stock A, *The hydrides of boron and silicon*, New York: Cornell University Press, 1933.
59. Wiberg E, Alfred Stock 1876–1946, *Chem. Ber.*, 1950, Vol. 83, XIX.
60. Midgley T and Henne A L, Organic fluorides as refrigerants, *Industrial & Engineering Chemistry*, 1930, Vol. 22, 542.
61. Stock A, Pohland E and Borwasserstoffe, VIII. ZurKenntnis des B_2H_6 und des B_5H_{11}, [Boric acid solution, VIII Regarding knowledge of B_2H_6 and B_5H_{11}], *Berichte*, 1926, Vol. 59, 2210.
62. Sidgwick N V, *The electronic theory of valency*, Oxford: Clarendon, 1927, 178.
63. Fiske C H and Subbarow Y, Phosphorus compounds of muscle and liver, *Science*, 1929, Vol. 70, 381.
64. Taylor H J, The disintegration of boron by neutrons, *Proc. Phys. Soc.*, 1935, Vol. 47, 873.
65. Plunkett R J, The History of polytetrafluoroethylene: Discovery and development, *High performance polymers: Their origin and development*, Seymour RB and Kirshenbaum GS, Eds, Elsevier, New York, 1986, 261–266.
66. Longuet-Higgins H C and Bell R P, 64. The structure of the boron hydrides, *J. Chem. Soc., (Resumed)*, 1943, 250.
67. Rochow E G, *Silicon and silicones*, Springer-Verlag, Berlin, 1987.
68. Rupprecht G, Untersuchungen der elektrischen und licht elektrischen Leitfähigkeit dünner Indiumoxyd schichten, *Z. Phys.*, 1954, Vol. 139, 504.
69. Gillespie R J and Nyholm R S, Inorganic stereochemistry, *Quart. Rev. Chem. Soc.*, 1957, Vol. 11, 339.
70. Weisz P B, FriletteV J, Intracrystalline and molecular-shape-selective catalysis by zeolite salts, *J. Phys. Chem.*, 1960, Vol. 64 , 382.
71. Pitochelli A R and Hawthorne F M, The isolation of icosahedral $B_{12}H_{12}^{2-}$ ion, *J. Am. Chem. Soc.*, 1960, Vol. 82, 3228.
72. Bartlett, Xenon hexafluoroplatinate (V) $Xe^+[PtF_6]^-$, *Proceedings of the Chemical Society. London: Chemical Society*, 1962, Vol. 6, 218.
73. Heying T L, Ager J W, Clark S L, Mangold D J, Goldstein H L, Hillman M, Polak R J and Szymanski J W, A new series of organoboranes, I. Carboranes from the reaction of decaborane with acetylenic compounds, *Inorg. Chem.*, 1963, Vol. 2, 1089.
74. Hawthorne M F, Young D C and Wegner P A, Carbametallic boron hydride derivatives, I. Apparent analogues of ferrocene and ferricinium ion, *J. Am. Chem. Soc.*, 1965, Vol. 87, 1818.
75. McCormick L W, *A short history of the airbag*, Consumers Unified LLC, 2006.
76. Olah G A and Schlosberg R H, Chemistry in super acids, I. Hydrogen exchange and polycondensation of methane and alkanes in FSO_3H-SbF_5 ("magic acid") solution. Protonation of alkanes and the intermediacy of CH_5^+ and related hydrocarbon ions. The high chemical reactivity of "paraffins" in ionic solution reactions, *J. Am. Chem. Soc.*, 1968, Vol. 90, 2726.

77. Wade K, The structural significance of the number of skeletal bonding electron-pairs in carboranes, the higher boranes and borane anions, and various transition-metal carbonyl cluster compounds, *J. Chem. Soc. D.*, 1971, 792.
78. Stadtman T, Selenium biochemistry, *Science*, 1974, Vol. 183, 915.
79. Raju T N, The Nobel chronicles, 1998: Robert Francis Furchgott (b 1911), Louis J Ignarro (b 1941), and Ferid Murad (b 1936), *Lancet*, 2000, Vol. 356, 346.
80. West R, Chemistry of the silicon–silicon double bond, *Angew Chem. Int. Ed.*, 1987, Vol. 26, 1201.
81. Christe K O, Chemical synthesis of elemental fluorine, *Inorg. Chem.*, 1986, Vol. 25, 3721.
82. Arduengo J, Harlow R L and Kline M, A stable crystalline carbene, *J. Am. Chem. Soc.*, 1991, Vol. 113, 361.
83. Herzog A, Liu F Q, Roesky H W, Demsar A, Keller K, Noltemeyer M and Pauer F, Trimethyltinfluoride: A new fluorinating reagent for the preparation of organometallic fluorides, *Organometallics*, 1994, Vol. 13, 1251.
84. Dagani R, Gallium 'triple bonds' under fire, *Chemical & Engineering News*, 1998, Vol. 76, 31.
85. Christe K O, Wilson W W, Sheehy J A and Boatz J A, N_5^+: A novel homoleptic polynitrogen ion as a high energy density material, *Angew Chem. Int. Ed.*, 1999, Vol. 38, 2004.
86. Cui C, Roesky H W, Schmidt H-G, Noltemeyer M, Hao H and Cimpoesu F, Synthesis and structure of a monomeric Al(I) compound: A stable analogue of a carbine, *Angew Chem. Int. Ed.*, 2000, Vol. 39, 4274.
87. Khriachtchev L, Pettersson M, Runeberg N, Lundell J and Räsänen M, A stable argon compound, *Nature*, 2000, Vol. 406, 874.
88. Seidel S and Seppelt K, Xenon as a complex ligand: The tetra xenonogold(II) cation in $AuXe_4^+(Sb_2F_{11}^-)_2$, *Science*, 2000, Vol. 290, 117.
89. Jemmis E D, Balakrishnarajan M M and Pancharatna P D, A unifying electron-counting rule for macropolyhedral boranes, metallaboranes, and metallocenes, *J. Am. Chem. Soc.*, 2001, Vol. 123, 4313.
90. Sekiguchi A, Kinjo R and Ichinohe M, A stable compound containing a silicon–silicon triple bond, *Science*, 2004, Vol. 305, 1755.
91. Welch G C, Juan R R S, Masuda J D and Stephan D W, Reversible, metal-free hydrogen activation, *Science*, 2006, Vol. 314, 1124.
92. Wang Y, Quillian B, Wei P, Wannere C S, Xie Y, King R B, Schaefer H F, III, Schleyer P V R and Robinson G H, A stable neutral diborene containing a B=B double bond, *J. Am. Chem. Soc.*, 2007, Vol. 129, 12412.
93. Wang Y, Xie Y, Wei P, King R B, Schaefer H F, III, Schleyer P V R and Robinson G H, A stable silicone(0) compound with Si=Si double bond, *Science*, 2008, Vol. 321, 1069.
94. Dyker C, Lavallo V, Donnadieu B and Bertrand G, Synthesis of an extremely bent acyclic allene (carbodicarbene): A strong donor ligand, *Angew Chem. Int. Ed.*, 2008, Vol. 47, 3206.
95. Geim A, Graphene: Status and prospects, *Science*, 2009, Vol. 324, 1530.
96. Li L, Fukawa T, Matsuo T, Hashizume D, Fueno H, Tanaka K and Tamao K, A stable germanone as the first isolated heavy ketone with a terminal oxygen atom, *Nature Chem.*, 2012, Vol. 4, 361.
97. auf der Gruenne J S, Mangst L M and Kraus F, Occurrence of difluorine F_2 in nature-In situ proof and quantification by NMR spectroscopy, *Angew. Chem. Int. Ed.*, 2012, Vol. 51, 7847.
98. Braunschweig H, Dewhurst R D, Hammond K, Mies J, Radacki K and Vargas A, Ambient-Temperature isolation of a compound with a boron–boron triple bond, *Science*, 2012, Vol. 336, 1420.
99. Mondal K C, Roesky H W, Schwarzer M C, Frenking G, Niepotter B, Wolf H, Herbst-Irmer R and Stalke D, A stable singlet biradicaloid siladicarbene, $(L:)_2Si$, *Angew. Chem. Int. Ed.*, 2013, Vol. 52, 2963.

100. Li Y, Mondal K C, Roesky H W, Zhu P, Stollberg P, Herbst-Irmer R, Stalke D and Andrada D M, Acyclic germylones: Congeners of allenes with a central germanium atom, *J. Am. Chem. Soc.*, 2013, Vol. 135, 12422.
101. Savage N, LEDs light the future, *Technology Review*, 2000, Vol. 103, 38.
102. M, Stoyanova I V, Cummings S, Stoyanov E S and Reed C A, The strongest Brønsted acid: Protonation of alkanes by $H(CHB_{11}F_{11})$ at room temperature, *Angew. Chem. Int. Ed.*, 2014, Vol. 53, 1131.
103. Sen S S, A stable silanone with a three-coordinate silicon atom: A century-long wait is over, *Angew. Chem. Int. Ed.*, 2014, Vol. 53, 8820.
104. Yang Z, Zhong M, Ma X, De S, Anusha C, Parameswaran P and Roesky H W, An aluminium hydride that functions like a transition metal catalyst. *Angew. Chem. Int. Ed.,* 2015, Vol. 54, 10225.
105. Yang Z, Zhong M, Ma X, Nijesh K, De S, Parameswaran P and Roesky H W, An aluminium hydride working as a catalyst in hydroboration and dehydrocoupling, *J. Am. Chem. Soc.*, 2016, Vol.138, 2548.
106. Drozdov A P, Eremets M I, Troyan I A, Ksenofontov V and Shylin S I, Conventional superconductivity at 203 kelvin at high pressures in the sulfur hydride system, *Nature*, 2015, Vol. 525, 73.
107. Khan S B J, Lewis R D, Chen K and Arnold F H, Directed evolution of cytochrome C for C–Si bond formation. Bringing silicon to life, *Science*, 2016, Vol. 354, 1048.
108. Zhang C, Sun C, Hu B, Yu C and Lu M, Synthesis and characterisation of the pentazolate anion cyclo-N_5^- in $(N_5)_6(H_3O)_3(NH_4)_4Cl$, *Science*, 2017, Vol. 355, 374.
109. Christe K O, Polynitrogen chemistry enters the ring, *Science*, 2017, Vol. 355, 351.
110. Xu Y, Wang Q, Shen C, Lin Q, Wang P and Lu M, A series of energetic pentazolate hydrates, *Nature*, 2017, Vol. 549, 78.

STRUCTURE AND BONDING OF *p*-BLOCK COMPOUNDS: UNIQUE ASPECTS

CHAPTER 2

2.1 SHAPES OF MOLECULES—VSEPR RULES

Each element of the *p*-block is unique with respect to the structures and shapes of its compounds. Therefore, knowing the possible shapes a molecule can have and the actual shape the molecule possesses is very significant for understanding its chemical properties.

A simple set of rules called the Valence Shell Electron Pair Repulsion (VSEPR) theory was proposed by Ronald Gillespie and Ronald Sydney Nyholm in 1957. This has been quite useful in arriving at the shapes of many molecules of the *p*-block elements.[1] The basic ideas of these rules were made from suggestions made first by N V Sidgwick and H E Powell in 1940.[2] Although these rules seem empirical, as shown by Gillespie, they do have a quantum mechanical basis – the Pauli exclusion principle – and predict molecular geometry in a reliable manner.[3] The crux of these rules is that valence electron pairs surrounding an atom will repel each other mutually and therefore will take on an arrangement around the central atom which minimises these repulsions and maximises their distances apart thus determining the shape of the molecule. Electrons show this type of behaviour as a result of the Pauli exclusion principle and not as a result of their electrostatic repulsion. The rules are based on the *points on a sphere* model as it envisages the arrangement of electron pairs as points on a spherical core. For the same reason, the rules have been quite successful in predicting the shape of compounds whose central atom is a *p*-block element, and not very useful in the case of compounds of transition metals and inner transition metals where the core is not always spherical. In this context, it is of interest to note that the concept of hybridisation, especially involving *d* orbitals has come under intense scrutiny recently with even suggestions as to whether 'is it time to retire the hybrid orbitals!'.[4] However, it does not affect the VSEPR rules as these rules and the *points on a sphere* model on which it is built, does not depend upon hybridisation and hybrid orbitals at all.[5]

The *AXE method* of electron counting is generally used when applying the VSEPR rules. *A* represents the central atom and always has an implied subscript one. *X* represents the number of ligands (atoms bonded to A) and *E* represents the number of lone pairs surrounding the central atom. Before applying the rules, one has to first determine a reasonable Lewis structure of the compound based on the electronic configuration of the elements and identify the central atom. The electrons are then arranged in pairs that are either bonding (shared) pairs [X] or non-bonding (unshared) pairs [E]. In the usual σ–π treatment, this usually means ignoring the π bonds temporarily since they will follow the σ

bonds and therefore the two bonding pairs of a double bond or three bonding pairs of a triple bond are all primarily considered as a single point on the sphere. The sum of the number of bonded atoms to a central atom (X) plus the number of lone pairs of its non-bonding valance electrons (E) is called its *steric number* or occupancy. The following rules are applied, one following the other to arrive at the final shape of a given molecule.[6]

The pairs of electrons in the valence shell adopt that arrangement which maximises their distances apart. In other words, it means that the bonding and non-bonding electron pairs are arranged as far apart from each other as possible on the surface of a sphere with the central atom A at the centre of the sphere.

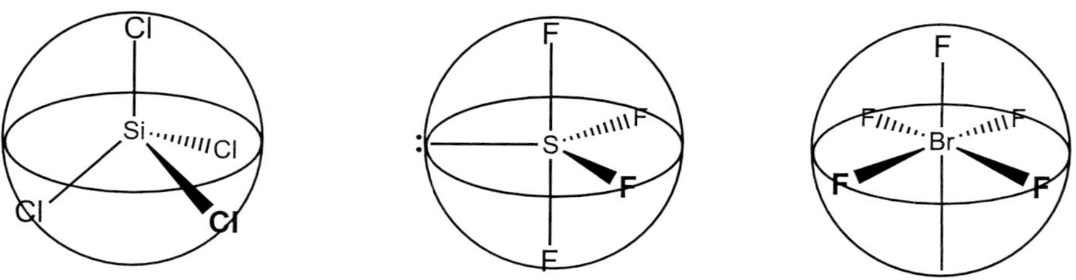

Fig. 2.1 Schematic representation of molecules based on points on a sphere model

While the ideal geometry can be arrived at by adopting this model, in reality, for many molecules the bond angles have been found to deviate significantly from the perfect polyhedral shapes predicted. Gillespie and Nyholm in 1957 accounted for these deviations by using three simple rules.

Rule 1: A non-bonding lone pair [E] occupies more space on the surface of the central atom than a bonding pair [X]. In other words, lone pair–lone pair repulsion is greater than lone pair–bond pair repulsion which in turn is greater than bond pair–bond pair repulsion. Lone pairs choose the largest site—equatorial in trigonal bipyramid and axial in pentagonal bipyramid geometries. If all sites are equal, lone pairs will be better placed *trans* to each other. On going down a group, the size of the central atom increases and there is a marginal increase in the effect of the lone pairs on the bonding pairs.

This accounts for the observation that the H–N–H angle of NH_3 is 107.3° while the H–O–H angle for water is 104.5°. The same also accounts for the observed difference in bond angles in comparison to what is expected based on the idealised shapes predicted (as shown in Fig. 2.2) for molecules such as SF_4 and ClF_3. For example the F–Cl–F angle of ClF_3 is 86.6° (Fig. 2.2).

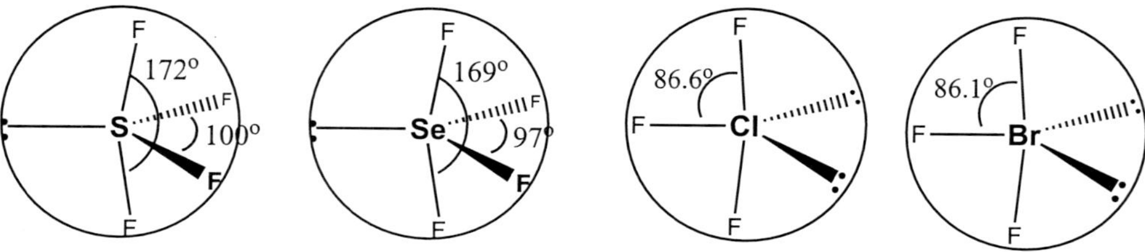

Fig. 2.2 Variations in bond angles based on lone pair–bond pair repulsions

Rule 2: The strengths of the repulsions between bonds having different multiplicities decreases in the order: triple–double > double–double > double–single > single–single.

Multiple bonds, since they are composed of two or three electron pair domains, take up more angular space than a bonding electron pair domain. This rule is useful to arrive at the shapes of molecules and ions such as SO_2Cl_2, XeO_2F_2, NSF_3, POF_3 where multiple bonds are present. As illustrated in Fig. 2.3, bond angles decrease in the order:

multiple bond–multiple bond > multiple bond–single bond > single bond–single bond.

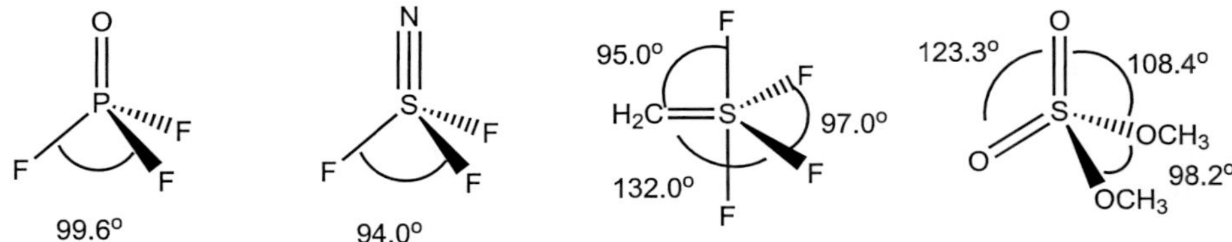

Fig. 2.3 Variations in bond angles due to the presence of multiple bonds

Rule 3: Bonding pairs to electronegative substituents occupy less space than those to more electropositive substituents. Therefore, angles between the single bonds decrease with increasing electronegativity of the substituent. This decrease in domain size with increasing electronegativity of the substituent occurs as more of the bonding electron density is shifted to the electronegative substituent. That is, the repulsion between single bonds is observed to decrease with increasing electronegativity of the substituent and/or decreasing electronegativity of the central atom.

For example, for triangular pyramidal phosphorus halides, the bond angles decrease in the order PI_3 102°, PBr_3 101°, PCl_3 100°, PF_3 98° as the electronegativity of the halogen increases. Similar variations are also observed in carbonyl halides and nitrogen trihalides (Fig. 2.4).

Fig. 2.4 Variations in bond angles based on the electronegativity of substituents

A valence shell containing seven or eight electron pair domains is very crowded and the idealised geometries predicted by VSEPR rules appear to hold only for main group compounds under two conditions: (i) where the ligands are quite electronegative, for example fluorine, as a result of which the binding domains in the valence shell of the central atom are small; (ii) where the central atom has a large valence shell especially if it is a fifth period element such as iodine or xenon. Although one can envisage the possibility of a capped octahedron (1:3:3) or monocapped trigonal prism (1:4:2) for the seven electron pair domain, the idealised pentagonal bipyramidal geometry forms the basis for AX_7, AX_6E and AX_5E_2 type of structures. For eight electron pairs (AX_8) the lowest energy geometry is for the square antiprism which has been observed in the structure of IF_8^-. The only known example of an AX_8E molecule is Xe F_8^{2-} which also has been reported to possess a square antiprism structure with very slight distortion. The observed distortion appears incompatible with a lone pair repulsion model and therefore it is also considered as an example of a stericially inactive lone pair which is not surprising due to the high coordination number.

2.1.1 Shapes of Compounds Having an Odd Number of Electrons

The VSEPR theory has also been extended to a few molecules with an odd number of electrons by treating the unpaired electron as half of a lone pair. In effect, the odd electron also has an influence on the molecular geometry similar to a lone pair but is of relatively lesser strength. As a result, the geometry will be midway between the molecule with a full electron pair and the molecule with one less electron pair on the central atom. Nitrogen dioxide (NO_2) has an unpaired electron ($AX_2E_{0.5}$) and it has been found to have a bent shape with an ONO angle of 134°. NO_2^- which is AX_2E is also bent but has an ONO angle approximately of 120° while NO_2^+ (AX_2) is linear.

Similarly ClO_2 ($AX_2E_{1.5}$) with a lone pair and an unpaired electron has OClO angle of 117.6° while ClO_2^- with two lone pairs (AX_2E_2) has an OClO angle of 111°.

2.1.2 Exceptions to the VSEPR Rules

Metal complexes and organometallic compounds based on transition elements have been found not to obey the VSEPR rules. The *points on a sphere model* is not applicable here as the cores of polarisable transition metal atoms and ions can be non-spherical because they have partially filled *d* orbitals and include non-bonding electrons or they are distorted by the ligands. Therefore, shapes of metal complexes are best described by using conventional valence bond theory based on hybridisation; deformation from the ideal shape is explained using Jahn–Teller distortion. Many triatomic alkaline earth metal dihalides, although expected to be linear have been found to be bent (approximate X–M–X angles: CaF_2, 145°; SrF_2, 120°; BaF_2, 108°; $SrCl_2$, 130°; $BaCl_2$, 115°; $BaBr_2$, 115° and BaI_2, 105°). Gillespie has proposed that the reason for this is due to loss of spherical symmetry of the inner shell of the metal atom due to the polarising effect of the relatively more electronegative halide substituents. Li_2O also is linear although it is expected to be bent and this has been ascribed to the bonding being more ionic resulting in repulsion between the lithium atoms.

The silyl ether $O(SiH_3)_2$ has a relatively larger Si–O–Si angle (144.1°), while similar bond angles of Cl_2O (110.9°) and $(CH_3)_2O$ (111.7°) are in the expected range. Gillespie's rationalisation for this observation is that the localisation of the lone pairs, and therefore their ability to repel other electron pairs, is highest when the ligand has an electronegativity similar to or greater than that of the central atom. If the central atom is more electronegative, as oxygen in $O(SiH_3)_2$, the lone pairs are less well-localised and have therefore a weaker repulsive effect. This fact, combined with the stronger ligand–ligand repulsion (SiH_3 is a relatively large substituent), gives the larger-than-expected Si–O–Si bond angle.

Certain molecules in which the central atom is from periods 3 and 4 and in which the ligands are less electronegative, do not have sufficient space in their valence shell to

Arrangement	Bonding pairs X	Lone pairs E	Occupancy/ Steric number	Shape of the molecule	Spatial arrangement	Examples
Linear	2	0	2	Linear AX_2	X—A—X	CO_2, CS_2, HCN, $BeCl_2$, N_2O, NO_2^+, NCF, NCBr
Triangular	3	0	3	Triangular planar AX_3	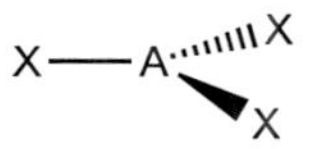	SO_3, BF_3, $COCl_2$, NO_3^-, CO_3^{2-}, BO_3^{3-}
	2	1	3	Bent (V-shape) AX_2E		SO_2, O_3, NOCl, $SnCl_2$, PbF_2, NOF
Tetrahedral	4	0	4	Tetrahedral AX_4	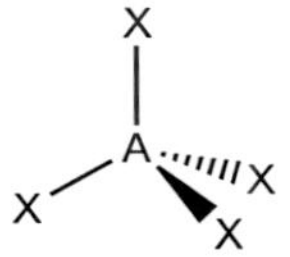	CH_4, $SiCl_4$, SO_2Cl_2, $POCl_3$, XeO_4, NSF_3, ClO_4^-, PO_4^{3-}
	3	1	4	Triangular Pyramidal AX_3E	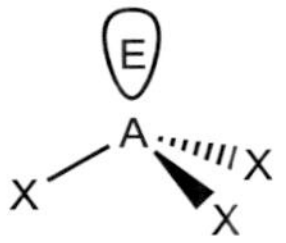	NH_3, PCl_3, $SOCl_2$, IO_3^-
	2	2	4	Bent AX_2E_2	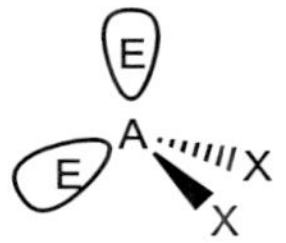	H_2O, SCl_2, NSF, ICl_2^+, $SeCl_2$, $[I(C_6H_5)_2]^+$, $Kr_2F_3^+$
Trigonal bipyramid	5	0	5	Trigonal bipyramidal AX_5	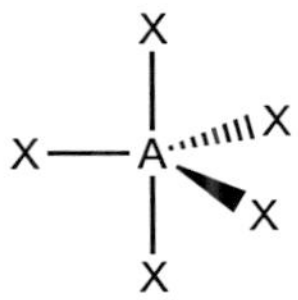	PF_5, SF_4O
	4	1	5	See saw AX_4E	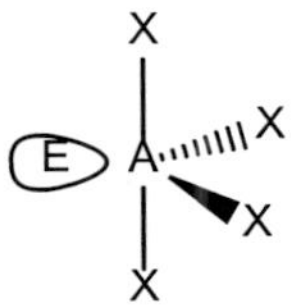	SF_4, XeO_2F_2, IF_4^+, IOF_3, $IO_2F_2^-$
	3	2	5	Bent T-shape AX_3E_2	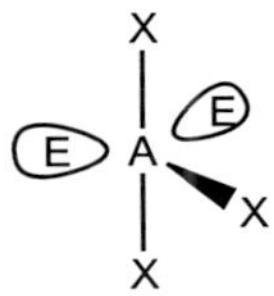	ClF_3, BrF_3, $XeOF_2$, $I(Ph)[(CF_3C(O)O)]_2$

Cont'd

Table Cont'd

Arrangement	Bonding pairs X	Lone pairs E	Occupancy/ Steric number	Shape of the molecule	Spatial arrangement	Examples
	2	3	5	Linear AX_2E_3	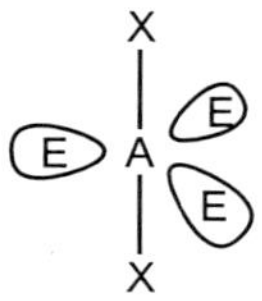	XeF_2, I_3^-, ICl_2^-, BrF_2^-
Octahedral	6	0	6	Octahedral AX_6	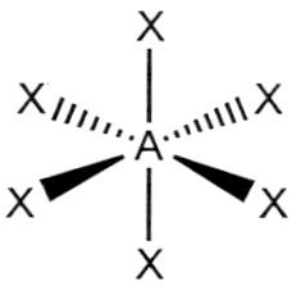	SF_6, PF_6^-, SiF_6^{2-}, AlF_6^{3-}, IF_6^+
	5	1	6	Square pyramidal AX_5E	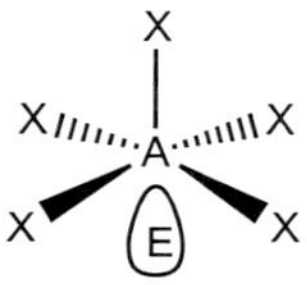	BrF_5, XeF_5^+, IOF_4^-, TeF_5^-, SbF_5^{2-}
	4	2	6	Square planar AX_4E_2	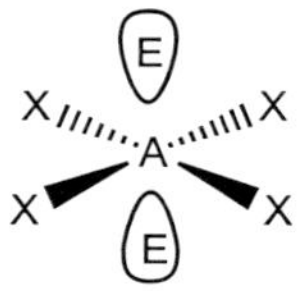	XeF_4, IF_4^-, $XeOF_3^-$
Pentagonal bipyramid	7	0	7	Pentagonal bipyramid AX_7	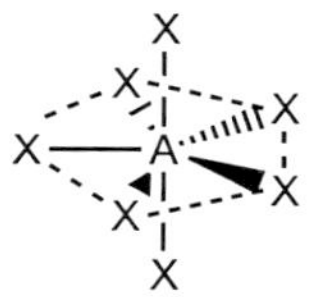	IF_7, $IO_2F_5^{2-}$, IOF_6^-, TeF_7^-, $CH_3OTeF_6^-$, $(CH_3O)_2TeF_5^-$, BiF_7^{2-}, SbF_7^{2-}
	6	1	7	Pentagonal pyramid AX_6E	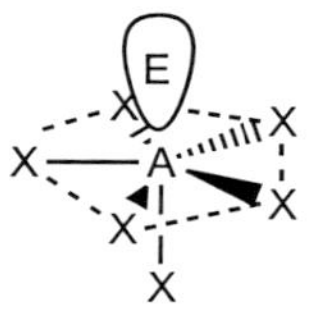	$XeOF_5^-$, IOF_5^{2-} $[IF_5(C_6H_5)]^-$
	5	2	7	Pentagonal planar AX_5E_2	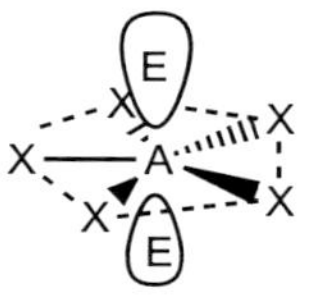	XeF_5^-, IF_5^{2-}

accommodate six bonding domains in addition to a large lone pair domain. In such molecules the lone pair is squeezed into a spherical domain surrounding the core and inside the bonding domains (stereochemically inactive *s* orbital) and therefore have an octahedral arrangement. Thus some AX_6E molecules such as $SeCl_6^{2-}$ and $TeBr_6^{2-}$ have a regular octahedral shape but with longer than normal bonds. An alternative explanation of their shapes is offered by the 3-centre-4-electron bond model. Another rationalisation is the inert pair effect.

2.1.3 Physical Basis of the VSEPR Rules, Criticisms and Modifications on the Original Model

Most of the bonding models which were proposed in the 1950–1970 period were not subject to rigorous theoretical calculations to substantiate and verify their validity as computational chemistry was still in its infancy and it was difficult to do theoretical studies on even moderately large molecules. However, with the advent of fast computing software and modern computing machines, such studies have now become routine and many of the old concepts and theories have been questioned and some are being discarded. Two such concepts at the centre of discussion are the participation of *d* orbitals in the bonding of main group compounds and the original concept of hybridisation itself. Interestingly, the VSEPR model has stood the test of time as it is not based on the concept of hybridisation and is still accepted as the simplest model to predict the shape of molecules. Attempts to provide more theoretical basis for the VSEPR model by Gillespie based on electron density studies on bonds has also had some criticism which will be discussed later.[7]

2.1.4 The Pauli Principle

While many teachers and practitioners of the VSEPR model think that its basis is electrostatics, the proponents of the VSEPR model have time and again insisted that these rules are based on the Pauli principle. Many questions such as, are electrons paired in lone pairs and bonding pairs, does the pairing happen on the free central atom, and so on, are explained based on the Pauli principle. The Pauli principle in its general form can be stated as follows.

1. The wave function for an electronic system must be antisymmetric to electron interchange.

   This statement is difficult to comprehend at the beginner's level without a sound knowledge of quantum mechanics and refers to the fact that one cannot determine the path of an electron but only the probability of finding an electron in a specific volume in space. The consequence of the Pauli principle can be explained in the following statements which are easier to understand.

2. Electrons with the same spin have a zero probability of being found simultaneously at the same point in space, a low probability of being found close together and are most probably placed as far apart from each other as possible.

3. No more than two electrons, which must be of opposite spin, can occupy an orbital.
4. No two electrons can have the same four quantum numbers.

Consider the case of a valence shell containing eight electrons such as O^{2-}. Four of the electrons will be of α spin and the remaining four of the opposite β spin. Since same-spin electrons keep as far from each other as possible, the most probable relative location of these four α electrons will be at the vertices of a tetrahedron which in a free atom can have any orientation in space. Similarly, the most probable relative location of the β electrons will be at the vertices of another tetrahedron which may take an orientation with respect to the α electrons so that that there is an equal probability of finding an electron anywhere in the valence shell of the ion or atom having octet under consideration. Thus any free atom or ion having an octet of electrons will have a spherical total valence electron density. However, in a molecule where the central atom is forming bonds to two or more atoms/ligands, the attraction of the nuclei of the combining atoms causes the two tetrahedra made up of α and β electrons to come into approximate coincidence thereby increasing the probability that a pair of electrons will be found in the bonding regions and also in the remaining tetrahedral directions (say for water molecule, two pairs in the bonding regions and two lone pairs in the remaining tetrahedral directions). Consequently the electron density is no more spherical and there are now four regions of enhanced electron density corresponding to the four electron pairs. It should be taken into consideration that these pairs, each having electrons of opposite spin are not formed due to any attraction between the electrons since all electrons repel each other electrostatically. Also in a free atom or ion there is no pairing of electrons of opposite spin formed by the attraction of the cores of the ligand atoms. This attraction overcomes the repulsion between the electrons and pulls the electrons of opposite spin to form bonding regions. The Pauli principle prevents the formation of pairs of electrons of the same spin.

2.1.5 The Electron Pair Domain Model

While the original rules of VSEPR were easy to apply to real molecules, it was subject to criticism and a few modifications were brought in by Gillespie and coworkers for better acceptability with respect to theory as well as to explain and reduce the number of molecules considered as violating the VSEPR rules. The first modification among these is the electron pair domain model proposed by Gillespie and Harigittai in 1991 followed by Gillespie and Robinson in 1996. This model was based not on pairs of electrons but on a region in space where there is more probability of finding a pair of electrons (bonding or lone pair). An electron pair domain is a region in space where there is an enhanced probability of finding an electron pair. An electron pair domain extends around the most probable position of finding an electron pair as determined by the Pauli principle and the probability decreases on increasing the distance from the most probable position. A lone pair domain can in general be thought of as more spread out and larger than a bonding pair domain as the electrons of a lone pair are subject to the attraction of only one atomic nucleus while those of a bonding pair are subject to the attraction of two nuclei taking part in the bond formation. A bonding pair domain therefore takes up less space around a nucleus compared to a lone pair domain because the former is stretched out more towards the ligand/ substituent of the central atom. Why a lone pair occupies more space than a bonding pair is well explained by the domain concept thereby explaining the variation observed in bond angles more meaningfully for molecules having one or more lone pairs on the central atom.[8]

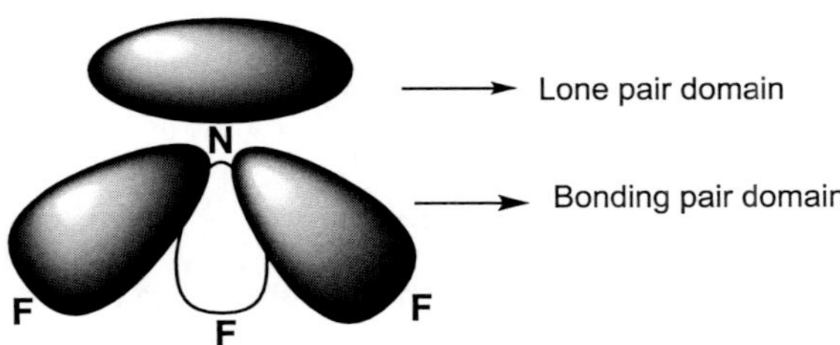

Fig. 2.5 Schematic representation indicating lone pair domain and bonding pair domain on NF_3

2.1.6 Ligand Close Packing Model

The basis of this model is that the bond angles are determined by the packing of the ligands or substituents around a central atom. Irrespective of the presence of lone pairs, the distance maintained between the substituents on a central atom will be almost same while the bond lengths and bond angles will be different as shown in the following examples. Comparing NF_3 and NF_4^+, the distance maintained between F substituents are the same (212 pm) (Fig. 2.6).

In molecules having lone pairs, the lone pairs will push the ligands together until the ligands reach the minimum inter-ligand distance which will be given by the sum of the ligand radii. It is also important to note that VSEPR rules are strictly followed when the ligands /substituents have high electronegativities comparable to or greater than that of the central atom. If the ligands are only weakly electronegative compared to the central atom, the electrons of the central atom will only be weakly localised into pairs and the central atom will be more or less like a spherical anion with charge delocalised all over. Such poorly delocalised lone pairs will hardly have any effect on the geometry of the molecule and its shape will be affected more by the ligand–ligand interactions. Two examples of such compounds whose actual structures violate the VSEPR rule are $H_3SiOSiH_3$ and $N(SiH_3)_3$.[9]

The two unusual structural features of a disiloxane – the large Si–O–Si bond angle and the short Si–O bond length – have been qualitatively explained by the concept of $p\pi$–$d\pi$ back-bonding. According to this earlier prevalent view, silicon with its unoccupied 3d orbitals, function as the acceptor in the donor–acceptor $p\pi$–$d\pi$ bond. As a consequence of this partial π bond, the σ bond shortens the Si–O distance, and the partial delocalisation of the oxygen lone pairs on to the silicon allows an increase of the Si–O–Si bond angle which

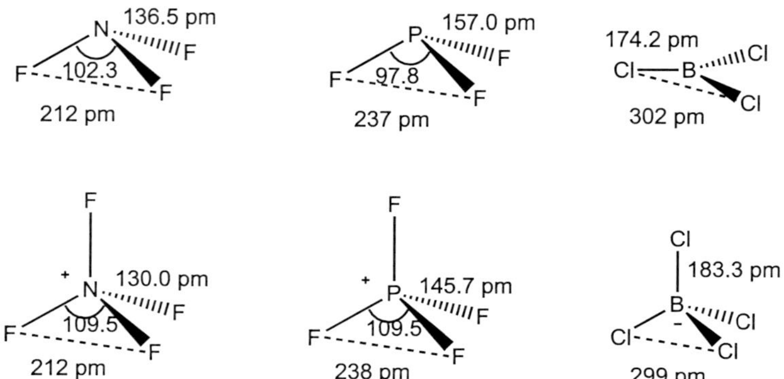

Fig. 2.6 Schematic representation indicating ligand close packing

results in the reduced basicity of the oxygen. However, semi-empirical MO calculations in the CNDO/2 approximation for a series of compounds containing the Si–O bond have shown that the oxygen bond angles in these compounds can be reproduced reasonably well without the use of Si (3*d*) functions. In addition, *ab initio* calculations carried out on such molecules, which also reproduced the geometry and dipole moments of Si–O–Si containing molecules very well, suggest that $p\pi$–$d\pi$ bonding is of minor importance in explaining the Si–O bond distance and Si–O–Si bond angle. A $p\pi$–$d\pi$ bond order of 0.4 only has been deduced from experimental bond lengths. Rather, the small value of the Si–O bond length is a consequence of the ionic character of the bond which is considerably higher than that estimated from electronegativity differences. Thus, it was concluded that the increase in the oxygen bond angles in siloxanes does not require $p\pi$–$d\pi$ contribution. A better explanation for the violation of the VSEPR model of Si–O–Si compounds is thus provided by the LCP model.[10,11]

2.2 The Concept of Hypervalency in *p*-Block Compounds

2.2.1 History and Current Status

A hypervalent molecule can be defined as a molecule in which there are more than four pairs of electrons surrounding the central atom in the conventional Lewis depiction of the molecule. J I Musher in 1969 originally defined hypervalent molecules as those formed by the nonmetals of groups 15–18 in any of their stable valence states higher than 3, 2, 1 and 0, respectively. As described in one of his first articles on the topic '*We refer to these molecules as* ***hypervalent*** *(or HV) since they involve atoms, called donor atoms, which exceed the number of valences allowed them by the traditional theory, and thus utilise more electron-pairs of bonding than provide stability in the Lewis–Langmuir theory. As hypervalent molecules have chemical formulas, and often the molecular* *structure, of the addition product of a stable molecule with two monovalent ligands or a single divalent ligand they could also be called hypermolecules or "molecules made out of molecules".'*[12]

The *p*-block abounds with examples of such hypervalent compounds which are neutral, ionic and even radicals. The N–X–L designation is used to describe hypervalent molecules where X is the central atom, N is the number of formally assignable valence electrons to the central atom and L the number of ligands/substituents directly bound to the central atom. Such compounds can have coordination numbers varying from two to six. All the known compounds with rare gases as the central atom come under the category of hypervalent molecules. Most hypervalent compounds have their structure derived from a trigonal bipyramid or octahedral geometry.[13]

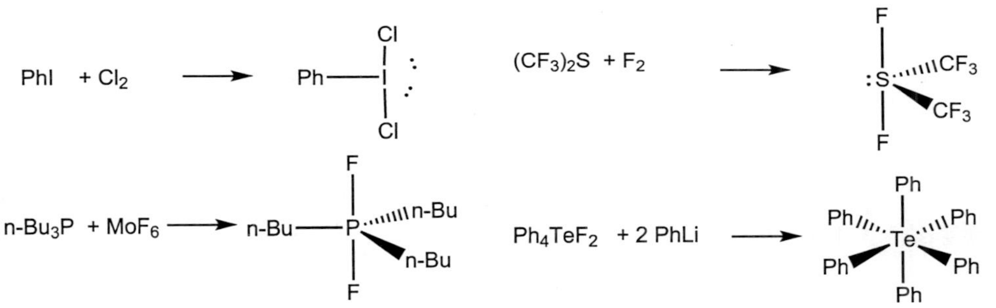

$PhI + Cl_2 \longrightarrow PhICl_2$

$(CF_3)_2S + F_2 \longrightarrow (CF_3)_2SF_2$

$n\text{-}Bu_3P + MoF_6 \longrightarrow n\text{-}Bu_3PF_2$

$Ph_4TeF_2 + 2\,PhLi \longrightarrow Ph_6Te$

Scheme 2.1 Synthesis of some hypervalent molecules

Initially, the bonding in hypervalent compounds was described in terms of sp^3d^n hybrid orbitals as opposed to sp^n orbitals used to describe the bonding as per the octet rule. However, in recent times theoretical calculations have advanced exponentially due to fast computing techniques and opposing the earlier belief, it has now been shown that the contribution of *d* orbitals in hybridisation with *s* and *p* orbitals is almost negligible for *p*-block elements of the second row (Al–Ar) and below due to the large energy difference between the relevant *p* (filled) and *d* (empty) orbitals. Also a distinction needs to be made between '*d* orbitals' in the valence bond sense and '*d* functions' that are included in the quantum mechanical calculation as polarisation functions. Theoretical calculations (*ab initio*) have clearly shown that *d* orbitals play only a very small role in bonding in the hypervalent molecules. Accordingly, other bonding models have been suggested that are in accordance with the octet rule, such as the 3-centre-4-electron (3c-4e) bond model (also known as the Rundle–Pimentel model)[7,8] or Pauling's proposal of combinations of resonance structures involving four covalent bonds and additional ionic bonds.[9] These descriptions were arrived upon in accordance with the octet rule, considers the polarity of the bonds and does not require *d* orbitals.[14-15]

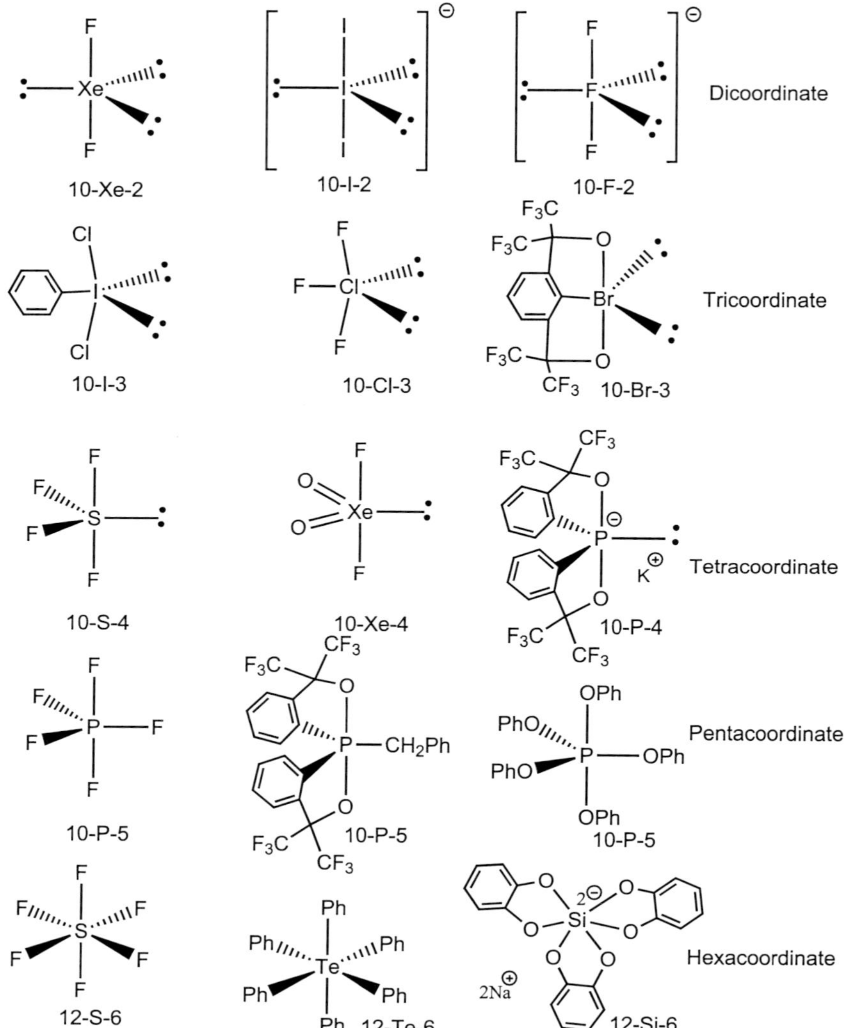

Fig. 2.7 Examples of hypervalent molecules and their N–X–L notations

2.2.2 Pauling's Resonance Model for Hypervalent Molecules

Pauling introduced a type of bonding model different from his conventional valence bond model for explaining hypervalent molecules. He introduced resonance structures in which there are only four covalent bonds and the rest are ionic bonds. Such structures imply that the bonds are polar which is in agreement with substituents which are more electronegative than the central atom. In the case of hypervalent compounds for which the substituents are more electronegative than the central atom, resonance structures can be drawn with no more than four covalent electron pair bonds and the rest completed with ionic bonds so that the octet rule is obeyed. Considering phosphorus pentafluoride (PF_5), one can write five resonance structures each with four covalent bonds and with one ionic bond with greater weight in the structures placing ionic character in the axial bonds. This satisfies the octet rule and also explains the geometry (TBP) and the variation in the axial (158 pm) and equatorial (154 pm) bond distances. In trying to explain the structure of SF_6 having six comparable S–F bonds, one can see the requirement of 15 resonance structures similar to the one given to describe six equally polar bonds such that the ionic character is distributed equally across each of the sulfur–fluorine bonds.[16]

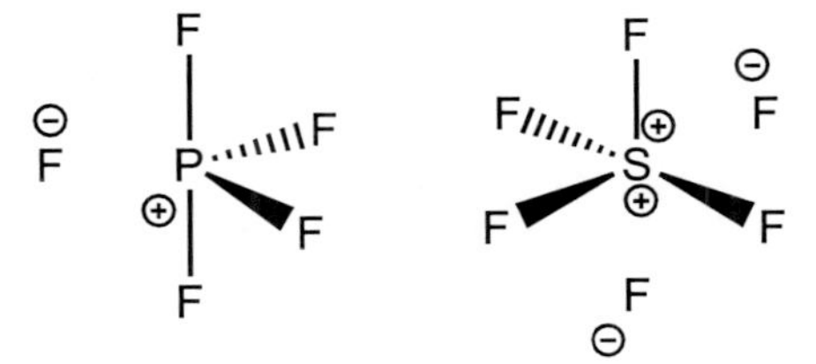

Hypervalency can therefore be considered as not a consequence of some special type of bonding. The bonds in a hypervalent molecule are similar to those in any other molecule and it can range from mostly covalent to mostly ionic.

2.2.3 Molecular Orbital Theory Applied to Hypervalent Molecules: The 3-Centre-4-Electron Bonding

Hypervalency is best explained by the 3-centre-4-electron bond (Rundle–Pimentel model).[17,18] The discovery of F_3^- which has a structure that is the same as that of I_3^- has been one of the biggest deathblows to the use of *d* orbitals in explaining the structure of hypervalent molecules as conventionally the central atom of I_3^- was assumed to have a trigonal bypyramid sp^3d geometry while the same fails for F_3^-. This is since fluorine is a first row *p*-block element and is expected not to have *d* orbitals for invoking hybridisation.[19,20] The 3-centre-4-electron bonding explains its structure in a better way.

The model considers the axial bonds of a trigonal bypyramid unit which is basically three atoms bonding in a linear fashion. The linear F–Xe–F subunit of XeF_2 or the F–F–F unit of F_3^- can be described by a set of three molecular orbitals (MOs) derived from co-linear *p*-orbitals on each atom. The Xe–F bonds result from the combination of a filled *p*-orbital in the central atom (Xe) with two half-filled *p* orbitals on the axial atoms (F). As a result we have three MOs—a bonding orbital (filled), a non-bonding orbital (filled), and an antibonding orbital (empty). The bond order for each Xe–F bond is 1/2. A similar bonding can be envisaged for F_3^- or I_3^- and the axial bonds of any hypervalent trigonal bypyramid or octahedral compounds.

In the case of the trigonal bypyramid PF_5, the three shorter equatorial bonds (153.4 pm) are described by localised orbitals formed by 2-centre-2-electron bonds. The longer axial bonds (157.7 pm) can be described by two of the three molecular orbitals formed by a single *p* orbital on phosphorus and a *p* orbital on each fluorine. The two non-bonding electrons are localised on the fluorine atoms and do not contribute to the bonding and the bonding is due to the two electrons in the bonding orbital and therefore each axial P–F bond is effectively a half bond.

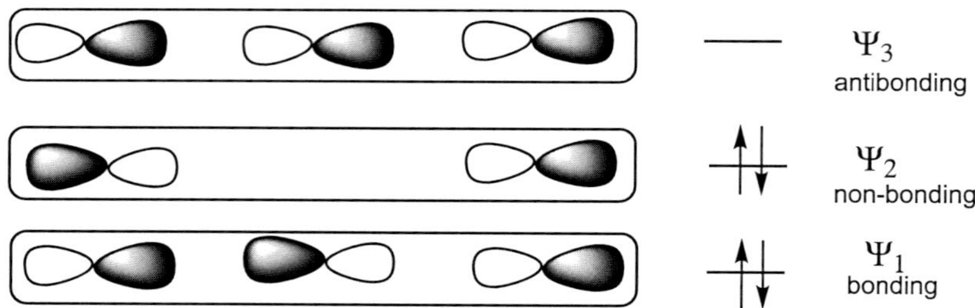

Fig. 2.8 The 3-centre 4-electron bond of F_3^- is stabilised by 46 kJ/mol compared to a state where an F_2 molecule and an F^- ion exist separately.

It should be stated that the Rundle–Pimentel model is not always completely successful and has been shown to be oversimplified, for example while accounting for the greater stability of XeF_2 (m.p. 140°C). Charge shift bonding is a type of bonding wherein the large covalent–ionic interaction energy is also a major stabilising factor in bond formation. The high stability of XeF_2 indicated by its large atomisation energy cannot be ascribed to any of its individual VB structures but mostly from an exceptionally large resonance energy arising from the mixing of several VB structures which are covalent and ionic. This model also supports the requirement that the electron must be transferred from the central atom to the ligand for hypervalency to take place. The model also stresses the importance of low ionisation energies for the central atom and high electron affinity for the ligands. For many strongly bonded hypervalent compounds such as XeF_2, the charge–shift bonding along with the 3-centre-4-electron bonding model is necessary to explain the higher stability.[21,22]

$$XeF_2 = F\bullet\!\!-\!\!-\!\!\bullet\ Xe^+F^- \longleftrightarrow F^-\ Xe^+\ \bullet\!\!-\!\!-\!\!\bullet F \longleftrightarrow F^-\ Xe^{2+}\ F^-$$

Fig. 2.9 Charge shift bonding based resonance structures of XeF_2

Extending the 3-centre-4-electron bonding, one can arrive at qualitative molecular orbital diagrams for hypervalent molecules such as SF_6 and PF_5 as explained below, by applying the ligand group orbital method.[23,24]

The SF_6 molecule belongs to the point group O_h and the *x*, *y* and *z* axes coincide with the S–F bonds. Applying point group symmetry and using the O_h character table, the valence orbitals of the sulfur atom of SF_6 can be classified as, 3*s* orbital with a_{1g} symmetry and 3*p* orbitals ($3p_x$, $3p_y$ and $3p_z$ which are degenerate) with t_{1u} symmetry. The ligand group orbitals for the F_6 part of SF_6 can be constructed from the 2*s* and 2*p* orbitals of fluorine. For a qualitative picture of the bonding, one can assume that the *s–p* separation in fluorine is large enough so that there is no notable *s–p* mixing of the orbitals. Separate sets of ligand group orbitals (LGOs) can therefore be formed for the 2*s* and 2*p* orbitals of fluorine. The 2*p* orbitals also fall under two categories, (a) the radial orbitals or those that point their lobes towards the S atom and (b) those that are tangential to the S atom. Since we are considering only σ bonds, we use only the radial orbitals.

Using the group theoretical treatment and the O_h character table, one can arrive at a reducible representation and reduce this to a sum of irreducible representations for obtaining the symmetry matching orbitals. The ligand group orbitals thus obtained for the F_6 part of SF_6 are a_{1g}, t_{1u} and e_g. The qualitative molecular orbital diagram of SF_6 can be constructed by matching the symmetry of the S valence orbitals and the LGOs of the F_6 fragment. While orbital overlap occurs for the a_{1g} and t_{1u} orbitals, the e_g set remains non-bonding. There is a total of 48 valence electrons in SF_6. In addition to the molecular orbitals shown in the MO diagram (Fig. 2.10) there are 18 more MOs accommodating electrons which are mostly of fluorine in character. The qualitative MO picture is consistent with six equivalent S–F bonds. Since the e_g orbitals are non-bonding, there are only 4 bonding pairs of electrons and therefore the bond order for SF_6 is 2/3.

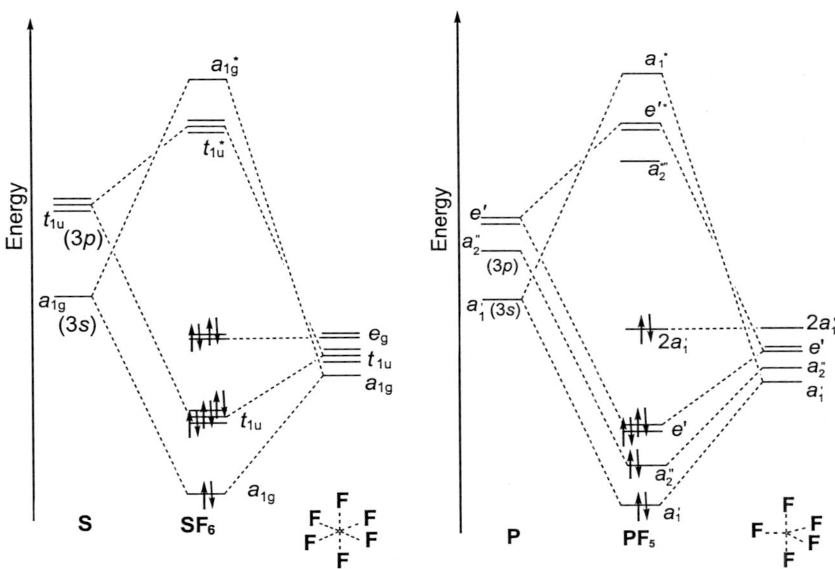

Fig. 2.10 The molecular orbital diagrams (qualitative, partial) for the formation of the hypervalent octahedral SF_6 and trigonal bypyramid PF_5 using the ligand group orbital approach with a basis set for sulfur and phosphorus composed of the 3s and 3p orbitals.

Extending a similar approach one can make the qualitative MO picture of PF_5. PF_5 belongs to the D_{3h} point group and by applying point group symmetry and using the D_{3h} character table, the valence orbitals of the phosphorus atom of PF_5 can be determined as a 3*s* orbital with a_1' symmetry, $3p_z$ orbital with a_2'' symmetry and $3p_x$, $3p_y$ orbitals with e' symmetry. The ligand group orbitals obtained are a_1', a_2'', e' and $2a_1'$. While orbital overlap occurs for the a_1', a_2'' and e' orbitals, the $2a_1'$ orbital remains non-bonding. Since there are only four bonding pairs, the bond order in PF_5 is 4/5.

2.2.4 Significance and Consequences of Hypervalency

Apicophilicity and pseudorotation

Apicophilicity is observed in hypervalent compounds which are pentacoordinate and tetracoordinate with an unshared electron pair. In such compounds, the apical bonds are the 3c-4e bonds and the apical substituents hold excess electrons. As a result, substituents which are more electronegative (electron withdrawing) and sterically less hindered preferentially occupy the axial positions. Steric effect is felt more at the axial positions than at the equatorial positions in TBP molecules. The experimentally determined order of apicophilicity of substituents in pentacoordinate phosphorus is F > H > OR ~ OH > Cl > NR_2 > Ph > O > CH_3.

Bents rule and berry pseudorotation

For trigonal bipyramidal molecules such as PCl_2F_3, the empirical Bents rule provides guidance for assigning the axial and equatorial substituents. Bents rule says that *more electronegative substituents prefer hybrid orbitals having less* s *character.* An sp^3d hybrid orbital set may be considered as a combination of $p_zd_z^2$ and sp_xp_y hybrid orbitals. As a consequence one can assign the fluorines to the axial positions, and the chlorines (which are less electronegative) to the equatorial positions.

One of the major differences between organic and inorganic chemistry is the relative emphasis placed on structure and reactivity. Structural organic chemistry is relatively simple as it is based on digonal, trigonal or tetrahedral carbon atoms leading to rather simple ways of looking at reaction mechanisms. In contrast, inorganic chemistry has a wider variety of structural types to consider and even for a single element there are many factors to consider. With the advent of multinuclear and variable temperature NMR spectroscopy and similar analytical tools, it has become possible to look into the dynamic nature of the structures of many inorganic compounds. The fluxionality shown by simple molecules is one such example. Inversion of ammonia is one such case. A more picturesque example is shown in the Berry pseudorotation by trigonal bipyramidal phosphorus compounds.[26] Berry pseudorotation is the facile interconversion of axial and equatorial substituents in AX_5 type of molecules without any bond-breaking. The trigonal bipyramid molecule converts to another trigonal bipyramid unit through a square pyramid intermediate. In this dynamic process, the axial bonds of the TBP become equatorial and two of the equatorial bonds become axial and the process keeps repeating. In certain molecules this process can be slowed down by cooling and can be monitored by NMR spectroscopy.

For a trigonal bipyramidal shaped molecule such as PF_5, since the environment for the axial and equatorial fluorines are different, one would expect different chemical shifts in the ^{19}F-NMR spectra of these molecules. However, in reality, it is observed that the axial and equatorial fluorines are indistinguishable by ^{19}F-NMR (what is observed is a simple doublet for all the five fluorines which are indistinguishable and which gets split due to coupling with the NMR active phosphorus core). In contrast, for PCl_2F_3, one observes that in the fluorine NMR spectra taken at room temperature, there is only one kind of fluorine but when measured at –143°C one gets two sets of peaks (a doublet of a doublet for the two axial fluorines and a doublet of a triplet for the equatorial fluorine at –143°C, Fig. 2.11). This shows that the axial and equatorial fluorines are exchanging positions at room temperature faster than the NMR technique can follow. One envisages the intermediate situation as a square pyramidal structure.

A similar fluxionality is shown by the T-shaped molecule ClF_3 (Fig. 2.12). At 60°C, it shows a single peak for the axial and equatorial fluorines in its ^{19}F-NMR indicating Berry pseudorotation. On measuring the NMR spectra at –60°C, the singlet splits into a doublet (–4 ppm) and a triplet (116 ppm).[27,28] Berry pseudorotation has also been reported for SF_4.[29]

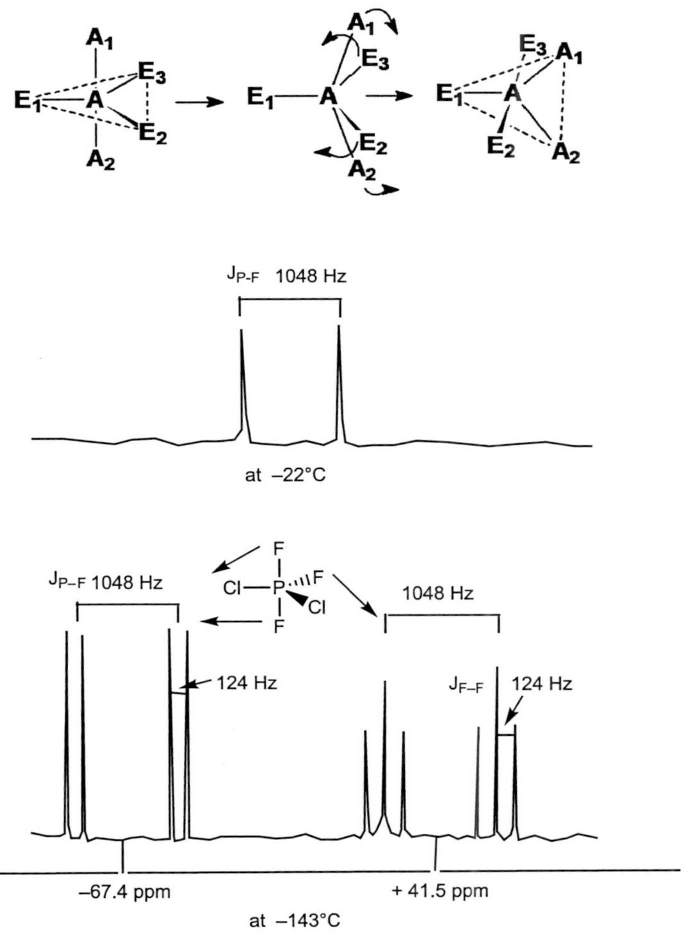

Fig. 2.11 Variable temperature ^{19}F-NMR spectra of PCl_2F_3

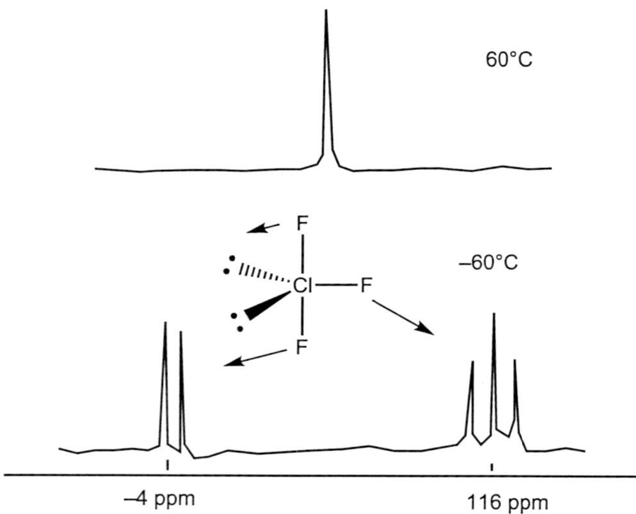

Fig. 2.12 Variable temperature ^{19}F-NMR spectra of ClF_3

Increased reactivity of hypervalent silicon compounds

Unlike tetrahedral silicon compounds, penta-coordinate hypervalent compounds display increased reactivity and this property is made use of in reactions such as Hiyama coupling of aryl halides with organosilicon compounds, and those reactions involving desilylation in the presence of fluoride ions or a strong base.[30,31]

Fig. 2.13 Comparing the hydrolytic stability of normal and hypervalent silicon compounds

Isolation of intermediates

The main difference between an S_N2 reaction on a carbon centre and an S_N2 reaction of a silicon or phosphorus centre is that the nucleophile added intermediate is often stable and even isolable in the latter while it is only a short lived transition state in the former.[32]

Stabilisation of main group elements in unusually low oxidation states

By making use of tailor-made multi-donor sites having anionic ligands, it is possible to stabilise main group elements in unusually low oxidation states. A fine example has been shown by Anthony J Arduengo who stabilised phosphorus and arsenic in the +1 oxidation state by using a suitable substituent having multiple donor sites which stabilises the central atom through hypervalent bonding.[33]

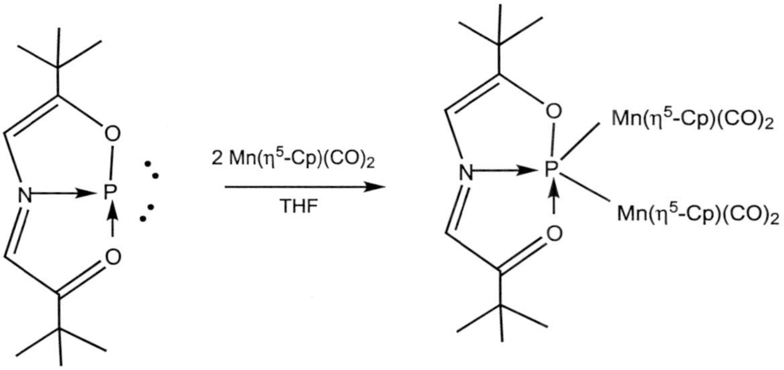

2 Mn(η⁵-Cp)(CO)₂

THF

Mn(η⁵-Cp)(CO)₂

Mn(η⁵-Cp)(CO)₂

Fig. 2.14 Stabilisation of P(I) by using a tailored substituent and hypervalency

2.3 Participation of *d* Orbitals in Bonding of *p*-Block Compounds: Current Status

The participation of *d* orbitals in describing the structures of many *p*-block element based compounds is a topic which has been the point of disagreement between theoretical chemists and experimentalists. The topic arose initially out of a particular qualitative intricacy in explaining the structure and bonding of hypervalent molecules. For example, to explain the bonding of SF_6 a very simple model in which one requires six equivalent bonds and the six sulfur σ bonds originating from six equivalent sp^3d^2 hybrid orbitals was envisaged. In this, the *d* functions originating from unoccupied outer *d* orbitals were possibly thought to result from the expansion of the valence shell of S by the promotion of electrons into these low lying *d* orbitals.[34] A similar approach was used for invoking sp^3d hybrid orbitals to explain the bonding of trigonal bypyramidal hypervalent molecules such as PF_5. The participation of *d* orbitals was also used to describe the π-type bonding phenomenon observed in phosphorus–oxygen and phosphorus–sulfur bonds of phosphine oxides, sulfides and phosphates, and the phosphorus ylide bond in the well known Wittig reagents. In addition, for explaining the VSEPR violations in the actual structures of silicon compounds such as $N(SiH_3)_3$, H_3SiNCS and so on, $d\pi$–$p\pi$ bonding was used. A comparison of earlier explanations based on *d* orbital participation with current theory supported explanations of bonding in some selected molecules is given in Table 2.1.

In a related explanation, the presence of available low lying *d* orbitals was quoted as the reason for the difference in chemistry between the first row (B to Ne) and the *p*-block elements of the lower rows, the classic case often summoned being the hydrolytic stability of CCl_4 versus the instability of $SiCl_4$. The hybridisation theory of bonding proposed by Pauling could also predict the shape of molecules (molecular stereochemistry) since bond directions are determined by the relative orientations of the lobes of the hybrid orbitals. One of the reasons why the bonding model of main group compounds invoking full participation of *d* orbitals was accepted without hesitation by the chemistry community till recent years, was its simplicity assisted by the extensive use of *sp* to sp^3 hybridisation by organic chemists to explain the structure and properties of organic compounds and the universal use of *d* orbitals in explaining colour, distortion and magnetic properties of transition metal complexes. Moreover, at the time this was proposed (1940–1950), detailed computational studies were almost nonexistent and there was no way of verifying the models proposed by detailed calculations using computers.[35,36]

Table 2.1 Explaining the bonding of hypervalent molecules with and without having *d* orbital participation

***p*-block compound and its property**	**Earlier explanation suggested based of participation of empty *d* orbitals**	**Modern explanation supported by theoretical studies where *d* orbitals are not used.**
PF_5 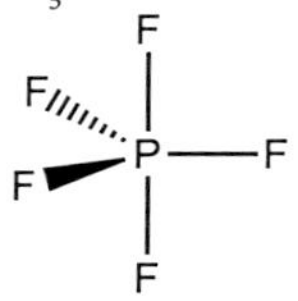	Assuming promotion of electrons to vacant higher energy *d* orbitals resulting in sp^3d hybridisation	Invoking 3c-4e bonding of the axial bonds: wavefunctions described as superpositions of resonance structures, some of which obey the octet rule (ionic) and some break the octet rule (covalent)
SF_6 	Assuming promotion of electrons to vacant higher energy *d* orbitals resulting in sp^3d^2 hybridisation	Invoking two 3c-4e bonds involving delocalisation of one of the two bonding pairs on to the substituents and resulting in a charge distribution of +1 on the central atom and −0.5 on each of the substituents
XeF_2 Structure and unusual stability 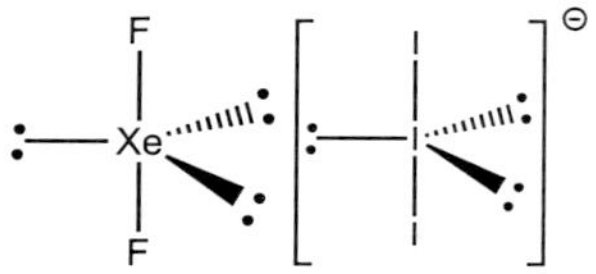	sp^3d hybridisation with equatorial positions of the trigonal bypyramid occupied by lone pairs	3-centre-4-electron bonding along with charge shift bonding. Extra stability due to charge shift bonding which are electron pair bonds with very large ionic–covalent resonance energies

Cont'd

Table 2.1 Cont'd

p-block compound and its property	Earlier explanation suggested based of participation of empty *d* orbitals	Modern explanation supported by theoretical studies where *d* orbitals are not used.
Phosphine oxide, sulfur dioxide 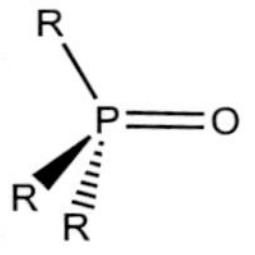 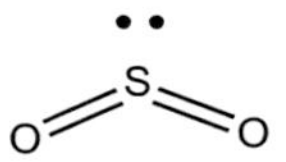 (R, P=O, R, R; S, O, O)	Covalent compounds having $p\pi$–$d\pi$ bonds resulting in shorter P–O and S–O bonds	Covalent bonds but not involving *d* orbitals. The back bonding of electron density to the σ^* acceptor orbitals of phosphorus or sulfur which are the LUMOs. An example of negative hyperconjugation effect (*n* to σ^*)
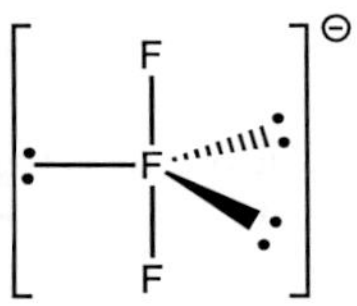 [F, F, F]⊖	No explanation possible based on *d* orbitals since fluorine, a first row *p*-block element has no low lying *d* orbitals	3-centre-4-electron bonding
General lack of hypervalency among the first row *p*-block elements	Absence of *d* orbitals of lower energy	Higher electronegativity of the first row *p*-block elements and smaller size of these elements
$H_3SiOSiH_3$ and $N(SiH_3)_3$ structures have large bond angles around oxygen/nitrogen, structures not predicted by VSEPR rules	$p\pi$– $d\pi$ bonds between silicon and oxygen/nitrogen reduces the effect of lone pair on geometry	VSEPR rules are applicable only if substituents have higher or comparable electronegativities as the central atom. Large bond angles are due to weak pairing of electrons on oxygen as a result of weakly electronegative SiH_3 group as well as electrostatic repulsions between bulky SiH_3 groups

Before the 1970s, theoretical discussions on bonding in hypervalent *p*-block compounds were mostly speculative in nature as sufficiently accurate wavefunctions could not be calculated. Recent years have witnessed an exponential growth in fast computing and computational software which have made computational analysis easier. The wavefunctions thus obtained are of sufficient quality to reproduce molecular geometries and to provide reliable bonding interpretations. The outcome of a series of such theoretical analyses on hypervalent compounds, which includes electron density analysis, natural population and natural bond orbital analysis, shared electron number and occupation number analysis, is that the traditional dsp^3 and d^2sp^3/sp^3d^2 models of bonding of the main group compounds have been found to be no more valid. The calculations indicate that the total *d* orbital occupancies are at the most 0.3e. The *d* orbitals can act as acceptor orbitals for back bonding from substituents. These studies also lay emphasis on the partial ionic character of these bonds. Also, by a combination of covalent and ionic bonding, the electronic octet rule is not violated.[37]

The 3-centre-4-electron bond theory gave a reasonably acceptable bonding picture for hypervalency without using the *d* orbitals. Rather than having a valence role, the *d* orbitals seem to have a role only as 'polarisation functions' by compensating for the limitations imposed on the molecular wave functions when it is restricted only to *s* and *p* functions centred on the nuclei. The *d* orbitals can be used as angular momentum functions to perturb the shape of the *p* orbitals.

The π acceptor ability of phosphines and silyl systems has been better explained by bringing in the negative hyperconjugation mechanism. This utilises backbonding from the filled *p* orbitals of the substituents to the empty σ^* orbitals on the phosphorus or silicon. The multiple bonds associated with traditionally hypervalent molecules such as $R_3P{=}O$, $O{=}S{=}O$, ClO_4^- are explained by bringing in alternate ionic valence bond structures such as $R_3P^+O^-$ and by negative hyperconjugation.

The absence or reluctance for forming hypervalent compounds by elements such as boron, nitrogen, oxygen and fluorine (first row of the *p*-block) has been one of the major arguments used in favour of *d* orbital participation as these elements do not have low lying *d* orbitals (they have only 2*s* and 2*p* orbitals). Two major factors disfavour the formation of hypervalent compounds by this row of *p*-block elements. One is the significantly small size of these elements resulting in high electron density on the atom. The second is the relatively high electronegativity of these atoms compared to those of the lower rows which also indicates increased electron density.

2.4 The Inert Pair Effect

The inert pair effect, first proposed by Sir Nevil Sidgwick in his book ***The electronic theory of valency*** tries to explain why the most stable oxidation states of some of the heaviest main group elements, especially of groups 13, 14 and 15, are two units less than the common oxidation state shown by the earlier elements of the same group. For example, for thallium, the most stable oxidation state is +1 unlike +3 shown by boron, aluminum, and so on. The same is true for Pb (+2) and Bi (+3). In other words, unlike the elements of the first few rows, which usually use all their valence electrons in compound formation, the heavier elements of group 13, 14 and 15 form more stable compounds when two of the valence electrons are not taking part in bonding (for example, TlCl, $PbCl_2$ and $BiCl_3$). Sidgwick postulated the existence of an *inert pair* of electrons in the heavier elements, thallium, lead and bismuth. The relevant part of the discussion in his 1927 book wherein he recognises 'two valence electrons to have become inert' reads as follows.[38]

'There is however one considerable class of abnormal valencies which can be recognised as being due to some common cause, that, namely, in which two of the valency electrons appear to have become inert, as though they had been absorbed into the core. This phenomenon is first to be noticed in group III B in InCl, and more markedly in the thallous compounds: in the typical and B elements of group IV it is especially prominent, from the divalent compounds of carbon to those of germanium, tin, and lead. In group V B we find it in antimony and still more in bismuth, in VI B in sulfur, selenium, and tellurium, and in VII B in iodine and bromine and possibly in chlorine. Thus it is most marked in the fourth group, and is generally prevalent among the heavier B elements. It is especially evident in the series of elements following gold:

mercury, thallium, lead, and bismuth. The new development of the Bohr theory due to Stoner and Main Smith indicates a possible reason for this peculiarity. We now realise that the first two electrons in any group, forming the grouplet N_{11}, correspond to the pair of l_{11} electrons in helium, and can have a certain completeness of their own, approaching that which they have in helium, where they form a complete group. We might thus anticipate that under some conditions the first two valency electrons of an element could become more like core-electrons, and refuse either to ionise, or to form covalencies, or both. How their presence in this state would react on the rest of the valency group whether they would have to be counted as part of it, or would change the conditions of its stability, or finally should be regarded as having no connection with it and genuinely forming a separate group of the core cannot at present be predicted; but something may be learnt by examining the chemical facts. Why this inertness of the N_{11} pair should appear precisely where it does in the periodic table, we cannot say: the spectral evidence supports the chemical, but no reason has been assigned for either.' Sidgwick N V, *The electronic theory of valency*, Oxford, Clarendon Press, 1927, p 178.

Grimm and Sommerfeld had attributed the inertness to stabilisation resulting from the completion of the *s* shell of electrons, while others have accredited it to stabilisation as a consequence of the *lanthanide contraction*. Although the inert pair effect is readily invoked to correlate the unusual behaviour of these elements, an adequate theoretical explanation has not yet been provided. For example, one should expect the ionisation potentials for these stabilised elements to indicate that a larger amount of energy is necessary to remove these inert pair electrons from the heavier atoms than from the lighter ones. The inert pair effect was thermodynamically evaluated by Drago in 1957 and a qualitative, quantum mechanical study for this phenomenon was carried out.[39] Drago noted from the values of ionisation potentials (IP) available for these elements, that *s* electrons are stabilised to a larger extent for fourth row elements than for sixth row elements. As shown in the graph (Fig. 2.15), although there is a slight increase in IP from the 5th to the 6th period of the periodic table, there is even more increase in IP for germanium and gallium. Inertness of the 6*s* electrons should be reflected in noticeably larger 6*s* ionisation potentials. Since the inert pair effect is not encountered in the compounds of gallium or germanium, the explanation advanced by Grimm and Sommerfeld that stabilisation of *s* electrons occurs is not adequate.

From a study of thermodynamic data available for group 13 and 14 chlorides, it is observed that one can account for the relative instability of $TlCl_3$ and $PbCl_4$ from the fact that in a particular family, the strength of the covalent bonds decreases as the atomic number increases. This decrease in covalent bond forming ability of such elements with increase of atomic number can be attributed to the following effects: (i) The radial part of the bond orbitals gets changed in such a way that the valence electrons are spread over a larger area resulting in less overlap. (ii) When the number of inner shell electrons increases, the coulombic repulsion between these electrons and the inner electrons of the other bonded atoms also increases. The higher solvation energy of gallium(III) has been attributed as the reason for the greater stability of gallium(III) compounds compared to thallium(III) compounds.

Sanderson explored the effect of inert pair on electronegativity by using polar covalence and calculated the electronegativity of elements in oxidation states lower than their maximum, from experimental bond energies. He observed that by reducing the effective nuclear charge, lone pair electrons significantly reduce the electronegativity for the low-valent forms.[41] For example, the value on the relative compactness scale for thallium(I) is only 0.99 compared to 2.25 for thallium(III); for lead(II) it is 1.92 compared to 2.29 for lead(IV). This results in the bond polarity being higher than normal in compounds showing the lower oxidation states. This explains the increased stability as well as changes in physical and chemical properties.

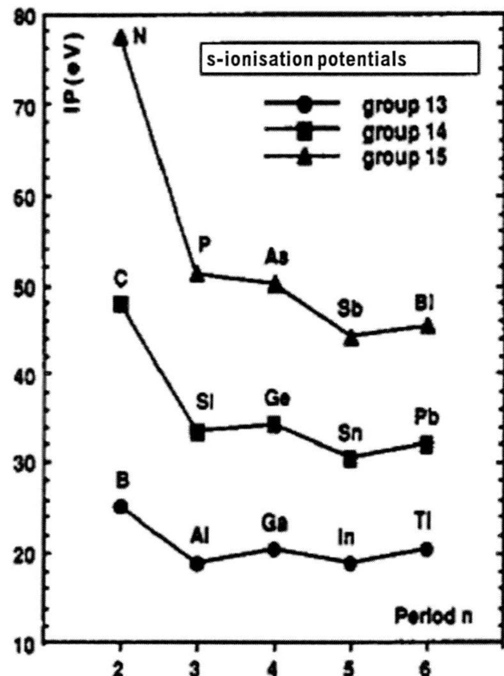

Fig. 2.15 Experimental first ionisation potentials for the group 13, 14 and 15 elements. Reprinted with permission from: Peter Schwerdtfeger, Graham A Heath, Michael Dolg et al., *J. Am. Chem. Soc.*, Vol. 114, 1992, 7518–7527. Copyright (2012) American Chemical Society [40]

Extensive ab initio theoretical computations were carried out by Schwerdtfeger, Bennet and coworkers in 1992 in order to study the occurrence of low valencies in the chemistry of heavy elements. They calculated a number of molecular properties, such as bond lengths, dipole moments, and dissociation energies, for a range of inorganic molecules. Their work further supported the views given by Drago in 1957.[40] Their study indicated that low valences in heavy main group elements arise naturally as a consequence of the periodic trend toward lower M–X bond strength with increasing atomic number. They concluded that although relativistic contributions play an important role in the chemistry of the 6th period elements, there is no evidence that the 6*s* electrons are more inert than the *s* electrons of lighter elements. Therefore Schwerdtfeger, Bennet are of the opinion that there is no special *inert pair effect* for the 6th period elements, suggesting that it is inappropriate to use this term to designate the low valencies of heavier elements.

2.5 Resurgence of Dative Bonds in *p*-Block Compounds

To use or not to use dative bonds (A →B) in describing the structures of compounds based on *p*-block elements has been a topic under considerable discussion recently, especially after the discovery of low valent and zerovalent compounds of main group elements that are stabilised using strong σ-donors such as N-heterocyclic carbenes (NHCs) and cyclic alkyl amino carbenes (cAACs).[42–44] Historically in 1929, it was Sidgwick who introduced the notation of the arrow for dative bonds or donor acceptor bonds, A→B to differentiate them from electron-sharing bonds A–B (often termed covalent bonds). It was widely used in depicting metal–ligand interaction of early coordination compounds. However, Linus Pauling, in his classical book ***The nature of the chemical bond***, discouraged the use of the dative bonds because it did not fit into his scheme of convenience as he preferred to depict compounds by resonance structures and use of formal charges on the atoms, wherever required. Pauling's method of using resonance structures had a dominating influence on the way structures were depicted in chemical literature until recently. The difference between dative bonds and electron sharing (covalent) bonds were only rarely displayed and was mostly restricted to the depiction of classical Lewis acid–base adducts. The strength of a dative bond between two atoms A→B is considered less compared to an electron sharing bond A–B between the same atoms. In the case of main group metal compounds, it has been suggested that a dative bond breaks heterolytically while an electron-sharing bond breaks homolytically.[45]

Recent interest in the necessity of using dative bonds came up after the discovery that many main group elements can be stabilised at low valency and low oxidation states by strong σ-donors such as N-heterocyclic carbenes. This also led to a re-look at some of the earlier known analogous compounds that were classically depicted as allenes or heteroallenes and therefore expected to be linear around the central atom. Instead, they were bent and their reaction chemistry was better explained by considering the central atom as being in the zero oxidation state. The early examples include carbon suboxide (O=C=C=C=O versus OC→C←CO), which was linear in the solid state but bent in the vapour phase with an angle of 156° (as determined by infrared spectral studies and agreeing with theoretical calculations). A second earlier example is carbodiphosphorane ($Ph_3P{=}C{=}PPh_3$ versus $Ph_3P{\rightarrow}C{\leftarrow}PPh_3$), which is best described as a complex of carbon atom in the excited ^1D state with two phosphine ligands and was found to possess a bent structure with an angle of 131.7°. Some of the recently prepared molecules whose structures need to be depicted using dative bonds, are given in Fig. 2.16. Compounds having a central carbon flanked by two carbenes are described as carbones (also called carbodicarbenes); these have divalent carbon in the zero oxidation state with two lone pairs of electrons on the carbon. These compounds show bent geometry at the central carbon while a classical description which considers the octet rule and electronegativity differences suggests a linear structure. The carbone (carbodicarbene) shown in Fig. 2.16 was called a bent allene initially, but the central carbon atom has been found to donate a pair of electrons to $RhCl(CO)_2$ acting as an η^1-ligand whereas regular allenes bind to metals by the η^2-mode. The two NCN planes of the carbene units of this carbone are also not mutually perpendicular as expected for classical allenes. These features forced a better nomenclature for this compound—carbone possessing a divalent carbon(0) with two NHC ligands.[46] Silylones, germylones and analogous tin compounds with Si, Ge and Sn in the zero oxidation state having a bent geometry and similar features to the carbone have also been prepared and structurally characterised, thus indicating that such compounds are better depicted by the use of dative bonds. Indeed, the reaction chemistry shown by such compounds further supports the fact that the central atom is in the zero oxidation state and possess two pairs of electrons (Eq. 2.1).[45]

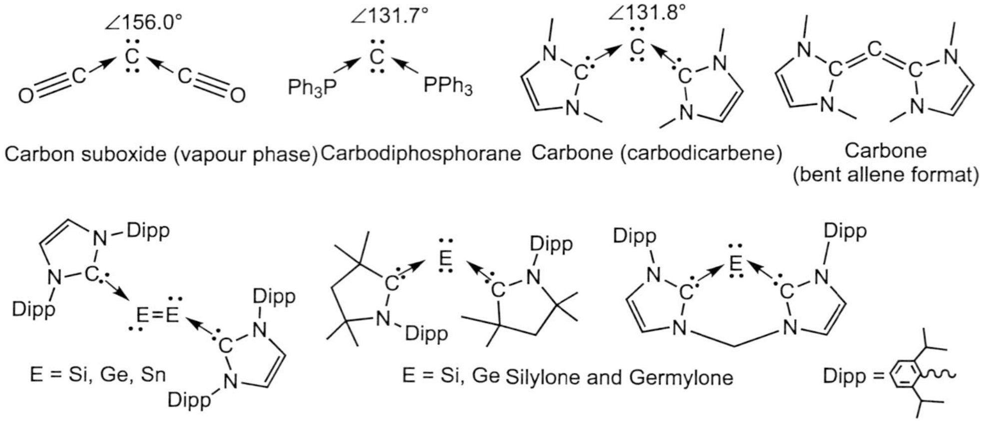

∠156.0° ∠131.7° ∠131.8°

Carbon suboxide (vapour phase) Carbodiphosphorane Carbone (carbodicarbene) Carbone (bent allene format)

E = Si, Ge, Sn E = Si, Ge Silylone and Germylone Dipp =

Fig. 2.16 Compounds of carbon, silicon, germanium and tin in zero oxidation state depicted using dative bonds

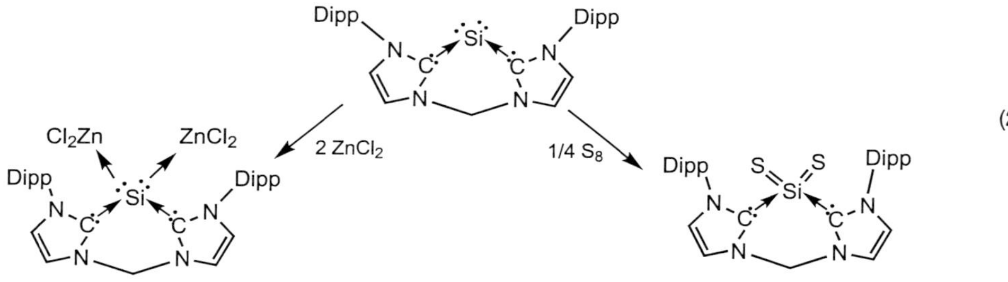

2 $ZnCl_2$ 1/4 S_8 (2.1)

Questions for self study (multiple choice)

2.1. In which of the following is the bond angle maximum?
(a) NH_3 (b) PCl_3 (c) SCl_2 (d) NH_4^+

2.2. Which of the following pairs are isostructural?
(a) XeF_2, IF_2^- (b) NH_3, BF_3 (c) CO_3^{2-}, SO_3^{2-} (d) PCl_5, ICl_5

2.3. Which of the following pairs are isoelectronic?
(a) CO_2, NO_2 (b) NO_2^-, CO_2 (c) CN^-, CO (d) CO_2, SO_2

2.4. In NO_3^-, the number of bond pairs and lone pairs on the nitrogen are:
(a) 2, 2 (b) 3, 1 (c) 1, 3 (d) 4, 0

2.5. Which of the following molecules has the maximum dipole moment?
(a) CO_2 (b) CH_4 (c) NH_3 (d) NF_3

2.6. Which of the following has a linear shape?
(a) O_3 (b) NO_2^- (c) SO_2 (d) NO_2^+

2.7. Which of the following is a polar molecule?
(a) SiF_4 (b) XeF_4 (c) SF_4 (d) BF_3

2.8. Which among the following is not a hypervalent molecule?
(a) SF_4 (b) XeF_4 (c) PCl_4^+ (d) ICl_4^+

2.9. The extra stability and high melting point of XeF_2 is best explained by
(a) sp^3d hybridisation (b) ionic bonding
(c) 3c-4e bonding (d) large ionic–covalent resonance energies

Solved concept-based questions

2.10. Why is the concept of lone pairs being replaced by lone pair domains in the VSEPR postulates?

An electron pair domain is defined as a region in space where there is an enhanced probability of finding an electron pair. The electron pair domain extends around the most probable position of finding an electron pair as determined by the Pauli principle. The probability decreases on increasing the distance from the most probable position.

A lone pair domain can in general be thought of as more spread out and larger than a bonding pair domain as the electrons of a lone pair are subject to the attraction of only one atomic nucleus while those of a bonding pair are subject to the attraction of two nuclei taking part in the bond formation. Why a lone pair occupies more space than a bond pair was poorly explained by the earlier model. The electron pair domain model also is in conformity with probability which is in tune with modern chemical bonding theories.

2.11. Many properties of main group compounds explained by using $d\pi$–$p\pi$ bonding or participation of empty *d* orbitals in bonding are not acceptable any more. Why?

The participation of *d* orbitals was brought in to provide an explanation to structures and bonding of hypervalent main group compounds. This was just a model with no theoretical support when it was introduced. Theoretical discussions were speculative in nature as sufficiently accurate wavefunctions could not be calculated. Recent years have witnessed an exponential growth in fast computing and modern computational software which have made computational analysis easier. Wavefunctions obtained are of sufficient quality to reproduce molecular geometries and to provide reliable bonding interpretations. The outcome of a series of such theoretical analyses on hypervalent compounds, which include electron density analysis, natural population and natural bond orbital analysis, shared electron number and occupation number analysis is that the traditional dsp^3 and d^2sp^3/sp^3d^2 models of bonding of the main group compounds have been found to be invalid or couldn't be substantiated by theory.

Problems and exercises

2.12. Arrange the following in the increasing order of F–X–F bond angles based on VSEPR rules.

ClF_3, BrF_3, $PhIF_2$

2.13. Arrange the following in the increasing order of X–S–X (X = halogen) bond angles based on VSEPR rules.

$SOCl_2$, SOF_2, $SOBr_2$

2.14. Arrive at the shapes of the following molecules using VSEPR theory.

NSF_3, IOF_4^-, $IO_2F_5^{2-}$, $Kr_2F_3^+$

2.15. Predict and draw the ^{19}F-NMR spectrum of SF_4 at low temperature and high temperature assuming it shows Berry pseudorotation.

2.16. What is the difference between lone pairs and lone pair domains?

2.17. Which among the following compounds are hypervalent? Indicate the NX L notation for the same.

NH_4^+, ICl, IF_7, TeF_6

2.18. What is charge shift bonding? Give a molecule on which stability is better explained using charge shift bonding.

2.19. How is the bonding explained for the hypervalent molecule PF_5 and SF_6 according to molecular orbital model and Pauling's valence bond model?

2.20. How does one account for the large Si–O–Si bond angles in silyl ethers without invoking dπ–pπ bonding?

Answer key for self study questions

2.1. (d) 2.2. (a) 2.3. (c) 2.4. (d) 2.5. (d) 2.6. (d) 2.7. (c) 2.8. (c) 2.9. (d)

Supplementary reading

1. Gillespie R J and Nyholm R S, Inorganic stereochemistry, *Quart. Rev. Chem. Soc.*, 1957, Vol. 11, 339.
2. Sidgwick N V and Powell H M, Bakerian Lecture, Stereochemical types and valency groups, *Proc. Roy. Soc. A*, 1940, Vol. 176, 153.
3. Gillespie R J, The electron-pair repulsion model for molecular geometry, *J. Chem. Educ.*, 1970, Vol. 47, 18.
4. Grushow A, Is it time to retire the hybrid atomic orbital? *J. Chem. Educ.*, 2011, Vol. 88, 860.
5. Grushow A, In response to those who wish to retain hybrid atomic orbitals in the curriculum, *J Chem. Educ.*, 2012, Vol. 89, 578.
6. Gillespie R J, *Molecular geometry*, Van Nostrand, London, 1972.
7. Gillespie R J, Fifty years of the VSEPR model, *Coord. Chem. Rev.*, 2008, Vol. 252, 1315.
8. Gillespie R J and Robinson E A, Electron domains and the VSEPR model of molecular geometry, *Angew Chem. Int. Ed.*, 1996, Vol. 35, 495.

9. Gillespie R J and Robinson E A, Models of molecular geometry, *Chem. Soc. Rev.*, 2005, Vol. 34, 396.
10. Gillespie R J and Poplier P L A, *Chemical bonding and molecular geometry: From Lewis to electron densities*, OUP, NY, 2001.
11. Gillespie R J, Teaching molecular geometry with the VSEPR model, *J. Chem. Educ.*, 2004, Vol. 81, 298.
12. Musher J I, The chemistry of hypervalent molecules, *Angew Chem. Int. Ed.*, 1969, Vol. 8, 54.
13. Jenson W B, The Origin of the term hypervalent, *J. Chem. Educ.*, 2006, Vol. 83, 1751.
14. Curnow O J, A simple qualitative molecular orbital/Valence bond description of bonding in main group hypervalent molecules, *J. Chem. Educ.*, 1998, Vol. 75, 910.
15. Gillespie R J, The octet rule and hypervalence: Two misunderstood concepts, *Coord. Chem. Rev.*, 2002, Vol. 233–234, 53.
16. Durrant M C, A quantitative definition of hypervalency, *Chem. Sci.*, 2015, Vol. 6, 6614.
17. Pimentel G C, The bonding of trihalide and bifluoride ions by the molecular orbital method, *J. Chem. Phys.*, 1951, Vol. 19, 446.
18. Rundle R E, Electron deficient compounds II Relative energies of half bonds, *J. Chem. Phys.*, 1949, Vol. 17, 671.
19. Ault B S and Andrews L, Infrared and Raman spectra of the $M^+F_3^-$ ion pairs and their mixed chlorine–fluorine counterparts, *Inorg. Chem.*, 1977, Vol. 16, 2024.
20. Braida B and Hilberty P C, What makes the trifluoride anion F_3^- so special? A breathing orbital valence bond ab initio study, *J. Am. Chem. Soc.*, 2004, Vol. 126, 14890.
21. Shaik S, Danovich D, Wu W and Hilberty P C, Charge–shift bonding and its manifestations in chemistry *Nature Chem.*, 2013, Vol. 1, 443.
22. Braida B and Hilberty P C, The essential role of charge shift bonding in hypervalent prototype XeF_2, *Nature Chem.*, 2009, Vol. 5, 417.
23. Housecroft C E and Sharpe A G, *Inorganic Chemistry*, 4th Ed., Pearson, 2012.
24. Hoffmann R, Howell J M and Muetterties, E L, Molecular orbital theory of pentacoordinated phosphorus, *J. Am. Chem. Soc.*, 1972, Vol. 94, 3047.
25. Berry R S, Correlation of rates of intermolecular tunneling processes with application to some group V compounds, *J. Chem. Phys.*, 1960, Vol. 32, 933.
26. Holmes R R, Cartel R P and Peterson G E, Molecular structures of PCl_4F, PCl_3F_2 and PCl_2F_3. Pure chlorine nuclear quadrupole resonance and low temperature ^{19}F-nuclear magnetic resonance spectra, *Inorg. Chem.*, 1964, Vol. 3, 1748.
27. Alexakos L G and Cornwell C D, Nuclear magnetic resonance (NMR) spectra of ClF_3 and ClF: gaseous spectra and gas to liquid shifts, *J. Chem. Phys.*, 1964, Vol. 41, 2098.
28. Minayev R M, Pseudorotation in ClF_3, *Chem. Phys Lett.*, 1992, Vol. 196, 203.
29. Bacon J, Gillespie R J and Quail J W, The nuclear magnetic resonance spectrum of sulfur tetrafluoride, *Can J. Chem.*, 1963, Vol. 41, 1016.
30. Johnson S E, Deiters J A, Day R O and Holmes R R, Pentacoordinated molecules. 76. Novel hydrolysis pathways of dimesityldifluorosilane via an anionic five-coordinated silicate and a hydrogen-bonded bisilonate. Model intermediates in the sol-gel process, *J. Am. Chem. Soc.*, 1989, Vol. 111, 3250.
31. Dieters J A and Holmes R R, Enhanced reactivity of penta-coordinated silicon species. An ab initio approach, *J. Am. Chem. Soc.*, 1990, Vol. 112, 7197.
32. Corriu R J P, Guerin C, Henner B J L and Wong Chi Man W W C, Pentacoordinated silicon anions: reactivity towards strong nucleophiles, *Organometallics*, Vol. 1988, 7, 237.
33. Arduengo A J (III) and Stewart C A, Low coordinate hypervalent phosphorus, *Chem. Rev.*, 1994, Vol. 94, 1215.

34. Reed A E and Weinhold F, On the role of *d* orbitals in SF_6, *J. Am. Chem. Soc.*, 1986, Vol. 108, 3586.
35. Gilheany D G, No *d* orbitals but Walsh diagrams and may be banana bonds: Chemical bonding in phosphines, phosphines oxides and phosphonium ylides, *Chem. Rev.*, 1994, Vol. 94, 1339.
36. Reed A E and Schleyer P V R, Chemical bonding in hypervalent molecules: The dominance of ionic bonding an negative hyperconjugation over *d* orbital participation, *J. Am. Chem. Soc.*, 1990, Vol. 112, 1434.
37. Magnusson E, Hypercoordinate molecules of second row elements: *d* functions or *d* orbitals?, *J. Am. Chem. Soc.*, 1990, Vol. 112, 7940.
38. Sidwick N C, *The electronic theory of valency*, Oxford: Clarendon Press, 1927, 178.
39. Drago R S L, Thermodynamic evaluation of the inert pair effect, *J. Phys Chem.*, 1968, Vol. 62, 353.
40. Schwerdtfeger P, Heath G A, Dolg, M and Bennett M A, Low valencies and periodic trends in heavy element chemistry. A theoretical study of relativistic effects and electron correlation effects in group 13 and period 6 hydrides and halides, *J. Am. Chem. Soc.*, 1992, Vol. 114, 7518.
41. Sanderson R T, The inert pair effect on electronegativity, *Inorg. Chem.*, 1986, Vol. 25, 1856.
42. Himmel D, Krossing I and Schnepf A, Dative bonds in main-group compounds: A case of fewer arrows, *Angew. Chem. Int. Ed.*, 2014, Vol. 53, 370.
43. Frenking G, Dative bonds in main-group compounds: A case of more arrows, *Angew. Chem. Int. Ed.*, 2014, Vol. 53, 6040.
44. Himmel D, Krossing I and Schnepf A, Dative or not dative, *Angew. Chem. Int. Ed.*, 2014, Vol. 53, 6047.
45. Yao S, Zhiong Y and Driess M, A new era in main group chemistry: Zerovalent monoatomic silicon compounds and their analogues, *Acc. Chem. Res.*, 2017, Vol. 50, 2026.
46. Zao L, Hermann M, Holzmann N and Frenking G, Dative bonding in main group compounds, *Coord. Chem. Rev.*, 2017, Vol. 344, 163.

The Chemistry of Boron

CHAPTER

3

Unique Properties of Boron and its Compounds

- *Boron has many allotropes and most of them have the icosahedral B_{12} as the basic structural unit.*
- *Boron and its compounds on burning produce a green coloured flame.*
- *Boron-10 isotope has a high propensity to absorb thermal neutrons. It has potential applications in nuclear reactors and boron neutron capture therapy, a non-invasive therapeutic method for cancer treatment.*
- *Borosilicate glass (SiO_2 80% + B_2O_3 13%) has very low coefficient of linear thermal expansion (~3.3×10^{-6}/°C). It can withstand thermal shock better than other glasses [for example, soda glass (SiO_2 73% + Na_2O 14% ~ 8.9×10^{-6}/°C)].*
- *Boron forms trivalent electron deficient compounds which are excellent Lewis acids.*
- *Boron halides, unlike other halides of the same group do not dimerise due to the smaller size and higher electronegativity of boron.*
- *Boron tribromide is one of the most widely used and well known reagents for the cleavage of ethers.*
- *Boron is a minor nutrient for plants.*
- *Boron compounds do not show perceptible toxicity and are therefore sought after intermediates in the pharmaceutical and cosmetic industries.*
- *Sodium perborate $Na_2H_4B_2O_8$, with the perborate anion $[B_2O_4(OH)_4]^{2-}$ releases oxygen at around 60°C. Hydrolysis of this salt gives H_2O_2. This property makes it a useful oxidiser for many organic reactions and as a chlorine-free mild bleach.*
- *Hydroboration is a well known addition reaction of the B–H bond to double and triple bonds. This occurs in the anti-Markovnikov manner.*
- *Aryl boronic acids $ArB(OH)_2$, and boronate esters are excellent reagents for palladium catalysed cross-coupling reactions (Suzuki coupling).*

- *Carboranes (clusters composed of C, B and H) have higher air stability to form useful and stable organo-derivatives much easier than borane clusters. They form strong superacids due to the extra high stability of the anionic carborane clusters.*
- *Boron carbide and boron nitride have hardness, close to that of diamond. They have high melting points making them useful materials in nuclear (control rods) and defence (bullet proof vests) applications.*
- *Sterically bulky Lewis acids such as $B(C_6F_5)_3$ form Frustrated Lewis Pairs (FLPs) with sterically bulky bases such as $P(Mes)_3$ which are potential metal-free catalysts especially for hydrogenation and other small molecule activation.*
- *Magnesium diboride (MgB_2), a simple binary ionic compound, is a superconductor with critical temperature (T_c) of –234°C (39 K). This is the highest T_c among conventional superconductors at atmospheric pressure.*
- *Borophene is a two dimensional boron allotrope, analogous to graphene showing metallic properties. The smallest unit has 32 boron atoms having a hexagonal hole in the centre and the whole unit is slightly curved from planarity.*

3.1 Isolation and Purification of the Element

Although boron compounds have been known for centuries, pure elemental boron was isolated only in 1808 by the efforts Humphry Davy, Gay-Lussac and Thenard (1777–1857). This was achieved by carrying out the reduction of boric acid (H_3BO_3) with potassium metal. Elemental boron does not have any bulk applications except as a dopant in the semiconductor industry.[1,2]

Boron is naturally present mainly as

- Borax (hydrated sodium borate) $Na_2B_4O_7.10H_2O$ (Na_2O, 16.2%; B_2O_3, 36.6%; H_2O, 47.2%).
- Kernite (hydrated sodium borate) $Na_2B_4O_7.4H_2O$ (Na_2O, 22.6%; B_2O_3, 51.1%; H_2O, 26.3%)
- Colemanite (hydrated calcium borate) $2CaO.3B_2O_3.5H_2O$ (CaO, 27.2%; B_2O_3, 50.9%; H_2O, 21.9%).
- Ulexite (hydrated sodium calcium borate) $Na_2O.2CaO.5\ B_2O_3.16H_2O$ (Na_2O, 7.7%; CaO, 13.8%; B_2O_3, 43.0%; H_2O, 35.5%).

Crystalline boron is a black solid not found in the elemental form on earth. It is chemically inert and is resistant to attack by boiling HF or HCl. It has two isotopes B-11 (80%) and B-10 (20%). When finely divided, it is attacked slowly by hot concentrated nitric acid. Boron can be obtained starting from borax as follows.

$$Na_2B_4O_7.10H_2O + 2\ HCl \longrightarrow 4\ B(OH)_3\ [\text{ or } H_3BO_3\ ] + 2\ NaCl + 5\ H_2O \quad (3.1)$$

Orthoboric acid when heated above 170°C, dehydrates to give metaboric acid (HBO_2):

$$H_3BO_3 \longrightarrow HBO_2 + H_2O \quad (3.2)$$

Metaboric acid is a white solid sparingly soluble in water. Metaboric acid melts around 236°C, and when heated above 300°C it further dehydrates, forming tetraboric acid or pyroboric acid ($H_2B_4O_7$):

$$4\ HBO_2 \longrightarrow H_2B_4O_7 + H_2O \quad (3.3)$$

The term boric acid may sometimes refer to any of these compounds. Further heating leads to boron trioxide.

$$H_2B_4O_7 \longrightarrow 2\ B_2O_3 + H_2O \quad (3.4)$$

Boron trioxide on treatment with magnesium gives pure boron.

$$B_2O_3 + 3\ Mg \longrightarrow 2B + 3\ MgO \quad (3.5)$$

Scheme 3.1 consolidates the steps involved in the preparation of elemental boron from borax.

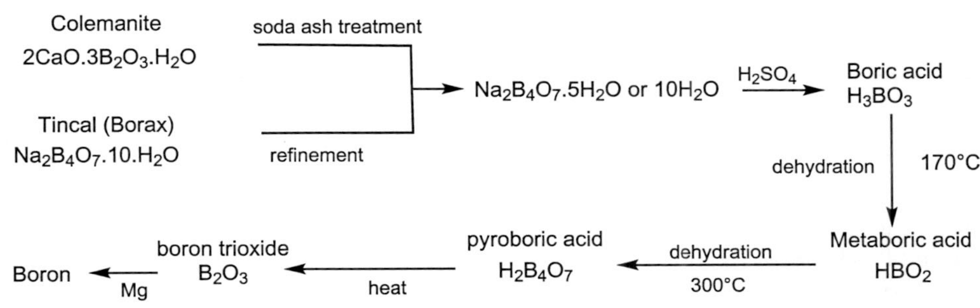

Scheme 3.1 Preparation of elemental boron from borax

Boron of high purity can also be made by the thermal decomposition of BBr_3 in the presence of hydrogen gas over a heated tantalum wire. Better results are obtained at temperatures of over 1000°C.

Table 3.1 Physical properties of boron

Boron Symbol: B At. no. 5	*Atomic mass*	*Electronegativity (Pauling scale)*	*1st ionisation energy*	*Electronic configuration*	*van der Waals radius*
	10.813 g.mol^{-1}	2.04	800.6 kJ.mol^{-1}	[He]$2s^2 2p^1$	0.192 nm
	Atomic radius	*Melting point*	*Boiling point*	*Electron affinity*	*Density*
	0.90 Å	2076°C	3927°C	26.989 kJ.mol^{-1}	2.08 g.cm^{-3} at m.p.

3.2 Allotropes of Boron

There are claims of at least 16 allotropes of boron till 2013—some of which are impure, poorly characterised, having insufficient evidence and can also be mixtures. The most pure and well characterised among these allotropes, α-rhombohedral boron (α-B_{12}), β-rhombohedral boron (β-B_{106}), γ-boron (a partially ionic high-pressure form of boron, γ-B_{28}, which can be quenched to ambient conditions), and β-tetragonal (β-B_{192} or T-192), have been confirmed to be thermodynamically stable (number indicates the number of boron atoms in a unit cell of the allotrope).[3,4] There is also a general understanding that as the temperature increases, α-B_{12} transforms to β-B_{106}. However, based on extrapolation of phase boundary, α-B_{12} has been claimed to be stable at normal temperatures and pressures. α-B_{12} can also be transformed to γ-B_{28} at high temperatures and pressures. All these allotropes of boron and many other impure forms contain icosahedral B_{12} clusters, with three-centre bonds within the icosahedra and covalent two- and three-centre bonds between the icosahedra.

Two novel allotropes have been reported in 2014 which do not contain B_{12} icosahedra—borophene, a two dimensional boron analogue of graphene (see Section 3.12.2) and borosphere, a quasi spherical allotrope having 40 boron atoms (see Section 3.12.3).

3.3 Boron Compounds of Industrial Importance

3.3.1 Borax

Borax (or tincal as its ore is known) is basically decahydrated sodium borate. The word tincal originates from *Sanskrit* as the early discoveries of borax were from the dry lakebeds of Tibet. It is the precursor for all other well-known boron–oxygen containing compounds and has by itself diverse applications.

Borax is a component of many detergents, enamel glazes and cosmetics. It has fire retardant properties and is mildly alkaline. Borox is added in fertilisers as boron is a micronutrient for plants. It is used as a flux and forms a glass in metallurgy. It is also used for *timborising*, a process by which wood is inserted into hot borax solution to make a protective impregnation against wood boring insets. In a traditional method of gold mining, borax has been used as a substitute for toxic mercury in the gold extraction process and this method is nowadays reintroduced due to mercury poisoning of traditional gold miners. Heating concentrated ore with borax gets the molten gold sink to the bottom aiding its separation. In the analytical laboratory, the name is related to the *borax bead test* introduced by Berzelius in 1812, which is a convenient flame test for many transition metal ions. In the borax bead test, a small amount of borax $Na_2B_4O_7.10H_2O$ is taken on a red hot nichrome or platinum loop and is heated to remove water of hydration converting it to a glass like bead (mostly sodium borate). By dipping this bead in a metal salt, the sodium ions are exchanged for the metal ions being analysed generating the corresponding metal borate and a flame test of this bead will now show a characteristic flame colour for that metal.[2]

$$\underset{\text{Borax}}{Na_2B_4O_7.10H_2O} \xrightarrow{\Delta} \underset{\text{Anhydrous borax}}{Na_2B_4O_7} \xrightarrow{\Delta} \underset{\text{Sodium borate}}{2\,NaBO_2} + \underset{\text{Boron trioxide}}{B_2O_3} \qquad (3.6)$$

Two structural forms are commonly used to depict borax $Na_2B_4O_7.10H_2O$. The actual form from structural studies is $Na_2[B_4O_5(OH)_4].8H_2O$ having two anionic tetra-coordinate boron atoms.

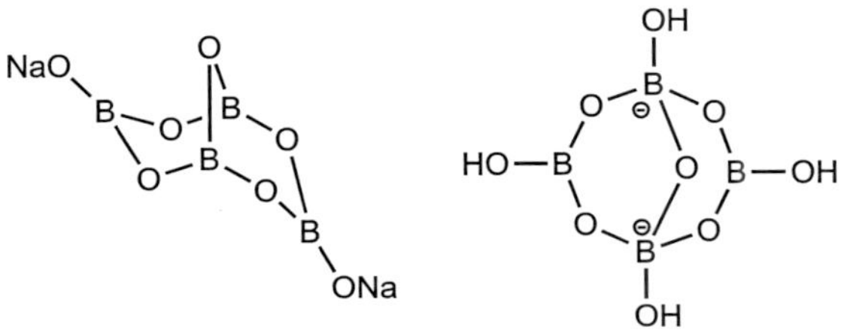

Fig. 3.1 Structural forms used for depicting borax

3.3.2 Sodium Perborate

Sodium perborate $NaBO_3.nH_2O$, is prepared from borax by treatment with NaOH (to convert it to sodium borate, $NaBO_2$) and further reaction with hydrogen peroxide. Two forms of this compound are available—one with $n = 1$ and the other with $n = 4$. The structurally characterised monohydrate showed it to be a disodium salt of 1,4 diboratetroxane dianion, $B_2O_4(OH)_4^{2-}$. It is highly water soluble and is a source of active oxygen used in detergents and bleaches and is also a disinfectant. It releases oxygen rapidly above 60°C. Some teeth brightening gels contain this as the active ingredient, sodium perborate which forms a very low concentration of hydrogen peroxide on the teeth (0.035% where the legal limit is 0.1%). This gives great results without any irritation of the gums or creating teeth sensitivity.

$$Na_2B_4O_7.10.H_2O \xrightarrow{NaOH} NaBO_2 \xrightarrow{H_2O_2} NaBO_3.H_2O \qquad (3.7)$$

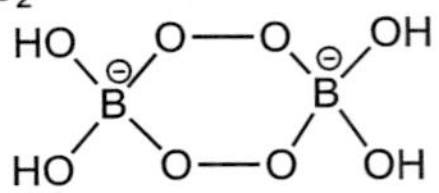

Sodium perborate is used also as an inexpensive safe and versatile oxidiser in many organic functional group modifications such as oxidation of aldehydes to acids, ketones to esters, aryliodides to diacetoxyiodoarenes, nitriles to amides and arylamines to nitroaromatics (Scheme 3.2). Reactions are often carried out in acetic acid, and mixing of sodium perborate with acetic acid is perceived to generate the percarboxylic acid in situ, which is the active oxidant in these reactions.[5]

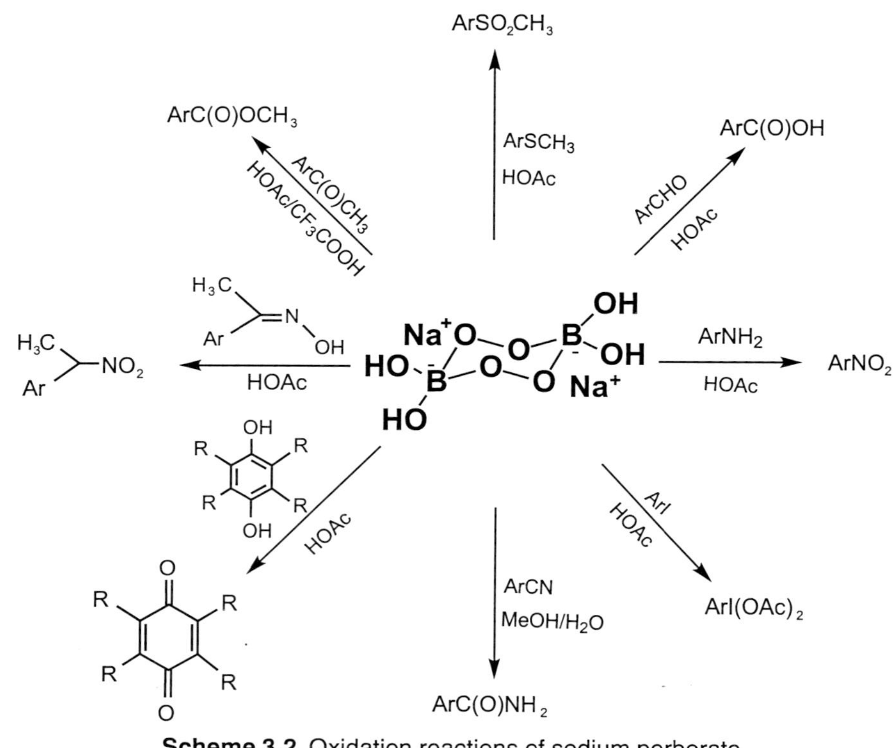

Scheme 3.2 Oxidation reactions of sodium perborate

3.3.3 Orthoboric Acid, H_3BO_3

Boric acid in the crystalline form consists of layers of $B(OH)_3$ molecules held together by hydrogen bonding. The distance between two such adjacent layers is 3.18 Å. It is readily prepared by reacting borax with a mineral acid such as HCl or H_2SO_4. As an acid, boric acid does not dissociate releasing H^+ but effectively donates a proton to another molecule of water itself acting as a Lewis acid (Eq. 3.8). So, although it is not a protonic acid it is a weak monobasic acid with pKa 9.14, comparable in acid strength to the ammonium ion. Among the common uses of orthoboric acid are for controlling of pests and treating eye infections.[6]

Fig. 3.2 Hydrogen bonding in orthoboric acid

$$B(OH)_3 + 2\,H_2O \rightleftharpoons B(OH)_4^- + H_3O^+ \qquad (3.8)$$

3.3.4 Metaboric Acid, HBO_2

Metaboric acid is obtained by dehydration of orthoboric acid. There are basically two types or modifications of metaboric acid – orthorhombic and monoclinic – both with the empirical formula HBO_2.

Boric acid on heating at 80–100°C releases one equivalent of water to form orthorhombic metaboric acid. This molecule is a trimer that is planar. On heating this in a sealed ampoule further at 130–140°C, the orthorhombic form converts to the monoclinic form which is polymeric and has one of the boron atoms of the individual trimeric unit being tetrahedral.

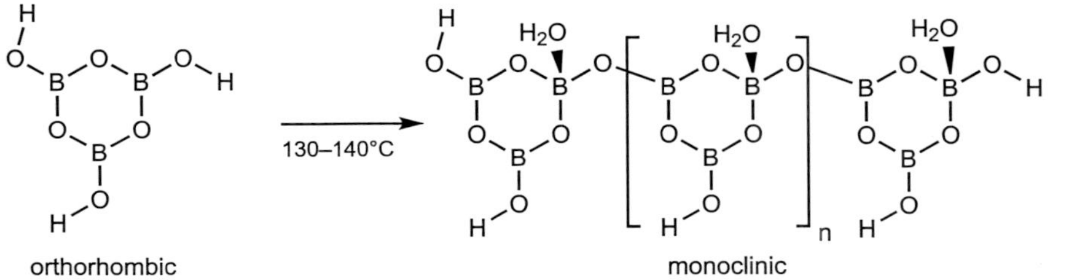

orthorhombic $\xrightarrow{130\text{–}140°C}$ monoclinic (3.9)

3.3.5 Boron Trioxide, B_2O_3

Boron trioxide is prepared by reacting borax with sulfuric acid in a fusion furnace. At temperatures above 750°C, the molten boron trioxide layer separates out from the sodium sulfate layer which can be separated by decanting.

Also, by heating boric acid above 300°C one can obtain boron trioxide. This goes through metaboric acid and pyroboric acid to the network of tricoordinate boron surrounded by three oxygens. The structure of B_2O_3 is best described as an infinite chain or ribbon of BO_3 triangles having both six-membered boroxol ring and non-ring units.

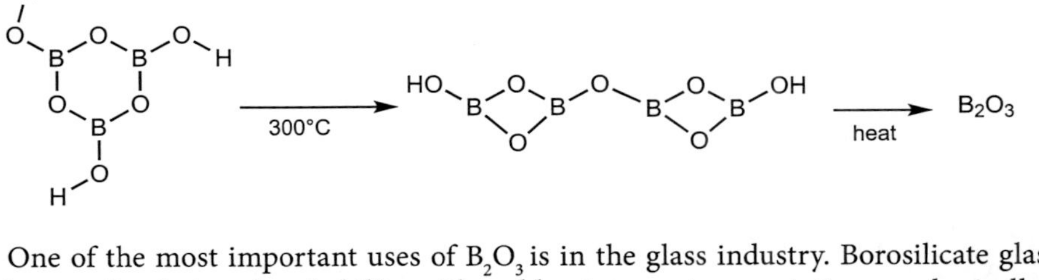

$\xrightarrow{300°C}$ HO–B(O)2B–O–B(O)2B–OH $\xrightarrow{heat}$ B_2O_3 (3.10)

One of the most important uses of B_2O_3 is in the glass industry. Borosilicate glasses, well known for their unbreakability with sudden temperature variations are basically vitreous solutions of SiO_2 (80%), B_2O_3 (13%), Al_2O_3 (4%) and Na_2O (4%). In contrast, soda lime glass, which breaks with sudden temperature variations has a higher coefficient of linear thermal expansion and is used for making window panes. The major components of sodalime glass are SiO_2 (73%), Na_2O (14%), CaO (9%) and MgO (4%). Boron trioxide is also an essential ingredient in high quality vitreous enamels. Modern day TVs, laptops and mobile phones require specialty glasses which do not contain sodium ions. Boric oxide is a major component of such alkali-free LCD glass. Boric oxide acts as a connecting network in the glass structure, improving optical properties and reducing the melting temperature.[7]

Borosilicate glass also finds use as a material for display panels because of its durability, transparency, thermal characteristics, and chemical resistance. While the cover glass panel of a mobile phone is made of alkali-aluminosilicate glass which is crack and scratch resistant, toughened and reinforced (for example, gorilla glass of Corning), the rear glass panel of a mobile phone or TFT screen is made of special borosilicate glass since the critically sensitive thin film transistor (TFT) is to be deposited on it in layers without any risk of contamination. Fabrication involves several cycles of heating which deposit thin films. To host these layers, the glass substrate must have very high chemical and heat resistance and should have a coefficient of thermal expansion which agrees with that of silicon. One of the basic requirements of such a glass is that it must be free of alkali metals. The diffusion of alkali ions to the film or the liquid crystal leads to breakdown of the TFT and sodium, if present is particularly harmful. Corning glassware has developed special alkaline earth boroaluminosilicate glasses for TFT deposition which has 69% SiO_2, 11.5% Al_2O_3 and 7.3% B_2O_3 and the rest alkaline earth metal oxides of Ca, Ba, Mg and Sr. Sodium is present only below 0.06 wt%.

3.3.6 Boron Carbide, B_4C

Henri Moissan, the father of fluorine chemistry was also the person who prepared boron carbide for the first time by the reduction of boron trioxide with magnesium in the presence of carbon in an electric arc furnace. It can also be prepared by the reaction of carbon on B_2O_3 at high temperature with the release of CO as side product.

$$2B_2O_3 + 7C \xrightarrow[\text{furnace}]{\text{electric arc}} B_4C + 6CO \tag{3.11}$$

With the more accurate chemical formula of $B_{12}C_3$, boron carbide consists of B_{12} icosahedra surrounding a carbon–boron–carbon chain at the centre. It is one of the hardest known materials with a hardness of 9.5 mhos which is slightly less than that of cubic boron nitride and diamond. This ceramic material is used for making bullet-proof vests, tank armour, and so on. Boron carbide has immense use in nuclear reactors as control rods and shutdown pellets due to its ability to absorb thermal neutrons and the very high melting point of 2763°C and boiling point of 3500°C.[8, 9]

3.3.7 Boron Halides

The three widely used boron halides are BF_3, BCl_3 and BBr_3. All of them are well known Lewis acids and form stable adducts with Lewis bases. Based on the relative exothermicities of these adduct-forming reactions, their relative Lewis acidity has been determined and follows the order $BF_3 < BCl_3 < BBr_3$.

Although the electronegativity of fluorine is high and one would expect a reverse order in Lewis acidity, the B–F bonds of BF_3 are found to have a partial double bond character from a possible $p\pi$–$p\pi$ bonding and resulting in a lower Lewis acidity for BF_3.[10, 11]

Synthesis of the boron halides is carried out as per the following equations.

$$B_2O_3 + 6\ HF \longrightarrow 2\ BF_3 + 3\ H_2O \tag{3.12}$$

$$B_2O_3 + 2\ C + 3\ Cl_2 \longrightarrow 2\ BCl_3 + 3CO \tag{3.13}$$

$$B_2O_3 + 2\ C + 3\ Br_2 \xrightarrow[HT]{} 2\ BBr_3 + 3CO \tag{3.14}$$

BBr_3 is also prepared by the following reaction

$$2B + 3Br_2 \longrightarrow 2BBr_3 \tag{3.15}$$

Alternatively BF_3 can be synthesised from $NaBF_4$

$$2\ NaBF_4 + B_2O_3 + 6\ H_2SO_4 \longrightarrow 8\ BF_3 + 3\ H_2O + 6\ NaHSO_4 \tag{3.16}$$

BF_3 and BCl_3 are gases at room temperature with boiling points of –100.3°C and 12.6°C and both are corrosive. A convenient way of handling them for chemical reactions are as adducts, $BF_3.OEt_2$ (b.p. 126°C), and $BCl_3.SMe_2$ (m.p. 88–90°C). While hydrolysis of BF_3 results in HBF_4 and boric acid, hydrolysis of BCl_3 gives HCl and boric acid.

Because of it Lewis acidity and its relatively higher resistance to hydrolysis, in comparison to other boron halides, BF_3 is used as a catalyst in many organic reactions such as:

- polymerisation of olefins, vinyl ethers, indenes, terpenes, styrene, unsaturated acids, and so on,
- alkylation of aromatic, phenols with olefins and alcohols,
- isomerisation and disproportionation of paraffins and unsaturated hydrocarbons,
- synthesis of acetals, steroids, hydroxy fatty acids, and,
- For promoting reactions of acetylenes with acids and addition of organic acids to olefins.

BF_3 forms Lewis acid base adducts with a range of inorganic compounds such as SO_2, H_2S, NH_3 and is also a precursor for the synthesis of boranes and aryl boron compounds.

BCl_3 has uses as precursor for boron and as a soldering flux for alloys of Al, Zn, Fe and W. It is also used in plasma etching as the gas etches metal oxides by the formation of $BOCl_x$ type compounds. BCl_3 and even its SMe_2 adduct react aggressively with water, releasing HCl.

Unlike BF_3 and BCl_3, BBr_3 is a liquid with a boiling point of 91°C and is therefore more convenient to handle although it reacts vigorously with moisture. The most important use of BBr_3 is for the cleavage of C–O bonds of alkyl ethers. The mechanism of this reaction goes through a complex formation of the boron centre with the ether oxygen followed by alkyl bromide elimination to yield a dibromo organoborane. This on further hydrolysis results in the alcohol.

$$ROR + BBr_3 \longrightarrow ROBBr_2 + RBr \tag{3.17}$$

$$ROBBr_2 + H_2O \longrightarrow ROH + B(OH)_3 + 2\,HBr \qquad (3.18)$$

BBr_3 also finds application in olefin polymerisation and dealkylation of aryl alkylethers. BCl_3 also can cleave C–O bonds of ethers; however, BBr_3 is preferred for this reaction due to higher reactivity and ease of handling as a liquid reagent.[12,13]

3.3.8 Boron Hydrides

Boron hydrides (also known as boranes) unlike their more familiar and stable oxygen containing counter parts – borax and boric acid – are a very interesting class of molecules from the perspective of structure and bonding. During the 1950s, nations such as the USA and USSR spent considerable amounts on exploring their chemistry to develop them as rocket fuels. Many of them, especially lower boranes, are very difficult to handle and some are so reactive that they catch fire when exposed to air, burning with boron's characteristic green flame. The chemistry of this unique family of boron compounds was first explored in 1912 by Alfred Stock, the father of boron chemistry who prepared some of these highly reactive molecules. The initial reaction of Stock was carried out between magnesium boride and phosphoric acid.[14]

$$Mg_3B_2 + H_3PO_{4\,(aq)} \longrightarrow B_4H_{10},\ B_5H_9,\ B_6H_{10} \qquad (3.19)$$

By using a sophisticated vacuum line apparatus for handling reactive gases, Stock separated the lower boranes and characterised them. Diborane B_2H_6, can now be prepared from $NaBH_4$ or BF_3 by many reagents as shown.

$$NaBH_4 \xrightarrow{I_2 \text{ or } H_3PO_4 \text{ or } BF_3} B_2H_6 \qquad (3.20)$$

$$BF_3 \xrightarrow{NaH \text{ or } LiH \text{ or } LiAlH_4} B_2H_6 \qquad (3.21)$$

Most of the higher boranes could be prepared from diborane by controlled pyrolysis reactions.

$$B_2H_6 \xrightarrow{\text{pyrolysis } (100\text{–}120°C)} B_4H_{10};\quad B_2H_6 \xrightarrow{\text{pyrolysis } (180\text{–}200°C)} B_5H_9;\quad B_2H_6 \xrightarrow{\text{pyrolysis } (Me_2O,\ 100\text{–}200°C)} B_{10}H_{14} \qquad (3.22)$$

W N Lipscomb, the 1976 Nobel Prize recipient for chemistry, was the first to classify boranes using the now redundant *styx* numbering. The currently accepted and widely used nomenclature based on the shape of boranes began in 1963 when a closed borane polyhedron was first classified as *clovo* which in Latin meant closed. In 1968 the American Chemical Society modified this to *closo* and also named another open polyhedra as *nido*. The words *arachno* and *hypho* were introduced later by Williams in 1970 and 1974 for further sub-classifying open polyhedra. The meaning of these names and examples of common lower boranes with electron count and structures are given below.

- *Closo* boranes—closed polyhedra, for example, $B_{12}H_{12}^{2-}$.
- *Nido* boranes (nest like)—B atoms occupy *n* corners of an *n*+1 cornered polyhedron, for example, B_2H_6, $B_{10}H_{14}$.
- *Arachno* (web like)—more open than nido with B atoms occupying *n* contiguous corners of an *n*+2 cornered polyhedron, for example, B_5H_{11}, B_6H_{12}.
- *Hypho* (net like)—open polyhedra with B atoms occupying *n* corners of an *n*+3 polyhedron, for example, B_8H_{16}.
- *Conjucto* (linked)—formed by linking two or more of the earlier type of clusters together, for example, $B_{10}H_{16}$, that is, $(B_5H_8)_2$.

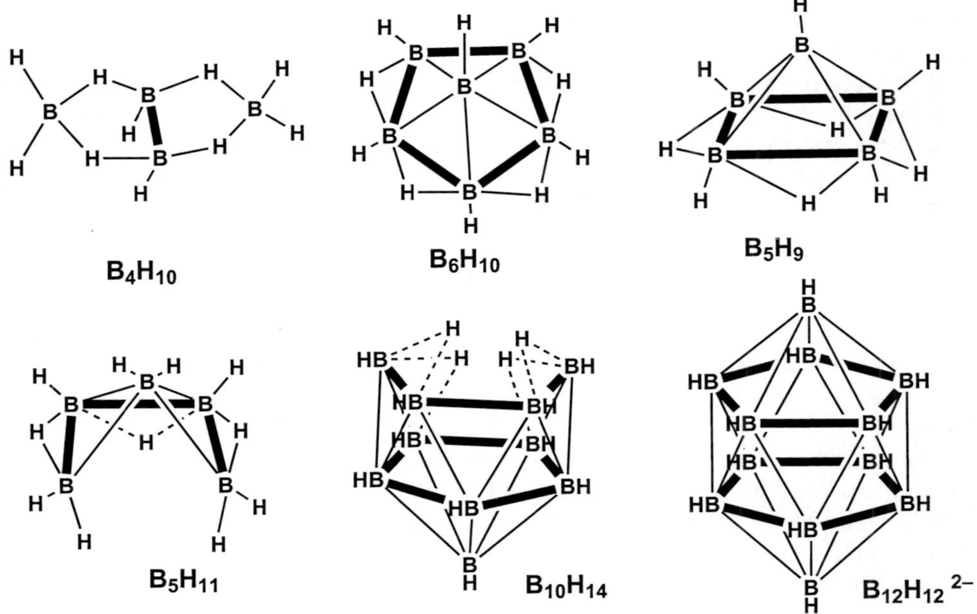

Fig. 3.3 Shapes of some boranes

It is interesting to note that most of the lower boron hydrides are gases and liquids. Decaborane ($B_{10}H_{14}$) and its higher homologues are found to be solids. The credit for separation and characterisation of almost all lower molecular weight boron hydrides (Table 3.2) goes to Alfred Stock who also designed the sophisticated vacuum lines of his times to make the separation of the volatile compounds effective. Since these vacuum lines used mercury and mercury manometers, it is recorded that he was identified with mercury poisoning. After finding out about the mercury poisoning, he took sufficient care in the latter part of his life and is also credited with about 50 research papers on the topic of mercury poisoning.[15, 16]

Table 3.2 Classification of small size boranes

Formula	B_2H_6	B_4H_{10}	B_5H_9	B_5H_{11}	B_6H_{10}	$B_{10}H_{14}$	$B_{12}H_{12}^{2-}$
Name	Diborane	Tetraborane	Penta borane-9	Penta borane-11	Hexaborane	Decaborane	Dodecaborate anion
b.p. (°C)	−92.5	16	58.4	65	108	213 m.p. 99.7	m.p. 650 (Cs salt)
Electron count	B_nH_{n+4}	B_nH_{n+6}	B_nH_{n+4}	B_nH_{n+6}	B_nH_{n+4}	B_nH_{n+4}	B_nH_{n+2}
Shape	*nido*	*arachno*	*nido*	*arachno*	*nido*	*nido*	*closo*

Stability of dodecaborate anion

Dodecaborate $B_{12}H_{12}^{2-}$ is quite stable, and survives heating with sodium hydroxide or hydrochloric acid. It is one of the most stable boron compounds and possibly the most stable boron hydride. The stability is attributed to the icosahedral structure. This is an accidental coincidence of the direction of the B–H bonds in the solid geometry of icosahedron versus what is required by the overlap requirements. In other words, if we take a pentagonal pyramidal $B_6H_6^{4-}$, it requires a B–H bending towards the cap. This angle of bending is close to the angle available in $B_{12}H_{12}^{2-}$ by virtue of the solid geometry (symmetry) of the icosahedron and does not happen for other polyhedra. Interestingly, although elemental boron exists in different polymorphic forms, the icosahedral B_{12} unit is the primary building block of all these forms. The most stable β-rhombohedral has maximum number of boron atoms forming a part of B_{12} icosahedra. The icosahedral unit is also the reason for the high stability of carboranes of the type $C_2B_{10}H_{12}$ and $(CHB_{11}X_{11})^-$ (X = Cl, F). The former is the most well known stable carborane with the maximum number of derivatives and the latter is the stable counter ion for the strongest solo superacids based on carboranes.[17]

Diborane

Diborane adopts a structure containing four terminal and two bridging hydrogen atoms. Initially it was thought to have an ethane-like structure which was proved wrong later. As per molecular orbital theory, the bonds between boron and the terminal hydrogens are 2-centre-2-electron bonds while the bonds between the boron and the bridging hydrogen atoms are different. Having used two electrons in bonding to the terminal hydrogen atoms, the bridging hydrogen atoms provide one electron each. Thus the B_2H_2 ring is held together by four electrons, an example of 3-centre-2-electron bonding. The lengths of the B–H_{bridge} bonds and the B–$H_{terminal}$ bonds are 1.31 and 1.19 Å respectively, and this difference in the lengths of these bonds reflects the difference in their strengths, the B–H_{bridge} bonds being relatively weaker.

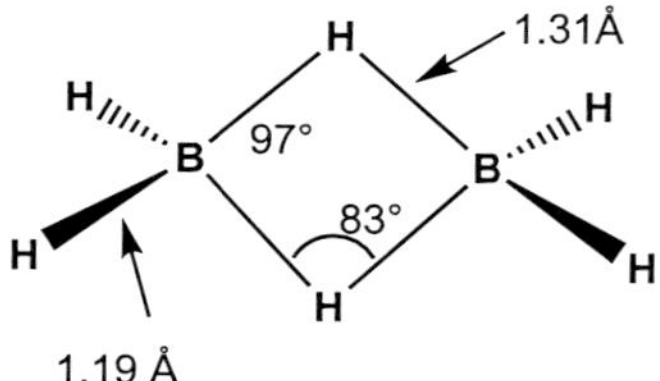

Fig. 3.4 Structure of B_2H_6

In search of a boron based superfuel

During the arms race between the USA and the USSR in 1950s, there was an intense effort to develop powerful jet and rocket fuels which could deliver weapons at distant continents. In the 1930s, boron's potential began to emerge when it was found that boron-based materials produced more heat than the corresponding carbon-based compounds. It was estimated that boron hydrides could produce up to 30,000 BTUs/lb of energy upon combustion, compared to ~18,000 BTUs/lb for comparable hydrocarbon fuel. With the rumour spreading that the USSR used boron hydride-based fuel for launching the sputnik satellite in 1957, the USA started many secret government programmes to develop boron hydride-based fuels. The three boranes initially selected were diborane, pentaborane-9 and decaborane. Out of these the most stable was decaborane and the heat of combustion of these boranes were very attractive to try them out as jet and rocket fuels.

Fuel	Heat released in btu/lb
Decaborane	27850
Diborane	31078
Pentaborane-9	29070
JP-4 (kerosene + gasoline 1:1)	18400
Hydrogen	51571
HEF-3	~25000

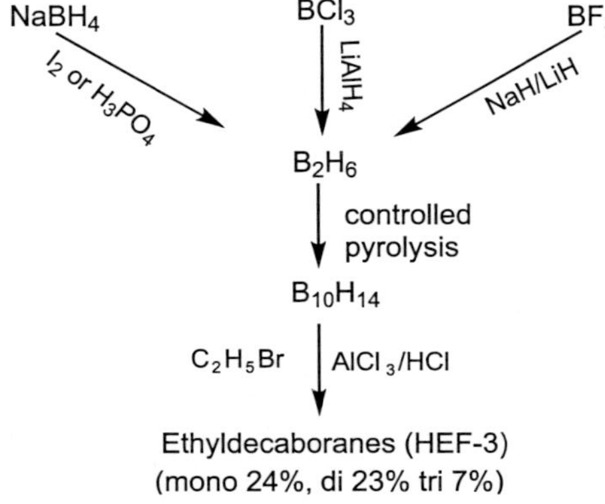

The ethyl derivatives of decaborane, made by Friedel–Crafts reaction of decaborane with ethylbromide was found to be more easy to store and handle and was code named HEF-3 (High Energy Fuel-3). This was considered for use as fuel in XB-70 Valkyrie, the prototype version of the B-70 nuclear armed deep penetration strategic bomber which could fly at three times the speed of sound similar to inter-continental ballistic missiles (ICBMs) and reach distant continents. But the programme was cancelled after spending millions of dollars. One reason was, the borane-based fuels are highly corrosive and could spontaneously ignite. Also when burned, they produced solid combustion byproducts especially hard boron carbide and boron oxides that got attached to the engine parts leading to mechanical faults and corrosion. In addition, the sooty exhaust was highly toxic, poisoning populated areas and had high visibility in the sky making it unfit for defence use. Adding to this was the exorbitant cost of making and storing such fuels. Hence the programme for making borane-based fuels was shelved. Nevertheless the billions of dollars spent on this programme generated many new compounds of boron such as carboranes which found their use elsewhere, especially in nuclear applications.

The diborane story

While the story about Neil Bartlett and the discovery of the first rare gas compound has found a place even in high school chemistry textbooks (see Chapter 15 on rare gases for a detailed description), a similar story on the discovery of the actual structure of diborane has gone rather unnoticed. In both the stories, Nobel Prize recipient, Linus Pauling is a character. The take-home message of both the stories also seems to be the same: never fully believe the words of greatly accomplished scientists. They, like the rest of the lot can also go wrong in their assumptions and conclusions.

There are two well accomplished chemists in this story—Alfred Stock, the father of borane chemistry and Linus Pauling, as mentioned earlier. Both strangely believed that the structure of B_2H_6 had to be similar to that of its carbon analogue, ethane.

The story begins with Alfred Stock who painstakingly prepared and separated the boron hydrides and without much thought proposed the ethane structure for B_2H_6. In 1937 Simon H Bauer, a former student of Linus Pauling at Caltech, while working at Cornell university did an electron diffraction study of diborane and reported a structure analogous to ethane and published his results in the Journal of American Chemical Society. In 1941, H I Schlesinger, a professor at the University of Chicago, who worked extensively on boron compounds wrote to Linus Pauling indicating that he thought the structure of diborane should have two hydrogens in a bridging fashion (see Fig. 3.4) based on his work on boron hydrides and after reading some literature on the topic by some Russian scientists. Pauling wrote back in four days the following in a letter.

'I do not feel very friendly towards the structure which you mention in your letter for the diborane molecule. So long as the suggested structure remains vague and indefinite, it is not easy to say that it is eliminated by electron diffraction data or other data. Also the force constant for the B–B vibration is I think, much stronger than that would be expected for the proposed structure in which there is no direct B–B bond'.

Pauling thus disagreed with the structure proposed and also discredited Schlesinger of the recognition for proposing the actual bridged structure. Two years after Schlesinger wrote to Pauling, a young undergraduate student of Oxford, H Christopher Longuet Higgins, after examining the physical evidence and after carrying out a proper and adequate theoretical study, came out with a paper in Journal of Chemical Society supporting the bridging hydrogen structure. This structure was further confirmed by infrared, NMR evidences and by 1946–47 three well known theoreticians, among them the famous Mulliken and Walsh, gave molecular orbital and valence bond descriptions of the bridged structures. The electron diffraction study which Pauling believed of his former student was redone which was then found to be consistent with the bridged structure proposed by Longuet Higgins.[18]

Diborane is a highly reactive reagent with many applications. Common reactions include formation of adducts with Lewis bases. Such Lewis acid–base adducts often react further to give new products. Its reaction in a 1:2 ratio with ammonia leads to borazine—inorganic benzene. It reacts with alcohols to give borate esters and undergoes hydrolysis to give boric acid. Diborane also reacts readily with alkynes to form substituted alkene products which will readily undergo further addition reactions. Diborane and its analogous hydrides are common reagents for hydroboration reactions, where alkenes add across the B–H bonds resulting in trialkylboranes. Diborane is the starting point for a host of neutral and anionic boranes and heteroboranes. It is used as a reducing agent whose reaction is almost complementary to the reactivity of lithium aluminium hydride. Diborane readily reduces carboxylic acids to the corresponding alcohols.[19, 20]

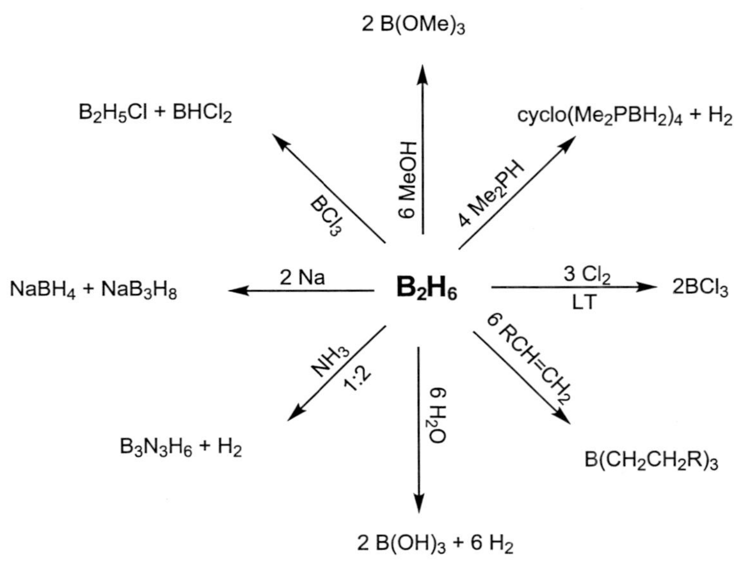

Scheme 3.3 Reactions of diborane

$$B_2H_6 + 2\,Et_3N \longrightarrow 2\,Et_3N.BH_3 \xrightarrow[-H_2]{B_4H_{10},\ 190°C} [Et_3NH]_2\,[B_{12}H_{12}] \qquad (3.23)$$

3.4 Hydroboration

Hydroboration, a well known reaction in organic chemistry deals with the addition of a hydrogen–boron bond to double bonds (C–C, C–N and C–O), as well as C–C triple bonds. The development of this reaction and its applications in organic chemistry resulted in the Nobel Prize for H C Brown of Purdue University, USA. In his Nobel lecture, Prof. Brown had given a lot of credit to his Indian coworker, B C Subba Rao for the discovery of hydroboration and for some of its uses.[21]

Hydroboration reactions follow anti-Markovnikov addition, where the hydrogen adds to the more substituted carbon of the double bond. The regiochemistry is opposite to that of a typical HX addition and it reflects the polarity of the $B^{\delta+}$–$H^{\delta-}$ bonds. The mechanism of hydroboration indicates that it proceeds through a four-membered transition state where the hydrogen and the boron atoms are added on the same face of the double bond. Since this mechanism is concerted, the C–B bond formation proceeds slightly faster than the C–H bond formation. As a consequence, boron develops a partially negative charge while the more substituted carbon bears a partial positive charge in the transition state. The partial positive charge is supported better by the carbon atom which is more substituted.

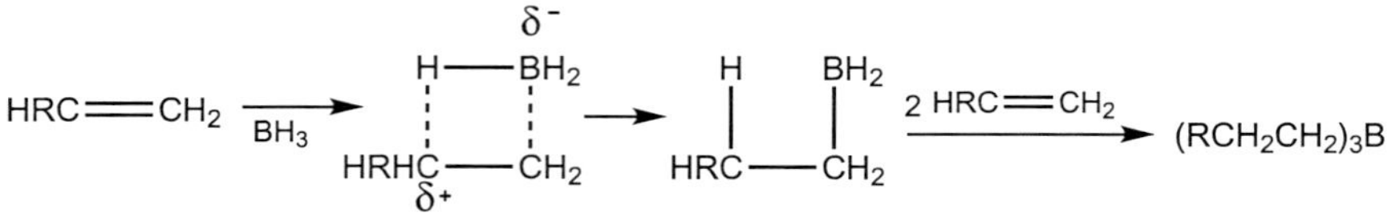

$$HRC{=}CH_2 \xrightarrow{BH_3} \left[\begin{array}{c} \overset{\delta^-}{H\text{---}BH_2} \\ \vdots \qquad \vdots \\ HRH\underset{\delta^+}{C}\text{---}CH_2 \end{array}\right] \longrightarrow \begin{array}{cc} H & BH_2 \\ | & | \\ HRC & \text{---}CH_2 \end{array} \xrightarrow{2\ HRC{=}CH_2} (RCH_2CH_2)_3B \qquad (3.24)$$

BH_3 + alkene (R_1, H / R_3, R_2) $\xrightarrow{25°C}$ H_2B–CR_1H–CR_3HR_2; with NH_3 → H_2N–CR_1H–CR_3HR_2; with X_2 → X–CR_1H–CR_3HR_2; with H_2O_2 → HO–CR_1H–CR_3HR_2; with H^+ → H–CR_1H–CR_3HR_2

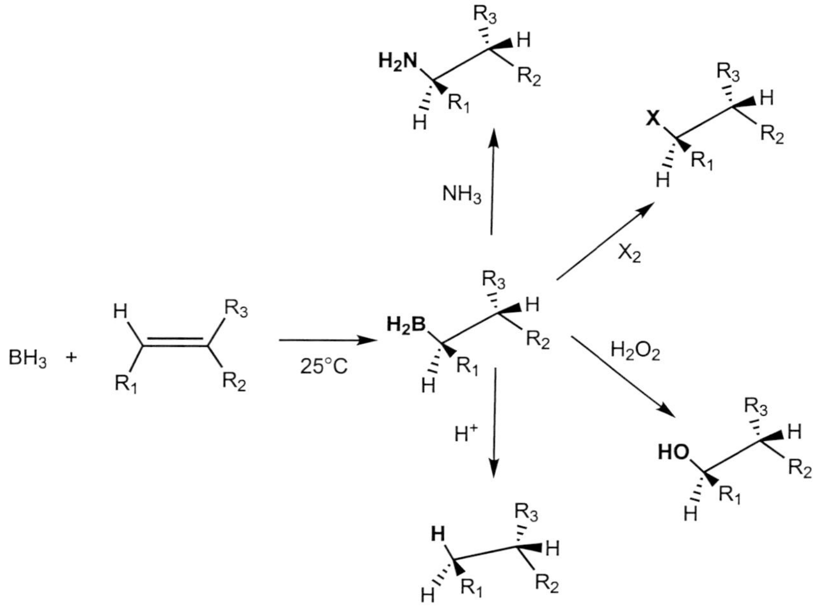

Scheme 3.4 Some products of hydroboration

Hydroboration results in organoborane compounds that can further react with a variety of reagents to produce compounds, such as amines, alcohols and alkyl halides. The most well known reaction of the organoboranes is oxidation leading to alcohols, by using hydrogen peroxide as the reagent. This reaction has promoted hydroboration due to its mild reaction conditions with a wide scope of olefins as substrates.

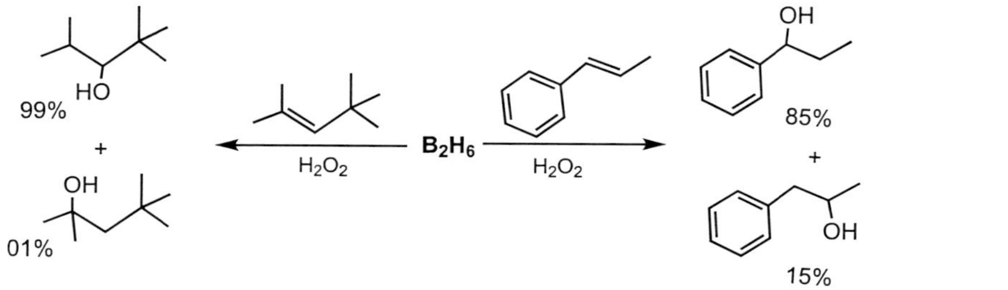

99% (HO) + 01% (OH) $\xleftarrow{H_2O_2}$ [alkene] B_2H_6 [styrene derivative] $\xrightarrow{H_2O_2}$ (OH) 85% + (OH) 15% (3.25)

When BH_3 is used as the hydroboration reagent, reactions often proceed beyond the monoalkylborane derivatives, especially for olefins which are less hindered. Trisubstituted olefins produce dialkylboranes, but further alkylation is slowed down due to steric hindrance. This significant rate difference in producing di- and tri-alkyl boranes is useful in the synthesis of bulky boranes which helps to enhance regioselectivity.[22]

The addition of the H–B bond to olefins is considered to be under electronic control rather than under steric control. In the case of trisubstituted alkenes, boron is predominantly placed on the less substituted carbon. For such 1,2-disubstituted olefins, regioselectivity can be observed only when one of the two substituents is a group such as a phenyl ring. There are many reagents for hydroboration which are more convenient to handle than diborane. One of the well known examples is 9-BBN.

9-BBN (9-Borabicyclononane)

Highly versatile among dialkylboranes is 9-borabicyclononane, 9-BBN. This compound is found to be a dimer which is distilled without decomposition at 195°C (12mm Hg). 9-BBN reacts normally at 60–80°C. Tetrasubstituted alkenes undergo hydroboration with 9-BBN at elevated temperatures. 9-BBN also shows very good regioselectivity in the hydroboration of alkenes.[23]

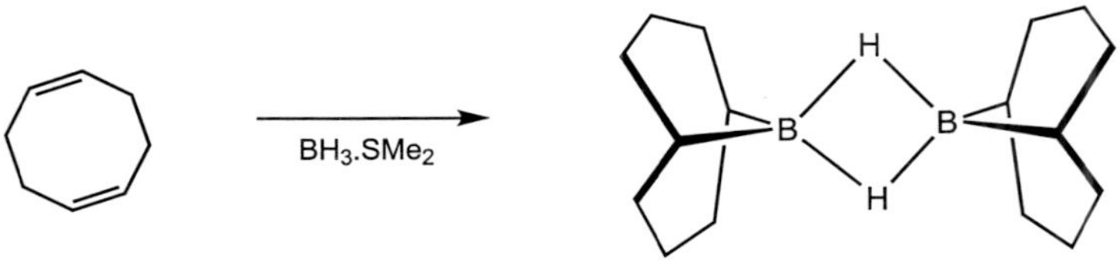

$BH_3.SMe_2$ (3.26)

Some of the other common widely used hydroboration reagents are (a) pinacolborane, (b) catecholborane and (c) disiamylborane (abbreviated Sia_2BH).

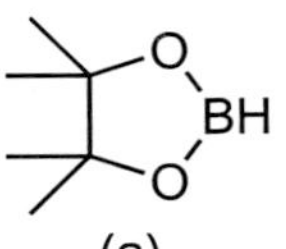

(a) 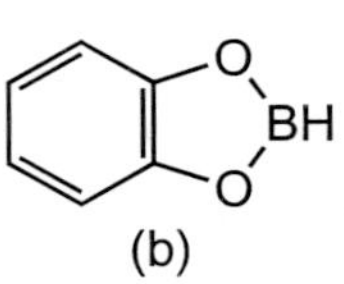 (b) 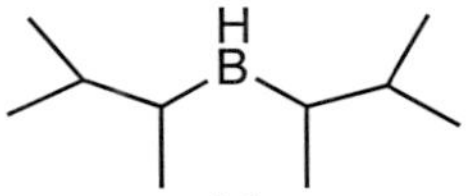 (c)

3.4.1 Sodium Borohydride

Among the most well known boron-based reagents used regularly in organic synthesis and reactions is sodium borohydride which is a mild reducing agent compared to lithium aluminium hydride. It was first prepared by Schlesinger from the methyl ester of boric acid (Eq. 3.27).

$$B(OMe)_3 + 4\,NaH \xrightarrow[250\text{–}270°C]{} NaBH_4 + 3\,NaOCH_3 \quad (3.27)$$

Sodium borohydride is used widely for the reduction of aldehydes and ketones to alcohols. However, it is not useful for reduction of acids, amides and esters as these reactions are either very slow at room temperature or do not occur at all. In the case of carboxylic acids, it forms the sodium carboxylate. Scheme 3.5 summarises the reactions of $NaBH_4$ with common organic functional groups.

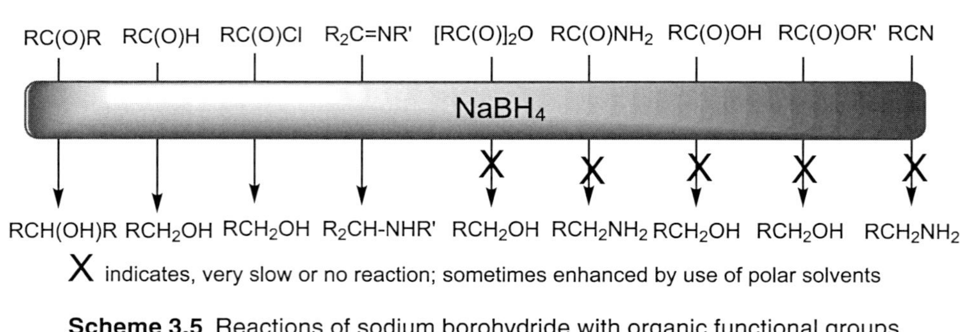

Scheme 3.5 Reactions of sodium borohydride with organic functional groups

3.5 Carboranes

Carboranes are borane clusters having carbon atoms replacing the framework boron atoms. They have also been classified based on the structure as, *closo, nido, arachno* and *hypho*. The 1,2-closo-dicarba-dodecaborane, $C_2B_{10}H_{12}$ (simply called carborane), was reported by groups at Olin Corporation and the Reaction Motors Division of Thiokol Chemical Corporation working under the US Air Force (work published in 1963). Before its discovery, decaborane derivatives were generally thought to be unstable with air and water. These research groups showed the unusual stability of the 1,2-closo-dodecaborane group, came out with a general synthesis and also showed transformation of substituents without destroying the framework of the carborane cluster. The most well-studied carborane is $C_2B_{10}H_{12}$ with a m.p. of 320°C. It is prepared by the reaction of acetylene with decaborane (Eqs 3.29, 3.30). A variation on this method involves the use of dimethylacetylene dicarboxylate to give $C_2B_{10}H_{10}(CO_2CH_3)_2$, which can be degraded to $C_2B_{10}H_{12}$.[24]

$$B_2H_6 \xrightarrow[\text{100–200°C}]{\text{Pyrolysis}} B_{10}H_{14} \tag{3.28}$$

$$B_{10}H_{14} + 2\ Et_2S \xrightarrow[-H_2]{n\text{-}Pr_2O} B_{10}H_{12}(Et_2S)_2 \tag{3.29}$$

$$B_{10}H_{12}(Et_2S)_2 \xrightarrow[-H_2,\ -2Et_2S]{HC\equiv CH} \underset{\text{Ortho carborane}}{C_2B_{10}H_{12}} \tag{3.30}$$

At 450°C *ortho*-carborane converts to the *meta*-isomer and at 620°C it converts to the *para*-isomer. In comparison, benzene requires >1000°C to induce skeletal rearrangement. Lipscomb has proposed a mechanism for this skeletal movement of carbon atoms from the *ortho* to *meta* to *para* as a diamond–square–diamond (DSD) mechanism. Like arenes, carboranes also undergo electrophilic aromatic substitution reactions.[25, 26]

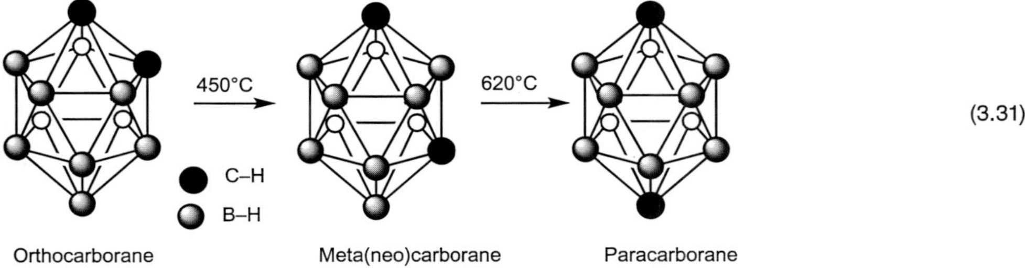

(3.31)

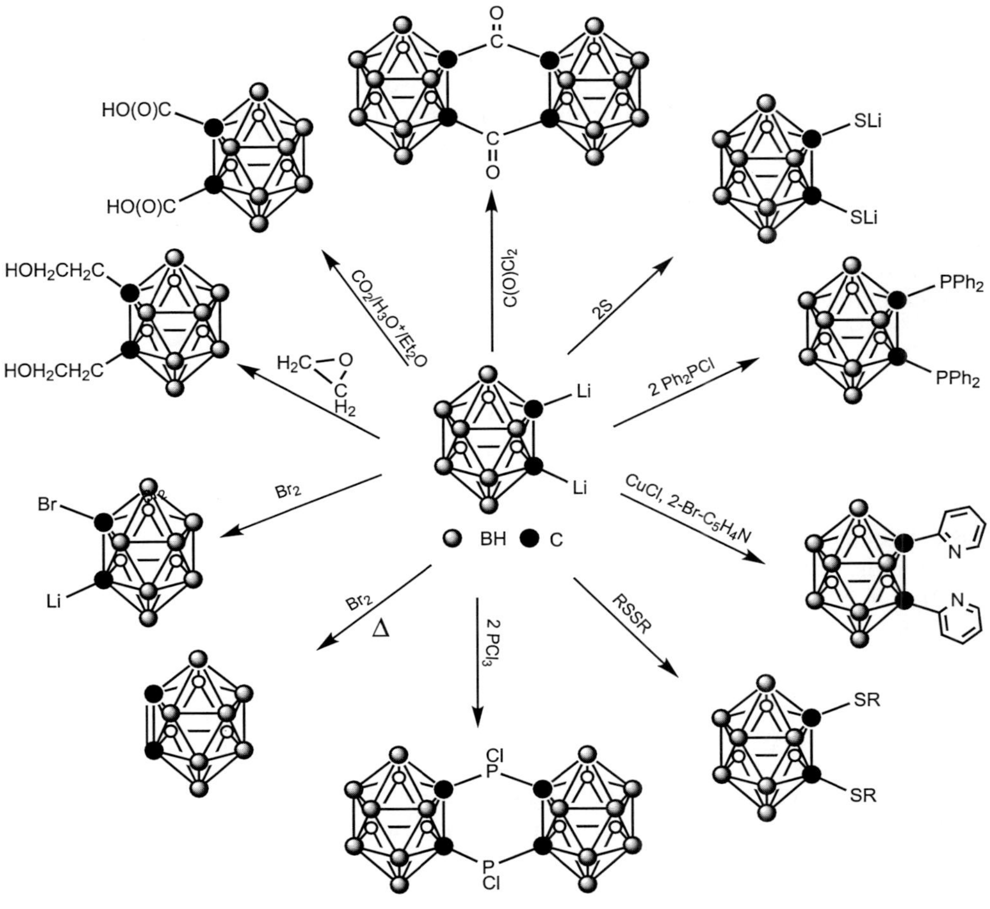

Scheme 3.6 Reactions of dilithiated carboranes

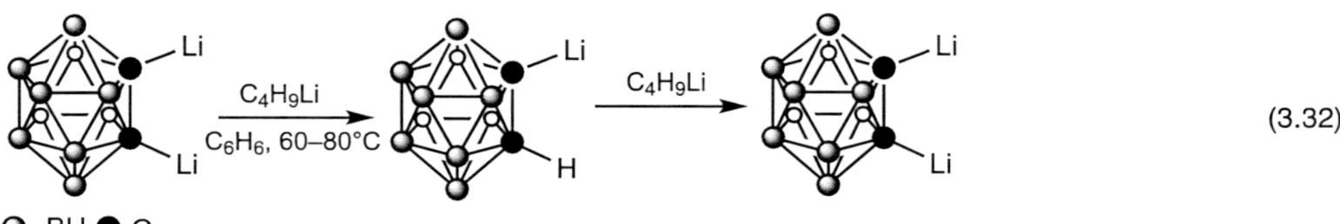

(3.32)

C-mono or dilithiated *o*-carboranes are versatile reagents which are easily prepared by the reaction of *n*-butyllithium on the carborane in benzene medium. Unlike many other organolithiated reagents, they show good functional group tolerance to groups such as ketones, esters and nitriles. A wide range of products which themselves are precursors to many other derivatives have been prepared from the dilithiated *o*-carborane, $C_2B_{10}H_{10}Li_2$.[24]

3.5.1 Carboranes as Ligands

Numerous studies have been carried out on derivatives of the so-called dicarbollide anion, $[B_9C_2H_{11}]^{2-}$ (the word '*olla*' from Spanish for pot). The first metal dicarbollide complexes were prepared by M F Hawthorne and coworkers in 1965. This anion forms sandwich compounds, called bis(dicarbollides), with many transition metal ions (Scheme 3.7). Some of them exist in otherwise unusual oxidation states. The dianion is a nido cluster obtained from the parent dicarborane by removal of a B and H.[27]

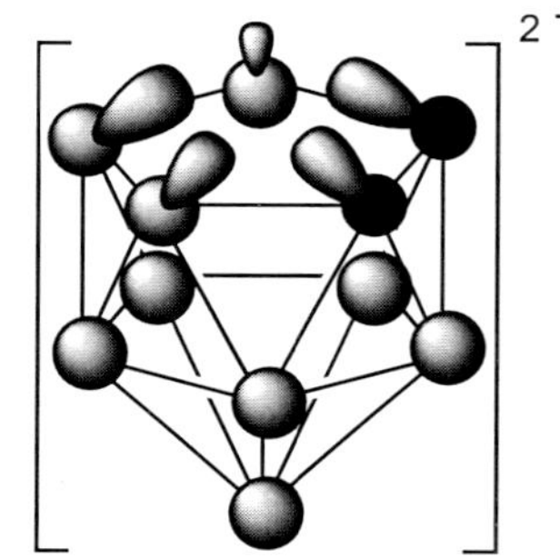

$$C_2B_{10}H_{12} + MeO^- + 2\,MeOH \xrightarrow{85°C} [C_2B_9H_{12}]^{2-} + H_2 + B(OMe)_3 \quad (3.33)$$

$$[C_2B_9H_{12}]^{2-} + NaH \longrightarrow [C_2B_9H_{11}]^{2-} + H_2 + Na^+ \quad (3.34)$$

Bis(dicarbollides) often exhibit properties very different from traditional metal sandwich compounds. For example, Ni-based bis(dicarbollide) cluster can be observed for the rare Ni(III) and Ni(IV) oxidation state of nickel. Hawthorne has shown the potential of these compounds in applications such as catalysis, ion-exchange materials for radioactive waste management and inert redox shuttles for dye-sensitised solar cells (DSSCs).

3.5.2 Carborane-Based Superacids

The idea of making carborane based superacids possibly originated from the fact that among polyhedral borane clusters, icosahedral borane clusters have been found to have extraordinary stability. This stability is the result of an accidental coincidence of the direction of the B–H bonds in the solid geometry of icosahedron versus what is required

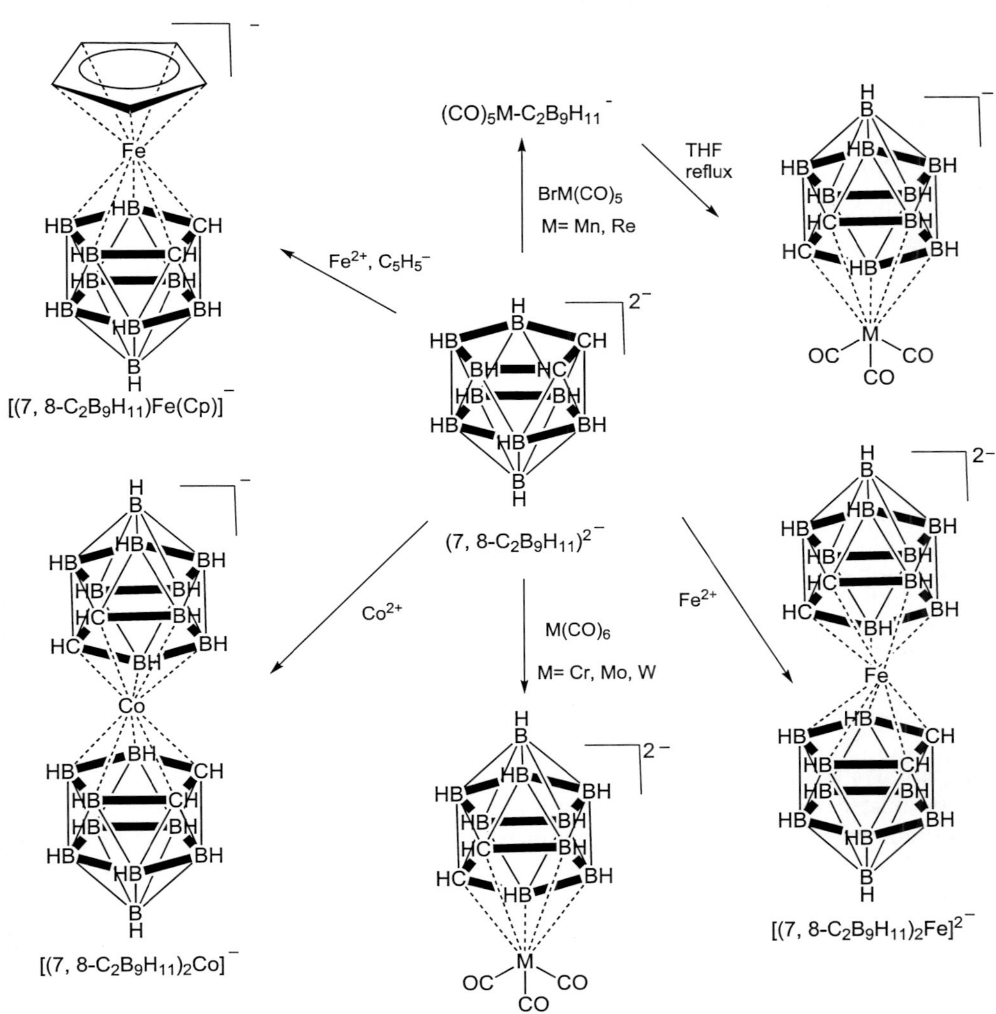

Scheme 3.7 Formation of metallacarboranes from carboranes

by the overlap requirements. The B_{12} icosahedral unit is present even in elemental boron. So if a mono-anion based on the icosahedral carborane is made, due to its extraordinary stability H^+ as its counterion will be very weakly bound. Therefore it will provide H^+ readily, making it a very strong acid. The carborane superacid $H(CHB_{11}Cl_{11})$, first prepared by Christopher Reed in 2004, is one million times stronger than sulfuric acid. The reason for this high acidity is that the acid anion $CH\,B_{11}Cl_{11}^-$ is highly stable and is substituted with electronegative chlorines. $H(CHB_{11}Cl_{11})$ and related carborane acids are known to protonate C_{60} fullerene without decomposing it. Additionally, it also forms a stable, isolable salt with protonated benzene, $C_6H_7^+$.[28]

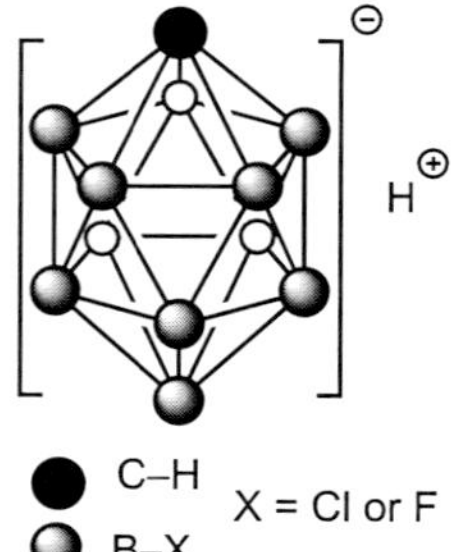

Though the boron-based solid carborane superacid is not as strong as fluoroantimonic acid and has a pH around −18, it is also noncorrosive and is a solo superacid unlike many other superacids which are mixtures of a Lewis acid and a Brønsted acid. In 2014, Reed and coworkers reported an even stronger carborane-based solo superacid—$H(CHB_{11}F_{11})$. This superacid was found to even protonate and isomerise butane, and protonate CO_2 as well.[29]

$$+\ H(CHB_{11}F_{11}) \longrightarrow \quad (CHB_{11}F_{11})^- \ + H_2 \tag{3.35}$$

$$CO_2\ +\ H(CHB_{11}F_{11}) \longrightarrow [H(CO_2)_2]^+\ (CHB_{11}F_{11})^- \tag{3.36}$$

3.6 Wade's and Jemmis rules

3.6.1 Wade's Rules

Kenneth Wade (1932–2014) was a faculty at Durham University UK. In the early 1970s he formulated the Wade's rules which provided a major breakthrough in the qualitative understanding of the electron deficient multicentre bonding of boron hydrides and their shape-based classification. Wade's rules define the electronic requirements of a *closo* polyhedron to have n+1 skeletal electron pairs where n is the total number of vertices of the polyhedron. Wade's rules have been very popular as they could be extended to other systems such as *nido* and *arachno* to rationalise the diverse structural patterns shown by polyhedral boranes.[30]

According to Wade, his rule correlates skeletal structures of boranes, carboranes, heteroboranes and their anions (*closo, nido, arachno, hypho*) with the number of skeletal electron pairs they contain. The rule states that a cluster having n skeletal atoms (vertices) will adopt a *closo* structure if it is held together by n+1 skeletal bonding electron pairs, *nido* if held together by n+2 skeletal electron pairs, *arachno* if held together by n+3 skeletal electron pairs, *hypho* if held together by n+4 skeletal electron pairs and *klado* if held together by n+5 skeletal electron pairs.

To apply this rule, one has to first determine the number of skeletal electron pairs in a cluster. Each BH unit furnishes 2 skeletal bonding electrons, each B as such gives three skeletal electrons, each C–H unit of a carborane furnishes 3 skeletal bonding electrons and each additional H· furnishes 1 skeletal bonding electron. Ionic charges must be included in the electron count. Extending this to borane clusters with other hetero-elements, one may replace C, Si, Ge and Sn of a cluster with a BH unit; N, P and As with a BH_2 unit and S and Se with a BH_3 unit for counting purposes.[31] Some examples are given below.

B_5H_9	$B_2H_7^-$	$C_2B_{10}H_{12}$	$C_2B_9H_{11}^{2-}$
5BH = 10e 4H = 4e 14e = 7 pairs $n = 5$ $n+2$ = *nido*	2BH = 4e 5H = 5e 1 (–) =1e 10e = 5 pairs $n = 2$ $n+3$ = *arachno*	2CH = 6e 10BH = 20e 26e = 13 pairs $n = 12$ $n+1$ = *closo*	2CH = 6e 9BH = 18e 2 (–) =2e 26e = 13 pairs $n = 11$ $n+2$ = *nido*
$(MeSi)_2B_{10}H_{10}$	**$CPB_{10}H_{11}$**	**SB_9H_{11}**	
Considering Si≡BH $Me_2B_{12}H_{12}$ 2Me = 2e 12BH = 24e 26e = 13 pairs $n = 12$ $n+1$= *closo*	Considering C≡BH $PB_{11}H_{12}$ Considering P≡BH_2 $B_{12}H_{14}$ 12BH = 24e 2H = 2e 26e = 13 pairs $n = 12$ $n+1$= *closo*	Considering S≡BH_3 $B_{10}H_{14}$ 4H= 4e 10BH = 20e 24e = 12 pairs $n = 10$ $n+2$= *nido*	Considering S has 6 valence electrons, 2 used for vertices; remaining 4 electrons for skeletal B_9H_9 = 18 2H= 2e S= 4e 24e = 12 pairs $n = 10$ $n+2$= *nido*

3.6.2 Jemmis *mno* Rules

Eluvathingal D Jemmis and his coworkers came out with a set of rules named *mno* rules, which can be used as an extension of Wade's rules for polyhedral boranes to macropolyhedral boranes and the Huckel rule to three dimensions. As stated by a fellow cluster chemist Thomas P Fehlner, '*Jemmis draws together a number of corollaries to Wade's rules into a single, easily employed general rule.*'[32]

Jemmis mno rule states that m+n+o skeletal electron pairs are necessary for a closed macropolyhedral system to be stable. [$(m + n + o + p)$ where p is also present.]

Here, m = number of individual polyhedra, n = number of vertices, o = number of single atom bridges between two polyhedra and p = number of vertices missing for open polyhedra if present. For example, for *nido*, $p = 1$ and *arachno*, $p = 2$; for a benzene ring or cyclopentadienyl ring, $p = 2$ and if it is already bound to an metal as η^6 or η^5, $p = 1$.

For transition metals, one needs to know the oxidation state and take the number of electrons for the electron count the same as the electrons lost by the neutral metal atom to reach that oxidation state. For example, Fe^{3+} gives 3 electrons. Conversely one can also use compliance with Jemmis rules for stability to arrive at the favoured oxidation state of a transition metal in that polyhedron.

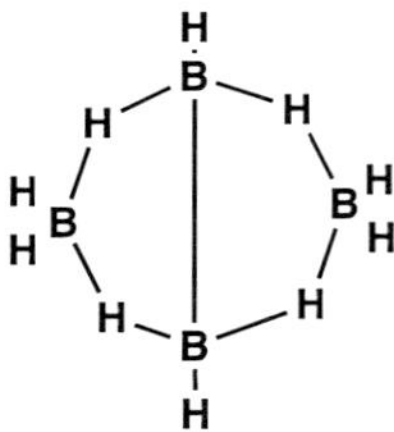

$m = 1, n = 4, o = 0, p = 2$: 7 pairs

skeletal electron count
4 BH = 4 pairs
6 H = 3 pairs
total 7 pairs
STABLE

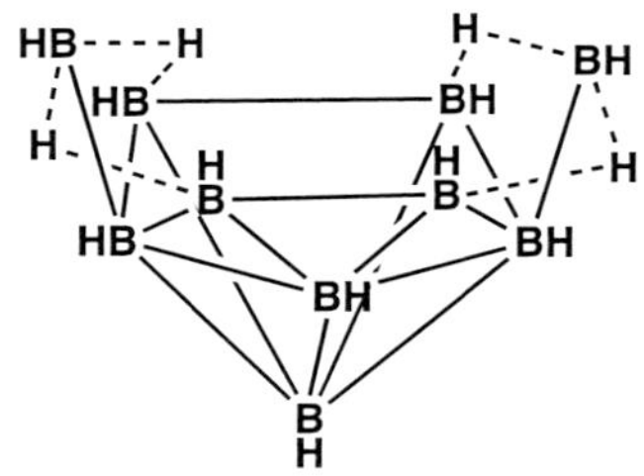

$m = 1, n = 10, o = 0, p = 1$: 12 pairs

skeletal electron counting
10 BH = 10 pairs
4 H = 2 pairs
total 12 pairs
STABLE

It is not necessary that the bridging hydrogen atoms should be around. Instead it could just have C replacing B so that $B_{12}H_{12}$ (–2) could have the corresponding *nido* structure as one of the many options from the list: $B_{11}H_{11}$ (–4), $C_2B_9H_{11}$ (–2), $C_2B_9H_{13}$, $B_{11}H_{15}$ and so on. $C_2B_9H_{11}$ (–2) is one of the *nido* structures that is known.

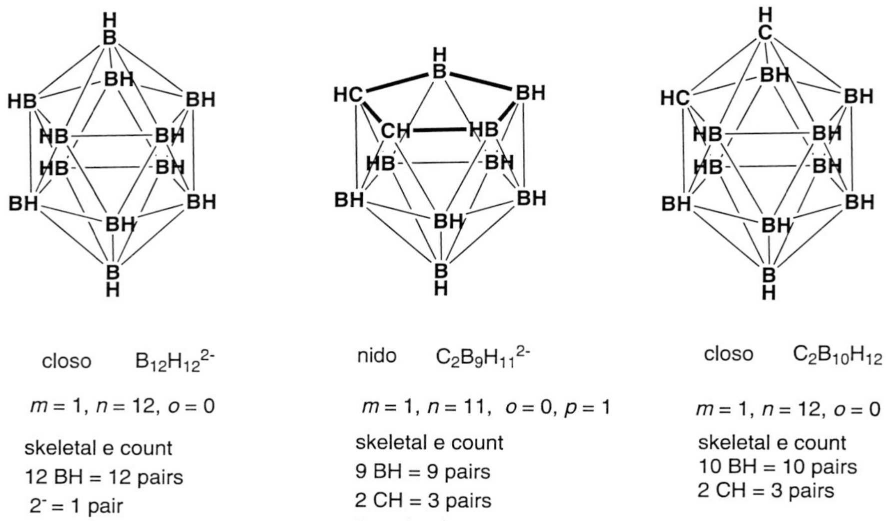

Jemmis *mno* rule is more universal as it can also be applied to metallocenes, polyhedron having metallocenyl fragments and even cyclic organic and inorganic compounds. The important difference is that Wade's rules are applicable to a single polyhedron while Jemmis rule is applicable to systems with any number of polyhedra condensed with each other. It reduces to Wade's rules when $m = 1$ (one polyhedra) and $o = 0$ (which is obvious when there

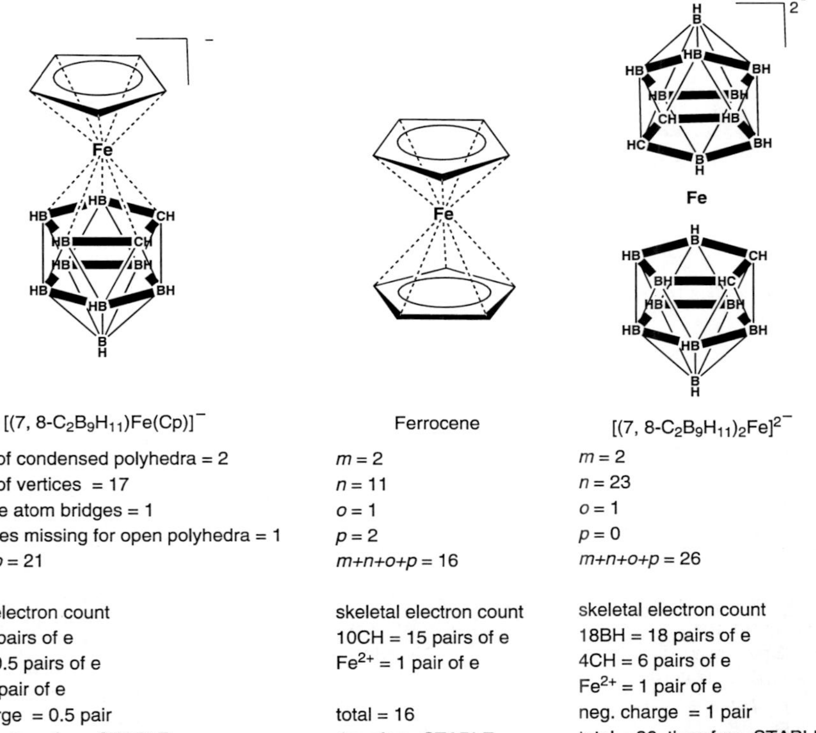

is only one polyhedron). With more than one polyhedron, Wade's rule cannot be used in a general way.[33] Worked out examples of Jemmis rules are provided on pages 79 and 80.

Wade's rule can also be used in principle for cyclic aromatic systems, but only for single ring systems, for example, benzene but not naphthalene. The Jemmis mno rule can be used for condensed systems. It is easily seen that the *mno* rule is the same as the Hückel rule if we add the sigma electrons to the Hückel π-electron count. One can do this for naphthalene and check. $4n+2$ where $n = 2$; so 10 π electrons. There are 11 C–C bonds, so, 22 σ electrons. *mno* rule for naphthalene gives the same.

3.7 Trispentafluorophenylborane: $B(C_6F_5)_3$

$B(C_6F_5)_3$ is a strong Lewis acid which is relatively inert and is hydrolytically more stable than BF_3. It is well known as an activator in Zeigler Natta and metallocene-based olefin polymerisation reactions, it is also a catalyst and reactant in organic and organometallic transformations. The strong Lewis acidity, which has been determined to be between that of BF_3 and BCl_3, and substantive steric bulkiness makes its chemistry useful in diverse

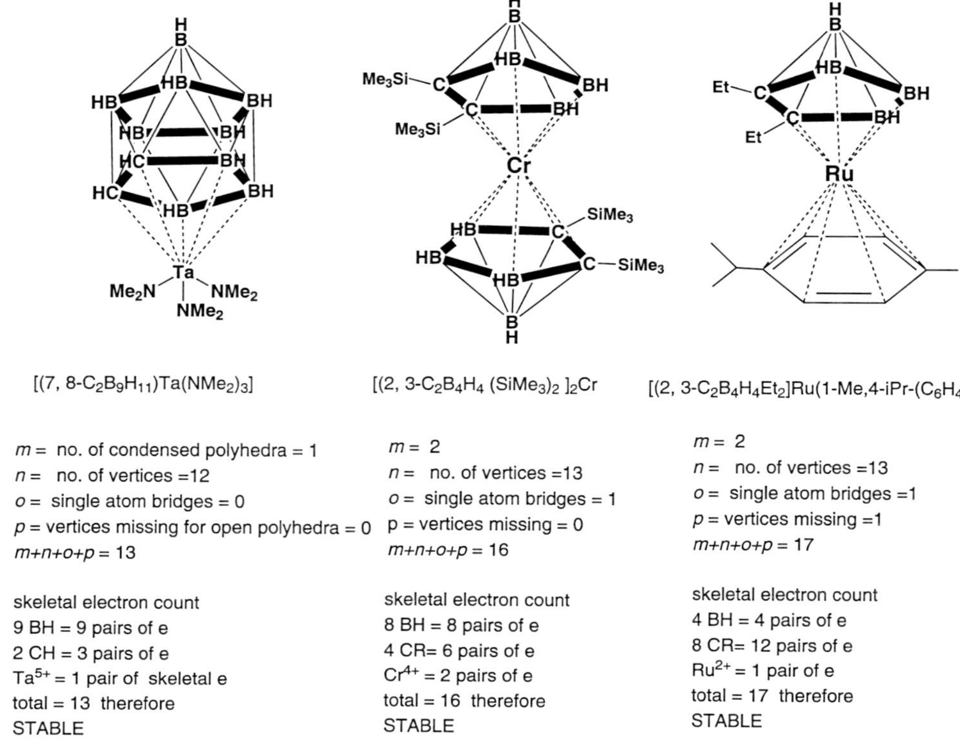

applications, the most well-known being the hydride and alkyl group abstraction from titanium group sandwich compounds.[34,35]

Tris(pentafluorophenyl)borane is prepared using a Grignard reagent derived from bromopentafluorobenzene.[36]

$$3\ C_6F_5MgBr + BCl_3 \longrightarrow (C_6F_5)_3B + 3\ MgCl \qquad (3.37)$$

Bromopentafluorobenzene can in turn be made from benzene as per the following sequence for reactions.

$$C_6H_6 \xrightarrow[\text{2. KF/HF}]{\text{1. } Cl_2/AlCl_3 \text{ cold, dark}} C_6F_6 \xrightarrow[\text{Mg, } PMe_3]{Cp_2ZrCl_2} C_6F_5H \xrightarrow{Br_2/AlCl_3} C_6F_5Br \qquad (3.38)$$

Because of its versatility and the relative inertness of the B–C bonds, $B(C_6F_5)_3$ is considered as an ideal Lewis acid. Related fluoro-substituted boron compounds, such as those containing B–CF_3 groups, decompose with the formation of B–F bonds. $B(C_6F_5)_3$ is well known as:

(i) an activator in Zeigler Natta and metallocene-based olefin polymerisation reactions,
(ii) for preparing frustrated Lewis acid base pairs (FLPs) for activation of H_2 and other small molecules, and,
(iii) for preparing Xe–C bonded compounds.

Details of (ii) and (iii) reactions are described elsewhere in this book.

3.8 Boronic Acids

A boronic acid is an alkyl or aryl substituted boric acid having the general formula $RB(OH)_2$ belonging to the larger class of organoboranes. Like the rest of the boron compounds, boronic acids act as Lewis acids. These compounds came to prominence after the landmark discovery of the Suzuki coupling reaction by Akio Suzuki in 1976 which is one of the most sought after aryl halide cross coupling reactions having excellent functional group tolerance using palladium as catalyst. Boronic acids are also capable of forming reversible covalent complexes with sugars, amino acids, hydroxamic acids, and so on. The pK_a of a boronic acid is ~9.[37]

Boronic acids are mainly used in organic chemistry as building blocks in palladium catalysed Suzuki coupling reactions. The key step of this reaction is transmetallation of its organic residue to palladium or nickel. The non-toxicity, ease of synthesis and low cost of boronic acids provides an edge to Suzuki coupling in the pharmaceutical industry over similar cross coupling reactions.

The most common synthetic route to boronic acids is the reaction of organolithium and organomagnesium compounds with borate esters. For example, phenylboronic acid is produced by the reaction of phenylmagnesium bromide and trimethyl borate followed by hydrolysis.

$$PhMgBr + B(OMe)_3 \longrightarrow PhB(OMe)_2 + MeOMgBr \tag{3.39}$$

$$PhB(OMe)_2 + H_2O \longrightarrow PhB(OH)_2 + MeOH \tag{3.40}$$

Another method is reaction of an arylsilane ($RSiR_3$) with boron tribromide (BBr_3) in a transmetallation to $RBBr_2$ followed by acidic hydrolysis.

$$RSiR_3 + BBr_3 \longrightarrow RBBr_2 \xrightarrow[H_2O/H^+]{} RB(OH)_2 \tag{3.41}$$

3.9 Boronate Esters, MIDA Boronates and Organotrifluoroborates

Boronic acids have become very important for Suzuki coupling reactions and a wide range of acids were prepared. However it was observed that they do have some shortcomings and to address these shortcomings, new modified organoboron based reagents were prepared.[38]

Boronate ester MIDA boronate Organotrifluoroborate

Boron compound	Advantages	Disadvantages
Boronic acid	Inexpensive. Wide range of reagents. Pharma-friendly since they are non-toxic and ecofriendly. Aryl boronic acids are very stable and easy to handle.	Instability of alkyl, alkenyl and alkynyl boronic acids—difficulty in purification and for extended storage. 2-pyridyl boronic acid is unstable—undergoes protodeborylation. Uncertain stoichiometry—due to presence of anhydrides of boronic acid (boroximes), polar solvents required. Relatively harsh reaction conditions needed.
Boronate ester[38]	Higher stability of aliphatics. No solvent specificity. No stoichiometry related problems.	Lesser reactivity than acids. Bispinacolato diboron and catechols are expensive. Cumbersome as many additional steps in synthesis compared to boronic acids.
MIDA boronate[39]	Higher reactivity due to sp^3 boron. Extended periods of bench-top storage possible, good solubility. In situ slow release of boronic acid. MIDA=N-methyliminodiacetic acid.	Expensive. An effective protecting group for boronic acids, facilitating various reactions such as oxidation, cross-coupling reactions and aldol reaction.
Organotrifluoro-borate[40]	Higher reactivity due to sp^3 boron. Easy to make with KHF_2. Less expensive, not hygroscopic. Extended storage and high stability for alkyl, alkenyl and alkynyl groups. More reactive than boronic acids.	Polar solvents required. Insoluble in hydrocarbons, ether and dichloromethane. Electron donating group.

The synthesis of boronate esters, MIDA boronates and organotrifluoroborates are shown in Scheme 3.8.

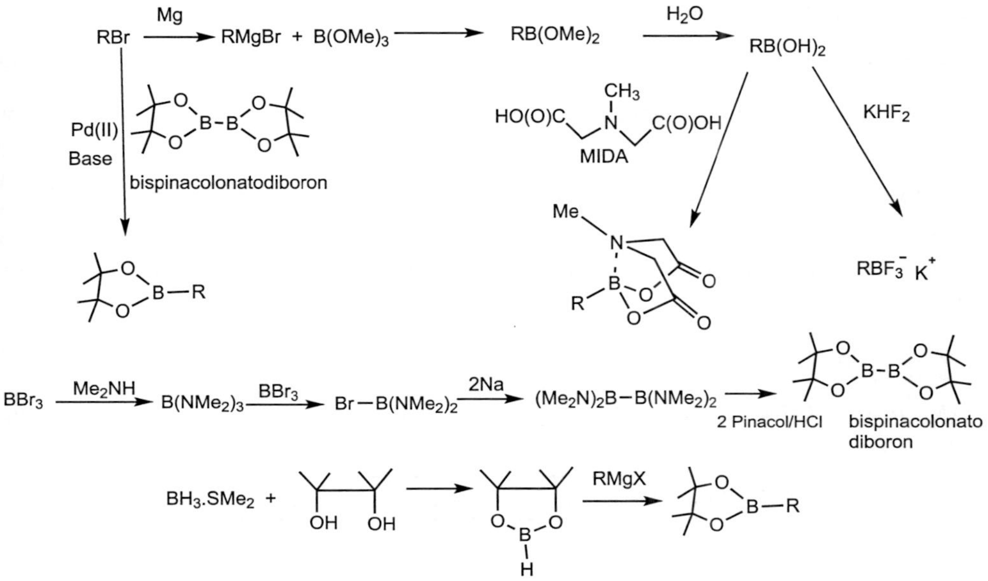

Scheme 3.8 Synthesis of boronate esters, MIDA boronates and organotrifluoroborates

3.10 Boron-Nitrogen Compounds

3.10.1 Borazine

Borazine is a six-membered inorganic heterocyclic compound which shows many physical properties similar to benzene and has therefore been called inorganic benzene. Borazine is isostructural with benzene and structural studies indicate that the bond length of the six B–N bonds is 1.436 Å. The carbon–carbon bond in benzene is slightly shorter than a double bond at 1.42 Å. The B–N bond distance is between that of the B–N bond (1.51Å) and the B=N bond (1.31 Å). This suggests partial delocalisation of the nitrogen lone pair. It is a colourless liquid with an aromatic smell. In water it hydrolyses to H_3BO_3, ammonia and hydrogen. Borazine is thermally very stable with $\Delta H_f = -531$ kJ/mol.[41]

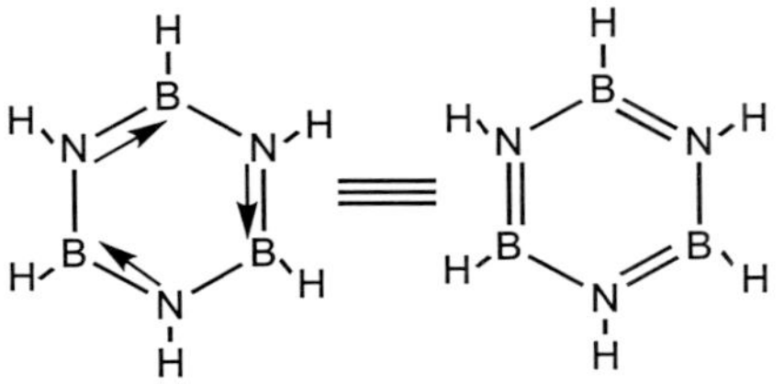

Borazine was first reported in 1926 by Alfred Stock and Erich Pohland by a reaction of diborane with ammonia at 250–300°C with a conversion of 50%.

$$3\ B_2H_6 + 6\ NH_3 \longrightarrow 2\ B_3N_3H_6 + 12\ H_2 \quad (3.42)$$

An alternative and more efficient route uses lithium borohydride and ammonium chloride.

$$3\ LiBH_4 + 3\ NH_4Cl \longrightarrow B_3H_6N_3 + 3\ LiCl + 9\ H_2 \quad (3.43)$$

In another two-step process for borazine, boron trichloride is first converted to trichloroborazine. The B–Cl bonds are subsequently converted to B–H bonds by using sodium borohydride.

$$3\ BCl_3 + 3\ NH_4Cl \longrightarrow Cl_3B_3H_3N_3 + 9\ HCl \quad (3.44)$$

$$2\ Cl_3B_3H_3N_3 + 6\ NaBH_4 \longrightarrow 2\ B_3H_6N_3 + 3\ B_2H_6 + 6\ NaCl \quad (3.45)$$

Properties	**Borazine**	**Benzene**
Mol. wt	80.5	78.1
b.p.	55°C	80°C
Density	0.81	0.81
Trouton const.	89.5	88.2
Critical temp.	252°C	288°C
Planarity	Planar	Planar
B–N/C–C Distance Å	1.436	1.42

Borazine has the same colligative properties as benzene, however the two are very different chemically. The electron density along the boron–nitrogen bond is not distributed evenly, due to the difference in electronegativities between the two atoms. Therefore, the molecular orbitals of the system are lumpy in appearance. This uneven distribution makes borazine susceptible to addition reactions.

$$B_3N_3H_6 \xrightarrow{+Br_2} B_3N_3H_6Br_6 \xrightarrow{-HBr} B_3N_3H_3Br_3 \quad (3.46)$$

$$C_6H_6 \xrightarrow[FeBr_3]{Br_2} C_6H_5Br \quad (3.47)$$

Heating of $B_3N_3H_6$ at 70°C results in the removal of H_2 with the formation of a borazinyl polymer or polyborazylene, in which the monomer units are coupled in a para fashion by new boron–nitrogen bonds. On heating polyborazylene to 1000°C, boron nitride is formed. Borazine can also be used to grow boron nitride thin films on to surfaces.[42]

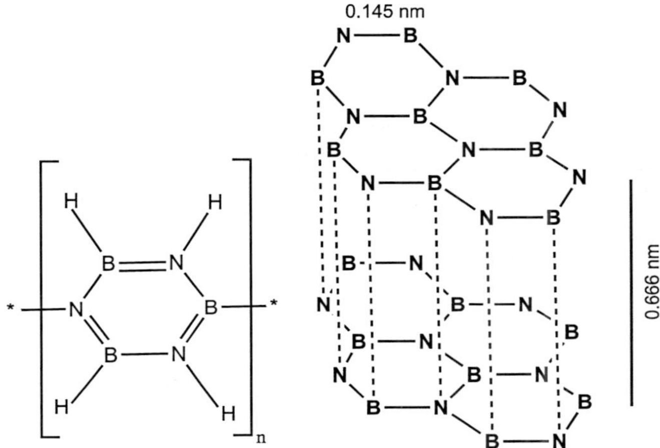

3.10.2 Boron Nitride

Boron nitride is a solid with the chemical formula BN, isoelectronic to a similarly structured carbon lattice but having both boron and nitrogen in equal numbers. Quite interestingly, BN, similar to allotropes of carbon exists in three different crystalline forms—hexagonal boron nitride (h-BN), cubic boron nitride (c-BN) and wurtzite boron nitride. The hexagonal form has considerable similarities to graphite; it is the most stable and softest among the BN polymorphs while the cubic variety has a structure and properties analogous to diamond. Its hardness is inferior only to diamond, but its thermal and chemical stability is superior. The wurtzite BN is rarely observed but its hardness is predicted to be even higher than the cubic form. Quite interestingly, similar to carbon nanotubes, boron nitride nanotubes have been prepared and characterised.[43, 44]

Boron nitride is produced synthetically. Hexagonal boron nitride is obtained by reacting boron trioxide (B_2O_3) or boric acid [$B(OH)_3$] with ammonia (NH_3) or urea [$CO(NH_2)_2$] in a nitrogen atmosphere. It is also obtained from calcium boride, CaB_6 which by itself is obtained by the reaction of CaO, B_2O_3 and Mg.

$$B_2O_3 + 2\,NH_3 \longrightarrow 2\,BN + 3\,H_2O\ (T = 900^\circ C) \tag{3.48}$$

$$B(OH)_3 + NH_3 \longrightarrow BN + 3\,H_2O\ (T = 900^\circ C) \tag{3.49}$$

$$B_2O_3 + CO(NH_2)_2 \longrightarrow 2\,BN + CO_2 + 2\,H_2O\ (T > 1000^\circ C) \tag{3.50}$$

$$B_2O_3 + 3\ CaB_6 + 10\ N_2 \longrightarrow 20\ BN + 3\ CaO\ (T > 1500°C) \quad (3.51)$$

Boron nitride is also prepared from borazine by pyrolysis.

The methods used for the artificial synthesis of diamond are also used for the synthesis of c-BN. Hexagonal boron nitride when subjected to high pressure and temperature (pressures between 5 and 18 GPa and temperatures between 1730 and 3230°C), gives c-BN.[44,45]

Table 3.4 Comparing hexagonal boron nitride with graphite

Property	Hexagonal boron nitride	Graphite
Stability to heat	1000°C (air), 2800°C (inert atm)	700°C (air)
Hardness (Mohs scale)	1–2	1–2
Colour	White (also called white graphite)	Black
Density	~2.1	2.09–2.23
Bandgap (eV)	5.2	−0.04
B–N/C–C distance	1.45 Å	1.42 Å
Interlayer distance	3.30–3.33 Å	3.33–3.35 Å
Interlayer interaction	More	Less
Electron delocalisation	Less	More
Electrical conductivity	Insulator	Conductor (in the direction of the plane)
Chemical reactivity	Inert	Reactive
Intercalation	Difficult (K, NH_3)	Easy
Lubricant properties	Both high and low temperature up to 900°C in oxidising atmosphere. Can be used in vacuum also (space applications)	Water vapour necessary. Not effective in vacuum. Useful till 450°C in oxidising atmosphere

Table 3.5 Comparing cubic boron nitride with diamond

Property	Cubic boron nitride	Diamond
Stability to heat	1300°C (air), 1525°C (inert atm., conversion to h-BN) 2973°C sublimes	720–800°C ignition point in oxygen, 1700°C in argon. Above 1700°C converts to graphite in vacuum
Hardness (Mohs scale)	9.5–10	10
Thermal conductivity	13 W/(cm·K) (predicted)	High, 4 times that of copper 22 W/(cm·K)

Cont'd

Table 3.5 Cont'd

Property	Cubic boron nitride	Diamond
Thermal expansion		Very low
Colour	White (also called white graphite)	White with high optical dispersion
Density	~2.1	3.52
Bandgap (eV)	6.4	5.47
B–N/C–C distance	1.45 Å	1.42 Å

Boron nitride nanotubes were predicted in 1994 and experimentally made in 1995. Although structurally similar to carbon nanotubes, their properties are found to be different. While carbon nanotubes can be metallic or semiconducting depending on the direction of rolling, a boron nitride nanotube is an electrical insulator with a bandgap of ~ 5.5 eV. They are much more stable than carbon nanotubes thermally and chemically.[45]

3.11 Boron-Based Materials for Nuclear Applications

Boron-10 finds use in two nuclear related applications. These are (i) as control rods/shutoff material in the nuclear industry, as sensors for neutron counting and as materials for making neutron shields, (ii) in boron neutron capture therapy, a cancer treatment therapy in which boron-10 based compounds after getting concentrated in cancer affected cells are irradiated with thermal (slow) neutrons from outside resulting in the release of high linear energy α-particles and recoiling lithium-7 nuclei whose combined path length is only around 9–12 μm which is approximately the diameter of a cell. The basic reaction for both these applications is the same (Fig. 3.5).[46]

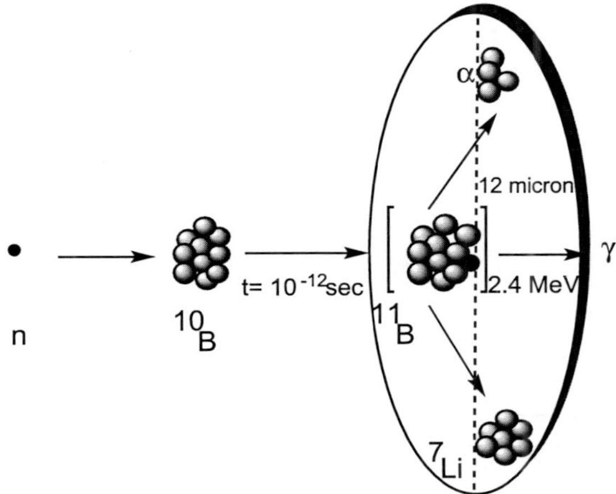

Fig. 3.5 Pictorial depiction of boron neutron capture therapy

3.11.1 Boron-Based Control Rods in Reactors

Boron compounds such as boric acid, boron carbide, rare earth and refractory metal borides find use as neutron sensors, instrument shielding against neutrons, control shutoff rods and in storage of nuclear fuel. The thermal neutron cross section for ^{10}B and ^{11}B are 3835 barn and 0.005 barn respectively. The reaction products – α particles and lithium – are formed as stable, non-radioactive isotopes. Boric acid is used as poison in moderator/coolant of research reactors, boric acid and boron carbide powder are used in concrete while constructing nuclear reactor buildings, and ^{10}B enriched boron is used in neutron detectors. Boron carbide powder and pellets made of enriched boron are used as control rod materials. In addition, polymer material such as bocarsil (boron carbide in silicone rubber) polyboron (boric acid in polyethylene) are used in parts for neutron shields.[47]

3.11.2 Boron Neutron Capture Therapy (BNCT)

This cancer treatment method has been primarily attempted for the treatment of *Glioblastoma multiforme* (GBM), which is the most common and most aggressive malignant primary brain *tumour found* in humans which affects the glial supportive tissues of the central nervous system. The method relies on targeting the tumour cells by an appropriate boron-10 containing chemical compound which preferentially concentrates in the cancer cells. There are many nucleids such as gadolinium-157 (2,59,000 barns) cadmium-113 (20,600 barns), boron-10 (3835 barns) and indium-115 (200 barns) that have a high propensity for absorbing low energy neutrons (thermal neutron cross section). However ^{10}B is the most attractive as (i) it has 20% naturally occurring boron, (ii) the particles emitted after it is hit by thermal neutrons are harmless and non-radioactive (He and ^{7}Li) and Li has high linear energy transfer, (iii) the combined path-length of the daughter nuclei is only 12 microns which is approximately the diameter of a single living animal cell, and (iv) boron has no toxicity and its chemistry is well understood to make suitable water soluble derivatives with high boron content.[48,49]

For BNCT to be successful, the required criteria are (i) selective uptake of the boron-containing compound by the tumour cells over the normal cells, (ii) concentration of boron should be of the required level for releasing a therapeutic dose (~10^9 atoms of ^{10}B per cell), (iii) retention of ^{10}B in the tumour cell alone and not in normal tissue and blood, and (iv) easy and cost effective synthesis of the boron compound. Three such compounds that have been approved by federal drug approving agency (FDA) of the USA are shown in Fig. 3.6.

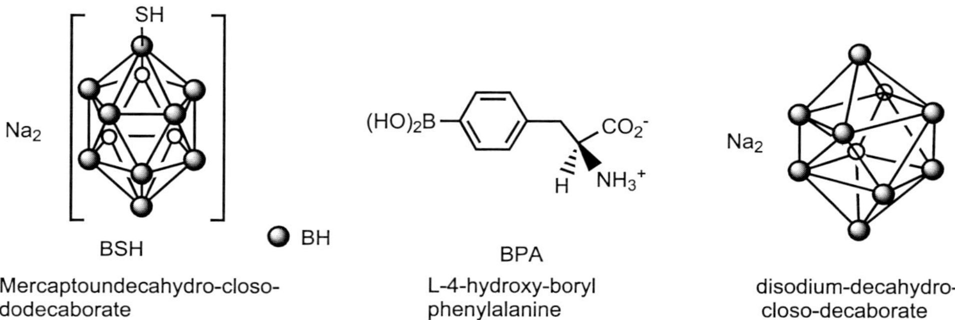

Fig. 3.6 Some examples of boron rich compounds tried for BNCT

The use of these compounds in BNCT against GBM, melanoma, and head and neck cancers have met with limited success. Efforts have been directed to make liposomes as carrier agents for boron-rich compounds since liposomes have been found to accumulate in most tumour cells with enhanced permeability and retention. The following boron-rich anionic borane, TAC and carborane MAC have been found to be very effective for concentrating boron in cancer cells (Fig. 3.7). Liposomal delivery of TAC, a trisodium salt of a dimeric decaborane $Na_3[1\text{-}(2'\text{-}B_{10}H_9)\text{-}2\text{-}(NH_3B_{10}H_8)]$ has been found to provide large therapeutic concentration of boron in the tumour cell with long periods of its retention. After BNCT, diminished tumour growth was observed with TAC and MAC introduced through liposomes in mouse cancer cells.[50–53]

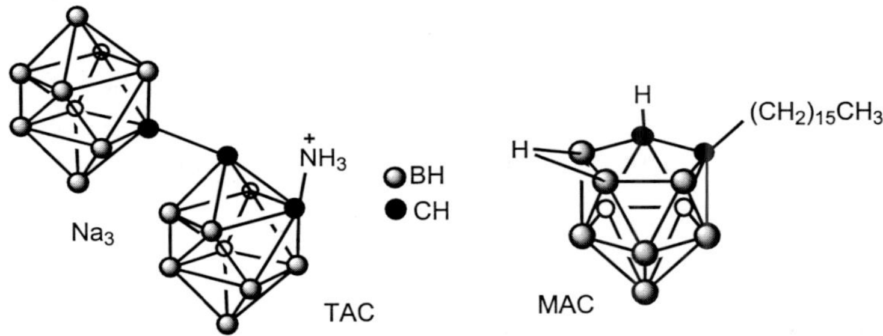

Fig. 3.7 TAC and MAC: boron-rich compounds tried for BNCT

Table 3.6 gives in a nutshell the major industrial applications of boron-based compounds.

Table 3.6 Industrial applications of boron based compounds in a nutshell

Boron compound	Property	Application
Borax (sodium tetraborate)	Replaceable sodium	Detergent additive
	Micronutrient	Fertiliser
	Registered poison	Pesticide for cockroach control
	Mild alkalinity	Detergent additive for softening water and neutralise acidity
	Fire retardant property	Flame proofing of cotton mattresses and to make glass wool
	Phyto-toxicity	Additive in herbicides
	Corrosion resistance	For making corrosion resistant alloys
Sodium perborate	Release of H_2O_2 on hydrolysis	Bleach for washing and reagent for oxidation reactions
Boric oxide	Additive in glass industry Break resistance of borosilicate glass	Laboratory glassware, unbreakable cover and domestic glass, Pyrex, Borosil and so on
	Additive in ceramic industry for making vitreous enamels	Ceramic industry

Cont'd

Table 3.6 Cont'd

Boron compound	Property	Application
Boric acid	Reagent for oxidation	Petrochemical industry: oxidation of cyclohexane to cyclohexanil and cyclohexanone
Aryl boronic acid	High stability and non-toxicity	In pharmaceuticals for introducing aryl, alkenyl units by Suzuki cross-coupling
Carborane	High stability	Boron rich compounds for BNCT
Boron carbide	Hardness	Bullet proof vests, armour
	High melting and boiling points	As control rods in nuclear reactors
Boranes	High heat of combustion	Propellants and explosives
Boron nitride	Hardness (cubic form) Layer (hexagonal form)	Materials, ceramics, abrasives, intercalation, lubricant in vacuum

3.12 Recent Developments

3.12.1 Compounds with Multiple Bonds Between Boron Atoms

One of the recent developments in boron chemistry has been on making hitherto unknown multiple bonds between boron atoms. The first example of a neutral compound having a B=B bond was prepared in 2007 by G H Robinson and coworkers using KC_8 as a reducing and dehalogenating agent on a BBr_3 adduct of an N-heterocyclic carbene (NHC). The compound was structurally characterised and indicated two NHCs to coordinate to the HB=BH unit in a trans manner with a B=B distance of 1.560 Å. This distance is shorter than the B–B single bond distance of an analogous NHC stabilised compound (1.828 Å).[54]

$$\text{NHC(Dipp)}\rightarrow BBr_3 \xrightarrow[\text{rt, 6 d}]{KC_8,\ Et_2O} \text{(NHC)HB=BH(NHC)} \qquad (3.52)$$

Dipp = 2,6 diisopropylphenyl

The first example of a stable diboryne having a B≡B and stabilised by N-heterocyclic carbenes was prepared by Holger Braunschweig in 2012 (Eq. 3.53). Emerald-green crystals were obtained for this stable compound containing a triple bond between two boron atoms, a feat that had previously been limited to only two other non-metal first row elements—carbon and nitrogen.[55] Carbon monoxide was found to add to this triple bond at room temperature and atmospheric pressure resulting in a bis(boralactone) in which four molecules of CO were found to be incorporated thus indicting a metal free binding and coupling of CO on a boron–boron triple bond.[56]

$Na^{+}C_{10}H_8^{-}$, –78°C

Dipp = 2,6 diisopropylphenyl (3.53)

CO, CO (3.54)

3.12.2 Borophene, an Analogue of Graphene

In 2015, the research group at Brown University led by Lai-Sheng Wang, prepared a boron analogue of graphene having a slightly curved two dimensional array of 36 boron atoms. Neutral B_{36} is a boron cluster which is almost planar and two dimensional possessing a six-fold symmetry with a hexagonal vacancy at the centre. Borophene is predicted to show metallic character typical of anisotropic 2-D metals, unlike graphene which is a semi-metal. The B–B bonds in borophene are nearly as strong as the C–C bonds in graphene. This small cluster provides the potential basis for extended two dimensional boron sheets. More examples of borophenes have been prepared since then as a single layer especially on metallic surfaces.[57]

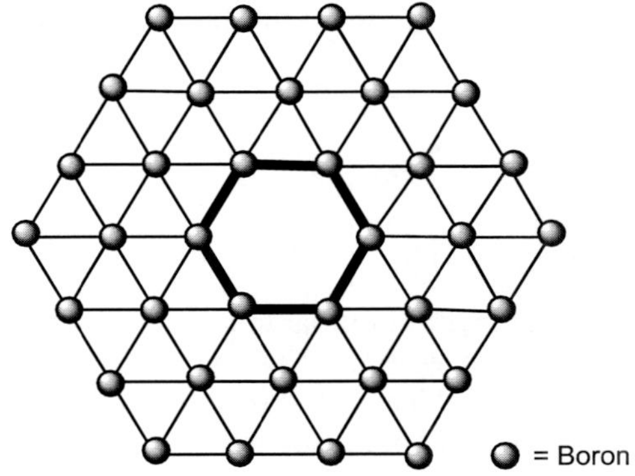

Fig. 3.8 The first two dimensional array of boron (borophene)

3.12.3 Borospherene, an Analogue of Fullerene

Borospherene is a fullerene-like cage-compound having 40 boron atoms, discovered by Lai-Sheng Wang in 2014 with a very low electron binding energy.[58] The cage-compound consists of 48 triangles, four 7-membered rings and two 6-membered rings connecting the 40 boron atoms (Fig. 3.9). Unlike fullerene, the surface of borospherene is less smooth as several boron atoms stick out from the doughnut-like shape and exhibit unusual heptagonal faces.

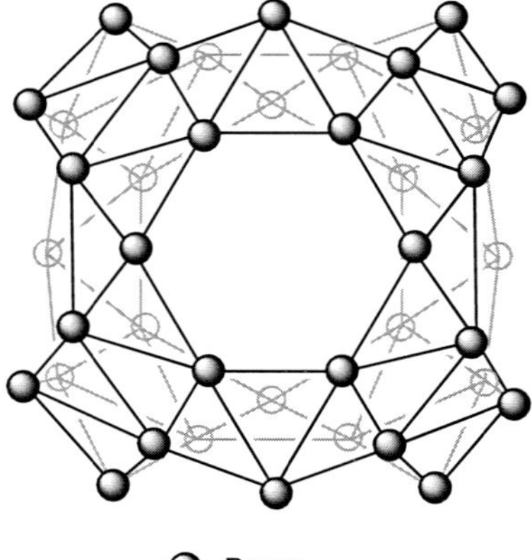

Fig. 3.9 Borospherene cage showing heptagonal, hexagonal and triangular faces

3.12.4 Boron-Based Lewis Bases

Generally boron compounds, such as boranes, are electrophilic in nature and are well known Lewis acids. During a reaction with a carbonyl compound, the boron centre in a borane binds to the oxygen atom (electron pair donor) thus making the carbon centre more susceptible to nucleophilic attack. The chemistry of tricoordinate boron is thus dominated by its electrophilic character. Segawa and coworkers prepared a $LiB(NR_2)_2$ compound in which they showed (with evidence) that the boron centre can be made nucleophilic in nature as well.

Using a bulky substituent on the nitrogen atoms of the group RNCH=CHNR (where R = 2,6-*i*-$PR_2C_6H_3$) the five-membered boron containing cyclic compound (RNCH=CHNR)BBr was prepared and it was converted to the lithium salt of its anion, (RNCH=CHNR)BLi (Eq. 3.55). The large aryl substituents on the nitrogen atoms prevent the close approach of another similar boron centre, which would have lead to the formation of a dimer during reduction process. Segawa and coworkers showed that this compound, which is isoelectronic with N-heterocyclic carbenes, serves as a boron nucleophile or Lewis base.

$$\mathrm{ArN{=}CH{-}CH{=}NAr} \xrightarrow{\text{Mg, BBr}_3} \mathrm{(ArNCH{=}CHNAr)B{-}Br} \xrightarrow[\text{Naphthalene}]{\text{Li}} \mathrm{(ArNCH{=}CHNAr)\ddot{B}^{-}\;Li^{+}} \qquad (3.55)$$

Ar = 2, 6- $(iPr)_2C_6H_3$

The cyclic diamino substituted boryllithium was found to react with a range of electrophiles. Reactions with carbonyl groups proceeded in the same way as carbanions react. For example the boryl moiety reacted with benzaldehyde followed by protonation to give α-borylbenzyl alcohol. A similar reaction with carbon dioxide followed by protonation yielded the boryl carboxylic acid. Scheme 3.9 consolidates such reactions carried out with the boryllithium compound.[59]

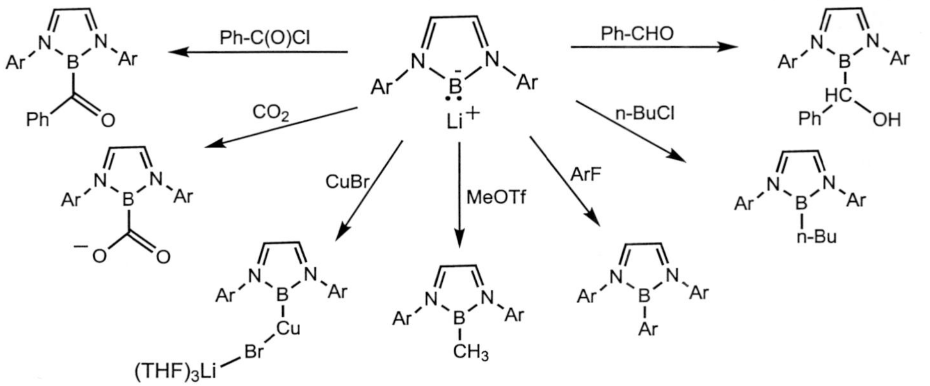
Ph-C(O)Cl
Ph-CHO
CO2
n-BuCl
CuBr
MeOTf
ArF
Li+
Ar
N
B
Ph
O
HC
OH
n-Bu
Cu
Br
(THF)3Li
CH3

Scheme 3.9 Reactions of diamino derived boryllithium with electrophiles

3.12.5 Borylenes

Borylenes are compounds of boron which are analogous to carbenes which can be depicted by the general structure of R–B: having R (an organic residue) and a pair of electrons reside on the boron. The first example of such a compound was the borylene–metal complex, $(CO)_5WBN(SiMe_3)_2$. Another way to stabilise a borylene is by using strong σ-donors such as carbenes. A recent addition to such borylenes possessing a nucleophilic centre is the bis-cyclic alkylamino carbene (cAAC) stabilised borylene (Eq. 3.56). This is prepared by the reduction of cAAC stabilised BBr_3 using KC_8, a well-known dehalogenating and reducing agent.[60]

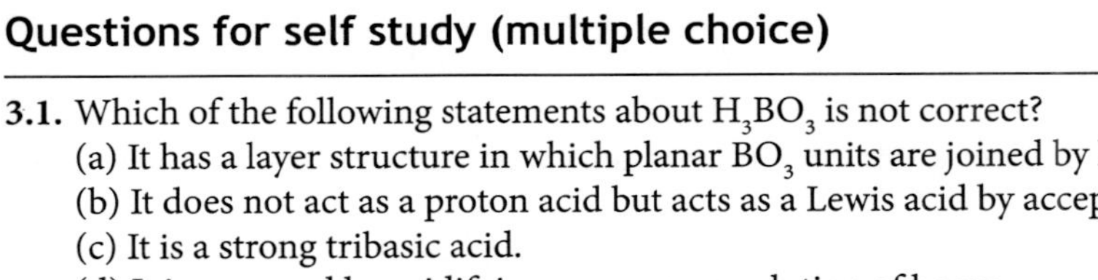
R
N
BBr3
KC8, Toluene
N–R
H
B
R = 2,6-(i-Pr)2C6H3

(3.56)

Questions for self study (multiple choice)

3.1. Which of the following statements about H_3BO_3 is not correct?
(a) It has a layer structure in which planar BO_3 units are joined by hydrogen bonds.
(b) It does not act as a proton acid but acts as a Lewis acid by accepting a hydroxyl ion.
(c) It is a strong tribasic acid.
(d) It is prepared by acidifying an aqueous solution of borax.

3.2. The type of hybridisation of boron in diborane is:
(a) sp^3 (b) sp^2 (c) sp (d) sp^3d

3.3. The tendency of BF_3, BCl_3 and BBr_3 to behave as Lewis acids decreases in the order:
(a) BCl_3>BF_3>BBr_3 (b) BBr_3>BCl_3>BF_3
(c) BBr_3> BF_3 >BCl_3 (d) BF_3>BCl_3> BBr_3

3.4. Which of the following has a structure similar to graphite?
(a) B_4C (b) B_2H_6 (c) BN (d) B_{12}

3.5. The compound added during the manufacture of laboratory glassware to make it withstand thermal shock is:
(a) BN (b) B_2O_3 (c) $NaBO_3$ (d) B_4C

3.6. Which among the following compounds is used for breaking the C–O bonds of alkyl ethers?
(a) BF_3 (b) BBr_3 (c) B_2H_6 (d) $B(OMe)_3$

3.7. Which among the following applications does not relate to boron compounds?
(a) Control rods in nuclear reactors
(b) Treatment of cancer using on-spot mild radiation
(c) Manufacture of laboratory glassware
(d) Light emitting diodes (LED)

3.8. The reason why carboranes act as superacids is:
(a) Lewis acidity of boron
(b) Acidity of boric acid released
(c) High stability of the carborane anion
(d) Electron-withdrawing nature of carborane

3.9. The statement which is not true with respect to boron nitride is:
(a) Hexagonal boron nitride shows similarity in structure and properties to graphite.
(b) Cubic boron nitride shows similarity in structure and properties to diamond.
(c) Boron nitride is white in colour.
(d) Boron nitride nanotubes show similarity in structure and properties to carbon nanotubes.

3.10. Borazine is called inorganic benzene. Which among the following statements is not true for both borazine and benzene?
(a) Both are liquids and have the same density.
(b) Both react with Br_2 forming tribromo derivatives.
(c) Both are planar molecules.
(d) Both have uniform C–C and B–N bond distances.

3.11. Among the given allotropes of boron, one that does not contain B_{12} icosahedral unit is:
(a) α-Rhombohedral (b) β-Rhombohedral
(c) γ-Boron (B_{28}) (d) Borospherene

3.12. The cross-coupling method in organic synthesis which involves boronic acids and boronate esters is:
(a) Suzuki–Miyaura Coupling (b) Negishi coupling
(c) Stille coupling (d) Sonogashira coupling

3.13. Among the given boron based compounds, the oxidising agent is:
(a) B_2O_3 (b) $NaBO_2$
(c) $Na_2B_4O_7$ (d) $Na_2B_2O_4(OH)_4$

Solved concept-based questions

3.14. BF_6^{3-} does not exist while Al_6^{3-} exists. Why?

Boron being a first row element cannot expand its valency beyond 4 due to the non-availability of suitable *d* orbitals of comparable energy. Aluminium being in the second row can expand its valency beyond 4 due to the availability of *d* orbitals.

3.15. $AlCl_3$ dimerises but BCl_3 does not dimerise. Why?

This is due to the large size difference between Al and B. The relative size of chlorine being large, boron is unable to accommodate more than 3 Cl units. (Although there are claims of BCl_4^- there are no examples which are structurally characterised by X-ray diffraction studies.) The reason why BF_3 does not dimerise is entirely different as covalently bound fluorine shows very little intermolecular interaction due to high electronegativity and also there is considerable $p\pi$–$p\pi$ interaction between B and F evident from the shorter B–F bond distance of 1.30 Å. BF_4^- also exists.

3.16. The Lewis acidity of BF_3 is less than that of BBr_3. Why?

The F atom is small compared to the larger Cl and Br atoms, and the lone pair electrons in the p_z orbital of F readily overlaps with the empty p_z orbital of boron. As a result, the π donation of F to B is greater than that of Cl or Br resulting in lower Lewis acidity of BF_3 compared to BBr_3.

3.17. BCl_3 and BF_3 hydrolyse readily but CCl_4 does not. Why?

BCl_3 and BF_3 are electron deficient and possess an empty *p* orbital; they are also triangular planar in shape. They can easily accommodate the lone pair of a water molecule becoming tetrahedral in the process. CCl_4 does not have empty *p* orbital and unlike silicon cannot use *d* orbitals to expand its valency.

3.18. Why are control rods used in nuclear reactors made of boron?

Boron has two naturally occurring and stable isotopes, ^{11}B (80.1%) and ^{10}B (19.9%). The ^{10}B isotope has one of the highest nuclear cross sections for thermal neutrons (propensity to absorb thermal neutrons). In a nuclear chain reaction, the rate is determined by the number of thermal neutrons produced in successive nuclear reactions. In fact, Gd has even better propensity for thermal neutrons but its natural abundance is less than that of boron.

Problems and exercises

3.19. Arrange BF_3, BCl_3 and BBr_3 in (a) increasing order of Lewis acidity and (b) resistance to hydrolysis of the B–X bond.

3.20. Using equations show two significant differences in the chemical reactivity of benzene and borazine.

3.21. For each of the following applications/chemical reactions, write/draw against each the most appropriate boron containing compound as reagent/catalyst from the list provided

(a) Preparing $C_6H_7^+$ (b) Olefin polymerisation (using metallocenes)

(c) Alkyl ether cleavage (d) Control rods in nuclear reactors

3.22. Using Wade's rules arrive at the skeletal electron count and shapes of the following compounds:

(a) $C_2B_7H_{15}$ (b) $C_2B_5H_7$ (c) $CB_7H_{10}Me^{2-}$ (d) $SeC_2B_8H_{10}$

(e) $C_2B_7H_{13}$ (f) $GeC_2B_9H_{11}$ (g) $SB_9H_{12}^-$

3.23. Given that the dicobaltacarborane $(Cp_2Co)_2(C_2B_8H_8)$ (macropolyhedral cluster) has two cobalt units at the 1 and 12 positions of the central icosahedrons, is overall neutral and highly stable, determine the oxidation state of the cobalt ions using Jemmis rule and skeletal electron count (same oxidation state for both cobalt units) (no marks without calculations).

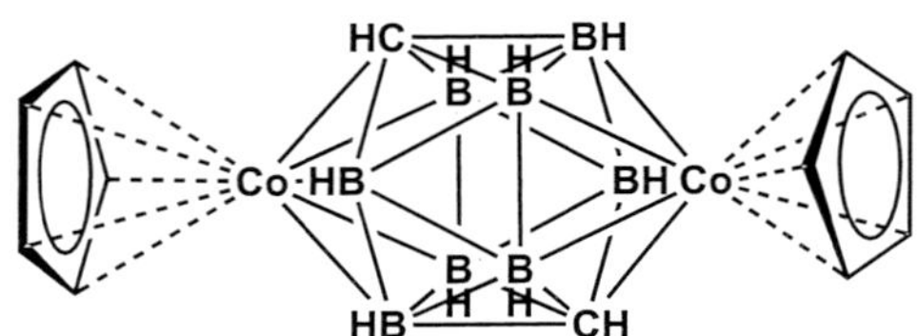

3.24. Is orthoboric acid, $B(OH)_3$ described best as a Lewis acid or a Brønsted acid? Justify your answer in a line or two. Why are borosilicate glasses unbreakable with sudden temperature variations?

3.25. In each case, between the given boron compounds A or B, choose the most appropriate reactant for preparing the given product by Suzuki coupling with aryl iodides. Your selection should consider the relative stability of the boron compounds and then minimum number of steps required.

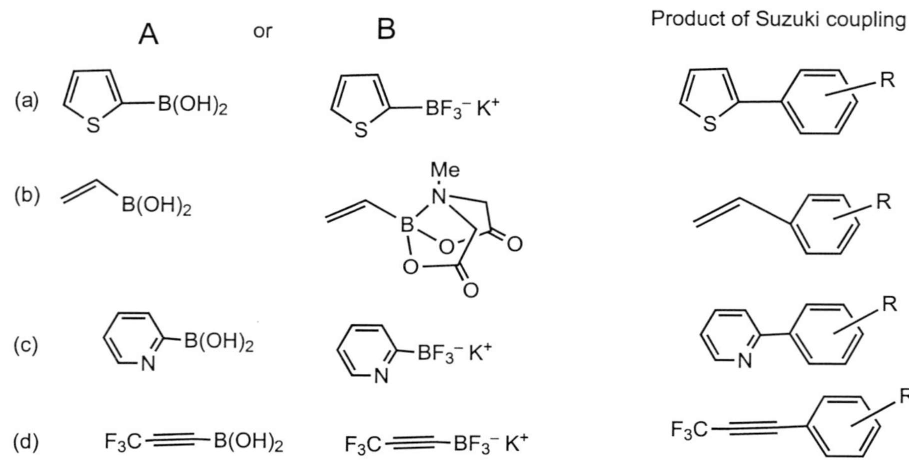

3.26. Starting from BF_3 show reactions with schemes for the preparation of para $C_2B_{10}H_{12}$? Compared to boranes, what are the advantages of carboranes?

3.27. Which is the world's strongest solo acid discovered so far (till 2015)? Write one unique reaction which supports its superior acid strength.

3.28. Vinyl and cyclopropyl boronic acids are highly unstable. Suggest and draw the structure of a possible alternative boron-based compound which is stable, easily prepared and soluble in non-polar solvents for Suzuki coupling involving vinyl and cyclopropyl groups.

3.29. Although gadolinium-157 and cadmium-113 have better nuclear cross section for thermal neutrons, they are less favoured compared to boron-10 for BNCT. Indicate four reasons for the same.

3.30. What is white graphite? What is the major difference in property between graphite and white graphite?

Answer key for self study questions

3.1. (c) 3.2. (a) 3.3. (b) 3.4. (c) 3.5. (b) 3.6. (b) 3.7. (d) 3.8. (c) 3.9. (d) 3.10. (b) 3.11. (d) 3.12. (a) 3.13. (d)

Supplementary reading

1. Wisniak J, Borax, Boric Acid, and Boron—From exotic to commodity, *Indian Journal of Chemical Technology, CSIR, New Delhi*, 2005, Vol. 12, 4.
2. Brotherton R J, Weber C J, Guibert C R and Little J L, *Boron compounds, Ullmann's Encyclopedia of Industrial Chemistry*, Weinheim: Wiley-VCH, 2005.
3. Ogitsu T, Schwegler E and Galli G, β-Rhombohedral boron: At the crossroads of the chemistry of boron and the physics of frustration, *Chem. Rev.*, 2013, Vol. 113, 3425.
4. Oganov A R, Chen J, Gatti C, Ma Y, Glass C W, Liu Z, Yu T, Kurakevych O O and Solozhenko V L, Ionic high-pressure form of elemental boron, *Nature*, 2009, Vol. 457, 863.
5. McKillop A and Sanderson W R, Sodium perborate and sodium percarbonate: Cheap, safe and versatile oxidising agents for organic synthesis, *Tetrahedron*, 1995, Vol. 51, 6145.
6. Perelygin Y P and Chistyakov D Y, Boric acid, *Russian Journal of Applied Chemistry, Pleiades Publishing*, 2006, Vol. 79, 2041.
7. Gurr G E, Montgomery P W, Knutson C D and Gorres B T, The crystal structure of trigonal diboron trioxide, *Acta Crystallographica B.*, 1970, Vol. 26, 906.
8. Balakrishnarajan M M, Pancharatna P D and Hoffmann R, Structure and bonding in boron carbide: The invincibility of imperfections, *New J. Chem.*, 2007, Vol. 31, 473.
9. Domnich V, Reynaud S, Haber R A and Chhowalla M, Boron carbide: Structure, properties, and stability under stress, *J. Am. Chem. Soc.*, 2011, Vol. 94, 3605.
10. Sowa F J, Hennion G F and Nieuwland J A, Organic reactions with boron fluoride. IX. The alkylation of phenol with alcohols, *J. Am. Chem. Soc.*, 1935, Vol. 57, 709.
11. Greenwood N N and Martin R L, Boron trifluoride coordination compounds, *Q. Rev. Chem. Soc.*, 1954, Vol. 8, 1.
12. McOmie J F W, Watts M L and West D E, Demethylation of aryl methyl ethers by boron tribromide, *Tetrahedron*, 1968, Vol. 24, 2289.
13. Doyagüez E G, Boron tribromide, *Synlett*, 2005, Vol.10, 1636.
14. Lipscomb W N, *Boron hydrides* (Dover books on chemistry), Dover Publications, 2012.
15. Bauer S H, Structures and physical properties of the hydrides of boron and of their derivatives, *Chem. Rev.*, 1942, Vol. 31, 43.

16. Schlesinger H I and Burg A B, Recent developments in the chemistry of the boron hydrides, *Chem. Rev.*, 1942, Vol. 31, 1.
17. Longuet-Higgins H C and Bell R P, The structure of the boron hydrides, *J. Chem. Soc.*, 1943, 250.
18. Laszlo P, A Diborane story, *Angew. Chem. Int. Ed.*, 2000, Vol. 39, 2071.
19. Norman A D, Jolly W L, Saturnino D and Shore S G, Diborane, *Inorg. Synth.*, 1968, Vol. 11, 15.
20. Mikhailov B M, The chemistry of diborane, *Russ. Chem. Rev.*, 1962, Vol. 31, 207.
21. Brown H C, Mead E J and Shoaf C J, Convenient procedures for the preparation of alkyl borate esters, *J. Am. Chem. Soc.*, 1956, Vol. 78, 3613.
22. Brown H C, Hydroboration—A powerful synthetic tool, *Tetrahedron*, 1961, Vol. 12, 117. Ramachandran P V and Brown H C, Organoboranes for syntheses, *ACS Symposium Series*, 2001, 783.
23. Dhillon R S, *Hydroboration and organic synthesis: 9-Borabicyclo [3.3.1] Nonane (9-BBN)*, Springer, 2007.
24. Grimes R N, *Carboranes*, 3rd Ed., Academic Press, 2016.
25. Hosmane N S, *Boron science: New technologies and applications*, CRC Press, 2011.
26. Ott J J, Brown C A and Gimarc B M, Diamond–Square–Diamond isomerisation of $C_2B_5H_7$, *Inorg. Chem.*, 1989, Vol. 28, 4269.
27. Hawthorne M F, The chemistry of the polyhedral species derived from transition metals and carboranes, *Acc. Chem. Res.*, 1968, Vol. 1, 281.
28. Cummings S, Hratchian H P and Reed C A, The strongest acid; Protonation of CO_2, *Angew. Chem. Int. Ed.*, 2016, Vol. 55, 1182.
29. Nava M, Stoyanova I V, Cummings S, Stoyanov E S and Reed C A, The strongest Brønsted acid: Protonation of alkanes by $H(CHB_{11}F_{11})$ at room temperature, *Angew. Chem. Int. Ed.*, 2014, Vol. 53, 1131.
30. Wade K, The structural significance of the number of skeletal bonding electron-pairs in carboranes, the higher boranes and borane anions, and various transition-metal carbonyl cluster compounds, *J. Chem. Soc. D.*, 1971, 792.
31. Welch A J, The significance and impact of Wade's rules. *Chem. Commun.*, 2013, Vol. 49, 3615.
32. Jemmis E D, Balakrishnarajan M M and Pancharatna P D, A unifying electron-counting rule for macropolyhedralboranes, metallaboranes, and metallocenes, *J. Am. Chem. Soc.*, 2001, Vol. 123, 4313.
33. Jemmis E D, Balakrishnarajan M M and Pancharatna P D, Electronic requirements for macropolyhedralboranes, *Chem. Rev.*, 2002, Vol. 102, 93.
34. Piers W E and Chivers T, Pentafluorophenylboranes: From obscurity to applications, *Chem. Soc. Rev.*, 1997, Vol. 26, 345.
35. Erker G, Tris(pentafluorophenyl)borane: A special boron Lewis acid for special reactions, *Dalton Trans.*, 2005, 1883.
36. Chivers T, Pentafluorophenylboron halides: 40 years later, *J. Fluor. Chem.*, 2002, Vol. 115, 1.
37. Hall D G, *Boronic acids*, WILEY-VCH Verlag GmbH & Co. KGaA, Weinheim, 2005.
38. Ishiyama T, Murata M, Ahiko T-aki and Miyaura N, Bis(pinacolato)diboron, *Org. Synth.*, 2004, Vol., 10, 115.
39. Knapp D M, Gillis E P and Burke M D, A general solution for unstable boronic acids: slow-release cross-coupling from air-stable MIDA boronates, *J. Am. Chem. Soc.*, 2009, Vol. 131, 6961.
40. Molander G A and Figueroa R, Organotrifluoroborates: expanding organoboron chemistry, *Aldrichim. Acta*, 2005, Vol. 38, 49.
41. Paetzold P, New perspectives in boron-nitrogen chemistry-I, *Pure Appl. Chem.*, 1991, Vol. 63, 345.

42. Sneddon L G, Mirabelli M G L, LynchA T, Fazen P J, Su K and Beck J S, Polymeric precursors to boron based ceramics, *Pure Appl. Chem.*, 1991, Vol. 63, 407.
43. Rudolph S, Boron Nitride (BN), *American Ceramic Society Bulletin*, 2000, Vol. 79, 50.
44. Paine R T and Narula C K, Synthetic routes to boron nitride, *Chem. Rev.*, 1990, Vol. 90, 73.
45. Wentorf Jr. R H, Synthesis of the cubic form of boron nitride, *J. Chem. Phys.*, 1961, Vol. 34, 809.
46. Greim J and Schwetz K A, *Boron carbide, boron nitride, and metal borides, Ullmann's Encyclopedia of Industrial Chemistry*, Weinheim: Wiley-VCH, 2005.
47. Chopra N G, Luyken R J, Cherrey K, Crespi V H, Cohen M L, Louie S G and Zettl A, Boron nitride nanotubes, *Science*, 1995, Vol. 269, 966.
48. Barth R F, Soloway A H and Fairchild R G, Boron neutron capture therapy for cancer, *Scientific American*, 1990, Vol. 263, 100 and106.
49. Barth R F, Vicente M G H, Harling OK, Kiger W S, Riley K J, Binns P J, Wagner F M, Suzuki M, Aihara T, Kato I and Kawabata S, Current status of boron neutron capture therapy of high grade gliomas and recurrent head and neck cancer, *Radiat. Oncol.*, 2012, Vol. 7, 146.
50. Moss R L, Critical review with an optimistic outlook on boron neutron capture therapy (BNCT), *Appl. Rad. Isotopes*, 2014, Vol. 88, 2.
51. Farr L E, Sweet W H, Robertson J S, Foster C G, Locksley H B, Sutherland D L, Mendelsohn M L and Stickley E E, Neutron capture therapy with boron in the treatment of glioblasto mamultiforme, *Am. J. Roentgenol, Rad. Therapy, and Nuclear Med.*,1954, Vol. 71, 279.
52. Soloway A H, Tjarks W, Barnum B A, Rong F G, Barth R F, Codogni I M and Wilson J G, The chemistry of neutron capture therapy, *Chem. Rev.*, 1998, Vol. 98, 1515.
53. Barth R F S F, Boron neutron capture therapy of cancer, *Cancer Research*, 1990, Vol. 50, 1061.
54. Wang Y, Quillian B, Wei P, Wannere C S, Xie Y, King R B, Schaefer, III, H F, Schleyer P V R and Robinson G H, A stable neutral diborene containing a B=B double bond, *J. Am. Chem. Soc.*, 2007, Vol. 129, 12412.
55. Braunschweig H, Dewhurst R D, Hammond K, Mies J, Radacki K and Vargas A, Ambient-temperature isolation of a compound with a boron-boron triple bond, *Science*, 2012, Vol. 336, 1420.
56. Brauschweig H, Dellermann T, Dewhurst R D, Ewing W C, Hammond K, Oscar J, Jimmenz-Halla C, Kramer T, Krummenacher I, Mies J, Phukan A K and Vargas A, Metal-free binding and coupling of carbon monoxide at a boron-boron triple bond, *Nature Chem.*, 2013, Vol. 5, 1025.
57. Piazza Z A, Hu H S, Li W L, Zhao Y F, Li J and Wang L S, Planar hexagonal B36 as a potential basis for extended single-atom layer boron sheets, *Nature Commun.*, 2015, Vol. 5, 3113.
58. Zhai H-J, Zhao Y-F, Li W-L, Chen Q, Bai H, Hu H-S, Piazza Z A, Tian W-J, Lu H-G, Wu Y-B, Mu Y-W, Wei G-F, Liu Z-P, Li J, Li S-D and Wang L-S, Observation of an all—boron fullerene, *Nature Chem.*, 2014, Vol. 6, 727.
59. Segawa Y, Yamashita M and Nozaki K, Boryllithium: Isolation, characterisation and reactivity as a boryl anion, *Science*, 2006, Vol. 314, 113.
60. Soleilhavoup M and Bertrand G, Borylenes: An emerging class of compounds, *Angew. Chem. Int. Ed.*, 2017, Vol. 56, 10282.

CHAPTER

4

THE CHEMISTRY OF ALUMINIUM, GALLIUM, INDIUM AND THALLIUM

UNIQUE PROPERTIES AND APPLICATIONS

Aluminium

- *Aluminium is the third most abundant metal on the earth's crust and it is found to be present in as many as 270 minerals.*
- *Aluminosilicates form a major component of the earth's crust.*
- *Aluminium is one of the lightest metals with low density and resists reactivity by forming a thin passivation layer of aluminium oxide (about 4 nm thickness) on any exposed aluminium surface.*
- *Although pure aluminium has a very low yield strength (the stress at which a material begins to deform plastically) its alloys with metals such as copper, magnesium and zinc have very high yield strength comparable to that of steel. Many of its alloys have very good strength to weight ratio and therefore find applications where a strong light metal is required such as in aeroplanes, modern trains and racing cars.*
- *It is a good thermal and electrical conductor, possessing 59% of the electrical conductivity of copper, while its density is only 30% that of copper.*
- $AlCl_3$ *is a very good Lewis acid and a reagent suited specifically for alkylation and acylation reactions of aromatics (Friedel–Crafts reaction). Sodium salts of long chain aryl sulfonates (for example, dodecyl benzenesulfonate), whose aryl–alkyl unit is made using* $AlCl_3$ *catalyst, are used as surfactants in laundry detergents.*
- *Aluminium sulfate is an important flocculating agent in the purification of water and a sizing agent of paper, replacing potassium alum. Positively charged aluminium cations neutralise the negative charge on colloidal particles in water leading to mutual adsorption and flocculation.*

- *Triethylaluminium and trimethylaluminium and a partially hydrolysed form of Me_3Al – methylalumoxane (MAO) – are important alkylation reagents and cocatalysts especially in olefin polymerisation (for example, Ziegler–Natta and Kaminsky catalysts).*
- *Aluminium oxide (alumina, α-Al_2O_3) constitutes the mineral corundum, varieties of which form the precious gemstones ruby and sapphire. Claus process (recovery of sulfur from H_2S) and alcohol dehydrogenation are industrial processes using alumina as the sole catalyst. Aluminium oxide is an amphoteric substance as it reacts with both acids and bases, such as hydrofluoric acid and sodium hydroxide, acting as an acid with a base and as a base with an acid, neutralising the other and producing a salt.*
- *$LiAlH_4$ is a strong reducing agent which is extremely effective in the reduction of acids, esters and amides to alcohols.*
- *Red-Al is a reducing agent comparable to lithium aluminium hydride in activity but not as sensitive to oxygen, nor pyrophoric and has good solubility in organic solvents.*
- *Partial reduction of esters and nitriles to aldehydes is achieved using DIBAL-H, an aluminium hydride reagent. DIBAL-H is also an excellent reagent for hydroalumination reactions.*
- *The stepwise growth of aluminium alkyls through alkene insertion (Aufbau reaction) is a very important step in the use of aluminium alkyls in Ziegler–Natta and metallocene catalysed polymerisation of alkenes.*
- *The thermite process, in which aluminium powder reacts with Fe_3O_4, releases very high temperatures (2500°C) and generates molten iron which has been used for spot welding, especially of railway tracks.*
- *The first aluminium beverage cans (which took over from tinplate cans) were manufactured by Reynolds Metals Company in 1963 and were used to package a diet cola called 'Slenderella' and by 1967 Pepsi and Coke entered the area. According to an estimate, on the average, more than 125,000 aluminium cans are recycled every minute of every day. Since 1972, more than 700 billion beverage cans have been used and recycled.*

Gallium

- *Gallium is the only liquid element of the group (m.p. 29.7°C). Unlike mercury (m.p. 38.8°C) it expands 3.1% on freezing and also wets glass. It also does not react with water till about 100°C.*
- *Gallium arsenide, when compared to silicon, has higher saturated electron velocity, higher electron mobility and is insensitive to overheating. GaAs has therefore been used to make solar cells at a maximum conversion efficiency of 28.8%. It also tends to create less noise in electronic circuits than silicon devices, especially at high frequencies.*

- *Unlike silicon which is a poor emitter of light due to an indirect and small band gap, GaAs has a direct and wider band gap, which translates to absorbing and emitting light efficiently. As it shows resistance to radiation damage, GaAs finds use in outer space electronics.*
- *Gallium nitride, unlike silicon, due to its wider band gap, emits brilliant blue light when used as an LED. By varying the band gap, light of any colour can be produced by adding more of indium to gallium nitride or one can produce ultraviolet light by adding aluminium to gallium nitride.*
- *GaN/InGaN based light emitting diodes (LEDs) are widely used and if one coats a blue GaN LED with a yellow-emitting phosphor such as Ce-YAG (yttrium aluminium garnet), a cool white light is produced which is replacing regular incandescent and CFL light bulbs.*
- *GaN transistors have the ability to operate at relatively higher temperatures and higher voltages than normal transistors. This makes them ideal power amplifiers at microwave frequencies.*

Indium

- *Indium being air stable, reactions mediated by it (such as Barbier type allylation) can be carried out in water medium and the organo-indium compounds that are formed are tolerant to water.*
- *Indium tin oxide (ITO) is one of the most widely used transparent conducting oxides (TCO). Compared to other TCOs its optical transparency and electrical conductivity are quite high, in addition to the ease with which it can be deposited as a thin film on surfaces. Touch screen mobile phones, ATM screens and similar devices use ITO. It is also used for heating of glass for removing frost formation especially of aircraft glass windows.*
- *$InCl_3$ is a mild, air and moisture stable Lewis acid which has been found to be useful in acid catalysed organic reactions. Unlike many other Lewis acids, it has excellent functional group tolerance (for epoxides, acetals, alkynes and alkenes) and gives better yield and selectivity compared to other mild Lewis acids such as $ZnCl_2$.*

Thallium

- *Soluble thallium salts such as thallium acetate and thallium sulfate (which are almost tasteless) are highly toxic, notably resulting in hair loss and often death. They were historically used in rat poisons, hair removing creams and insecticides. Use of thallium compounds has been banned in many countries and it has gained notoriety as 'the poisoner's poison' since it does not leave any trail to follow up and is well known historically as a chemical for carrying out murder without evidence.*

Table 4.1 Physical properties of aluminium

Aluminium	*Atomic mass*	*Electronegativity (Pauling scale)*	*1st ionisation energy*	*Electronic Configuration*	*van der Waals radius*
Symbol: Al	26.981 g.mol^{-1}	1.61	577.5 kJ.mol^{-1}	[Ne]$3s^2 3p^1$	0.184 nm
At. no. 13	*Atomic radius*	*Melting point*	*Boiling point*	*Electron affinity*	*Density*
	1.43 Å	660.32°C	2470°C	41.762 kJ.mol^{-1}	2.375 g.cm^{-3} at m.p.

Table 4.2 Physical properties of gallium

Gallium	*Atomic mass*	*Electronegativity (Pauling scale)*	*1st ionisation energy*	*Electronic Configuration*	*van der Waals radius*
Symbol: Ga	69.723 g.mol^{-1}	1.81	578.8 kJ.mol^{-1}	[Ar]$3d^{10} 4s^2 4p^1$	0.187 nm
At. no. 31	*Atomic radius*	*Melting point*	*Boiling point*	*Electron affinity*	*Density*
	1.35 Å	29.76°C	2400°C	41 kJ.mol^{-1}	6.095 g.cm^{-3} at m.p.

Table 4.3 Physical properties of indium

Indium	*Atomic mass*	*Electronegativity (Pauling scale)*	*1st ionisation energy*	*Electronic Configuration*	*van der Waals radius*
Symbol: In	114.818 g.mol^{-1}	1.78	558.3 kJ.mol^{-1}	[Kr]$4d^{10} 5s^2 5p^1$	0.193 nm
At. no. 49	*Atomic radius*	*Melting point*	*Boiling point*	*Electron affinity*	*Density*
	1.67 Å	156.6°C	2072°C	37.043 kJ.mol^{-1}	7.02 g.cm^{-3} at m.p.

Table 4.4 Physical properties of thallium

Thallium	*Atomic mass*	*Electronegativity (Pauling scale)*	*1st ionisation energy*	*Electronic Configuration*	*van der Waals radius*
Symbol: Tl	204.383 g.mol^{-1}	1.62	589.4 kJ.mol^{-1}	[Xe]$4f^{14} 5d^{10} 6s^2 6p^1$	0.196 nm
At. no. 81	*Atomic radius*	*Melting point*	*Boiling point*	*Electron affinity*	*Density*
	1.70 Å	304°C	1473°C	36.4 kJ.mol^{-1}	11.22 g.cm^{--3} at m.p.

4.1 About the Elements

4.1.1 Aluminium

Aluminium and aluminum are two words used to name the same element. Sir Humphry Davy named this new element, first spelling it *alumium* (1807) then changed it to *aluminum*, and finally settled for *aluminium* in 1812. It should be noted that North America still uses the word aluminum while the rest of the world uses *aluminium* (the officially approved word for the element by IUPAC in 1990). However, IUPAC also recognised *aluminum* as an acceptable variant in 1993.

Looking back historically, in 1855, this metal was displayed along with the crown jewels at the Paris world exhibition while it is known that Napoleon III used aluminium cutlery at state dinners. Being the most abundant metal in the earth's crust and the major constituent of many common minerals such as zeolites, rocks, soil and clays, it has a wide range of applications. The remarkable properties of this metal are its ability to resist corrosion through passivation of its outer surface and a very good strength to weight ratio of its alloys making it the most sought after metal for construction of different modes of transportation where high strength, corrosion resistance, durability and low weight are desired.[1]

Aluminium is a soft, lightweight metal with a silvery grey appearance. It is non-magnetic and does not easily ignite. The yield strength of pure aluminium is 7–11 MPa, while aluminium alloys have yield strengths ranging from 200 MPa to 683 MPa, close to that of steel. Aluminium has around one-third of the density and stiffness of steel. It is a good conductor, possessing around 59% the conductivity of copper, both thermal and electrical, while its density is only 30% of that of copper.

Aluminium has good corrosion resistance as a very thin layer of Al_2O_3 (~4 nm thickness) forms on its surface when the freshly cut metal is exposed to air which effectively prevents further oxidation. The aluminium alloys containing copper are less corrosion-resistant as a result of galvanic reactions with copper. In highly acidic solutions, aluminium reacts with water forming hydrogen, and in highly alkaline solutions it forms aluminates. Under such conditions, the passivation layer gets removed. Due to its general resistance to corrosion, aluminium retains its silver coloured reflectance in finely powdered form. This makes aluminium useful as a component of silver coloured paints. Aluminium is oxidised by water at temperatures below 280°C to produce hydrogen, aluminium hydroxide and heat:

$$2\,Al + 6\,H_2O \longrightarrow 2\,Al(OH)_3 + 3H_2 \tag{4.1}$$

This conversion is of interest for the production of hydrogen. However, commercial application of this fact has challenges in circumventing the passivating oxide layer, which inhibits the reaction.[2]

4.1.2 Gallium

Gallium was discovered in 1875 by a French chemist Paul-Émile Lecoq de Boisbaudran. It is a soft and silvery-blue metal having hardness of 1.5 moh, and specific gravity of 5.904 and

6.905 g/cm^3 for the solid and liquid states, respectively. It exhibits an unusually large liquid range due to its low melting point of 29.76°C (302.98 K) and high boiling point of 2400°C (2673 K). Its compounds give a blue colour in the flame test. A solid piece of gallium will liquefy when placed in one's hand and a spoon made of gallium will melt if used to stir hot water. It is stable to air and water till about 100°C. Upon solidification, gallium metal expands by 3.1% of its volume. Gallium can exist in the form of six isotopes of which only two, ^{69}Ga (60.4%) and ^{71}Ga (39.6%), are stable. It rapidly dissolves in concentrated sodium hydroxide and also reacts with strong oxidants. Gallium is suited for making low-melting alloys.

In material-based applications such as LEDs, laser diodes, solar cells and digital IC chips, gallium based compounds with other main group elements such as nitrogen, arsenic, phosphorus and antimony are being used. These materials possess qualities such as high processing speeds, radiation and temperature tolerance, compatibility with other optical devices and lesser power requirement.

Since the 1970s, gallium metal has attracted a lot of interest especially in making materials which compete with silicon in electronic and solar cell applications. Gallium combined with arsenic and nitrogen display semiconducting properties forming gallium arsenide (GaAs) and gallium nitride (GaN). The latter is one of the most sought after compounds for making white LED light devices which are fast replacing traditional devices of lighting. GaAs and GaN are valuable materials employed in advanced semiconductors for microwave transceivers, laser diodes in compact discs and DVDs, and many other electronic applications. After white LED started replacing fluorescent lights, the overall demand for gallium has increased exponentially.[3]

4.1.3 Indium

The discovery of indium is credited to Reich and Richter in 1863 while they were analysing sphalerite ores at Freiburg in Germany but commercial production and use were not started until 1934. The name *indium* is not directly related to India but attributed to the indigo-blue colour of its spectrum. Indium is a soft, low melting silvery-white crystalline metal with a brilliant lusture. It consists of a mixture of two isotopes; 96% indium-115 and 4% indium-113. With a specific gravity of 7.31 it can be subjected to deformation and is ductile and malleable. In addition to its relatively low melting point of 156.6°C (429.8 K) and high boiling point (2353.2 K) it becomes a superconducting material at 3.37 K. It is interesting to note that an alloy *galinstan*, containing indium (5%–22%), gallium (62%–95%) and tin (0%–16%) remains liquid at room temperature melting at –19°C. Due to the low toxicity and low reactivity, galinstan finds use as a replacement for many applications replacing toxic mercury or the reactive sodium–potassium alloy. Indium, like gallium and aluminium, is self passivating and forms an oxide layer of 30–40 Å thickness when freshly cut and exposed to air. This oxide layer becomes 80–100 Å thick over 2–3 days and is easily removed by washing with dil. HCl. Indium metal will burn at red heat to form yellow coloured indium oxide (In_2O_3).

The major end use of indium metal is in the preparation of a transparent conducting oxide namely indium tin oxide (ITO). Approximately 70% of indium has been practically utilised to produce indium tin oxide thin films which is an In–Sn mixed metal oxide consisting of indium oxide (In_2O_3) and tin oxide (SnO_2) at a rough mass ratio of 9:1. The transparent and conducting ITO film is an important material used in liquid crystal displays (LCD) and touch screen devices. Indium, like gallium, wets clean glass. Indium semiconductor compounds are used in infrared detectors, high-efficiency photovoltaic devices, photo conductors, high-speed transistors and so on. The largest end use for indium thin-film coatings are in LCD screens, flat panel displays, solar cells with high conversion efficiency (for example, copper indium diselenide), and electroluminescent lamps. Indium trichloride is also a very useful Lewis acid which is not sensitive to air and moisture and has good functional group tolerance.

Sensing its huge potential, USA had reportedly reserved indium for national strategic use and stopped producing indium from the 1990s. China is currently the major producer. The concentration of indium in the earth's crust has been estimated at around 50–200 ppb, which is one sixth the amount of gold present in the earth's crust. Metal market prices indicate increased price of indium every year. In addition, some data analyses predict that the indium reserve in the earth's crust is likely to be used up by the year 2025. Hence, currently, methods are being developed to recycle the indium from waste LCDs.

Table 4.5 Comparison of some properties and abundance of group 13 elements

Element	Melting point (°C)	Flame colour	Abundance on earth's crust
Boron	2763	Green	0.00086%
Aluminium	660.3	White	8.10%
Gallium	29.7	Blue	0.0019%
Indium	156.6	Indigo	0.000016%
Thallium	303.8	Green	0.000053%

4.1.4 Thallium

In 1861, the British chemist William Crookes and the French chemist Claude-Auguste Lamy independently discovered the element thallium. Crookes while investigating a batch of sulfuric acid which was contaminated saw a very unique green colour when a flame test was conducted. This green colour (535 nm) was different from the green colour shown by flame tests of copper and barium salts. Lemy, on the other hand observed the same green colour while carrying out flame tests of a deposit found on the lead chambers of the lead chamber process of making sulfuric acid. He was able to extract this new element and make even an ingot of it from the sidelines of sulfuric acid production. The sulfuric acid connection of thallium is possibly linked to the lead chamber process (wherein chambers made of lead were used to contain the heat liberated during the synthesis of sulfuric acid) as thallium is found along with ores of lead and zinc. William Crookes named the new element *thallium*, which is related to the Greek word *thallos*, meaning 'green shoot or twig' due to the green spectral emission lines that are related to this element.

Thallium was used to treat tuberculosis, gonorrhea, syphilis and ringworm; it was also used as a chemical for removal of excess hair. During the last century, a product known as *Koremlu cream or Celio cream* containing thallium salts were marketed in the United States for the treatment of ringworm as well as for removing unwanted hair. By 1934, around 700 cases of thallium poisoning were reported with more than 30 deaths. Thallium was also widely used earlier as a rodenticide and was banned in the United States in 1965.

4.2 Isolation and Purification of the Elements

4.2.1 Aluminium

Historically it is believed that Persians made the best pottery out of clay containing aluminium oxide about 7,000 years ago. Since aluminium never occurs in its metallic form in nature, it proved quite difficult to isolate the pure metal. By 1825, the Hans Christian Ørsted from Denmark prepared a sample of aluminium by heating a potassium amalgam and alum-based mixture. Friedrich Wöhler, the father of synthetic organic chemistry, improved this process by using metallic potassium. For a long time, the only way to prepare aluminium metal was by the cumbersome method of reacting metallic sodium or potassium with aluminium chloride. When the Washington monument, built to commemorate the American president George Washington was made in 1884, a piece of pure aluminium weighing six pounds (quite expensive at that time) was placed at its very top as an ornament and also to serve as a lightning arresting unit, due to the high electrical conductivity and corrosion resistance of aluminium.[4]

On February 23, 1886, in his woodshed laboratory at the family home at Oberlin, Ohio, USA, Charles Martin Hall was successful in producing aluminium metal. He was able to make pure aluminium by passing electricity through aluminium oxide dispersed in molten cryolite, Na_3AlF_6. He has observed that fluoride salts do not absorb water from the air unlike chlorides. He prepared cryolite, and showed that it would dissolve more than 25% by weight of Al_2O_3. The melting point of cryolite is 1000°C while that of alumina is 2072°C. In his clay reaction vessel, he made a lining of graphite to make it conduct electricity. He was able to lower the melting point of the alumina/cryolite solution by adding aluminium fluoride to it and passed electric current for several hours. The melt thus obtained was cooled and in the presence of his three sisters and father as witnesses, he broke open the clay reactor. They found several small silvery globules, which when tested with hydrochloric acid was found to be pure aluminium. He took them to his supervisor, Professor Jewett, who had earlier obtained a sample of pure aluminium from Friedrich Wöhler of Göttingen and confirmed that it was indeed aluminium.[5]

The Bayer process

Bauxite, a reddish-brown mineral, first discovered in *Les Baux*, France, in 1821 is the most important ore of aluminium. The Bayer process is the main industrial method of refining bauxite to produce alumina (aluminium oxide). Bauxite contains only 30%–54% aluminium oxide (Al_2O_3), the rest being a mixture of silica, various iron oxides and titanium dioxide. The aluminium oxide Al_2O_3, has to be purified before it can be refined to aluminium metal. In this process, bauxite, under pressure is washed with a hot sodium hydroxide solution at 175°C, which converts the ore to soluble sodium aluminate, $NaAlO_2$.

$$Al_2O_3 + 2\ NaOH \longrightarrow 2\ NaAlO_2 + H_2O \tag{4.2}$$

Lime, $Ca(OH)_2$ is added at this point to precipitate the silica as calcium silicate. The solution is clarified by filtering off the solid impurities. The alkaline solution is then cooled and treated by bubbling carbon dioxide into it, upon which aluminium hydroxide precipitates. Another method is by seeding the supersaturated solution with high-purity $Al(OH)_3$ crystals.

$$2\ NaAlO_2 + CO_2 \longrightarrow 2\ Al(OH)_3 + Na_2CO_3 + H_2O \tag{4.3}$$

When heated to 980°C, aluminium hydroxide decomposes to aluminium oxide.

$$2\ Al(OH)_3 \longrightarrow Al_2O_3 + 3\ H_2O \tag{4.4}$$

The leftover NaOH solution (also called the Bayer liquor) is then recycled. This, however, allows gallium and vanadium impurities to build up in the liquor. Bayer liquor is the main source for extracting gallium metal industrially.

Hall–Hérault process

Pure aluminium is produced using the Hall–Héroult electrolytic process, a method independently developed and patented by Charles Martin Hall of the USA, and by Paul L T Héroult of France who was granted a French patent for the same process. It is carried out in large electrolytic cells called pots with a direct electric current input of up to 280,000 amperes and about 5 volts. Carbon blocks are used as cathode. Collector bars made of steel are inserted into the cathode blocks to carry current away from the pot.

Molten cryolite (Na_3AlF_6) is kept in a cavity formed by the cathode carbon blocks. Anodes, also of carbon, are inserted in the cryolite to complete the circuit. Alumina (Al_2O_3) is taken in the powder form and is dissolved in the cryolite bath. Molten aluminium is produced while the anode is consumed liberating CO and CO_2.[5]

$$Al_2O_3 + 3/2\ C \longrightarrow 2\ Al + 3/2\ CO_2 \tag{4.5}$$

$$Al_2O_3 + 3\ C \longrightarrow 2\ Al + 3\ CO \tag{4.6}$$

4.2.2 Gallium

Gallium is most often associated with bauxite ore to the extent of 0.002% to 0.1% and when bauxite is digested with alkali, it dissolves along with aluminium. After extracting the aluminium as aluminium hydroxide, the remaining solution is called Bayer liquor. By successive leaching of the Bayer liquor, its gallium concentration can be made to reach 100–250 mg/dm^3. World commercial production of gallium comes mostly from this Bayer liquor and it is the main raw material resource for gallium production. Many methods have been developed to recover gallium from Bayer liquor such as electrochemical deposition by amalgamation, cementation with aluminium, fractional precipitation, solvent extraction and ion exchange.

During the Bayer process, when bauxite is digested with alkali, gallium is converted to the sodium salt of gallate, $NaGa(OH)_4$. When Bayer liquor is subjected to electrolysis using a mercury cathode and nickel anode, sodium metal gets deposited in the mercury forming an amalgam and later gallium (III) in the solution gets reduced using this sodium and passes into the amalgam. This Hg–Ga amalgam when re-oxidised along with NaOH, gives pure $NaGa(OH)_4$ which on electrolysis gives pure gallium. However, due to toxicity of mercury this method is now less preferred.

$$Na_3Hg_x + NaGa(OH)_4 + 2\,H_2O \longrightarrow GaHg_x + 4\,Na^+ + 4\,OH^- \quad (4.7)$$

The cementation process with aluminium is a non-toxic method which is basically displacement of gallium by aluminium of sodium gallate. There is a difference in potential of about 300 mV between Al and Ga in concentrated alkaline medium which makes this process feasible. Gallium is obtained with a purity of 99%.

$$Al + NaGa(OH)_4 \xrightarrow[H_2O]{NaOH} Ga + 2\,NaAl(OH)_4 \quad (4.8)$$

4.2.3 Indium

It is a minor component in zinc sulfide ores and hence it is made as a byproduct of zinc production. The major ore of zinc, sphalerite [(Zn,Fe)S] is also the mineral used for indium production. The traditional method for indium extraction from In–Zn concentrates is as follows. Initially ferric iron is removed by the method of jarosite [a basic hydrous sulfate of potassium and iron with a chemical formula of $KFe_3(OH)_6(SO_4)_2$] and indium goes into jarosite. To recover indium, the jarosite residue is dissolved in hot sulfuric acid solution and indium is directly extracted from this leach solution by D_2EHPA [di-(2-ethylhexyl)phosphoric acid]. This method of solvent extraction is suitable for solutions having high concentration of indium.

While part of the indium present in zinc is recovered by aluminium cementation, electrolysis is the main method for producing metallic indium in zinc, lead and tin refining. Indium is the 68th most abundant element in the earth's crust, approximately as abundant as cadmium. Due to the exponential growth and use of smartphones and other hand held devices using ITO based touch screen technology, there is an increased concern whether the indium available in the earth's crust will be used up completely in the near future.

4.2.4 Thallium

Thallium is mostly obtained as a byproduct from the processing of zinc and lead. World production of thallium compounds is very low as it does not have any major industrial use. The high toxicity comes as a deterrent in the attempts made to find some use. Thallium oxide is used in a relatively small scale for the production of highly refracting lenses. Thallium sulfate is used even now in some less developed countries as a rat poison. Thallium present in the soil is quite often absorbed by many plants due to its similarity in size to that of potassium; more the thallium present in the soil, more of it is absorbed. This often results in some leafy vegetables like cabbage becoming toxic if they have been grown in an area with high thallium content in the soil.

4.3 The Chemistry of Aluminium

4.3.1 Aluminium Metal-Based Alloys

Aluminium alloys are widely used in fabrication of engineering structures and components where light weight or more precisely, good strength to weight ratio and corrosion resistance, are required. The typical alloying elements are copper, magnesium, manganese, silicon and zinc. The discovery of an alloy of aluminium and copper (duralumin) in 1906 by the German scientist Alfred Wilm, was the path breaking discovery which increased the strength to weight ratio of aluminium. There are broadly two types, namely casting alloys and wrought alloys, both of which are further subdivided into the categories heat-treatable and non-heat-treatable. About 85% of aluminium is used for wrought products, for example rolled plates, sheets, foils and extrusions. The most important cast aluminium alloy is silumin, where the high levels of silicon (4%–13%) contribute to give good casting characteristics. Industry uses a four digit number for classifying aluminium alloys. The first digit generally helps to distinguish the additional alloy element present in a specific aluminium alloy.[6]

Aluminium alloy 7068 contains additional elements Zn (7.3%–8.3%), Mg (2.2%–3.0%), Cu (1.6%–2.4%) and Zr (0.05%–0.15%) is the strongest commercially available aluminium alloy, and is comparable to that of some steels. 7068-T6511 has a typical ultimate tensile strength of 710 MPa. The typical yield strength for alloy 7068-T6511 is 683 MPa versus 590 MPa for a similar product produced from 7075-T6511 and of high steel (690 MPa).

Table 4.6 Classification of aluminium alloys

Four digit number	Alloying elements (major)	Specific applications
1XXX	Al (99%)	
2XXX	Al and Cu	Duralumin type; aircraft (fuselage skin, flaps and so on)
3XXX	Al and Mn	Work hardened
4XXX	Al and Si	Silumin, castability and corrosion resistance
5XXX	Al and Mg	Boat and ship building (sea water corrosion resistant), automobile frames
6XXX	Al, Mg and Si	Weldable, cycling alloy, automotive body parts, scuba diving cylinder
7XXX	Al, Zn and Mg	Aircraft frames, skin
8XXX	Al, Li and so on (miscellaneous)	Space shuttle external tank, jetliner and helicopter frames

Aluminium and aeroplanes

Not many are aware about the role played by aluminium in making 'aeroplanes made of metal' a reality. Although the Wright brothers in 1903 were the first to fly a mechanical device, their 'flying machine' had no future or sustainability to expand as it was made of spruce, a lightweight wood. So, the search for a suitable metal was on. The initial attempts in this direction were made by a professor of mechanical engineering, and aircraft enthusiast from Aachen university, Germany—Hugo Junkers. He had attempted to assemble an all-metal aeroplane many times without success. On 15 December 1915, he tested a small aeroplane (J1 glider) made of a 1 mm thick sheet iron (J standing for Junkers). But the German military administration representatives who came to inspect the aeroplane rejected it, calling it 'a tin donkey' as it was too heavy, and its maneuverability and climbing capacity were not up to the level of expectations for military use. Junkers realised that the main culprit for the poor performance of the aircraft was the metal used. He needed an alternative metal to replace the iron sheets.

Pure aluminium, although available at this time, was not a suitable metal to make an aeroplane as it has only about one-third the density (2.7 g/cc) and stiffness of steel. The yield strength of pure aluminium is 7–11 MPa while the yield strength of high strength steel is 690 MPa (*yield strength is defined as the stress at which a material begins to deform plastically without returning to its original shape).* A breakthrough came in 1906 when Dr Alfred Wilm, a German metallurgist, discovered that pure aluminium, when alloyed with copper and properly heat-treated, could have a much higher yield strength. This alloy of aluminium with copper was called duralumin and had low specific gravity and high yield strength (450 MPa). Prof. Junkers realised that the alternative he was looking for the iron sheets had arrived. Duralumin met all the requirements for making an efficient aeroplane—high strength, forgeability, incredible lightness and good strength to weight ratio. By 1917, the D1 fighter entirely built of the 'light' alloy duralumin, was made and tested. After seeing the success, more D Junker aeroplanes were ordered by Germany for participation in World War I. During the military campaign, duralumin completely proved Junkers expectations and protected the pilot from bullets and shells. Junker's duralumin aeroplanes were nicknamed 'flying tanks'. There is a recorded case when a duralumin made aeroplane sustained 480 bullet shots on its wings and fuselage, and it not only completed the mission, but also successfully landed at the airbase. Germans however, kept the composition of duralumin a military secret.

The enemies of Germany had no idea how the Germans made these aeroplanes. In the spring of 1922, a Junkers duralumin fighter plane was shot down by the Russian army—a priceless trophy, to learn the secret of the German aircrafts. This was delivered to the Central Aerohydrodynamics Institute (CAHI) established in Russia. A separate 'Material Testing Division' group studied the composition of the metal used for making the aeroplane. The Russian researchers did not just determine the formula of the alloy but were successful in developing a stronger alloy modification, able to compete with duralumin. By late 1922, the Russians started production of 'kolchoug-aluminium'—the first Soviet high-strength aluminium alloy. The following year the first Russian-made aluminium alloy aeroplane Antonov AN-2 was launched, thus announcing the first competitor for Junker's duralumin aeroplanes. The study expanded and now there is a wide range of aluminium alloys suitable for different applications of transport.

The use of aluminium alloys was expanded for making commercial aircrafts. They dominated the scene till the 1980s as Boeing 747s were made with 81% aluminium alloys.

Thereafter, composites (carbon fibre reinforced plastic) and titanium slowly started replacing aluminium. The following table shows how slowly materials with better strength to weight ratio is replacing aluminium in aircraft manufacture.

Table 4.7 Commercial aeroplanes and their aluminium content

Aircraft	**Boeing 747**	**Boeing 777**	**Boeing 787 Dream liner**	**Airbus A350 XWB**
Year of introduction	1970	1995	2011	2015
Aluminium content	81%	70%	20%	19%
Titanium content	4%	7%	15%	14%
Carbon fibre (composites)	1%	11%	50%	53%
Steel content	13%	11%	10%	6%

4.3.2 The Thermite Process

The German chemist Hans Goldschmidt discovered and patented the *thermite reaction* (in 1895) which is also known as the *Goldschmidt reaction*. Goldschmidt's attempt to avoid carbon in smelting for making pure metals resulted in this extremely useful reaction involving metallic aluminium in the welding of iron at remote places without any external heating and extreme convenience in handling. It is no surprise that railway engineers all over the world use the thermite process for welding of railway tracks on the spot even in difficult terrain. This application was first used in the welding of tram tracks in the city of Essen in Germany in 1899. The actual chemical reaction uses aluminium powder to reduce iron oxide along with production of very high temperatures (up to 2400°C) which melts the metallic iron and welds the metal parts it comes in contact with. Often, strips of magnesium metal are used for ignition of the thermite. Since metals like magnesium burn without releasing cooling gases, they can easily attain the high temperatures needed for thermite ignition.[7]

$$3\,Fe_3O_4 + 8\,Al \longrightarrow 4\,Al_2O_3 + 9Fe + 3341\text{ kJ (upto 2400°C)} \quad (4.9)$$

The factors which helped the widespread applicability of this process are (i) aluminium is the least expensive of highly reactive metals, (ii) the passivation layer on aluminium makes it easy to store and handle without special care unlike other reactive metals such as sodium, (iii) the relatively low melting point (660°C) of aluminium makes it easy to melt and reaction proceeds quickly in the liquid phase, (iv) the high boiling point (2519°C) of aluminium enables the reaction to reach temperatures of 2400°C (common for transition metals but rare for metals such as Na and Mg), and (v) low density of Al_2O_3 makes it float above the molten iron preventing oxidation of iron and contamination.

While the iron thermite is the most important thermite considering the perspective of large scale applications, other metal oxides such as chromium oxide and copper oxide can also be reduced to the respective individual metals.

$$3\,CuO + 2\,Al \longrightarrow 3Cu + Al_2O_3 \tag{4.10}$$

The reaction between $KMnO_4$ and glycerol or ethylene glycol is also used as an alternative to the magnesium method for initiating the burning of the thermite. When they mix, a spontaneous reaction will begin, slowly increasing the temperature of the mixture until it starts burning. The heat released is sufficient to initiate a thermite reaction.

4.3.3 Aluminium Oxide (Alumina)

One of the very early references to alumina is by Greville in 1798 as a mineral from India that had the composition of Al_2O_3 and it was named corundum, from the native name of this stone. This possibly originates from the ancient Sanskrit language word *Kuruvinda* meaning ruby (gem stone). Red and blue varieties of gemstones are called ruby and sapphire, respectively. Since early times, rubies and sapphires in India have ranked among the *Maharatnani* (five great gemstones—the other three being diamonds, emeralds and pearls).

Corundum is a common aluminium mineral found in igneous and metamorphic rocks. The lattice of corundum is composed of hexagonally close packed oxygen ions. Two-thirds of the octahedral interstices are occupied by aluminium ions. The structure may be described as consisting of alternating layers of aluminium and oxygen ions.[8]

Alumina is produced by calcination of aluminium hydroxide above 1380 K.

$$2\,Al(OH)_3 \xrightarrow[1380\,K]{} Al_2O_3 + 3\,H_2O \tag{4.11}$$

There are different crystalline polymorphic phases for alumina. The most stable and naturally occurring phase is α-alumina which also forms the gemstones such as ruby and sapphire and has the highest density. The β-form has sodium peroxide and the γ-form is the one with the maximum applications as catalysts due to its high surface area. Alumina also exists in many more phases such as η, χ, κ, θ and so on, each possessing a unique crystal structure and properties. A comparison of α, β and γ alumina is given in Table 4.8.

Table 4.8 Comparison of properties of α, β and γ alumina

α-Alumina	β-Alumina	γ-Alumina
Most stable	$Na_2O.11\,Al_2O_3$	Highest surface area
Naturally occurring	Shows conductivity due to sodium ions	Prepared from Bayerite
Density 3.97 g/cc	Used as battery electrolyte	Density 3.65 g/cc
Surface area 5 sq.m/g		Surface area 100 sq.m/cc
Hexagonal close packed		Used as catalyst and in column chromatography

γ-Alumina is one of the inorganic chemicals produced in great volume. A good amount of it is being used for preparing adsorbents, ceramics, refractory materials and catalysts. Activated alumina is obtained by thermal dehydration of different aluminium hydroxides in the range of 520–1070 K. It is an amphoteric substance and thus reacts with both bases and acids.

$$Al_2O_3 + 2\,NaOH + 3\,H_2O \longrightarrow 2\,NaAl(OH)_4 \quad (4.12)$$

$$Al_2O_3 + 6\,HF \longrightarrow 2\,AlF_3 + 3\,H_2O \quad (4.13)$$

There are many catalytic applications of alumina and many industrial processes use alumina as a catalyst. The major ones are alcohol dehydration and the Claus process. Typical reaction conditions for olefin production by dehydration of alcohols are temperatures in the range of 360–500°C and pressures of 0.1–0.2 MPa. Lower temperatures can result in ether formation. While pure γ-alumina gives only conversion in the range of 80%, modified alumina such as γ-Al_2O_3-ZnO (Phillips) or γ-Al_2O_3-MgO-SiO_2 (SynDol) or γ-Al_2O_3-TiO_2 gives almost 100% conversion.

$$C_2H_5OH \longrightarrow C_2H_5OC_2H_5 \longrightarrow H_2C{=}CH_2$$
$$C_2H_5OH \xrightarrow{\gamma\text{-}Al_2O_3,\ 450°C} H_2C{=}CH_2 \quad (4.14)$$

The Claus process which produces elemental sulfur from H_2S is currently the most well-known natural gas desulfurising process. Carl Friedrich Claus patented this process in 1883 and it is now the industry standard overtaking the Frasch process for making elemental sulfur. A major portion of sulfur produced worldwide in 2011 was byproduct sulfur from refineries and other hydrocarbon processing plants. The Claus process reaction has two steps—partial oxidation of H_2S to SO_2 followed by a self-reduction process (Eqs 4.15, 4.16).

$$2H_2S + 3\,O_2 \longrightarrow 2\,SO_2 + 2H_2O \quad (4.15)$$

$$2H_2S + SO_2 \longrightarrow 3\,S + 2H_2O \quad (4.16)$$

The feed gas is first burned in a furnace using air to oxidise one-third of the H_2S it contains to SO_2 (1.5 bar and 1000°C). At these conditions, the Claus reaction sets in thermally in the reaction furnace. There is no requirement of catalyst for this step. About 70% of the H_2S in the feed gas is thus thermally converted into sulfur. The remaining unreacted gas is then separated, reheated and allowed to enter catalytic reactors having Al_2O_3 catalyst one after the other, maintained at temperatures of about 305°C, 225°C and 200°C where almost all of the remaining H_2S gets converted to sulfur. The catalyst used is γ-alumina with a high surface area.

4.3.4 Aluminium Trichloride

Anhydrous $AlCl_3$ can be prepared by the reaction of chlorine or HCl with aluminium in the temperature range of 650–750°C. It is also possible to prepare it by the displacement reaction of $CuCl_2$ with Al.

$$2\,Al + 3\,Cl_2 \xrightarrow[650\text{–}750°C]{} 2\,AlCl_3 \tag{4.17}$$

$$2\,Al + 6\,HCl \xrightarrow[650\text{–}750°C]{} 2\,AlCl_3 + 3H_2 \tag{4.18}$$

$$2\,Al + 3\,CuCl_2 \longrightarrow 2\,AlCl_3 + 3Cu \tag{4.19}$$

Anhydrous $AlCl_3$ is known to adopt different structures, depending on the temperature and the physical state. Solid $AlCl_3$ (density 2.48 g/cm^3) has a sheet-like layered cubic close packed structure in which each Al centre exhibits octahedral coordination geometry. In the molten form (density 1.78 g/cm^3), aluminium trichloride exists as the dimer Al_2Cl_6, with two bridging chlorine atoms and a tetracoordinate tetrahedral aluminium. Al_2Cl_6 dimers are also found in the vapour phase. At higher temperatures, the Al_2Cl_6 dimers dissociate into trigonal planar monomeric $AlCl_3$, which is structurally similar to BF_3.[9]

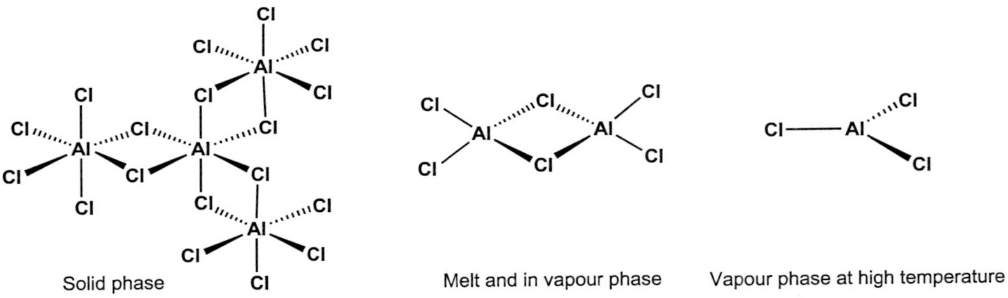

Solid phase

Melt and in vapour phase

Vapour phase at high temperature

Fig. 4.1 Structure of $AlCl_3$ in the solid, melt and vapour phase

$AlCl_3$ is most well known as the catalyst for Friedel–Crafts reactions. The Friedel–Crafts reactions were developed by Charles Friedel and James Crafts in 1877 basically to attach substituents to an aromatic ring. There are three main types of Friedel–Crafts reactions—alkylation, acylation and chlorination reactions, all of which proceed by electrophilic aromatic substitution (Scheme 4.1).

The most well known industrial application of Friedel–Crafts alkylation is the alkylation of benzene by alkyl halides to form alkylbenzenes that are used in the production of synthetic detergents, such as alkylbenzene sulfonates. Aluminium chloride also catalyses the ethylation of benzene with ethylene to yield ethylbenzene (which is a precursor in the production of styrene). This reaction is now performed better with zeolite catalysts. Ethyl chloride is manufactured by the reaction of hydrochloric acid and ethylene with aluminium chloride as catalyst but the declining demand for tetraethyl lead has however reduced the consumption of ethyl chloride significantly. $AlCl_3$ is still used in reforming hydrocarbons, as a polymerisation catalyst in the production of hydrocarbon resins, and butyl rubber. It is a catalyst also for the production of many organic compounds, such as aromatic aldehydes, ketones, and 2-phenyl-ethanol, the latter being used as artificial rose fragrance.[10]

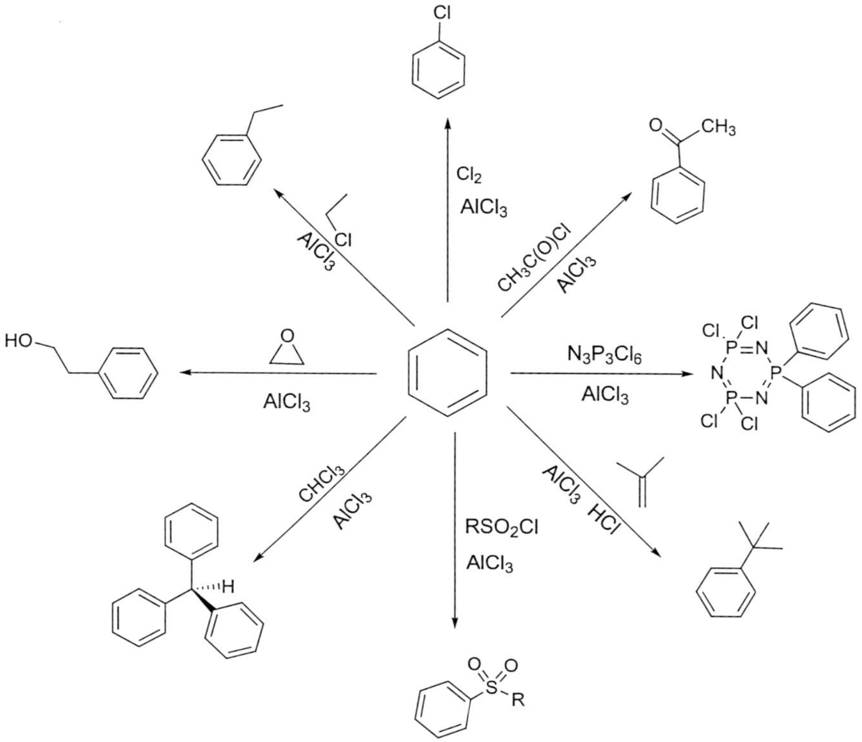

Scheme 4.1 Friedel–Crafts reactions

One of the drawbacks of the Friedel–Crafts reaction is that, often isomerisation also occurs, giving rise to a mixture of products.

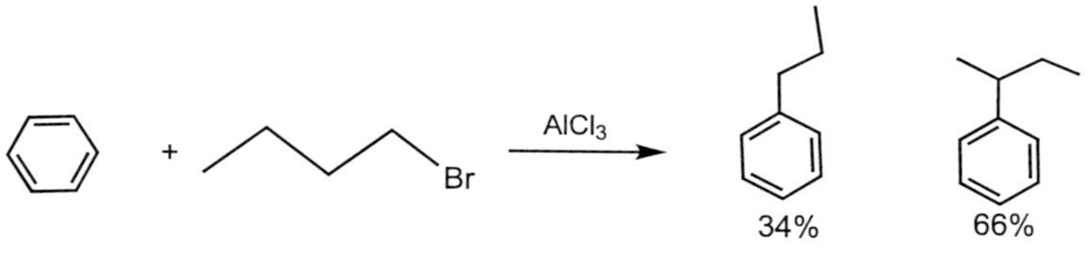

$$\text{C}_6\text{H}_6 + \text{CH}_3\text{CH}_2\text{CH}_2\text{CH}_2\text{Br} \xrightarrow{AlCl_3} \text{C}_6\text{H}_5\text{CH}_2\text{CH}_2\text{CH}_2\text{CH}_3\ (34\%) + \text{C}_6\text{H}_5\text{CH}(\text{CH}_3)\text{CH}_2\text{CH}_3\ (66\%) \qquad (4.20)$$

The aliphatic Friedel–Crafts reaction is not used as often as its aromatic counterpart since it does not always result in simple substitution, and gives a mixture of products. However, some substrates such as vinyltrimethylsilanes give a good yield of acylated products with acylation occurring on the carbon possessing the trimethylsilyl group.[11]

$$\text{(4-}t\text{-butylcyclohexenyl)SiMe}_3 \xrightarrow[AlCl_3]{CH_3C(O)Cl} \text{(4-}t\text{-butylcyclohexenyl)C(O)CH}_3 \quad 77\% \tag{4.21}$$

4.3.5 Aluminium Sulfate

Aluminium sulfate, $Al_2(SO_4)_3$ is the second-most important compound of aluminium industrially, after aluminium oxide. $Al_2(SO_4)_3$ is insoluble in anhydrous alcohol but is highly soluble in water and the aqueous solutions are acidic in nature. Aluminium sulfate with high purity is produced on a large scale from aluminium hydroxide and sulfuric acid (Giulini process).

$$2\ Al(OH)_3 \xrightarrow[170°C,\ 5\ atm]{60\%\ H_2SO_4} Al_2(SO_4)_3 \tag{4.22}$$

It is an important flocculating agent for purifying water. The method by which it works is as follows: positively charged aluminium ions get hydrated or hydroxylated when dispersed in water and neutralise the negative charge on the colloidal material present in water. As a result of mutual adsorption, the material flocculates/sediments out and settles as a sludge. A major portion of the aluminium sulfate production is used for water treatment. Another part of the production goes into the paper industry for paper sizing, pH adjustment and wastewater purification. It may be noted that the industrial production of alums ceased to have importance when it became possible to produce aluminium sulfate of high purity economically, as only the aluminium content is of practical importance in the applications of both the compounds. Alum production methods now have only limited, partly only historical, interest.[9]

4.3.6 Aluminium Hydrides [$LiAlH_4$, AlH_3, Red-Al and DIBAL-H]

After the report of Schlesinger and coworkers on the synthesis of lithium aluminium hydride ($LiAlH_4$, LAH) and alane (AlH_3) in 1947, aluminium hydrides have become indispensable reducing agents in chemical reactions. Unlike LAH and AlH_3, the ease of handling and high solubility of ionic or neutral aluminium compounds such as diisobutylaluminium hydride (DIBAL-H) and $Na[H_2Al(OCH_2CH_2OCH_3)_2]$ (Red-Al) have made aluminium hydrides the reagents of choice for the reduction of many types of unsaturated organic moieties, including those having carbon–carbon double and triple bonds.

$LiAlH_4$

Lithium aluminium hydride (LAH) is one of the most well known reducing agents having aluminium hydride units. It can be prepared from sodium aluminium hydride or from aluminium chloride as shown.

$$Na + Al + 2\,H_2 \xrightarrow[\text{THF, 150°C, 350 bar}]{} NaAlH_4 \quad (4.23)$$

$$NaAlH_4 + LiCl \longrightarrow LiAlH_4 + NaCl \quad (4.24)$$

$$AlCl_3 + 4\,LiH \longrightarrow LiAlH_4 + LiCl \quad (4.25)$$

Lithium aluminium hydride is well known as a reducing agent especially for the reduction of esters, carboxylic acids and amides. The solid in the pure form is colourless and is grey when not so pure. It reacts very vigorously with water, releasing hydrogen gas. It is not very soluble in non-polar solvents and has good solubility only in ethers such as diethyl ether (5.9 mol/L) and THF (3.0 mol/L) at 25°C. It is metastable at room temperature and on prolonged storage converts to Li_3AlH_6.

$$3\,LiAlH_4 \longrightarrow Li_3AlH_6 + 2\,Al + 3H_2 \quad (4.26)$$

Due to the weak nature of the Al–H bond compared to the B–H bond, it is a more powerful reducing agent than the analogous tetrahydride, $NaBH_4$. It has been found to convert esters, carboxylic acids, acyl chlorides, aldehydes and ketones into the corresponding alcohols. In addition, it converts amides, nitriles, imines, oximes, azides and aliphatic nitro compounds into amines. It reduces quaternary ammonium cations to the corresponding tertiary amines and alkyl halides to alkanes. It is most commonly used for the reduction of esters and carboxylic acids to primary alcohols (which borohydrides do not do), and was a difficult conversion involving sodium in boiling ethanol (Bouveault–Blanc reduction). When epoxides are reduced using LAH, the reagent attacks the less hindered end of the epoxide.[12]

Reaction of $LiAlH_4$ with $AlCl_3$ results in the formation of a polymeric aluminium hydride $(AlH_3)_n$. This compound shows very similar reducing activity to that of $LiAlH_4$ but is even more reactive undergoing violent reaction with water and air. Unlike $LiAlH_4$, it does not reduce nitro groups and organohalogens providing selectivity in ester reduction.[13]

O_2N–substituted ethyl ester (OEt, C=O) $\xrightarrow{AlH_3}$ O_2N–substituted alcohol (OH) (4.27)

Cl–chain carboxylic acid (OH, C=O) $\xrightarrow{AlH_3}$ Cl–chain alcohol (OH) (4.28)

Red-Al [$NaAlH_2(OC_2H_4OCH_3)_2$]

Although it is an excellent reducing agent, $LiAlH_4$ does have some problems especially in storage and handling. It is pyrophoric in nature and often ignites when exposed to moist air. It is metastable and has a relatively short shelf-life, and has limited solubility in many organic solvents especially non-polar solvents. Red-Al [$NaAlH_2(OC_2H_4OCH_3)_2$], also known as Vitride or Synhydrid is a commercially available reducing agent which is the best alternative for LAH. The trade name Red-Al means a reducing aluminium compound. The compound has a tetrahedral aluminium centre attached to two hydride and two alkoxide groups, the latter belonging to 2-methoxyethanol. Sodium bis(2-methoxyethoxy) aluminium hydride is comparable in reactivity and is a safer substitute for LAH and related hydrides. It does not have the inconvenient pyrophoric nature, short shelf-life, or limited solubility of LAH. Upon contact with air and moisture, sodium bis(2-methoxyethoxy) aluminium hydride reacts exothermically but does not ignite. It tolerates temperatures up to 200°C. LAH on the other hand decomposes around 150°C. Red-Al has an unlimited shelf life under dry conditions and is soluble in many aromatic solvents, while LAH is only soluble in ethers. Therefore Red-Al has the convenience which LAH does not have but retains most of the reactivity of LAH. Red-Al is synthesised by reacting sodium aluminium tetrahydride ($NaAlH_4$) and 2-methoxyethanol.[14]

$$NaAlH_4 + 2\ CH_3OCH_2CH_2OH \xrightarrow[-H_2]{} \quad (4.29)$$

Red-Al converts ketones, aldehydes, esters, carboxylic acids, acyl halides and acid anhydrides to the corresponding primary alcohols. Nitrogen-based compounds such as nitriles, amides and imines are reduced to the corresponding amines. Lactones and epoxides are reduced to diols. Unlike $LiAlH_4$, Red-Al can be used to reduce aliphatic *p*-toluenesulfonamides to the corresponding free amines and is one of the few reagents that can carry out this challenging reduction under normal conditions.

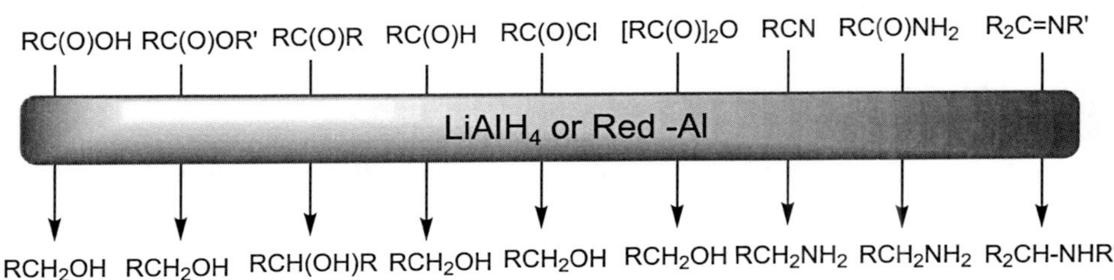

Scheme 4.2 Reactions common for lithium aluminium hydride and Red-Al

DIBAL

Diisobutylaluminium hydride, DIBAL-H is another useful aluminium hydride for a variety of reductions, including converting esters and nitriles to aldehydes. In contrast, $LiAlH_4$ reduces esters and acyl chlorides to primary alcohols, and nitriles to primary amines. DIBAL-H also reduces α–β unsaturated esters to the corresponding allylic alcohols. DIBAL-H reacts slowly with electron-poor compounds, and faster with electron-rich compounds. Triisobutylaluminium is readily prepared by the reaction of activated aluminium powder with isobutene and hydrogen (Eq. 4.30).[15]

$$3\,(CH_3)_2CH{=}CH_2 + \underset{\text{activated by } I_2 \text{ or by milling}}{\text{Al powder}} + H_2 \xrightarrow[\substack{140-150°C \\ 40–60\ atm \\ 4–5\ h}]{(i\text{-}Bu)_3Al\ (optional)} (i\text{-}Bu)_3Al \qquad (4.30)$$

Triisobutyl aluminium has been found to be 40% dimeric and 60% monomeric at 10°C. At higher temperatures, the monomeric form dominates. DIBAL-H is prepared by heating triisobutylaluminium to induce β-hydride elimination (Eq. 4.31).

$$2\,[(i\text{-}Bu)_3Al]\ (\text{b.p. } 86°C) \xrightarrow[\Delta\quad -2\ H_2C{=}C(CH_3)CH_2]{\beta\text{–H elimination}} [(i\text{-}Bu)_2Al(\mu\text{-}H)]_2\ (\text{b.p. } 116°C) \qquad (4.31)$$

Although DIBAL-H is an aluminium hydride, its reactivity, especially capability in reduction, is more similar to that of sodium borohydride ($NaBH_4$). It is most useful for the reduction of esters to aldehydes and nitriles to aldehydes (via imines). DIBAL-H reduces esters, but not to alcohols —it stops at the aldehyde stage. The temperature needs to be kept low and only one equivalent of DIBAL-H should be added for obtaining exclusively the aldehyde as the product. It will also reduce other carbonyl compounds such as amides, aldehydes and ketones. In addition, DIBAL-H can do all the reductions that $NaBH_4$ does. DIBAL-H will also do partial reductions of nitriles to imines which are then hydrolysed to aldehydes with water. It may be noted that $LiAlH_4$ reduces nitriles all the way to the amines. The mechanism of reduction by DIBAL-H is slightly unusual compared to the mechanism of reduction with $NaBH_4$. While $NaBH_4$ is considered a nucleophilic reductant, that is, it delivers the hydride (H^-) directly to a carbonyl carbon, DIBAL is considered an electrophilic reductant. That is, the first step is coordination of a lone pair from the carbonyl oxygen (a nucleophile) to the aluminium (electrophile). As shown in Eq. 4.32, only after binding to its carbonyl host, does DIBAL transfer its hydride to the carbonyl carbon. DIBAL-H is extensively used in hydroalumination reactions.

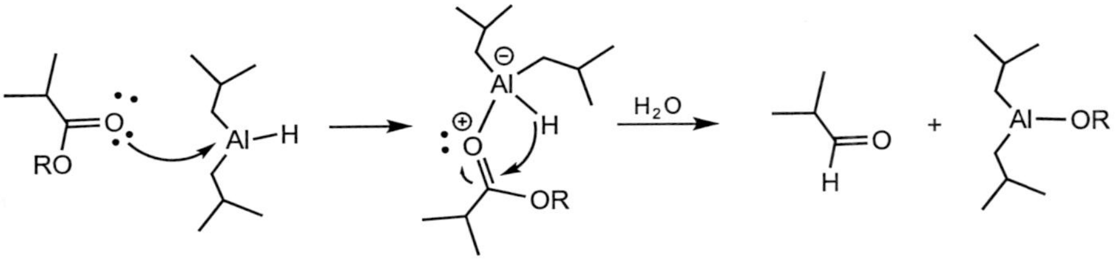

(4.32)

A comparison of the reactivity of three Al–H reagents with nitriles is shown in Eq. 4.33.

$$RCHO \xleftarrow[\text{Low temp, controlled stoichiometry}]{\text{DIBAL}} RCN \xrightarrow{\text{LAH or } AlH_3} RCH_2NH_2 \quad (4.33)$$

4.3.7 Hydroalumination

Hydroalumination is the addition of Al–H bonds to the double or triple bonds of unsaturated organic compounds. This reaction finds application as a useful method for the reduction of specific substrates in organic syntheses. Commercially available DIBAL-H is a convenient and efficient hydroaluminating agent for linear and cyclic alkenes and also shows functional group tolerance. While high temperatures are usually required for alkenes, by using zirconium compounds such as Cp_2ZrCl_2 or $ZrCl_4$ as catalysts, reactions can be made to proceed at room temperatures.[16]

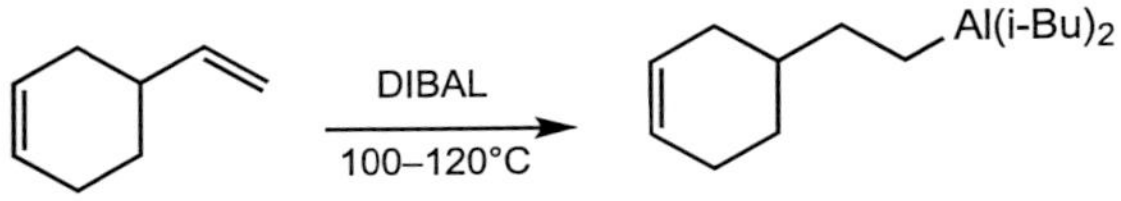

(4.34)

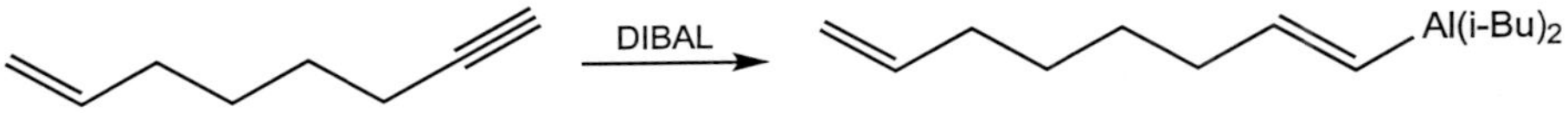

(4.35)

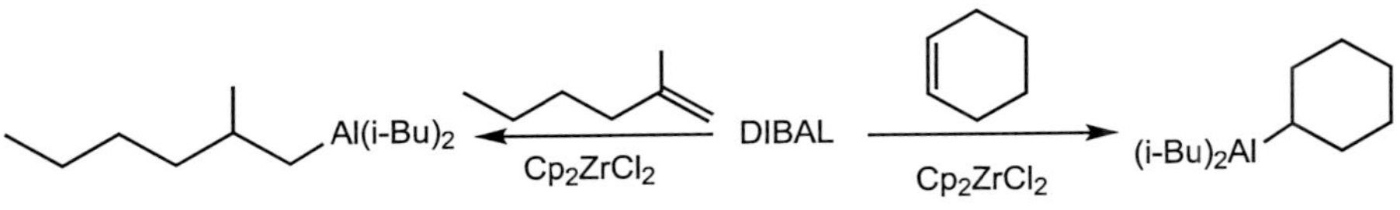

(4.36)

The rate of addition of the Al–H bond to olefins is greatest for ethylene, and decreases with 1-alkenes, 2-alkyl-1-alkenes, and 1,2-dialkylethylenes. Owing to the high reactivity of the Al–H bonds, the reactions with alkynes proceed under relatively mild conditions. There is also an exceptionally high regioselectivity which is caused by the specific polarity of the Al–H bond with the negative charge localised at the hydrogen atoms.

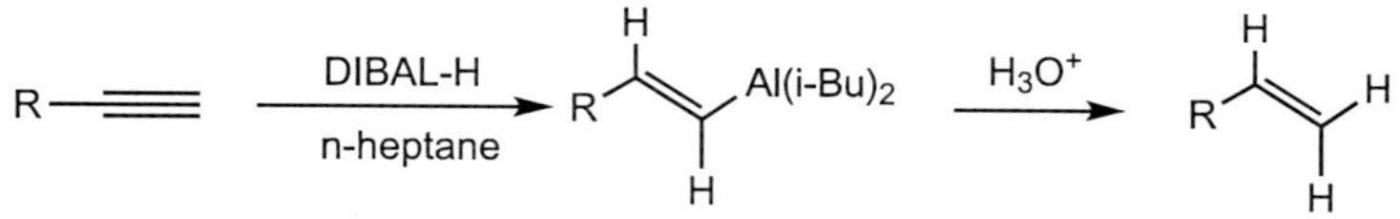

$$R-C\equiv CH \xrightarrow[\text{n-heptane}]{\text{DIBAL-H}} RCH=CH\text{-Al(i-Bu)}_2 \xrightarrow{H_3O^+} RCH=CH_2 \quad (4.37)$$

Temperature and solvent play a significant role in deciding the stereochemistry of hydroalumination. When a dialkylalane such as DIBAL-H is used, the hydrogen and aluminium delivered from the reagent end up *cis* in the resulting alkenylalane. Analogous to hydroboration reactions with R_2BH, hydroalumination with R_2AlH follows anti-Markovnikov selectivity. In general, mono-substituted alkynes can afford a mixture of alkyl, vinyl and alkynyl alanes. Disubstituted alkynes rapidly undergo hydroalumination depending on the reaction temperature and solvent used. Upon hydrolysis they afford a mixture of *cis* and *trans* olefins, *trans* conjugated dienes or hexasubstituted benzenes.

R—C≡C—R

50°C, R_2AlH, H_3O^+ → Z/E isomers Olefins

100°C, R_2AlH, H_3O^+ → E isomers Trans- conjugated dienes

150°C, R_2AlH, H_3O^+ → Hexasubstituted benzene

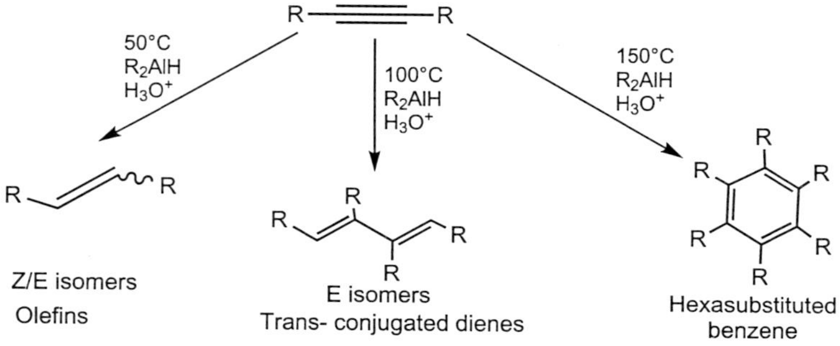

Scheme 4.3 Hydroalumination of alkynes

Selectivity between *cis* and *trans* alkene products can be achieved by using selected solvents as shown in Scheme 4.4.

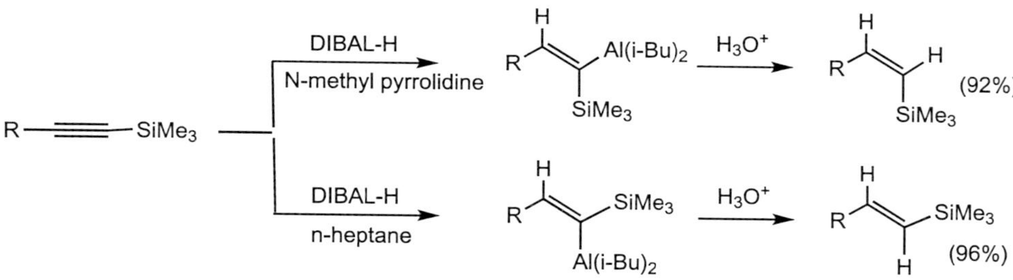

$R-C\equiv C-SiMe_3$

DIBAL-H, N-methyl pyrrolidine → RCH=C(SiMe₃)Al(i-Bu)₂ → H_3O^+ → RCH=CH(SiMe₃) (92%)

DIBAL-H, n-heptane → RCH=C(SiMe₃)Al(i-Bu)₂ → H_3O^+ → RCH=CH(SiMe₃) (96%)

Scheme 4.4 Selectivity of DIBAL-H reactions by solvent polarity

4.3.8 Trialkyl Aluminium Compounds

Compounds containing an aluminium–carbon bond figure in the highest volume, highest value organometallic compounds used in chemical industry. Important products made using organoaluminium compounds include polyethylene, polypropylene, alcohols, pharmaceuticals, agricultural chemicals, ceramics and electronic materials. The value of these products runs into billions of dollars every year.

Organoaluminium compounds began to attract interest with the discoveries of Ziegler and coworkers, commencing in the early 1950s, from Germany. Of technical importance was the discovery of the inexpensive direct synthesis of trialkylaluminium compounds, particularly triethylaluminium and triisobutylaluminium, from metallic aluminium, H_2 and the olefins, ethene and isobutene (Eq. 4.30). The two steps of this reaction are given below which indicates the formation of diethylaluminium hydride intermediate followed by hydroalumination.[17]

$$Al + 1.5\,H_2 + 2\,Et_3Al \xrightarrow[100\text{–}200\text{ bar}]{80\text{–}160°C} 3\,Et_2AlH \qquad (4.38)$$

$$6\,Et_2AlH + 6\,C_2H_4 \xrightarrow[1\text{–}10\text{ bar}]{80\text{–}110°C} \underset{\text{b.p }128\text{–}130°C}{3\,Et_6Al_2} \qquad (4.39)$$

Triethylaluminium can also be prepared by the following method

$$3\,C_2H_5Cl + 2\,Al \xrightarrow[\text{absence of }O_2/H_2O]{\text{traces of Hg or }I_2} Et_2Al(\mu\text{-}Cl)_2AlClEt \qquad (4.40)$$

$$3\,Et_2Al(\mu\text{-}Cl)_2AlClEt + 9Na \longrightarrow Et_2Al(\mu\text{-}Et)_2AlEt_2 \qquad (4.41)$$

Carboalumination of olefins, especially ethylene chain growth, helped in the production of polymers. The discovery by Kaminsky in 1976 that zirconocene–methylaluminoxane (MAO) catalyst systems were highly active in ethylene and propylene polymerisation has led to a major expansion of research in polyolefin chemistry.[17]

Organoaluminium compounds are generally soluble in hydrocarbon solvents. Normal trialkylaluminium compounds exist largely as associated dimers at 25°C. Trialkyls branched at the β-carbon, such as triisobutylaluminium, are partly monomeric. Dimeric trialkylaluminium compounds are often in equilibrium with their monomers. With increasing temperature, increased dissociation of the dimers occurs. A breakthrough in organoaluminium chemistry stems from the discovery of the alkene insertion reaction discovered by Ziegler and the low-pressure polymerisation of ethene and propene in the presence of organometallic AlR_3 and $TiCl_4$ catalysts discovered by Ziegler and Natta.

Among trialkylaluminium compounds, trimethyl- and triethylaluminium dominate as molecules of interest in syntheses as reagents for reactions such as alkylation and those involving removal of small molecules such as methane. Triethylaluminium ignites and decomposes on contact with air or water, or with any other oxidiser. It is one of the few substances that can ignite on contact with liquid oxygen thus making it desirable for rocket engine ignition. A mixture of triethylaluminium and triethylborane is used as a first-stage igniter in many modern rockets.

Synthesis

The industrial process for preparation of trimethylaluminium is as follows. The reactivity of aluminium can be improved by alloying it with 0.01%–2% of titanium.

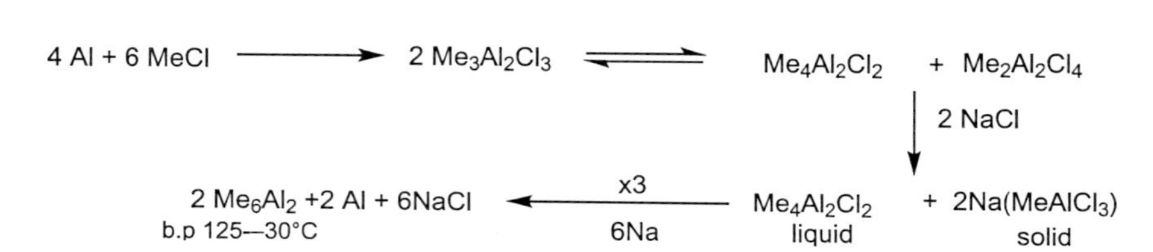

Scheme 4.5 Industrial synthesis of trimethylaluminium

Hydroalumination reactions also are used for synthesis of many trialkyl- and dialkylaluminium compounds by a series of successive β-H eliminations and alkene 1,2-migratory insertions. In the hydroalumination reaction, the affinity for the Al–H bond increases with decreasing steric hindrance in the order $CH_2{=}CR_2 < CH_2{=}CHR < CH_2{=}CH_2$.

$$(i\text{-}Bu)_3Al \xrightarrow[-3\ H_2C=CMe_2]{3\,H_2C=CHMe} (n\text{-}Pr)_3Al \xrightarrow[-3H_2C=CHMe]{3\ H_2C=CH_2} Et_3Al \tag{4.42}$$

Most of the sterically unhindered trialkylaluminium and triphenylaluminium compounds have been found to be dimeric in nature. The gas electron diffraction structure of Al_2Me_6 clearly shows two different Al–C bond distances with the bridging Al–C bond being longer.

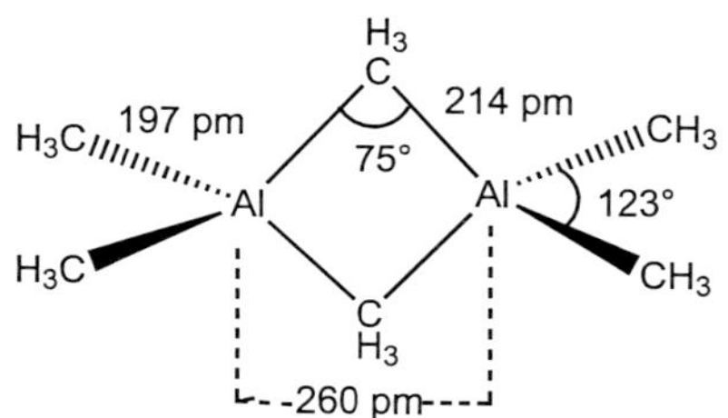

Fig. 4.2 Structure of Al_2Me_6

The general reactions of trimethylaluminium are shown in Scheme 4.6. One can see its important utility as a methylating agent for main group elements, as a Lewis acid to form adducts with amine bases and as a source for making methyl alumoxane, MAO.

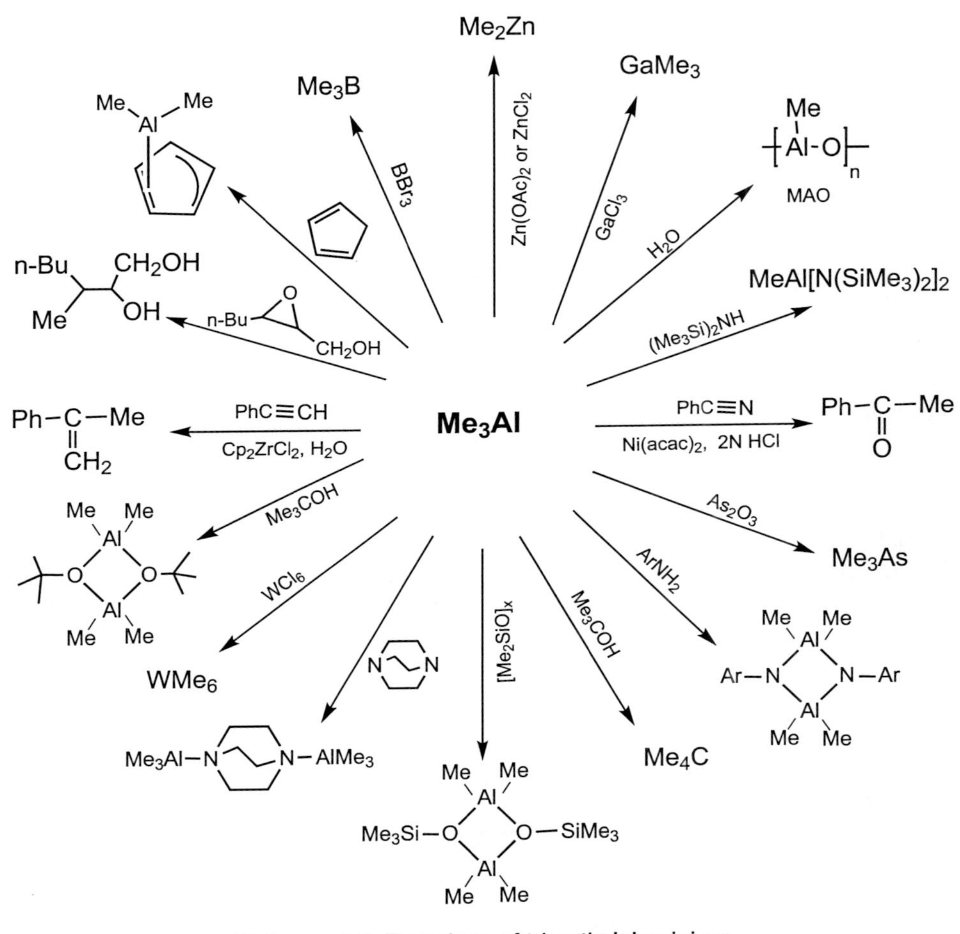

Scheme 4.6 Reactions of trimethylaluminium

4.3.9 Methyl Alumoxane (MAO)

Methyl alumoxane is a very reactive cocatalyst which is used in the olefin polymerisation catalysed by zirconocene-derived sandwich compounds such as Cp_2ZrCl_2. Its role consists of methylation of the zirconium site followed by demethylation to generate a cationic active site for polymerisation reactions (Scheme 4.7).

To overcome the difficulties to spectroscopically study the details of the heterogeneous olefin polymerisation reactions, especially Ziegler–Natta catalysts at the molecular level, their homogeneous analogues were developed. Detailed interest was directed to alkylaluminium-

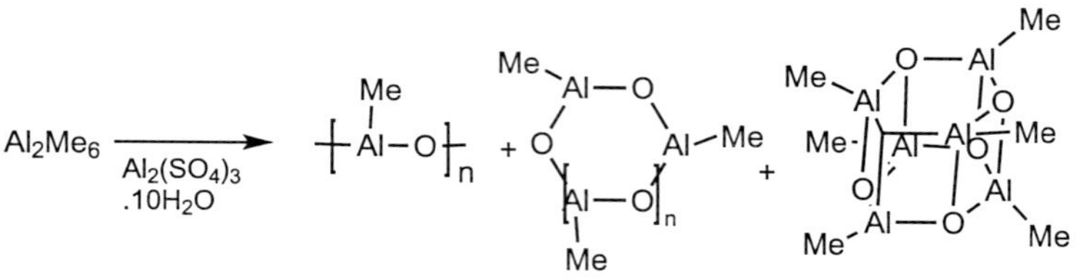

$$Al_2Me_6 \xrightarrow[.10H_2O]{Al_2(SO_4)_3} [\text{Al(Me)}-O]_n + [\text{cyclic (MeAlO)}_n] + [\text{cage (MeAlO)}_n] \quad (4.43)$$

activated group 4 metallocene dichlorides, Cp_2MCl_2 (M = Ti, Zr, Hf). Their unreactive, bulky cyclopentadienyl rings function as spectator ligands, leaving only two available reactive sites unlike in the tetrachlorides. In 1980, Walter Kaminsky and coworkers were studying the halogen-free Cp_2ZrMe_2/Me_3Al system in the presence of ethylene by low-temperature ^{1}H-NMR spectroscopy. In an attempt to simplify the procedure of preparing the NMR samples, a student replaced the sealed NMR tube with a regular plastic-cap-stoppered tube. Careful analysis of the obtained spectrum showed a small new NMR signal for CH_2, which indicated polyethylene formation. Performing a larger-scale experiment revealed the presence of a highly active catalyst system. Further experiments, with and without traces of oxygen and chloride as the enhancer, showed that water was the necessary component to obtain this highly active catalyst system. In fact, better activities were obtained when water was knowingly added to Me_3Al in a little less than equimolar ratio. Analysis of the resulting mixture showed partial hydrolysis of Me_3Al releasing methane and a product with an approximate composition of $Me_{1.5}AlO_{0.75}$. This partially hydrolysed Me_3Al was named methylalumoxane (MAO). The role of MAO in generating an active cationic catalytic site is shown in Scheme 4.7.[18–20]

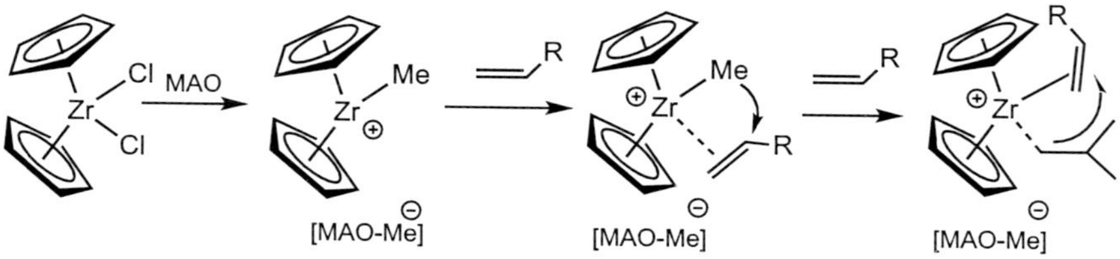

Scheme 4.7 Generation of the active catalyst for olefin polymerisation by MAO

4.3.10 Carboalumination

Carboalumination is a unique reaction shown by aluminium alkyls especially triethylaluminium which has a very important industrial use. It is used for oligomerisation of ethylene to make long chain alcohols which are used in the preparation of detergents. The first step involves the oligomerisation of ethylene by the reaction well known as the *aufbau* reaction (meaning building up), which gives a mixture of 'trialkylaluminium' compounds where the ethylene keeps on adding to the ethyl groups of triethylaluminium, forming long chains by the carbalumination reaction. The formation of alkyl chains by the *aufbau* reaction is shown in Eq. 4.44.

$$Al(CH_2CH_3)_3 \xrightarrow{n\ CH_2=CH_2} Al[(CH_2CH_2)_x CH_2CH_3][(CH_2CH_2)_y CH_2CH_3][(CH_2CH_2)_z CH_2CH_3] \quad (4.44)$$

Subsequently, by making use of oxygen, the Al–C bonds (which are highly susceptible to oxidation) get oxidised to aluminium alkoxides, which are then hydrolysed to long chain alcohols along with $Al(OH)_3$.

$$Al_2(CH_2CH_3)_6 \xrightarrow[110°C,\ 100\ bar]{6n\ H_2C{=}CH_2} Al_2[(CH_2CH_2)n(CH_2CH_3)]_6 \quad (4.45)$$

$$Al_2[(CH_2CH_2)n(CH_2CH_3)]_6 \xrightarrow{O_2} Al_2[(OCH_2CH_2)n(CH_2CH_3)]_6 \quad (4.46)$$

$$Al_2[(OCH_2CH_2)n(CH_2CH_3)]_6 \xrightarrow{H_2O} 6\ CH_3CH_2(CH_2CH_2)_nOH + 2\ Al(OH)_3 \quad (4.47)$$

This stepwise growth of aluminium alkyls through ethylene insertion was first reported by Ziegler in the early 1950s. For over half a century, this reaction formed the basis of large-scale industrial production of ethylene oligomers, with the advantage that a relatively narrow and controllable distribution of chain length is possible. The sequence of ethylene insertions can proceed up to chain lengths of C_{200}. The reaction also can be used for the generation of long chain terminal alkenes by making use of the β-H elimination reaction, which is incidentally a limiting factor of this reaction, and is often referred to as the displacement reaction. If propene or another bulky 1-alkene is used, only a single insertion to the Al–C bond takes place. This can be used for catalytic dimerisation of propene using $(n\text{-Pr})_3Al$. The analogous *aufbau* reaction with acetylene was explored by Ziegler and Wilke, but met with limited success as the reaction was found to stop after one insertion, leading to $Et_2Al(CH{=}CHEt)$. While the carboalumination reaction discovered by Ziegler requires high temperature and pressure and also has limitations in making long chain polymers, a combination $AlEt_3$ and $TiCl_4$ discovered by Ziegler in 1953 was very efficient for room temperature and atmospheric pressure polymerisation of ethylene. In 1954, Guilo Natta found that a similar catalyst using $TiCl_3$ and $AlEt_2Cl$ works very well for polymerisation of propene producing isotactic polypropylene. These landmark discoveries (Ziegler–Natta catalysts for olefin polymerisation) fetched them the Nobel prize.

Ziegler's work

$$Al(Et)_3 + TiCl_4 \xrightarrow[1\ atm\quad 20–70°C]{=} \text{linear polyethylene} \quad (4.48)$$

linear polyethylene
Mw = 10,000–2,000,000

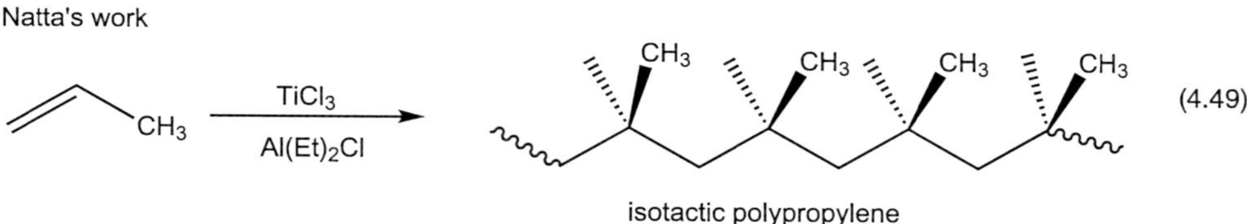

4.4 Al(I) Compounds

Extensive studies have been carried out by Schnökel on making Al(I) compounds which resulted in tetrahedral aggregates. The tetraalumina tetrahedrane, $Cp^*_4Al_4$ is the first Al(I) organometallic compound to be prepared which is quite stable against hydrolysis and oxidation.[21]

$$2Al + 2HCl \xrightarrow[(2)\ 55K]{(1)\ 1200\ K,} 2\ AlCl + H_2 \xrightarrow[Toluene]{Et_2O} AlCl.\ OEt_2 \qquad (4.50)$$

$$4AlCl.\ OEt_2 + 2\ (C_5Me_5)_2Mg \xrightarrow[-MgCl_2]{} (Cp^*Al)_4 \qquad (4.51)$$

Aluminium monochloride is prepared by a high temperature reaction of Al with HCl followed by rapid chilling. It forms adducts with diethyl ether which forms metastable solutions stable at –50°C for weeks. This adduct has been quite useful for the synthesis of many Al(I) organo derivatives.

The halides of composition $[Cp^*AlX(\mu\text{-}X)]_2$ (X = Cl, Br or I) are prepared from AlX_3 and Cp^*SiMe_3. The reduction of $[Cp^*AlX(\mu\text{-}X)]_2$ with Na/K alloy also results in the formation of $(Cp^*Al)_4$ with maximum yields when X = I. More examples of structurally characterised compounds of the type $(RAl)_4$, having aluminium in the +1 oxidation state with the ligands tris(trimethylsilyl)methyl, tris(trimethylsilyl)silyl, and tri *tert*-butylsilyl were also prepared. However the steric bulk of these ligands was not good enough to stabilise Al–R in the monomeric form.[22]

One of the early successes in isolating Al(I) compounds in the monomeric form was achieved by Herbert Roesky and coworkers. A room temperature stable monomeric compound of composition $[HC(CMeNAr)_2]Al$ (Ar = 2,6-i-$Pr_2C_6H_3$) was synthesised by reduction of the aluminium diiodide $LAlI_2$ [L = $HC(CMeNAr)_2$] with potassium metal. Using the diketimine unit, the cyclic six-membered compound was obtained as a red crystalline solid, stable in organic solvents such as toluene, benzene and hexane. It is stable both thermally and in an inert atmosphere, but decomposes above 150°C.[23]

$$LAlI_2 + 2\,K \xrightarrow[-2\,KI]{} \text{LAl:} \qquad \text{Ar} = 2,6\text{-}{}^iPr_2C_6H_3 \tag{4.52}$$

This compound, also named NacNacAl, is monomeric and is the first dicoordinate Al(I) compound to be prepared and characterised. This compound has been described as an aromatically stabilised carbenoid, featuring an Al-based lone pair and an accessible antibonding π* (Al–N) orbital. This Al(I) centre has been found to undergo oxidative addition reactions with a range of molecules of the type HX where X = H, Si, B, Al , C, N, P and O.[24]

$$\text{LAl:} \xrightarrow{HX} \text{LAl(H)(X)} \tag{4.53}$$

HX	HX
H–H	H–BPin
H–SiH_2Ph	H–Oi-Pr
H–SiHMePh	H–NHPh
H–PPh_2	H–Cp*

4.5 The Chemistry of Gallium and Indium

Gallium and indium show oxidation states of +1 and +3. Thallium mostly forms compounds with the +1 oxidation state. Among the many compounds of these elements, a few have become important due to the potential applications of binary compounds with other main group elements resulting in semiconductors with a wider band gap. Synthesis of some of these compounds and their physical properties are described in this section.

Gallium trichloride and indium trichloride are the most important precursors for making organometallic derivatives of gallium and indium. Solid gallium trichloride exists as a dimer with the formula Ga_2Cl_6. It is similar to $AlBr_3$ which is different from the solid structures of $AlCl_3$ and $InCl_3$ which have octahedral metal centres in the solid phase. Ga_2Cl_6 is colourless and soluble in most solvents, which is truly unusual for a metal halide. It is the major precursor for making derivatives of gallium and is also used as a reagent in organic synthesis.

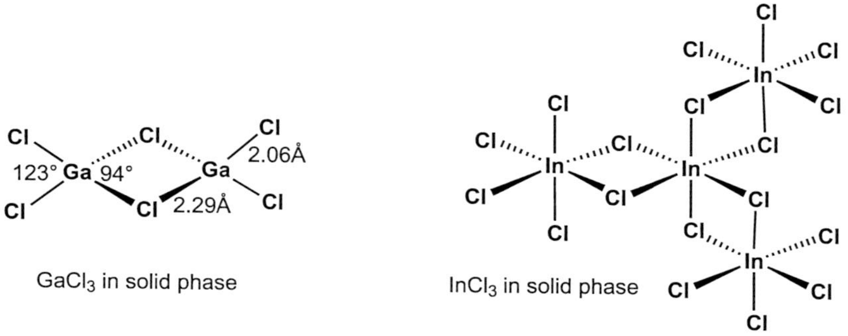

$GaCl_3$ in solid phase

$InCl_3$ in solid phase

Fig. 4.3 Solid phase structures of $GaCl_3$ and $InCl_3$

As a Lewis acid, $GaCl_3$ is milder than aluminium trichloride. Gallium(III) is easier to reduce than aluminium(III).

Gallium trichloride can be prepared by heating elemental gallium in a supply of chlorine gas or from the reaction of gallium oxide with thionylchloride and purifying the product by sublimation under vacuum.

$$2\,Ga + 3\,Cl_2 \longrightarrow 2\,GaCl_3 \tag{4.54}$$

$$Ga_2O_3 + 3\,SOCl_2 \longrightarrow \underset{\text{m.p. } 78^\circ C}{2GaCl_3} + 3SO_2 \tag{4.55}$$

In a similar manner, indium trichloride can also be synthesised.

$$2\,In + 3\,Cl_2 \longrightarrow \underset{\text{m.p. } 586^\circ C}{2\,InCl_3} \tag{4.56}$$

Gallium compounds, especially volatile ones, have been used for the preparation of gallium arsenide and gallium nitride thin layers. Two of the most common gallium compounds used in chemical vapour deposition are trimethyl gallium and triethyl gallium. The synthesis of these air and moisture sensitive compounds and their boiling/melting points are given in the following equations. It is important that the synthesis should not involve a coordinating solvent since the product is a Lewis acid and can form an adduct. Grignard reagents are therefore not suitable and reactions with alkylzinc and alkylaluminium compounds are preferred. A solid air-stable adduct which liberates Me_3Ga on heating in vacuum has also been designed for preparing pure Me_3Ga.[25]

$$Me_2Zn + GaCl_3 \longrightarrow \underset{\text{b.p. } 55.7^\circ C/\ 762 \text{ mm Hg}}{Me_3Ga} + ZnCl_2 \tag{4.57}$$

$$MeMgI + GaCl_3 \xrightarrow[Et_2O]{} \underset{\text{b.p.} 98.3^\circ C}{Me_3Ga.OEt_2} + MgClI \tag{4.58}$$

$$Me_3Ga.OEt_2 + \text{Liq. } NH_3 \longrightarrow \underset{\text{m.p. 31°C}}{Me_3Ga.NH_3} \quad (4.59)$$

$$6MeLi + 2GaCl_3 + \underset{\text{dppe}}{Ph_2P(CH_2)_2PPh_2} \longrightarrow \underset{\text{m.p.162°C}}{(Me_3Ga)_2.(Ph_2PCH_2)_2} \quad (4.60)$$

$$(Me_3Ga)_2.(Ph_2PCH_2)_2 \xrightarrow{\text{heat, vacuum}} 2\ \underset{\text{b.p.56°C}}{Me_3Ga} + (Ph_2PCH_2)_2 \quad (4.61)$$

$$(GaCl_3)_2 + 3\ (Et_3Al)_2 \xrightarrow{\text{pentane}} \underset{\text{b.p.143°C}}{2\ Et_3Ga} + 3\ Et_2AlCl \quad (4.62)$$

$$3\ Et_2AlCl + 6\ KCl \longrightarrow 6\ K^+\ AlEt_2Cl_2^- \quad (4.63)$$

Similary trimethylindium, $InMe_3$, is used for the chemical vapour deposition of indium in the production of indium tin oxide as well as in the synthesis of white LED lights developed by the Nobel Prize recipient S Nakamura. Nakamura achieved exceptional brightness for his white LED by placing an active layer of Zn doped InGaN sandwiched between *p*-type and *n*-type GaN layers from where the emission happens.[25]

Using diethyl ether in the synthesis as a solvent resulted in the formation of ether adducts of trimethyl indium.

$$InCl_3 + 3\ MeLi \xrightarrow{Et_2O} Me_3In.\ OEt_2 + 3\ LiCl \quad (4.64)$$

$$InCl_3 + 3\ EtMgBr \xrightarrow{Et_2O} Et_3In.\ OEt_2 + 3\ MgBrCl \quad (4.65)$$

The ether was removed by applying vacuum. There are many methods which can result in the formation of ether-free trimethyl indium. To get pure Me_3In, one can also use ligands such as dppe or mbda (N,N,N,N-tetramethyl-4,4-methylene-bis-aniline) whose adducts with Me_3In can be easily separated by applying heat and vacuum.

$$6\ MeLi + 2\ InCl_3 + \underset{\text{dppe}}{Ph_2P(CH_2)_2PPh_2} \longrightarrow \underset{\text{m.p.150°C}}{(Me_3In.Ph_2PCH_2)_2} \quad (4.66)$$

$$(Me_3In.Ph_2PCH_2)_2 \xrightarrow{\text{heat, vacuum}} \underset{\text{m.p.88°C}}{2\ Me_3In} + (Ph_2PCH_2)_2 \quad (4.67)$$

$$3MeLi + InCl_3 \xrightarrow{Et_2O} Me_3In.OEt_2 \xrightarrow{\text{mbda}} Me_3In.mbda \quad (4.68)$$

$$Me_3In.mbda \xrightarrow[\text{80-130°C, vacuum}]{} 2\ \underset{\text{m.p.88°C}}{Me_3In} + \text{mbda} \quad (4.69)$$

$$\text{mbda} = Me_2N{-}C_6H_4{-}\underset{H_2}{C}{-}C_6H_4{-}NMe_2$$

$$4MeLi + InCl_3 \xrightarrow{Et_2O} Me_4InLi.(OEt_2)_3 \xrightarrow{vacuum} Me_4InLi \quad (4.70)$$

$$3Me_4InLi \xrightarrow{InCl_3} 4\ Me_3In + 3\ LiCl \quad (4.71)$$

Indium in organic synthesis

Indium metal is air-stable and non-toxic with a first ionisation potential of 5.8 eV, which is close to that of Na and Li, and is therefore an efficient single electron transfer agent.

Element	Zn	Mg	Sn	In	Li	Na
1st IP (eV)	9.39	7.65	7.43	5.79	5.39	5.12

Allyl indium halides are formed without formation of Wurtz type coupling products by mixing indium powder and allyl halides (useful for allylation of carbonyl compounds). Indium mediated Barbier type allylations can be done in water medium. Organoindium compounds are tolerant to water and reactions do not need protection of OH or similar protonated functional groups.

$$RCHO + CH_2{=}CHCH_2Br \xrightarrow[H_2O]{In} RCH(OH)CH_2CH{=}CH_2 \quad (4.72)$$

60–97%

Indium has a tendency to coordinate to heteroatoms especially oxygen of OH or C=O making reactions regio-, stereo and chemo-selective.

$$MeOCH_2CH(OMe)CH_2COCH_3 + CH_2{=}CHCH_2Br \xrightarrow[H_2O]{M} MeOCH_2CH(OMe)CH_2C(OH)(CH_3)CH_2CH{=}CH_2 \quad (4.73)$$

M
In 70% yield
Sn 10% yield
Zn 0% yield

When the allylation was carried out in water medium using indium catalyst, the reaction proceeded smoothly at room temperature without the requirement of a co-solvent or promoter. Zinc and tin requires heating, sonication or additional acid catalysts. The mild conditions make it possible to allylate methyl ketones in the presence of acid-sensitive acetal functional groups which does not proceed with Sn and Zn.

$InCl_3$ is a well known mild Lewis acid which is stable in an aqueous medium, and it effectively catalyses the aldol, Diels–Alder, and Friedel–Crafts reactions. Unlike $AlCl_3$ it is not moisture sensitive and reactions involving $InCl_3$ as catalyst can be carried out in water or in organic solvents such as acetonitrile.

4.6 Wide Band Gap Semiconductors Based on Gallium and Indium

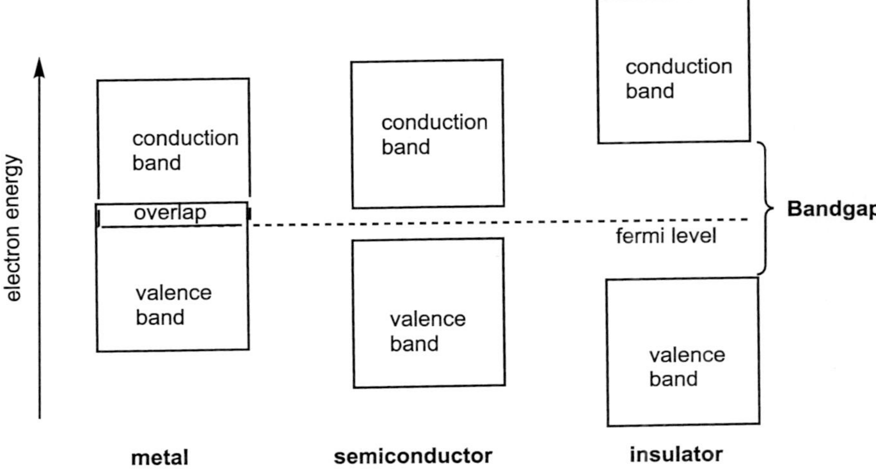

Fig. 4.4 Band gaps of metals, semiconductors and insulators

Wide band gap semiconductors are semiconductors that possess band gaps significantly greater than that of silicon. Germanium, silicon and gallium arsenide have band gaps of the order of 0.7–1.4 eV, a relatively small amount of energy. The low band gap limits silicon based devices to operational temperatures below 100°C, beyond which the temperature will cause the electrons to switch on their own to the conduction band, thereby losing the semiconductivity. On the other hand, materials with wide band gaps have band gaps in the order of 2–4 eV (see Table 4.9), allowing them to operate at much higher temperatures (up to 300°C). In addition, semiconductors with a wide band gap permit devices also to operate at much higher voltages and frequencies, making the power electronic modules using these materials significantly more powerful and energy efficient than those made from conventional silicon-based materials. The overall consequence of this wider band gap is that it changes the way we were looking at semiconductors compared to traditional silicon and germanium with many new applications that were not possible with semiconductors possessing narrow band gaps.

Table 4.9 Band gaps of well known semiconductors

Semi-conductor	Ge	InN	Si	GaAs	ZnSe	SiC	ZnO	GaN	Diamond
Band gap (eV)	0.7	0.7	1.12	1.43	2.82	3.3	3.4	3.45	5.45

4.7 Advantages of Wide Band Gap Semiconductors

Many materials with a wide band gap emit light in the visible colour range, an optical property useful for applications in solid state lighting. For example, gallium nitride (GaN)

with a band gap of 3.45 eV, is the material behind the highly efficient white light emitting diodes (LEDs) which are replacing traditional incandescent and CFL lighting. In contrast, due to a smaller band gap (1.43 eV), GaAs emits only in the infrared region. Electronic devices with wide band gap are expected to cut down energy losses. Electronic devices based on them are expected to eliminate up to 90% of the power losses that currently occur during electricity conversion from direct to alternating current and vice versa. They can be used for voltages more than 10 times higher than Si-based devices, greatly enhancing performance in high-power applications. They are also suitable for higher-temperature operation as they can operate at temperatures over 300°C (which is almost twice the maximum temperature of Si-based devices). This tolerance for higher operating temperature enables smaller and lighter systems and results in better overall system reliability, and creates openings for new types of applications.

Semiconductors with wide band gap can operate at frequencies at least 10 times higher than Si-based devices, making possible more compact, less costly product designs providing scope for new applications, such as radio frequency (RF) amplifiers. LEDs made of GaN or InGaN which have wide band gaps, have greater lighting efficiency: LEDs produce 10 times more light per watt of input energy when compared to incandescent bulbs and extend service life by 30 times or more. The deep ultraviolet LEDs, mostly based on GaN, have potential applications not only in white lighting (~255 nm) but also in UV-based air and water purifiers (~265 nm) and biomedical sensors (~280 nm). The utility of a semiconductor in LED applications is also dependent upon whether it has a direct band gap or an indirect band gap. The band gap is called 'direct' if the momentum of electrons and holes is the same in both the conduction band and the valence band. As a consequence an electron can directly emit a photon. In an 'indirect' band gap, the momentum of electrons and holes is not the same in both the valence conduction bands. A photon cannot be emitted because the electron must first pass through an intermediate state and transfer momentum to the crystal lattice. Silicon and germanium have indirect band gaps while GaAs and GaN have direct band gaps.

Some of the devices which are likely to result from the use of semiconductors that possess wide band gap are (i) variable speed drives for motors (used in pumps, fans etc), (ii) large, high-efficiency data centres and compact power supplies (for consumer electronics), (iii) inverters which can convert the DC electricity generated from solar and wind energy into AC electricity while reducing losses by around 50%, (iv) transformers with reduced size providing greater efficiency in converting AC to DC, (v) LED lighting that saves billions of dollars in cumulative energy costs, (vi) rugged sensors that can withstand harsh military applications, and (vii) electronics that tolerate higher operating temperatures leading to considerably reduced size of cooling systems in such devices.

4.8 Gallium Nitride

Gallium nitride is possibly the most important new semiconductor discovered since silicon and is a material with a wide band gap (direct band gap of 3.4 eV). Additional properties of GaN include high breakdown field, high thermal and chemical stability and high saturation drift velocity. The combination of these properties makes it an ideal material for extended use in laser diodes (LDs), light emitting diodes (LEDs), photonics and high-power and high-temperature operation devices. Unlike silicon, gallium nitride *p–n* junction diode emits brilliant blue light when a small electric current is passed through it. In fact one can produce light of any colour by adding different amounts of indium to gallium nitride. On the other hand, if we add aluminium to gallium nitride, it produces UV light. Gallium nitride based light emitting diodes (LEDs) and laser diodes are already in widespread use, from street lights to Blu-ray laser diodes for DVD players. If we coat a blue GaN LED with a yellow-emitting phosphor Ce-YAG (yttrium aluminium garnet) then a cool white light is produced. Compared to traditional lighting sources, gallium nitride is quite efficient with a long life and lower power consumption.

Methods of making GaN thin layers

Since the applications require a fine layer of GaN, there has been considerable growth in the development of epitaxial growth techniques. The main methods used are metal organic chemical vapour deposition (MOCVD), hydride vapour phase epitaxy (HVPE) and molecular beam epitaxy (MBE). MOCVD uses volatile organogallium compounds such as trimethylgallium and triethylgallium to transport gallium atoms to the substrate to be deposited and react with nitrogen sources such as NH_3 at a convenient deposition temperature. For example, GaN growth can be carried out on a sapphire substrate by keeping it in a reaction chamber at 800–1000°C and introducing trimethyl gallium or triethyl gallium and ammonia simultaneously. Different chemical routes reported for the synthesis of GaN are given in Scheme 4.8. The commonly used precursors for gallium are triethylgallium, trimethylgallium and triisobutylgallium. The nitrogen precursors used are ammonia, dimethylhydrazine, *t*-butylamine or phenylhydrazine. MOCVD method provides very high quality film, is of intermediate cost and has a high growth rate and throughput rate. However, one disadvantage is the need for large quantities of ammonia. The GaN layers grown using triethylgallium as a precursor exhibited superior electrical and optical properties and a lower concentration of carbon impurity. There are two possible reaction mechanisms for the reaction of triethylgallium with ammonia in growing GaN: (i) β-hydride elimination (C_2H_4 + H) and (ii) direct desorption of C_2H_6 (similar to the trimethylgallium growth mechanism).[26]

In the HVPE method, the reactor consists of two zones namely the source zone and the growth zone. In the source zone, metallic gallium is reacted with HCl at 860°C to make GaCl. The gallium monochloride formed is injected into the growth zone using a carrier gas such as hydrogen or ammonia where the substrate to be coated is kept at around 1000°C and allowed to react with ammonia resulting in the coating of GaN. HVPE method requires high temperatures, shows a high growth rate and provides reasonably good quality films. In the MBE method, elements such as gallium and arsenic, in ultra-pure form, are heated in separate outpouring cells or electron beam evaporators until they start to slowly sublime. The gaseous elements then condense on the wafer, where they react with each other forming the desired semiconductor. In the case of gallium and arsenic, single-crystal gallium arsenide is formed.

Table 4.10 Comparison of silicon, GaAs and GaN

Material	Density g/cc and m.p.	Band gap eV (300 K) and type	LED activity	Electron Mobility $Cm^2/V.s$ (300 K)	Crystal lattice	Cost	Thermal conductivity
Silicon	2.33 1414°C	1.12 Indirect	Nil	1500	Diamond	Inexpensive	149 W/(m.K)
Gallium arsenide	5.32 1238°C	1.42 Direct	Infrared emission	8500	Zinc blend	Expensive	0.55W/(cm.K)
Gallium nitride	6.15 2530°C	3.39 Direct	Blue and UV (InGaN)	440	Wurtzite/ zinc blend	Expensive	1.30 W/(cm.K)

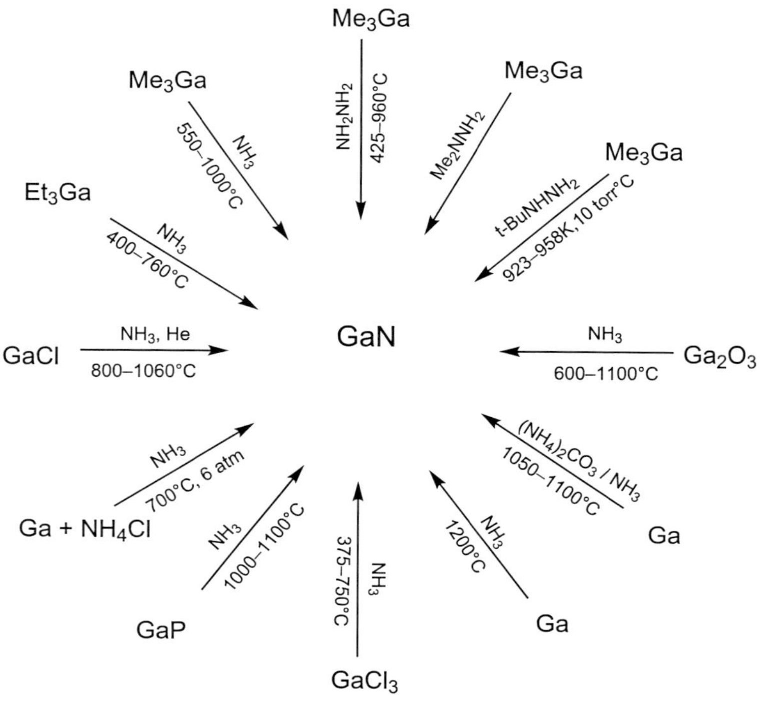

Scheme 4.8 Different methods for synthesis of gallium nitride

4.9 Gallium Nitride-Based LED for White Light

The band gap determines the wavelength at which LEDs can emit light and the wavelength at which photovoltaics operate most efficiently. The GaAs band gap of 1.43 eV, corresponds to a wavelength of 890 nm, which falls in the infrared range (the wavelength is determined by dividing the constant 1240 nm-eV by the energy in eV, so 1240 nm-eV/1.4 eV = 886 nm). Therefore, GaAs photovoltaics are not ideal for converting shorter wavelength visible light into electricity. Silicon at 1.1 eV (1100 nm) is also not good. For LED applications, the ability to emit visible light rather than IR radiation is necessary. The rise of LEDs for lighting applications depends particularly on the development of nitride semiconductors with wide band gap.[27]

The Nobel prize for physics in 2014 was shared by three scientists—Isamu Akasaki and Hiroshi Amano of Nagoya University in Japan, and Shuji Nakamura of the University of California, Santa Barbara, for work spanning three decades that eventually led to blue and white light emitting diodes.[22,23] Efficient white light emitting diodes are currently replacing traditional incandescent bulbs, fluorescent tubes and compact fluorescent lamps (CFL) at a very fast rate. This changeover is expected to generate 40% energy savings in the USA alone by 2030 and eliminate the need for more than thirty 1000 W power plants. The changeover prevents generating millions of tons of carbon dioxide.[28]

A light emitting diode is basically a *p–n* junction diode which when activated with a small voltage, emits radiation in the form of visible light. For example, the GaN blue LED devised by Nakamura uses 2.8 V power and has a GaN *p*-type and GaN *n*-type layer sandwiching a small layer of InGaN. When electricity is applied, electrons combine with holes within the device releasing energy in the form of photons (the process is called electroluminescence). Interestingly, the wavelength of the light emitted, corresponding to the energy of the photon is decided by the energy band gap of the semiconductor unit which is sandwiched between the *p* and *n* type layers (InGaN in this case) providing scope for generating lights of different wavelengths.[29, 30]

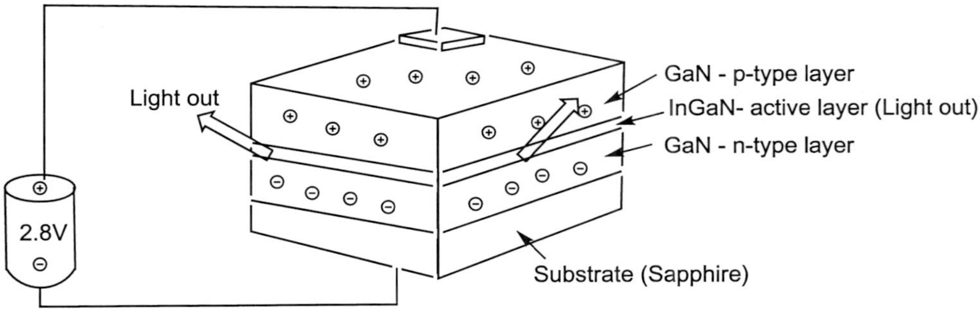

Fig. 4.5 Basic structure of the blue LED developed by Nakamura

Cerium(III)-doped YAG (Ce:YAG or YAG:Ce) is a phosphor (scintillator). It emits yellow light when irradiated with blue or ultraviolet light. It is used in white light-emitting diodes,

How the sole desire to get a Ph. D led to a Nobel prize winning discovery

The discovery of efficient blue light emitting diodes of InGaN in 1993 by Nakamura et al. is one of the greatest useful discoveries of the modern world to mankind as this blue LED was easily converted to white LED by embedding phosphors in a plastic cap surrounding it. Also this blue light has been combined with red and green light to get quality white light. Since then, white LEDs have been replacing other common sources of white light all over the world as it is more efficient, long lasting, non-polluting and consuming much less power thereby saving electricity. For comparison, while fluorescent light (CFL) generates 70 lm/W, white LEDs generate 300 lm/W and do not contain toxic mercury. An estimate says that by 2030 the power saved by using white LEDs will be 261 TWh* which corresponds to electricity savings of over 40%. This energy saving will also reduce CO_2 emission by 185 million tons by 2030 as less number of fossil fuel using power plants need to be made. Governments of energy demanding countries like India have even started freely distributing two white LED bulbs per household keeping energy savings in mind.

Shuji Nakamura along with Isamu Akasaki and Hiroshi Amano were conferred the 2014 Nobel Prize in physics for this astounding discovery. In his Nobel lecture, Nakumura explained how he decided to work in the field of gallium nitride which was a less trodden path unlike the area of zinc selenide. His major contribution and breakthrough which fetched him the Nobel Prize was designing a metal organic chemical vapour deposition (MOCVD) reactor that could produce uniform highest quality 2 inch GaN layers needed for LED making. Very interestingly, to learn the technique of MOCVD, he as a young researcher went to the University of Florida, USA. He recounts how he felt ashamed when his lab colleagues asked him if he has published any research paper or has a Ph.D, both of which he did not have. The following is an excerpt from his Nobel lecture.

'My entry into the field started in April of 1988, when I went to the University of Florida as a visiting researcher. The main purpose of my visit was to learn how to use a MOCVD (Metal Organic Chemical Vapour Deposition) system to grow GaAs crystals on a silicon substrate, as I had no experience in how to use a MOCVD. During my stay there, I worked together with graduate students and they all asked me if I had a Ph.D, I said no. At the time, I only had a Masters. Next, they asked me if I had published any scientific papers. Again, I said no, I had never published a single paper. Consequently, they treated me as a technician. In the U.S., this meant one has to help the researcher and one's name would not appear on papers or patents. Gradually, I became very frustrated with this arrangement.

One year later, in March of 1989, I came back to Japan. It was my dream to get a Ph.D. degree. In Japan, at the time, it was possible to be awarded a Ph.D if one published five scientific papers. This type of degree was called a paper degree and one did not need to go to the university to get the degree. It was therefore my ultimate dream to publish at least five papers and get a Ph.D. With this in mind, I noted that the ZnSe field was publishing lots of papers. As I had never published a paper, I had no confidence in publishing a paper. In the GaN field, only very few papers had been published, mainly from Professor Isamu Akasaki and Hiroshi Amano. I was therefore confident that I could publish lots of papers, though had no confidence that I could actually invent the blue LED. My only objective was to get a Ph.D. That's it.'

* 1 Terawatt hour = 1 000 000 000 kilowatt hour (1 kWh is the energy used by a 1kW (1000 watt) electric heater for 1 hour)

Excerpt reprinted from '*Background story of the invention of efficient blue InGaN light emitting diodes*' Nobel Lecture, December 8, 2014 by *Shuji Nakamura;* with permission from the Nobel foundation, Stockholm, Sweden© The Nobel Foundation.

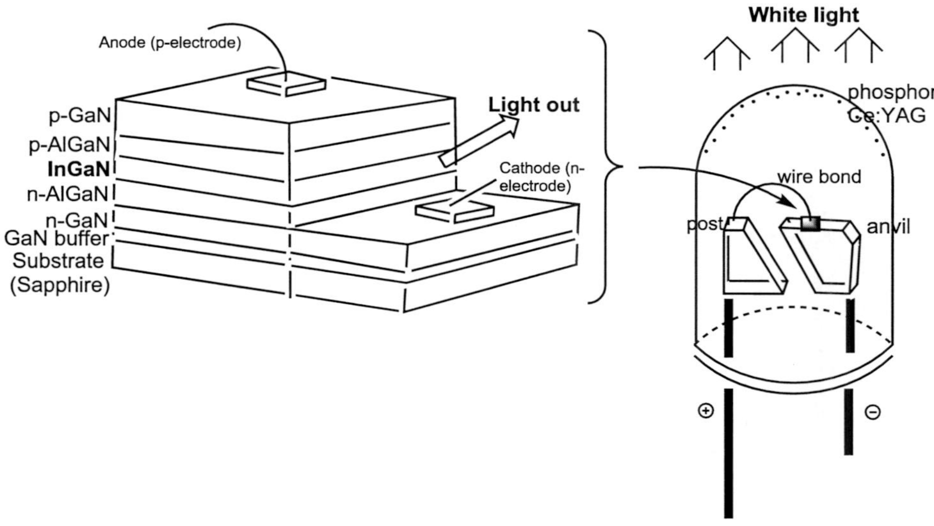

Fig. 4.6 Detailed design and structure of a white light LED developed by Nakamura

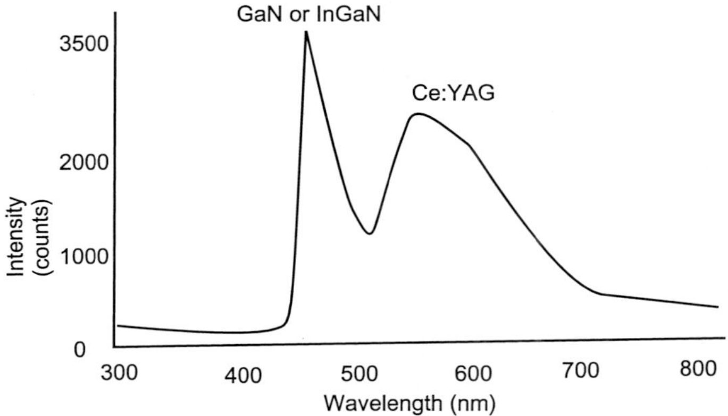

Fig. 4.7 Spectra of white light LED resulting from using a Ce:YAG phosphor on GaN blue LED

as a coating on a high-brightness blue InGaN diode, converting part of the blue light into yellow, which along with the blue then appears as white.

InGaN, is one candidate as the active layer for the blue emission because its band gap can be varied depending on the indium mole fraction. Band gap of InGaN can be

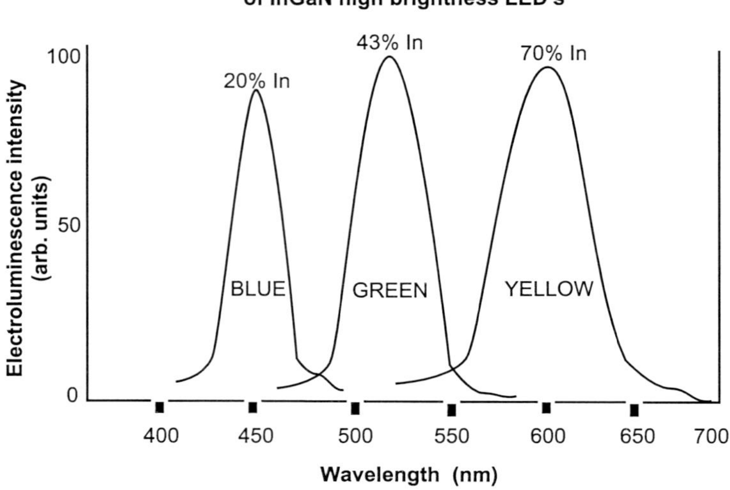

Fig. 4.8 Output colour of InGaN when amount of indium is varied

tuned by varying the amount of indium in the alloy. The band gap of $In_xGa_{1-x}N$ spans from the infrared (0.70 eV) for InN to the ultraviolet (3.4 eV) of GaN. The ratio of In/Ga is usually 0.02/0.98 and 0.3/0.7. Indium gallium nitride is the light-emitting layer in modern blue and green LEDs, often grown on a GaN buffer on a transparent substrate such as sapphire or silicon carbide. InGaN films can be grown by the two-flow metal-organic chemical vapour deposition (MOCVD) method using trimethylindium and trimethylgallium and a nitrogen source.[31]

4.10 Gallium Arsenide

As an infrared emitter it has only minimal applications in producing visible radiation such as the radiation needed for remote control units. However, the utility of GaAs become significant in the design and use of high efficiency solar cells.[31]

4.10.1 Synthesis of Gallium Arsenide

Some of the well known chemical methods of synthesising gallium arsenide are given in Scheme 4.9.

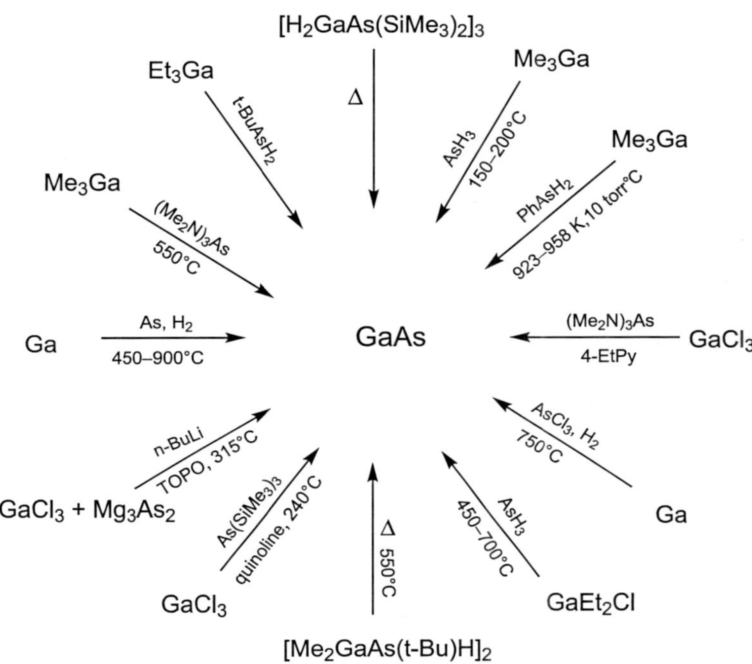

Scheme 4.9 Reagents used for synthesis of GaN for chemical vapour deposition method

4.10.2 GaAs Advantages

The GaAs-based infrared LED was discovered and patented by Robert Biard and Gary Pittman working at Texas Instruments. The uses of GaAs are diverse and include diodes, field effect transistors (a transistor that uses an electric field to control the electrical behaviour of the device), integrated circuits and most importantly solar cells. Components made of gallium arsenide are useful in ultrahigh radio frequencies and fast electronic switches. The benefits include lesser noise than other semiconductors. Therefore, it is useful for weak signal amplification. Gallium arsenide LEDs work in the infrared region and are useful in optical communications and control systems. Compared to silicon, gallium arsenide units are insensitive to heat, resistant to radiation damage and are highly efficient, thus making them perfect for space-related applications. Almost all spacecrafts, space stations and satellites use gallium arsenide based solar cells to the maximum extent. The main drawback compared to silicon is the cost factor.

High saturated electron velocity and higher electron mobility are properties of gallium arsenide which are superior to those of silicon. This allows transistors made of gallium arsenide to function at frequencies higher than 250 GHz. GaAs devices tend to create less noise in electronic circuits than silicon devices, especially at high frequencies.

Relative to silicon, GaAs is not affected by overheating. This is an outcome of higher carrier mobilities and lower resistive device parasitics (parasitics are some circuit elements such as resistance, inductance or capacitance possessed by an electrical component which are not desirable for its intended purpose). These superior properties result in the use of GaAs circuitry in mobile phones and satellite communications and also in higher frequency radar systems. It is also used in the manufacture of Gunn diodes, a form of diode with negative resistance used for the generation of microwaves.

GaAs, unlike silicon and germanium has a direct band gap. GaAs-based solar cells show one of the highest efficiencies among single-junction solar cells (28.8%). When sunlight falls on a gallium arsenide solar cell, which is basically a *p–n* junction, part of it is absorbed which excites electrons and allows them to flow freely. Solar cells also have one or more electric fields that act as a mediator which forces the electrons liberated to flow in a certain direction resulting in an electric current. This electricity is taken out by placing metal contacts on the top and bottom of the cell.

The high resistivity combined with the high dielectric constant of pure GaAs, makes it a very good substrate for integrated circuits which provides natural isolation between devices and circuits (unlike silicon).[32]

Table 4.11 Comparison of silicon and gallium arsenide transistors and solar cells

Property	Silicon	Gallium arsenide	Outcome
Band gap (eV)	1.1	1.4	GaAs transistors and solar cells are insensitive to overheating. GaAs is resistant to radiation damage.
Saturated electron velocity (maximum electron velocity attained in the presence of high electric fields)	1.0×10^7 cm/s	1.2×10^7 cm/s	Allows gallium arsenide transistors to work at frequencies higher than 250 Hz.
Electron mobility (how quickly an electron will move through a semiconductor when pulled by an electric field)	1400 cm^2/V.s	8500 cm^2/V.s	Less noise in electronic circuits of GaAs; suitable for mobile phones and satellite communication devices.
Temperature coefficient (performance loss of a solar cell versus temperature relative to 25°C)	−0.41%/°C	−0.0%/°C	No loss of solar cell efficiency for GaAs solar cells with increase in day temperature.

What semiconductors are ideal for making solar cells?

One of the interesting differences between GaAs and GaN is that the former has been used for making solar cells as well as LEDs while the latter which is an excellent material for making LEDs is not useful for making solar cells. This has been linked to the band gap of the two materials. So how does sunlight generate electricity? The spectrum of sunlight which covers from infrared to ultraviolet region occurs in the range of approximately 0.5 eV to 2.9 eV. When a photon from sunlight enters a solar cell, it makes loose a valence electron and allows the electron to travel freely. By leaving the organised structure of the parent atom, the electron leaves a 'hole' behind. Thus an electron–hole pair is generated. Once the electron is free, the electric field pulls it towards the *p*-side and the hole is pulled towards the *n*-side of the semiconductor. All the free electrons forced to one side of the cell form the current. And the current combined with the voltage of the electric field gives power.

The main reason why solar cells are not 100% efficient is due to the fact that a single semiconductor does not respond to the entire spectrum of sunlight. Only those photons with sufficient energy to knock an electron out of a semiconductor has the ability to create an electron–hole pair, which is directly linked to the band gap of the material. For example, the amount of energy required to knock an electron off from gallium arsenide is 1.43 eV (which is its band gap). If a photon has 2.0 eV energy and if it falls on a 1.43 eV GaAs solar cell, the excess energy (0.57 eV) will be lost in the form of heat. Thus, for a material to be useful for making a solar cell, first of all its band gap has to fall within the solar spectrum, secondly there should be sufficient photons of that energy, and thirdly there should be minimum loss of energy as heat. The band gap range between 1.0 and 1.7 eV has been found to make the best solar cells. In this range, less heat is produced during the functioning of the cell. A low band gap is desirable for a solar cell so that it can absorb as much of the solar spectrum as possible and for the formation of a large built in voltage, a slightly larger band gap is desired. The Shockley–Queisser limit places maximum solar conversion efficiency as around 33.7% assuming a single *p*–*n* junction with a band gap of 1.34 eV. This means that of all the power contained in sunlight falling on an ideal solar cell (about 1000 W/m^2), only 33.7% of that could ever be turned into electricity (337 W/m^2). One of the reasons why multi-junction solar cells give more than 40% conversion efficiency is because they utilise the maximum possible radiation from sunlight by using a combination of semiconductors with different band gaps such as Ge (0.7 eV), InGaAs (1.4 eV) and InGaP (1.86 eV).

4.11 Indium Tin Oxide

Transparent conductors (TCs) have a wide variety of uses. Most well known are the TCOs or transparent conductive oxides. Thin films of metal oxides like cadmium oxide (CdO), tin oxide (SnO_2) and indium oxide (In_2O_3), are the classical examples for optically transparent and electrically conducting materials. Among the several types of TCOs, SnO_2, ZnO, In_2O_3, Ga_2O_3, and CdO are the well-known binary compounds. A doping element is generally introduced to improve their electrical conductivity, and the frequently used doped TCOs are In_2O_3:Sn (ITO), In_2O_3:F, SnO_2:F, SnO_2:Sb, ZnO:Al (AZO), a variety of ternary and even more complex TCO materials, such as Zn_2SnO_4, Cd_2SnO_4 and $In_4Sn_3O_{12}$.

Table 4.12 Comparison of common transparent conducting oxides

TCO	Transmittance of visible light (%)	Electrical resistivity (ohm.cm)	Deposition method	band gap (eV)
SnO_2	95	6.1×10^{-3}	Sputtering	4.1
In_2O_3	86	2.0×10^{-4}	PLD	–
ITO (In_2O_3:SnO_2 9:1)	95	2.4×10^{-4}	Sputtering	4.0
Cd_2SnO_4	93	5.0×10^{-4}	Sputtering	2.7
$CdIn_2O_4$	90	2.7×10^{-4}	Sputtering	3.2
$In_4Sn_3O_2$	> 80	3.5×10^{-4}	Sputtering	3.5
ZnO	> 80	2.0×10^{-3}	Sputtering	3.3
ZnO:Al (AZO)	90	10^{-2}	Sputtering	1.5

Aluminium zinc oxide (AZO) used to be a common choice of TCO as it is relatively inexpensive and has good optical transmission performance in the visible spectrum. However, indium tin oxide is the most widely used TCO now because of its two main properties—better electrical conductivity and optical transparency, as well as the ease with which it can be deposited as a thin film. ITO defeats AZO in all performances including chemical resistance to moisture and if moisture does penetrate, ITO has been found to degrade less than AZO. ITO has been found to remain intact in rooftop solar cells for up to 30 years. ITO can also be precisely etched into fine patterns unlike AZO. The only drawback is the cost factor, as ITO is priced at more than AZO.

This compound, also known as tin doped indium oxide, is a 90: 10 percent mixture by weight of indium(III) oxide (In_2O_3) and tin(IV) oxide SnO_2 or more specifically 74% In, 18% O and 8% Sn by weight. It is yellow-green in colour in powder form but is colourless and transparent when deposited as a thin film with thickness of 1000–3000 Å. It can be deposited both on glass and clear plastic such as polyester. These thin films are stable up to 150°C.

Other than the exponential rise in their use in touch screen devices such as smart phones and ATMs, the ability of ITO to reflect thermal infrared radiation (90% reflectance) is exploited in making energy-conserving low emissivity windows. When ITO is incorporated into oven windows it helps to conserve energy and to maintain a lower outside temperature. The front-surface electrodes for solar cells and flat-panel displays (FPDs) also utilise ITO. Electric current is passed through TCOs such as ITO to defrost windows in vehicles and aeroplanes and to keep display cases frost-free. The conductivity also helps to dissipate static electricity from the windows.

Indium tin oxide films are transparent conductors that have tremendous applications in the design of optoelectronic devices. These films combine many technologically useful properties such as good electrical conductivity, high transparency for visible light, excellent substrate adherence on both glass and plastic, stability against oxidation by air and moisture and infrared reflectivity. Typical use of these films include LCD, OLED, plasma flat panel displays, touch screen devices such as mobile smart phones and ATM machines, electroluminescent lamps and displays, electromagnetic interference shields (EMI shields), antistatic coatings, solar panels, energy efficient windows, aircraft and automobile self-demisting and deicing. It is also used to a lesser extent as infrared reflecting coating to reflect heat energy from devices and allows visible and UV radiation to pass through.

Advantages of using ITO

- At a given thickness, it has the highly desired combination of both better electrical conductivity and higher optical transparency compared to any other doped metal oxide.
- It can be easily deposited as a thin film (sputtering).
- It has more resistance against moisture penetration.
- It has better chemical stability compared to other TCOs.
- It can be easily mass produced.
- ITO can be precisely etched into fine patterns unlike other transparent conducting oxides (such as aluminium zinc oxide) which are less expensive.
- ITO shows almost 90% reflectance of infrared radiation.

4.12 Thallium

Thallium(I) sulfate, which is the most common precursor for thallium compounds, is made by the reaction of thallium metal with sulfuric acid. It is colourless, odourless, tasteless and is highly toxic. Its toxic effects are derived from the thallium(I) cation's high solubility in water and similarity to K^+ (effective ionic radius: K^+ 138 pm; Tl^+ 150 pm). The lethal dose of thallium(I) sulfate for an adult is about 1 gram. CpTl or cyclopentadienyl thallium used to be a convenient organothallium compound for synthesising metal sandwich compounds. CpTl which is highly air stable has been prepared by the reaction of Tl_2SO_4, NaOH and cyclopentadiene. However the use of these compounds has become almost non-existent due to the thallium toxicity issue.

4.12.1 Toxicology of Thallium

Some of the major toxicological effects of thallium are (i) disruption of potassium-dependent processes, (ii) interference with cysteine residues (iii) riboflavin sequestration (iv) ribosomal inhibition and (v) injury to myelin sheath.

Thallium mimics the nutritionally essential element potassium and easily passes through the stomach wall into the blood stream. The body, after a while realises that it is not potassium and sends it back to the intestines. However, further down the intestines it is again mistaken for potassium and reabsorbed. This cycle needs to be broken if thallium has to be removed from the body. It accumulates in organs and tissues where potassium is normally found in high concentrations such as heart, muscle and nerve tissues. Pyruvate kinase and succinate dehydrogenase are two major enzymes involved in thallium toxicity. Inhibition of these enzymes results in impaired glucose metabolism and disruption of the Kreb's cycle which eventually leads to diminished ATP production. In addition, the sodium–potassium pump also gets affected, resulting in injury to the cell membrane. All these lead to mitochondrial and cell death. Thallium can disrupt disulfide bonds. This affects cross-linking of cysteine residues thus affecting the formation of keratin. This is the reason of hair loss during thallium poisoning and the reason why it was used in hair removing creams in the early days before its toxicity was detected.

Elimination of thallium from the body is slow with a half-life for elimination of 3–30 days. Because of this reason, thallium may act as a cumulative poison. Many complexing agents and chelating ligands were attempted to selectively remove thallium from the system and break the cycle of excretion and absorption without success and thallium poisoning was therefore thought to be incurable. In 1969 the German pharmacist Heydlauf suggested the well known dark blue complex, Prussian blue [$KFe(Fe(CN)_6$] as an antidote. The soluble form of Prussian blue exchanges its potassium for thallium to which it is found to bind strongly and gets eliminated from the body.[33]

4.13 Recent Developments

4.13.1 Hydroboration Using Aluminium Hydride Catalysts

Hydroboration reaction catalysed by transition metal complexes is well known from 1985 when Wilkinson's catalyst $RhCl(PPh_3)_3$ was effectively used as a catalyst for hydroboration of alkenes with catacholborane. Following this, a variety of transition metal catalysts using metals such as rhodium and iridium were developed, and used for hydroboration and asymmetric hydroboration. However till 2015, no example of main group metal compounds acting as catalysts for hydroboration was known. Yang, Parameswaran, Roesky and coworkers showed for the first time that the sterically hindered aluminium dihydrides $LAlH_2$ [L = $HC(CMeNAr)_2$, Ar = 2,6-$Et_2C_6H_3$] and LAlH(OTf) [L = $HC(CMeNAr)_2$, Ar = 2,6-$iPr_2C_6H_3$] can effectively bring about hydroboration of alkynes, aldehydes and ketones in almost quantitative yields.[34,35]

H$_3$Al.NMe$_3$, toluene, 0°C or LiAlH$_4$, toluene, reflux → A (4.74)

R–C≡CH + H–Bpin —A, 3 mol%, 30°C→ R–CH=CH–Bpin (4.75)

Following this study, commercially available aluminium hydrides and aluminium alkyls such as DIBAL-H and Et_3Al were also found to be effective as catalysts for hydroborylation of alkynes but in lesser yields. Further, hydroboration of alkenes with HBPin was found to be effectively catalysed by well known and easily available aluminium hydrides such as $LiAlH_4$ and Red-Al in yields of up to 95%.[36]

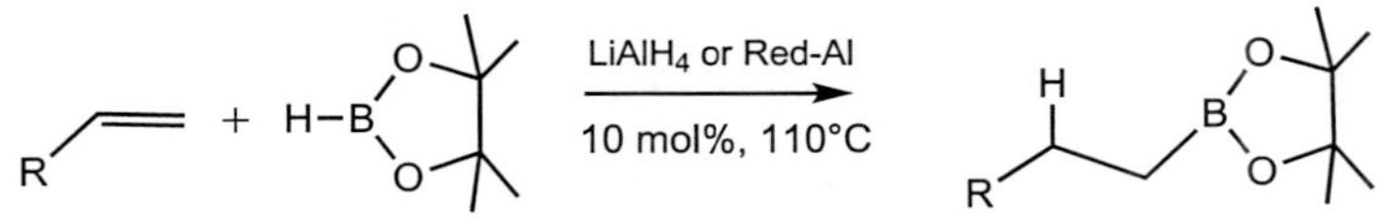
LiAlH4 or Red-Al
10 mol%, 110°C
H–B
R
H
B
O

(4.76)

4.13.2 Al=Al Compounds

Homo-dinuclear double bonded neutral aluminium compounds, which are aluminium analogues of alkenes, have been a very difficult synthetic target over several decades. Till recently only masked Al=Al compounds or charged species were known. The first neutral compound was recently prepared by using the moderately bulky N-heterocyclic carbene, 1,3-diisopropyl-4,5-dimethyl-imidazolin-2-ylidene (Scheme 4.10). The Al=Al bond length is 2.394 Å units which is the shortest distance reported for any molecular aluminium species.[37] The compound was found to readily form four-membered rings with ethylene as well as phenylacetylene, further confirming the presence of a double bond. It is noteworthy that attempts to make double bonds between gallium, indium and thallium using the sterically demanding terphenyl ligand systems resulted in compounds of the general formula ArMMAr where Ar = 2,6-dipp$_2$C$_6$H$_3$, M = Ga, In and Tl. However, the M–M bonds in all these compounds turned out to be longer than the corresponding single bonds, showed considerable trans bending geometry and dissociated into their monomers in hydrocarbon solvents (Scheme 4.11).[38–40]

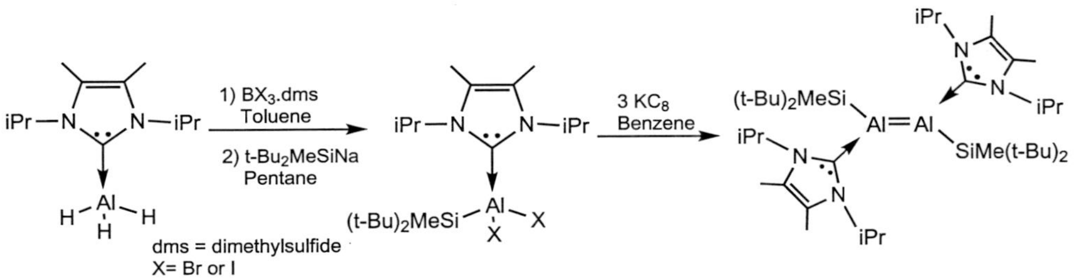
1) BX3.dms
Toluene
2) t-Bu2MeSiNa
Pentane
3 KC8
Benzene
iPr—N
N—iPr
H–Al–H
H
(t-Bu)2MeSi–Al–X
X
(t-Bu)2MeSi
Al=Al
SiMe(t-Bu)2
iPr
N
dms = dimethylsulfide
X= Br or I

Scheme 4.10 Synthesis of the first neutral Al=Al bonded compound

123.2° 121.2° 119.7°

Ga—Ga 2.627 Å; In—In 2.979 Å; Tl—Tl 3.094 Å

M–M distances longer than correspondng single bonds

Scheme 4.11 Comparison of bond parameters of ArMMAr (M = Ga, In, Tl)

Questions for self study (multiple choice)

4.1. In Friedel–Crafts reaction, toluene can be prepared by
(a) $C_6H_6 + CH_3Cl$ (b) $C_6H_5Cl + CH_4$
(c) $C_6H_6 + CH_4$ (d) $C_6H_6 + CH_3COCl$

4.2. The stability of the +1 oxidation state increases in the sequence
(a) Tl < In < Ga < Al (b) In < Tl < Ga < Al
(c) Ga < In < Al < Tl (d) Al < Ga < In < Tl

4.3. Duralumin is mainly an alloy of aluminium with
(a) Zn (b) Cu (c) Mg (d) Mn

4.4. The reason why thallium is toxic is that its ionic size and oxidation state are very close to that of
(a) K^+ (b) Na^+ (c) Li^+ (d) Cu^+

4.5. The increasing order of band gap for the following semiconductors is
(a) Si < GaN < GaAs < InN (b) Si < InN < GaAs < GaN
(c) InN < Si < GaAs < GaN (d) InN < Si < GaAs < GaN

4.6. Molten cryolite is used in the electrolysis of Al_2O_3 because it
(a) has lower melting point than Al_2O_3
(b) has lower melting point; does not absorb moisture from air
(c) does not absorb moisture from air
(d) lower melting point and has ease of ionisation

4.7. Which of the following statements is not true for $LiAlH_4$?
(a) It is pyrophoric and often ignites in moist air
(b) It is poorly soluble in non-polar solvents
(c) On prolonged storage at room temperature it can release H_2
(d) It is used for converting nitriles to aldehydes

4.8. MAO is
(a) a cocatalyst in olefin polymerisation
(b) a methylating agent
(c) obtained by partial hydrolysis of Me_3Al
(d) Has a linear polymeric structure

4.9. Which of the following statements is not true for GaAs?
(a) GaAs has a band gap of 3.4 eV
(b) GaAs can work as an infrared LED
(c) GaAs can be made from Et_3Ga and AsH_3
(d) GaAs solar cells have better efficiency than silicon solar cells

4.10. Which of the following properties is not true for indium tin oxide?
(a) High optical transparency
(b) High electrical conductivity
(c) Ease of deposition as a thin film
(d) High thermal conductivity

Solved concept-based questions

4.11. $AlCl_3$ dimerises but BCl_3 does not dimerise. Why?

This is due to the large size difference between Al and B. The relative size of chlorine being large, boron in unable to accommodate more than three Cl units. (Although there are claims of BCl_4^- there are no examples which are structurally characterised by X-ray diffraction studies.) The reason why BF_3 does not dimerise is entirely different as covalently bound fluorine shows very little intermolecular interaction due to high electronegativity. There is also considerable $p\pi$–$p\pi$ interaction between B and F evident from the shorter B–F bond distance of 1.30 Å. BF_4^- also exists.

4.12. Solid $TlCl_3$ is unstable and decomposes at 40°C, losing chlorine to give TlCl. Why?

This is due to inert pair effect. Inert pair effect occurs when electrons are pulled closer to the nucleus, making them more stable and more difficult to ionise. The *s*-orbital electrons are affected in this way since they have a greater penetrating power. In thallium the *s* electrons in the outer shell are not as well shielded as those in the inner core. They are therefore pulled into the inner core of electrons and thus become inert giving rise to an oxidation state two units less than the group oxidation state.

4.13. Why do $AlCl_3$ and $AlMe_3$ dimerise?

$AlCl_3$ and $AlMe_3$ are electron deficient. By dimerisation $AlCl_3$ can get one more pair of electrons (lone pair on Cl) to complete its octet. Dimerisation observed of $AlMe_3$ is akin that of B_2H_6 (electron deficient).

4.14. Gallium is not stored in glass and metal containers. Why?

Upon solidification, gallium metal expands by 3.1% of its volume. Therefore, storage in either glass or metal containers has to be avoided, due to the possibility of container rupturing with freezing.

4.15. Why is gallium arsenide a better semi conducting material than silicon?

GaAs has higher saturated electron velocity and higher electron mobility, which permits transistors made of gallium arsenide to function at higher than normal frequencies. GaAs solar cells and devices are insensitive to heat; a consequence of the wider band gap of GaAs. Also, at higher frequencies, GaAs devices have lesser noise compared to silicon devices. These properties make GaAs circuitry the preferred one in mobile phones, satellite communications, and so on. Another advantage of GaAs is that it has a direct band gap, which means that it can be used to emit light efficiently (infrared LED). Silicon has an indirect band gap and so is very poor at emitting light.

4.16. Hydrated aluminium trichloride, $AlCl_3.6H_2O$, does not bring about Friedel–Crafts reactions. Why?

Friedel–Crafts reaction requires the formation of $AlCl_4^-$ thereby converting an alkyl or acyl halide to an electrophile such as R^+ or $RC(O)^{+\cdot}$. This is possible only if $AlCl_3$ retains its Lewis acidity. By forming $AlCl_3.6H_2O$ which is hexa-coordinate $Al(H_2O)_6$, (as seen from crystal structure) its Lewis acidity is lost.

Answer key for self study questions

4.1. (a) 4.2. (d) 4.3. (b) 4.4. (a) 4.5. (c) 4.6. (b) 4.7.(d) 4.8. (d) 4.9. (a) 4.10. (d)

Problems and exercises

4.17. What is the role of aluminium compounds in the production of elemental sulfur? Explain with equations.

4.18. Phenylethyl alcohol is present in rose flowers and is the compound which gives the characteristic fragrance for rose water. Starting with benzene, suggest a method for making phenylethyl alcohol (without using Grignard reagents).

4.19. Suggest a method using an aluminium compound for making ethylbenzene from benzene. What is the industrial use of ethylbenzene?

4.20. How is Red-Al synthesised? In what ways it is better than $LiAlH_4$? What reaction of Red-Al is different from that of LAH?

4.21. What alloy of aluminium is closest in strength to that of steel?

4.22. Using $(i\text{-}Bu)_3Al$ as a precursor, show how ethylene can be converted to the long chain alcohol, $CH_3CH_2(CH_2CH_2)_nOH$ and terminal alkene $CH_3CH_2(CH_2CH_2)_nCH_2{=}CH_2$.

4.23. GaInN can indeed produce all the colours of the rainbow. Explain the principle behind this.

4.24. Et_3Ga has been found to be a better reagent for deposition of Ga by MOCVD than Me_3Ga. Why?

4.25. With reference to their intrinsic properties, give three advantages of GaAs over silicon as a solar cell semiconductor.

4.26. Compare GaAs with Si with reference to band gap, temperature coefficient of solar cells and electron mobility.

4.27. A *p*-block element A on reaction with anhydrous HCl gave a compound B. Compound B on reaction with excess of a metal hydride, MH, resulted in compound C. Compound C was found to convert acetonitrile to an amine D. Identify A to D.

4.28. What are the most suitable compounds used for the MOCVD of GaN? Starting from the individual elements N_2 and Ga, how are these compounds prepared?

4.29. Give four reasons why indium tin oxide is the most sought after transparent conducting oxide.

4.30. While GaAs can work both as a solar cell material and as a red LED, GaN is used only as an LED material. What specific modification of GaN can make it suitable for working as a solar semiconductor?

4.31. Represent the thermite process using an equation. What properties of aluminium makes it the most suitable material for the thermite process?

4.32. What were the elements additional to aluminium present in the original duralumin alloy prepared by Alfred Wilm? Which among these are present in the 2XXX and 7XXX types of alloys used in aircraft manufacture nowadays? Which additional element(s) make(s) aluminium salt-water-corrosion free?

4.33. Identify compounds A to F and write down their chemical formula.

$$M + Na + H_2 \longrightarrow A$$
$$A + LiCl \longrightarrow B$$
$$B + CH_2{=}CH{-}CH_3 \longrightarrow C$$
$$C + AlCl_3 \longrightarrow D + LiCl$$
$$D + 3\ CH_2{=}CH_2 \longrightarrow E$$
$$E + O_2 + H_2O \longrightarrow F + EtOH$$

4.34. In a room with room temperature above 30°C, you are provided with three elements, all of which are silvery white liquids and all of which are stored under kerosene oil. Given that these elements are mercury, cesium and gallium how will you identify each of them if you are provided with only a beaker of ice cold water?

Supplementary reading

1. Aldridge S and Downs A J, *The group 13 metals aluminium, gallium, indium and thallium*, Wiley 2011.
2. Emsley J, *Aluminium—nature's building blocks: An A-Z guide to the elements*, Oxford University Press, 2001.
3. Greber J F, *Gallium and gallium compounds* in *Ullmann's encyclopedia of industrial chemistry*, Wiley-VCH, Weinheim, 2012.
4. Binczewski G J, The point of a monument: a history of the aluminium cap of the Washington monument, *J. Mater.*, 1995, Vol. 47, 20.
5. *Production and commercialisation of aluminium: The Hall–Herault process*, National history chemical landmarks, ACS, 1997.
6. Lyle J P, Granger D A and Sanders R E, *Aluminium alloys* in *Ullmann's encyclopedia of industrial chemistry*, 2005, Wiley-VCH, Weinheim.
7. Frank W B, *Aluminium* in *Ullmann's encyclopedia of industrial chemistry*, 2009, Wiley-VCH, Weinheim.

8. Hudson L K, Misra C, Perrotta A J, Wefers K and William F S, *Aluminium oxide* in *Ullmann's encyclopedia of industrial chemistry*, , Wiley-VCH, Weinheim, 2012.
9. Helmboldt, O, Hudson L K, Misra C, Wefers K, Heck W, Stark H, Danner M and Rosch N, *Aluminium compounds, Inorganic* in *Ullmann's encyclopedia of industrial chemistry*, 2005, Wiley-VCH, Weinheim, 2007.
10. Olah G A, *Friedel–Crafts and related reactions*, 1963, Vol.1, Interscience, NY.
11. Fleming I and Pierce A, Aliphatic Friedel–Crafts acylation of vinyl silanes, *Chem. Comm.*, 1975, 633.
12. Finholt A E, Bond A C and Schlesinger H I, Lithium aluminium hydride, aluminium hydride and lithium gallium hydride, and some of their applications in organic and inorganic chemistry, *J. Am. Chem. Soc.*, 1947, Vol. 69, 1199.
13. Brown H C, Reduction by lithium aluminium hydride, *Org. React.*, 1951, Vol. 6, 469.
14. Gugelchuk M, Silva L F, Vasconcelos R S and Quintiliano S A P, *Sodium bis(2-methoxyethoxy) aluminium hydride, Encyclopedia of reagents for organic synthesis*, John Wiley & sons, NY, 2007.
15. Galatsis P, *Diisobutyl aluminium hydride* in *Encyclopedia of reagents for organic synthesis*, John Wiley & sons, NY, 2001.
16. Ashby E S and Noding S A, Hydrometallation 3. Hydroalumination of alkene and dienes catalysed by transition metal halides, *J. Org. Chem.*, 1979, Vol. 44, 4364.
17. Ziegler K, Martin H and Krupp F, Metallorganishe Verbindungen XXVII: Aluminium trialkyle und dialkyl aluminium hydride aus aluminium isobutyl verbindungen, *Liebigs Ann. Chem.*, 1960, Vol. 629, 14.
18. Zijlstra H S and Harder S, Methyl alumoxane-History, production, properties and applications, *Eur. J. Inorg. Chem.*, 2015,19.
19. Kaminsky W and Sinn H, *Methyl alumoxane: Key component for new polymerisation catalysts*, Kaminsky W, Ed., Springer Verlag, Berlin–Heidelberg, 2013, Vol. 1, 1.
20. Kaminsky W, Discovery of methyl alumoxane as cocatalyst for olefin polymerisation, *Macromolecules*, 2012, Vol. 45, 3289.
21. Dohmeier C, Loos D and Schnockel H, Aluminium(I) and gallium(I) compounds: Synthesis, structure and reactions, *Angew Chem. Int. Ed*, 1996, Vol. 35, 129.
22. Cui C, Roesky H W, Schmidt H-G, Noltemeyer M, Hao H and Cimpoesu F, Synthesis and structure of an monomeric Al(I) compound: A stable analogue of a carbine, *Angew Chem. Int. Ed.*, 2000, Vol. 39, 4274.
23. Roesky H W, The renaissance of aluminium chemistry, *Inorg. Chem.*, 2004, Vol. 43, 7284.
24. Chu T, Korobkov I and Nikonov G I, Oxidative addition of sigma bonds to an Al(I) center, *J. Am. Chem. Soc.*, 2014, Vol. 136, 9195.
25. Yamaguchi M and Shibasaki M, *Gallium trichloride* in *Encyclopedia of reagents for organic synthesis*, John Wiley & sons, NY, 2005.
26. Akasaki I, Blue Light: A Fascinating journey (Nobel Lecture), *Angew. Chem. Int. Ed.*, 2015, Vol. 54, 7750.
27. Nakamura S and Krames M R, History of gallium-nitride-based light-emitting diodes for illumination, *Proc. IEEE*, 2013, Vol. 101, 2211.
28. Zheludev N, The life and times of the LED, A 100 year history, *Nature Photonics*, 2007, Vol.1, 189.
29. Pimputkar S, Speck J S, Den Baara S P and Nakamura A, Prospects for LED lighting, *Nature Photonics*, 2009, Vol. 3, 180.
30. Ohkawa K, Blue LEDs: Helping to Save Earth, *AAPPS Bulletin*, 2014, Vol.24, 11.
31. Dollen P V, Pimputkar S and Speck J S, Let there be light—with gallium nitride: The 2014 Nobel prize in Physics, *Angew. Chem. Int. Ed.*, 2014, Vol. 53, 13978.

32. Brozel M R and Stillman G E, *Properties of gallium arsenide*, EMIS datareviews series, 3rd Ed., INSPEC, 1996.
33. Heydlauf H, Ferric cyanoferrate (II): An effective antidote in thallium poisoning, *Eur. J. Pharmacol*, 1969, Vol.6, 340.
34. Yang Z, Zhong M, Ma X, De S, Anusha K, Parameshwaran P and Roesky H W, An aluminium hydride that functions as a transition metal catalyst, *Angew. Chem. Int. Ed.*, 2015, Vol. 54, 10225.
35. Yang Z, Zhong M, Ma X, Nijesh K, De S, Parameshwaran P and Roesky H W, An aluminium dihydride working as a catalyst in hydroboration and dehydrocoupling, *J. Am. Chem. Soc.*, 2016, Vol. 138, 2548.
36. Bismuto A, Cowley M J and Thomas S P, Aluminium catalysed hydroboration of alkenes, *ACS Catal*, 2018, Vol. 8, 2001.
37. Bag P, Porzelt P, Altmann P J and Inoue S, A stable neutral compound with an aluminium–aluminium double bond, *J. Am. Chem. Soc.*, 2017, Vol. 139, 14384.
38. Hardman N J, Wright R J, Phillips A D and Power P P, Synthesis and characterisation of the neutral digallene, ArGaGaAr and its reduction to Na_2ArGaGAAr, *Angew. Chem. Int. Ed.*, 2002, Vol. 41, 2842.
39. Wright R J, Phillips A D, Hardman N J and Power P P, The diindene ArInInAr: Dimeric versus monomeric In(I) aryls; para substituent effects in terphenyl ligands, *J. Am. Chem. Soc.*, 2002, Vol. 124, 8538.
40. Wright R J, Phillips A D, Hino S and Power P P, Synthesis and reactivity of ArTlTlAr and trimeric $(ArTl)_3$ thallium (I) derivatives; Tl(I)–Tl(I) bonding in species ligated by monodentate ligands, *J. Am. Chem. Soc.*, 2005, Vol. 127, 4794.

The Chemistry of Carbon

CHAPTER 5

The Inorganic Chemistry of Carbon

As organic chemistry, also referred to as the chemistry of carbon compounds, is by itself a well studied and separate branch of chemistry, in this chapter the discussion will be restricted only to compounds of carbon other than those which are considered traditionally as organic. The diverse allotropes of carbon such as fullerenes, nanotubes, graphene, and so on, as well as new useful materials based on carbon such as carbon molecular sieves, carbon fibre reinforced polymers and persistent carbenes form the content of this chapter.

Unique Properties of Carbon

- *The electronegativity of carbon (2.55) is too low to allow carbon to form C^{4-} ions with most metals and at the same time too large for carbon to form C^{4+} ions when it reacts with nonmetals. Carbon therefore forms covalent bonds with other elements.*
- *Although carbon was initially known to have diamond and graphite as the only allotropes, over the years the number of its allotropes have increased and now it has the maximum number of allotropes known for any element.*
- *The newly discovered allotropes include different types of fullerenes (C_{60} and C_{70} being the most well known) along with carbon nanotubes, nanorods, nanowires, and so on, and graphene.*
- *Fullerenes provide a robust three dimensional core on which a wide range of organic and inorganic functionalisation is possible. The most well known among them is C_{60} and the high stability of C_{60} has been attributed primarily to the absence of double bonds inside the pentagon rings.*
- *The most promising industrial application of carbon nanotubes so far has utilised its electrical conductivity as a reinforcement material in polymer composites* in *order to prevent static charge formation on aeroplane and automotive parts. In*

aeroplanes, this can result in additional weight reduction, as antistatic paint coating, and aluminum in fabrication can be avoided.

- *Carbon molecular sieves due to their unique pore size and sieving capability are widely used for separation of gases, especially separating pure nitrogen from other gases.*
- *Graphite is a substrate on which intercalation chemistry has been carried out resulting in many new compounds with applications (for example, KC_8, a very strong reductive dehalogenating agent).*
- *Carbon monoxide is a versatile ligand in organometallic complexes and sensitivity of the carbonyl stretching frequency to electronic effects is ably monitored by IR spectroscopy.*
- *Many metal carbonyls are excellent catalysts in industrial organic synthesis, such as in the methanol to acetic acid process.*
- *Mond's process, where $Ni(CO)_4$ is utilised for purification of impure nickel is a well established and energy efficient industrial process in the metallurgy of nickel.*
- *Carbon dioxide is one of the primary greenhouse gases of the earth's atmosphere. CO_2 has an atmospheric concentration of 400 ppm (as of 2013) and contributes up to 26% of the green house effect by such gases. It is the reference standard for the determination of the global warming potential of other gases.*
- *Supercritical CO_2 (CO_2 maintained above its critical point; the critical parametres of carbon dioxide being T_{crit} 31°C, p_{crit} 7.4 MPa) has properties midway between a gas and a liquid. It is an important commercial and industrial solvent due to its role in chemical extraction, in addition to its non-toxicity (for example, caffeine extraction from coffee beans).*
- *Carbon fibre reinforced polymer (CFRP), is an extremely strong and light fibre-reinforced plastic which contains woven carbon fibres made stronger with a binding polymer such as an epoxy resin. CFRPs are commonly used wherever high strength-to-weight ratio and rigidity are required, such as aerospace, automotive, civil engineering and sports goods. Upto 53% CFRP is used in the construction of modern passenger aircrafts replacing traditional aluminum alloys.*
- *Persistent carbenes such as N-heterocyclic carbenes (NHCs) and cyclic alkylamino carbenes (cAACs) are some of the strongest σ-donor ligands in transition metal chemistry and catalysis. They are capable of stabilising main group elements in low and zero oxidation states and also act as organocatalysts.*

5.1 The Formation of Carbon in the Universe

Carbon is present in interstellar space and has been found in all of its main allotropic forms—diamond, graphite and fullerene. The discovery of fullerene by Harry Kroto was linked to the study of carbon in interstellar space. The formation of carbon in the universe is associated with the death of stars like our sun. The death of a star begins with the complete conversion of hydrogen to helium. When a dying star begins to run out of hydrogen for fusion, its core begins to collapse until the temperature rises to about 10^8 K which is about six times hotter than the sun's core. Next, contraction and the mass added to the core during hydrogen burning will result in burning of helium and its fusion at such temperatures. The triple-alpha process is a set of nuclear fusion reactions by which three helium nuclei combine to form carbon-12. The carbon later gets ejected into interstellar and intergalactic space during stellar collapse and supernova explosions. As a side reaction of this process, some carbon nuclei can fuse with extra helium to produce a stable isotope of oxygen with the release of energy. Astronomical observations in the last decade have shown that carbonaceous compounds are omnipresent in our own as well as in distant galaxies. Large carbon abundances are detected from observations of the strong C(II) and CO lines in the spectral analysis of the stars.[1,2]

5.2 Graphite

Graphite is a naturally occurring form of carbon which is a crystalline semimetal, has excellent lubricating properties and is the most stable form of carbon under standard conditions. Until 1779, graphite was thought to be a form of lead, when Carl Wilhelm Scheele demonstrated experimentally that graphite produces the same amount of carbon dioxide per gram as amorphous carbon does. Graphite occurs in metamorphic rocks as a result of the reduction of sedimentary carbon compounds during metamorphism. Minerals associated with graphite include quartz, calcite, micas and tourmaline. World production of natural graphite in 2012 was 1,100,000 tonnes. Graphite electrodes have many applications such as in the manufacture of steel and in the electrolytic cells used for making aluminium, and so on. The electric and thermal conductivity, thermal shock resistance, mechanical strength and machinability of graphite makes it the choicest material for electrodes.

Three principal types of natural graphite have been found to be present in different types of ore deposits—(i) crystalline flake graphite (flake graphite) which occurs as isolated, flat, plate-like particles with hexagonal edges, (ii) amorphous graphite which is basically very fine flake graphite, and (iii) lump graphite (also called vein graphite). India is a major global producer of crystalline flake graphite with reserves of nearly 11 m tonnes.

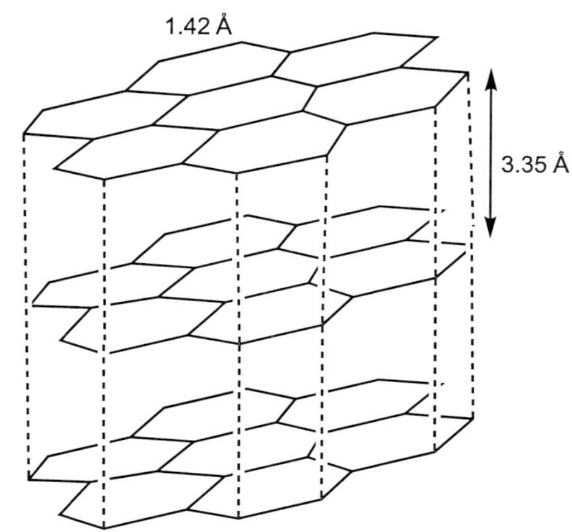

Fig. 5.1 Structure of graphite

Graphite has a layered, planar structure. In each layer, the carbon atoms are arranged in a hexagonal honeycomb lattice with a separation of 1.42 Å, and the distance between the layers is 3.35 Å. Each carbon atom is sp^2-hybridised and is covalently bound to three other carbon atoms in a trigonal geometric arrangement. The bond angle in graphite is 120°. Three out of four valance electrons of each carbon atom are used in bond formation while the fourth electron is free to move around in the plane, resulting in graphite becoming electrically conductive. Interestingly, it does not conduct in a direction at right angles to the plane and only in the direction of the plane. Weak bonding between the layers of graphite is by van der Waals forces which allows layers of graphite to be easily separated, or to slide past each other providing lubricating properties.[3]

Natural graphite is formed when carbon is subjected to heat and pressure in the earth's crust and in the upper mantle. To produce graphite, synthetically high pressures (~75,000 pounds per square inch) and high temperatures (~750°C) are needed. Synthetic graphite has a purity of over 99% carbon, and it is used in manufactured products where an extremely pure sample of graphite is required.[4]

5.2.1 Graphite Intercalation Compounds

Some of the most practically useful products based on graphite are the graphite intercalation compounds. The most well known example of such a compound is KC_8, which is obtained by melting elemental potassium along with graphite.

KC_8 is currently one of the most well known dehalogenative reducing agents. Its use has resulted in many metal–metal multiple bonds and metals and nonmetals in unusually low oxidation states. KC_8 is prepared by melting potassium over oven dried graphite powder. The potassium is absorbed into the graphite increasing the interlayer distance from 335 to 540 nm and one can visibly see that the material changes colour from grey to bronze. The resulting solid is highly pyrophoric in nature. The bond between anionic graphite layers and potassium cations is ionic. The electrical conductivity of the material is greater than that of α-graphite. KC_8 is a superconductor but with a very low critical temperature (T_c = 0.14 K).[5]

5.3 Diamond

Diamond is an allotrope of carbon with sp^3 hybridised carbon atoms. In diamond, the C–C–C bond angle is 109.5°. These tetrahedral units unite with one another and produce a cubic unit cell. The C–C bond length in diamond is 1.54 Å. The C–C bond energy is 347 kJ/mol. It is less stable when compared to graphite but under standard conditions the rate of conversion from diamond to graphite is negligible. The excellent physical properties of diamond basically originate from the strong covalent bonds between the carbon atoms. Diamond also has the highest hardness (10 in the mohs scale) and thermal conductivity (the thermal conductivity of natural diamond was measured as

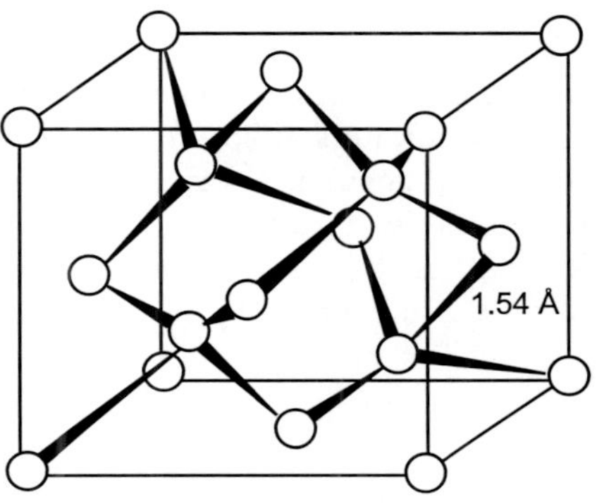

Fig. 5.2 Structure of diamond

22 W/cm.K, which is five times more than copper) for any bulk material. Single crystals of synthetic diamond enriched with isotopically pure diamond (C-12 = 99.9%), have one of the highest thermal conductivities reported for any material, 30 W/cm·K at room temperature, which is 7.5 times higher than copper. The major industrial application of diamond in cutting and polishing tools results from its high hardness. While pure diamond is colourless, small amounts of defects or impurities (about one per million of lattice atoms) colour the diamond blue (boron), yellow (nitrogen), brown (lattice defects), and so on. Diamond also has relatively high optical dispersion which is the ability to disperse light of different colours.[6]

Since diamond is a less stable form of carbon, it gets oxidised in air if heated over 700°C. In 1772, Antoine Lavoisier showed that diamonds are a form of carbon by burning carefully weighed diamond and carbon samples and showed that both substances produced no water vapour and the same amount of carbon dioxide gas was produced per gram. In the absence of oxygen, such as under a flow of high-purity argon gas, diamond can be heated up to about 1700°C. At high pressure (~20 GPa) diamond can be heated up to 2500°C. Natural diamonds are formed at high temperature and pressure at depths varying from 140–190 kilometres inside the earth's mantle. The carbon source is carbon-based materials, and the growth occurs over periods ranging from 1–3.3 billion years (25% to 75% of the age of the earth). Diamonds are brought close to the earth's surface by lava coming from deep volcanic eruptions. Diamonds can also be produced synthetically in a high pressure–high temperature method which approximately simulates the conditions inside the earth's mantle.

Although it is the hardest, due to structural weaknesses, its toughness is only moderate. The exact tensile strength of diamond is not known. Diamond has a high refractive index (2.417) and moderate dispersion (0.044) properties which give cut diamonds their brilliance. Most diamonds are electrical insulators but extremely efficient thermal conductors.

5.4 Fullerenes

The term fullerene is often used to describe a family of related molecules of carbon such as spherical fullerenes (C_{60}, buckyball), ellipses (C_{70}), hollow spheres (carbon nanotubes) and other variations. The whole story specifically started with the discovery of C_{60} (buckminster fullerene) which resembles a soccer ball and has a truncated icosahedral shape with isolated pentagons of carbon surrounded by hexagons.

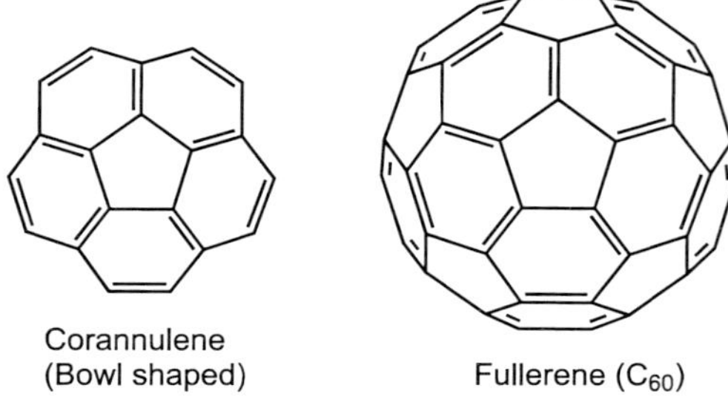

Fig. 5.3 Corannulene and fullerene (C_{60})

5.4.1 The Discovery

In 1970, Osawa who was stimulated by the synthesis of corannulene, proposed a spherical icosahedral symmetric structure of C_{60}. Subsequently few theoretical studies on C_{60} were reported along with other three dimensional aromatic systems. Hückel calculations of C_{60}

were reported. In 1984 it was observed that when graphite is vaporised with laser, large all-carbon clusters, C_n with n = 30–190, can be produced. Interestingly, only ions with even numbers of carbon atoms were observed in the mass spectra of these carbon clusters, among which C_{60} and C_{70} were also present but their identity was not recognised at that time. The breakthrough in fullerene discovery arrived in 1985 when Prof. Harry Kroto visited Rice University in Houston, Texas where Prof. Richard Smalley and coworkers had developed a technique for generating carbon clusters by focusing a pulsed laser on solid graphite and analysing the products by mass spectrometric techniques. Kroto and Smalley were trying to make cyanopolyynes [HC_nN (n = 7,9)] (which are predicted to be present in interstellar space) under laboratory conditions, in which experiments simulating conditions under which carbon nucleates in the atmosphere of red giant stars were carried out. Incidentally in 2010, the presence of C_{60} was discovered in a cloud of cosmic dust surrounding a distant star 6500 light years away by identifying the molecule's unique infrared bands using NASA's infrared telescope. Under specific conditions of cluster formation, a peak attributed to C_{60} (m/z = 720) and a less intense peak attributed to C_{70} (m/z = 840) showed pronounced intensity in the spectra. Kroto and Smalley were able to modify conditions under which the mass spectra were completely dominated by the C_{60} signal. They concluded that the stability of C_{60} is due to its almost spherical, truncated icosahedral structure. They proposed that it can be the football shaped C_{60} and ellipsoidal $C_{70.}$ The second major breakthrough was by Krätschmer and Huffman who vaporised graphite in a helium atmosphere. The soot collected on extraction with benzene or by sublimation, resulted in fullerenes.[7]

5.4.2 Preparation and Purification of Fullerenes, C_{60} and C_{70}

There are at least three different heating methods of graphite which result in the formation of fullerenes. The resistive heating of graphite was the first method used for producing macroscopic quantities of fullerenes (10%–15%). The apparatus first used by Krätschmer and coworkers for the production of C_{60} was a bell jar connected to a pump and gas inlet. In the jar, two graphite rods – one sharpened to a conical point and the other with a flat end – were kept in contact by a spring. After evacuating the apparatus, it was filled with helium at 14 kPa, and an electric current was passed through the rods. This heated the point of contact to 2500–3000°C and resulted in the formation of a black smoke, which settled inside the bell jar. The soot was scrapped out and was extracted with toluene to get the fullerenes.

Arc vaporisation (where an electric arc is used to vaporise material present on a cathode target) of graphite was first developed by Smalley and coworkers. This method allowed efficient evaporation of carbon and the yield of fullerenes obtained was about 15%. The disadvantage of this method was the intense UV radiation originating from the arc plasma which resulted in the formation of some non-vaporisable insoluble side products. Later on, the problem of intense UV radiation has been avoided by the use of solar furnaces as fullerene generators, which led to far lower exposure of the fullerenes to radiation than arc vaporisation or resistive heating. Although fullerenes can be obtained in this way, the efficiency of the solar method is not very promising. Laser ablation of graphite in an

Table 5.1 Comparison of general properties of graphite, diamond and C_{60}

Properties	Graphite	Diamond	C_{60}
m.p.	Sublimes at 4300 K	Pressure dependant (4500 K)	Sublimes at ~ 600°C
Density	2.09–2.23 g/cm^3 1.3–1.95	3.15–3.53 g/cm^3	1.65 g/cm^3
Hardness	1–2 mohs scale	10 mohs scale	NA
Heat conductivity	25–470 (W/m.K)	900–2320 (W/m.K)	~0.4 (W/m K) (single crystal)
Standard heats of formation (k cal mol^{-1})	0.4	0.0	9.08
Electrical conductivity	Conducts along the plane	Good electrical insulator	–
Solubility	Soluble in molten nickel, warm chlorosulfuric acid	Resistant to acids, but dissolves irreversibly in hot steel	Insoluble in water soluble in many organic solvents, for example, 1,2 dichlorobenzene
Burning temperature in oxygen	700°C	720–800°C in oxygen and 850–1000°C in air	-
C–C bond distance	1.42 Å 3.35Å (interlayer)	1.55 Å	5,6 bond =1.46 Å 6,6 bond =1.38 Å
Band gap (eV)	−0.04	5.5 (5.47)	1.9 (1.86)

argon atmosphere at room temperature is another method for the generation of fullerenes. Fullerenes have also been obtained by pyrolysis of naphthalene at 1000°C in an argon stream with yields of 0.5% and also from many other high carbon containing organic compounds such as the insoluble soot obtained by the burning of camphor.[8]

Other than C_{70}, the important higher fullerenes include carbon clusters having 76, 78, 84, 90 and 96 atoms. It was again Kroto who separated C_{60} and C_{70} for the first time in the pure form. The soot formed in the graphite heating reactors were scraped and dispersed in benzene whereupon a wine-red solution was obtained. This was filtered from the insoluble solids and concentrated. This mixture of C_{60} and C_{70} was then separated on an alumina column using hexane as eluant. The magenta coloured C_{60} elutes out first followed by the port wine coloured C_{70}. Currently, several modifications have been made on the original separation procedure. Toluene has entirely replaced benzene and a combination method called soxhlet chromatography has been introduced to speed up the separation of fullerenes from the soot. Neutral alumina is widely used as the column material but better separation has been achieved by a 1:2 mixture of charcoal and silica gel, or better, by using a polystyrene gel column. Pure solid C_{60} is mustard coloured [UV 596, 604, 625nm] and C_{70} is red (UV 600, 617, 644 nm).

5.4.3 Structure of C_{60} and C_{70}

The structure of C_{60} belongs to the truncated icosahedron (I_h point group) while C_{70} has a prolate ellipsoid structure (D_{5h} point group). Generally Euler's theorem can be used to find out the number of hexagons in a fullerene if each fullerene contains 12 pentagons. For a fullerene C_n, the number of hexagons $m = (C_n - 20)/2 : m = 20$ for C_{60}, $m = 25$ for C_{70}, $m = 32$ for C_{84} and so on.

The structure of C_{60} can be described as having 12 pentagons and 20 hexagons with the pentagons sharing no common edges and the hexagons sharing edges with another hexagon or a pentagon. All the carbons are tri-coordinate, pyramidal and all pentagons and hexagons are planar. Two types of C–C bonds are present in the molecule with differing bond lengths, 1.38 Å for a 6,6 bond (common for two hexagons) and 1.46 Å for a 5,6 bond (common for a hexagon and a pentagon). With regard to numbering, no common approach seems to be available, however one widely accepted method is by assigning numbers in a spiral form from one pentagonal cap to its antipode.

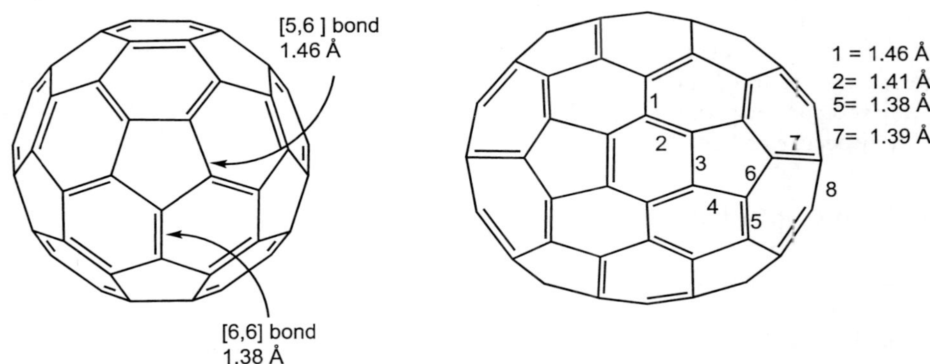

Fig. 5.4 Types of bonds and their distances of C_{60} and C_{70}

In C_{60} all carbon atoms are identical, whereas C_{70} has five different types of carbon atoms depending on the carbon environment. These are easily differentiated by ^{13}C-NMR spectroscopy. In C_{70}, there are eight different types of bonds with distances varying from 1.38–1.46 Å. The bonds at the two end caps of C_{70} resemble those of C_{60} and reactions of C_{70} happen mostly at the short [6.6] end cap bonds. Another fullerene C_{76}, whose X-ray structure has been solved belongs to the D_2 point group, is chiral and occurs as a racemic mixture.

5.4.4 Stability of C_{60}

The unusual stability of C_{60} compared to other higher/lower fullerenes has been explained. In the structure of C_{60}, all the twelve pentagons are isolated from each other. Again, only in C_{60} are double bonds arranged in such a way that they are located only in the six-membered rings and none in the five-membered rings. This is favoured as there will be less strain on the already strained five-membered rings. The need to avoid double bonds in pentagons largely governs the stability of fullerenes since the five-membered rings with five planar hexagons around are already strained, and unsaturation will only increase the strain further. C_{60} has only arrangement (a) in its structure while C_{70} and C_{84} have five and six of the (b) arrangements in their structures respectively.

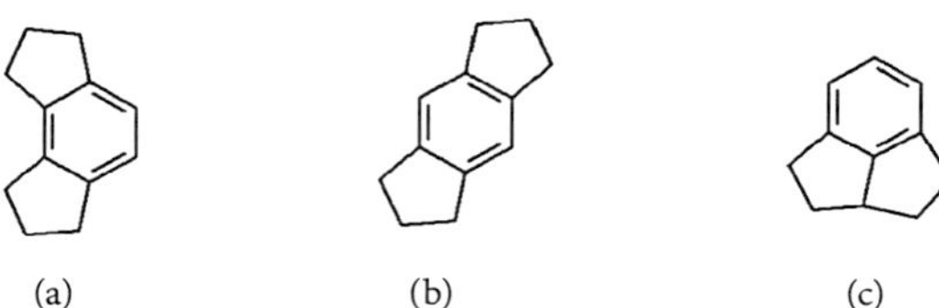

Fig. 5.5 Types of arrangements possible between two pentagons and a hexagon

Fullerenes were also found to decompose slowly in presence of UV radiation and on reaction with oxygen and ozone. Early indications of these were observed by Kroto who observed that during separation on the alumina column, if elution was stopped and the column exposed, the colour on the bands originally due to C_{70} and C_{60} slowly disappeared. $C_{60}O$ and $C_{70}O$ have also been prepared by different methods. This also explains why fullerenes are not naturally occurring.

5.4.5 Concept of Aromaticity in C_{60}

Initially C_{60} was predicted to be extremely stable and aromatic. More than 12,500 resonance structures were proposed for the molecule. The terms superaromaticity and three dimensional aromaticity were freely used to describe its structure. However, the presence of strained five-membered rings adjacent to benzenoid rings were overlooked in such discussions. Only one structure exists for C_{60} which avoids having any double bonds in the pentagons. Another important feature which was overlooked was the fact that the π bonds on the hexagonal faces were displaced about 10° away from the plane of the hexagon. The two important consequences of this finding were—(i) the delocalisation of electrons in C_{60} is poor and so it is more reactive than expected, and (ii) C_{60} is not so much an aromatic compound and the double bonds in it are isolated. This is clearly indicated by the η^2 coordination shown by C_{60} with many metallic and organometallic reagents.

$$(Ph_3P)_2Pt(\eta^2\text{-}C_2H_4) + C_{60} \longrightarrow (Ph_3P)_2Pt(\eta^2\text{-}C_{60}) + C_2H_4 \qquad (5.1)$$

5.4.6 C_{60} as an Electron Acceptor

C_{60} shows a tendency to take up electrons to form anions (even up to 6 electrons). Close examination of a ball and stick model reveals that there are six pyracyclene units on C_{60} and C_{70} and on the basis of the evidence that pyracyclene is a 4n–π system (considering only the electrons around the periphery of the pyracyclene) it was hypothesised that each unit will provide a driving force for the capture of up to two electrons.

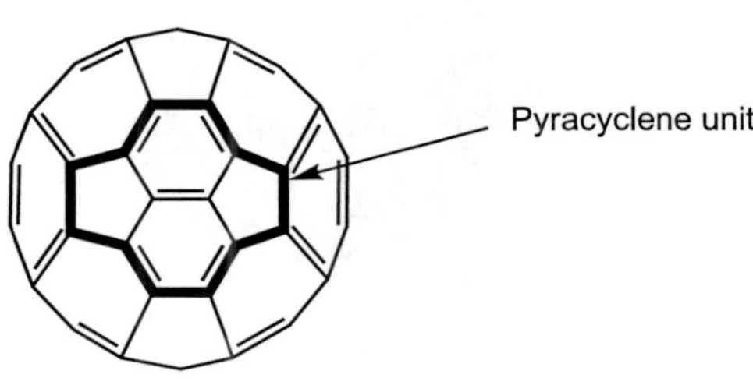

5.4.7 Spectral Characterisation and Comparison of C_{60} and C_{70}

Although vibrational and UV-visible spectra of C_{60} and C_{70} are different, conclusive evidence for their structural differences are obtained from mass spectral and ^{13}C-NMR spectral data. The calculated and the observed IR spectra of C_{60} agree well as shown below.

Calculated (cm^{-1})	**Observed (cm^{-1})**
472	528
618	577
1119	1183
1434	1429

The UV-vis spectra of C_{60} show peaks at 217, 257 and 329 nm which are different from that of C_{70}. The region 400–700 nm also shows slight differences (596, 604 and 625 nm for C_{60} and 600, 617 and 644 nm for C_{70}). The FAB mass spectra of a mixture of C_{60} and C_{70} will clearly show the peaks at 720 amu and 840 amu as the strongest peaks. Also evidence for other even-numbered species like C_{58}, C_{68} and C_{76} are obtained from high resolution mass spectra. The ^{13}C-NMR spectrum of C_{60} gives just one signal at 142.8 ppm indicating that all the 60 carbon atoms have an identical environment. However, C_{70} gives five different lines in the range 150.7 to 130.3 ppm as expected for the D_{5h} molecule in the ratio 10:10:20:20:10 (A:B:C:D:E) (Fig. 5.6).

5.4.8 Types of Fullerene Derivatives

Osmylation of C_{60}

The first ever successful reaction on C_{60} which led to the isolation and X-ray structural characterisation of a C_{60} derivative was osmylation. Hawkins and coworkers reacted C_{60} with OsO_4 in the presence of a ligand, 4-*t*-butyl pyridine. The impetus for this reaction came from the fact that OsO_4 is a powerful yet selective oxidant and the osmylation of polycyclic aromatic hydrocarbons have been known for many years. Anthracene adds 2 equivalents of OsO_4 in the presence of pyridine. The resulting compound with C_{60} has C–O bonds and

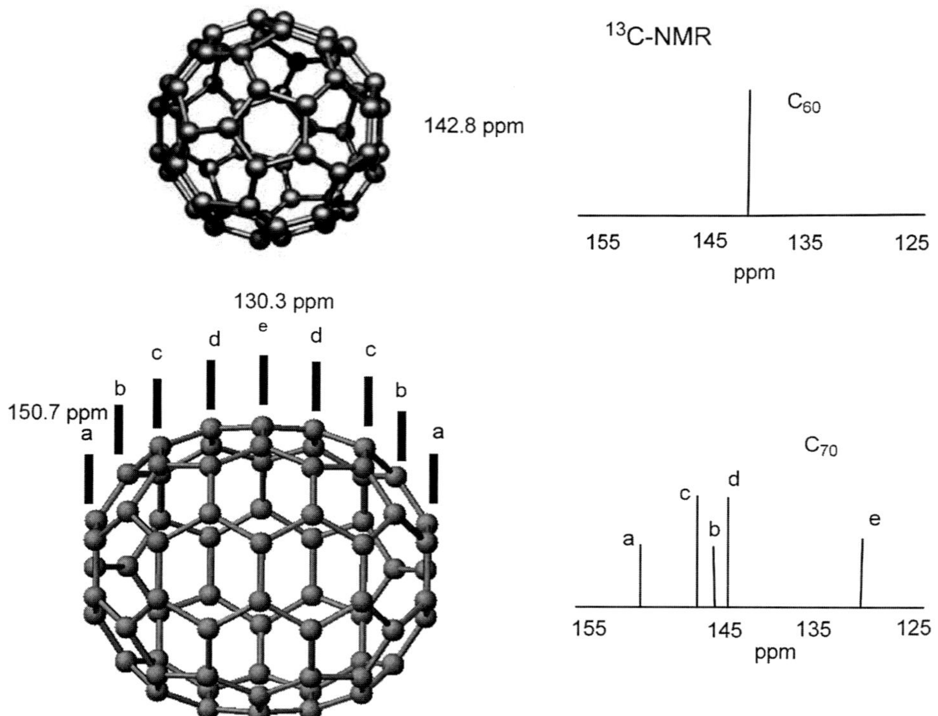

Fig. 5.6 Comparison of the ^{13}C-NMR spectra of C_{60} and C_{70}

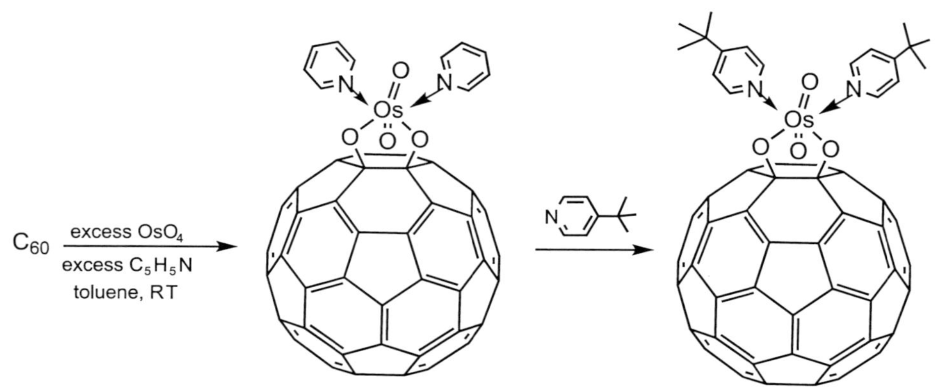

Scheme 5.2 Synthesis of the first structurally characterised C_{60} derivative

the two *t*-butyl pyridine units provide the handles to solubilise it in organic solvents. There are no metal–carbon bonds present in the system. Two different adducts of C_{60}, 1:1 and 1:2, were prepared and the 1:1 adduct was structurally characterised. The differences in bond distances of the 5,6 and the 6,6 bonds were first obtained from this study.

Early attempts to obtain an X-ray crystal structure of the un-derivatised C_{60} were unsuccessful due to crystal disorder and poor diffraction. While the buckyball molecules pack in an ordered fashion, their nearly spherical symmetry was found to result in orientational disorder. Hawkins and coworkers realised that if C_{60} could be derivatised to get rid of its high (icosahedral) symmetry, it might crystallise with orientational order and one may be able to carry out a detailed crystallographic analysis. Initially they attempted a reaction with OsO_4 and pyridine which gave the compound $C_{60}(OsO_4)(py)_2$ which was chromatographically identified as a single regioisomer, rather than a mixture of the two regioisomers which are possible from the soccer-ball structure for C_{60}. To get better solubility in organic solvents, pyridine was replaced with 4-*t*-butyl pyridine. This gave a complex with better solubility and formed crystals which diffracted well. The crystal structure showed that the O–Os–O unit has added across the fusion of two six-membered rings. The tricoordinate carbons C_3–C_{60} all lie within a spherical shell of radius 3.46–3.56 Å, with an average distance of 3.51 Å from the centre of the cluster. The tetracoordinate oxygen-bound carbons C1 and C2 lie significantly outside of this shell at distances of 3.80 and 3.81 Å from the centre. They have approximately tetrahedral geometry with sums of C–C–C angles equal to 330°, slightly more than 328° for ideal tetrahedral arrangement.[10]

While some of the early derivatives of fullerenes were mainly metal complexes, soon organic reactions were made possible on the C_{60} skeleton. C_{60} also undergoes six reversible, one-electron reductions to $[C_{60}]^{6-}$. The midpoint potentials obtained for these species using cyclic voltammetry against ferrocenium/ferrocene couple as standard at 213 K are given in Eq. 5.2.

$$C_{60} \underset{}{\overset{-0.81V}{\rightleftharpoons}} [C_{60}]^{\ominus} \overset{-1.24V}{\rightleftharpoons} [C_{60}]^{2\ominus} \overset{-1.77V}{\rightleftharpoons} [C_{60}]^{3\ominus} \overset{-2.22V}{\rightleftharpoons} [C_{60}]^{4\ominus} \overset{-2.71V}{\rightleftharpoons} [C_{60}]^{5\ominus} \overset{-3.12V}{\rightleftharpoons} [C_{60}]^{6\ominus} \tag{5.2}$$

Over the years, a large number of fullerenes and diverse derivatives have been prepared which can be classified under different categories. These include:

- Fullerene salts where the fullerenes have been reduced. A large number of fullerene salts were made following these studies. Cesium-doped fullerene Cs_3C_{60}, shows superconductivity at 38 K, but only under applied pressure.
- Exohedral fullerenes where one or more carbon atoms of the surface have been substituted with organic and inorganic substitutents.
- Open cage fullerenes where the cage has been partially opened by what is now well known as molecular surgical synthesis, small molecules deposited and the cage closed to realise endohedral fullerenes.

- Quasi fullerenes which are closed cage fullerenes containing rings other than 5 and 6 membered rings obtained as a result of chemical modification.
- Heterofullerenes where at least one carbon atom is replaced by heteroatoms such as nitrogen, oxygen, boron, and so on, resulting in azafullerenes, oxafullerenes and borafullerenes.

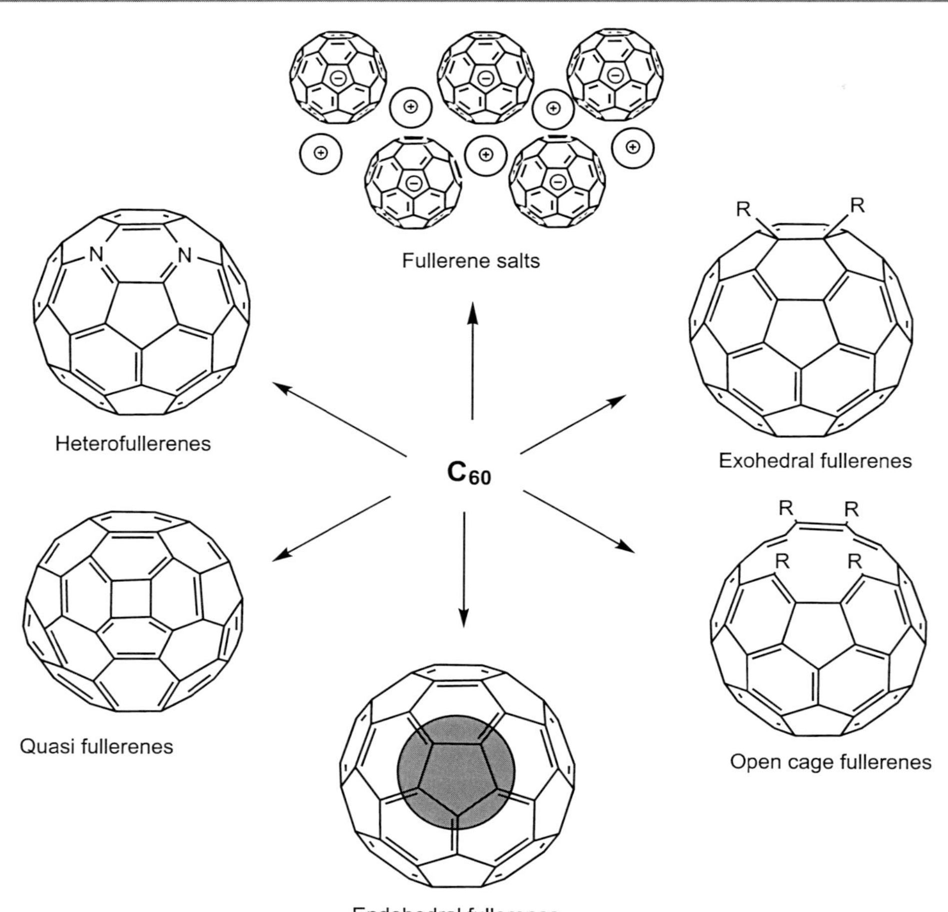

Scheme 5.3 General types of derivatives known for C_{60}

The most reactive bond of the C_{60} is the 6,6-bond which forms an edge between two hexagons. As shown in Scheme 5.4, most of the fullerene derivatives have been obtained by carrying out different types of addition reactions on the 6,6-bond. These include hydrogenation,

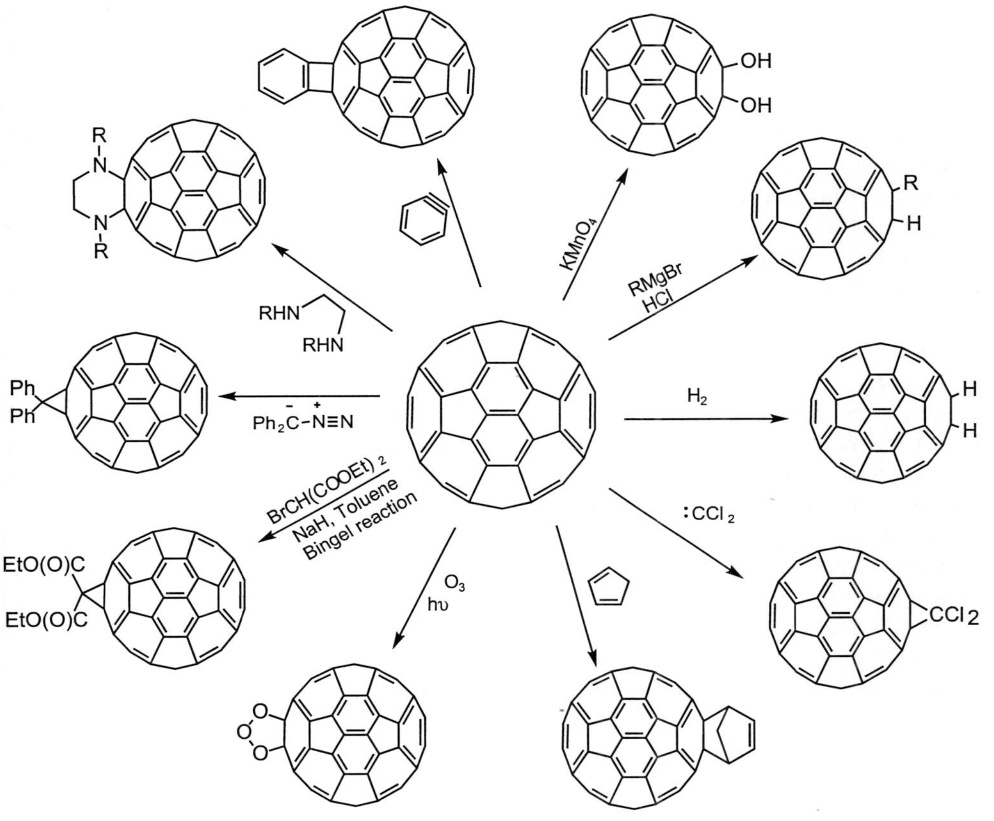

Scheme 5.4 Organic derivatives of C_{60}

ozonide formation, cycloaddition reactions, reactions of carbenes and benzyne. Among the cyclopropanation reactions, the Bingel reaction first discovered by C Bingel in 1993, with the bromo-derivative of diethyl malonate in the presence of a base such as sodium hydride or DBU is important in fullerene chemistry as it allows the introduction of useful extensions to the fullerene sphere which can alter properties of fullerenes, such as, solubility.

5.4.9 Organometallic Chemistry of Fullerenes

Fullerenes can in principle form η^1, η^2, η^3, η^5 and η^6 complexes with metals. However, research in this area has resulted in the synthesis of mostly η^2-complexes, a few examples of η^1- and some η^5-complexes. The types of reactions shown by fullerenes leading to the formation of inorganic derivatives mostly involve its electron acceptor capability; the electron deficiency of fullerenes has been well established. Single crystal and spectroscopic analysis of various fullerene derivatives show that the reactions of metals with fullerene proceed in a similar fashion to those with electron deficient olefins in a dihapto manner to one or more [6,6]-bonds. Packing effects as well as supramolecular π–π interactions also control to some extent, the regioselectivity of substitution on the fullerene ring. Fullerenes form simple charge transfer complexes, charge separated systems (fullerides), halides with varying number of halogens, oxides, complexes where C_{60} and C_{70} act as η^2-ligands towards transition metal complexes or complexes wherein a stable C–O–M bond is formed between the fullerene and reactive TM oxides (for example, OsO_4).

The reaction of Schwartz's reagent, $Cp_2ZrCl(H)$ with C_{60} resulted in a red solution in contrast to the usual green solution of η^2-complexes. Proton NMR studies confirmed the formation of the air-sensitive η^1-complex with the hydrogen being transferred from the Schwartz's reagent to C_{60} giving a peak at δ 6.09 ppm which is the typical proton chemical shift for hydrogenated fullerenes. Hydrolysis of the metal complex with aq. HCl resulted in $C_{60}H_2$ as an isomerically pure 1,2-addition product.

While some theoretical studies still predict that the hexagonal face of the fullerenes will form η^6-complexes, practical studies have clearly shown that between η^6- and η^2-hapticities, C_{60} prefers to form η^2-complexes. Considering the geometry of the carbon *p*-orbitals within a hexagon of the C_{60} framework, which is tilted away from the centre of the ring, it is clear that binding of C_{60} in a hexahapto manner is not favoured. However, coordination of more than one double bonds of a hexagonal face to multi-metal clusters has been observed and with three such bonds from a single hexagonal face is the closest which has been achieved to an η^6-binding (see Section 5.4.11). Peng and his group identified an intermediate $[C_{60}RuC_p^*]^+$, similar to $[C_6H_6RuC_p^*]^+$ and by analysing mass spectroscopic data, proposed that the coordination in this compound is η^6. However, there was no structural confirmation for this suggestion. Due to the curvature of C_{60} the exohedral π-orbitals are displaced perpendicularly about 10° away from the face of a hexagon. Thus sufficient overlap with one η^6-bound metal is difficult.[12]

5.4.10 η^2-Complexes of C_{60}

A wide variety of metal complexes with C_{60} and C_{70} binding in the η^2-mode have been reported and many of these have been structurally characterised. Further, complexes with more metals symmetrically substituted on the fullerene (up to 6) and two fullerenes bound to the same metal, all in the η^2-mode were reported. It was always the 6,6-bond which was involved in the bonding. Nickel, palladium and platinum with phosphines were found to give symmetrically hexa-substituted derivatives with C_{60}. From the structural analysis of these derivatives, it was seen that there is a gain in aromaticity of the unsubstituted hexagonal rings of the fullerene with such a symmetrical hexa substitution indicated by the diminished difference in bond lengths between alternate bonds of the hexagons.

5.4.11 μ^3-η^2-η^2-η^2 Complexes of Fullerenes

Although theoretical studies predict that η^6-coordination to one hexagon of C_{60} is possible, no examples of such compounds have been prepared so far. However, a face capping of the

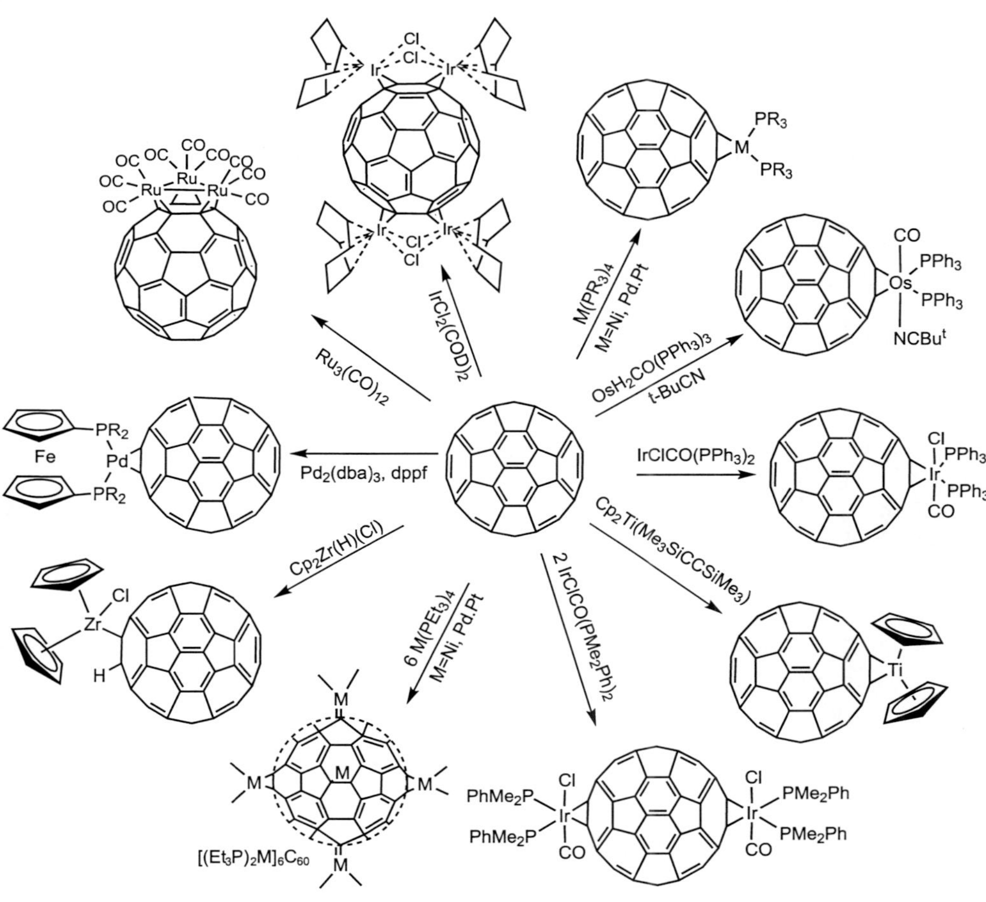

Scheme 5.5 Transition metal η^2-derivatives of C_{60}

hexagonal face of C_{60} has been achieved by a μ^3-η^2-η^2-η^2-bonding with trinuclear metal carbonyls but orbital overlap of π-orbitals with the metal triangular arrangement of metal clusters can be quite effective. The metal–carbon bonds of $Ru_3(CO)_9(\mu^3\text{-}\eta^2\text{-}\eta^2\text{-}\eta^2\text{-}C_{60})$ are found to be similar and even stronger than that of $Ru_3(CO)_9(\mu^3\text{-}\eta^2\text{-}\eta^2\text{-}\eta^2\text{-}C_6H_6)$. It is also possible to get η^2–η^2 substitution on the same hexagonal face as shown by the iridium complex obtained from the reaction of $IrCl_2(COD)_2$.

5.4.12 Halogenation of Fullerenes

Except iodination which has not been successful, possibly due to the weakness of the C_{60}–iodine bond and the bulky nature of iodine, all other halogens have produced derivatives with

C_{60} and some with C_{70} with a varying number of halogen substituents. While fluorination by F_2 gives the highest substitution namely $C_{60}F_{60}$, the maximum substitution achieved with chlorine and bromine are only 30 and 24 respectively. Bromination and chlorination of fullerenes are some of the very few reactions that proceed with remarkably high selectivity. Chlorination of C_{60} or C_{70} using iodine monochloride produces individual isomers of $C_{60}Cl_6$ or $C_{70}Cl_{10}$ in almost quantitative yields. Bromination follows two major patterns—in $C_{60}Br_6$ two bromines are adjacent to each other, whereas in $C_{60}Br_8$ the bromines display a 1,3- or 1,4-relationship. Although the bromo-compounds are interesting from the structural viewpoint, they are highly insoluble in most organic solvents, which limits their use as starting materials. By contrast, the fluoro- and chlorofullerenes are quite soluble in organic solvents. $C_{60}Cl_6$ is highly soluble in benzene, carbon disulfide and tetrachloromethane, moderately soluble in chloroform, dichloromethane and toluene, and slightly soluble in pentane, hexane, diethyl ether and acetone.[13]

Cl Cl Cl Cl Cl Cl ← ICl in C_6H_6, 298 K — C_{60} — Br_2 in C_6H_6 or CCl_4, 298 K → Br Br Br Br Br Br (5.3)

The reaction of $C_{60}Cl_6$ with organic cyanides was shown to produce salts of the stable pentacyanofullerene anion $[C_{60}(CN)_5]^-$. It has also been shown that hexachlorofullerene $C_{60}Cl_6$ reacts readily with thiols producing a novel family of fullerene derivatives $C_{60}[SR]_5H$.

5.4.13 η^5-Complexes of Fullerene Derivatives

The preparation of the first η^5-fullerene compound happened rather serendipitously from a one step fivefold addition of a phenyl copper(I) reagent to C_{60} in 1996 by Nakamura and coworkers. Although obtained initially in very poor yield, the methodology was improved to yield η^5-fullerene compounds in very good yields. The reaction has been found to be general for various aryl, methyl and alkenyl groups. Pentamethyl and pentaphenyl derived C_{60} were prepared using the corresponding Grignard reagents along with CuBr and SMe_2. This way of pentaalkylation and arylation increased considerably the electron cloud present in the five-membered ring in whose vicinity they are present, converting it almost to a cyclopentadienyl system. A very interesting outcome of this study was the preparation of the first ferrocene–fullerene hybrid molecule (called buckyferrocene) prepared in relatively good yields by the reaction of iron cyclopentadienyl dicarbonyl dimer and $Me_6HC_{60.}$ The C_{70} analogue was also prepared from a trimethyl substituted C_{70} derivative but in lesser yields.[14–19]

The study was extended by Nakamura and coworkers to double decker buckyferrocenes, and to a range of η^5-fullerene containing organometallic compounds with metals such as Ru, Rh, Ir, Co, Ni, W, Mo and Pd. The synthesis of these compounds is shown in Scheme 5.6.

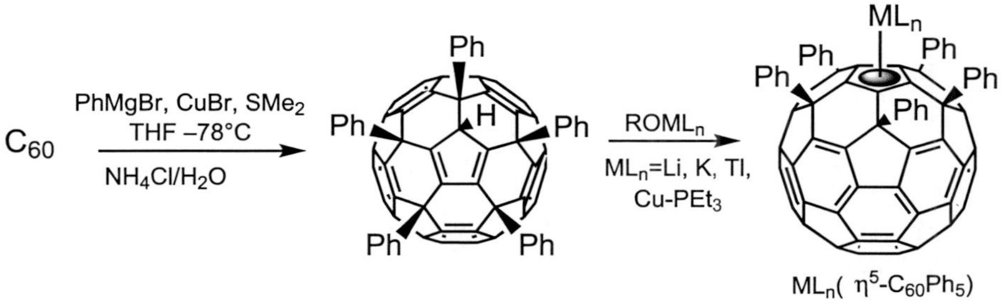

C_{60} —PhMgBr, CuBr, SMe_2 / THF −78°C / NH_4Cl/H_2O→ $C_{60}Ph_5H$ —$ROML_n$ / ML_n=Li, K, Tl, Cu-PEt_3→ $ML_n(\eta^5\text{-}C_{60}Ph_5)$ (5.4)

$C_{60}Me_5H$ —$[CpFe(CO)_2]_2$ / PhCN, 180°C, 8h→ Bucky ferrocene —MeMgBr / $CuBr.SMe_2$/ 1,2–$Cl_2C_6H_4$, THF, Py, 25°C / $[CpFe(CO_2]_2$/ PhCN, 185°C→ Double decker bucky ferrocene (5.5)

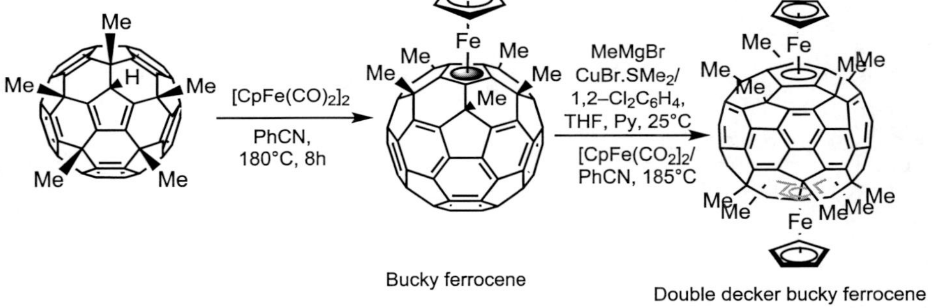

5.4.14 Complexes of C_{70}

The well known Vaska's complex forms η^2-complex readily with C_{70}. The complex $Ir(CO)Cl(PPhMe_2)_2$ is more reactive than the original Vaska's complex in oxidative addition reactions. The binding constant of $Ir(CO)Cl(PPhMe_2)_2$ is 200 times higher than that of the PPh_3 substituted Ir complex. Such modified Vaska's complexes yield multiple substitution on the $C_{60.}$ The bis-adduct of C_{70} is obtained when a six to twelve fold excess of this iridium complex is used in a reaction with C_{70} in benzene.

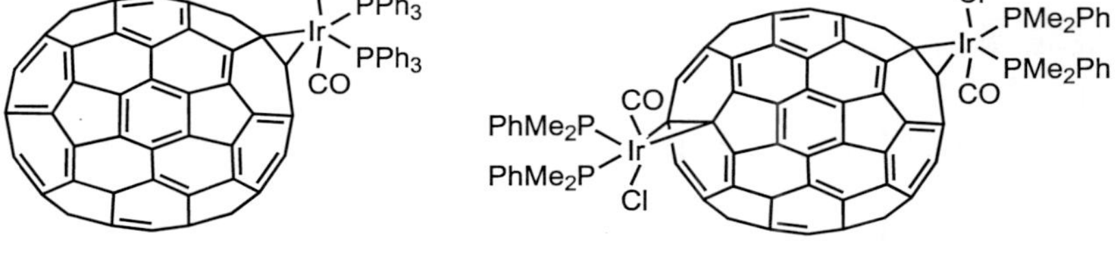

Fig. 5.7 Iridium complexes of C_{70}

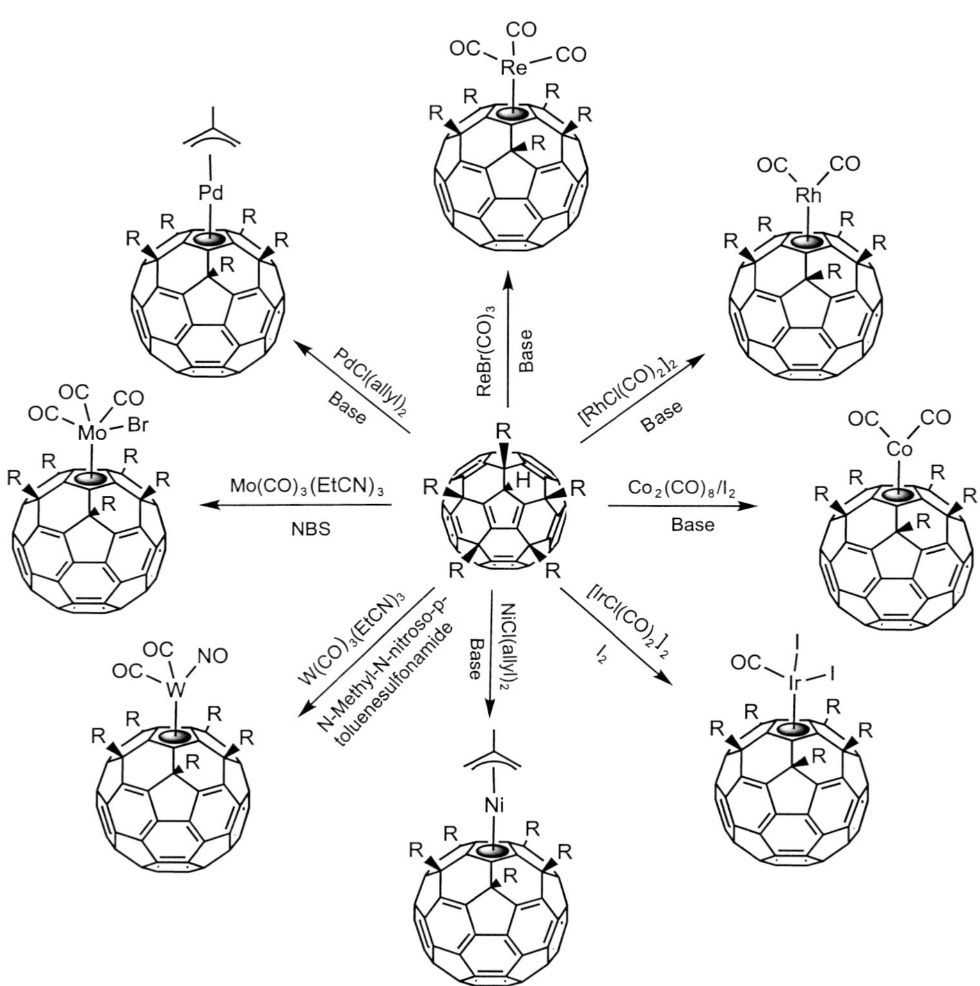

Scheme 5.6 Transition metal η^5-derivatives of C_{60}

5.4.15 Molecular Surgical Synthesis of Endohedral Fullerenes

Endohedral fullerenes are fullerenes having additional atoms, ions or small clusters enclosed or entrapped inside the fullerene cage. These were initially prepared by direct synthesis where the molecule to be entrapped was present during the fullerene synthesis or by using pressure. Saunders and coworkers in 1993 produced endohedral fullerenes $He@C_{60}$ and $Ne@C_{60}$ by pressurising C_{60} to approximately 3 bar pressure with the corresponding noble

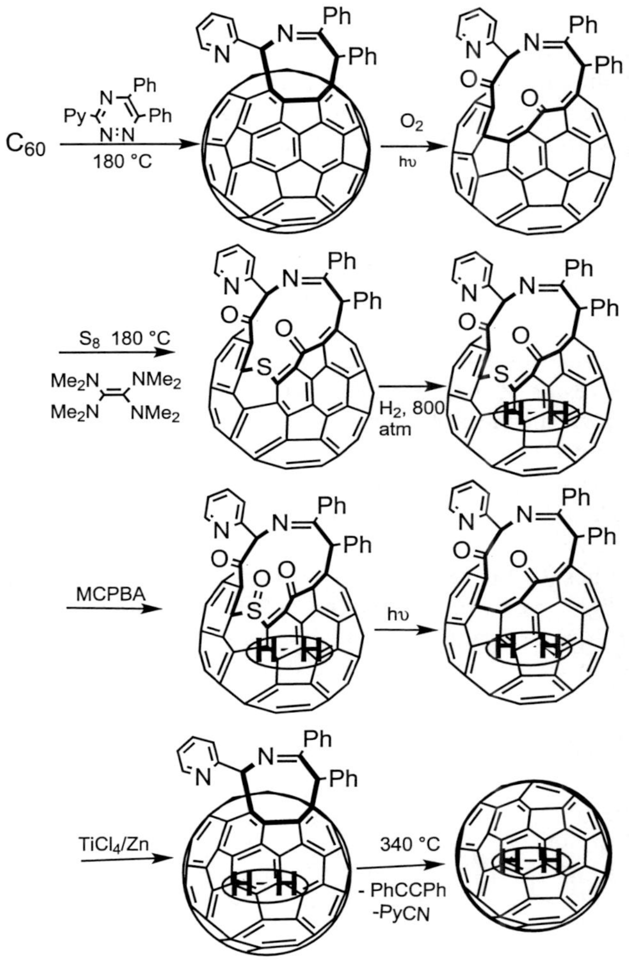

Scheme 5.7 Surgical synthesis of C_{60} leading to $H_2@C_{60}$

gas. The formation of endohedral complexes with neon, argon, krypton and xenon were also demonstrated. In 1999 the synthesis of the $Sc_3N@C_{80}$ by H Dorn and coworkers, showed for the first time that molecular fragments can be incorporated in a fullerene cage. This compound has been prepared by the arc vaporisation method at temperatures up to 1100°C of graphite rods packed with scandium(III) oxide iron nitride and graphite powder in a nitrogen atmosphere at 300 torr. However, the yields of the endohedral fullerenes were very poor in these types of syntheses.

Closed fullerenes encapsulating small molecules have been synthesised by a sequence of organic reactions opening up the fullerene cage, a methodology which has been named as molecular surgical synthesis of endohedral fullerenes. The first among these was the synthesis of the dihydrogen endofullerene $H_2@C_{60}$, by Komatsu. This was followed by water endofullerene $H_2O@C_{60}$, and hydrogen fluoride endofullerene $HF@C_{60}$, by Murata and Whitby respectively. The encapsulated molecules display unusual physical properties which have been studied by a variety of physical methods. The methodology adopted for the molecular surgical synthesis of $H_2@C_{60}$ by Komatsu is shown in Scheme 5.7. The process can be described as making a hole on the C_{60} surface by a chemical reaction, widening up more of the hole by further reactions, insertion of a guest gaseous species under pressure inside the cavity created and closure of the hole without loss of the encapsulated guest. The process involves three reactions to open and widen the hole and four reactions to gradually close the hole for making $H_2@C_{60}$. The yield obtained based on consumed C_{60} of $H_2@C_{60}$ was 9%. An 8-membered ring hole was formed initially by the use of 1,2,4-triazine, which was further widened to a 12-membered hole. This hole was not wide enough for a H_2 molecule to enter. Further it was made into a 13-membered hole upon which H_2 molecules could enter the cavity. The reduction in the hole size was brought about by oxidation of the S with MCPBA and McMurry coupling of the two C=O units. The steps involved are shown in Scheme 5.7.[20]

In 2010, fullerenes (C_{60}) have been discovered in a cloud of cosmic dust surrounding a distant star 6500 light years away. Using NASA's Spitzer infrared telescope, infrared bands corresponding to C_{60} were identified in this cosmic dust. This finding provides evidence that fullerenes existed since time immemorial in the distant regions of our galaxy.

5.5 Carbon Nanotubes

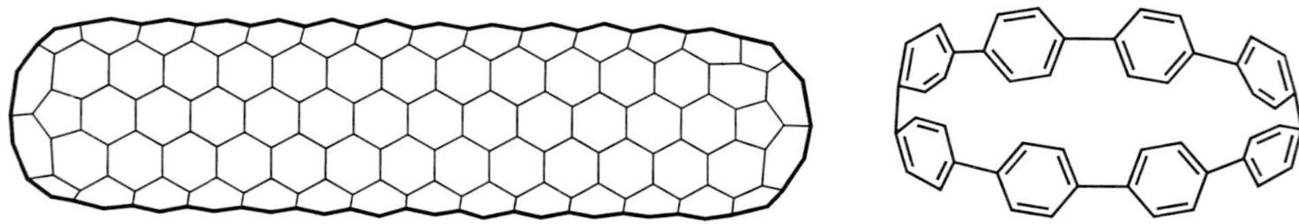

Carbon nanotubes (CNTs) are allotropes of carbon which resemble cylindrically extended fullerenes. They can be considered formally as sheets of graphite rolled into a cylinder and sealed at the ends by a hemifullerene end cap. These molecules have interesting properties that make them potentially useful for applications in nanotechnology, electronics and optics. They exhibit extraordinary strength and unique electrical properties and are efficient conductors of heat. Carbon nanotubes have diameters as small as 1 nm and lengths up to several centimetres and they can be open or closed. The shortest carbon nanotube can be considered as cycloparaphenylene having eight phenyl rings forming a ring connected through their para positions. In general, the diameter of single-walled carbon nanotube is less than 2 nm, while diameters of multi-walled nanotubes range between 2 and 100 nm,

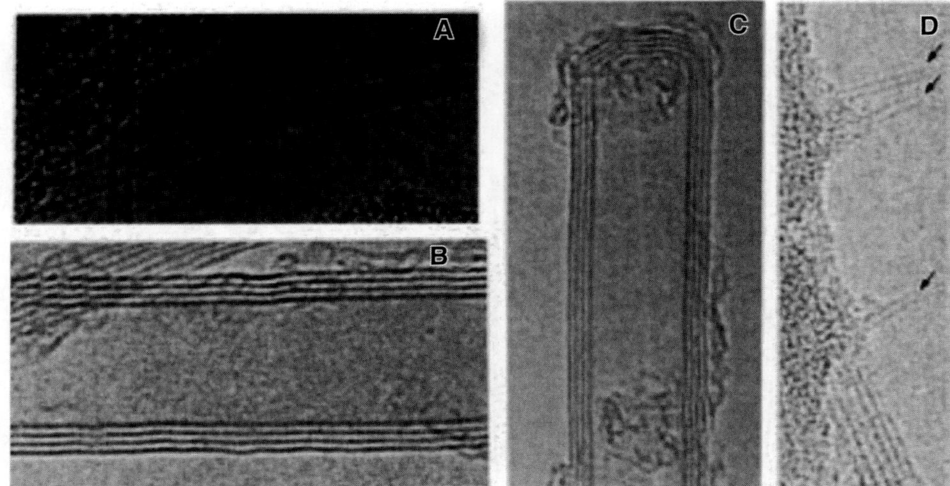

Fig. 5.8 High resolution transmission electron microscopy images of (A) SWNT (B) MWNT (C) MWNT tip and (D) SWNT tips

Source: Reprinted with permission from (Ajayan P M, Nanotubes from carbon, *Chem. Rev.*, 1999, Vol. 99, 1787), Copyright 1999, American Chemical Society.

also depending on the number of shells present. CNTs are typically only a few microns long but, tubes up to 4 cm in length have also been reported.[21]

The excitement in the scientific community on carbon nanotubes began after Sumio Iijima, while investigating the formation of fullerene soot by the electric arc method, observed the formation of multilayer carbon nanotubes. He published his observations with the title 'On helical microtubules of graphitic carbon' in *Nature*, 1991, followed by another paper in the same journal in 1993 on single shell carbon nanotubes. However, the actual discovery of carbon nanotubes has been claimed by many groups much earlier. The earliest among these are by the Soviet scientists L V Radushkevich and V M Lukyanovich, who in 1952 published clear images of tubes of 50 nm diameter made of carbon in the *Soviet Journal of Physical Chemistry* (Vol. 26, pp 88–95) in Russian.

Carbon nanotubes can in general be a single walled (SWNT), double walled (DWNT) or multiple walled (MWNT) type with coaxially nested tubes having varying diameters. The outer diameter ranges between 2 and 50 nm. In the case of MWNT, the interlayer spacing of individual walls is constantly 0.34 nm, very close to the interlayer spacing between graphite sheets. CNT has the highest strength of all known materials with a tensile strength almost 100 times as that of steel. In general, CNTs are synthesised using arc discharge between graphite electrodes, laser vaporisation of graphite or chemical vapour deposition methods. The arc discharge method favours formation of MWNT. However, if a catalyst like Ni or Fe is present in the graphite, single walled carbon nanotubes (SWNT) can be preferentially produced. Gram scale synthesis of SWNT can be carried out by the laser vaporisation technique. In chemical vapour deposition method, use of precursors like acetylene or ethylene with Fe, Co or Ni catalysts facilitate the formation of SWNT.[22, 23]

Three classes of carbon nanotubes are known, based on the direction of roll formation considering a single sheet of graphite. These are the armchair, zigzag and chiral types of CNT. The armchair type nanotubes show metallic electrical properties, the chiral type shows semiconducting properties with a small band gap and the zigzag type shows semiconducting behaviour.

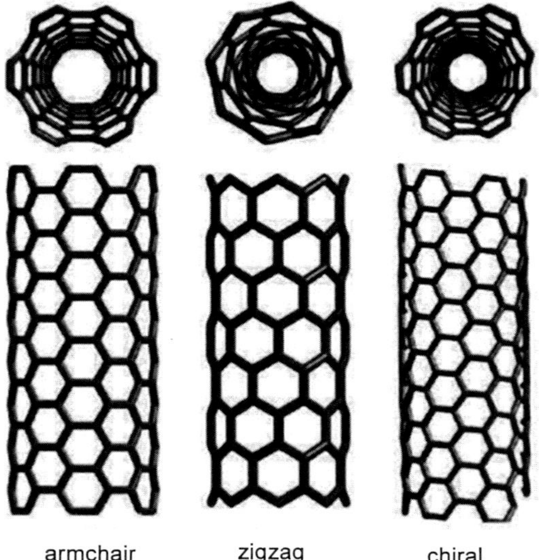

Table 5.2 Comparing properties of carbon nanotubes with that of stainless steel

Property	Armchair	Zigzag	Chiral	Stainless steel
Electrical conductivity	Metallic	Semiconducting	Semiconducting	Metallic
Young's modulus (TPa)	0.94	0.94	0.92	~ –0.02
Tensile strength	126.2	94.5	–	~0.65–1.00

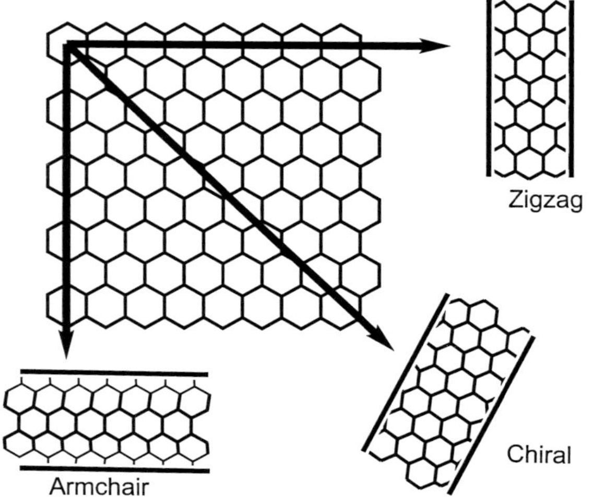

Fig. 5.9 The three types of carbon nanotubes based on the direction of roll formation

5.5.1 Unique Properties of Carbon Nanotubes[24]

(i) **High electrical conductivity**

CNTs can be either metallic or semi-conducting in their electrical behaviour. Armchair structured CNTs appear to conduct better than other CNTs. The conductivity and resistivity of ropes of SWNTs has been measured. The resistivity of the SWNT ropes was of the order of 10^{-4} ohm-cm at 27°C. This indicates that SWNT ropes have the highest conductivity among all known carbon fibres. The achieved current density was 10^7 A/cm^2.

(ii) **High thermal conductivity**

Ultra-small SWNTs have even been shown to exhibit *superconductivity* below 20 K. Studies show that an individual SWNT has thermal conductivity along its axis of about 3500 Wm^{-1}K^{-1} at room temperature. When compared to this, copper metal transmits 385 Wm^{-1}K^{-1}. Across its axis, in the radial direction, an individual SWNT has thermal conductivity of about 1.52 Wm^{-1}K^{-1} at room temperature. Macroscopic assemblies of nanotubes such as films or fibres have shown thermal conductivity up to 1500 Wm^{-1}K^{-1} so far. The carbon nanotubes have been found to be stable up to 2800°C in vacuum and upto 750°C in air.

Table 5.3 Comparing thermal and electrical conductivities of carbon nanotubes with copper

Material	Thermal conductivity [W/(m·K)]	Electrical conductivity [S/m]
Carbon nanotubes	~ 3,500	10^6–10^7
Copper	385	6×10^7

(iii) **High tensile strength**

Carbon nanotubes are one of the strongest and stiffest materials in terms of tensile strength and elastic modulus respectively. This strength is a consequence of the covalent sp^2 bonds formed between the individual carbon atoms. A multi-walled carbon nanotube was tested to have a tensile strength of 63 giga pascals (9,100,000 psi) which means that it has, in principle, the ability to endure tension of a weight equivalent to 6,422 kilograms-force on a cable with a cross-section of 1 square millimetre. Further studies indicate that individual CNT shells have strengths of up to ≈100 gigapascals (15,000,000 psi). It may be noted that under excessive strain, the tubes will undergo plastic deformation.

(iv) **Good electron field emitting ability and sharpness as a probe**

Field emission is the emission of electrons from the surface of a conductor under the influence of a strong electrostatic field. Their high electrical conductivity and the incredible sharpness of their tip (the smaller the radius of curvature of the tip, the more concentrated will be the electric field), leads to increased field emission when CNTS are used. When a carbon nanotube is put into an electric field, it will emit electrons from the end of the nanotube like a micro gun. If those electrons are allowed to bombard a phosphor screen, one can create an image. Some television manufacturers are using this technology to replace the existing bulky electron guns of television sets with these significantly smaller carbon nanotube electron guns. In CNT-based picture displays, there is a separate electron gun (or even many of them) for each individual pixel in the display. Their high current density, low turn-on and operating voltages, and steady, long-lived behaviour make CNTs very attractive field emitters in this application.

Atomic force microscopy (AFM) also called scanning force microscopy (SFM) is a type of very high resolution scanning probe microscopy, with demonstrated resolution of the order of fractions of a nanometre. On an experimental basis, single-walled carbon nanotubes have been attached to the tip of an AFM probe to make the tip sharper. This allows much higher atomic resolution of the surface being investigated. In addition, since the nanotubes are flexible and not rigid, it prevents damage to the sample surface and the probe tip if the probe tip happens to rub into the surface.

(v) **High aspect ratio for making materials conducting**

While plastics and composites such as carbon fibre reinforced polymer (CFRP) are nowadays replacing the traditional role played by metals due to their high strength to weight ratio, one major difference between them and the metals is the electrical conductivity. The electrically insulating property of plastics can sometimes be a hindrance when one wants to introduce conductivity in non-conducting polymers. This deficiency is overcome by loading these polymers with conducting fillers, such as larger graphite fibres. It is known that, higher the aspect ratio of filler particles, the lower is the loading required to achieve a given level of conductivity. CNTs have the highest aspect ratio known for any carbon fibre. Since they also have a tendency to form ropes, they provide long conductive pathways with low loadings of the nanotube.

Carbon nanotubes (arm chair type) can carry an electrical current density of 4×10^9 A/cm^2 which is greater than the electrical conductivity of metals such as copper. For industrial-scale applications of carbon nanotubes, the most promising area seems to be the reinforcement of polymer composites and introduction of CNT in polymer composites, especially in carbon fibre reinforced polymers in order to avoid static charges especially on the outer surface of aeroplanes. Modern day aeroplanes use 50%–56% CFRP replacing aluminum in their manufacture. One drawback of CFRP is relatively lower electrical conductivity especially for dissipating lightning strikes; aluminum strips are being used to ensure the electrical continuity of the fuselage. The high electrical conductivity of CNT reinforced in the CFRP polymer composite helps to replace the aluminium strips. For the aeronautical industry, this may lead to enormous weight reductions because antistatic painting, aluminum framework and sheet covering of aeroplanes can be reduced.

In addition to the above properties, CNTs are highly flexible and can be bent considerably without damage. They are also highly elastic and have a low thermal expansion coefficient as well.

5.5.2 Functionalising of Carbon Nanotubes

CNTs are considered as attractive materials in diverse nanotechnological applications, such as fillers in polymer matrices, molecular tanks, biosensors, and so on. However, very

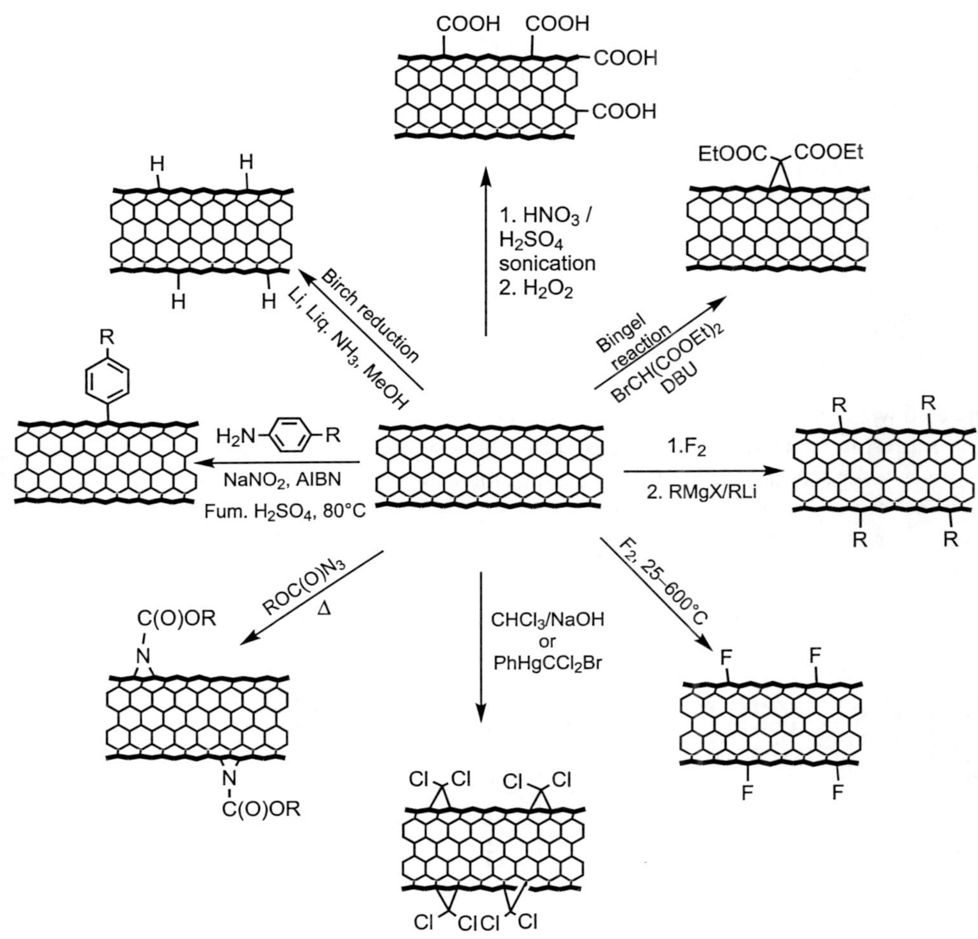

Scheme 5.8 Reactions leading to functionalisation of carbon nanotubes

poor solubility and the difficulty in manipulation in any solvent have imposed considerable limitations on the use of CNT. Indeed, pure carbon nanotubes are insoluble in all organic solvents and aqueous solutions. By sonication, they can be dispersed in some solvents, but precipitation occurs when this procedure is stopped. However, CNTs can undergo chemical reactions that make them more soluble for their integration into inorganic, organic and biological systems. CNTs are functionalised by many reactions, especially oxidation, that result in reactive groups at their end-caps and/or sites of defect, or by modifying their sidewalls. Covalent attachment to nanotubes involves the direct addition of a functionality to the CNT through the formation of chemical bonds, whereas non-covalent attachment involves interactions such as electrostatic, van der Waals or similar weak interactions. It may be noted that significant covalent functionalisation, which alters its C–C bonds from sp^2 to sp^3

structure, can result in substantial loss of electrical conductivity of the functionalised SWNTs. Many biomolecules such as carbohydrates, oligonucleotides, proteins, enzymes and DNA have been attached to CNTs using oxidative methods that introduce carbonyl and carboxylic acid groups on the open ends of the CNTs and at defect sites along the CNT sidewalls.[25]

Early studies on oxidation of carbon nanotubes involved reactions with nitric acid vapour in air, which led to indiscriminate functionalising of the carbon nanotubes with carboxylic, carbonyl or hydroxyl groups. In the liquid phase, oxidising solutions of nitric acid or a combination of nitric and sulfuric acid were effective, but over oxidation often occurred causing the carbon nanotube to break up. Later it was observed that sonication assisted oxidation with sulfuric and nitric acid produced carbonyl and carboxyl groups which when followed by treatment with hydrogen peroxide, reduced the extent of structural damage of the carbon nanotubes.

The presence of carboxylic acid groups is helpful for esterification and amidation reactions. The carboxylic acid group can be converted to an acyl chloride which is then reacted with the desired amine, amide or alcohol.

Tethering or connecting ordered assemblies of carbon nanotubes to metal surfaces is a very important step for making nanotube-based materials which can have potential applications especially in electronic devices. Equation 5.6 shows how carboxylic acid derived carbon nanotubes can be chemically modified to connect on to gold surfaces by using thiol groups as termini.[26]

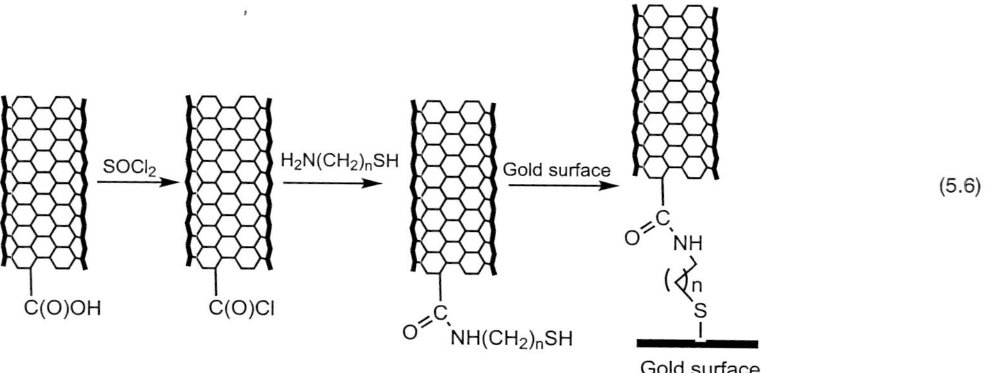

(5.6)

5.6 Graphene

Graphene, is basically a thin layer of pure sp^2 hybridised carbon or more precisely, a single layer of the well known carbon allotrope, graphite. It is a single, tightly packed layer of carbon atoms that are bound together in a hexagonal honeycomb lattice with a C–C bond length of 0.142 nm. Layers of graphene stacked on top of each other form graphite, with an interplanar spacing of 0.335 nm. One of the ways of obtaining graphene is by exfoliation of graphite.

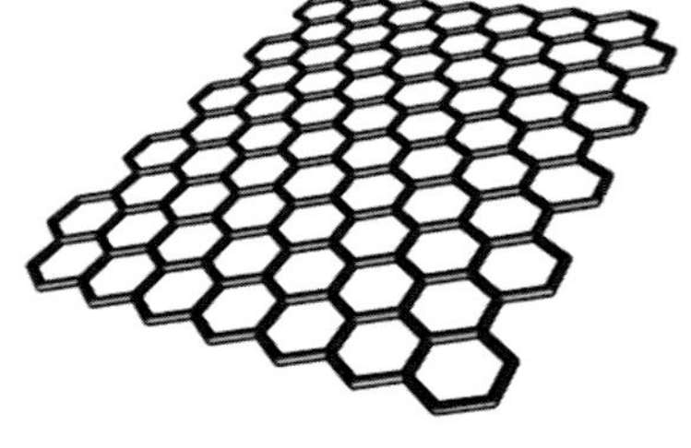

The term graphene for the first time was used in 1987, mainly to describe layers of graphite separated from each other by inserting other molecules and ions in between (a process termed as intercalation). Also in 1984, it was theoretically predicted that electric current would be carried effectively by massless charge carriers in graphene. In fact carbon nanotubes are considered as rolled up graphene sheets in different directions.

The scientists who discovered graphene are Andre Geim and his colleague Kostya Novoselov of the University of Manchester. Carbon nanotubes were a major area of materials research during that time and Geim got his inspiration from the possibilities found with carbon nanotubes. He tried to put into practice an idea to polish down a graphite block to just 10 or 100 layers thick and then study the properties of the resulting material. One of his students produced a speck of graphite roughly 1000 layers thick. That is when Geim thought that simple Scotch tape can be tried to peel away the top layer from this graphite piece. Graphite flakes came off onto the tape, and the process was repeated several times to achieve progressively thinner flakes attaching to the tape. When the scotch tape was dissolved in a solution, it left behind ultra-thin flakes of graphite: just 10 layers thick. Soon the team was able to fabricate rudimentary transistors with this material. Subsequent modifications of the scotch tape technique finally resulted in the first graphene sheets which are basically a single two dimensional layer of graphite.[27]

In 2004, Geim published his discovery announcing the achievement of graphene sheets in the journal *Science* (Vol. 306, 2004, pp 666), with a paper entitled 'Electric field effect in atomically thin carbon films'. By 2005, researchers had succeeded in isolating graphene sheets by other methods as well. Graphene is a mere one atom thick – perhaps the thinnest material and two dimensional material in the universe – and forms a high-quality crystal lattice, with no vacancies or dislocations in the structure. This structure gives it intriguing and unique properties.[28,29]

The unique properties of graphene are

- It is the thinnest compound known to man (at one atom thick) and having a high level of transparency. The optical absorption of a single graphene layer is shown to be ~2.3% over the visible spectrum. The transparency and high electrical conductivity makes it a potential replacement for indium tin oxide as transparent conducting material for touch screen devices.
- It is the lightest material known (the weight of 1 sq.m is approximately 0.77 mg).
- It is the strongest compound discovered (100–300 times stronger than steel and with an intrinsic tensile strength of 130 GPa and a Young's modulus (stiffness) of 1 TPa (150000000 psi).
- It is the best conductor of heat at room temperature with conductivity $(4.84\pm0.44) \times 10^3$ to $(5.30\pm0.48) \times 10^3$ $Wm^{-1}K^{-1}$.
- It is the best conductor of electricity known (studies have shown electron mobility at values of more than 15,000 $cm^2V^{-1}s^{-1}$). The resistivity of graphene sheets is 10^{-6} Ωcm which is less than the resistivity of silver, the otherwise known material of lowest resistivity at room temperature.

- It can act as a filter and barrier. Even helium cannot pass through it. Water purification is being attempted by allowing selective permeation through structural defects of single-layer graphene or graphene oxide.
- Graphene has a theoretical specific surface area (SSA) of 2630 m^2/g. This is much larger than that reported to date for carbon black (typically smaller than 900 m^2/g).
- Graphene can self-repair holes in its sheets, when exposed to molecules containing carbon. Bombarded with pure carbon atoms, the atoms perfectly align into hexagons, completely filling the holes.
- Graphene is the only form of carbon (or solid material) in which every atom is available for chemical reaction from two sides (due to the 2D structure).

5.6.1 Methods of Producing Graphene

Exfoliation from graphite produces graphene with the lowest number of defects and highest electron mobility. Alternatively, a sharp single crystal diamond wedge can be used to cleave layers from a graphite source. Graphite oxide when rapidly heated and exfoliated yields highly dispersed carbon powder with a few per cent of graphene flakes. Another method involves the reduction of graphite oxide monolayer films by reagents such as hydrazine.

5.6.2 Applications of Graphene

3-D printing also called additive manufacturing, is the process of creating a three dimensional, solid object from a digital file. 3-D printers have the capability to create models from polymers, metals and many other materials. 3-D printing is achieved using an additive process in which successive layers of material are laid down to create nearly any imaginable shape and it differs from the well known method of subtractive manufacturing as there is minimum material waste. Graphene produced from natural graphite is a useful starting material for the production of graphene infused polymers which are now used to make 3-D printing filaments. Addition of graphene to polymers which are traditionally used in 3-D printing improves the properties of the polymer especially mechanical strength, electrical and thermal conductivity.

Experiments are being conducted by some smartphone manufacturers to use graphene as a replacement for transparent conducting oxides (such as indium tin oxide) in modern touch screen devices.

5.7 Carbon Fibre Reinforced Polymers (CFRP)

Carbon fibre reinforced polymers (CFRP) are composite materials where a carbon fibre as the reinforcement provides strength and the matrix is a polymer resin, often an epoxy resin, which binds the carbon fibres together. Because CFRP consists of two distinct constituents, the material properties depend on these two ingredients. Carbon fibre reinforcement provides the CFRP its strength and rigidity which is measured in stress and elastic modulus.

Unlike materials such as aluminum and steel, CFRP has directional strength properties. Its properties depend on the direction of layouts of the carbon fibre and the proportion of the carbon fibres relative to the polymer content. The most important property however is the strength-to-weight ratio or specific strength which is a material's strength (force per unit area) divided by its density. This property is especially useful in deciding applications where strength cannot be compromised while weight needs to be reduced (for example, as replacement for aluminum alloys in modern aircraft and sports car parts).

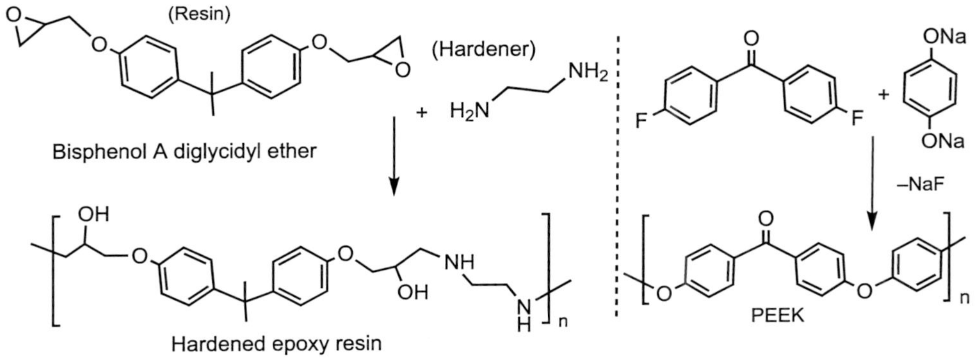

Scheme 5.9 Synthesis of polymeric resins used for CFRP manufacture

The primary constituent of CFRP is a carbon filament which is produced from a precursor polymer such as polyacrylonitrile (PAN), rayon or petroleum pitch. When PAN or rayon is used, the polymer is first spun into filament yarns, using chemical and mechanical processes to initially align the polymer chains in a way to enhance the final physical properties of the completed carbon fibre. Afterwards, the polymer filament yarns are heated to drive off non-carbon atoms (carbonisation), producing the final carbon fibre. In three successive stages of heating, all elements other than carbon are mostly eliminated. The fibre is black in colour and is essentially unaffected by UV exposure. The honeycomb shaped crystals organise themselves in long flattened ribbons. This crystal alignment makes the ribbon strong in the long axis. In turn, these ribbons align themselves within fibres.

The binding polymer is often a thermosetting resin such as epoxy, but other thermoset or thermoplastic polymers, such as polyester, vinyl ester or nylon, are sometimes used. The composite may contain other fibres, such as an aromatic amide polymers (such as Kevlar), aluminium, ultra-high-molecular-weight polyethylene (UHMWPE) or glass fibres. Recent efforts to toughen CFRPs include modifying the existing epoxy material and finding alternative polymer matrices. One such material with high promise is PEEK (polyether ether ketone), which exhibits an order of magnitude of higher toughness than epoxy with similar elastic modulus and tensile strength. However, PEEK is much more difficult to process and more expensive. While the carbon fibres themselves are not affected by the moisture diffusing into the material, the moisture can plasticise the polymer matrix to some extent.

Polyether ether ketone PEEK has a T_g of around 143°C and melts around 343°C. Some grades have a useful operating temperature of up to 250°C. The possible ways by which CFRP gets formed by heat treatment of polyacrylonitrile is shown in Scheme 5.10.

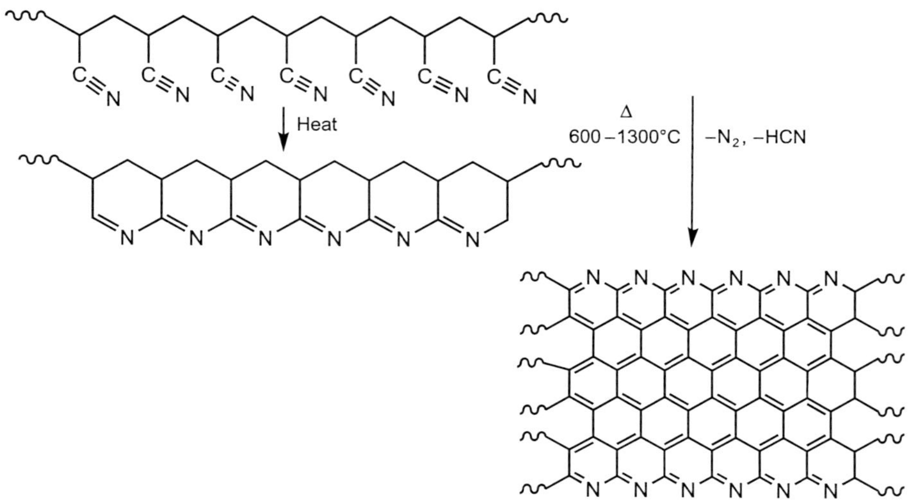

Heat

Δ
600 – 1300°C
$-N_2$, –HCN

Scheme 5.10 Possible way by which carbon fibres are formed upon heating of polyacrylonitrile

5.7.1 Applications of Carbon Fibre Reinforced Polymers

It is interesting to note that commercial aeroplane manufacturers such as Boeing and Airbus have, over the years, replaced aluminum partly with CFRP in aircraft manufacture. While a Boeing 747 (introduced in 1970) had 1% composites (mainly CFRP), 81% aluminum and 4% titanium used in its structure, Boeing 787 Dreamliner (introduced in 2011), has 50% CFRP, 20% aluminum and 15% titanium. The Airbus A350 XWB (introduced in 2015) is built with 52% CFRP including wing spars and fuselage components, overtaking the Boeing 787. SpaceX, an American aerospace manufacturer and space transport services company, uses carbon fibre for the entire primary structure of their super heavy-lift ITS launch vehicle. 'Solar Impulse', the first aeroplane to go around the world in 2016 with 0% fuel, and energy from only solar cells, had its structure made almost entirely with CFRP.[31]

CFRPs are used extensively in the making of high-end automobile racing cars. The high cost of carbon fibre is eased by the material's unsurpassed strength-to-weight ratio and low weight is essential for high-performance automobile racing. Such car manufacturers have also developed methods to give carbon fibre pieces strength in the load-bearing directions. The use of carbon fibre in a vehicle can appreciably reduce the weight and hence the size of its frame. This will also facilitate designers' and engineers' creativity and allow more in-cabin space.

CFRP has become a notable material in structural engineering applications. It has proved to be cost effective in a number of field applications strengthening masonry, steel, concrete, cast iron and timber structures. CFRP is now widely used in sports equipment such as in squash, tennis and badminton racquets, sport kite spars, hockey sticks, fishing rods, bicycle frames, and so on. Amputee athletes use carbon fibre blades for running.[32]

5.8 Carbon Molecular Sieves

A carbon molecular sieve (CMS) is the porous carbon skeletal framework that remains after oxygen-free pyrolysis of a polymeric precursor such as polyvinylidene chloride, polyacrylonitrile, phenol formaldehyde or coal. It can be considered as a special type of activated carbon. Unlike activated carbon which has micropores of 10–500 Å diameter, CMS has uniform super-micropores of less than 10 Å (1 nm) pore diameter. In addition, activated carbon separates molecules through differences in their adsorption equilibrium constants while carbon molecular sieves bring about molecular separations based on the rate of adsorption rather than the differences in adsorption capacity. This behaviour is clearly evident in the working of pressure swing adsorbers (PSA). The most important practical application of CMS is the separation of nitrogen from oxygen. It is because the difference in molecular sizes is very small between O_2 (0.28 nm × 0.39 nm) and N_2 (0.30 nm × 0.43 nm). The best molecular sieving carbon for N_2/O_2 separation is precisely controlled to have slightly larger pores than N_2. The difference in pore size results in the fact that the N_2 is harder to adsorb and O_2 is easier to adsorb. On application of pressure, the O_2 remains caught up in the pores of the CMS while nitrogen passes out. Once all the pores get saturated with O_2, the pressure is brought down to atmospheric pressure upon which the O_2 molecules are released from the CMS pores.

5.9 Common Compounds of Carbon

5.9.1 Carbon Monoxide

Carbon monoxide is a toxic, odourless, tasteless and colourless gas generated by the incomplete combustion of carbon-containing compounds. It was first prepared in 1776 by J M F Delassone by heating zinc oxide with charcoal. Its chemical composition was recognised 20 years later by W Cruikshank. According to research done in the 1970s, less than 10% of CO comes from man-made emissions and 80% of the estimated 3.8×10^9 tons originates from oxidation of methane which is given off by decaying organic matter. Industrial production of CO is basically from gasification of coal or by steam reforming of natural gas or petroleum. The main industrial applications of CO are as a reducing agent for the production of metals, in the production of hydrogen by the water gas shift reaction, and for the carbonylation of many organic substrates.[33]

Carbon monoxide has the strongest chemical bond known so far with a bond energy of 1072 kJ/mol and a short bond distance of 113 pm while the isoelectronic molecule N_2 has a bond energy of 942 kJ/mol and a distance of 110 pm. However in stark contrast to N_2 which is

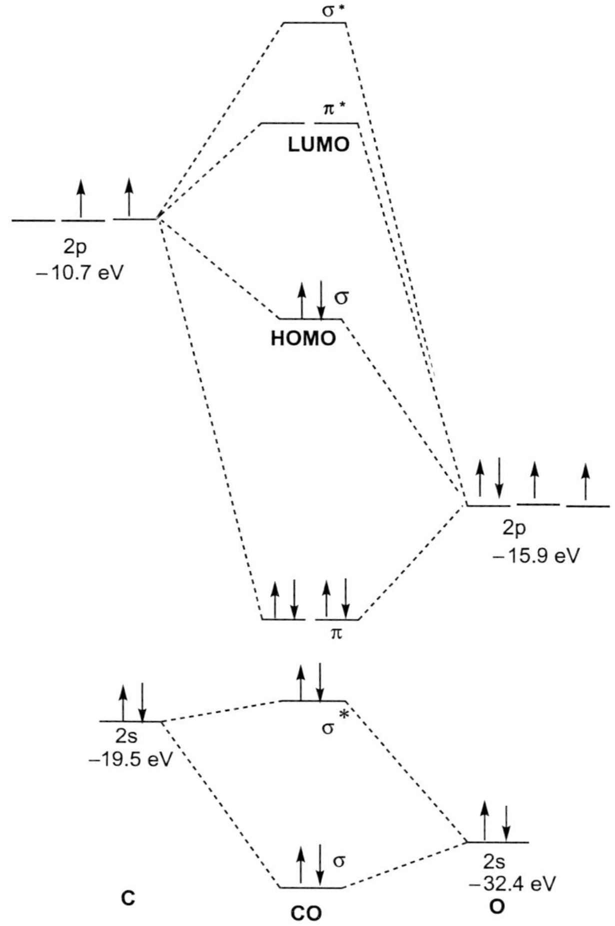

Fig. 5.9 Molecular orbital picture of carbon monoxide

relatively quite unreactive, CO, in spite of the high bond strength, is a very reactive molecule especially as a ligand in bioinorganic and organometallic chemistry. This difference can be explained by viewing the molecular orbital diagram of CO. The highest occupied molecular orbital (HOMO) of CO is weakly antibonding (compared to the O atomic orbitals) and is an MO which is carbon based. Secondly, the π^* antibonding orbital which is the lowest unoccupied molecular orbital (LUMO) is also comparatively of lower energy which makes it possible for it to interact with metal t_{2g} orbitals for π bonding. There exists a strong back bonding of metal electrons to the π^* antibonding orbitals of CO.

The site at which CO poisoning occurs in human beings is the principal oxygen-carrying biomolecule hemoglobin. When CO binds to hemoglobin, it does so irreversibly and produces a compound known as carboxyhemoglobin. The carbon monoxide toxicity is due to the formation of carboxyhemoglobin, which reduces the oxygen-carrying capacity of the hemoglobin and inhibits the transport, delivery and utilisation of oxygen by the body. The affinity between hemoglobin and carbon monoxide is approximately 230 times stronger than the affinity between hemoglobin and oxygen so hemoglobin binds to carbon monoxide in preference to oxygen.

Methods of CO production

The three industrial methods by which CO is produced in large quantities are (i) coal gasification, (ii) steam reforming, and (iii) partial oxidation of hydrocarbons.

$$C + H_2O \xrightleftharpoons{900\text{–}1200\ K} CO + H_2 \quad \Delta H^\circ_{298} = +131.30\ kJ/mol \tag{5.7}$$

$$CH_4 + H_2O \xrightleftharpoons{1000\text{–}1100\ K,\ 1.5\text{–}4.0\ MPa} CO + 3H_2 \quad \Delta H^\circ_{298} = +206.15\ kJ/mol \tag{5.8}$$

$$CH_4 + 1.5\ O_2 \xrightleftharpoons{1500\text{–}1800\ K,\ 1.5\text{–}4.0\ MPa} CO + 2H_2O \quad \Delta H^\circ_{298} = -519.33\ kJ/mol \tag{5.9}$$

Dehydration of formic acid with conc. H_2SO_4 is a convenient method for preparing CO in the laboratory.

Reactions of carbon monoxide

The chemistry of carbon monoxide is dominated by its reducing properties and its ability to combine with a large variety of molecules. As early as 1921, Hopcalite, a mixture of the oxides of manganese, copper, cobalt and silver [MnO_2 (50%), CuO (30%), Co_2O_3 (15%), and Ag_2O (5%)], has been used in gasmasks to oxidise CO to CO_2 with atmospheric air at room temperature. Also important are ceramic-supported palladium, platinum and rhodium which are heterogeneous catalysts used in catalytic converters to reduce carbon monoxide emissions in vehicle exhaust at high temperatures.

Solutions of copper(I) salts are capable of reversibly absorbing carbon monoxide to form carbonyl complexes. This property is used in many industrial processes to selectively separate carbon monoxide from a mixture of gases. The copper–ammonium salt wash method was developed by the German company BASF, by making use of aqueous ammonia solutions of copper carbonate, acetate, lactate, and so on. The absorption–desorption reaction is highly dependent on factors such as the temperature, pressure and acidity of the solution.

$$[Cu(NH_3)_2]^+ + CO + NH_3 \xrightleftharpoons[350\ K]{283\text{–}303\ K,\ 20\ MPa} [Cu(NH_3)_3CO]^+ \tag{5.10}$$

The gas mixture containing carbon monoxide is made to come in contact with the solution in countercurrent at 283–303 K and a pressure of up to 20 MPa in a tower with good surface area. Carbon monoxide is recovered by heating the CO rich solution to 350 K, usually at atmospheric pressure.

Some portable analysers use wet oxidation with I_2O_5, whereby a reactive adsorbent turns greenish brown as it adsorbs the iodine formed (Chapter 14; Eq. 14.40).

Metal carbonyls form a major class of carbon monoxide based compounds. While $Ni(CO)_4$ and $Fe(CO)_5$ can be made by direct carbonylation of the metal with CO, all other carbonyls have been prepared by reductive carbonylation methods. In the case of metal halides, a reducing agent such as an alkali metal or Al is used. When metal oxides are used, CO itself will act as the reducing agent to bring the metal to the zero oxidation state. Except in the case of $Ni(CO)_4$, all other syntheses require high temperatures and pressures.

The reaction of carbon monoxide with chlorine catalysed by activated charcoal, yields phosgene, an intermediate in the synthesis of isocyanates which are used for making polyurethanes.

$$CO + Cl_2 \xrightarrow[\text{act. C}]{} COCl_2 \tag{5.11}$$

Methanol to acetic acid is an industrially important process for which cobalt, rhodium or iridium based homogeneous catalysts have been used. The well known processes include the Monsanto and Cativa processes.

$$CH_3OH + CO \xrightarrow[\text{Co/Rh/Ir}]{} CH_3COOH \tag{5.12}$$

Other uses of pure carbon monoxide include the production of formic acid and methyl formate in the presence of a basic catalyst (350–420 K, 1–20 MPa).

$$CO + ROH \xrightarrow[\text{basic catalyst}]{} HCOOR \quad R = H, CH_3 \tag{5.13}$$

In a similar manner, dimethyl formamide is also produced.

$$(CH_3)_2NH + CO \xrightarrow[\substack{120°C,\ 20\ atm \\ NaOR\ catalyst}]{} HCON(CH_3)_2 \tag{5.14}$$

$$HC{\equiv}CH + CO + H_2O \xrightarrow[\text{Ni, 470 K, 5 MPa}]{\text{Reppe carbonylation}} H_2C{=}\underset{H}{C}{-}COOH \tag{5.15}$$

5.9.2 Carbon Dioxide

Carbon dioxide was known from the time of Pliny in the first century AD and was recognised as a distinct gas by van Helmont during the early 1600s. In 1757 it was shown that during respiration, part of the atmospheric air was changed and the new gas formed can have lethal effect on life. Thereafter CO_2 was called *fixed air*. The French chemist Lavoisier was the first to prove the chemical composition of fixed air by showing that it was formed when carbon is heated in oxygen. In the initial days, the interest in CO_2 was mainly to manufacture artificial mineral water or soda.[34]

The amount of carbon dioxide present in the earth's atmosphere is about 0.03% by volume. The total amount of carbon dioxide in the atmosphere is about 2.3×10^{12} tons. Combustion of carbon-based materials, especially dried plants and trees, respiration of plants and animals, decay of organic material, and industrial processes, such as cement production and the production of hydrogen for ammonia synthesis, all generate carbon dioxide. There are also processes which remove carbon dioxide from the air, such as photosynthesis, weathering of rocks and chemosynthesis by certain bacteria. The total CO_2 content of the seas and oceans is even more than that in the atmosphere and is about 1.4×10^{14} tons. In the oceans, carbon dioxide exists as carbonates, hydrogencarbonates, carbonic acid and as the dissolved gas itself.

Table 5.11 Comparison of CO_2 emission from natural and human sources

Carbon dioxide emission			
Natural source		**Human source**	
Ocean–atmosphere exchange	43%	Burning of fossil fuels	87%
Plant and animal respiration	28.5%	Clearing of forests and land use	9%
Soil decomposition and degradation	28.5%	Cement production (using $CaCO_3$)	4%

The industrial development of CO_2 began with the experiments of Faraday who was successful in liquefying CO_2 in a bent glass tube. Thilorier who repeated Faraday's experiments on a large scale also made solid carbon dioxide (dry ice). He made the observation that it undergoes sublimation.

Carbon dioxide is an essential substance in the life cycle of plants and animals and therefore is basic to all life on the earth. It is absorbed from the atmosphere by the leaves of plants and by utilising solar energy and using chlorophyll, it reacts with water to form glucose, the most basic form of carbohydrates.

$$6\,CO_2 + 6H_2O \xrightarrow[\text{chlorophyll}]{\text{sunlight}} C_6H_{12}O_6 + 6O_2 \quad \Delta H = 2803\ \text{kJ/mol} \tag{5.16}$$

Glucose is converted by the plant into sugars and starch. Animals convert these carbon compounds built up by plants into carbon dioxide and water using oxidative phosphorylation whereby ATP molecules (energy stores) are synthesised. This reaction is the reverse of photosynthesis. Thus, carbon is being continually exchanged between the bodies of plants and animals and atmospheric carbon dioxide is the medium of this exchange.

$$C_6H_{12}O_6 + 6O_2 \xrightarrow[\text{Oxidative phosphorylation}]{} 6\,CO_2 + 6H_2O \quad \Delta H = -2803\ \text{kJ/mol} \tag{5.17}$$

From 1850 to 1998, there has been an increase in the CO_2 concentration in the atmosphere from 280 ppm to 364 ppm. This is mainly due to human activities during and after the industrial revolution which started around 1850. By the burning of fossil fuels, by land clearing and forest/crop residue burning and by producing cement, humans have been increasing the amount of CO_2 in the air by about 2 ppm per year. Since CO_2 is a potent greenhouse gas, efforts are being introduced across the world to control its increase.

Radiocarbon dating

A very useful application of atmospheric carbon dioxide is the determination of the age of carbon-containing fossils. Naturally occurring carbon contains 98.89% ^{12}C, 1.11% ^{13}C, and traces of ^{14}C. The ^{14}C is generated in the lower stratosphere and upper troposphere by cosmic rays, which generate neutrons which in turn create ^{14}C when they strike nitrogen atoms. The ^{14}C is termed as radiocarbon as it undergoes radioactive decay to form nitrogen, releasing a β-particle. This process occurs with a half life of 5730±40 years. The method of radiocarbon dating developed by Willard Libby (who received the Nobel Prize for this work in 1960) became a standard tool for archaeologists to determine the age of fossils. It is based on the fact that radiocarbon combines with atmospheric oxygen to form radioactive carbon dioxide, which is incorporated into plants by photosynthesis; hebrivorous animals then acquire ^{14}C by eating the plants and carnivorus animals get the ^{14}C by eating the herbivorous animals. This is an equilibrium process, therefore, as long as the plant or animal is alive the relative amount of ^{14}C remains unchanged. When the animal or plant dies, the exchange of carbon with its environment stops, and from that time onwards the amount of ^{14}C it contains begins to decrease since it undergoes radioactive decay. Measurement of the amount of ^{14}C in a fossil sample provides information that can be used to calculate when the animal or plant died. By comparing this quantity with the amount of radioactive carbon present in a live organism, the age of the fossil sample can be determined with an accuracy of ±5%. The older a sample is, the less ^{14}C there is to be detected. As the half-life of ^{14}C is about 5,730 years, the oldest dates that can be reliably measured by this process is around 50,000 years.

Industrial production and purification of CO_2

Steam reforming of hydrocarbons gives a gaseous mixture of hydrogen, carbon dioxide and carbon monoxide. The carbon monoxide is further converted to CO_2 by the water gas shift reaction. While pressure swing absorption is a method by which CO_2 can be separated, most industries still use chemical methods of separating CO_2. Few methods are available based on the absorption of carbon dioxide into a solution, under pressure. The absorbed gas is then liberated in a separate vessel by raising the temperature and lowering the pressure. Under normal conditions, the solubility of carbon dioxide in water is too low for this to be used as the absorbent. The solubility of CO_2 is increased by adding a solute to the water so that chemical combination of carbon dioxide with the solution is made to happen. Alkali metal carbonates such as K_2CO_3 and ethanolamines have been efficiently used for this purpose.

The potassium carbonate method works by the following reaction. The carbon dioxide absorption increases with lowering of temperature, increase of partial pressure and increase in concentration of K_2CO_3 solution.

$$K_2CO_3 + CO_2 + H_2O \rightleftharpoons 2\ KHCO_3 \quad \text{`(5.18)}$$

Another method for CO_2 capture employs aqueous amine solutions. Currently, monoethanolamine (MEA) is the one most commonly used, at concentrations of 30 wt%, and the predominant product of the reaction of MEA with CO_2 is known to be the carbamate.

$$HOCH_2CH_2NH_2 + CO_2 + H_2O \rightleftharpoons [(HOCH_2CH_2NH_3)_2]\ CO_3 \quad (5.19)$$

Liquefaction and solidification of CO_2

The critical parametres of carbon dioxide are $T_{crit} = 31°C$, $p_{crit} = 7.4$ MPa. This shows that CO_2 can be liquefied at any temperature between 31°C and its triple point (–56.6°C) by compression to the required liquefaction pressure at that particular temperature and removal of the heat of condensation. It is, therefore, possible to liquefy CO_2 near the critical temperature by using a pressure of 7.4 MPa and water as the coolant, but because water supply at temperatures below 20°C is not easy to achieve, this method is less favoured and instead a refrigerant is used.

Supercritical CO_2 in the liquid form has been found to be an extremely useful solvent for many lipophilic organic compounds. Carbon dioxide has attracted attention in the pharmaceutical and other chemical processing industries as a non-toxic, green alternative to more traditional solvents such as organochlorides. The applications include decaffenation of coffee beans and powder, and as a reaction medium for selected reactions. It is also used in the preparation of some aerogels.

When liquid carbon dioxide is allowed to expand to atmospheric pressure, dry ice is formed. The loose snow-like powder is shaped by a hydraulic press to solid blocks of density 1.5–1.6 kg/L.

Uses of carbon dioxide

- Synthesis of urea and methanol.
- For enhancing oil recovery from oil wells.
- As a nontoxic and no-residue refrigerant which can be in contact with food stuff.
- As an inert atmosphere for food to prevent bacterial growth.
- As a fire extinguisher for electrical and oil fires.
- For growth promotion of plants in green houses.
- Heat transfer medium in some nuclear reactors.
- As a supercritical fluid for extraction of lipophilic compounds.

Reactions of CO_2

Carbon dioxide as the end product of combustion of hydrocarbons is a very stable molecule with a standard Gibbs free energy of formation of -394.4 kJ/mol. Therefore, most of the conversions of carbon dioxide are endothermic reactions and require substantial amounts of energy input or co-reactants which have high Gibbs free energy (for example, methane or hydrogen). Also, for many reactions, a high kinetic barrier has to be overcome and to achieve this, efficient catalysts which can promote the reaction are required. Consequently, carbon dioxide can be utilised to produce value added products, which is of interest to the chemical industry.

The largest use of carbon dioxide is urea synthesis using CO_2 and NH_3 at high pressure (180–200 atm) and high temperature (185–190°C) conditions. The 2011 annual production of urea amounted to 1.54×10^8 tonnes which used 1.13×10^8 tonnes of carbon dioxide. The urea production involves two separate reactions. First, ammonia reacts with carbon dioxide under pressure forming ammonium carbamate. Decomposition of ammonium carbamate then gives urea and water.

$$CO_2 + 2NH_3 \xrightarrow[180\text{–}200\ atm]{185\text{–}190°C} H_2NC(O)ONH_4 \quad \Delta H = -37.4\ Kcal \tag{5.20}$$

$$H_2NC(O)ONH_4 \longrightarrow H_2NC(O)NH_2 + H_2O \quad \Delta H = +6.3\ Kcal \tag{5.21}$$

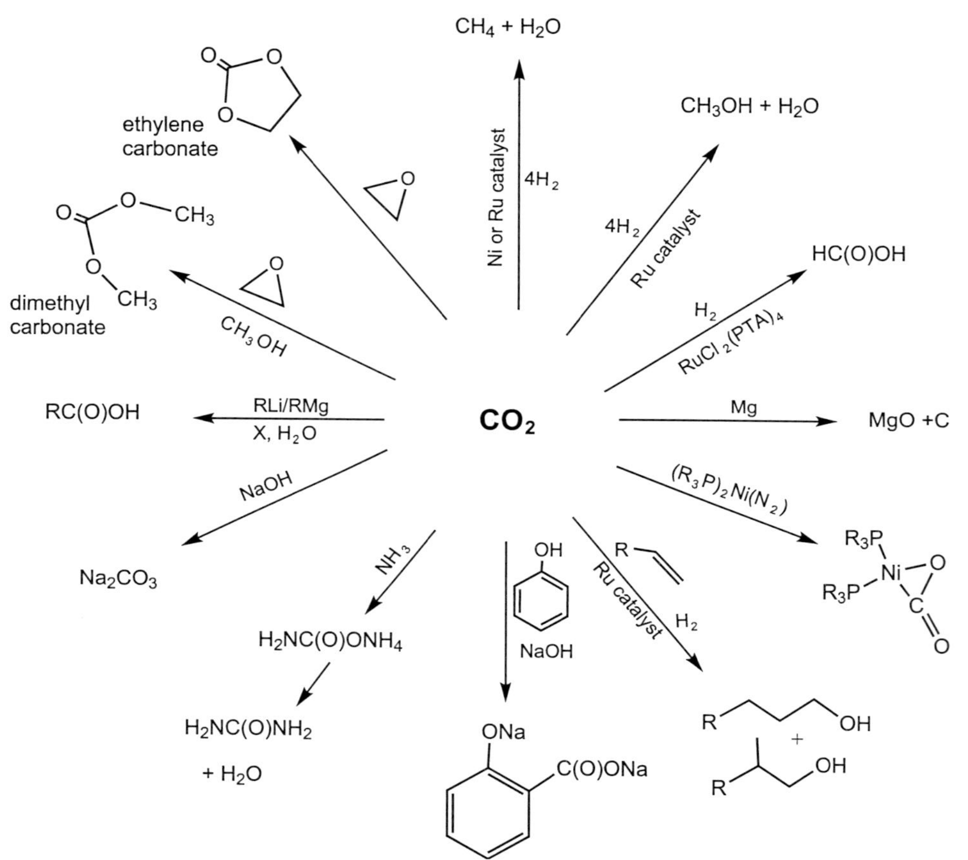

Scheme 5.11 Common reactions of carbon dioxide

NH_3 and CO_2 are separately compressed and fed into a high pressure (180 atm) autoclave. A mixture of urea, ammonium carbamate and water is formed. The liquid mixture is brought down to 27 atm pressure and fed to a special flash-evaporator containing a gas–liquid separator and condenser. Unreacted NH_3, CO_2 and water are removed and these are recycled. The aqueous solution of carbamate–urea is passed to a flash drum at atmospheric pressure where further decomposition of carbamate gives urea. The aqueous urea solution is often used as it is, or transferred to a vacuum evaporator to obtain urea containing less water.

Industrial production of methanol also involves carbon dioxide but only to a lesser degree. Generally, methanol is produced from synthesis gas or by partial oxidation of heavier feedstocks such as coal. Synthesis gas comprises carbon dioxide, hydrogen and carbon monoxide, and is converted to methanol. It may be noted that there are other methods of synthesising methanol and many examples of homogenous catalysts for the same have also evolved.

Other industrially important reactions involving CO_2 are synthesis of formic acid and methane. In addition, there are many other reactions which are important in the laboratory scale. The reaction of CO_2 with NaOH is the method used for estimating the amount of carbon in an organic compound after the carbon is oxidised to CO_2. Other well known reactions of carbon dioxide are shown in Scheme 5.11.[35]

As CO_2 is a green house gas and is produced on a large scale by human activities, it has a significant impact on the climate of the world. During the period 1956–2005, the estimate is that the increase in the global average temperature is ~0.65 K. For minimising its effect on climate, the maximum temperature rise has to be limited to 2 K with reference to preindustrial levels. Therefore, the maximum allowable atmospheric CO_2 concentration should not be allowed to exceed 450 ppm, and the CO_2 production has to be reduced significantly.

5.9.3 Persistent Carbenes

Carbenes are divalent carbon compounds which are generally highly reactive organic intermediates possessing six valence electrons with the general formula R_2C or R_1R_2C. Carbenes traditionally were so reactive that they were only considered as reaction intermediates or transition states and were only indirectly studied, often by trapping them in the presence of suitable reagents. The first attempt to prepare carbenes dates back to the times of the French chemists Henri Victor Regnault (1810–1878) and Jean Baptiste André Dumas (1800–1884), when they tried to prepare methylene by dehydrating methanol. Dichlorocarbene as a reactive intermediate was proposed as early as 1862 when Anthon Geuther viewed chloroform as $Cl_2C.HCl$. Another proponent of carbenes or reactive methylenes was J U Nef who proposed a *general methylene theory*, which proposed that many substitution reactions proceed by the sequence of α-elimination and addition. However, only by the 1950s did the modern era of carbene chemistry begin. The initial emphasis was also in making stable metal complexes of carbenes such as Fischer and Schrock carbenes.[36]

Carbenes are generally classified as belonging to two types—triplet and singlet. In singlet carbenes the two electrons are paired and the molecule is diamagnetic, while for triplet carbenes the two electrons are unpaired and the molecule is paramagnetic. Carbenes can be either linear or bent. Linear geometry indicates *sp* hybridisation with two nonbonding and degenerate p_x and p_y orbitals while bent geometry indicates sp^2 hybridisation where the p_y is unchanged and is called $p\pi$ and the p_x gains stability and some *s* character and is termed σ. Most carbenes are bent and their frontier orbitals are σ and $p\pi$. The four possible electronic states for carbenes are given in Fig. 5.10.[37]

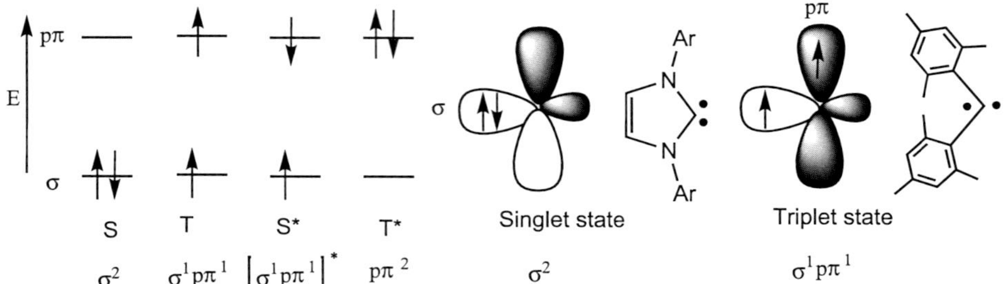

Fig. 5.10 Comparison of singlet and triplet states of carbenes

Persistent carbenes are those carbenes with some amount of stability, that can be isolated in the free form and studied using spectral methods, and can be used as reactants or ligands. Such carbenes are relatively recent. The tendency of some types of carbenes to undergo dimerisation has been overcome by introducing sterically hindered groups on the carbene carbon. Between singlet and triplet carbenes, singlet carbenes are relatively more stable and some of them have been isolated and stored under inert atmosphere for indefinite periods of time.

Persistent triplet carbenes

Diazo compounds that are prone to undergoing degradation under mild conditions are usually used as precursors for preparing triplet carbenes by photolysis with UV radiation or by using a XeCl excimer laser. The diradical nature of these compounds makes them extremely reactive and quite difficult to isolate. They can be trapped with oxygen to convert them to ketones. Different ways by which triplet carbenes decompose or degrade include dimerisation, C–H insertion, cyclopropanation, H abstraction from an alkene, and so on. Triplet carbenes are generated at low temperature and are characterised by spectral techniques which identify radicals such as ESR spectroscopy and UV-vis spectroscopy. They are generally very short-lived with lifetimes of a few seconds especially in solutions, but some of them are stable for extended periods in the solid phase. One example of a relatively long lived triplet carbene is shown in Eq. 5.22, which is stable for more than a week in a dilute benzene solution at room temperature.[38] Unlike singlet carbenes, these species do not react with carbon–halogen bonds and their stability was found to increase when they have bromo or trifluoromethyl groups in the vicinity of the carbene carbon.

N_2 ; hν, $-N_2$ (5.22)

Persistent singlet carbenes

N-Heterocyclic carbenes

The best examples of persistent singlet carbenes are N-heterocyclic carbenes (NHCs) based on imidazole rings (both saturated and unsaturated). In 1962, Wanzlick first investigated the reactivity and stability of N-heterocyclic carbenes. Shortly afterwards, he reported the first application of NHCs as ligands for metal complexes. However, isolation of these carbenes remained elusive. The breakthrough was made by Anthony J Arduengo, who in 1991 prepared an isolable and bottleable NHC. 1,3-Diadamantyl imidazol-2-ylidene was the first in a series of such stable N-heterocyclic carbenes. This crystalline compound was obtained in near quantitative yield by the deprotonation of 1,3-di-1-adamantylimidazolium chloride with sodium or potassium hydride in the presence of catalytic amounts of a base such as potassium tert-butoxide or the dimethyl sulfoxide anion.

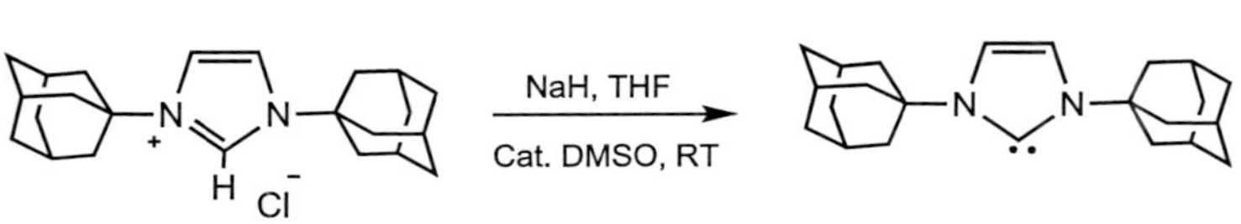

NaH, THF
Cat. DMSO, RT (5.23)

NHCs are in general defined as heterocycles containing a carbene carbon with at least one nitrogen atom as part of the ring. They generally have sterically hindered substituents on the atom adjacent to the carbene carbon which helps to provide kinetic stability. In the traditional NHCs, such as the one prepared by Arduengo, the HOMO has been described as a formally sp^2 hybridised lone pair and the LUMO has been described as an unoccupied p orbital on the carbene carbon. The nitrogen atoms of this heterocycle are σ-electron withdrawing and π-electron donating thus providing inductive and mesomeric stabilisation for the carbene. The cyclic structure also stabilises and favours a singlet ground state by forcing the carbene carbon into a bent sp^2-like arrangement. The stability of singlet carbenes results mostly from electronic effects [mesomeric (+M) as well as inductive (–I) effects], while steric bulkiness of the substituents also contributes significantly.[39] NHCs have been described as better ligands than phosphines (Table 5.12). Some well known examples of NHCs are given in Fig. 5.11. In their abbreviated names, S stands for saturated and I stands for imidazolyl.

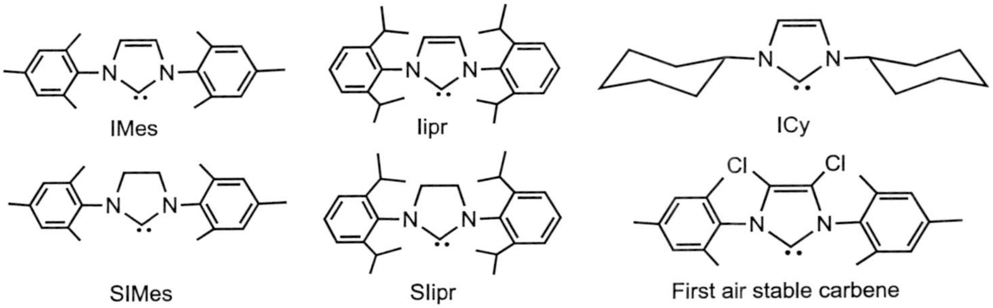

IMes Iipr ICy

SIMes SIipr First air stable carbene

Fig. 5.11 Some well known examples of N-heterocyclic carbenes

Table 5.12 Comparison of properties of N-heterocyclic carbenes with phosphines

Property	N-Heterocyclic carbenes	Phosphines
Stability towards water	Hydrolyses rapidly	Not sensitive to degradation in deoxygenated water, especially aryl phosphines
Stability towards oxygen	Does not react directly with O_2; requires special reagents for oxidation	Highly reactive towards oxidation especially with O_2
σ Donation of electrons	Better σ-donors than even alkyl phosphines	Some alkyl phosphines are good σ-donors
Range of electron donor ability	Narrow range	Wide range
Steric bulkiness based on $\%V_{bur}$	Generally sterically hindered	Steric bulkiness depends on substituents on phosphorus
Shape as ligand in the metal-bound form	Fan like	Cone shaped
Metal–ligand bond strength	Stronger bonds, not easily displaced	Not as strong as NHCs; can be displaced during reactions
Ligand stability in oxidation reactions	Stable	Unstable—phosphine ligands are prone to oxidation

The two most well known types of diamino carbenes are—those with unsaturation in the five-membered ring (imidazol-ylidenes) and those with a saturated five-membered ring (imidazolidin-ylidenes).

imidazolidinylidene imidazolylidene

While NHCs with saturated rings (imidazolidinylidenes) are easily dimerised in the absence of stercially hindered substituents, those with unsaturation in the ring (imidazolylidenes) do not undergo dimerisation even in the absence of sterically hindered groups (Scheme 5.12).

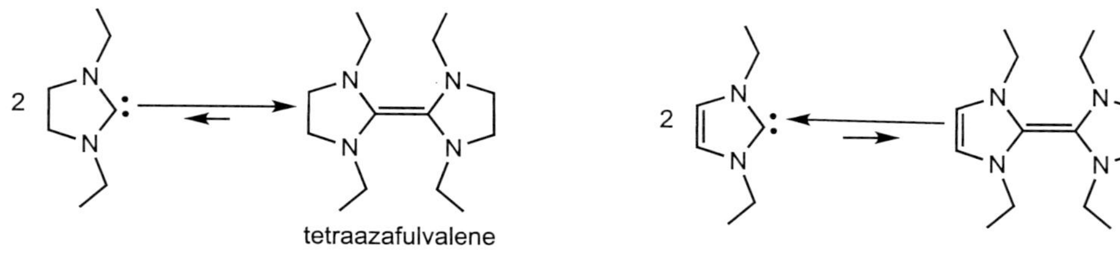

tetraazafulvalene

Scheme 5.12 Dimerisation of N-heterocyclic carbenes

NHCs are excellent spectator ligands in transition metal and organometallic chemistry and they are even better σ-donors than traditional phosphines and sterically more bulky. Percent buried volume (%V_{bur}; Chapter 9) has been effectively used for classifying carbenes based on their steric bulk. The efficiency of many existing catalysts was improved by introducing NHCs as ligands. For example, Grubbs 2nd generation catalyst for olefin metathesis and PEPPSI catalyst for cross coupling reactions are better than other related catalysts due to the fact that the better σ-donor ability of the NHCs over phosphines and amines helps in the labilisation of ligands *trans* to them (*trans effect*). In addition, the steric bulk of the NHCs provide kinetic stability to the metal centre making it easy to handle the complex under air and moisture. Many examples of Ag(I) and Au(I) NHC complexes have also shown promise as antibacterial and anticancer agents.

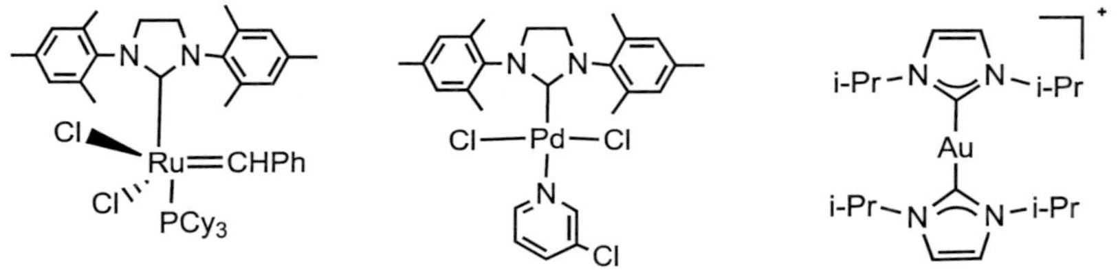

Grubbs 2nd gen. catalyst

PEPPSI catalyst

Fig. 5.12 Well known NHC-based catalysts and anticancer agent

NHCs are found to be extremely useful in stabilising unusually low oxidation states of many main group elements such as phosphorus, silicon and boron. They are also used for stabilising hitherto unknown multiple bonds between *p*-block elements. They are also found to be useful as organocatalysts.[39]

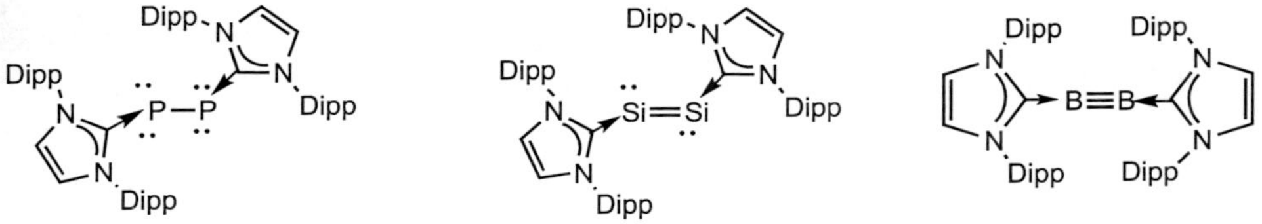

Dipp = 2,6 diisopropylphenyl

Fig. 5.13 Stabilising low oxidation states and multiple bonds of *p*-block elements with NHCs

5.10 Recent Developments

New types of singlet carbenes

After the development of saturated and unsaturated NHCs possessing two ring nitrogens, chemists expanded the scope of NHCs by introducing more nitrogens, other main group elements and carbon as part of the five-membered heterocycles as well as by bringing in acyclic carbenes. While many of them had comparable properties to that of the NHCs, a few of them were found to be better than the NHCs with regard to their donor capabilities, nucleophilicity as well as electrophilicity. Some of these new carbenes are given in Fig. 5.14.

Cyclic alkyl amino carbene cAAC or CAAC | Acyclic alkyl amino carbene aAAC | Cyclic diamido carbene cDAC | Abnormal N-heterocyclic carbene aNHC | Bicyclic alkyl amino carbene BICAAC

Fig. 5.14 Different types of novel carbenes

Among the new type of carbenes, cyclic alkyl amino carbenes (cAACs) introduced first by Bertrand in 2005 have shown better σ donor properties than NHCs. cAACs are carbenes where one of the σ-withdrawing and π-donating nitrogen atoms of the NHC has been replaced by a σ-donating quaternary carbon atom with a σ-donating alkyl group.[40] By this replacement, the carbene centre has been made relatively more nucleophilic (σ-donating) and electrophilic (π-accepting) compared to the NHCs (Fig. 5.15).[41] The presence of a quaternary carbon in the position α to the carbene carbon in the heterocycle provides steric environments that differentiate the cAACs from other carbene ligands.

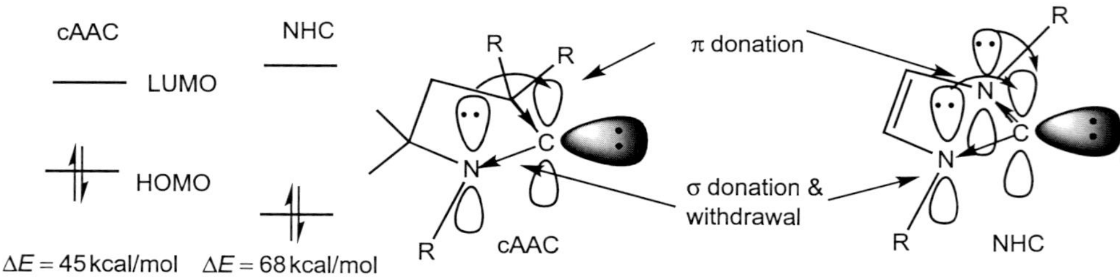

Fig. 5.15 Comparison of electron donation and electron withdrawal of cAACs and NHCs

A comparison of the ν_{CO} stretching frequency of the rhodium carbonyl complexes of (NHC)Rh(CO)$_2$Cl (2039 cm^{-1}) and (cAAC)Rh(CO)$_2$Cl (2036 cm^{-1}) indicate the relatively better σ-donating capability of the cAACs. Computational studies have also indicated that the HOMOs of the cAACs are slightly higher in energy and the singlet–triplet gap is smaller than that of the NHCs. Similar properties have been reported for acyclic alkyl amino carbenes (aAACs) as well. Abnormal NHCs (aNHC) where the carbene carbon is no longer located between two nitrogens have also been predicted to be more basic than normal NHCs based on computational studies.[42] Bicyclic alkyl amino carbenes (BICAAC), first prepared in 2017, are found to be more nucleophilic and electrophilic than cAACs.[43]

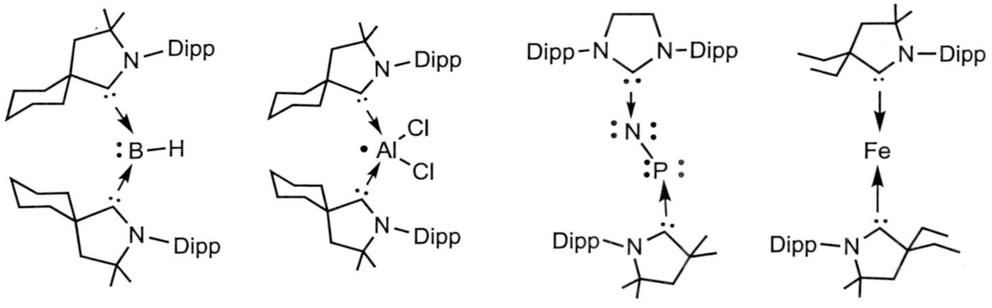

Fig. 5.16 Stabilisation of low oxidation states and radicals by cAACs

cAACs have been found to be useful for stabilising low valent main group and transition metal species such as boron(I), antimony(0), iron(0), and so on. They have also been found to be efficient for the stabilisation of paramagnetic complexes, in which the metal is often in a formal zero oxidation state (Fig. 5.16). In addition, their utility in modifying the Grubbs catalyst for olefin metathesis reactions has also been evaluated. Especially for cross metathesis reactions of methyl oleate and ethylene gas, very good catalytic activity was observed with cAAC substituted form of Grubbs–Hoveyda catalyst with a turnover number (TON) of 340,000. The catalyst turnover numbers for the same reaction using commercially available Grubbs olefin metathesis catalysts were in the range of 2000–5000 only.[39]

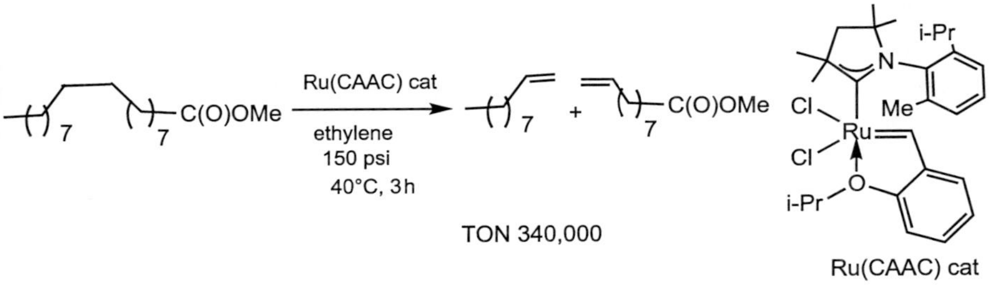

Ru(CAAC) cat; ethylene 150 psi 40°C, 3h; TON 340,000; Ru(CAAC) cat (5.24)

Another unique and unprecedented property shown by both cAACs and aAACs is their ability to mimic transition metal centres in activating small molecules such as hydrogen, carbon monoxide and ammonia by themselves at ambient conditions.[44] In contrast, NHCs or acyclic NHCs do not react with H_2, CO or NH_3. The mode of activation of H_2 by cAACs and transition metals are entirely different. For H_2 activation, computational studies have indicated that the initial step is the electron donation from the carbene centre to the antibonding σ^* orbital of H_2 leading to the elongation (1.071 Å versus 0.77 Å for H_2) and polarisation of the H–H bond. Although cyclic diamido carbenes (cDACs) do not activate H_2, they were found to activate NH_3 and CO.[45]

i-Pr; N; Dipp — H_2 (1 atm, 35°C) or NH_3 (−78°C to RT) → i-Pr; H; H (or NH_2); N; Dipp (5.25)

N — H_2 (1 atm, 35°C) or NH_3 (−78°C to RT) → H; H (or NH_2); N (5.26)

O; Mes; N; N; O; Mes — NH_3(−78°C to RT), Mes = Mesityl → O; Mes; N; H; NH_2; N; O; Mes (5.27)

Questions for self study (multiple choice)

5.1. Which of the following statements is not true for graphite?
(a) The carbon atoms are sp^2 hybridised
(b) It has a layer structure with a distance between the layers of 0.142 nm.
(c) It conducts electricity in the direction of the plane.
(d) Binding between the layers is by weak van der Waals forces.

5.2. Which of the following statements is not true for KC_8?
(a) It is made by melting potassium over powdered graphite.
(b) It is a very strong reducing agent especially for dehalogenation reactions.
(c) Unlike graphite it does not conduct electricity.
(d) The potassium increases the interlayer distances of graphite layers.

5.3. Among the given the properties, which does not belong to diamond?
(a) Highest hardness for any compound or element.
(b) sp^3 hybridised carbons.
(c) Highest thermal conductivity for any naturally available bulk material.
(d) Very high electrical conductivity.

5.4. The major types of metal complexes formed by C_{60} are of the type having:
(a) η^2 and η^1 hapticities
(b) η^6 and η^1 hapticities
(c) η^2 and η^5 hapticities
(d) η^2 and η^6 hapticities

5.5. Among carbon nanotubes, the ones which are electrically conducting in nature are:
(a) Arm chair and zigzag
(b) Zigzag
(c) Arm chair
(d) Zigzag and chiral

5.6. Which of the following statements is not true about single walled carbon nanotubes (SWNT)?
(a) They have electrical conductivity close to that of copper.
(b) Their thermal conductivity is about 10 times more than at of copper.
(c) They show superconductivity at liq. N_2 temperatures.
(d) The use of acetylene and ethylene with metal catalysts in chemical vapour deposition favours their formation.

5.7. The property of carbon fibre reinforced polymer (CFRP) which has fetched excellent applications is:
(a) High electrical conductivity
(b) High thermal conductivity
(c) Specific strength (strength to weight ratio)
(d) Very high melting point

5.8. The statement which is not true for carbon molecular sieves is:
(a) It is most suitable for the separation of CO_2 from O_2.
(b) It is most suitable for the separation of N_2 from O_2.
(c) It has pore sizes around 10 Å diameter.
(d) It is made from oxygen-free pyrolysis of some organic polymers.

5.9. Hopcalite, the material used for oxidation of CO to CO_2 at room temperature in gas masks, contains:
(a) MnO_2
(b) CuO
(c) Co_2O_3
(d) All of the above

5.10. The chemical reaction used for quantitative estimation of carbon in an organic compound during elemental analysis is:
(a) Reaction of CO_2 with monoethanolamine.
(b) Reaction of CO_2 with NaOH.
(c) Reaction of CO_2 with K_2CO_3.
(d) Reaction of CO_2 with Mg.

Solved concept-based questions

5.11. Among fullerenes why is C_{60} the most common and stable?

In the structure of C_{60}, the twelve pentagons are isolated from each other. Only in C_{60} the double bonds are arranged in such a way that they are located only in six-membered rings and none in five-membered rings. As a result there is less strain on the already strained five-membered rings.

5.12. Why are fullerenes not naturally occurring on earth?

Fullerenes have been found to decompose slowly in the presence of UV radiation and by reaction with oxygen and ozone. $C_{60}O$ and $C_{70}O$ have been prepared and characterised too. Therefore fullerenes undergo oxidative degradation on earth.

5.13. Why have Boeing and Airbus replaced aluminium partially with CFRP in aircraft manufacture?

For an aeroplane, less weight without compromising on strength means more passenger capability and less wastage of fuel. Among metals, aluminium alloys are the best and has been used for a long time in aeroplane manufacture. However, the strength-to-weight ratio of CFRP is much better than that of aluminium. So in the modern commercial planes made by these major players, more of CFRP is being used instead of aluminium.

5.14. Carbon monoxide has the strongest bond known. Still it is highly reactive and is a better ligand in organometallic chemistry unlike N_2. Why?

The highest occupied molecular orbital of CO is weakly antibonding (compared with the O atomic orbitals) and is an MO which is carbon-based. Secondly, the π^* antibonding orbital which is the lowest unoccupied molecular orbital is also of comparatively lower energy which makes it possible to interact with metal orbitals for π bonding. The fact that it is a heteronuclear compound also brings in more polarity unlike N_2.

5.15. Why are cyclic alkylamino carbenes better σ-donors than N-heterocyclic carbenes?

In CAACs, one of the ring nitrogen atoms has been replaced by a quaternary carbon atom with two σ-donating alkyl groups. This reduces the electron withdrawing effect which was due to the nitrogen atom and increases electron density on the carbene carbon thereby making the carbene a better σ-donor.

Problems and exercises

5.16. How is C_{60} differentiated spectroscopically from C_{70}?

5.17. Why was osmylation a success for the derivatisation of C_{60}?

5.18. What is Bingel reaction and why is it important?

5.19. How is bucky ferrocene prepared from C_{60}?

5.20. What is the important industrial use of carbon nanotubes?

5.21. What is hopcalite and how does it help in removing carbon monoxide?

5.22. Suggest a method to remove carbon monoxide from air.

5.23. How does one test if a given gas is carbon monoxide or not?

5.24. Suggest two reagents which can reversibly bind carbon dioxide.

5.25. Why is Grubbs 2nd generation catalyst better in activity than Grubbs 1st generation catalyst?

5.26. Give two reactions which are shown by cAACs but not by NHCs.

Answer key for self study questions

5.1. (b) 5.2. (c) 5.3. (d) 5.4. (c) 5.5. (c) 5.6. (c) 5.7. (c) 5.8. (a) 5.9. (d) 5.10. (b)

Supplementary reading

1. Salpeter E E, Nuclear reactions in stars without hydrogen, *The Astrophysical Journal*, 1952, Vol. 115, 326.
2. Wilson R, *The stars, their birth, life and death; astronomy through the ages; the story of the human attempt to understand the universe*, 1997, Basingstoke, Taylor and Francis.
3. Cadek M, Vostrovsky O and Hirsch A, *Carbon 7, Fullerenes and carbon nanomaterials* in *Ullmann's encyclopedia of industrial chemistry*, 2012, Wiley-VCH, Weinheim.
4. Stocks A R, A new structure of carbon, *Nature*, 1942, Vol. 149, 328.
5. Chakraborty S, Chattopadhyay J, Guo W and Billups W E, Functionalization of potassium graphite, *Angew Chem., Int. Ed.*, 2007, Vol. 46, 4486.
6. Lipson H, Ottmers D M and Rase H F, Potassium graphites prepared by mixed-reaction technique, *Carbon*, 1966, Vol. 4, 125.
7. Harlow G E, *The nature of diamonds*, Cambridge University Press, 1998.
8. Kroto H W, Heath J R, O'Brien S C, Curl R F and Smalley R E, C_{60}: Buckminsterfullerene, *Nature*, 1985, Vol. 318, 162.
9. Fagan P J, Calabrase J C and Malone B, The chemical nature of buckminsterfullerene and the characterisation of a platinum derivative, *Science*, 1991, Vol. 252, 1160.
10. Hawkins J M, Meyer A and Lewis T A, Crystal structure of osmylated C_{60}. Confirmation of soccerball framework, *Science*, 1991, Vol. 252, 312.
11. Balch A L and Olmstead M M, Reactions of transition metal complexes with fullerenes ($C_{60,}$ C_{70} etc.) and related materials, *Chem. Rev.*, 1998, Vol. 98, 2123.
12. Soto D and Salcedo R, Coordination modes and different hapticites for fullerene organometallic complexes, *Molecules*, 2012, Vol. 17, 7151.
13. Khakina E A, Yurkova A A, Peregudov A S, Troyanov S I, Trush V V, Vovk A I, Mumyatov A V, Martynenko V M, Balzarini J and Troshin P A, Highly selective reactions of $C_{60}Cl_6$ with thiols for the synthesis of functionalised [60] fullerene derivatives, *Chem Comm.*, 2012, Vol. 48, 7158.
14. Sawamura M, Kuninobu Y, Toganoh M, Matsuo Y, Yamanaka M and Nakamura E, Hybrid of ferrocene and fullerene, *J. Am. Chem. Soc.*, 2002, Vol. 124, 9354.
15. Nakamura E and Sawamura M, Chemistry of η^5-fullerene metal complexes, *Pure Appl. Chem.*, 2001, Vol. 73, 355.
16. Nakamura E, Bucky ferrocene and ruthenocene: serendipity and discoveries, *J. Organomet. Chem.*, 2004, Vol. 689, 4630.
17. Marczak R, Wielopolski M, Gayathri S,S, Guldi D,M, Matsuo Y, Matsuo K, Tahara K and Nakamura E, Uniquely shaped double decker buckyferrocenes-distinct donor-acceptor interactions, *J. Am. Chem. Soc.*, 2008, 130, 16207.
18. Matsuo Y, Tahara K and Nakamura E, Synthesis and electrochemistry of double-decker buckyferrocenes, *J. Am. Chem. Soc.*, 2006, 128, 7154.
19. Matsuo Y and Nakamura E, Selective multiaddition of organocopper reagents to fullerenes, *Chem Rev.*, 2008, Vol. 108, 3016.
20. Komatsu K, Murata M and Murata Y, Encapsulation of molecular hydrogen in fullerene C_{60} by organic synthesis, *Science*, 2005, Vol. 307, 238.
21. Hudson J L, Casavant M J and Tour J M, Water soluble, exfoliated non roping single wall carbon nanotubes, *J. Am. Chem. Soc.*, 2004, Vol. 126, 11158.

22. Ajayan P M, Nanotubes from carbon, *Chem. Rev*, 1999, Vol. 99, 1787.
23. Iijima S, Helical microtubules of graphitic carbon, *Nature*, 1991, Vol. 354, 55.
24. Ebbesen T W, *Carbon nanotubes—preparation and properties*, CRC Press, Boca Raton, Florida, 1997.
25. Karousis N, Tagmatrachis N and Dimitros T, Current progress on the chemical modification of carbon nanotubes, *Chem. Rev.*, 2010, Vol. 110, 5366.
26. Sun Y-P, Fu K, Lin Y and Huang W, Functionalised carbon nanotubes: Properties and applications, *Acc. Chem. Res.*, 2002, Vol. 35, 1096.
27. Novoselov S, Geim A K, Morozov S V, Jiang D, Zhang Y, Dubonos S V, Grigorieva I V and Firsov A A, Electric field effect in atomically thin carbon films, *Science*, 2004, Vol. 306, 666.
28. Geim A, Graphene: Status and Prospects, *Science*, 2009, Vol. 324, 1530.
29. Geim A K and MacDonald A H, Graphene: Exploring carbon flatland, *Physics Today*, 2007, Vol. 60, 35.
30. Liu Y, Zwingmann B and Schlaich M, Carbon fibre reinforced polymer for cable structures—A review, *Polymers*, 2015, Vol. 5, 2078.
31. Hiremath N, Mays J and Bhat G, Recent developments in carbon fibres and carbon nanotube based fibres, A review, *Polymer Rev.*, 2017, Vol. 57, 339.
32. Liu Y and Kumar S, Polymer/carbon nanotube nano composite fibres—A review, *ACS Appl. Mater. Interfaces*, 2014, Vol. 6, 6069.
33. Bierhals J, *Carbon monoxide* in *Ullmann's encyclopedia of industrial chemistry*, 2012, Wiley-VCH, Weinheim.
34. Topham S, Bazzanella A, Schiebahn S, Luhr S, Zhao L, Otto A and Stolten D, *Carbon dioxide* in *Ullmann's encyclopedia of industrial chemistry*, 2012, Wiley-VCH, Weinheim.
35. Liu Q, Wu Li, Jackstell R and Beller M, Using carbon dioxide as a building block in organic synthesis, *Nature Commun.*, 2015, Vol. 6, 5933.
36. Bourissou D, Guerret O, Gabbai F P and Bertrand G, Stable carbenes, *Chem. Rev.*, 2000, Vol. 100, 39.
37. Fremont d P, Marion N and Nolan S P, Carbenes: Synthesis, properties and organometallic chemistry, *Coord. Chem. Rev.*, 2009, Vol. 253, 862.
38. Iwamoto E, Hirai K and Tomioka H, A triplet carbene surviving a week in solution at room temperature, *J. Am. Chem. Soc.*, 2003, Vol. 123, 14664.
39. Hopkinson M N, Richter C, Schedler M and Glorius F, An overview of N-heterocyclic carbenes, *Nature*, 2014, Vol. 510, 485.
40. Melami M, Jazzar R, Soleilhavoup M and Bertrand G, Cyclic (alkyl)(amino)carbenes (CAACs): Recent developments, *Angew. Chem. Int. Ed.*, 2017, Vol. 56, 10046.
41. Roy S, Mondal K C and Roesky H W, Cyclic alkyl(amino) carbene stabilised complexes with low coordinate metals of enduring nature, *Acc. Chem. Res.*, 2016, Vol. 49, 357.
42. Aldeco-Perez E, Rosenthal A J, Donnadieu B, Parameswaran P, Frenking G and Bertrand G, Isolation of an C5-deprotonated imidazolium, a crystalline abnormal N-heterocyclic carbene, *Science*, 2009, Vol. 326, 556.
43. Tomas-Mendivil E, Hansmann M M, Weinstien C M, Jazzar R, Melaimi M and Bertrand G, Bicyclic (alkyl) (amino) carbenes (BICAACs): Stable carbenes more ambiphilic than CAACs, *J. Am. Chem. Soc.*, 2017, Vol. 139, 7753.
44. Martin D, Soleilhavoup M and Bertrand G, Stable singlet carbenes as mimics for transition metal centres, *Chem. Sci.*, 2011, Vol. 2, 389.
45. Frey G D, Lavallo V, Donnadieu B, Schoeller W W and Bertrand G, Facile splitting of hydrogen and ammonia by nucleophilic activation at a single carbon centre, *Science*, 2007, Vol. 316, 439.

The Chemistry of Silicon

CHAPTER

6

Unique Properties of Silicon and its Compounds

- *Silicon mostly forms tetravalent tetracoordinate compounds. However, divalent, monovalent and zerovalent silicon compounds have also been prepared by kinetic stabilisation using sterically hindered substituents.*
- *Silicon is a semiconductor (bandgap 1.14 eV at 273 K) and is used in solar cells and electronics.*
- *Organosilicon compounds are mostly nontoxic and are therefore used in a wide range of applications such as cosmetics, pharmaceuticals and prosthetics.*
- *Ferrosilicon, an impure form of silicon containing iron, is used by the steel industry for removing dissolved oxygen from steel melt.*
- *Quartz (silicon dioxide) shows piezoelectric properties.*
- *The Si–O and Si–F bond dissociation energies are much higher compared to that of C–O and C–F bonds and the formation of these two bonds is the driving force for many reactions of silicon compounds.*
- *Silicon–silicon multiple bonds are rare and to stabilise such bonds, sterically bulky groups are required.*
- *Silicones, polymers with a Si–O–Si backbone, are the most sought after inorganic polymers having a wide range of properties and unique applications.*
- *Polydimethylsiloxane (PDMS) has the lowest glass transition temperature (temperature below which a polymer loses its flexibility) for any polymer discovered so far (–135°C). It is hydrophobic, has oxygen and water vapour permeability, is nontoxic, has low surface tension and is stable to UV radiation.*
- *Silicates, aluminosilicates and related compounds are environmentally friendly, mostly naturally occurring compounds with a wide range of applications (talc, clays for pottery, solid acid catalysts and building materials).*

- *Aluminosilicates, wherein a tetrahedral silicon is replaced by a tetrahedral aluminum, have replaceable counter ions, often sodium which make zeolites excellent water softeners.*
- *Zeolites which are open framework aluminosilicates having pore and cavities of specific dimensions can act as acid catalysts, especially as shape selective catalysts. Many zeolites can act as dehydrating agents due to the presence of multiple replaceable water molecules in their channels and cavities.*
- *Silica gel is used as a drying agent and as a stationary phase for chromatographic separations and as a support for catalysts.*
- *Trialkylsilyl groups are well known protecting groups for alcohols and phenols in organic chemistry and as terminal groups for silicone polymers.*
- *Silicon carbide with a hardness close to 9 on the mho scale is a semiconductor with an indirect band gap and is used in light emitting diodes.*
- *Unlike carbon, silicon does not form double bonds readily with oxygen. A few silanones, with tricoordinate Si=O bonds prepared recently, have been found to be highly sensitive to moisture.*

6.1 About the Element

Silicon is a grey coloured, air and moisture-stable brittle solid having high melting and boiling points (Table 6.1). It is a semiconductor and has a negative temperature coefficient of resistance as the number of charge carriers increases with temperature. Naturally occurring silicon is composed of three stable isotopes, silicon-28, silicon-29 and silicon-30, with silicon-28 being the most abundant (92% natural abundance).[1]

Table 6.1 Physical properties of silicon

Silicon	*Atomic mass*	*Electronegativity (Pauling scale)*	*1st ionisation energy*	*Electronic configuration*	*van der Waals radius*
Symbol: Si	28.0855 g.mol^{-1}	1.8	786.3 kJ.mol^{-1}	[Ne]3s^{2}3p^2	0.132 nm
At. no. 14	*Atomic radius*	*Melting point*	*Boiling point*	*Electron affinity*	*Density*
	2.02 Å	1414°C	3265°C	134.068 kJ.mol^{-1}	2.33 g.cm^{-3} at 20°C

6.2 The Important Ores of Silicon

Silicon is the second most abundant element in the earth's crust after oxygen and is the eighth most common element in the universe by mass. In nature, silicon rarely occurs as the pure free element. Over 90% of the earth's crust is composed of silicate minerals. Silicon dioxide, commonly called silica is a major component of common sand. Silica occurs in three different crystalline forms—quartz, tridymite and cristobalite and is the source for realising pure silicon.[1]

6.3 Isolation and Purification of the Element

The first step in making pure silicon is reduction of silica to an impure form of silicon known as ferrosilicon. The isolation of ferrosilicon and later amorphous silicon for the first time has been credited to Berzelius who wrote that '*silica is reduced by carbon to form a substance that is capable of combining with iron and by its nature should be a metal, because it does not abolish the iron forgeability*' indicating his discovery of ferrosilicon.[2]

Ferrosilicon is an alloy that contains elemental silicon and iron in varying ratios (15%–90% silicon). Ferrosilicon is commercially prepared by the reaction of high-purity sand often with some scrap iron, wood, charcoal or coal in an electric arc furnace using carbon electrodes. At temperatures over 1900°C, carbon and silicon undergo a reduction reaction. Liquid ferrosilicon collects at the bottom of the furnace, which is then drained and cooled. The silicon produced in this manner is also called metallurgical grade silicon and is around 98% pure.

$$SiO_2 \text{ (Sand)} + 2\,C \text{ (Coke)} \xrightarrow[2000^\circ C]{} \underset{\text{(Ferrosilicon)}}{Si(Fe)} + CO_2 \qquad (6.1)$$

Ferrosilicon accounts for about 80% of the world's production of elemental silicon. Ferrosilicon is primarily used by the steel industry to remove dissolved oxygen present in the steel melt.[1]

6.3.1 Ultrapure Silicon for Solar Cells and Electronics

Pure amorphous silicon was first prepared by Berzelius in 1823 as a dark brown powder by the reaction of potassium metal with silicon tetrafluoride. However, it was Wöhler in 1843 who first prepared crystals of pure silicon by the high temperature reaction of silicon tetrachloride vapours with hydrogen gas.[2] Pure elemental silicon when doped with elements such as boron and phosphorus is one of the best semiconductor materials. Silicon semiconductors are used in electronics and in solar cells.

For solar cell fabrication, one requires silicon of very high purity (+99.9%). For electronic grade, the purity required is even higher (99.9999999%; known as 9 nines). Although molten salt electrolysis of SiO_2 or zone refining of metallurgy grade silicon can be carried out for purification, a better known and less expensive method involves conversion of ferrosilicon to $SiCl_4$ or $HSiCl_3$. These two relatively low boiling liquids can be purified to a very high level of purity by repeated distillation and then converted back to silicon while the impurity, which is mostly iron chloride is not distillable since it is a solid. Another precursor, especially for amorphous silicon and chemical vapour deposition of silicon is silane, SiH_4 which is a gas at room temperature (b.p. –112°C). Trichlorosilane $HSiCl_3$, is most commonly used as the silicon precursor, although silicon tetrachloride and silane are also used. When these compounds in the gaseous form are blown over silicon at high temperature, they decompose to deposit high-purity silicon.[1]

$$Si(Fe) + Cl_2 \longrightarrow SiCl_4 \text{ (Liquid, b.p. 58°C)} + FeCl_3 \text{ (solid)} \quad (6.2)$$

Trichlorosilane is produced by treating powdered ferrosilicon with hydrogen chloride gas at 300°C.

$$Si(Fe) + 3\ HCl \longrightarrow HSiCl_3 \text{ (Liquid, b.p. 31.8°C)} + FeCl_3 \text{ (solid)} + H_2 \quad (6.3)$$

Trichlorosilane when boiled on a resinous bed containing $AlCl_3$ as catalyst forms silane (SiH_4) and silicon tetrachloride by disproportionation.

$$4\ HSiCl_3 \xrightarrow{AlCl_3} SiH_4 \text{ (gas, b. p. −112°C)} + 3\ SiCl_4 \quad (6.4)$$

One of the initial processes for preparing ultrapure silicon was by treating silicon tetrachloride with Zn vapours at high temperature (950°C). Due to practical problems, especially in dealing with large quantities of $ZnCl_2$, this process was replaced by the '*Siemens process*'. Reduction of $SiCl_4$ with hydrogen gas at around 1250°C also gives polycrystalline silicon.

$$SiCl_4 + 2\ Zn \xrightarrow{950^oC} Si + 2\ ZnCl_2 \quad (6.5)$$

The Siemens process is a convenient method in which highly pure silicon rods are exposed at high temperatures to trichlorosilane vapours. Polycrystalline silicon gets deposited on the silicon rod. This is a batch process.

$$2\ HSiCl_3 \xrightarrow[1150°C]{\text{high purity Si rod}} Si + 2\ HCl + SiCl_4 \quad (6.6)$$

In a third method which is a fluidised bed continuous process, silane gas is generated from the disproportionation of trichlorosilane which on pyrolysis gives amorphous silicon or polysilicon.

$$3\ SiCl_4 + Si + 2\ H_2 \longrightarrow 4\ HSiCl_3 \quad (6.7)$$

$$4\ HSiCl_3 \xrightarrow{AlCl_3} SiH_4 + 3\ SiCl_4 \quad (6.8)$$

$$SiH_4 \xrightarrow{>600°C} Si + 2\ H_2 \quad (6.9)$$

To obtain monocrystalline silicon, one often uses the method of Czochralski that involves introducing a crystal of silicon as seed into melted polysilicon, slowly and slightly lowering the temperature so that it crystallises and then very slowly pulling the crystal out by rotation which results in a cylindrical rod of crystalline silicon. These rods are then mechanically cut into thin wafers, polished, etched, cleaned and further polished before they are fabricated as solar cells or integrated circuits.

6.4 Silicon in Alloys

6.4.1 Ferrosilicon

Ferrosilicon is produced by the reduction of silica with coke (carbon) in the presence of scrap iron. It is primarily used by the steel industry to remove dissolved oxygen in steel melt, as a source of silicon to reduce metals from their oxides and also to deoxidise steel and other ferrous alloys. Carbon is present in carbon steel (not the same as stainless steel) up to 2% which improves hardness and strength, and reduces the melting point of steel. Removing dissolved oxygen prevents the loss of carbon as carbon monoxide and dioxide from the molten steel. Treatment of ferrosilicon with hydrogen chloride is the basis of the industrial synthesis of $HSiCl_3$.[1]

Ferrosilicon has a military use—to produce hydrogen gas for balloons quickly. The simple chemical reaction uses ferrosilicon, sodium hydroxide and water (Eq. 6.10). The hydrogen generator which is small and convenient, can be fitted on a truck. Note that the reagents are stable, and they do not generate hydrogen till they are mixed.[3]

$$2\ NaOH + Si(Fe) + H_2O \longrightarrow Na_2SiO_3 + 2\ H_2 + Fe(OH)_3 \quad (6.10)$$

It may be noted that sodium silicate, also known as water glass, is the most water soluble form of silicate (solubility 22.2g/100 mL of water) unlike calcium silicate ($CaSiO_3$, solubility 0.01g/100 mL of water) which is one of the common slags obtained during metallurgical processes.

6.4.2 Aluminum-Silicon Alloy (Silumin)

The most important cast alloy of aluminum is Al–Si, where the high levels of silicon (4.0%–13.0%) contribute to give good casting characteristics. Cast aluminium is widely used in the automobile industry and forms the single largest industrial use (about 55% of the total) of pure silicon (here purified silicon is added to pure aluminium, whereas ferrosilicon need not be purified before being added to steel). This alloy, also called silumin is lightweight and has high-strength. The advantage of silumin is its high resistance to corrosion, making it useful in humid environments. The addition of silicon to aluminium also makes it less viscous when liquid, which together with its low cost (both component elements are relatively inexpensive), makes it a very good casting alloy. There are many additional uses for silumin such as fabrication of rifle scope mounts and camera mounts.[1]

6.5 Silicon Hydrides

Unlike hydrocarbons, silicon hydrides are sensitive to oxidation and hydrolysis. The first five members of the homologous series (SiH_4, Si_2H_6...) are known and all of them undergo decomposition to SiO_2 in the presence of O_2 and to $Si(OH)_4$ in the presence of water liberating hydrogen. Monosilane, SiH_4, a colourless gas with smell similar to that of acetic acid was first prepared by Wöhler in 1857 by the reaction of aluminium silicide and hydrochloride acid. A laboratory synthesis of SiH_4 can be carried out by the reaction of magnesium silicide with hydrochloric acid.

$$4\ HCl + Mg_2Si \longrightarrow SiH_4 + 2\ MgCl_2 \quad (6.11)$$

Reactions of silicon tetrahalides with metal hydrides such as $LiAlH_4$ also results in silicon hydride. Industrially, SiH_4 is prepared by the disproportionation of trichlorosilane in the presence of $AlCl_3$ catalyst. Silane is widely used in chemical vapour deposition of amorphous silicon on surfaces needed for photovoltaics and thin film transistors (TFT).

6.6 Comparison of Carbon and Silicon Bond Dissociation Energies

As Table 6.2 indicates, Si–O and Si–F bonds are much stronger than the analogous carbon-based bonds. The Si–Si and Si–H bonds are weaker than the C–C and C–H bonds and therefore are more reactive. Although the Si–Cl bond is stronger compared to the C–Cl bond, reactivity of the former is higher due to the larger size of silicon compared to carbon which provides less steric hindrance for the attacking nucleophile.[1]

Table 6.2 Comparing silicon and carbon bond strengths with other atoms

	Bond energy, kJ/mol	
	X = Si	**X = C**
H_3X–F	565	485
H_3X–OH	452	358
H_3X–Cl	381	327
H_3X--H	318	411
H_3X–CH_3	343	378
H_3X–SiH_3	310	343

6.7 Silicon Halides and Hydrohalides

The most well known halide of silicon, which is also easy to handle and purify is $SiCl_4$, which incidentally is one of the well utilised precursors for many other silicon compounds. Unlike CCl_4, $SiCl_4$ is highly sensitive to moisture and hydrolyses immediately on exposure to moisture.[1]

Table 6.3 Comparing properties of silicon tetrahalides

Silicon halide	**m.p. (°C)**	**b.p. (°C)**	**Si–X Bond enthalpy (kJ/mol)**	**Si–X Bond length (Å)**
SiF_4	−90.2	−98.6	582	1.55
$SiCl_4$	−70	57.6	391	2.02
$SiBr_4$	5	153	310	2.20
SiI_4	121	290	234	2.43

The reactivity of the Si–Cl bonds of $SiCl_4$ with other organosilicon halides and hydrohalides is comparable and they are highly reactive towards substitution reactions with propensity for forming Si–O bonds whenever oxygen is present in the reactant. Well known reactions of $SiCl_4$ are given in Scheme 6.1.

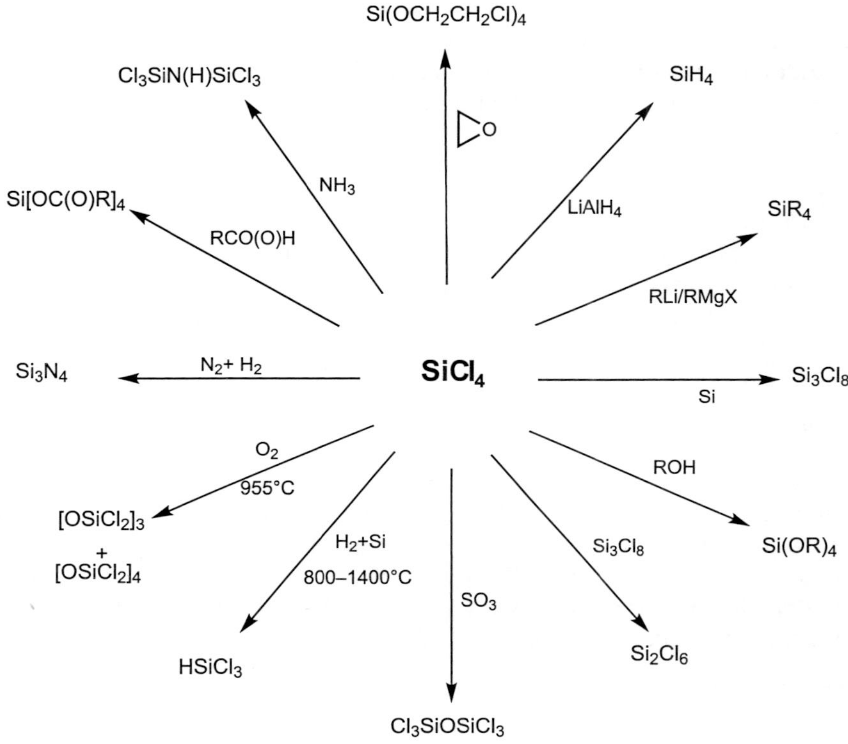

Scheme 6.1 Common reactions of silicon tetrachloride

6.8 Trichlorosilane

$HSiCl_3$ is the most useful silicon hydrohalide. This compound was prepared first by Wöhler from the reaction of ferrosilicon with HCl, which is the method followed even today for its preparation. The low boiling point of this compound makes it an even better choice than $SiCl_4$ as precursor for ultrapure silicon as it can be repeatedly purified by distillation at a much lower temperature (31.8°C) than $SiCl_4$, thereby saving energy. While the Si–Cl bond of this compound undergoes almost all the reactions known for the Si–Cl bonds of $SiCl_4$, the Si–H bond undergoes reactions unique to silanes, especially the hydrosilylation reaction (Eq. 6.12).[4]

$$HSiCl_3 + CH_2{=}CH_2 \xrightarrow[10^{-5}\ \text{mol\%}]{H_2PtCl_6} CH_3CH_2SiCl_3 \qquad (6.12)$$

6.9 Fumed Silica

Fumed silica is a very fine white powder of SiO_2 made by flame pyrolysis of silicon tetrachloride along with hydrogen and oxygen gases and is the silicon equivalent of carbon powder, especially for use as a filler in silicone based polymers.

$$SiCl_4 + 2\,H_2 + O_2 \xrightarrow{1500°C} SiO_2 + 4\,HCl \quad (6.13)$$

Fumed silica is used as a thickening agent and an anticaking agent (free-flow agent) in many products. Similar to silica gel, it can serve also as a desiccant. It is used in many cosmetics and has light-diffusing properties. It is used in products like toothpaste as a light harmless abrasive. Other uses include, as filler in silicone polymers and for adjustment of viscosity in paints, adhesives and resins.

6.10 Organosilanes and Halides

The first example of an organosilicon compound, tetraethylsilane was prepared by Friedel and Craft in 1863 by the reaction of diethylzinc and silicon tetrachloride. Two years later, tetramethylsilane, now the accepted internal standard for calibrating the chemical shift for proton, carbon and silicon NMR spectra, was also made by the same team using dimethylzinc. Dialkyl zinc reagents were also reacted with tetraethoxysilane in the presence of sodium metal to make silicon–carbon bonds.[4]

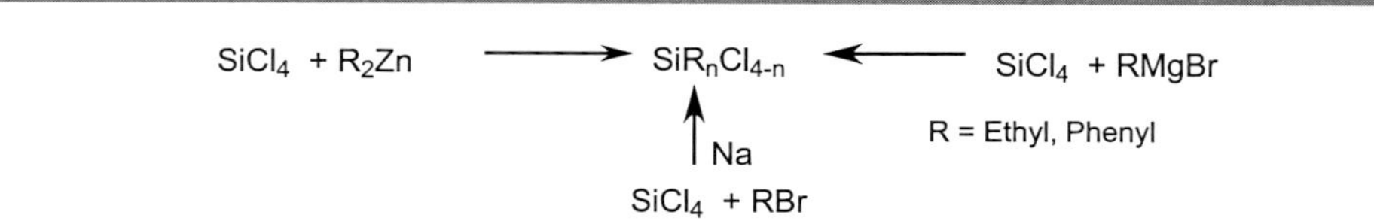

Scheme 6.2 Early methods for making organosilanes

Later on RMgX, $Me_3Al_2Cl_3$ and RLi reagents were used for the preparation of organosilicon compounds. F Stanley Kipping first used organozinc reagents, followed by Wurtz coupling and later Grignard reagents to make organosilicon compounds. In 57 research papers, Kipping created the current basis of organosilicon chemistry. He made mostly ethyl and phenyl organochlorosilanes as methyl halides were not easy to handle at that time. The first examples of silicone polymers (having ethyl and or phenyl groups) were reported by Kipping in 1904 and he described the products of his reaction as 'sticky messes of no particular use'! Kipping coined the word silicone. In 1937, after 30 years of research, on the occasion of his retirement (while delivering the Bakrerian lecture of the Royal Society) he said, '*We have considered all the known types of organic derivatives of silicon and we see how few are their number, in comparison with the purely organic compounds. Since the few which are known are very limited in their reactions, the prospect of any immediate and important advancement in this section of chemistry does not seem very hopeful*'. The exponential development in the chemistry of silicone polymers which began shortly after his retirement proved how wrong he was in making such a statement.[6] The world silicone demand in 2015 was for 15 billion US dollars.

In 1943, Dow Corning Company (USA) took up industrial production of silicone polymers. James Frank Hyde at Corning hydrolysed Grignard derived organosilanes to make silicones industrially. He used ethylphenyl silicone along with glass fibres to make an excellent insulating resin which during World War II helped aircrafts to fly around 35,000 feet for extended hours without having the phenomenon of corona discharge (which affected aircraft communications with ground control) and being flown directly from USA to UK and north Africa above the Atlantic (earlier such small airplanes were transported on ships). This discovery was described as 'the most propitious single wartime discovery in the entire area of resins and plastics'. Dow (magnesium expertise) and Corning (silicon expertise) jointly started the company Dow Corning which is now one of the world leaders in the manufacture of silicone based compounds and discovery of new applications. Another company which has made giant strides in diverse industrial applications of silicones is Wacker Chemie of Germany.

6.10.1 The Direct Synthesis of Methylchlorosilanes[7]

While many organosilicon compounds are still prepared by using organometallic precursors which are rather expensive for scaling up and for industrial production, a landmark discovery by Rochow and Müller, well known as *the direct synthesis*, changed the whole scenario. In this method, methyl chloride was efficiently converted to a mixture of organochlorosilanes by reaction with a silicon–copper (9:1) alloy at 300°C. This discovery made synthesis of methyl chlorosilanes quite easy and inexpensive, leading to the fast development of the related silicone polymer area for which the main monomer required Me_2SiCl_2 could be produced in high yield by the Rochow–Müller process.

$$CH_3Cl + Si/Cu\ (9:1) \xrightarrow{300°C} (CH_3)_nSiCl_{4-n} \qquad (6.14)$$

$$\underset{80\%}{Me_2SiCl_2} \quad \underset{3\%}{MeHSiCl_2} \quad \underset{8\%}{MeSiCl_3} \quad \underset{3\%}{Me_3SiCl}\ldots\ldots\ldots\ldots$$

6.10.2 The Mechanism of the Direct Synthesis[8]

The mechanism of the direct synthesis or the Rochow–Muller process has been subject to many investigations. Since this is a heterogeneous reaction, understanding the mechanism becomes quite cumbersome. However, a few interesting facts have evolved from the observations made by many research groups which provides some insight into the mechanism.

- Methyl chloride does not react with pure silicon directly.
- There is a lengthy induction period before the chlorosilane production begins which becomes faster at higher temperatures.
- The catalytic activity of the copper lies to a significant extent in the formation of bimetallic phases such as Cu_3Si, for which short induction periods of the reaction were observed.

- The products observed during the induction period are not chlorosilanes but methane, H_2 and C, possibly resulting from the initial cracking of CH_3Cl on copper (possibly also forming CuCl).
- Spectral evidence has been obtained for the formation of unstable silylenes ($SiCl_2$ and SiMeCl) by the chlorination of silicon with CuCl. Trapping studies of these silylenes using butadiene showed further evidence for the formation of both these silylenes on the surface of the alloy.

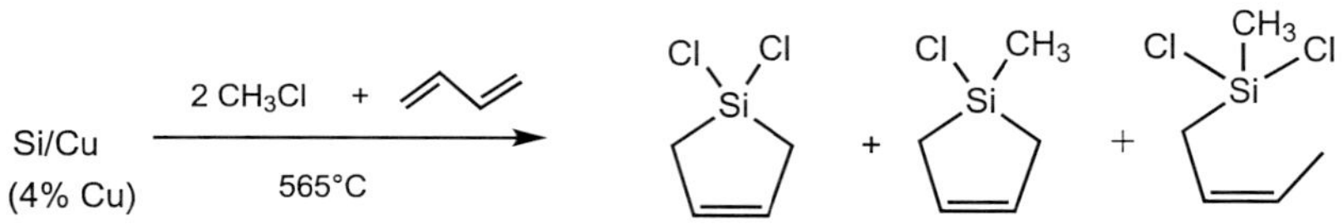

Si/Cu (4% Cu) + 2 CH_3Cl + butadiene —(565°C)→ 1,1-dichlorosilacyclopent-3-ene + 1-chloro-1-methylsilacyclopent-3-ene + CH_3SiCl_2(butenyl) (6.15)

- A few promoters (0.5% Zn, 0.01% Sn) have been found to help the reaction. These promoters possibly assist in the transfer of chlorine and methyl groups to silicon and facilitate diffusion of copper into silicon.
- The main role of copper is now believed to be in weakening in the Si–Si metallic bonds on the Si–Cu contact mass surface of the alloy. Owing to the Si–Cu bond, the electron density gets redistributed in such a way that the nearby Si–Si bond gets weakened.

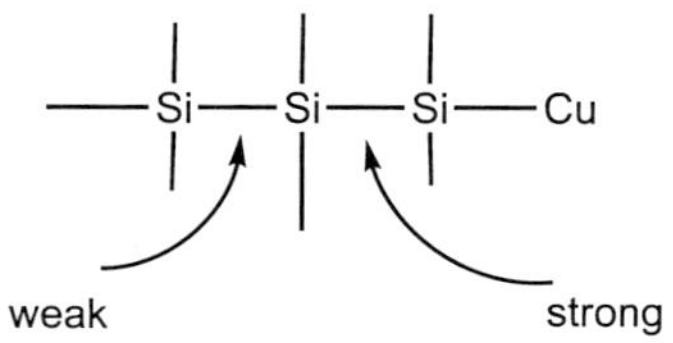

—Si—Si—Si—Cu (weak, strong) (6.16)

- Based on these factors, a plausible chain mechanism for the formation of dimethyldichlorosilane has been proposed by Michael P Clarke (Scheme 6.3).[7]

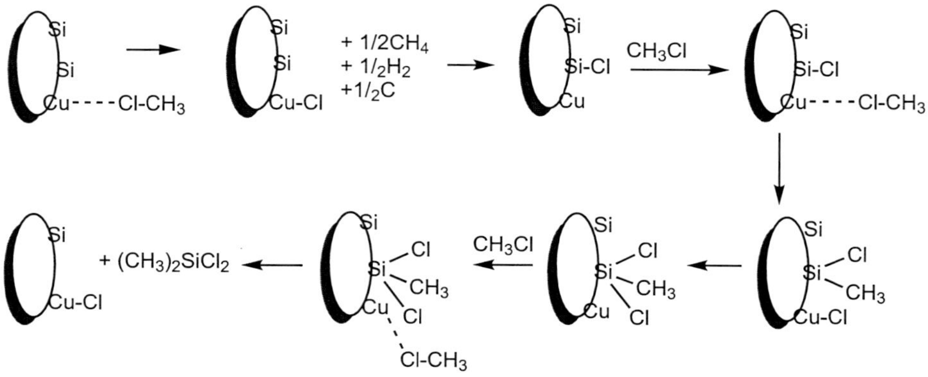

Si, Si, Cu----Cl-CH$_3$ → Si, Si, Cu-Cl + 1/2CH$_4$ + 1/2H$_2$ +1/2C → Si, Si-Cl, Cu —CH$_3$Cl→ Si, Si-Cl, Cu----Cl-CH$_3$ → Si, Si(Cl)(CH$_3$), Cu-Cl → Si, Si(Cl)(CH$_3$), Cu, Cl —CH$_3$Cl→ Si, Si(Cl)(CH$_3$), Cu, Cl----Cl-CH$_3$ → Si, Cu-Cl + (CH$_3$)$_2$SiCl$_2$

Scheme 6. 3 Proposed mechanism for the Rochow–Müller process

6.10.3 Chemistry of Methyl Chlorosilanes

Trimethylchlorosilane has been termed 'bulky hydrogen' as it quite often mimics the chemistry of a bulky hydrogen analogue in organic reactions. It is a well known protecting group for alcohols, amines and alkynes and it introduces hydrophobicity on organic compounds and polymers. It is also an alternative side product for HCl from many condensation reactions as it can be easily removed due to its high volatility and relatively low boiling point (57°C). As Scheme 6.4 indicates, it is a well known starting material for a host of organosilicon compounds and is also the terminating group for siloxane polymers. Interestingly, it is also prepared by Rochow–Müller process albeit in lesser yields compared to Me_2SiCl_2.

Unlike Me_3SiCl, the other chloromethylsilanes Me_2SiCl_2 and $MeSiCl_3$ are known mostly as propagating groups in polymer synthesis involving silicon-based monomers.

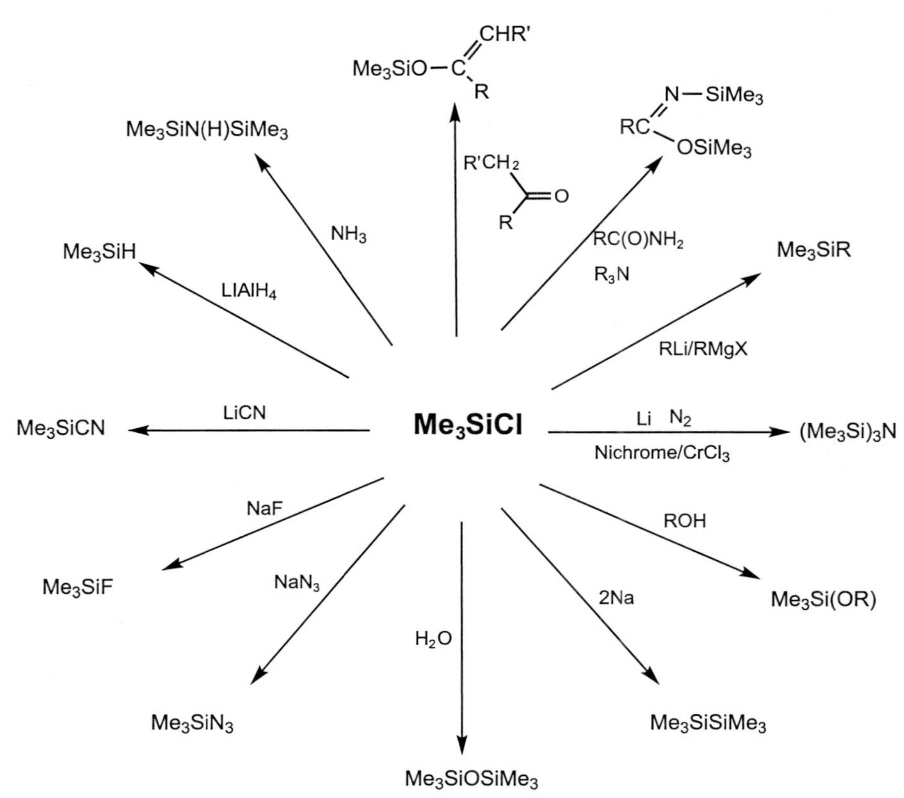

Scheme 6.4 Common reactions of trimethylchlorosilane

6.10.4 Bistrimethylsilylamine

Among the derivatives of Me_3SiCl, one of the most well known is bis-trimethylsilylamine, $(Me_3Si)_2NH$ (also called hexamethyl disilazane, HMDS) which is prepared by the reaction of NH_3 and Me_3SiCl. Due to the steric bulkiness of Me_3Si group, only two of the hydrogens of NH_3 are substituted under normal conditions and this liquid (b.p. 126°C) although undergoes slow hydrolysis in moist air can be easily distilled and stored. It is a very good reagent to convert alcohols to trimethylsilyl ethers (in the presence of some catalysts) as the side product of the reaction (NH_3) can be easily expelled from the reaction medium.[9,10]

$$(Me_3Si)_2NH + 2\,ROH \xrightarrow{\text{neat}} 2\,Me_3SiOR + NH_3\,(gas) \tag{6.17}$$

It is useful for silylating OH groups on laboratory glassware to make it hydrophobic and also for water-proofing window glass. The amide of this secondary amine can be easily made by displacing the hydrogen of the acidic N–H bond with alkali metals such as Li, Na and K leading to LiHMDS, NaHMDS and KHMDS. These bis-trimethylsilyl amides are non-nucleophilic bases for organic reactions and have also been some of the most sought after sterically hindered ligands for stabilising low oxidation states of transition and inner transition metals.[11] The metal amides prepared in this manner are also excellent precursors for other organometallic compounds due to the high reactivity of the M–N bond.

These metal bis(trimethylsilyl)amide complexes are lipophilic and sterically hindered and hence are soluble in a range of nonpolar organic solvents. This property is useful as it often makes them more reactive than the corresponding metal halides, which are otherwise difficult to solubilise. The steric bulkiness of HMDS ligands causes their complexes to be discrete and monomeric (with the exception of group 1 complexes).

6.11 Hydrosilylation

Hydrosilylation reaction is the addition of a Si–H bond across multiple bonds such as C=C, C≡C, C=O and C=N (the latter two being rare).[12] Hydrosilylation was first reported using transition metal catalysts mostly based on platinum. Later, rhodium, palladium, and recently even iron-based catalysts have been found to bring about hydrosilylation. Alkylsilanes are important monomers for many industrial applications and hydrosilylation plays an important role in their synthesis. Organosilicon compounds having at least one Si–H bond adds across double and triple bonds and often helps in crosslinking of polymers and expanding the size of dendrimers. For industrial synthesis, Spier's catalyst ($H_2PtCl_6.nH_2O$) has been used extensively while Karstedt's catalyst, Pt_2(tetramethyldivinylsiloxane)$_3$ is preferred in laboratory scale reactions since lesser induction period is required for the Karstedt's catalyst as platinum is in the zero oxidation state.[13]

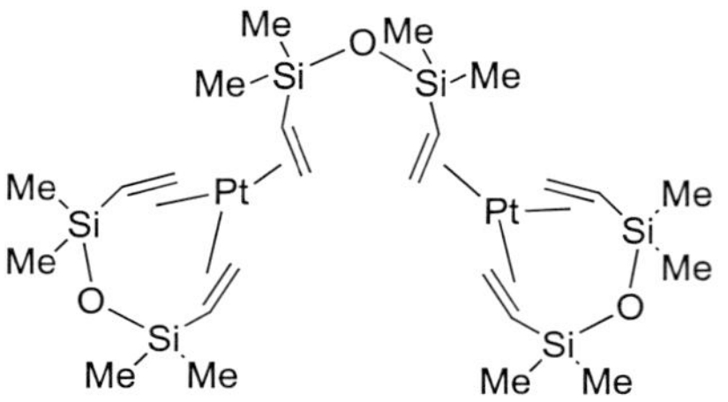

Fig. 6.2 Karstedt's catalyst

The hydrosilylation of alkenes produces terminal anti-Markonikoff products. Reactions of internal olefins produce the same product as that of terminal olefins as a result of isomerisation of secondary alkyl complexes to the terminal alkyl complexes. Hydrosilylation of alkynes produces vinyl silanes.

$$HSiCl_3 + CH_2{=}CH_2 \xrightarrow[H_2PtCl_6]{} CH_3CH_2SiCl_3 \quad (6.18)$$

$$\text{cyclohexene} + MeHSiCl_2 \xrightarrow[H_2PtCl_6]{10^{-5}\ \text{mol\%}} \text{cyclohexyl-}SiMeCl_2 \quad (6.19)$$

$$\text{isoprene} + HSiMe_2Ph \xrightarrow[H_2PtCl_6]{} (CH_3)_2C{=}CHCH_2SiMe_2Ph\ (72\%) \quad (6.20)$$

$$\text{1-hexyne} + HSiCl_3 \xrightarrow[H_2PtCl_6]{} \text{CH}_3(CH_2)_3CH{=}CHSiCl_3\ (78\%) + CH_3(CH_2)_3C(SiCl_3){=}CH_2\ (22\%) \quad (6.21)$$

72%

6.12 Reaction of Hydrosilanes and Carbon Dioxide

The thermodynamic and kinetic stability of CO_2 has been a challenge to achieve its chemical activation to realise useful organic products. The catalytic reduction of CO_2 to silylformates,

$$Et_3Si\text{-}H + CO_2 \xrightarrow[RuCl_3.n\ H_2O]{} Et_3Si\text{-}O\text{-}C(H){=}O$$

$$Ph_2SiH_2 + CO_2 \xrightarrow[\text{Mes = mesityl}]{\text{Mes-N}\ \text{(NHC)}\ \text{N-Mes}} Ph_2SiH\text{-}O\text{-}CH{=}O \xrightarrow{Ph_2SiH_2} Ph_2SiH\text{-}O\text{-}CH_2\text{-}O\text{-}SiHPh_2 \xrightarrow{Ph_2SiH_2} Ph_2SiH(OMe) + Ph_2SiH\text{-}O\text{-}SiHPh_2$$

$$Ph_2SiH(OMe) \xrightarrow{H_2O} CH_3OH + Ph_2SiH(OH)$$

Scheme 6.5 Reactions of hydrosilanes with carbon dioxide methoxysilanes and methane with silanes is favoured thermodynamically. Among the above mentioned compounds, silylformates are of great interest due to their potential application as synthons for the preparation of a range of valuable chemicals. Quite a few inexpensive and environmentally safe hydrosilanes are commercially available. Thus, catalytic CO_2 hydrosilylation to silylformates could present an alternative methodology for sustainable catalytic CO_2 transformation into useful chemicals.[14]

Many metal catalysts are useful for the synthesis of formic acid by hydrogenation of carbon dioxide. This is a thermodynamically unfavourable reaction and therefore, relatively high pressures are required. The early examples of catalysts used for CO_2 hydrosilylation employed late transition metals such as ruthenium and iridium. Presently, there are several new catalysts, including transition metal species acting alone or together with $B(C_6F_5)_3$, as well as metal-free frustrated Lewis pairs (FLPs) and organocatalysts, the most well known among them being N-heterocyclic carbenes.[15]

6.13 Silicone Polymers or Siloxanes

In his attempts to make keto analogues of silicon, Kipping tried out many reactions but never succeeded. His reactions always resulted in sticky viscous liquids. He called these silicones and first examples of silicone polymers (having mostly ethyl and or phenyl groups) were reported by him in 1904.[6] Over the years, silicone polymers, especially polymers based on polydimethylsiloxane (PDMS), have made exponential progress and is currently the most widely used and sought after inorganic polymer due to its unique and diverse properties leading to a plethora of applications. Some of the properties of silicone fluids and applications are listed in Table 6.4.[16,17]

Table 6.4 Properties of silicone polymers and related applications

Property	Application
High thermal stability	Heat transfer media
Low glass transition temperature (–135°C)	Refrigerant in cryostats
Low surface tension	Release agents, antifoams
Water repellency	Water proofing, paints, polishes
Antifriction	Lubricant
Gas permeability	Masks, contact lenses
Good dielectric property	Power transmission fluids
High water vapour permeability	Water repellent breathable materials
Nontoxicity	Medical and prosthetic devices
Tolerance to skin and physiology	Cosmetics, ointments
Stability to UV	Sunscreen, building protection

6.14 Industrial Production of Silicones[5]

Although simple hydrolysis of Me_2SiCl_2 can result in a silicone polymer, its molecular weight is over a wide range, there is considerable cross-linking and it results in a mixture of cyclic and linear oligomers. To obtain silicone polymers with a better control over molecular weights and unnecessary cross-linking, a different approach is followed. In this method, Me_2SiCl_2 is first hydrolysed leading to the formation of cyclic and acyclic oligomers out of which the major component is the tetramer $[Me_2SiO]_4$. This is called D_4 as per industry nomenclature.

Table 6.5 Types of silicon–oxygen units found in silicone polymers

Symbol	Functionality	Application	Symbol	Functionality	Application
M	$O_{0.5}$–Si(Me)(Me)(Me)	Chain ends in silicone polymers; TMS protecting groups (^{29}Si NMR δ + 7 ppm)	D	Me–Si(Me)($O_{0.5}$)($O_{0.5}$)	Propagating group, in silicon fluids, rubbers and elastomers (^{29}Si NMR δ –22 ppm)
T	Me–Si($O_{0.5}$)($O_{0.5}$)($O_{0.5}$)	Silicone resins, masonry protection, impregnating agents (^{29}Si NMR δ –66 ppm)	Q	Si($O_{0.5}$)($O_{0.5}$)($O_{0.5}$)($O_{0.5}$)	Silicone resins (^{29}Si NMR δ –110 ppm)

$$Me_2SiCl_2 \ (4\ L) \xrightarrow[12\ L]{H_2O} [Me_2SiO]_4 \ (42\%) + \text{Trimer (0.5\%), Pentamer (7\%), Hexamer (2\%)} \quad (6.22)$$

$$[Me_2SiO]_4 \xrightarrow[\text{Thermal and/or by acid or base catalyst (example, conc. } H_2SO_4)]{Me_3SiOSiMe_3} Me_3SiO\!\left[\!\begin{array}{c} Me \\ | \\ Si\!-\!O \\ | \\ Me \end{array}\!\right]_n\!SiMe_3 \quad (6.23)$$

Polydimethyl siloxane PDMS

The D_4 molecule can be polymerised in the presence of hexamethyldisiloxane resulting in linear chains of PDMS terminated at both ends by M units. Here, there is better control on molecular weights (which will depend on D_4: hexamethyldisiloxane ratio) and absence of any crosslinking. Polymerisation can be carried out under acidic or basic conditions.

Since around 60% of the hydrolysis products of Me_2SiCl_2 are not used in this method, a more comprehensive and inclusive method is adopted nowadays by the polymer industry (Scheme 6.6). Me_2SiCl_2 can be made to undergo either hydrolysis or methanolysis leading to both cyclic and acyclic silicone products. Methanolysis is a better method as MeCl formed during the reaction can be reused as starting material for the Rochow–Müller process. The acyclic products are cyclised using KOH. Finally the cyclised products, which are mostly a mixture of tetramer and pentamer (D_4 and D_5), are treated with required amounts of hexamethyldisiloxane or disproportionated with MD_yM to prepare PDMS of the required molecular weight. The base catalysed reaction is neutralised by phosphoric acid while the acid catalysed reaction is neutralised by Na_2CO_3/ZnO.

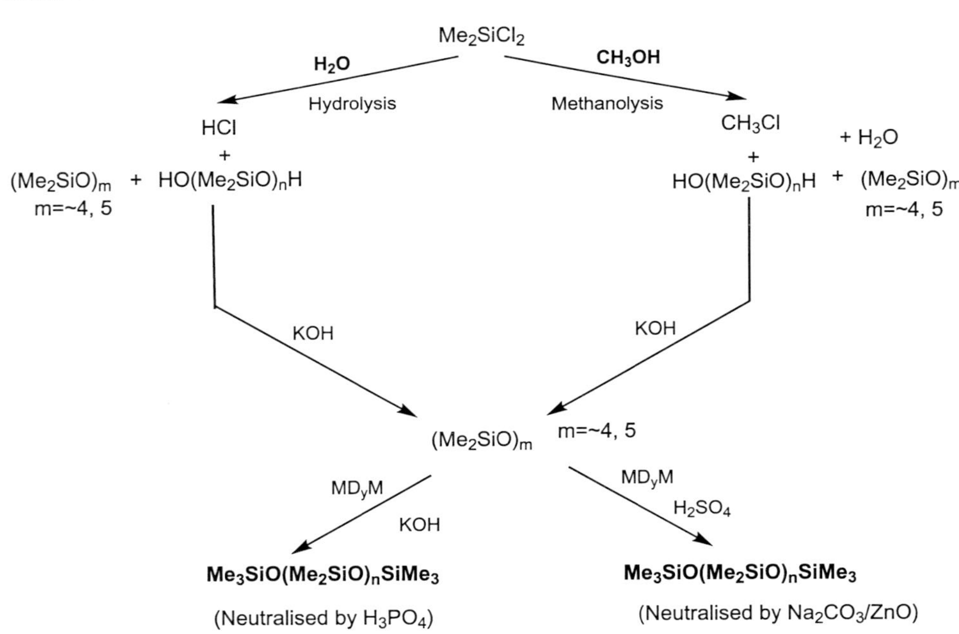

Scheme 6.6 Industrial methods used for production of PDMS

6.15 Crosslinking of Silicone Fluids to Resins and Rubbers

Silicone fluids can be crosslinked using free radical initiators or by using hydrosilylation methods (Scheme 6.7).

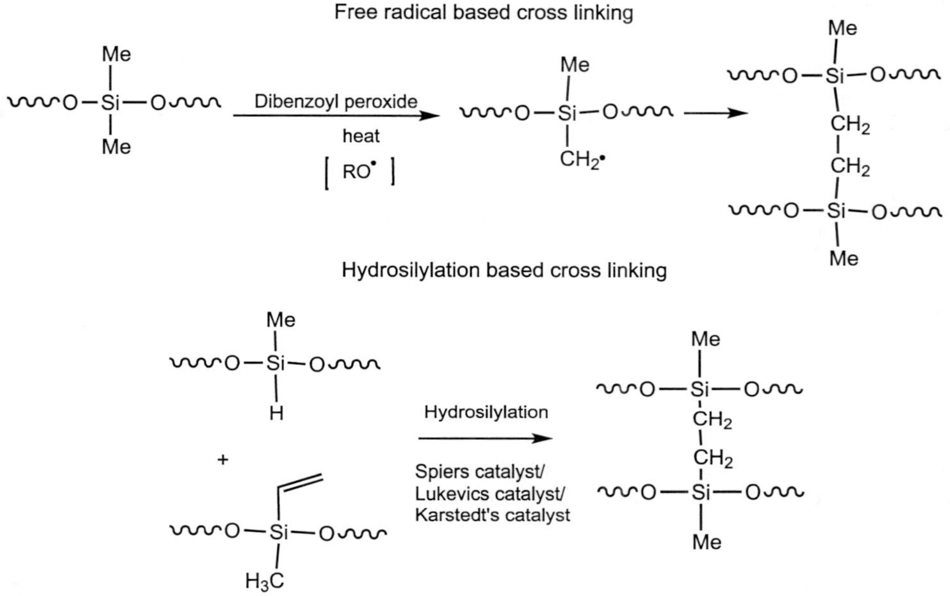
Free radical based cross linking
Me
O–Si–O
Me
Dibenzoyl peroxide
heat
[RO•]
Me
O–Si–O
CH2•
Me
O–Si–O
CH2
CH2
O–Si–O
Me
Hydrosilylation based cross linking
Me
O–Si–O
H
+
O–Si–O
H3C
Hydrosilylation
Spiers catalyst/
Lukevics catalyst/
Karstedt's catalyst
Me
O–Si–O
CH2
CH2
O–Si–O
Me

Scheme 6.7 Methods used for crosslinking of PDMS

6.16 Polysilanes

Polysilanes are inorganic polymers where the polymer backbone is entirely made of silicon–silicon single bonds. Structurally and property-wise, they are similar to organic polymers such as polyolefins. However, they exhibit some very unusual properties. The cumulative silicon–silicon bonds of these polymers allow extensive electron delocalisation of the σ electrons, resulting in some remarkable optical and electronic properties. This especially includes the photophysical properties of polysilanes which are not shown by other polymers except for the less known and more costly polygermanes and polystannanes.

These polymers can be obtained as homopolymers or copolymers having aryl or alkyl groups as pendant groups. The first example of a polysilane was described by C A Burkhard of General Electric Company in 1949. He combined one pound of sodium metal with 700 mL of Me_2SiCl_2 and heated the mixture in a steel autoclave to make $(Me_2Si)_n$. The polymer formed was a highly crystalline material which decomposed above 250°C and was found to be insoluble in organic solvents. However, studies carried out between 1949 and 1975 on linear polydimethyl silanes indicated some interesting properties. One of these was that they show intense UV bands which shifted to longer wavelengths and became more intense when the number of silicon atoms in the chain was increased. The electronic properties of cyclic peralkyl silanes with ring sizes of 4–6 also resembled aromatic hydrocarbons such as benzene. Similar to aromatic hydrocarbons such as naphthalene which can be reduced with alkali metals to deeply coloured radical anions, colourless cyclic polysilanes also undergo reduction with alkali metals to form coloured radical anions.

$$\underset{\text{colourless}}{C_{10}H_8} \xrightarrow[\text{THF}]{e^-} \underset{\text{green}}{[C_{10}H_8]^{\bullet -}} \tag{6.24}$$

$$\underset{\text{colourless}}{(Me_2Si)_6} \xrightarrow[-60°C\ \text{THF}]{e^-} \underset{\text{blue}}{[(Me_2Si)_6]^{\bullet -}} \tag{6.25}$$

One of the early soluble polysilanes to be prepared was the copolymer polymethylphenylsilane–dimethyl silane (Eq. 6.27).

$$Me_2SiCl_2 + PhMeSiCl_2 \xrightarrow[\text{Na, 110°C}]{\text{Toluene}} -(SiMe_2)_n-(SiPhMe)_m- \tag{6.26}$$

Interest in polysilanes was revived with the finding of Yazima and coworkers that $(Me_2Si)_n$ could be converted to polycarbosilanes by heating and further to silicon carbide fibres containing strength reinforcing β-SiC crystallites.

$$[Me_2Si]_n \xrightarrow[450°C]{\text{argon}} [SiH(Me)-CH_2]_n \xrightarrow[280°C]{\text{air}} \text{cross linked polycarbosilane} \xrightarrow[1300°C]{\text{nitrogen}} \text{SiC fibre} \tag{6.27}$$

The three major synthetic routes to polysilanes are (i) by the Wurtz coupling process, (ii) by the use of masked polysilanes and (iii) by catalytic dehydrogenation of silicon hydrides.

The Wurtz coupling process can be carried out with sodium metal (Eq. 6.27) which requires higher temperatures or by the use of KC_8 which can be carried out at below 0°C thereby obtaining better polymer properties such as higher molecular weights and better polydispersity index.

$$Cl-SiPhMe-Cl \xrightarrow[\text{THF, < 0°C}]{KC_8} [SiPhMe]_n \tag{6.28}$$

Masked disilene monomers can be prepared by the reaction of symmetrically and unsymmetrically substituted dichlorodisilanes with the biphenyl anion radical. These monomers undergo rapid polymerisation with alkyl or silyl metal initiators (for example, n-BuLi) to give high molecular weight polysilanes of highly ordered structure with concomitant extrusion of biphenyl.

$$Cl{-}\underset{Me}{\overset{Me}{Si}}{-}\underset{Me}{\overset{NEt_2}{Si}}{-}Cl \xrightarrow{\text{biphenyl anion radical}} \text{[Me}_2\text{Si–Si(Me)(NEt}_2\text{) bridged Ph-substituted bicyclic adduct]} \xrightarrow[\text{EtOH, }-\text{Ph–Ph}]{\text{n-BuLi}} n\text{-Bu}{-}\left[\underset{Me}{\overset{Me}{Si}}{-}\underset{Me}{\overset{NEt_2}{Si}}\right]_n{-}H \quad (6.29)$$

Bent sandwich compounds such as Cp_2TiMe_2 and Cp_2ZrMe_2 have been found to catalyse the conversion of primary aryl silanes such as $PhSiH_3$ at room temperatures giving short chain polysilanes $H\text{-}(PhHSi)_nH$ containing Si–H terminal groups.

$$PhSiH_3 \xrightarrow{Cp_2ZrMe_2} H{-}\left[\underset{H}{\overset{Ph}{Si}}\right]_n{-}H \quad n = \sim 20 \quad (6.30)$$

Properties of polysilanes

Most polysilanes are soluble in organic solvents with the solubility decreasing with increasing crystallinity of the polymer. Polysilanes show electronic delocalisation in the σ-bonded polymer back bone which is a consequence of lower ionisation potential and greater electron affinity of silicon rather than carbon. Due to this delocalisation of electrons, they show strong absorption in the region of 300–400 nm. Since σ–σ* transitions are allowed transitions, the extinction coefficients of these absorptions are also very high. These polymers are radiation sensitive and can degrade upon exposure to certain frequencies of UV light. The glass transition temperatures of polysilanes vary with the nature of the substituents on the silicon and are observed in the range (–75 to +120°C). Some of the polysilanes can be thermally converted to polycarbosilanes and further to silicon carbide fibres.

6.17 Tetraethyl Orthosilicate, $Si(OEt)_4$

Tetraethyl orthosilicate TEOS, can be considered to be the ethyl ester of orthosilicic acid, $Si(OH)_4$. It is prepared by the reaction of ethyl alcohol with silicon tetrachloride:[5]

$$SiCl_4 + 4\,CH_3CH_2OH \longrightarrow Si(OEt)_4 + 4\,HCl \quad (6.31)$$

TEOS has the remarkable property of easily converting into silicon dioxide. This reaction occurs upon the addition of water:

$$Si(OEt)_4 + 2\,H_2O \longrightarrow SiO_2 + 4\,EtOH \quad (6.32)$$

This hydrolysis reaction is an example of a sol–gel process. The reaction proceeds via a series of condensation reactions that convert the TEOS molecule into a mineral-like solid via the formation of Si–O–Si linkages. TEOS is mainly used as a crosslinking agent in silicone polymers and as a precursor to silicon dioxide in the semiconductor industry. TEOS is also used as the silica source for the synthesis of zeolites. At elevated temperatures (>600°C), TEOS converts to silicon dioxide with the release of ether as side product.

6.18 Silicon Carbide[1]

Silicon carbide was accidentally discovered during an attempt to produce artificial diamonds by E G Acheson in 1891. Acheson electrically heated a mixture of clay and powdered carbon in an iron bowl, the bowl and a carbon arc-light were used as the electrodes. He observed bright green crystals forming on the carbon electrode and believed that he had prepared a new compound consisting of carbon and alumina from the clay. He named the new compound carborundum since alumina is called corundum. He also noticed that the crystals obtained had hardness of diamond and applied for a patent. The early product was used for the polishing of gems.

$$SiO_2 + 3\,C \longrightarrow SiC + 2\,CO \tag{6.33}$$

Henri Moissan, the father of fluorine chemistry, while examining rock samples found in the Canyon Diablo meteorite in Arizona, USA, discovered a rare, naturally occurring SiC mineral. This mineral was later named in his honour as moissanite. Moissan also synthesised SiC by several routes, including dissolution of carbon in molten silicon and melting a mixture of calcium carbide and silica. Silicon carbide powder has been mass produced since 1893 for use as an abrasive. Very pure silicon carbide is white or colourless; crystals of it are used in semiconductors for high-temperature applications. Silicon carbide grains can be bonded together by sintering to form very hard ceramics that have potential in applications where high endurance is required, such as, vehicle brakes and clutches, ceramic plates in bulletproof vests, and so on. Applications of silicon carbide as light-emitting diodes (LEDs) were first demonstrated around 1907 by Henry Joseph Round and today SiC is used extensively in high-temperature/high-voltage semiconductor electronics. Silicon carbide has a melting point of 2730°C and density of 3.21 g/cc.

Till the advent of boron carbide and cubic boron nitride, silicon carbide was the hardest synthetic material known. It has a Mohs hardness rating of 8.5–9.0, close to that of diamond. Its hardness and fracture characteristics make it useful (sometimes as a strong composite material with Al_2O_3) in the manufacture of abrasive machining tools such as metal cutting and grinding wheels and abrasive metal polishing paper. Manufacture of yellow and blue LEDs made of SiC in the 1970s and 1980s was stopped because gallium nitride showed 10–100 times brighter emission. This difference in efficiency is due to the fact that SiC has an unfavourable indirect bandgap while GaN has a direct bandgap which favours light emission. It may be noted that SiC is still a major LED component as it is a popular substrate for growing GaN devices, and as a heat spreader in high-power LEDs.[18]

6.19 Silicates, Aluminosilicates and Zeolites[1]

A silicate is an anionic silicon compound having some or all of its valencies satisfied by Si–O bonds and some or all of the oxygens shared by other silicon atoms. Silicates comprise the major mineral component of the earth's crust, as well as other planets, moons and asteroids. Sand, clay, bricks, porcelain crockery, Portland cement, and thousands of minerals are examples of silicates and/or alumino silicates. 'Orthosilicate' is the anion SiO_4^{4-}. Related to orthosilicate are families of anions (and their compounds) with the formula $[SiO_{2+n}]^{2n}$. Silicates are the most widespread of all minerals. By themselves they make up over 90% of the weight of the earth's crust. Most rocks, soil and clays are composed mainly of this class of minerals.

Silicate compounds, including the minerals, consist of silicate anions whose charge is balanced by various cations. Numerous silicate anions can exist, and each can form compounds with many different cations. Hence this class of compounds is very large. Both minerals and synthetic materials fit in this class. Silicate minerals are classified according to the structure of their silicate and corner sharing of oxygen into the groups (Table 6.6).

In almost all of silicates, the silicon is in a tetrahedral environment, being surrounded by four oxygen units. These tetrahedra sometimes occur as isolated SiO_4^{4-} centres, but most often, they are joined together by corner sharing (sharing of one or more oxygens), such as $(Si_2O_7^{6-})$ and rings $(Si_6O_{18}^{12-})$. Commonly the silicate anions are chains, double chains, sheets and three dimensional frameworks. Classification of silicates is done on the basis of number of corners of these tetrahedra being shared by other SiO_4 tetrahedra.

From phyllosilicates onwards, often one finds one or more tetrahedral aluminium atoms replacing tetrahedral silicon atoms resulting in aluminosilicates. Aluminium, unlike silicon, can be present in minerals in the tetrahedral or octahedral geometry. Kaolinite, the main component of China clay (used for making porcelain crockery and bone china tea cups) is a two layer structure having the formula $Al_2(OH)_4Si_2O_5$ where the aluminum ions are in the octahedral sites and the OH groups are on the periphery. The peripheral OH groups make this silicate sticky as they can get involved in hydrogen bonding with O and OH groups. In contrast, talc another phyllosilicate which is smooth and non-sticky, has the composition $Mg_3(OH)_2Si_4O_{10}$. It is a three-layer silicate where the magnesium layer is flanked on both sides by the silicon layer and the OH groups are not in the periphery. In the phyllosilicate mica, $KAl_2(OH)_2[Si_3AlO_{10}]$, one of the tetrahedral silicons is replaced by a tetrahedral aluminium and among counter-ions, we have octahedral aluminium as well. Other than quartz, most of the tectosilicates are alumino silicates having one or more tetrahedral silicon atoms replaced by tetrahedral aluminium atoms. Since aluminium is trivalent, every such replacement results in an extra negative charge on the aluminosilicate unit per aluminium, and this is neutralised by a suitable counter ion such as Na^+.

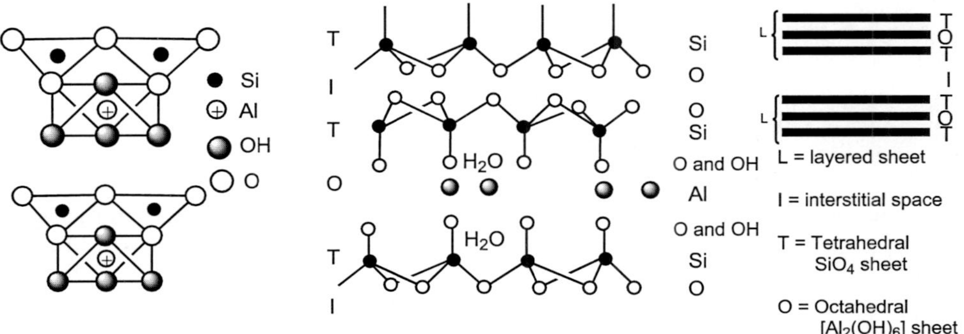

Fig. 6.3 Comparison of the structures of kaolinite and montmorillonite

Table 6.6 Classification of silicates based on corner sharing

Class	Corners shared	Formula	Example
Nesosilicates (Orthosilicates)	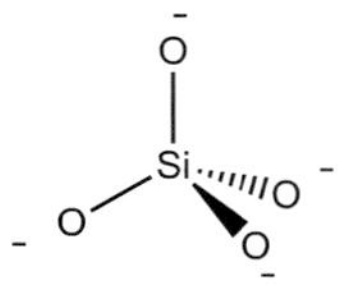 No corner oxygen shared	$[SiO_4]^{4-}$	Example olivine $(Mg^{+2}, Fe^{+2})_2SiO_4$
Sorosilicates (Disilicates)	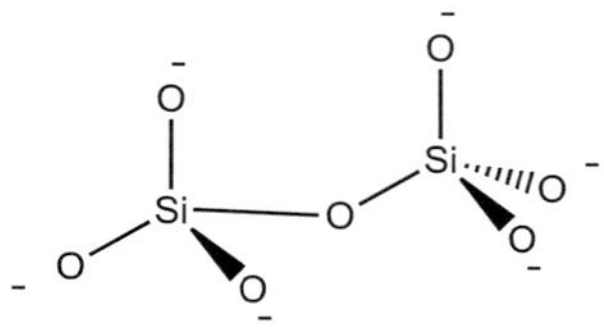 One corner oxygen shared	$[Si_2O_7]^{6-}$	Example thortvetite, $Sc_2(Si_2O_7)$
Cyclosilicates (ring) (Metasilicates)	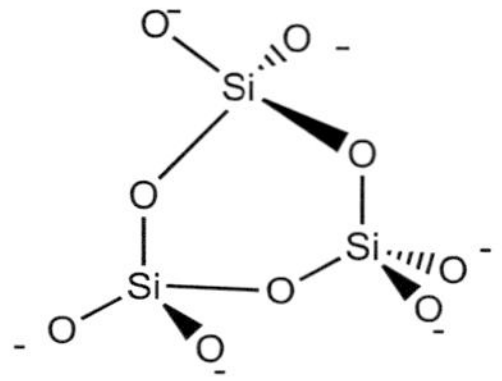 Two corners oxygen shared	$[Si_nO_{3n}]^{2n-}$	Example beryl, $Be_3Al_2(Si_6O_{18})$
Inosilicate (Chain)	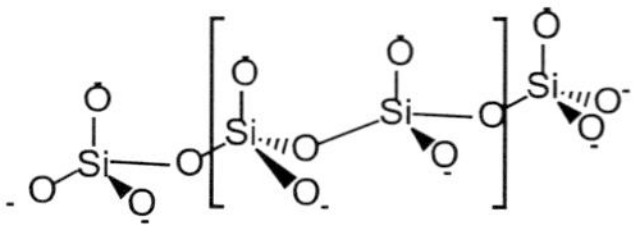 Two corners oxygen shared	$[SiO_3^{2-}]_n$	Example pyroxene group $(Na_2SiO_3)_n$
Inosilicates (double chain)	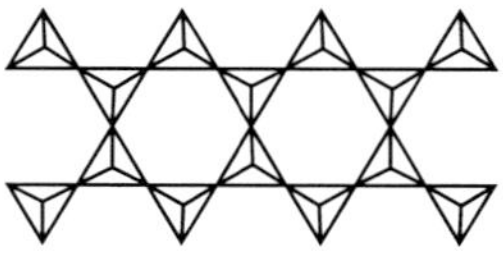 Three corners oxygen shared	$[Si_4O_{11}^{6-}]_n$	Example amphiboles, asbestos minerals tremolite: $Ca_2Mg_5(Si_4O_{11})_2(OH)_2$
Phyllosilicates (sheets)	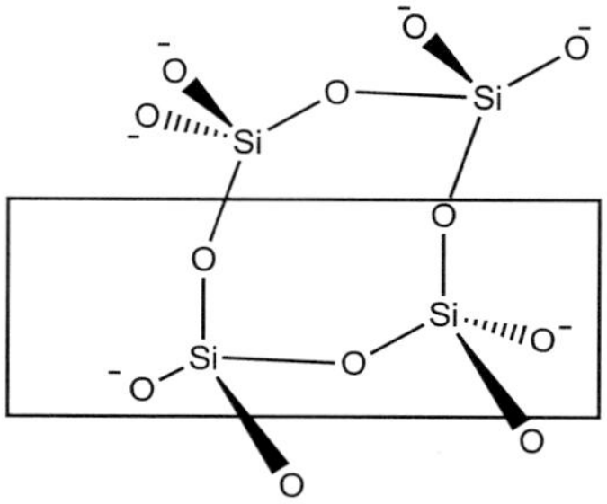 Three corners oxygen shared	$[Si_{2n}O_{5n}]^{2n-}$	Example talc $Mg_3(OH)_2Si_4O_{10}$, clay: kaolinite $Al_2(OH)_4Si_2O_5$, montmorillonite
Tectosilicates (3-D framework)	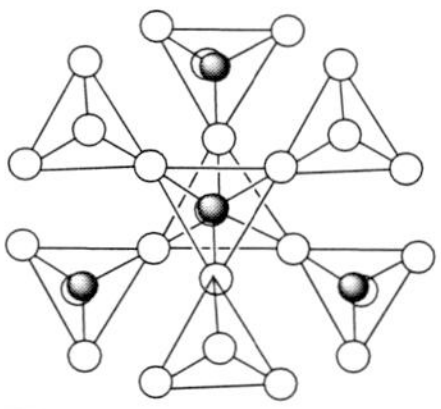 Four corners oxygen shared	$[SiO_2]^{4-}$	Example quartz, feldspars, ultramarines, zeolites

Montmorillonite, another soft phyllosilicate is present in Fuller's earth which is used as a skin cleansing agent and drilling mud. It has a three layer structure where octahedral $Al_2(OH)_6$ units are flanked on both sides by tetrahedral SiO_4^{4-} units. The presence of Na^+ ions in the interlayer space helps it to absorb water and swell in volume. Montmorillonite is also used as a catalyst in the synthesis of the gasoline additive methyl tertiary butyl ether (MTBE).

Silicon dioxide, commonly known as silica, is one of the most important tectosilicate compounds occurring in nature. This compound occurs in three major crystalline forms—quartz, tridymite and cristobalite. The last two are mostly found in volcanic rocks and do not have industrial applications. Quartz is a piezoelectric substance used to make crystal oscillators for watches and to provide clock signals for digital ICs. More than 2 billion quartz crystals are manufactured every year for use in wristwatches, clocks, cell phones and computers. They are also used to stabilise amplifier circuits in radio transmitters, to measure high electrical potentials or to measure high instantaneous pressures.

In a tectosilicate, for every tetrahedral silicon(IV) replaced by a tetrahedral aluminum(III), the framework attains an extra negative charge. This will be compensated by cations such as Na^+. Such compounds are termed framework aluminosilicates. The three well known types of tectoaluminosilicates are feldspars, ultramarines and zeolites. The best examples of feldspars are igneous rocks with the formula $K(AlSi_3O_8)$. It has been observed that over a long period of time feldspars undergo weathering to form kaolinite, by reaction with CO_2 and water.

$$K(AlSi_3O_8) + CO_2 + H_2O \longrightarrow Al_2(OH)_4(Si_2O_5) + 4\ SiO_2 + K_2CO_3 \tag{6.34}$$

Sodalite is an ultramarine and has structure very similar to Linde A molecular sieves. However, the truncated octahedral units are connected directly to each other through the square faces and therefore do not have large enough channels which can allow small molecules to pass through. Some ultramarines (for example lapis lazuli, which contains a blue cubic mineral called lazurite) are coloured due to the presence of polysulfur radical anions. The major component of lazurite is a complex sulfur-containing sodium-silicate ($Na_{8-10}Al_6Si_6O_{24}S_{2-4}$). Some chloride is often present in the crystal lattice as well. The blue colour of the pigment is due to the $S_3^{-\cdot}$ radical anion, which has an unpaired electron.

While feldspars and ultramarines are not open framework alumino silicates, zeolites are open framework structures, which basically means that their structural units have plenty of voids as channels and cavities at molecular dimensions. Zeolites are highly environmentally friendly and have a wide range of applications. For example, Linde-A, $Na_{12}(AlO_2)_{12}(SiO_2)_{12}.27H_2O$, has been used as a water softener (due to the presence of twelve replaceable Na^+ ions) and as a dehydrating agent (as most of the water of hydration can be expelled by heating above 100°C). Many zeolites, especially those having low Si/Al ratio, for example ZSM-5 ($Na_3Al_3Si_{93}O_{192}\cdot 16H_2O$) have been used as acid catalyst, especially in shape selective catalysis.

6.19.1 Structure of Zeolites and the Lowenstein's Rule[19]

Generally for zeolites, the AlO_4^- tetrahedra are always interspersed with SiO_4 tetrahedra, or Al–O–Al units do not occur in zeolites. This rule of thumb is called the *Lowenstein's rule.* The basic structural unit of zeolites Linde-A and Faujasite as well as sodalite are the same—a truncated octahedron. As depicted below, each edge of this truncated octahedron represents an Si–O–Si unit or an Si–O–Al unit. The counter ions are located close to the oxygen atoms present at the centre of these edges.

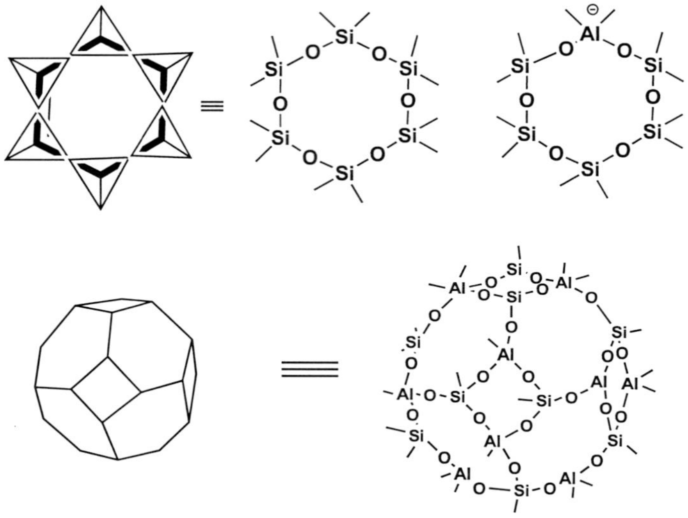

Fig. 6.4 Basic framework found in Linde Type A and Faujasite molecular sieves

One can describe the structure of Linde Type A (LTA, molecular sieves 4A) as truncated octahedra connected to other truncated octahedra through a cube (called D4R, 'double four ring' in zeolite nomenclature). Such an arrangement will result in the channels constituting an eight membered ring or having eight oxygen atoms forming a ring. The diameter of this ring has been determined to be 4.1 Å. In contrast, the structure of Faujasite has the truncated octahedron connected to other such truncated octahedra through a hexagonal prism (also called D6R, double six ring). Such an arrangement will result in channels having 12 oxygen atoms forming a ring.

6.19.2 Zeolite Classification[20]

Zeolites are basically classified under two different considerations. The first is pore or channel size and the second is silicon to aluminium ratio. The channel diameter of zeolite determines the possibility of a particular molecule to enter the porous network of a zeolite. Figure 6.5 indicates the channel sizes of Linde type (channel diameter 4.1 Å), Faujasite (channel diameter 7.4 Å) and ZSM-5 (channel diameter 5.1 and 5.5 Å).

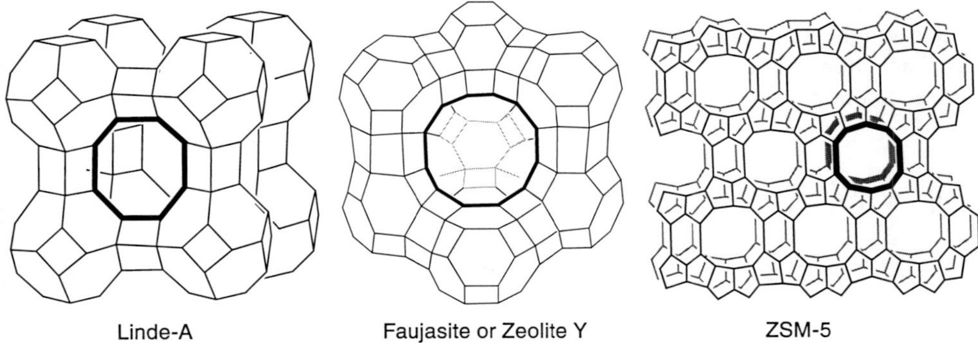

Fig. 6.5 Schematic showing structural units of Linde-A, Faujasite and ZSM-5

Linde type A (LTA) with a channel diameter of 4.1 Å allows small molecules such as H_2O, EtOH, H_2S, CO_2, C_2H_4, C_2H_6 and C_3H_8 to pass through while ZSM-5 with pore diameter 5.1 × 5.5 Å also allows molecules of larger size such as benzene and toluene. Faujasite (13X) with a pore diameter of 7.4 Å allows even large molecules such as di-n-butylamine to pass through. Faujasite is also suitable for making metal complexes trapped inside zeolite cavities for metal specific catalysis (a synthetic procedure known as 'ship in a bottle synthesis').

High concentrations of protons (from a low Si/Al) give a high acidity but low concentrations of protons yield strong acid sites. The acidic form of ZSM-5 shows high catalytic activity and has been used for olefin oligomerisation, methanol conversion and toluene alkylation.

6.19.3 Zeolite Synthesis

There are different methods for synthesising zeolites in the laboratory. Some of them involve the use of a template and some are done without any template.

Template synthesis: The cavity and channel size is controlled and maintained by building the framework around a specific template, often organo-ammonium hydroxide. These organo-ammonium hydroxide compounds can ultimately be converted to volatile products at about 500°C and the aluminosilicate cage retains its framework structure.

$$\underset{\text{Low Si/Al}}{Na_2SiO_3 + Al(OH)_3} \xrightarrow[100\text{–}200°C]{Me_4NOH} \xrightarrow[500\text{–}600°C]{} \text{Sodalite} \qquad (6.35)$$

$$\underset{\text{High Si/Al}}{Na_2SiO_3 + Al(OH)_3} \xrightarrow[100\text{–}200°C]{(CH_3CH_2CH_2)_4NOH} \xrightarrow[500\text{–}600°C]{} \text{ZSM-5} \qquad (6.36)$$

6.19.4 Zeolite Applications in Catalysis

Due to the presence of replaceable sodium ions, zeolites can be readily converted to solid Brønsted acids as well as Lewis acids, as shown. Washing with a strong mineral acid displaces the Na^+ ions with H^+ ions making the framework a solid Brønsted acid. Heating this Brønsted acid to a higher temp (~500°C) results in the expulsion of water molecules converting it to a Lewis acid having tricoordinated aluminium and silicon. Both these acidic forms have been utilised for carrying out heterogeneous acid catalysis involving organic substrates.

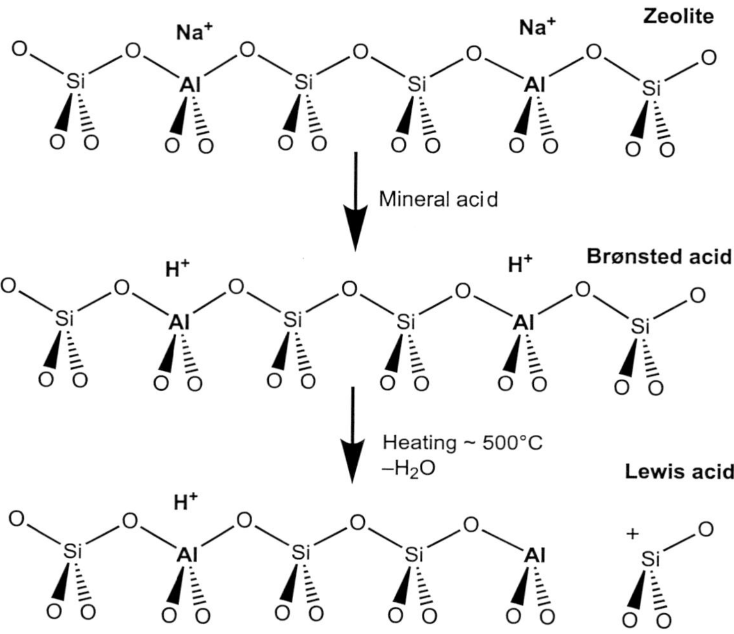

Fig. 6.6 Generation of Bronsted and Lewis acid sites in zeolites

Acid catalysis also is of two types. In some cases there is no shape selectivity but the method is very efficient and environmentally friendly. The Badger–Mobil process for the preparation of ethylbenzene from benzene and ethylene, (which is a precursor for styrene and thereby polystyrene) is one such ZSM-5 catalysed reaction by which one-fourths of the world's production of ethylbenzene is carried out.

$$C_6H_6 + CH_2{=}CH_2 \xrightarrow{\text{ZSM-5}} C_6H_5CH_2CH_3 \qquad (6.37)$$

6.19.5 Shape Selective Catalysis[21]

One of the defining features of zeolites is that the shape or topology of its internal pore structure can strongly affect the selectivity with which particular product molecules are formed in catalytic transformations. This unique property termed *shape selective catalysis*, is a type of heterogeneous catalysis where there is selectivity in the nature of products. This is a consequence of the catalyst showing preference either to the shape of the reactant, transition state or products. The classical definition of shape selectivity is that the product distribution will deviate from the ideal distribution if the formation of some of the molecules is inhibited by the constraints on molecular size and shape imposed by the channels and cavities. This selectivity in turn is dependent exclusively upon the diameter of the channels of the zeolite used as well as the size of the cavities formed at the junctions of these channels. The classic example of shape selective catalysis is alkylation of toluene using methanol in the presence of H-ZSM-5 catalyst. A normal catalyst will give a mixture of *o*, *m* and *p*-xylenes as products. But in the presence of H-ZSM-5, *p*-xylene is obtained as the major product. While the pore size permits both toluene and methanol to enter the cavities though the channels, *p*-xylene is able to come out preferentially from the honeycomb structure due to the restricted size of the channels. Therefore most of the *ortho* and *meta* isomers are trapped inside the cavities and they undergo further isomerisation. Whenever a *p*-xylene molecule is formed, it moves out of the catalyst cavities.[22]

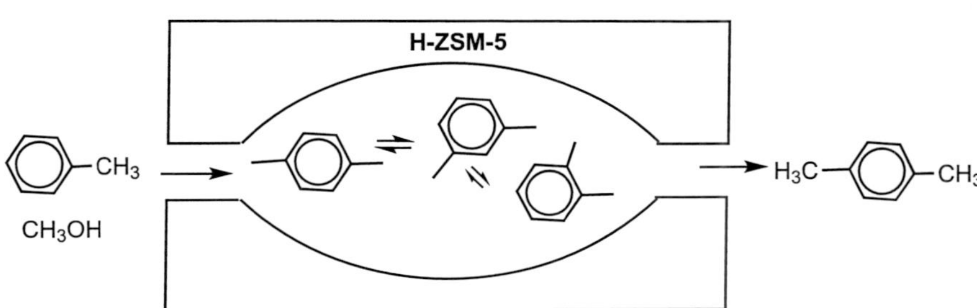

Scheme 6.9 Shape selective catalysis by H-ZSM-5 showing selectivity for *p*-xylene

Basically, shape selective catalysis is classified under three categories—(i) reactant selectivity, (ii) product selectivity and (iii) restricted transition state selectivity.

(i) Reactant selectivity

Fig. 6.7 Schematic indicating reactant selectivity

This occurs when some of the molecules in a reactant mixture are too large to diffuse through the catalyst pores. Catalysis happens mostly to those molecules which diffuse through. Examples are distillate and tubeoil dewaxing of petroleum products using H-ZSM-5. The zeolite allows conversion of only n-paraffins and some monomethyl paraffins to propane and butane. There is very little cracking of branched paraffins and cyclic molecules. Another example is the dehydration of alcohols over molecular sieves Ca-A. While n-butyl alcohol undergoes dehydration readily, secondary butyl alcohol shows great reluctance. The rate of dehydration of iso-butyl alcohol, a branched primary alcohol, is very low.

Table 6.9 Reactant selectivity observed for dehydration of alcohols

Reactant alcohol	Reaction temperature (ϒC)	Conversion Wt% with zeolite	Conversion Wt% without zeolite
n-butyl alcohol	260	60	64
sec-butyl alcohol	130	0	82
iso-butyl alcohol	260	<2	85

(ii) **Product selectivity**

This happens when some of the products formed within the pores and cavities are too bulky to diffuse out as products. They are either converted to less sterically hindered molecules (by equilibration, isomerisation or cracking) or such hindered molecules deactivate the catalyst by blocking the pores. Examples of reactions showing product selectivity using zeolite catalysts are: isomerisation of xylene where selectivity is present for p-xylene, reaction of methanol and ammonia to selectively give Me_2NH over the catalyst H-RHO and disproportionation of toluene to give benzene and p-xylene.

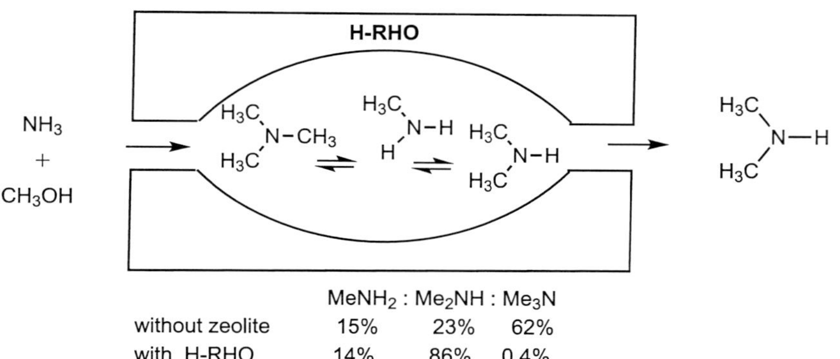

Scheme 6.10 Product selectivity for secondary amines over shape selective catalyst H-RHO

(iii) **Restricted transition state selectivity**

This type of selectivity is observed when certain reactions are prevented because the corresponding transition state requires more space than is available in the cavities/pores. Neither the reactant nor the product molecules are prevented from diffusing through the pores. Reactions requiring a smaller transition state will proceed unhindered. A well known example is the acid catalysed *trans*-alkylation of m-xylene over 'H-mordenite' zeolite catalyst. This is a bimolecular reaction and involves a diphenylmethane type transition state. The transition state leading to 1,3,5-trimethylbenzene does not form due to limited space in the cavities of mordenite. So the preferred product is 1,2,4-trimethylbenzene.

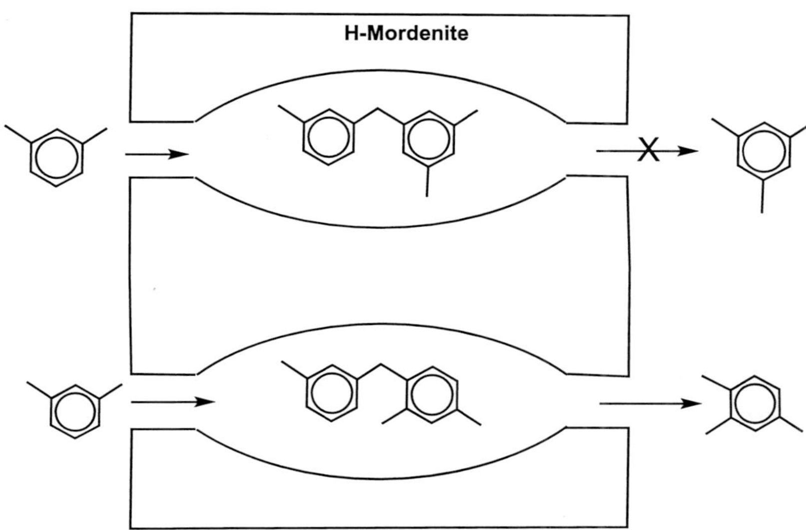

Scheme 6.11 Transition state selectivity of trimethylbenzenes over H-mordenite

6.20 Challenges and New Advancements in the Chemistry of Silicon

While general organic chemistry centred around silicon developed rapidly during 1950–1980 leading to major applications especially as silicones and polysilanes, there were still some major challenges to overcome. One among these was the absence of silicon–silicon multiple bonds and another was the inability to isolate compounds with a silicon–oxygen double bond. These challenges were taken up by synthetic chemists, especially bringing in steric bulkiness, to provide kinetic stability to highly reactive molecules. Another strategy was to use the high reactivity and steric bulk of N-heterocyclic carbenes to stabilise highly reactive species. The attempts directed to achieve these goals also resulted in other interesting discoveries.

6.20.1 Multiple Bonds Between Silicon Atoms

Disilenes

The first example of a compound having a Si=Si bond was a disilene made by Robert West in 1981.[23] This was achieved by photolytically removing $Me_3Si–SiMe_3$ from a sterically hindered silane having two mesityl groups. The divalent silene formed was found to dimerise, to yield the first disilene. The compound was found to readily undergo addition reactions with a wide range of reagents. Quite interestingly, a reaction with O_2 was found to ultimately result in the cleavage of the Si–Si bond as well.

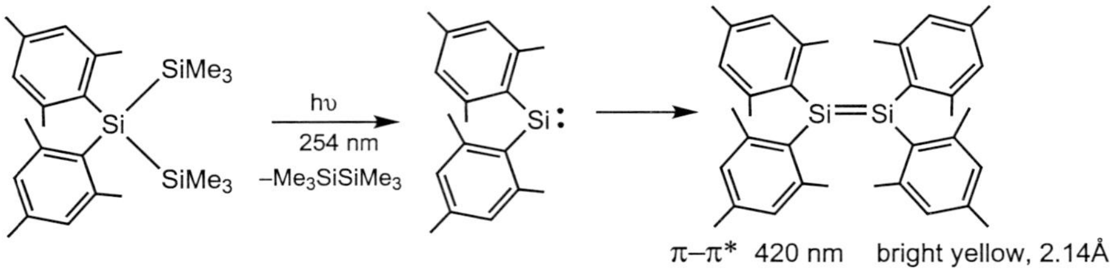

$$\mathrm{Mes_2Si(SiMe_3)_2} \xrightarrow[\;254\ \mathrm{nm},\ -\mathrm{Me_3SiSiMe_3}\;]{h\nu} \mathrm{Mes_2Si}: \longrightarrow \mathrm{Mes_2Si{=}SiMes_2} \quad (6.38)$$

π–π* 420 nm bright yellow, 2.14Å

Reactions of $Mes_2Si{=}SiMes_2$:
- ROH → $HO(Mes)_2Si–Si(Mes)_2R$
- $LiAlH_4$ → $H(Mes)_2Si–Si(Mes)_2H$
- Cl_2 → $Cl(Mes)_2Si–Si(Mes)_2Cl$
- R_2CO → four-membered ring O–CR_2–$Si(Mes)_2$–$Si(Mes)_2$
- PhC≡CH → four-membered ring PhC=CH–$Si(Mes)_2$–$Si(Mes)_2$
- $(Ph_3P)_2Pt(C_2H_4)$ → $(Ph_3P)_2Pt$ bound to $Si(Mes)_2$–$Si(Mes)_2$ (three-membered ring)
- O_2 → O–O–$Si(Mes)_2$–$Si(Mes)_2$ ring → $Mes_2Si(\mu\text{-}O)_2SiMes_2$
- $[PhC(O)]_2$ → six-membered ring $Si(Mes)_2$–O–C(Ph)=C(Ph)–O–$Si(Mes)_2$

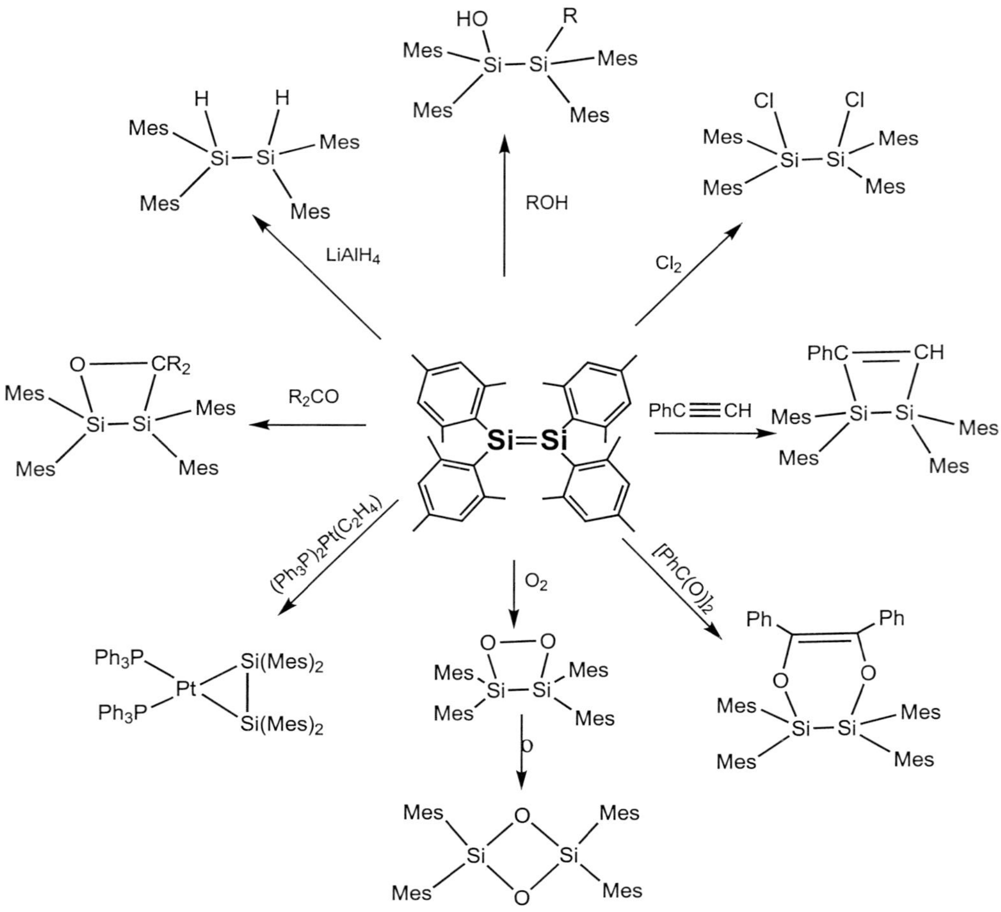

Scheme 6.12 General reactions of tetramesityl disilene

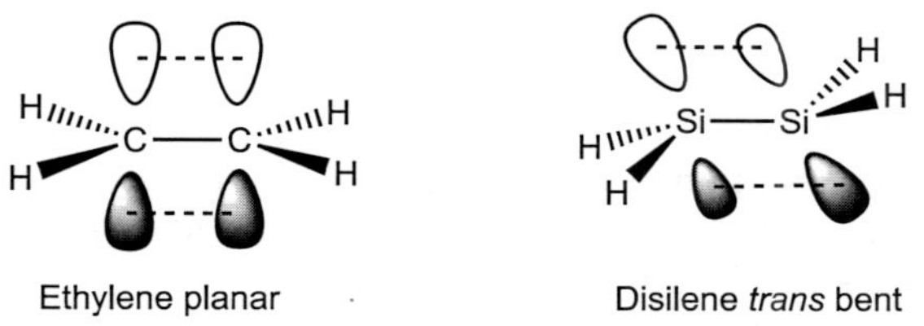

Fig. 6.8 Comparison of bonding between ethylene and silylene

The bonding in silicon–silicon multiple-bonded compounds has been a topic of broad interest. These compounds, and others in neighbouring groups marked a departure from the widely-assumed double bond rule that heavier main group elements do not form double or triple bonds with each other. The main development that enabled their stabilisation was the use of sterically hindered hydrocarbon substituents that blocked their decomposition via association or elimination reactions. In the numerous reviews and discussions of heavier main group multiple bonds, the focus has been almost exclusively on details of the interaction between the two heavier elements, that is, the orbitals likely to be involved in the multiple bonds, their relative energies, their likely bonding character, and the extent of the overlap. Unlike the planar structure of alkenes, the bonding in these compounds has been found to be of the *trans* bent type (Fig. 6.9). The bonding is also of the type named as double donor–acceptor type which involves donation of electron pairs instead of sharing of electrons.[24]

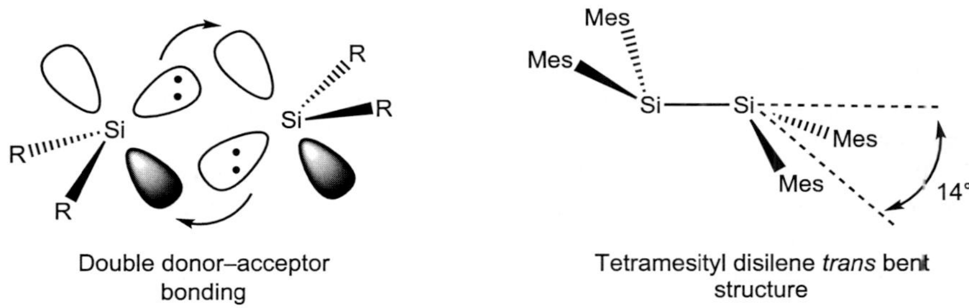

Fig. 6.9 Double donor acceptor bonding and *trans* bent structure of disilenes

Other types of silicon–silicon double bonded compounds include cyclic disilenes, spiro pentadisilenes, silabutadiene and trisila-allenes. More than 50 examples of disilenes have been prepared till 2017.

R= t-Bu_2MeSi, R' = t-$BuMe_2Si$ Tip = 2,4,6, triisopropylphenyl

Fig. 6.10 Different types of silicon–silicon double bonded compounds

A comparison of the energy levels of π and π* orbitals between ethylene and disilene indicates that the energy levels of the π and π* orbitals in disilene are higher and lower than those in the case of ethylene respectively (Fig. 6.11). Also, the π–π* splitting in disilene is calculated to be almost half as that in ethylene. These differences in energy levels suggest that the reactivity of disilene to both nucleophiles and electrophiles is higher than that of ethylene. Unlike alkenes which are colourless, the colour of disilenes ranges from pale yellow to blue and the longest absorption maxima of disilenes are assignable to the π–π* transitions. The λ_{max} values for tetraalkyl disilenes are in the range of 393–493 nm while that of tetra-aryl disilenes fall in the range of 400–460 nm. For tetrasilyldisilenes it falls in the range of 411–612 nm.[25]

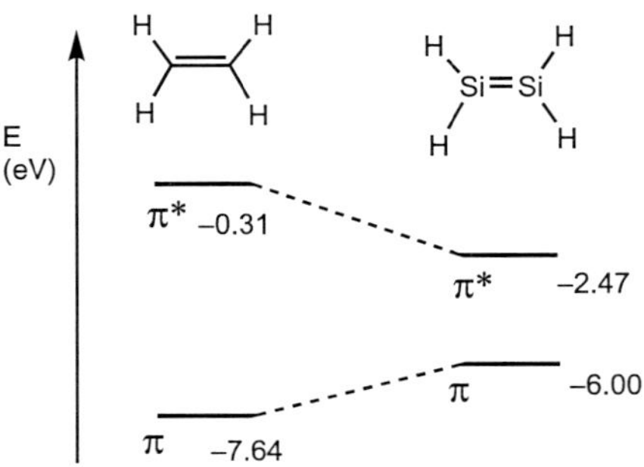

Fig. 6.11 Comparison of calculated energy levels of ethylene and disilene

Disilynes

The first example of a silicon–silicon triple bond was discovered by Akira Sekiguchi in 2004.[26] This was achieved by the KC_8 reduction of a tetrabromo linear tetrasilane having bulky groups at the periphery (Eq. 6.39). The emerald green crystalline compound formed had four Si atoms in the chain which are not collinear. In this respect, this compound differs from the related traditional carbon-based alkyne. The four silicon atoms in the chain are however perfectly coplanar, the first and fourth silicon atoms are *trans* to each another, and the angle between the triple bond and the adjacent silicon is 137.4°. The central triple bond is 206.2 pm [around 4% shorter than Si–Si double bonds (214 pm)] and the other Si–Si single bonds are 237 pm. The UV-vis spectrum of this compound gave two weak absorption peaks at 690 and 483 nm in the visible region and two strong bands (328 and 259 nm) in the UV region. Based on theoretical calculations, the colour is determined to be due to a weak π–π* transition from HOMO to LUMO and HOMO-1 to LUMO or HOMO to LUMO+1 (Fig. 6.13). The chemistry of this triple bond has also been explored leading to partial saturation and even complete cleavage of the silicon–silicon triple bond (Scheme 6.13).[27–31]

$$[(Me_3Si)_2HC]_2({}^iPr)Si{-}SiBr_2{-}SiBr_2{-}Si({}^iPr)[CH(SiMe_3)_2]_2 \xrightarrow[THF]{KC_8(4\ eq)} [(Me_3Si)_2HC]_2({}^iPr)Si{-}Si{\equiv}Si{-}Si({}^iPr)[CH(SiMe_3)_2]_2 \qquad (6.39)$$

π–π* 328 nm
Emerald green, 2.06 Å

While alkynes are linear, disilynes are found to be of *trans* bent geometry.[32] The *trans* bent angle in the case of the first disilyne was found to be 137.4° which is close to that predicted for the parent disilyne HSiSiH (124.5–127.4°). This observation further disproved the *double bond rule* which states that, multiple bonds between elements with a principal quantum number equal to or greater than 3 cannot be formed due to Pauli repulsion between the inner shells of these elements. The bonding in alkynes is compared with that in disilyne (Fig. 6.12) as well as the contributing resonance structures for such a *trans* bent geometry of disilynes (Fig. 6.13).

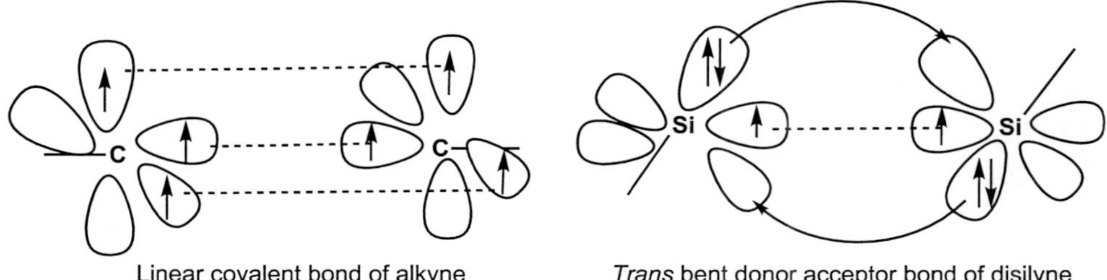

Fig. 6.12 Comparison of bonding of alkynes with disilynes

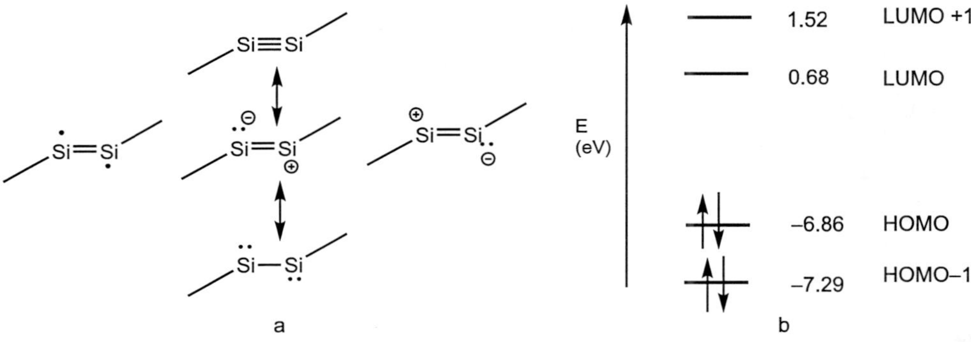

Fig. 6.13 (a) Resonance structures possible for *trans* bent disilynes and (b) frontier orbitals of disilynes and their energy levels

6.20.2 Extent of Triple Bond Formation by Group 14 Elements

The synthesis of the first silicon–silicon triple bond was indeed a landmark discovery and soon more examples were made, characterised and their reactivity explored. But this was not an easy task as attempts to realise triple bonds between heavier group 14 elements were not always successful. The most noted example was diplumbyne in which the bond between the two lead atoms remained a long single bond and the pair of remaining electrons did not form a multiple bond but was localised on each of the lead atoms. The use of highly stericially hindered aryl substituents was required to stabilise these so called triple bonds of Pb, Sn and Ge but these substituents were not successful with silicon. The significant trans-bending observed in all these compounds indicates that the bond order is less than three and the non-bonding lone pair character increases on going down the group.[25]

Scheme 6.13 General reactions of silicon–silicon triple bonds

E_2R_2	RSi–Si≡Si–SiR	Ar–Ge=Ge–Ar	Ar–Sn=Sn–Ar	Ar–Pb–Pb–Ar
Colour	Emerald green	Orange red	Purple	Dark green
Distance E–E (Å)	2.062	2.285	2.667	3.188
Angle E–E–Ar (°)	137.4	126.7	125.2	94.26

Fig. 6.14 Comparison of extent of triple bond formation by group 14 elements

6.21 Recent Developments

6.21.1 Silicon in +1 and 0 Oxidation States

Robinson and coworkers prepared a complex of a sterically hindered NHC with $SiCl_4$ and this compound was reduced with KC_8. Depending upon the amount of reductant used, dimeric compounds were obtained having silicon in +1 and 0 oxidation states.[33] Controlled oxidation of the dimeric Si(0) compound using N_2O or O_2 resulted in novel compounds having both Si–O and Si=O stabilised by N-heterocyclic carbenes (Scheme 6.14).[34]

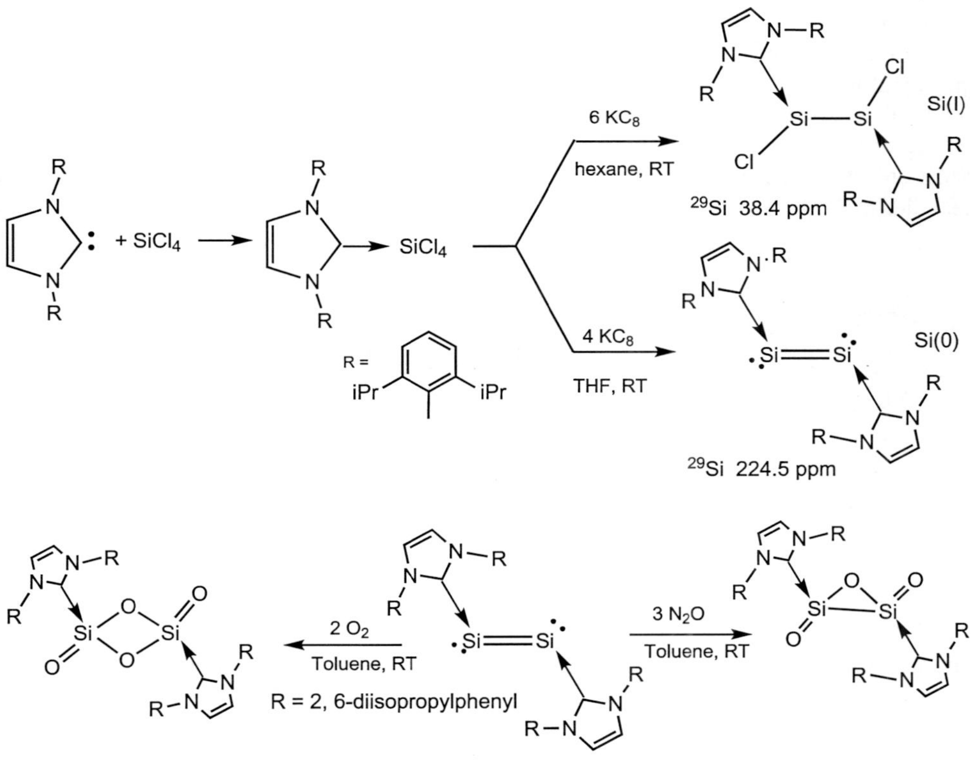

Scheme 6.14 Silicon in +1 and 0 oxidation states and controlled oxidation of the Si(0) compound

6.21.2 Stable Silylenes [Silicon (II) Compounds]

A significant difference between silylenes and germylenes on one hand and traditional methylenes on the other hand is the difference in the multiplicity of their ground states.

Methylene is a ground state triplet while silylene and germylene are ground state singlets. In the case of methylene its singlet state lies 9.0 kcal/mol higher in energy. This reversal in the multiplicity of the ground state on going from methylene to silylene has a significant effect on the chemistry of silylenes compared to methylenes. The HOMO–LUMO gap is also larger in SiH_2 (52 kcal/mol) than in the case of CH_2 (27 kcal/mol). For SiH_2, the energy gained in forming a triplet configuration is not sufficient to compensate for the large HOMO–LUMO gap and therefore SiH_2 favours singlet configuration.[35]

The search for stable silicon(II) compounds started even before the discovery of the first NHC. Peter Jutzi stabilised Si(II) by using pentamethylcyclopentadienyl groups in 1986.[36] The 1980s also witnessed some silylenes being stabilised in argon/hydrocarbon matrices at very low temperatures. The first NHC analogue of silicon N-heterocyclicsilylene (NHS) was prepared by Denk and West in 1994.[37] This silylene, crystalline and persistent at room temperature, was found to be highly stable, decomposing only at 220°C. The preparation of this unsaturated NHS was followed soon by its saturated analogue by West and coworkers, and then benzo and pyrido fused silylenes of a similar type followed. The first dialkylsilylene having carbon atoms flanking the Si(II) was made by Kira and coworkers in 1999.[38] The ^{29}Si-NMR of this compound gave a signal for the divalent silicon at 567.4 ppm, an unusually low field Si-NMR signal for a silicon compound.

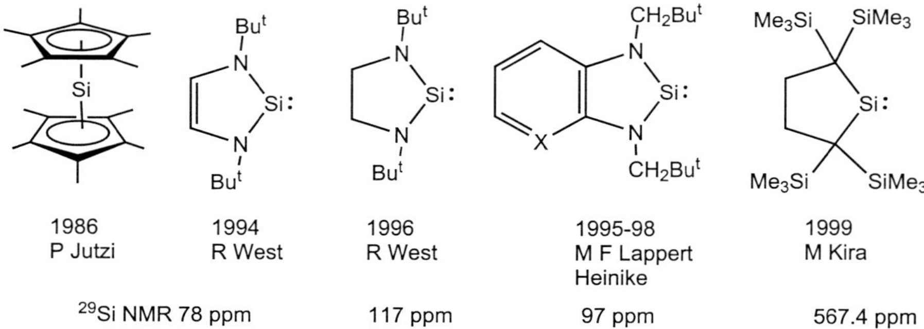

Fig. 6.15 Examples of stable silylenes with their ^{29}Si-NMR shifts

Glyoxal on reaction with t-butylamine gave the diimine which was reduced with lithium metal to get the dilithiated diimine. The dilithiated imine was reacted with $SiCl_4$. The dichlorosilyl compound was reduced with molten potassium in refluxing THF to make the silylene in good yields.[39]

$$\text{OHC–CHO} \xrightarrow{Bu^tNH_2} Bu^tN{=}CH{-}CH{=}NBu^t \xrightarrow{2.1\ eq.\ Li} Bu^tN(Li)CH{=}CHN(Li)Bu^t \xrightarrow{SiCl_4} (Bu^tNCH{=}CHNBu^t)SiCl_2 \xrightarrow{K,\ THF} (Bu^tNCH{=}CHNBu^t)Si: \quad (6.40)$$

A similar reduction was carried out with KC_8 on the five-membered cyclic dialkyldibromosilane to get the silylene having carbon atoms flanking the Si(II).

The compound was stable as a solid at 0°C but as solution in hexane, undergoes slow rearrangement to yield the silicon–carbon double bonded compound.

Me3Si SiMe3 SiBr2 Me3Si SiMe3 —KC8/THF, −50°C→ Me3Si SiMe3 Si Me3Si SiMe3 —R.T, hexane→ Me3Si Si-SiMe3 Me3Si SiMe3 (6.41)

Carbene stabilised $SiCl_2$

A donor stabilised $SiCl_2$ was prepared by Roesky and coworkers in 2009 by using an N-heterocyclic carbene according to the procedure given in Eq. 6.43.[40] The complex was reacted with a better σ-donating cyclic alkylamino carbene with a view to displace the NHC and form a more strongly bonded cAAC→$SiCl_2$ complex. Quantum chemical calculations have predicted that the Si–C bond in such a complex with cAAC will be shorter and stronger than the NHC→$SiCl_2$ complex.[41] However the reaction gave a completely different product, $SiCl_2(cAAC)_2$, with two CAAC units bound to the $SiCl_2$ and the Si–C bonds were even shorter than the predicted bond distances. Since $SiCl_2$ has only one coordination site, this experimental finding has been explained by a changeover from a dative bond in NHC→$SiCl_2$ to electron sharing (covalent) bond in $SiCl_2(cAAC)_2$ which means that the complex has converted from a singlet silylene to a stable biradical. The presence of a biradical was ascertained by magnetic studies and ESR spectroscopy. The $SiCl_2$ bound biradical on reduction with KC_8 was found to result in a silylone.[42]

NHC: + HSiCl3 ⟶ NHC→SiCl2 + [NHC-H]+ Cl− ; R = iPr–C6H3(Me)–iPr (6.42)

NHC→SiCl2 + 2 cAAC: ⟶ Triplet (biradical) —KC8→ Silylone (6.43)

6.21.3 Stable Silylones (Silicon 0 Compounds)

Silylones are two-coordinate Si(0) compounds of the type L:→Si(0)←:L where L is a strong donor such as an N-heterocyclic carbene, cyclic alkyl(amino) carbene, and so on. Theoretical calculations predicted that these compounds, which are similar to allenes will not be linear but would adopt a bent geometry around the silicon(0). This is in accordance with the VSEPR

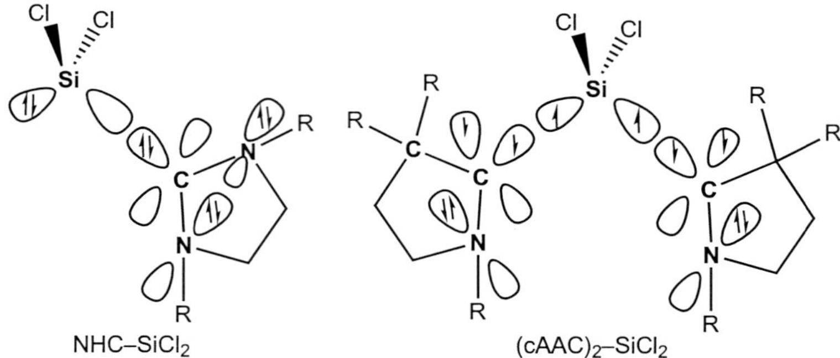

Fig. 6.16 Differences in the bonding and orbitals of NHC–$SiCl_2$ and $(cAAC)_2SiCl_2$

theory as silicon will have two lone pairs of electrons and two dative bonds from the donor ligands. A classic example of a silylone was described by Roesky and coworkers[42] where the ligands were cyclic alkylamino carbenes. This was followed by other examples where bis-NHC ligands were used.[43]

E = Si, Ge Silylone and germylone

Fig. 6.17 Examples of silylones and germylones

The zero oxidation state of the silicon in the silylone and the presence of two lone pairs have been experimentally proved by reactions with reagents such as elemental sulfur and zinc chloride.[44]

(6.44)

Silylium ions—trivalent tricoordinate silicon cations

Unlike carbocations which have a rich and diverse chemistry, isolation and characterisation of unassociated silyium cations turned out to be a difficult task. Reed and coworkers isolated such a silyium cation analogous to the triphenylmethyl (trityl) cation in 2002, which led to a debate on how free the silylium cation actually is. This debate clearly reflected the challenges in isolating and structurally characterising such electron deficient compounds in the free form. The difficulty in generating R_3Si^+, unlike R_3C^+, is the exceptional electrophilicity of the trivalent silicon atom. The electrophilicity as well as the larger size of the silicon atom result in C–Si bonds which are longer than the C–C bonds; the orbital overlap is also less effective. Orbitals of silicon show a considerable reluctance towards π overlap and hyperconjugation, both of which have been found to stabilise carbenium ions. As a consequence, silylium cations react readily with any nucleophiles (both σ and π donors) including solvents and counter ions.[45]

In the synthesis of $Mes_3Si^+[HCB_{11}Me_5Br_6]^-$ this long standing silyium ion problem was solved by using non-nuclophilic solvents, a weakly coordinating carborane counter ion and sterically hindered mesityl groups. The carborane anion is one of the most chemically robust and least coordinating anions known to date. The compound was spectroscopically and structurally characterised. Its ^{29}Si-NMR chemical shift is 225.5 ppm, downfield of TMS, which indicates that the cation is quite unattached to any other units.

$$Mes_3Si(CH_2CH{=}CH_2) + Et_3Si(HCB_{11}Me_5Br_6) \longrightarrow Mes_3Si^{\oplus}\,[HCB_{11}Me_5Br_6]^- + Et_3Si(CH_2CH{=}CH_2) \quad (6.45)$$

Silylium cations have been found to be extremely useful for the activation of C–F bonds which are otherwise quite unreactive and bring about hydrodefluorination reactions.

6.20.6 Silanones: A Reality?

Silanones are some of the most sought after compounds of silicon. Unlike ketones, silanones are very highly unstable and reactive. This is due to:

- a weak π bond having unfavourable overlap between pπ (Si) and pπ (O) orbitals, and
- a strongly polarised $Si^{\delta+}$–$O^{\delta-}$ bond with a significantly large bond energy difference between Si–O π (58 kcal/mol) and Si–O σ (109 kcal/mol) bonds.[34]

The strongly polarised bond gives rise to a highly electrophilic silicon centre. As a consequence, such bonds easily undergo oligomerisation to give polysiloxanes through exothermic reactions which are barrier-less. While many silanones were prepared, they could be stabilised only by using an extra base bound to the silicon resulting in a tetracoordinate silicon. Although the first example of an ideal and stable germanone having the Ge=O bond tricoordinate and bound to two carbons was prepared in 2012,[46] analogous silicon compounds having tricoordinate silicon and which are stable at room temperature were not made. However, three tricoordinate silanones are worth mention as they come very close to an ideal tricoordinate and room-temperature-stable silanones.

A C Filippou and coworkers prepared a tricoordinate silanone (Scheme 6.15) having a carbon from NHC and chromium flanking on both sides the Si=O bond.[47] N_2O was used as the oxidising agent. The very high reactivity of this compound was observed from the fact that even 1 ppm of moisture impurity present in argon gas was sufficient to convert the Si=O unit to $Si(OH)_2$. Raman spectra gave a band for the Si=O bond at 1157 cm^{-1}. Single crystal X-ray diffraction analysis proved the presence of a trigonal-planar coordinated silicon centre unambiguously (the sum of bonding angles at Si is 359.98).

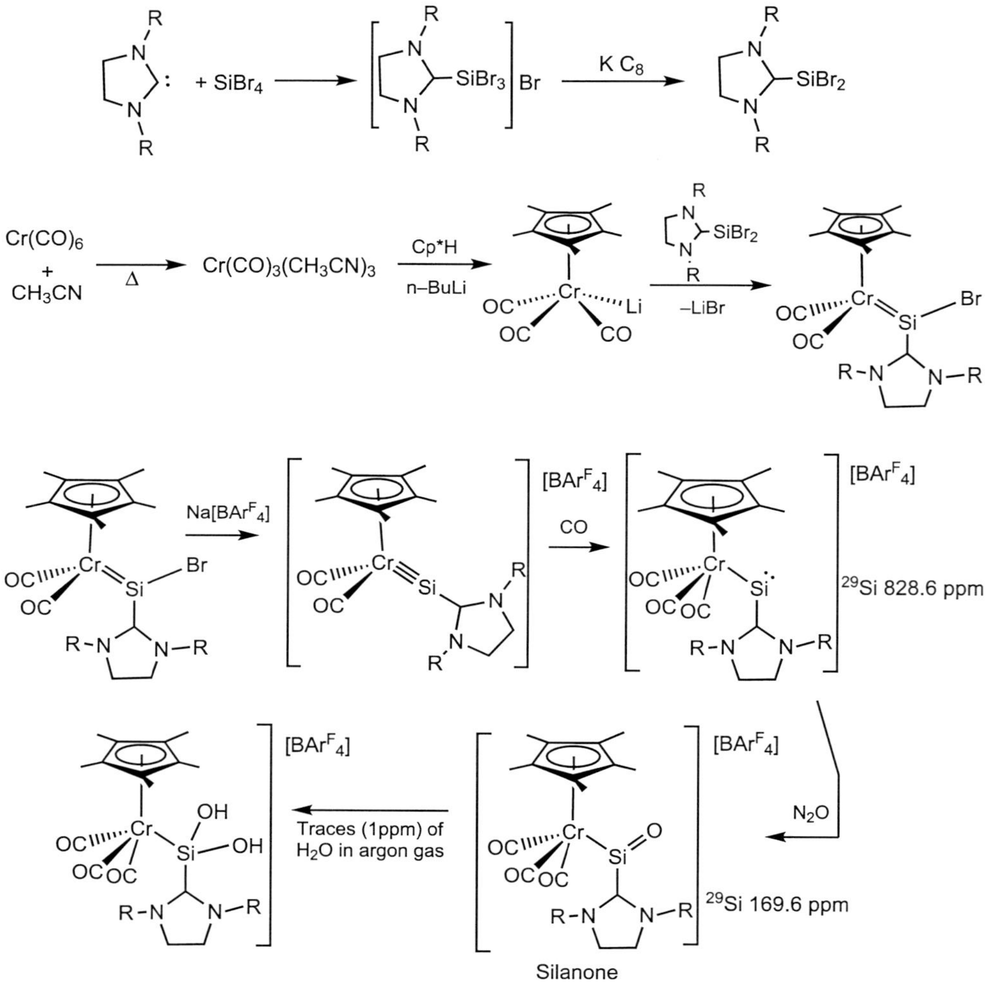

Scheme 6.15 Synthesis of the first stable tricoordinated silanone from $SiBr_4$

In another development, Kira and coworkers prepared a tricoordinate silanone flanked on both sides of the Si=O bond by carbon atoms. However this compound was stable only below –80°C, it rearranged to a new compound with a Si–O–Si bond above this temperature.[48]

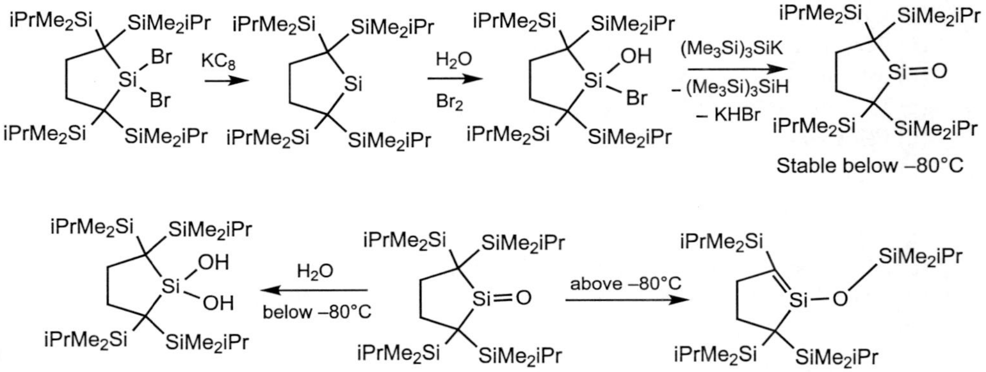

Scheme 6.16 Attempted synthesis of a tricoordinated silanone and its rearrangement

A third type of silanone was reported in 2017 which was stable below –50°C and was crystalline. A sterically hindered (amino)(ylide) silylene (Eq. 6.46) was found to readily react with N_2O to give the silanone which however underwent dimerisation at room temperature to yield the corresponding head to tail dimer.[49]

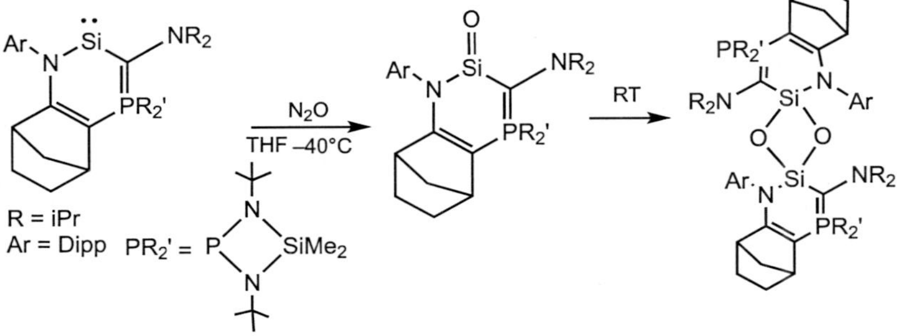

(6.46)

Questions for self study (multiple choice)

6.1. Ferrosilicon, the impure form of silicon metal is used in the steel industry. This is because:

(a) Silicon reacts with dissolved oxygen to prevent oxidation of iron to iron oxides.

(b) Silicon reacts with dissolved oxygen so that carbon added to steel is not converted to CO and CO_2 by reacting with the dissolved oxygen and escape from the melt.

(c) Silicon alloys with iron giving steel more strength and durability.
(d) Addition of silicon prevents rusting of steel by forming an alloy.

6.2. The basic structural unit of silicates is:
(a) SiO_3^{2-} (b) SiO_4^{2-}
(c) SiO^- (d) SiO_4^{4-}

6.3. Which one among the following statements is wrong?
(a) Beryl is an example of a cyclic silicate.
(b) Mg_2SiO_4 is an orthosilicate.
(c) The basic structural unit in silicates is the SiO_4 tetrahedron.
(d) Feldspars are not aluminosilicates.

6.4. A straight chain silicone polymer is formed by:
(a) Hydrolysis of CH_3SiCl_3 followed by condensation polymerisation.
(b) Hydrolysis of $(CH_3)_2SiCl_2$ followed by condensation polymerisation.
(c) Hydrolysis of $(CH_3)_3SiCl$ followed by condensation polymerisation.
(d) Hydrolysis of $(CH_3)_4Si$ followed by addition polymerisation.

6.5. Which of the following statements about zeolites is false?
(a) They are used as cation exchangers.
(b) They have open structures that enable them to take up small molecules.
(c) Zeolites are alumino silicates having a three dimensional network.
(d) Some of the SiO_4^{4-} units are replaced by AlO_4^{4-} and AlO_6^{4-} units.

6.6. Na^+, Mg^{2+}, Al^{3+} and Si^{4+} are isoelectronic. Their ionic sizes follow the order
(a) $Na^+ > Mg^{2+} < Al^{3+} < Si^{4+}$ (b) $Na^+ < Mg^{2+} < Al^{3+} < Si^{4+}$
(c) $Na^+ > Mg^{2+} > Al^{3+} > Si^{4+}$ (d) $Na^+ < Mg^{2+} > Al^{3+} < Si^{4+}$

6.7. Silicon of very high purity required for electronic and solar cells is obtained by:
(a) Zone refining of silicon.
(b) Crystallisation of silicon from a melt.
(c) Converting to $SiCl_4$ and $HSiCl_3$, followed by repeated distillation and reduction.
(d) Controlled melting and allowing melted silicon to move down a slanting surface.

6.8. The increasing order of bond energies is:
(a) Si–O < Si–F < Si–H < Si–C (b) Si–C < Si–H < Si–O < Si--F
(c) Si–H < Si–C < Si–F < Si–O (d) Si–C < Si–H < Si–F < Si–O

6.9. Which of the following statements is not true for the Rochow–Muller process?
(a) Me_2SiCl_2 is the major product of this process.
(b) Methyl chloride does not directly react with pure silicon.
(c) The reaction proceeds at room temperature.
(d) Me_3SiCl is a minor product of this reaction.

6.10. The Badger–Mobil process of synthesis of styrene using ZSM-5 catalyst is an example of:
(a) Reactant selectivity in shape selective catalysis.
(b) Product selectivity in shape selective catalysis.
(c) Restricted transition state selectivity in shape selective catalysis.
(d) No shape selectivity in catalysis.

Solved concept-based questions

6.11. Unlike CCl_4, $SiCl_4$ hydrolyses readily. Why?

Factors which need to be considered

(a) The C–Cl electronegativity difference is 0.5, while for Si–Cl, it is 1.2. This corresponds to an approximate ionic character of 6% for the C–Cl bond and 30% for the Si–Cl bond. (b) The covalent radius of carbon is 77 pm while that of silicon is 118 pm.

The *d*-orbitals of silicon are available to accept the lone pair of electrons from the attacking hydroxide ion/ water molecule and the Si–Cl bond does not have to break first; however, it is easier for the Si–Cl bond to break compared to the C–Cl given the much higher polarity of the former. (b) Since the silicon atom is larger in size than the carbon, there is less steric hindrance in $SiCl_4$ for the formation of the necessary 5-coordinate transition state. It may be also be possible that it is a true 5-coordinate intermediate as well, since stable 5-coordinate intermediates are well known in silicon chemistry.

6.12. Silicon forms double bonds with difficulty while carbon forms double bonds readily. Why?

The $3p$ orbitals of silicon which should involve in $p\pi$–$p\pi$ bonding are larger and more diffuse in nature compared to the $2p$ orbitals of oxygen. So $p\pi$–$p\pi$ bonding is less strong in Si–Si π bonds. Similarly $p\pi$–$d\pi$ or $d\pi$–$d\pi$ bonds which can form between silicon units are less compact and less strong compared to the $p\pi$–$p\pi$ bond of O_2.

6.13. $(SiH_3)_3N$ is almost planar while $(CH_3)_3N$ is pyramidal. Why?

The earlier explanation was that the lone pair on the nitrogen is involved in a $p\pi$–$d\pi$ bonding interaction with the empty *d* orbitals of silicon. However, detailed theoretical studies now rule out the concept of $p\pi$–$d\pi$ bonding having any role to play (rather, a vey insignificant role only). The modern reason focuses on the higher electronegativity of the N compared to the silicon. Large bond angles are due to weak pairing of electrons on oxygen as a result of weakly the electronegative SiH_3 group as well as electrostatic repulsions between bulky SiH_3 groups.

6.14. Silicones are hydrophobic while silicates are not. Why?

Methyl groups which are the substituents on the Si–O polymer back bone bring in the hydrophobicity. Silicates do not have organic non-polar groups and the termini are ionic sites which makes them hydrophilic.

6.15. Silane is a reducing agent under normal conditions while methane is not. Why?

Because of the greater electronegativity of hydrogen in comparison to silicon, the hydrogen atoms in silane have a partial negative charge and the silicon a partial positive charge. This Si–H bond polarity is the opposite of that observed in the C–H bonds of methane. Silicon (1.90) is less electronegative than hydrogen (2.20); as a result, the silicon–hydrogen bond possesses some hydridic character. In the presence of a strong electrophile, organosilanes containing a Si–H bond (hydrosilanes) serve as hydride donors or reducing agents.

6.16. Sodalite (molecular sieves 4A) and ZSM-5 are both zeolites. But only ZSM-5 brings about shape selective catalysis such as conversion of toluene to p-xylene selectively. Why?

Shape selective catalysis shown by zeolites depends upon the pore size of the channels found on a zeolite, its cavity size at the junctions of these channels, as well as the ratio of Si to Al of a zeolite. The pore size of zeolite 4A is only 4 Å which is too small for small organic molecules such as toluene to enter the channel and undergo catalysis. Pore size of ZSM-5 is between 5 and 6 Å. Also the ratio of Al : Si is 1:1 in Zeolite 4A while it is 1:31 in ZSM-5.

6.17. Why were silicone-based shoes and outfits used in the lunar expeditions of USA?

Poly dimethyl silicone has a glass transition temperature which is quite low (–125°C). Therefore, materials made of this polymer do not become hard and brittle till it reaches the lower limit of its glass transition temperature. The thermal decomposition of silicones also sets in above 200°C. The surface of the moon during a lunar day can vary from –153 to +107°C.

Problems and exercises

6.18. Ethyl benzene, the precursor for styrene, is industrially prepared by the reaction of benzene and ethylene in the presence of ZSM-5 catalyst (Badger–Mobil process, makes one-fourth of the world's requirement). Is this a shape selective catalysis? A small amount of diethylbenzene also has been reported to form in this reaction. Predict its most likely structure from among the structural isomers possible.

6.19. With reference to aluminum content what is the difference between kaolinite, muscovite mica and molecular sieves 4A? What is the stickiness of kalonite due to?

6.20. Given only pure silicon and MeCl as feedstock (starting materials) write the steps for preparing the linear polymer with specific structure $Me_3SiO(SiMe_2O)_{200}SiMe_3$ (PDMS). Indicate conditions/catalysts at each step.

6.21. The chromatographic analysis of the MTG process of Mobil corporation using ZSM-5 catalyst indicated that in addition to the major fractions of alkyl chain hydrocarbons, a small percentage of methyl substituted aryl hydrocarbons were also obtained resulting from ring formation and aromatisation. Given that the maximum substitution of methyl groups observed to form on a benzene ring was four from the chromatographic analysis, predict the structures of the MAJOR isomer of (a) di, (b) tri and (c) tetra methyl substituted benzene derivatives obtainable from this shape selective catalysis.

6.22. Starting from sand, show the steps for the preparation of hexamethyldisilazane, $(Me_3Si)_2NH$.

6.23. Three silicon compounds which are common precursors for making ultrapure silicon are provided. Hydrolysis of A (gas) gives SiO_2 and the water soluble residue does not answer the test for chloride. Hydrolysis of B (liquid) gives SiO_2, the residue gives a positive test for chloride and the proton-NMR of the compound indicates a singlet at 6.12 ppm. Hydrolysis of C (liquid) gives SiO_2 and the residue gives a positive test for chloride and the proton-NMR of the compound does not give any signals. Identify compounds A to C.

6.24. How is ferrosilicon prepared from pure sand? What is its most important use in industry and why it is important?

6.25. Match the type of silicon bond with its name for the following

(i)	$\rangle Si{-}O{-}Si\langle$	(a) silylone
(ii)	$\rangle Si{=}O$	(b) silylene
(iii)	$\rangle Si:$	(c) silicone
(iv)	$-Si{-}H$	(d) silanone
(v)	$: \rightarrow \ddot{\underset{\cdot\cdot}{Si}} \leftarrow :$	(e) silane

6.26. How is $Cl_3Si{-}SiCl_3$ prepared starting with silicon? Show its most important use by giving an example.

6.27. The ^{29}Si-NMR spectrum of a methyl silicone compound showed two peaks, one at –110 ppm and the other at +7 ppm. The compound has only five silicon atoms. Propose a suitable structure for this compound.

6.28. Separate hydrolysis of two methyl silicon compounds A and B, gave as major isolable products C and D respectively. The ^{29}Si-NMR spectrum of C gave a single peak around –22 ppm and that of D gave a single peak around +7 ppm. The compound C has four silicon atoms while compound D has two silicon atoms. Identify compounds A to D. An equimolar mixture of C and D with catalytic amounts of conc. H_2SO_4 gave a new compound E. Predict its structure.

6.29. Indicate how the following reaction can be carried out with a minimum number of steps (given ethylene and ethylene oxide; use a catalyst of your choice).

$$HSiCl_3 \xrightarrow{?} CH_3CH_2Si(OCH_2CH_2Cl)_3$$

Answer key for self study questions

6.1. (b) 6.2. (d) 6.3. (d) 6.4. (b) 6.5. (d) 6.6. (c) 6.7.(c) 6.8. (b) 6.9. (c) 6.10. (d)

Supplementary reading

1. Simmler W, *Silicon compounds, Inorganic.*in *Ulmann's encyclopedia of industrial chemistry,* Wiley VCH, 2012, Vol. 32, 615.
2. Voronkov M G, History of Chemistry: Silicon era, *Russ. J. Appl. Chem.,* 2007, Vol. 80, 2190.
3. Weaver E R, The generation of hydrogen by the reaction between ferrosilicon and a solution of sodium hydroxide, *Ind. Eng. Chem.,*1920, Vol. 12, 232.
4. Rӧsch L, John P and Reitmeier R, *Silicon compounds, Organic.*in *Ulmann's encyclopedia of industrial chemistry,* Wiley VCH, 2012,Vol. 32, 637.

5. Moretto HH, Schulze M and Wagner G, *Silicones* in *Ulmann's encyclopedia of industrial chemistry*, Wiley VCH, Vol. 32, 675, 2012.
6. Thomas N R, Frederic Stanley Kipping- Pioneer in silicon chemistry: His life and legacy, *Silicon*, 2010, Vol. 2, 187.
7. Rochow E G, *Silicon and silicones*, Springer Verlag, 1987.
8. Clarke M P, The direct synthesis of methylchlorosilanes, *J. Organomet. Chem.*, 1989, Vol. 376, 165.
9. Bruynes C A and Jurriens T K, Catalysts for silylation with 1,1,1,3,3,3-hexamethyldisilazane, *J. Org. Chem.*, 1982, Vol. 47, 3966.
10. Azisi N and Saidi M R, Novel and Efficient Method for the Silylation of Hydroxyl Groups with Hexamethyldisilazane (HMDS) under Solvent-Free and Neutral Conditions, *Organometallics*, 2004, Vol. 23, 1457.
11. Lappert M, Protchenko A, Power P and Seeber A, *Metal amide chemistry*, Wiley-VCH, 2009.
12. Troegel D and Stohrer J, Recent advance and actual challenges in late transition metal catalysed hydrosilylation of olefins from an industrial point of view, *Coord. Chem. Rev.*, 2011, Vol. 255, 1440.
13. Gupta B D and Elias A J, *Basic Organometallic Chemistry: Concepts, syntheses and applications*, 2nd Ed., Universities Press, Hyderabad, 2013.
14. Fernández-Alvarez F J, Aitani A M and Oro L A, Homogeneous catalytic reduction of CO_2 with hydrosilanes, *Catal. Sci. Technol.*, 2014, Vol. 4, 611.
15. Riduan S N, Zhang Y and Ying J Y, Conversion of carbon dioxide into methanol with silanes over N-Heterocyclic carbene catalysts, *Angew. Chem., Int. Ed.*, 2009, Vol. 48, 3322.
16. Mark J E, Allcock H R and West R, *Inorganic polymers*, Prentice Hall, Englewood, NJ, 1992.
17. Seethapathy S and Górecki T, Applications of polydimethylsiloxane in analytical chemistry: A review, *Analytica chimica acta*, 2012, Vol.750, 48.
18. Madar M, Materials science: Silicon carbide in contention, *Nature*, 2004, Vol.430, 974.
19. Ceika J, Corma A and Zones S, *Zeolites and catalysis: Synthesis, reactions and applications*, Wiley VCH, 2010.
20. Weitkamp J and Puppe L, *Catalysis and zeolites; Fundamentals and applications*, Springer, 1999.
21. Csicsery S M, Catalysis by shape selective zeolites-science and technology, *Pure Appl. Chem.*, 1986. Vol. 58, 841.
22. Csicsery S M, Shape selective catalysis in zeolites, *Zeolites*, 1984, Vol. 4, 202.
23. West R, Fink M J and Michl J, Tetramesityl disilene, a stable compound containing a silicon–silicon double bond, *Science*, 1981, Vol. 214, 1343.
24. Kira M and Iwamoto T, Progress in the chemistry of stable disilenes, *Adv. Organomet. Chem.*, 2006, Vol. 54, 73.
25. Kira M, Bonding and structure of disilenes and related unsaturated group 14 element compounds, *Proc. Jpn. Acad. Ser. B*, 2012, Vol.88, 167.
26. Sekiguchi A, Kinjo R and Ichinohe M, A stable compound containing a silicon-silicon triple bond, *Science*, 2004, Vol. 305, 1755.
27. Takeuchi K, Ichinohe M and Sekiguchi A, A new disilyne with π-accepting groups from the reaction of disilyne RSi^iSiR ($R = Si^iPr[CH(SiMe_3)_2]$) with isocyanides, *J. Am. Chem. Soc.*, 2012, Vol. 134, 2954.
28. Sekiguchi A, Ichinohe M and Kinjo R, The chemistry of disilyne with a genuine Si–Si triple bond: Synthesis, structure and reactivity, *Bull Chem. Soc. Jpn*, 2006, Vol. 79, 825.

29. Kinjo R, Ichinohe M, Sekiguchi A, Takagi N, Sumimoto M and Nagase S, Reactivity of disilene RSi^iSiR ($R = Si^iPr[CH(SiMe_3)_2]_2$) towards π-bonds: Stereospecific addition and a new route to an isolable 1,2-disilabenzene, *J. Am. Chem. Soc.*, 2007, Vol. 129, 7766.
30. Kinjo R, Ichinohe M and Sekiguchi A, An isolable disilyne anion radical and a new route to the disilenide ion upon reduction of a disilyne, *J. Am. Chem. Soc.*, 2007, Vol. 129, 26.
31. Takeuchi K, Ichinohe M and Sekiguchi A, Hydroboration of disilyne RSi^iSiR ($R = Si^iPr[CH(SiMe_3)_2]_2$), giving boryl-substituted disilenes, *Organometallics*, 2011, Vol. 30, 2044.
32. Landis C R and Weinhold F, Origin of the trans-bent geometries in maximally bonded transition metal and main group molecules, *J. Am. Chem. Soc.*, 2006, Vol. 128, 7335.
33. Wang Y, Xie Y, Wei P, Bruce King R, Schaefer III H , Schleyer P R and Robinson G H, A stable silicone(0) compound with Si=Si double bond, *Science*, 2008, Vol. 321, 1069.
34. Wang Y, Chen M, Xie Y, Wei P, Schaefer III H F, Schleyer P and Robinson G H, Stabilisation of elusive silicon oxides, *Nature Chem.*, 2015, Vol. 7, 509.
35. Apeloig Y, Pauncz R, Karni M, West R, Steiner W and Chapman D, Why is methylene a ground state triplet while silylene is ground state singlet? *Organometallics*, 2003, Vol. 22, 3250.
36. Haaf M, Schmedake T A and West R, Stable silylenes, *Acc. Chem. Res.*, 2000, Vol. 33, 704.
37. West R and Denk M, Stable silylenes: synthesis, structure, reactions, *Pure Appl. Chem.*, 1996, Vol. 68, 785.
38. Kira M, Ishida S, Iwamoto T and Kabuto C, The first isolable dialkylsilylene, *J. Am. Chem. Soc*, 1999, Vol. 121, 9722.
39. Gerhaus B, Lappert M F, HeinickeJ, Boese R and Blaser D, Synthesis, structure and reactions of new stable silylenes, *Chem. Commun.*, 1995, 1931, 9.
40. Mondal K C, Roesky H W, Schwarzer M C, Frenking G, Neudeck S, Tkach I, Wolf H, Kratzert D, Herbst-Irmer R, Niepotter H and Stalke D, Conversion of a singlet silylene to a stable biradical, *Angew Chem Int. Ed.*, 2013, Vol. 52, 1801.
41. Niepotter B, Herbst-Irmer R, Kratzert D, Samule P P, Mondal K C, Roesky H W, Jerabek P and Frenking G, Experimental charge density study of a silylone, *Angew. Chem. Int. Ed.*, 2014, Vol. 53, 2766.
42. Mondal K C, Roesky H W, Schwarzer M C, Frenking G, Niepotter H, Herbst-Irmer R and Stalke D, A stable singlet biradicaloid siladicarbene $(L:)_2Si$, *Angew Chem Int. Ed.*, 2013, Vol. 52, 2963.
43. Xiong Y, Yao S, Inoue S, Epping J D and Driess M, A cyclic silylone (siladicarbene) with an electron rich silicon(0) atom, *Angew. Chem. Int. Ed.*, 2013, Vol. 52, 7147.
44. Xiong Y, Yao S, Mueller R, Kaupp M and Driess M, From silylone to an isolable monomeric silicon disulfide complex, *Angew. Chem. Int. Ed.*, 2015, Vol. 54, 10254.
45. Klare H F T and Oestreich M, Silylium ions in catalysis, *Dalton Trans*, 2010, Vol. 39, 9146.
46. Li L, Fukawa T, Matsuo T, Hashisume D, Fueno H, Tanaka K and Tamao K, A stable germanone as the first isolated heavy ketone with a terminal oxygen atom, *Nature Chem.*, 2012, Vol. 4, 361.
47. Fillipou A C, Baars B, Chernov O, Lebdev Y N and Schankenburg G, Silicon–oxygen double bonds: a stable silanone with a trigonal planar coordinated silicon centre, *Angew. Chem. Int. Ed.*, 2014, Vol. 53, 565.
48. Ishida S, Abe T, Hirakwa F, Kosai T and Sato K, Persistent dialkylsilanone generated by dehydrobromination of dialkylbromosilanol, *Chem . Eur. J.*, 2015, Vol. 21, 15057.
49. Alvarado-Beltran I, Rosas-Sanchez A, Baceiredo A, Saffon-Merceron N, Branchadell V and Kato T, A fairly stable crystalline silanone, *Angew Chem. Int. Ed.*, 2017, Vol. 56, 10481.

THE CHEMISTRY OF GERMANIUM, TIN AND LEAD

CHAPTER 7

UNIQUE PROPERTIES OF Ge, Sn, Pb AND THEIR COMPOUNDS

GERMANIUM

- *The unique optical properties of germanium are, highest refractive index (4.003) for any glass forming material and low optical dispersion. These properties make it useful as an additive to reduce attenuation (reduction in power of light signal) of fibreoptic cables used in telecommunication.*
- *Germanium is the material from which the first transistors were made. A tin-doped germanium rectifier diode for solid state signal processing was found to be highly resistant to burnout by stray voltages during the second world war. It was extensively used as a semiconductor material (band gap 0.67 eV at 273 K) from 1940–1960.*
- *Modern day germanium-containing semiconductors find use in mobile phones and hand held devices as they consume much less power and are much faster than conventional semiconductors.*
- *Its use as a semiconductor declined considerably after the advent of silicon-based semiconductors as silicon was inexpensive; large scale and very large scale integrated circuits (VLSI) were made readily from silicon.*
- *Germanium and germanium-based glasses transmit near IR radiation more effectively than silicon. This property made it useful in the making of passive night vision lenses which use heat radiated by objects to view them in the dark.*
- GeO_2 *is extensively used as a catalyst for the manufacture of polyethylene terephthalate (PET) as unlike PET made using antimony-based catalysts, the mineral water bottles made out of PET, polymerised using* GeO_2*, retain transparency and do not turn yellow with time.*
- *Germanium (II) compounds, often prepared from* $GeCl_2$*.dioxane are more stable compared to Si(II) compounds.*
- *The first stable germanone (heavier analogue of a ketone),* $(Eind)_2Ge{=}O$ *with a tri-coordinate germanium was prepared in 2012 using sterically hindered substituents on the germanium.*

Tin

- *Bronze, an alloy of tin and copper was the first alloy to be used on a large scale since 3000 BC. After 600 BC, pure metallic tin was produced.*
- *An important application for tin is corrosion-resistant tin plating of steel especially for making cans for storing food without decomposition.*
- *Tin has the second lowest melting point (232°C) among air- and moisture-stable solid metals (comparable melting points: aluminium 660°C, zinc 419.5°C, lead 327.5°C, cadmium 321°C and indium 156.6°C). Lower melting points than tin are only for alkali metals which are moisture sensitive, also mercury, gallium and indium. Indium has a very low natural abundance and is an expensive metal.*
- *The low melting point, good natural abundance, air and moisture stability and ease of forming flexible alloys makes tin a sought after metal for making low melting alloys especially solders. All well known lead-containing and lead-free soldering alloys have tin as a major component, most notably tin/lead soft solders, which are typically 60% or more tin.*
- *In 1810 Peter Durand, a British merchant, received the first patent for preserving food using tin cans. Durand made containers out of tinplate which was made out of wrought iron sheets coated with tin to prevent rusting.*
- *Stannous chloride, easily prepared by the reaction of tin and HCl is a well known reducing agent for organic transformations such as, nitro to amino groups and nitrile to aldehyde groups.*
- *While tin and its inorganic salts are generally found to be nontoxic, organic tin compounds have a high level of toxicity. Some of the organotin compounds have been used as biocides especially bis-tributyltin oxide (TBT).*
- *Tributyltin hydride,* $n\text{-}Bu_3SnH$ *with a weak Sn–H bond (74 kcal/mol) is a very useful reagent in organic synthesis for the conversion of C–X bonds to C–H bonds by a free radical mechanism.*
- *Niobium–tin is a superconductor with a critical temperature of 18 K and critical magnetic field of 24.5 tesla. This is useful for constructing superconducting magnets which can withstand high current densities (200,000* A/cm^2*) at the highest working magnetic field of 15 T. Although it is brittle, magnetic coils are made by wrapping separate strands of niobium in tin and then fusing them to make the alloy. Another well-known intermetallic compound of tin is indium tin oxide (ITO) which is the most widely used transparent conducting oxide.*

Lead

- *The high density, low cost, ease to work with, makes lead the most suitable material for use in radiation shielding. Gamma radiation, being high-energy photons, is difficult to stop. However, this is achieved with lead blocks of 30 mm thickness.*

- *The most important application of lead is in lead–acid storage batteries for which more than 85% of the lead produced in the world is currently used. According to an estimate by the Environmental Defense and the Ecology Center of Ann Arbor, Michigan, USA, the batteries of vehicles on the road in 2003 contained an estimated 2,600,000 metric tons of lead. The auto industry uses over 1,000,000 metric tons every year, with 90% of it going to conventional lead–acid vehicle batteries.*
- *Tetraethyllead (TEL) was one of the most widely used antiknock additives of gasoline. The use of TEL has been stopped in the developed and developing countries due to lead poisoning. The use of catalytic converters and newer model cars to meet tighter emissions regulations, started a gradual phase-out of leaded gasoline. Currently TEL is used only in a few grades of aviation gasoline and in a handful of less developed Asian and African countries.*
- *Lead poisoning is one of the common problems associated with the use of lead compounds. The upper limit for blood lead for adults is 10 μg/dl (10 μg/100 g) and for children it is 5 μg/dl.*

7.1 About the Elements

7.1.1 Germanium

Like carbon and silicon, germanium crystallises in the cubic diamond structure. Its density is 5.32 g/cc and it melts at 937.4°C. Germanium metal is brittle and hard, with a mohs hardness of 6, close to that of silicon and much softer than diamond. Till 400°C germanium is stable in air, above which it will get oxidised. It is not attacked by mineral acids such as hydrochloric acid or hydrofluoric acid, unless an oxidiser is present with it; hot concentrated sulfuric acid attacks only slowly. However, it is readily attacked by nitric acid. Germanium dissolves readily in molten NaOH or KOH to form germanates but is not attacked by aqueous alkali solutions. A unique property of germanium is its transparency to near-IR radiation in the wavelength range of 1,800 to 23,000 nm. It has a high refractive index for IR radiation (4.0026 at 11,000 nm) which is the highest for any glass-forming material and also posses low chromatic dispersion. Its oxide glass also has a high refractive index and low dispersion. Germanium is an intrinsic *n*-type semiconductor with a small bandgap of 0.7 eV.

The main compounds of commercial importance of germanium are germanium tetrachloride and germanium dioxide. Unlike silicon, germanium forms stable divalent compounds such as $GeCl_2$ and GeO. Another difference with silicon is that it forms $GeCl_6^{2-}$ and $GeCl_3^-$. Germanium tetrachloride ($GeCl_4$), made from the reaction of zinc smelter residues with chlorine, is the feedstock for germanium metal and its compounds. Zone refined crystalline germanium typically is 99.9999 per cent pure and impurities are typically less than 100 ppb.[1]

Germanium is a dispersed element and in nature, it never exists as the native metal. It is present in trace amounts in many minerals, including a few of the common minerals used for zinc and copper extraction. Argyrodite, Ag_8GeS_6, is the mineral from which germanium metal was first isolated. Germanium is most often found in the zinc sulfide minerals, wurtzite and sphalerite (both ZnS) and, less commonly in the copper minerals, bornite (Cu_5FeS_4), enargite (Cu_3AsS_4), chalcopyrite ($CuFeS_2$), and tennantite ($Cu_{12}As_4S_{13}$). Most of the world's supply of germanium is a byproduct of zinc smelting and to a minor extent copper smelting, both as sulfide ores. In addition, germanium used in the manufacturing of electronic and optical devices is returned as scrap for recycling. Worldwide, up to 35 per cent of the total germanium used is derived from recycled scrap.

7.1.2 Tin

Tin was originally known as plumbum candidum or white lead in Latin. Berzelius proposed the symbol Sn from the Latin word *stannum*. It is a soft white metal which has a relatively low melting point of 231.9°C. Tin has played an important role in the history of the human race being an essential component of its alloy with copper, bronze which gave the name to the Bronze Age. Human beings first began mining and mixing tin with copper to create bronze around 5000 years ago thus marking the beginning of the Bronze Age. The first bronze artifacts appeared in Egyptian tombs during 650–750 BC. A bronze statuette named *dancing girl*, 10.5 cm high and about 4,500 years old was excavated from Mohenjo-daro in the Sind province of Indo-Pakistan in 1926 by Ernest Mackay. In the 19th century, Cornwall in England was the major producer of the metal. Deposits were later found in Bolivia and East Asia. Currently China is the leading producer, followed by Indonesia and Peru. In 2006, about half of all tin produced was used in making solder. The rest was divided between tin plating, tin chemicals, brass and bronze alloys.

In the periodic table, tin is a metalloid lying on the border between metals and nonmetals. The unique transformation of tin from the β- to the α- form is accompanied by a complete change of lattice structure, affecting the physical and mechanical properties. At 13.2°C pure tin transforms from the silvery, ductile metallic allotrope of the β-form (*white tin*) to the brittle, nonmetallic, α-form (*grey tin*) with a diamond structure. The transformation is slow to initiate due to high activation energy. A very low temperature of roughly –30°C is sometimes required for the initiation. This transformation is of practical importance, as it involves a volume increase of 21%. This transformation process is also termed as *tin pest*. One of the reasons attributed to the failed Russian campaign of Napolean in 1812 was the tin buttons used on the coats of his army men. Under the severe cold Russian winter, these tin buttons crumbled as they transformed from β- to the α-form, exposing the soldiers to the severity of the freezing cold leading to their death.

Table 7.1 Comparison and properties of α snd β forms of tin

Allotrope	α-form *grey tin*	β-form *white tin*
General properties	Brittle, nonmetallic fcc diamond type	Ductile, metallic Tetragonal
Density	5.77 g/cc	7.29 g/cc
Lattice constants	a = 648.92 pm	a = 583.16 pm c = 318.13 pm
Transformation temperature		13.2°C ; β-form to α-form

7.1.3 Lead

Lead is a relatively unreactive heavier main group metal. It tarnishes on exposure to air, has a melting point of 327°C and is soft and malleable. The properties which make it useful for applications are high density, low melting point, ductility and relative inertness to oxidation. Lead and lead oxides react with both acids and bases indicating its amphoteric nature. Compounds of lead are found in the +2 oxidation state more often than the +4 state. Organolead compounds are mostly covalent in nature and are of +4 oxidation state. It has a density exceeding that of most common main group elements (11.34 g/cc). Lead has one of the highest atomic numbers of the classically stable elements and also lies at the end of three major radioactive decay chains of heavier elements.

Lead is easily extracted from its ores which is mainly galena (PbS) and was known to prehistoric people in Western Asia. Since galena is associated with traces of silver, interest in silver helped to initiate lead extraction and use. Its natural abundance and inexpensive nature resulted in the use of lead in solders, plumbing, lead–acid batteries, bullets, weights, fusible alloys and radiation shielding. By the late 19th century, toxicity of lead was identified, leading to reduction in its use. As a neurotoxin that accumulates in soft tissues and bones, lead damages the nervous system and in mammals causes blood disorders.

7.2 Isolation and Industrial Production of Germanium

The relative natural abundance on the earth's crust of group 14 elements is as follows (indicates the rareness of germanium).

Carbon	0.18%
Silicon	27%
Germanium	0.00014%
Tin	0.00022%
Lead	0.00099%

Although John Newlands in 1864 predicted the existence of an element between silicon and tin in the carbon group (named ekasilicon by Mendeleev in 1871), the credit of discovering germanium goes to C A Winkler who in 1886 isolated germanium from the mineral Argyrodite (Ag_8GeS_6). He named this new element after his home country.

The industrial process of obtaining germanium starts from germanium sulfide or oxide. If starting with GeS_2 rather than GeO_2, the product is oxidised to GeO_2 with a reagent such as sodium chlorate. GeO_2 when dissolved in concentrated HCl forms germanium tetrachloride ($GeCl_4$) which is a fuming liquid similar to $SiCl_4$ with a boiling point of 86.5°C. The $GeCl_4$ is purified by fractional distillation and the purified $GeCl_4$ is hydrolysed with deionised water to yield GeO_2. The GeO_2 after drying, is reduced with hydrogen to form germanium metal in the powder form. The germanium powder is melted and cast into bars which are known as first-reduction bars. These bars are then zone-refined to a metal that typically contains less than 100 ppb of total impurities. The purified metal is used in the polycrystalline form or, for electronic uses, is converted to single crystal ingots by the Czochralski crystal pulling method similar to silicon single crystal making process.[2]

$$GeS_2 \xrightarrow[\text{Ex. NaClO}_3]{\text{Oxidiser}} GeO_2 \xrightarrow{\text{Conc. HCl or Cl}_2} GeCl_4 \text{ (b.p. 86.5°C)} \xrightarrow{\text{distilled}} \text{Pure } GeCl_4$$

(from ZnS or CuS ores)

$$\text{Pure } GeCl_4 \xrightarrow{\text{deionised water}} GeO_2 \xrightarrow[\text{760°C}]{H_2 \text{ reduction}} \text{Germanium Powder} \xrightarrow{\text{casting}} \text{First reduction bars} \xrightarrow[\text{m.p. 937.4°C}]{\text{zone refining}} \text{Ultrapure polycrystalline germanium}$$

Scheme 7.1 Preparation of ultrapure germanium from GeS_2

7.3 Isolation and Industrial Production of Tin

There are no historical evidences concerning the oldest methods of tin extraction. It is known that around 100 BC in Cornwall, England, tin was smelted from very pure ore using traditional methods. Till 1200 AD, Cornwall provided most of the tin needed for Europe.

Industrially, tin is derived primarily from the mineral cassiterite (SnO_2) which has a tin content of around 79% and is the economically important ore. Stannite $Cu_2(Fe,Zn)SnS_4$, is a sulfide ore with a tin content of up to 27.6%. World production of tin is two-thirds from oxidic raw materials and the rest from sulfidic ores. The problem in treating sulfidic ores is in their composition. No single metal sulfide ore for Sn has been found and most have two or more other metals along with tin sulfide making the separation tedious.

Cassiterite, is nonmagnetic and is not suitable for floatation, so mainly the gravimetric sorting process is used. Under practical reaction conditions, above 900°C, the reaction of cassiterite with CO becomes rate determining.

$$SnO_2 + 2CO \xrightleftharpoons{900°C} Sn + 2CO_2 \tag{7.1}$$

Among the uses of tin, tin plate which is a thin sheet of steel having a protective coating of tin on its surface is the most important. Tin plate is used for food cans as it does not react with the acids present in food. Alloys of tin include bronze (combination of tin and copper), pewter (combination of tin and lead), superconducting wire used in the manufacture of extremely powerful magnets, (combination of tin and niobium), babbitt metal used for the surface of bearings (combination of tin, copper and antimony), bell metal with a higher content of tin for making bells (combination of tin and copper) and solder (a combination of tin and lead).

Molten tin (m.p. 232°C) due to its low melting point, is used in the process of making float glass. To make glass with uniform thickness and a flat surface, molten glass is floated on the surface of molten tin and cooled. Tin sheets were also used for wrapping food, the product often known as *tin foil*, although it has been replaced by aluminium foil nowadays.

How tin cans ruled the world of preserved food for 150 years

Tin is a household name in English speaking countries (except North America where it is known as Can) associated with storing of perishable food and beverages without refrigeration and without losing nutrients (for example, biscuit tins, tins of salmon, tins of boiled peas, and so on). Its use has been spread over 150 years of human development (1812–1963) till aluminium took over. The unique properties of tin such as low melting point, ready availability in the pure form, air and moisture stability, ease of coating and non-toxicity of inorganic tin compounds are factors which made the food preservation industry turn to this element. By giving a coating of tin on wrought iron or steel sheets, the material 'tinplate' was created which was ideal for making food storing containers.

Historically, the demand for preserving food, especially fresh vegetables, fruits and meat, came from sailors who took long voyages. Napoleon famously noted that an army 'travels on its stomach'. Colonial expansion and long voyages at sea required that a way of keeping nutritious food unspoiled over distance and time be discovered. Napoleon even announced a reward for finding a method to preserve fresh food. The first invention was made by a Frenchman named Nicholas Appert who successfully preserved food by partially cooking it, sealing it in glass bottles with cork stoppers after immersing the bottles in boiling water. Appert found that, similar to wine, exposure to air spoiled food. He observed that food in an airtight container, with the air expelled through boiling, would stay fresh. It worked and some of Appert's preserved food was used in the ships belonging to Napoleon's troops. Meat, vegetables and gravy were among different items sealed in glass containers. 'Not a single substance had undergone the least change at sea', Appert wrote. He was rewarded for his discovery by Napoleon in 1810. However, the problem with Appert's method of preservation was that glass in those days was not of high quality and used to break quite frequently in addition to being quite heavy.

In 1810, an Englishman named Peter Durand discovered the use of tin coated metal cans instead of glass bottles for preserving freshly boiled food. A cylindrical canister and soldered lid was much easier to handle than a fragile bottle and an unreliable cork. Durand was granted a patent for his method by King George III. Made of iron coated with tin which prevented rusting and corrosion, tinplate could be sealed by molten tin and made airtight and was not fragile like glass. The story of tin cans began from there. Commercial canning factories were set up. The British Royal Navy used as many as 24,000 large cans of preserved food amounting to nearly 40,000 pounds on its ships each year by 1818. The canned vegetables with its nutrients preserved were a great relief to sailors who previously had depended on live cargo or salted meat and were affected by debilitating scurvy. (Scurvy, which used to affect sailors in large numbers, results in decreased red blood cells, gum disease and bleeding from skin leading to death and was caused by lack of vitamin C.) The salt-cured foods lost most of their vitamins and nutrients in the preservation process. The most reliable and interesting certificate for tin canned food came from Sir William Edward Parry who made two arctic expeditions to the Northwest Passage in the 1820s. He used to take tinned food on his journeys. One four-pound tin of roasted veal (tender beef), carried twice during his trips but never opened, was kept as an artifact of the expedition in a museum until it was opened out of curiosity in 1938; almost after 120 years. The contents were chemically analysed and were found to have kept most of their nutrients intact and to be in fairly perfect condition. It was fed to a cat, which ate it without any hesitation and survived!

The use of tin cans for food and drinks declined after the first aluminium beverage can was manufactured by Reynolds Metals Company in 1963 to pack a diet cola named 'Slenderella'. By 1967, Pepsi and Coke also changed from tin cans to aluminium cans. Tin cans are still made in large quantities for marketing viscous liquids such as paints, due to its inexpensive, non-corrosive and sturdy nature, however less of it is used for preserving food these days.

7.4 Isolation and Production of Lead

Some of the well known ores of lead are galena (PbS), lead ochre (PbO), cerrusite ($PbCO_3$) and angelesite ($PbSO_4$). Lead is mainly extracted from galena. Galena contains lead sulfide along with traces of silver. The galena ore is initially concentrated by the froth floatation process followed by roasting and smelting in a reverberatory furnace. During roasting, PbS gets converted to a mixture of PbO and $PbSO_4$.

Roasting

$$2\,PbS + 3\,O_2 \longrightarrow 2\,PbO + 2SO_2 \quad (7.2)$$

$$PbS + 2\,O_2 \longrightarrow PbSO_4 \quad (7.3)$$

Self reduction

$$PbS + 2\,PbO \longrightarrow 3\,Pb + SO_2 \quad (7.4)$$

$$PbS + PbSO_4 \longrightarrow 2\,Pb + 2\,SO_2 \quad (7.5)$$

During smelting, which is done after adding more of galena in the same furnace and at a different temperature, self reduction happens between the sulfide, sulfate and oxide resulting in metallic lead along with sulfur dioxide as the side product.

Lead extracted by the above method contains impurities such as silver, copper, tin, bismuth, gold and iron. By the process of liquation (heating on a sloping furnace) lead is allowed to melt and flow down the slope while impurities remain. Further, very pure lead is obtained by electrolytic refining using a mixture of $PbSiF_6$ and H_2SiF_6 as the electrolyte.

7.5 Properties of Germanium

Six salient properties of germanium which differ from that of silicon are the foundations for all its applications.[1] These are:

(i) intrinsic semiconductivity,
(ii) infrared transparency (ability to transmit near-IR radiation in the 1,600–18,000 nm range),
(iii) glass forming ability of Ge–O tetrahedra forming extended three dimensional structures,
(iv) highest refractive index (4.0026) which is the highest for any glass forming material,
(v) low chromatic dispersion (variation of refractive index with wavelength), and
(vi) ability of GeO_2 to catalyse polymerisation of terephthalic acid.

7.6 Application in Electronics

Germanium is an intrinsic semiconductor with a small band gap. It is particularly effective at high frequencies and low operating voltages. Although electronics was the major use of germanium from the 1950s to the 1970s, it had diminished by 2000 to around 5 per cent of the total germanium use. Initially there were no good methods to produce silicon with good purity to replace germanium, but after 1954, silicon began replacing germanium in electronic transistors. Most of the electronic parts used at the beginning of the 21st century are based on silicon as they are preferred to germanium devices due to better temperature stability, cost factor, ease of manufacture and higher maximum operating temperature (150°C versus 100°C).

Interest in germanium as an electronics material has been revived recently by the development of a successful method for growing exceedingly thin epitaxial layers of germanium on silicon. Transistors that use silicon–germanium are almost five times faster than silicon and use much less power. These are being used in chips of cell phones and computers and holds great promise in hand held devices which needs charged power. Because they start conducting at lower voltage [0.2 to 0.3 V versus 0.6 V] germanium diodes are preferred in some circuits. Germanium-based photo voltaic cells are still critical to many satellite operations.

7.7 Application in Night Vision Devices

Germanium is transparent to part of the infrared spectrum, both in the crystalline and glassy states. The passive devices which are thermal imagers are sometimes referred to as forward looking infrared (FLIR) devices and are different from active system IR devices. Active night vision systems emit infrared radiation and measure the reflected radiation differentially from objects in the field of view. In contrast, FLIR devices form images by sensing the thermal contrast between objects and their background. Therefore FLIR devices are suitable for working in complete darkness and are not affected by smoke. In military use, passive devices made of germanium are preferred as they are rugged and not locatable by the enemy while the radiation emitted from an active device makes it locatable. Germanium-based FLIR systems are used in night vision goggles, surveillance cameras, gunsights, and IR heat-seeking missiles. It also finds use in detecting overheated bearings, fire fighting under thick smoke, satellite mapping, detecting faults in structural materials, and so on.

7.8 Application in Fibreoptics

Until the 1980s, copper was used for long-distance land and submarine lines. These have now almost completely been replaced with fibreoptic lines with much greater information carrying capacities. The fibres used in long-distance cables are silica fibres of diameter 7–10 μm, which are doped with GeO_2. It is capable of transmitting digital information accurately at very high rates (~100 gigabits per second) with low signal attenuation. Worldwide, more than 600 million km of optical fibre is now deployed which indicates consumption of germanium in tons. The most important commercially used optical fibres are essentially silica glass doped with around 4 weight per cent GeO_2, which is useful as a glass component because of its transparency to near-IR radiation, its high refractive index and its low chromatic dispersion. This application of germanium has now surpassed all other germanium uses.

The role of germanium in the fibreoptic revolution

Fibreoptic cables support much of the world's internet, cable television and telephone systems today. They have replaced traditional copper communication cables as the amount of network bandwidth they can carry easily exceeds that of a copper cable with similar thickness. A fibreoptic cable consists of a bundle of glass threads, each of which is capable of transmitting digital information and data modulated onto light waves. Fibreoptics has several advantages over traditional metal cables. They have a much greater bandwidth which means they can carry more data. These cables are much thinner and lighter than metal wires thus reducing weight and volume. Fibreoptic cables are less susceptible than metal cables to electromagnetic interference. Digital transmission of data, which is the natural form for computer data is also possible.

The Indian born physicist, Narendra Singh Kapany coined the term fibreoptics in an article in *Scientific American* in 1960. He also wrote the first book about fibreoptics and was one of the field's most prominent researcher, writer and spokesperson. Kapany first demonstrated successfully that light can be transmitted through bent glass fibres during his doctoral work at the Imperial College of Science in London in the early 1950s, and published the findings in a paper in *Nature* in 1954.

The basic requirement of an optical data link is that the transmitted modulated light should reach the receiver with enough power so that it can be demodulated properly. The term attenuation is used to describe the reduction in power of a light signal as it is transmitted over long distances. The cause of attenuation is often related to components such as nature of cables, cable splices and connectors. Early samples of fibreoptic cables suffered from considerable attenuation. Reduction of attenuation is very significant as it will help to reduce the number of signal booster units needed in a long distance cable. Donald B Keck of Corning glass company experimented with different glass compositions and methods of heat in treating the glass to reduce attenuation. Corning had a target of achieving attenuation below 20 db/km. In August 1970, Keck prepared a new batch and took measurements of the fibres. He observed that light passed through a 65 foot glass fibre without any measurable loss, and exclaimed, 'Good grief, what do I have here?' More measurements of this optical fibre indicated that it had an attenuation of 16 db/km, exceeding the company's goal of 20 db/km. Upon this discovery, Keck wrote in his laboratory notebook, 'Attenuation equals 16 db it says. Eureka', followed by a 'Whoopee!'. Keck and coworkers had invented the first low-loss optical fibre which was composed of titanium-doped silica.

Keck continued to improve upon the optical fibre he had invented. He had a much better result when he replaced the titanium oxide doped glass of 1970 with germanium oxide doped glass, and eventually achieved a consistent and much lower 4 db/km attenuation by June 1972. By 1979, Corning was mass-producing the refined optical fibre invented by Keck with germanium oxide. Currently the highest use of germanium (more than 35%) is in the manufacture of fibreoptic cables.

7.9 Application as Catalyst in Polyethylene Terephthalate Synthesis

Another industrial application, involves the unique of ability of GeO_2 to catalyse the synthesis of the polymer, polyethylene terephthalate (PET), without undesirable colouring of the final product. PET is a plastic used extensively in the making of mineral water and beverage bottles, food packaging and many others where food safety is of prime concern. GeO_2 when used as catalyst yields a polymer with better transparency and clarity than other catalysts. Sb_2O_3 or antimony triacetate [$Sb(CH_3COO)_3$] were used initially but PET made with these were found to give an undesirable yellow colour, especially with ageing.[3]

7.10 Organogermanium Compounds

Due to the highly successful and versatile applications possible with silicone polymers, the possibility of making analogous compounds with germanium has been explored. Bond dissociation energies indicate that germanium–carbon bonds are much weaker than Si–C bonds and should be similar to Sn–C bonds (Table 7.2).

Table 7.2 Comparison of the bond dissociation energies of group 14 elements with carbon and oxygen

Bond	Bond dissociation energies (kJ/mol)	Bond	Bond dissociation energies (kJ/mol)
C–C	346	C–O	358
Si–C	318	Si–O	462
Ge–C	238	Ge–O	326
Sn–C	192	Sn–O	322
Pb–C	130	Pb–O	378

Interestingly one of the methods to prepare halogenated organogermanium compounds is by disproportionation reactions (similar to tin compounds)[4].

$$GeR_4 + GeX_4 \longrightarrow 3\ GeR_2X_2 \qquad X = Cl, Br \tag{7.6}$$

Similar to the Rochow–Müller process, one can make methylgermanium compounds by the reaction of germanium copper alloy with methyl chloride. However the major product of this reaction is Me_3GeCl.[5]

$$\underset{10:4 \text{ wt ratio}}{Ge + Cu} \xrightarrow[400°C]{CH_3Cl} Me_3GeCl\ (20\%) + Me_2GeCl_2\ (61\%) + MeGeCl_3\ (2.7\ \%) \tag{7.7}$$

Me_2GeCl_2 (b.p. 124°C) on hydrolysis in the presence of NaOH gives the colourless crystalline compound $(Me_2GeO)_4$ in 67% yield similar to $(Me_2SiO)_4$ (D_4) one of the silicone precursors.[6,7]

$$Me_2GeCl_2 \xrightarrow[NaOH]{H_2O} \underset{\text{m.p } 92°C}{[Me_2GeO]_4} \tag{7.8}$$

Unlike the silicon compound D_4, its germanium analogue is highly soluble in water. Cryoscopic measurements on a dilute aqueous solution gave a molecular weight corresponding to one Me_2GeO unit possibly resulting in the diol $Me_2Ge(OH)_2$. A white fibrous material was obtained when the aqueous solution of $(Me_2GeO)_4$ was evaporated possibly forming a high polymer $(Me_2GeO)_n$. This high polymer is insoluble in common organic solvents but dissolves in water. It is needless to mention that the water solubility and subsequent decomposition of the tetramer and high polymer was a dampener for those who were keen to develop a polymer chemistry akin to the chemistry of silicones.[8]

Starting from germanium or germanium dioxide, a chlorine-free protocol which is based on mechano-chemical synthesis has also been devised to make organogermanium compounds (Scheme 7.2).[9]

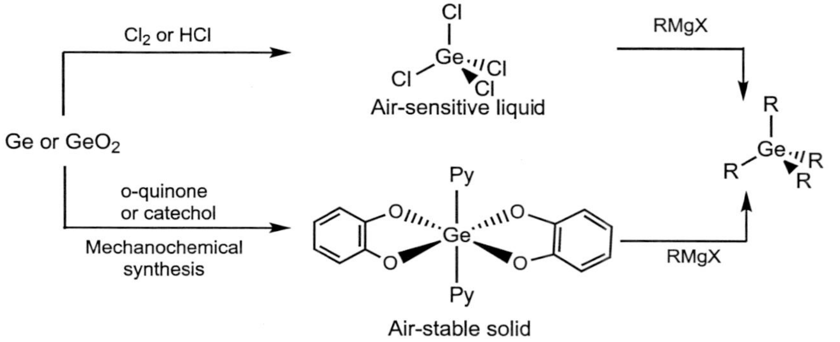

Ge or GeO_2 —(Cl_2 or HCl)→ $GeCl_4$ (Air-sensitive liquid) —(RMgX)→ GeR_4

Ge or GeO_2 —(o-quinone or catechol; Mechanochemical synthesis)→ $Ge(O_2C_6H_4)_2(Py)_2$ (Air-stable solid) —(RMgX)→ GeR_4

Scheme 7.2 Chlorine-free mechano-chemical synthesis of organogermanium compounds

7.11 Multiple Bonded Germanium Compounds

The first example of a compound with a germanium–germanium double bond (digermene) was prepared by the Grignard reaction of $MgClR(OEt)_2$ where R = $CH(SiMe_3)_2$ with $GeCl_2$.dioxane by Lappert and coworkers in 1984.[10] The compound with the double bond, stabilised by bis(trimethylsilyl)methyl groups was bright yellow in colour and showed a Ge=Ge distance of 2.347 Å when analysed by X-ray diffraction studies. However, this compound was found to dissociate to the corresponding singlet germylenes in solution. Following this work, bis(2,6-dimethylphenyl) substituted digermenes were prepared by Masamune by first preparing the cyclic trigermane and then converting it to the digermene by photo-conversion using a mercury lamp.[11] The analogous yellow coloured tetrakis(2,6-diethylphenyl) digermene was also prepared and structurally characterised with a Ge=Ge bond distance of 2.213 Å.[12] This digermane was found to retain its structural integrity in solution as a reaction with MeOH was found to add across the double bond.

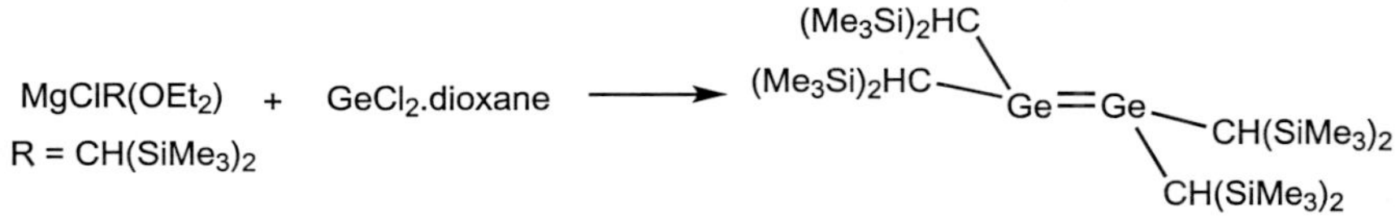

$$MgClR(OEt_2) + GeCl_2.dioxane \longrightarrow ((Me_3Si)_2HC)_2Ge{=}Ge(CH(SiMe_3)_2)_2 \qquad R = CH(SiMe_3)_2 \tag{7.9}$$

$$(Ar)_2Ge\text{–}Ge(Ar)_2\text{–}Ge(Ar)_2 \text{ (cyclotrigermane)} \xrightarrow{h\nu} Ar_2Ge{=}GeAr_2 \quad (7.10)$$

Ar = 2,6-dimethylphenyl (Me, Me) or 2,6-diethylphenyl (Et, Et)

Multiple bonded germanium, tin and lead compounds which are formal analogues of acetylenes have also been prepared by using sterically hindered substituents such as terphenyl (C_6H_3-2,6-Dipp$_2$; Dipp = C_6H_3-2,6-i-Pr$_2$).[13] These compounds have been found to show *trans* bent geometry with varying amounts of non-bonding electron density on the Ge, Sn and Pb atoms (see Chapter 6; Section 6.20.2). Structural parameters of selected compounds are given in Table 7.3.[14]

$$2\ MClAr^* + 2K \xrightarrow[M\ =\ Ge,\ Sn\ or\ Pb]{} Ar^*MMAr^* \quad (7.11)$$

Ar* (terphenyl bearing i-Pr groups)

Table 7.3 Structural parameters for alkyne analogues of germanium, tin and lead

Compound	M–M (Å)	M–C (Å)	M–M–C (°)	Torsion angle (°)	UV-Vis λ (nm)
Ar′GeGeAr′	2.285	1.996	128.67	0.4	500, 371
Ar′SnSnAr′	2.668	2.191	125.24	3.5	597, 410
Ar*PbPbAr*	3.188	2.303	94.26	3.4	750, 397

Ar′= (C_6H_3-2,6-Dipp$_2$; Dipp = C_6H_3-2,6-i-Pr$_2$); Ar*=(C_6H_3-2,6-Trip$_2$; Trip = C_6H_2-2,4,6-i-Pr$_3$)

The strength of the multiple bond is reduced due to the removal of electron density from the region between the two participating atoms. In fact the Ge–Ge distance of 2.285 Å of this compound is shorter than the Ge–Ge single bond distance (2.44 Å) and longer than the calculated distance of 2.197 Å of MeGe≡GeMe. On the other hand, the Pb–Pb bond distance is significantly longer than those found in structurally characterised single bonded diplumbanes which fall in the range 2.85–2.95 Å. However, the distance of 3.188 Å is significantly shorter than the interatomic distance of 3.49 Å of metallic lead. The M–M bonds are therefore not triple bonds but almost double bonds for Ge and Sn, and single bond for Pb.

It may be noted that germanium in such compounds can have either a double bond between the two germanium atoms with two radicals on each of the atoms or a single bond with two lone pairs. Examples of both these types have been observed. The singlet diradical character of Ar′GeGeAr′ (Ar′ = (C_6H_3-2,6-Dipp$_2$; Dipp = C_6H_3-2,6-i-Pr$_2$) is observed in its reactions with compounds such as 2,3-dimethyl-1,3-butadiene or N_3SiMe_3. With N_3SiMe_3, the reaction affords a cyclic N_2Ge_2 diradical which is a highly reactive purple solid and attacks solvents such as benzene and reacts with H_2 directly. The singlet diradical character of digermynes is also supported by theoretical (DFT) studies which indicates the presence of low lying triplet states near the *trans* bent singlet state.[13]

$$\text{Ar'GeGeAr'} + 2\,N_3SiMe_3 \xrightarrow[-2\,N_2]{} \text{cyclic } [\text{Ar'Ge}(\mu\text{-NSiMe}_3)_2\text{GeAr'}] \text{ diradical} \qquad (7.12)$$

The example of a digermyne having a single bond and two lone pairs was prepared by Jones and coworkers where the Ge–Ge distance of 2.709 was 0.4 Å longer than other digermynes and more than the sum of two germanium covalent radii (2.44 Å).[15] The molecule also undergoes addition with H_2 (Eq. 7.13) further supporting the single bond character of the Ge–Ge bond.

$$\text{Ar(Me}_3\text{Si)N–Ge–Ge–N(SiMe}_3\text{)Ar (Ge–Ge 2.709 Å)} \xrightarrow{H_2} \text{Ar(Me}_3\text{Si)N–Ge–GeH}_2\text{–N(SiMe}_3\text{)Ar} \qquad (7.13)$$

Ar = 2,6-bis(diphenylmethyl)-4-methylphenyl (Ph, Ph / Ph, Ph)

7.12 Low Valent Germanium Compounds

Going down the group from carbon to lead, it is observed that the ease of formation and stability of divalent compounds increases. While divalent carbon and silicon compounds require sterically hindered groups for stabilisation, low valent germanium compounds can be prepared more readily. The most well known precursor of germanium(II) compounds is $GeCl_2$.dioxane which is prepared by the comproportionation reaction of $GeCl_4$ and Ge or by heating $HGeCl_3$ at high temperature to remove HCl. A more practical method of preparing $GeCl_2$.dioxane is by the reduction of $GeCl_4$ by Bu_3SnH in the presence of dioxane.[16]

$$GeCl_4 + \text{Dioxane} + (n\text{-Bu})_3SnH \xrightarrow[\text{RT, 30 min}]{\text{hexane/Et}_2\text{O (1:1)}} GeCl_2.\text{dioxane} \; 84\% \qquad (7.14)$$

Comproportionation of GeO_2 and Ge at 1000°C gives GeO. GeS is also obtained in a similar manner.

There are many examples of germylenes reported and structurally characterised. Some of them dimerise to give the double bonded germenes as well.

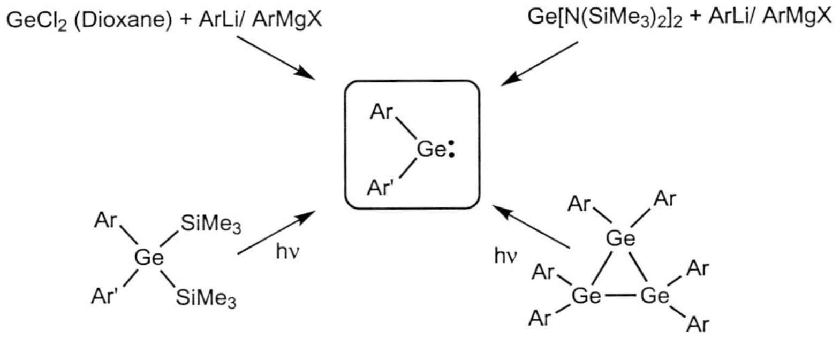

Scheme 7.3 Different methods of preparing heteroatom-bound germylenes

Using $GeCl_2$.dioxane, germanium(II) compounds having elements other than carbon bound to germanium have also been prepared.

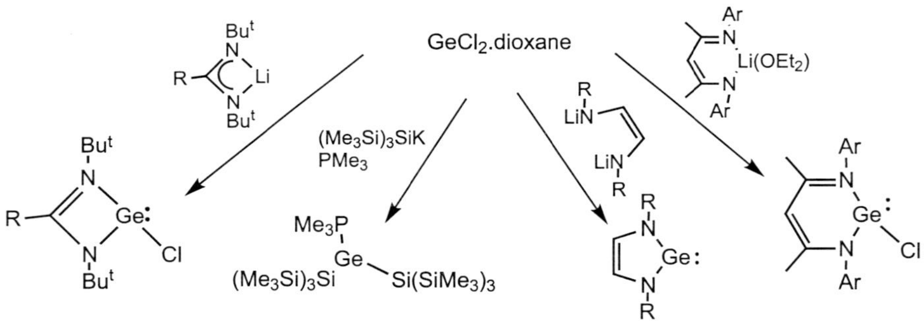

Scheme 7.4 Utility of $GeCl_2$.dioxane in germylene synthesis

By the use of a bis-N-heterocyclic carbene (bNHC), a new chlorogermyliumylidene complex was prepared by Driess and coworkers in 95% yield. This complex upon dechlorination with sodium naphthalenide, yielded a cyclic germadicarbene, also termed as germylone having germanium in the zero oxidation state. The structural analysis of this compound indicated a puckered boat shape conformation bent at the germanium indicating possibility of two lone pair domains on the germanium centre.[17] Germylones were first prepared by Roesky and coworkers using cyclic alkylamino carbene (cAAC) ligands (Fig. 6.17).

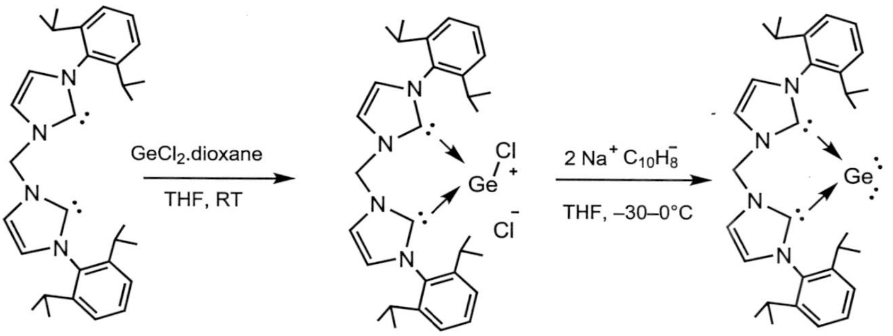

(7.15)

Germylone

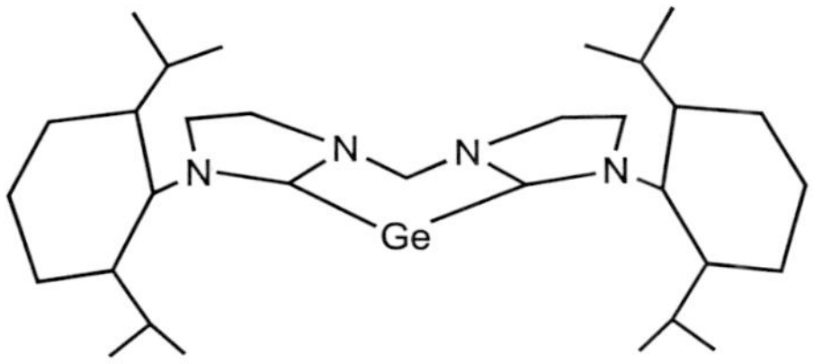

7.13 Stable Tricoordinate Germanone

Germanones, similar to silanones, are germanium analogues of ketones with a stable Ge=O bond. Similar to silicon, in general, germanium preferentially forms Ge–O single bonds. Making an ideal Ge=O bond was a synthetic challenge to chemists. Attempts to make germanones, heavier germanium analogues of ketones were made by several groups. However, most of the compounds made had a Ge=O unit with germanium in a tetracoordinate state.[18]

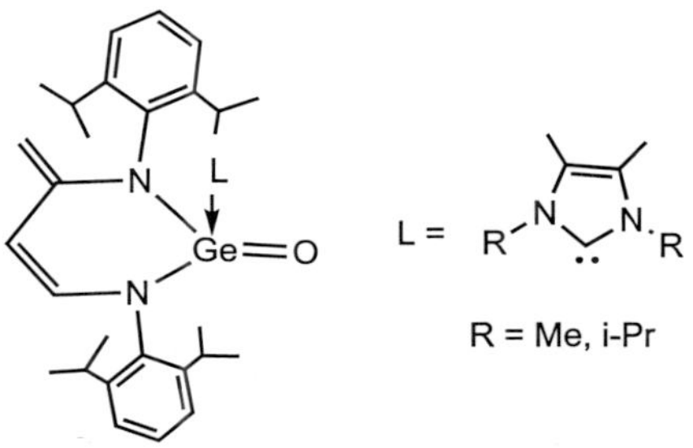

L = R = Me, i-Pr

Fig. 7.1 Example of a tetracoordinate germanone

The first isolation and characterisation of a stable monomeric germanone $(Eind)_2Ge{=}O$ (Eind = 1,1,3,3,5,5,7,7-octaethyl-s-hydrindacen-4-yl) having a planar tricoordinate germanium was achieved by K Tamao and coworkers in 2012. Reaction of the lithiated form of the sterically hindererd Eind substituent with $GeCl_2$. dioxane, resulted in the Ge(II) dicoordinate compound. Oxidation of the Ge(II) unit with Me_3NO gave $(Eind)_2Ge{=}O$ as a stable compound. The sum of the surrounding bond angles of germanium in $(Eind)_2Ge{=}O$ is 359.8 indicating planar tri-coordinate germanium. The Ge=O double bond length is 1.645 Å which is 6% shorter than the typical Ge–O single bond length of 1.76 Å and has a calculated bond order of 1.252. Computational studies and chemical reactions suggest that this Ge=O is highly polarised with contribution from a charge separated form, $(Eind)_2Ge^{+}{-}O^{-}$ as well. The Ge=O stretching frequency was observed at 916 cm^{-1} in the infrared spectrum. This germanone exhibits unique reactivity, not shown by ordinary ketones.[19]

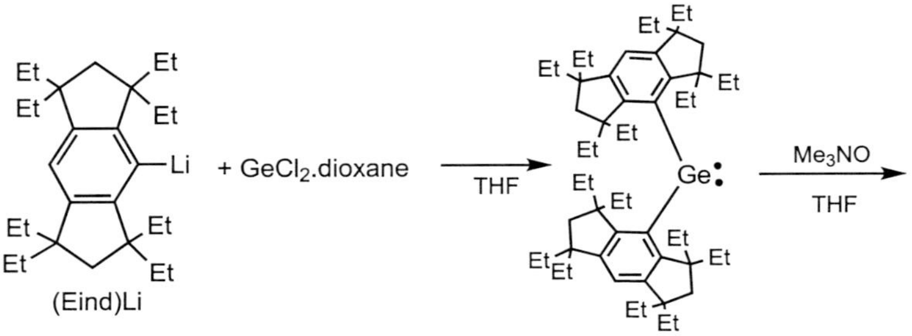

(Eind)Li + $GeCl_2$.dioxane →(THF) $(Eind)_2Ge$: →(Me_3NO, THF) $(Eind)_2Ge=O$ (7.16)

Tamao and coworkers also explored the basic chemistry of the Ge=O bond. The successful reactions of $(Eind)_2Ge=O$ are given in Scheme 7.5.[20]

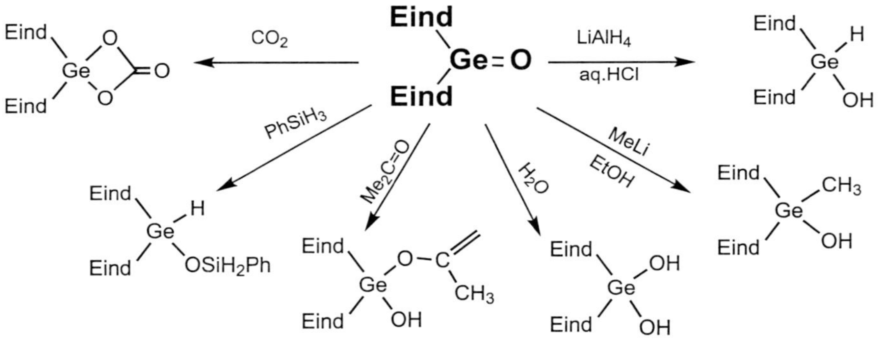

Scheme 7.5 Reactions of $(Eind)_2Ge=O$

7.13.1 Recent Developments

Stable germanone complexes

The first examples of metal complexes of germanones containing formal Ge=O→M bonds (M = Zn, Sn) were prepared by Nagendran and coworkers. These were prepared by a novel synthetic route using a germanium μ-oxo dimer as the starting material instead of a regular germanone.[21]

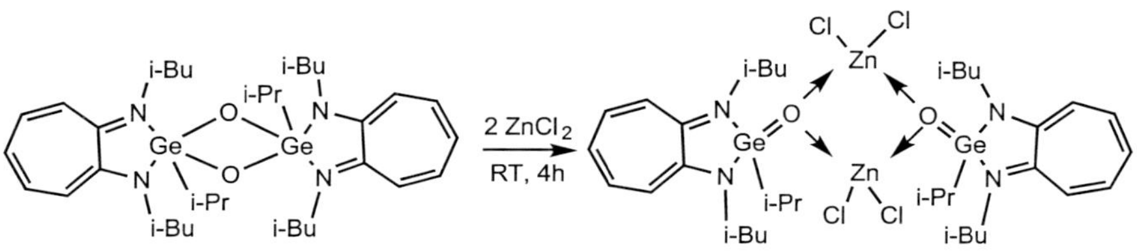

(7.17)

Stable germanium analogue of vinylidene

Vinylidene (CH_2=C) or 'methylidene carbene' is an isomer of acetylene but has only been observed transiently and has a very short life time close to 10^{-10} seconds. It has only been known to exist as a transient species under high vacuum and/or high temperature conditions. Difluorovinylidene, CF_2=C, the fluoro analogue of vinylidene is slightly more stable and has been generated and investigated spectroscopically at 15 K in an argon matrix. Free vinylidenes that are stable at room temperature were not known till the discovery of a germanium-based vinylidene that was prepared by A Rit et al. in 2016.[22]

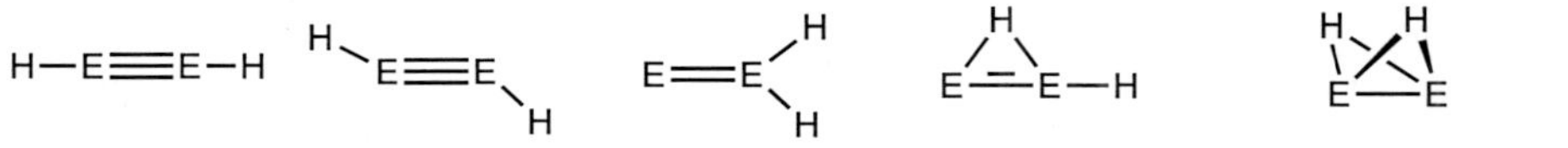

Fig. 7.2 Types of isomers possible for triple bonded species

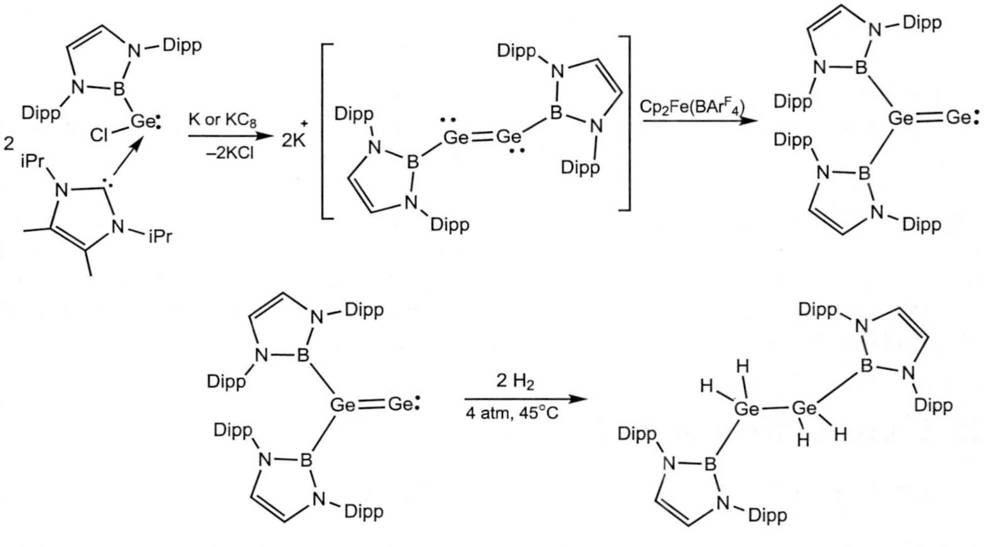

Scheme 7.6 Synthesis and hydrogenation studies of germanium analogue of vinylidene

The reaction of the NHC stabilised (boryl)germanium(II) chloride ($I^{Pr}{}_2{}^{Me}{}_2$) GeCl[B(NDippCH)$_2$] with potassium metal or KC_8 was found to result in a highly reduced formally Ge(0) complex K_2(HCDippN)$_2$B]GeGe[B(NDippCH)$_2$ with a formal Ge=Ge double bond (2.312 Å). The reaction of this compound with one-electron oxidants such as (Cp_2Fe) [$BAr^f{}_4$] or $Ph_3C[B(C_6F_5)_4]$ resulted in a 1,2 boryl migration yielding the vinylidene compound

$[(HCDippN)_2B]_2Ge{=}Ge$ as a crystalline solid which was structurally characterised [Ge=Ge double bond (2.312 Å)]. This novel vinylidene has been found to convert to a diboryl digermane by the uptake of two equivalents of dihydrogen after a 1,2-boryl shift (Scheme 7.6).

7.14 Properties of Tin

Tin is one of the important constituents of low-melting nonferrous alloys. The following important properties of tin have been utilised in its uses and applications (i) low melting point, (ii) low hardness, (iii) good wetting properties, (iv) ease of forming alloys, and (v) good compatibility with food. Tin cooking utensils have been known for over 5000 years. These utensils were made of tin alloys, as the pure metal is too soft to form a vessel. The common alloying elements are antimony, copper and lead. Some of the well known alloys of tin are listed below.

Babbit	Sn (86%); Cu (7%); Sb (7%)
Pewter	Sn (92%);Cu (2%); Sb (6%)
Pb–Sn solder	Sn (63%); Pb (37%)
Lead-free solder	Sn (95+%); Ag (3%); Cu (1%) (SAC)

Most solders are based on the tin–lead binary system which has a eutectic at ca. 63% tin and 183°C. Due to toxicity, lead solders are nowadays being phased out. However, tin remains one of the components of many new lead-free solders. Niobium–tin or more precisely triniobium–tin is an alloy used industrially as a type II superconductor. The material's ability to support high currents and magnetic fields has helped in large scale applications of superconductivity. Since Nb_3Sn is extremely brittle, it is difficult to draw a wire out of it, which is required for winding superconducting magnets. Wire manufacturers draw composite wires containing ductile precursors to overcome this problem. The high field modern NMR magnets are made incorporating niobium–tin wire.[23]

7.15 Useful Compounds of Tin

7.15.1 Stannous Chloride and Stannous Fluoride

Stannous chloride is a convenient, easily prepared and handled reducing agent. $SnCl_2.2H_2O$ is prepared by the reaction of hydrochloric acid on tin metal, while the anhydrous form is prepared by the action of HCl gas with tin.

$$Sn + 2\,HCl \longrightarrow SnCl_2 + H_2 \qquad (7.18)$$

$SnCl_2$ reduces many metal ions such as Ag(I) and Au(I) to the elemental form, Fe(III) to Fe(II) and Cu(II) to Cu(I).

For reduction of organic compounds, it is conveniently used for the conversion of aryl nitriles to aldehydes, aryl nitro to amino groups and quinine to hydroquinone.

$$ArCN \xrightarrow{HCl} Ar{-}C(=\overset{+}{N}H_2)Cl\ Cl^- \xrightarrow{SnCl_2} [Ar{-}C(=NH)H]_2^+\ SnCl_6^{2-} \xrightarrow{aq.HCl} ArCHO \quad (7.19)$$

Since the same conversion can also be brought about by DIBAL-H, conversion of nitriles to aldehydes using $SnCl_2$, also known as *Stephen reduction* is currently less used.

$$Cl{-}C_6H_4{-}NO_2 \xrightarrow[EtOH]{SnCl_2} Cl{-}C_6H_4{-}NH_2 \quad (7.20)$$

$$O{=}C_6H_4{=}O \xrightarrow[H_2O]{SnCl_2} HO{-}C_6H_4{-}OH \quad (7.21)$$

During electrolytic coating of tin metal to make tin cans, a solution of tin(II) chloride containing HCl is used. When electric potential is applied, tin metal is formed at the cathode.

Stannous fluoride (SnF_2) is an effective additive in toothpastes to make tooth enamel resistant to bacteria-generated acid attacks by converting hydroxyapatite to fluorapatite. It has been found to be more effective than NaF. This is because, in toothpastes containing calcium minerals, NaF becomes ineffective over time due to the formation of the insoluble $CaF_2.SnF_2$ is prepared by evaporating a solution of SnO in 40% HF.[24]

7.15.2 Trimethyltin Chloride

Trimethyltin chloride (b.p. 148°C) is one of the most widely used trialkyl tin halides. Unlike trimethylsilyl chloride, the Sn–C bond is less strong and it can therefore be prepared by a comproportionation reaction. It is the starting material for a host of organotin compounds.[25]

$$SnCl_4 + 3\ SnMe_4 \xrightarrow[\text{inert atm.}]{} 4\ Me_3SnCl \quad (7.22)$$

$$Me_3SnOH \xrightarrow[\text{HCl or } SOCl_2]{} Me_3SnCl \quad (7.23)$$

7.15.3 Trimethyltin Fluoride

Trimethyltin fluoride is the simplest member of the group of triorganyltin fluorides. It is easily prepared by the halogen exchange reaction of $(CH_3)_3SnCl$ with KF or NaF in a water–ethanol mixture. It is insoluble in most non-polar organic solvents but is soluble in hot alcohols. The relatively high melting point of 375°C is strongly indicative of its aggregation in the solid state. Although it was the first triorganyltin fluoride to be studied by X-ray crystallography

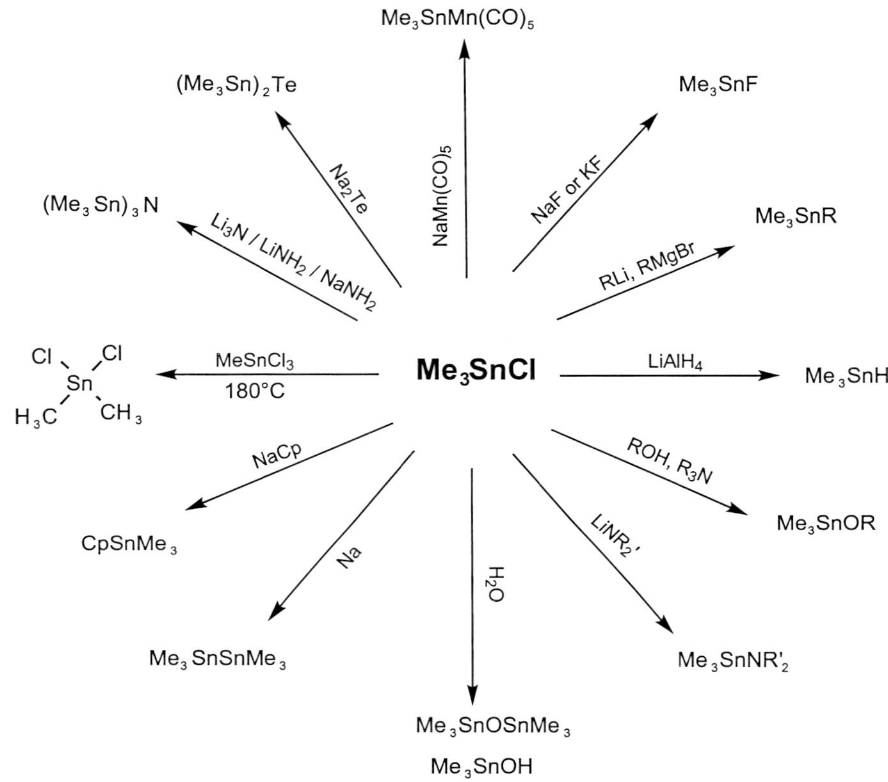

Scheme 7.7 Common reactions of Me_3SnCl

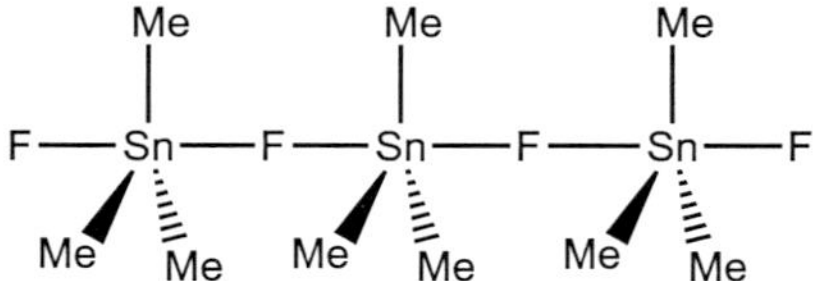

Fig. 7.3 Structure of Me_3SnF

in 1964, its structural characterisation had some anomalies. The 1964 studies proposed two models—(i) a pyramidalised trimethylstannyl moiety with disordered fluorine atoms and bent Sn–F...Sn bridges, and (ii) planar trimethylstannyl groups that are tilted in an alternating fashion along the chain axis and bent Sn–F...Sn bridges.[26] The structure was reinvestigated in 2015 which clearly indicated that the trimethyltin moiety was trigonal planar and the fluorine bridge is symmetric with a Sn–F distance of 2.162 Å and essentially linear (Fig. 7.3). Herbert Roesky and coworkers first showed the synthetic utility of this compound as a very convenient, highly efficient and recyclable fluorinating agent for group 4 to 6 organometallic compounds and some main group chlorides with reactions that are clean without impurities.[27]

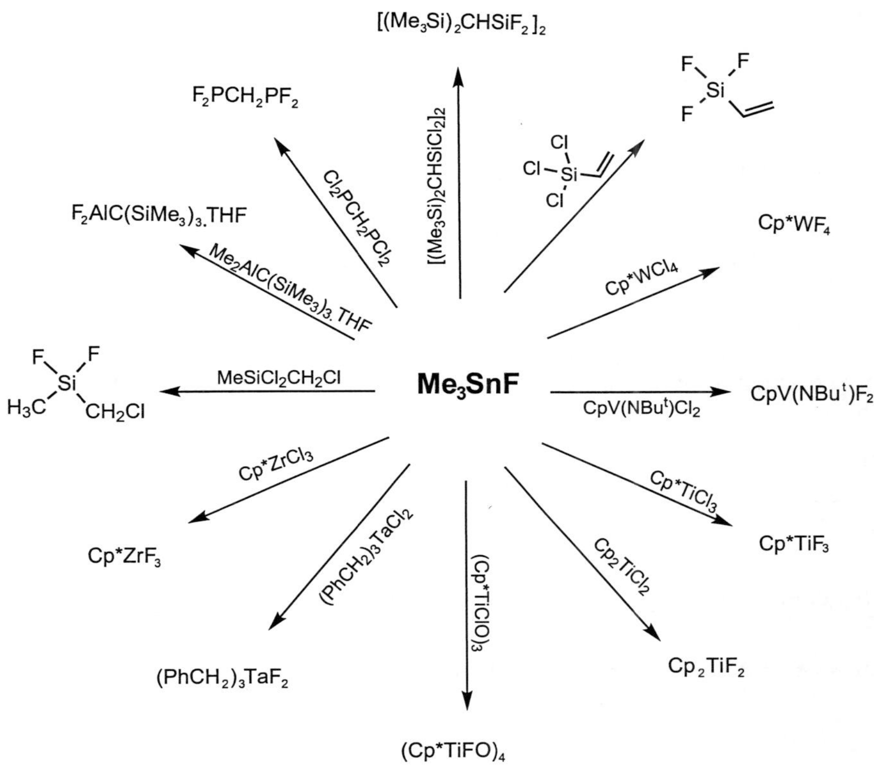

Scheme 7.8 Fluorination reactions of Me_3SnF

7.15.4 Tributyl Tin Oxide

$$(n\text{-Bu})_4Sn + SnCl_4 \xrightarrow{210°C} n\text{-}Bu_3SnCl \tag{7.24}$$

$$n\text{-}Bu_3SnCl \xrightarrow{aq.NaOH} (n\text{-}Bu_3)SnOSn(n\text{-}Bu)_3 + NaCl \tag{7.25}$$

Tributyl tin oxide is a well known biocide and was extensively used for removing biofouling from the submerged parts of ships and other sea vessels. However its use as a biocide had to be discontinued due to the environmental damage caused to sea crustaceans. It is also the starting material for making the well known reagent, tri-n-butyltin hydride.[28]

7.15.5 Tri n-Butyltin Hydride

Tri n-butyltin hydride is a very good radical reducing agent due to the relatively weak, nonionic bond between tin and hydrogen (Bu_3Sn-H 74 kcal/mol) that can cleave homolytically. It is a colourless liquid that is soluble in most organic solvents. The compound is used as a source of hydrogen atoms in organic syntheses. Reaction of silyl monohydrides such as Ph_3SiH, Bu_3SiH or polymethylhydrosilane with tributyltin oxide results in its formation.

$$x\,(n\text{-}Bu_3Sn)_2O + 2\,(MeSiHO)_x \longrightarrow 2\,(n\text{-}Bu)_3SnH + 2\,(MeSiO_{1.5})_x \quad (7.26)$$

$$(n\text{-}Bu_3Sn)_2O + Ph_3SiH \longrightarrow (n\text{-}Bu)_3SnH + Ph_3SiOSnBu_3 \quad (7.27)$$

Tributyltin hydride converts organic halides (and related groups) to the corresponding hydrocarbon often with AIBN or by shining UV light. The process happens by a radical chain mechanism involving the radical $Bu_3Sn\bullet$. Along with a free radical initiator such as AIBN, tributyltin hydride generates the tributyltin free radical. This radical can cleave an alkyl/aryl halide bond forming a free radical and Bu_3SnX. Another molecule of Bu_3SnH provides a hydrogen free radical for making an alkyl–aryl hydrogen bond and propagates the radical chain reaction.[29]

$$(CH_3)_2C(CN){-}N{=}N{-}C(CN)(CH_3)_2 \xrightarrow[\Delta]{-N_2} 2\,(H_3C)_2\dot{C}{-}CN \quad (7.28)$$

$$(n\text{-}C_4H_9)_3Sn\text{-}H + (CH_3)_2\dot{C}CN \longrightarrow (n\text{-}C_4H_9)_3Sn\bullet + (CH_3)_2CHCN \quad (7.29)$$

$$(n\text{-}C_4H_9)_3Sn\bullet + RBr \longrightarrow (n\text{-}C_4H_9)_3Sn\text{-}Br + R\bullet \quad (7.30)$$

$$(n\text{-}C_4H_9)_3Sn\text{-}H + R^\bullet \longrightarrow (n\text{-}C_4H_9)_3Sn^\bullet + RH \quad (7.31)$$

Tributyltin hydride TBTH, also finds use in related cyclisation reactions involving dehalogenation. However, TBTH has some drawbacks especially for use in the pharmaceutical industry due to the high toxicity and fat solubility of the organotin compound. Alternatives include the analogous germanium hydrides (Bu_3GeH) or other radical H-donors such as indium hydrides or silanes.[21]

OCH3 I O —TBTH / AIBN→ OCH3 O (7.32)

One of the large scale uses of molten tin is in the industrial manufacture of float glass which is basically flat glass sheets used extensively for making window panes and flat panel displays. Making large sheets of soda glass of uniform thickness and very flat surface was a very difficult task before the British glass manufacturer, Pilkington, discovered this method in 1950, now known as the *Pilkington process*. In this method, molten glass is allowed to spread uniformly on a bed of molten tin under an atmosphere of N2/H2 mixture. Upon cooling, perfectly flat glass sheets are formed which are easily separated from the tin surface. The density of molten tin is 6.99 g/cc while that of soda glass is 2.53 g/cc which makes the molten glass float on the former. Tin remains as a liquid without boiling in the range of 232°C (m.p.) and 2603°C (b.p.) while soda glass completely melts in the range of 1400–1600°C. The reducing N_2/H_2 atmosphere prevents any oxidation of the molten tin. The world production of quality float glass was 70 million metric tons in 2016.

7.16 Properties and Applications of Lead

One of the oldest known metals dating back to 3000 BC, lead was used by the Romans for making water pipes. Later it was used for glazing pottery and for making crystal glassware and chandeliers as PbO easily forms lead silicates with SiO_2. Its relatively high density among easily available metals coupled with the low cost as metal makes it an ideal material for making radiation shields to protect from γ-radiation.[30]

The use of lead and lead-based compounds has seen a sharp decline in the last 3–4 decades due to realisation that lead is highly toxic. Lead poisoning is one of the common problems associated with the use of metallic lead and lead compounds. The centre for disease control (USA) has set the upper limit for the blood lead for adults at 10 μg/dl (10 μg/100 g) and for children at 5 μg/dl. It causes almost 10% intellectual disability and can result in behavioural problems. Some of the effects are permanent. In severe cases, anemia, seizures, coma or even death may occur. Chelation therapy has been found to be useful for removing lead from the body.[31]

Table 7.4 Metals with relatively high densities

Metal	Density (kg/m³)	Metal	Density (kg/m³)
Aluminium	2712	Palladium	12160
Titanium	4500	Hafnium	13310
Zinc	7135	Gold	19320
Tin	7280	Tungsten	19600
Iron	7850	Platinum	21400
Copper	8940	Osmium	22610
Lead	**11340**	Iridium	22650

Lead was one of the major components of solders used for electrical soldering. Well known solders such as 60/40 Sn–Pb which melts at 188°C, and 63/37 Sn–Pb were used principally in electrical/electronic work. 63/37 is a eutectic alloy of these metals, which has the lowest and sharp melting point (183°C) of all the tin–lead alloys. With restrictions on the use of lead due to lead poisoning, lead-free solders are now being preferred. Most lead-free replacements for conventional 60/40 and 63/37 Sn–Pb solder have melting points from 5–20°C higher.

Lead is prohibited in modern plumbing applications. A higher proportion of lead was used earlier as it had the advantage of making the alloy solidify more slowly and prevent leaks. Although lead water pipes were displaced by copper and PVC pipes, lead-based solder was still used until the 1980s because it was thought that the amount of lead that could leach into water from the solder was negligible from a properly soldered joint.

By 2006, the European Union Waste Electrical and Electronic Equipment Directive (WEEE) and Restriction of Hazardous Substances Directive (RoHS) came into effect. These directives prohibit the inclusion of lead in most consumer electronics produced in the European union. In the US, manufacturers are given tax benefits if they reduce the use of lead-based solders. Lead-free solders contain mainly contain tin, copper and silver along with zinc, antimony, bismuth, indium and traces of other metals.

The most important use of lead currently is in lead–acid storage batteries for which no economically viable lead-free replacement has been found yet. Some of the common rechargeable batteries are listed in Table 7.5.[30]

Table 7.5 Properties of different types of rechargeable batteries

Rechargeable battery	Properties and applications
Lead acid	Highly economical for larger power applications where weight does not matter. Preferred choice for automobiles, for hospital equipment, emergency lighting and UPS systems. 85% efficiency in storing power. Works well even at –20°C for starting vehicles during winter. Lead is easily recycled. Inexpensive.
Nickel cadmium	Used where long life, high discharge rate and economical price are important. Main applications are in biomedical equipments, professional video cameras and power tools. Contains toxic cadmium. Relatively low in energy density. Expensive.
Ni metal hydride	Has a higher energy density compared to NiCd at the expense of reduced life. NiMH contains no toxic cadmium. Applications include mobile phones and laptop computers. Expensive.
Lithium ion Lithium cobalt oxide/ lithium iron phosphate	Fastest growing battery system. Used where high energy density and light weight is of prime importance. Better lifespan than lead–acid batteries. Can suffer from overheating and also severe power loss at subzero temperatures. 100% efficiency in storing power. $LiFePO_4$ is non-combustible unlike $LiCoO_2$. Applications include computers and cellular phones. Expensive.
Lithium polymer	Has the attributes of the Li ion but is smaller in size. Main application is in mobile phones. Expensive.

7.17 Oxides of Lead

The industrially useful compounds are mostly the oxides of lead. Lead forms three different oxides—PbO, Pb_3O_4 and PbO_2.

Lead(II) oxide, PbO is the product obtained on oxidation of galena (PbS) at about 1000°C. It is also obtained from the thermal decomposition of some lead salts such as $Pb(NO_3)_2$ or $PbCO_3$. PbO is an amphoteric oxide. It exists in two crystalline polymorphs—tetragonal and orthorhombic. The tetragonal form, also called litharge, has four Pb–O bonds of the same length while the orthorhombic form (also known as massicot) has two different Pb–O bond lengths. The colour varies from yellow to red. PbO is extensively used in making lead glass or crystal glassware which has exceptional reflecting properties when cut. It is also used for making ceramic glazes. Due to high toxicity its use is being discouraged.

Lead(II, IV) oxide, triplumbic tetroxide, is a red coloured oxide obtained by calcination of PbO in oxygen at around 480°C. It is also obtained by heating of lead carbonate in air. Its main use is as a red pigment for making primer paints.

Lead(IV) oxide, PbO_2 is a dark brown crystalline powder which is poorly soluble in water. It is a strong oxidant. It is prepared by the oxidation of Pb_3O_4 by oxidisers such as chlorine, hypochorite or nitric acid. It is also obtained by oxidation of $Pb(OAc)_2$ with calcium hypochorite. It reacts with acids such as H_2SO_4 and HNO_3 liberating oxygen gas. The most important use of PbO_2 is in lead–acid batteries. Heating of PbO_2 in air results in its conversion to other forms of lead oxide with the release of O_2 (Eq. 7.33).

$$PbO_2 \xrightarrow{290°C} Pb_{12}O_{19} \xrightarrow{350°C} Pb_{12}O_{17} \xrightarrow{375°C} Pb_3O_4 \xrightarrow{600°C} PbO \quad (7.33)$$

7.18 The Lead-Acid Battery

As of 2012, more than 85% of the world production of lead was used for making lead–acid batteries followed by 5% in pigments and related products. Discovered by the French physicist Gaston Plante in 1859, the lead–acid battery tops the list of inexpensive rechargeable batteries with widespread applications in automobiles and vehicles of all kinds, and as backup power supply units (inverters) in households, hospital power inverters and cell phone towers. Although when compared to lithium rechargeable battery and other such cells it has low energy to weight and energy to volume ratios, it has high power to weight ratio and ability to supply high surge currents needed for automobile starter motors. The two reactions occurring during discharge of electricity from the battery are the following.[32]

Negative plate reaction

$$Pb(s) + HSO_4^-(aq) \longrightarrow PbSO_4(s) + H^+(aq) + 2e^- \quad (7.34)$$

Positive plate reaction

$$PbO_2(s) + HSO_4^-(aq) + 3H^+ + 2e \longrightarrow PbSO_4(s) + 2H_2O(l) \quad (7.35)$$

Overall reaction

$$Pb(s) + PbO_2(s) + 2HSO_4^-(aq) + 2H^+ \longrightarrow 2PbSO_4(aq) + 2H_2O(l) \quad (7.36)$$

In the fully charged state, the negative plate consists of lead and the positive plate lead dioxide, with concentrated sulfuric acid as the electrolyte. In the fully discharged state, both the positive and negative plates become $PbSO_4$, and the electrolyte loses much of its dissolved H_2SO_4 and becomes water. The process of discharge is driven by the conduction of electrons from the negative plate back into the cell at the positive plate in the external circuit.

During charging of the battery, lead sulfate of the positive plate gets converted into lead peroxide; lead sulfate of negative plate gets converted to pure lead. Specific gravity of sulfuric acid increases potential of the cell increases.[33]

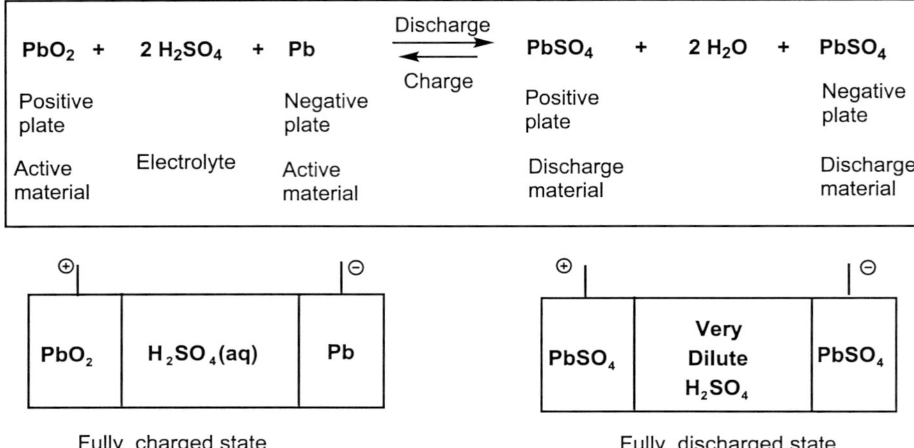

Fig. 7.4 Schematic diagram indicating the functioning of the lead–acid battery

Questions for self study (multiple choice)

7.1. Which among the following statements is not correct for tin which on cooling below 13.2°C undergoes a phase change known as tin pest.
(a) During this change it becomes brittle
(b)The lattice constants of tin change during the tin pest transformation
(c) The density of tin changes during the tin pest transformation
(d) This transformation can also be made by change of pressure

7.2. The relative thermal stability of lead oxides follows the order:
(a) $PbO_2 > PbO > Pb_3O_4$
(b) $PbO_2 > Pb_3O_4 > PbO$
(c) $Pb_3O_4 > PbO_2 > PbO$
(d) $Pb_3O_4 > PbO > PbO_2$

7.3. Which of the following statements is not true for stannous chloride?
(a) It is a reducing agent for the conversion of nitro to amino groups
(b) It is prepared by the reaction of Sn and HCl
(c) It is unstable in moisture
(d) It is used for making tin cans by electrolytic reduction

7.4. Which of the following statements is not true about Me_3SnF?
(a) It is added in toothpastes to prevent tooth decay
(b) It has a polymeric structure with bridging fluorine atoms
(c) It is prepared by fluorination of Me_3SnCl with NaF or KF
(d) It is used for the conversion of M–Cl bonds to M–F bonds in organometallics

7.5. The lead compound used in lead–acid batteries is:
(a) PbO
(b) Pb_3O_4
(c) PbO_2
(d) PbS

7.6. Which of the following statements is not true for GeO_2?
(a) It reacts with Cl_2 in dioxane giving $GeCl_2$.dioxane
(b) It is used in the manufacture of fibreoptical glasses to reduce attenuation
(c) It is reduced by hydrogen to elemental germanium
(d) It is used as a catalyst for making polyethylene terephthalate

Solved concept-based questions

7.7. PbO is much more stable than PbO_2 which decomposes readily to PbO. Why?

This is due to the inert pair effect. Inert pair effect occurs when electrons are pulled closer to the nucleus, making them more stable and more difficult to ionise. The *s*-orbital electrons are affected in this way since they have greater penetrating power. In thallium the *s* electrons in the outer shell are not as well shielded as those in the inner core. They are therefore pulled into the inner core of electrons and thus become inert rendering an oxidation state two units less than the group oxidation state.

7.8. Sn^{2+} ions in solution are good reducing agents. Why?

The stability of Sn^{2+} can be attributed to some amount of inert pair effect. Tin can easily get oxidised to +4 since the inert pair effect is not so strong compared to lead; also, +4 is the group valency.

7.9. What property of lead makes it useful as a protective shield in nuclear reactors and similar nuclear facilities?

Lead is able to moderate certain kinds of radiation because of its high density and high atomic number. It is effective against gamma and X-rays. High density is a result of its high atomic mass and the relatively small atomic radius. Due to its density and number of electrons, it is ideal for scattering X-rays and gamma rays. These rays form photons, which impart energy to electrons when they come into contact with them. In the absence of a lead shield, the electrons within the human body can get affected, leading to DNA damage and resulting in cancer.

Problems and exercises

7.10. Compound A, a well known mineral, on oxidation with O_2 at 1000°C gave a yellow-orange coloured compound B. B on calcination with O_2 at 480°C gave a red coloured compound C. C on reaction with dil. HNO_3, it yielded a brown coloured compound, D. Compound D is used extensively in the manufacture of a well known rechargeable battery. Predict compounds A to D.

7.11. What are the six salient properties of germanium which reflect in its applications? Which among these contribute to the use of germanium in fibreoptic cables?

7.12. Draw the structure of a stable tricordinate germanone and comment on its bond order.

7.13. Starting from $SnCl_4$, show the steps for the preparation of tri n-butyl tin hydride. What property of Bu_3SnH makes it useful for dehalogenation of alkyl and aryl halides?

7.14. How is Me_2GeCl_2 prepared from germanium? Why has a polymer chemistry similar to that of silicones not been realised in germanium chemistry?

7.15. Write three properties of lead which currently find scientific and industrial applications.

7.16. Why is GeO_2 the preferred polymerisation catalyst in the manufacture of PET?

7.17. What are the advantages and disadvantages of using germanium as a semiconductor in electronics and solar cell manufacture?

7.18. Me_3SnF is a recyclable fluorinating agent for organometallic chlorides. Explain how it is recycled.

7.19. What are the structural and geometrical isomers predicted as possible for triple bonded group 14 species? Which of these isomers are known with germanium compounds?

7.20. Indicate with examples the differences between a germylene, germylone and germanone.

Answer key for self study questions

7.1. (d) 7.2. (b) 7.3. (c) 7.4. (a) 7.5. (c) 7.6. (a)

Supplementary reading

1. Butterman W C and Jorgenson J D, Germanium open file report–Mineral commodity profiles, *USGS*, 2005.
2. Moskalyk R R, Review of germanium processing worldwide, *Minerals Engg.*, 2004, Vol. 17, 313.
3. Thiele U K, The current status of catalysis and catalyst development for the industrial process of poly(ethyleneterephthalate) polycondensation, *Int. J. Polym. Mater.*, 2001, Vol. 50, 387.
4. Johnson O H, Germanium and its inorganic compounds, *Chem. Rev.*, 1952, Vol. 51, 431.
5. Quane D and Bottei, R S, Organogermanium chemistry, *Chem. Rev.*, 1963, Vol. 63, 403.
6. Rochow E G, Organic compounds of germanium. The direct synthesis from elemental germanium, *J. Am. Chem. Soc.*, 1947, Vol. 69, 1729.
7. Rochow E G and Brown M P, Dimethylgermanium oxide and dimethylgermanium sulfide, *J. Am. Chem. Soc.*, 1960, Vol. 82, 4166.
8. Moedritzer K, Improved direct synthesis of methyl germanium halides, *J. Organomet. Chem.*, 1966, Vol. 6, 282.
9. Giavinovic M, Kruase M, Yang L, McLeod J A, Liu L, Baines K M, Friscic T and Lumb J-P, A chlorine-free protocol for processing germanium, *Science Advances*, 2017, Vol. 3, e 1700149.
10. Hitchcock P B, Lappert M.F, Miles S J and Thorne A J, *Chem. Commun.*, 1984, 480.
11. Masamune S, Hanzawa Y and Williams D J, Synthesis of a cyclotrigermane and its conversion to a digermene, *J. Am. Chem. Soc.*, 1982, Vol. 104, 6136.
12. Snow J T, Murakami S, Masamune S and Williams D J, Synthesis and characterisation of tetrakis(2,6- diethylphenyl)digermene, *Tetrahedron Lett.*, 1984, Vol. 25, 4191.
13. Power P P, Silicon, germanium, tin and lead analogues of alkynes, *Chem. Commun.*, 2003, 2091.
14. Power P P, Synthesis and some reactivity studies of germanium, tin and lead analogues of alkynes, *Appl. Organometal. Chem.*, 2005, Vol. 19, 488.
15. Li J, Schlenk C, Goedecke C, Frenking G and Jones C, A digeryne with a Ge–Ge single bond that activates dihydogen in the solid state, *J. Am. Chem. Soc.*, 2011, Vol. 133, 18622.
16. Roskamp C A and Roskamp E J, Germanium dioxide–dioxane complex, *Encyclopedia of reagents for organic synthesis*, 2001.

17. Xiong Y, Yao S, Tan G, Inoue S and Driess M, Acyclic germadicarbene (germylone) from germyliumyldene, *J. Am. Chem. Soc.*, 2013, Vol. 135, 5004.
18. Yao S, Xiong Y and Driess M, From NHC-germylenes to stable NHC-germanone complexes, *Chem. Commun.*, 2009, 6466.
19. Li L, Fukawa T, Matsuo T, Hashisume D, Fueno H, Tanaka K and Tamao K, *Nature Chem.*, 2012, Vol. 4, 361.
20. Power P P, Main group chemistry: a heavier element ketone at last, *Nature Chem.*, 2012, Vol. 4, 343.
21. Sinhababu S, Yadav D, Karwasara S, Sharma M K, Mukherjee G, Rajaraman R and Nagendran S, *Angew. Chem. Int. Ed.*, 2016, Vol. 128, 7873.
22. Rit A, Campos J, Niu H and Aldridge S, A stable heavier group 14 analogue of vinylidene, *Nature Chem.*, 2016, Vol. 8, 1022.
23. Graf G G, Tin, tin alloys and tin compounds, in *Ullmann's encyclopedia of industrial chemistry*, 2000.
24. Hattab F, The state of fluorides in toothpaste, *J. Dentistry*, 1989, Vol. 17, 47.
25. Thoonen S H L, Deelman B J and van Koten G, Synthetic aspects of tetraorganyltins and organotin(IV) halides, *J. Organometal. Chem.*, 2004, Vol. 689, 2145.
26. Chaudhary P, Bieringer M, Hazendonk P and Gerken M, The structure of trimethyltin fluoride, *Dalton Trans*, 2015, Vol. 44, 19651.
27. Herzog A, Liu F-Q, Roesky H W, Demsar A, Keller K, Noltemeyer M and Pauer F, Trimethyltin fluoride: A new fluorinating agent for the preparation of organometallic fluorides, *Organometallics*, 1994, Vol. 13, 1251.
28. Hayashi K, Iyoda J and Shiihara I, Reactions of organotin oxides, alkoxides and acyloxides with organotin hydrides, New preparative methods for organotin hydrides, *J. Orgaanometal. Chem.*, 1967, Vol. 10, 81.
29. Rajan Babu T V, Page P C B and Buckley B R, Tri n-butyl stannane in *Encyclopedia of reagents for organic synthesis*, John Wiley and sons, 2004.
30. Davidson A, Ryman J and Sutherland C A, Lead, in *Ullmann's encyclopedia of industrial chemistry*, 2014.
31. Reddy A and Braun C L, Lead and the Romans, *J. Chem. Educ*, 2010, Vol. 87, 1052.
32. Dell R, David A and James R, *Understanding batteries*, RSC, 2001.
33. Linden D and Reddy T B, Eds, *Handbook of batteries*, 3rd Ed., McGraw-Hill, New York, 2002.

CHAPTER

8

The Chemistry of Nitrogen

Unique Properties of Nitrogen and its Compounds

- *Nitrogen gas, N_2 forms 78.09% of the earth's atmosphere which amounts to an estimated 4,000 trillion tons.*
- *Unlike O_2, N_2 is quite stable in the atmosphere and does not get involved much in chemical reactions. Thus, over geological time, it has built up in the atmosphere to a much greater extent than oxygen.*
- *Nitrogen is the most widely used blanketing and purging gas for maintaining an inert atmosphere in metallurgy, chemical reactions, storage of flammable compounds and protecting compounds which are prone to air oxidation such as vegetable oils.*
- *Compounds of nitrogen have three major applications, namely, fertilisers (ammonia, urea and related compounds), explosives (organic nitro and polynitrogen compounds) and as an inhalation anesthetic (nitrous oxide).*
- *Since it is non-combustible, unreactive and available in plenty, N_2 is also used for generating high pressure to force crude oil to reach the earth's surface. It is slightly soluble in water (two litres in 100 litres of water).*
- *N≡N is the second strongest bond known; it has a bond dissociation enthalpy of 945 kJ/mol.*
- *Nitrogen is an essential element for life and is present basically in proteins as amino acid units.*
- *Nitrogen shows oxidation states of –3 (NH_3), –2 (N_2H_4),–1 (NH_2OH), 0 (N_2), +1 (N_2O), +2 (NO), +3(HNO_2), +4 (NO_2) and +5 (HNO_3) in its compounds.*
- *Ammonia is one of the most industrially important inorganic compounds and a precursor to almost all nitrogen-based compounds, such as nitric acid, nitrogen-based fertilisers and many explosives.*
- *Rhizobium bacteria present in root nodules of leguminous plants convert atmospheric nitrogen to ammonia which is further converted to soluble nitrates by nitrifying bacteria (nitrogen fixation).*

- *Nitric oxide (NO), which is produced by the endothelial cells of blood vessels, acts as a vasodilator, relaxing blood vessels thereby increasing blood flow and reducing blood pressure.*
- *Linear azide (N_3^-), bent pentazenium (N_5^+) and cyclic pentazolate (N_5^-) are polynitrogen compounds which have been stabilised with non-energetic counter ions and solvent molecules and are high energy density materials.*
- *The formation of the very strong N≡N bond is the main driving force behind the detonating property of all nitrogen-based explosives.*

8.1 About the Element

Nitrogen was first isolated in 1772 by Daniel Rutherford who called it *noxious air*. Since the phlogiston theory was prevalent during that time, Scheele, Cavendish and Priestly called the gas *phlogisticated air* which meant burnt air. Lavoisier found that animals died and flames went out in its presence and called it '*azote*', meaning lifeless.

At atmospheric pressure, N_2 liquefies at 77K (–195.79°C, density 1.25) to a colourless liquid and freezes at 63 K (–210°C). In contrast, O_2 liquefies at 90K (–183°C) to a pale blue liquid with a density of 1.42. By volume, dry air contains 78.09% N_2, 20.95% O_2, 0.93% Ar, 0.039% CO_2 and small amounts of other gases. Air also contains a variable amount of moisture, on an average 0.4% over the entire atmosphere.[1]

Table 8.1 Physical properties of nitrogen

Nitrogen	*Atomic mass*	*Electronegativity (Pauling scale)*	*1st ionisation energy*	*Electronic configuration*	*van der Waals radius*
Symbol: N	14.007	3.040	1402.328 (kJ/mol)	$1s^2 2s^2 2p^3$	155 pm
At. no. 7	*Atomic radius*	*Melting point*	*Boiling point*	*Electron affinity*	*Density*
	1.55 Å	-210.0°C	-195.795°C	7 kJ/mol	1.2504 g/L at 0°C

Humanity realised the importance of nitrogen-based compounds, especially for fertilisers and explosives during 1800–1900 and the only major source of nitrogen in those days was Chile saltpetre ($NaNO_3$) which was found in plenty in the Atacama desert (Chile/Peru) as an ore called *caliche*. Other than this, the only naturally available sources of nitrogen compounds in minor quantities were Peruvian guano (accumulated bird excrement) and sal ammoniac (a mineral containing ammonium chloride). In 1837, Liebig discovered the importance of nitrogen for plant growth and in 1867 Alfred Nobel discovered dynamite which was a mixture of nitroglycerine and diatomaceous earth. Nitric acid along with sulfuric acid was required for the nitration of glycerine for making nitroglycerine. The North Americans and the Europeans travelled to the Atacama desert to mine this industrially important nitrogen source.

During the World War I, Chile saltpetre became a highly sought after compound for making gunpowder and explosives and the Allies initiated a trade blockade against Germany to prevent them from receiving it. Germans discovered that ammonia could be oxidised to get NO which could be converted to NO_2 and further to nitric acid. This necessity resulted in attempts to make ammonia from atmospheric nitrogen and led to the discovery of the Haber–Bosch process. After the Haber–Bosch process became operative in 1909, the dependence on Chile saltpetre became less and less significant for making nitrogen-based compounds.

8.2 Isolation and Purification of the Element

There are basically three different methods for isolation and purification of nitrogen gas required for industrial use. The difference between these three methods depends basically on the purity of nitrogen required and the quantity of nitrogen gas required. These methods are (i) cryogenic process, (ii) pressure swing adsorption (PSA) using carbon molecular sieves and related processes, and (iii) membrane-based separation process.

The cryogenic process, which requires high amounts of energy input as electricity, produces N_2 gas with purity levels of 99.9999%. This purity is required for electronic and related industry. This process also is used for generating liquid nitrogen. The difference in the boiling points of liquid N_2 and liquid O_2 is utilised in the separation process.

Cryogenic fractional distillation of liquid air was first perfected by the German engineer Carl von Linde in 1895 and is the oldest method for nitrogen production. Even today, large commercial air separation plants work by this process and account for almost 60%–70% of the total nitrogen production of the world.[2]

Pressure swing adsorption (PSA), a process discovered in the 1970s, provides N_2 gas with purity levels of 95% to 99.9995%. The process which works on the principle of selective adsorption uses a special adsorbent which is carbon molecular sieve (CMS).[3] A carbon molecular sieve (CMS) is the porous skeletal framework of carbon that remains after the oxygen-free pyrolysis of a polymeric precursor such as polyvinylidene chloride, polyacrylonitrile, phenol formaldehyde or coal. It can be considered as a special type of activated carbon. Activated carbon, which is generally used in organic chemistry for adsorption of gases and liquids has micro and transitional pores of 10–500 Å pore diameter. In contrast, carbon molecular sieves have only uniform super-micro pores of less than 10 Å (1 nm) pore diameter. In addition, activated carbon separates molecules through differences in their adsorption equilibrium constants while carbon molecular sieves bring about separations based on rate of adsorption of different molecules rather than the differences in adsorption capacity. This behaviour is clearly observed in the functioning of pressure swing adsorbers (PSA). The separation of nitrogen from air by PSA is the single most important application of carbon molecular sieves.[3]

Pressure swing adsorption systems for nitrogen production operate typically over the pressure range of 1–7 atm. The CMS adsorbs O_2, CO_2 and moisture present in compressed air in a short period of time while N_2 gas, which is not adsorbed under pressure is available in the compressed form at the outlet. Afterwards, when the pressure is reduced to the atmospheric or vacuum level, CMS desorbs oxygen and other gases and the unit is regenerated for the next cycle. Two such adsorption columns can work side by side for a constant supply of pure nitrogen gas.[4]

The third process, membrane separation, operates on the principle of selective gas permeation through a membrane. Membrane systems have an efficiency to produce N_2 gas with purity levels of 95% to 99.5% only. The process unit contains several modules, each containing thousands of hollow, mostly polymeric fibres made of polymers such as polysulfones, polyamides and also CMS. Structurally, a hollow fibre membrane represents a cylindrical cartridge functioning as a spool with specifically reeled polymer fibres. Gas flow is supplied under pressure into a bundle of membrane fibres. Due to the difference in partial pressures on the external and internal membrane surface, gas flow separation is accomplished, The membrane allows nitrogen to pass and other gases (such as oxygen, water vapour and CO_2) to permeate. Compressed air enters at the inlet of the generator, nitrogen exits at the outlet and the other gases permeate out through the sides of the unit thus providing a constant supply of purified nitrogen.

Laboratory synthesis of small amounts of pure nitrogen can be achieved by the following reactions.

$$NH_4Cl_{(aq)} + NaNO_{2(aq)} \longrightarrow N_{2(g)} + NaCl_{(aq)} + 2\,H_2O_{(l)} \quad (8.1)$$

$$2\,Ba(N_3)_2 \xrightarrow[\Delta]{180°C} 2\,Ba + 3\,N_2 \quad (8.2)$$

$$2\,NaN_3 \xrightarrow[\Delta]{280°C} 2\,Na + 3\,N_2 \quad (8.3)$$

$$(NH_4)_2Cr_2O_7 \xrightarrow[burn]{} Cr_2O_3(s) + N_2 + 4\,H_2O \quad (8.4)$$

8.3 Reactions of Dinitrogen, N_2

Unlike other diatomic gases, reactivity of dinitrogen is very poor and the major reason for this can be ascribed to the very high bond strength of N≡N (945 kJ/mol) which is second only to C≡O (1072 kJ/mol). Unlike C≡O, the N≡N bond is non-polar and the HOMO–LUMO energy gap is significantly larger, contributing further to the poor reactivity. Although N_2 reacts with lithium at room temperature, better reactivity has been observed when lithium metal is dissolved in molten sodium giving Li_3N at around 400°C. Beryllium (at 700–1400°C) and magnesium (at 300–700°C) gives Be_3N_2 and Mg_3N_2 and burning of calcium and strontium in air generates calcium and magnesium nitrides along with their oxides. Li_3N melts to decomposition at 548°C, Be_3N_2 melts at 2200°C while Mg_3N_2 decomposes above 271°C. Reduction of metal oxides or halides in the presence of N_2 also results in some metal nitrides.

$$Al_2O_3 + 3\,C + N_2 \longrightarrow 2\,AlN + 3\,CO \quad (8.5)$$

From molecular orbital pictures one can see that N_2 and CO are isoelectronic species. The bond length and bond energies of the two species are compared in Table 8.2 for single, double and triple bond formation.

Table 8.2 Bond energy and bond length comparison of C–O and N–N bonds

	Bond energy **Bond length**		**Bond energy** **Bond length**		**Bond energy** **Bond length**
C–O	326 kJ/mol 0.143 nm	C=O	803 kJ/mol 0.120 nm	C≡O	1075 kJ/mol 0.113 nm
N–N	163 kJ/mol 0.145 nm	N=N	418 kJ/mol 0.125 nm	N≡N	945 kJ/mol 0.110 nm

Although not as reactive as CO, N_2 reacts with many transition metal complexes forming dinitrogen complexes. The first dinitrogen complex $[Ru(NH_3)_5N_2]^{2+}$ was accidentally discovered from the reaction of $RuCl_3$ and $N_2H_4.H_2O$.[5] This complex possessing terminal N_2 bonding was later prepared by the reaction of $[Ru(NH_3)_5H_2O]^{2+}$ with N_2 gas. The infrared spectrum of this compound indicated the $\nu_{N\equiv N}$ band at 2110 cm^{-1} (compared to free N_2 gas at 2331 cm^{-1}) and the N–N bond was found to be slightly lengthened to 1.12 Å (compared to free N_2 1.09 Å). The dimeric version of this complex, $[(NH_3)_5Ru\text{-}N_2\text{-}Ru(NH_3)_5]^{4+}$ was prepared by several routes including the reaction of $[Ru(NH_3)_5H_2O]^{2+}$ and $[Ru(NH_3)_5N_2]^{2+}$. Due to the high symmetry, the strong $\nu_{N\equiv N}$ band was seen only in the Raman spectrum at 2100 cm^{-1} for the dimer.

There are basically four different types of dinitrogen complexes known so far (Fig. 8.1). These are broadly classified under *end on* or *side on* categories. A couple of examples having both types of bonding are also observed.[6]

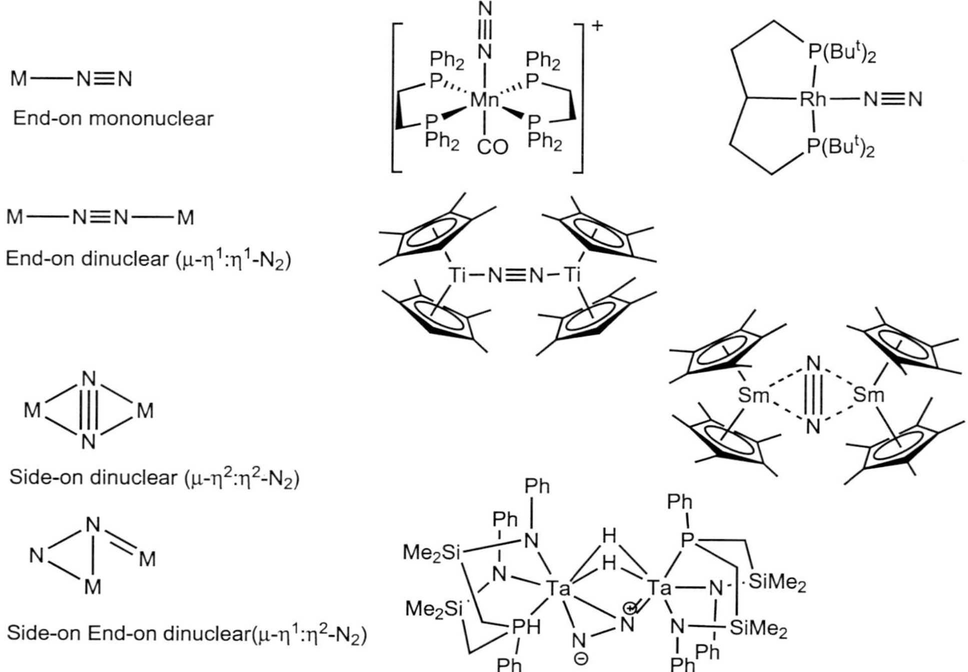

Fig. 8.1 Types of dinitrogen complexes with examples

8.4 Biological Nitrogen Fixation

Dinitrogen complexes are significant as they are considered as model systems for biological nitrogen fixation which involves the conversion of dinitrogen to ammonia. This is efficiently carried out by the nitrogenase enzymes present in the rhizobium bacteria which make colonies in the root nodules of leguminous plants (Scheme 8.1).[7]

$$N_2 + 8\,H^+ + 8e + 16\,MgATP \xrightarrow[\text{Fe–Mo nitrogenase}]{} 2\,NH_3 + H_2 + 16\,MgADP + 16\,P_i \quad (8.6)$$

The active site of the nitrogenase enzymes contains a metal–sulfido cluster. The three known types of nitrogenase enzymes have been found to have different combinations of metals in the active site—Fe–Mo, Fe–V and Fe alone.[7] The structure of the active site of the Fe–Mo nitrogenase has been elucidated by X-ray structural studies.[8]

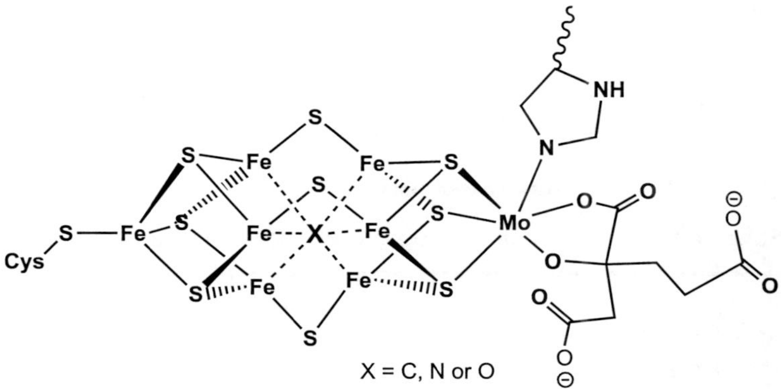

Fig. 8.2 Active site of the Fe–Mo nitrogenase enzyme

8.5 Industrial Processes for the Conversion of N_2 to Useful Products

Industrially, three processes were developed where N_2 gas was used as one of the reactants to make useful compounds. They are described below.

8.5.1 Birkeland-Eyde Arc Process for Nitric Acid (1903)

This process, developed by the Norwegian industrialist and scientist Birkeland along with business partner Eyde in 1903, is based on a method used by H Cavendish for making nitric acid before the discovery of the Ostwald process. The electric arc (~ 3000°C) converts nitrogen and oxygen to NO which gets converted to NO_2 and further to HNO_3.[10]

$$2N_2(g) + 5O_2(g) + 2H_2O(l) \xrightarrow[\text{electric arc furnace}]{} 4HNO_3(l) \quad (8.7)$$

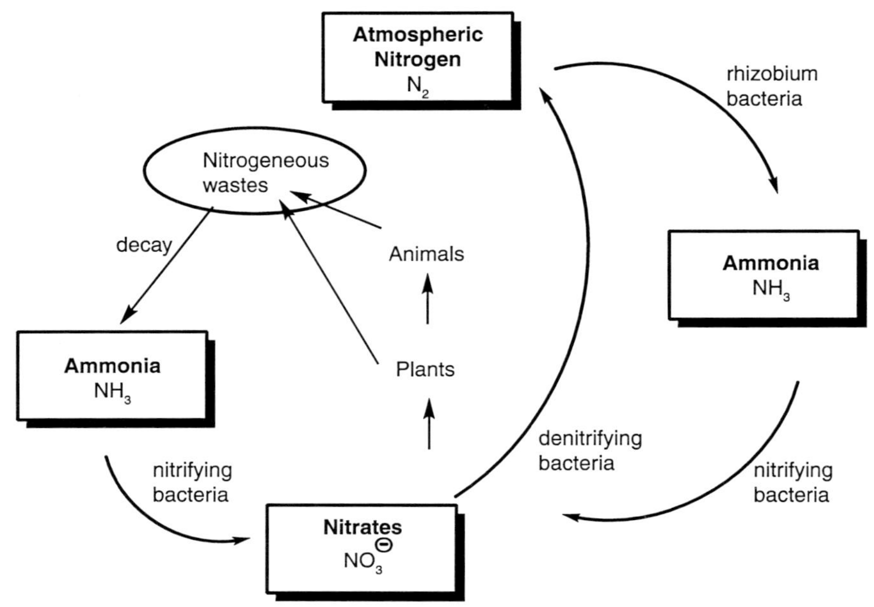

Scheme 8 .1 The biological nitrogen cycle indicating nitrogen fixation[9]

8.5.2 Frank-Caro Cyanamide Process for Calcium Cyanamide (Nitrolime)

Adolf Frank and Nikodemus Caro of Germany came out with this first industrial process for fixing atmospheric nitrogen in 1898 to convert calcium carbide to calcium cyanamide by heating the carbide with nitrogen at 1000°C.[11]

$$CaO + C \xrightarrow[\text{electric arc}]{2000°C} CaC_2 \xrightarrow[\text{Frank–Caro process}]{N_2\ \ 1000°C} \underset{\text{calcium cyanamide}}{CaCN_2 + C} \quad (8.8)$$

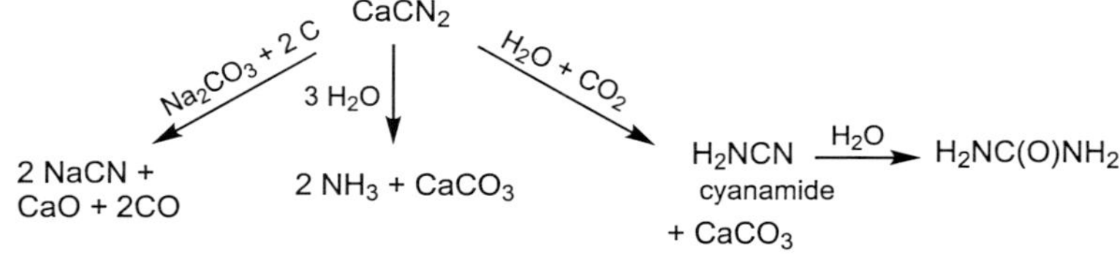

Scheme 8.2 Reactions of calcium cyanamide

Calcium cyanamide can be used for the synthesis of ammonia, NaCN and cyanamide (H_2NCN). Cyanamide has been found to give urea upon hydrolysis. With the development of the Haber–Bosch process and preparation of urea and HCN from ammonia, the route through cyanamide was discarded. Another reason for the termination of this method is the poor atom efficiency and energy efficiency of this process.

8.5.3 Haber-Bosch Process for Ammonia

Although C L Berthelot, a French chemist, discovered in 1784 that ammonia is made up of one nitrogen atom and three hydrogen atoms, there was no procedure to make ammonia from these two gases for the next 125 years. Fritz Haber, an outstanding chemist of German origin, who understood the laws governing such chemical reactions, together with an English scientist Robert Le Rossignol, solved the problem. They calculated that at 200°C and at a pressure of 200 atm, nitrogen and hydrogen would react. However, when the reaction was performed they found that it was extremely slow and a catalyst was required to speed it up—possibly a metal on whose surface hydrogen and nitrogen could combine. After much experimentation, they found that osmium accelerated the reaction very well (Eq. 8.9). In 1908, they were able to show that about 125 mL/h of ammonia could be made by this method, which is about 8% conversion, and obtained a patent.

$$N_2 + 3\,H_2 \xrightarrow[\text{200 atm, 200°C}]{\text{Os catalyst}} 2\,NH_3 \qquad (8.9)$$

This method is still the best method for converting atmospheric nitrogen to ammonia. It was procured by the BASF company and Carl Bosch of BASF converted the process into the industrial level using a promoted iron catalyst discovered by Allwyn Mittash. Both Haber and Bosch received the Nobel Prize for their path breaking discovery in 1918 and 1931 respectively.[11,12]

Over the years, processes other than the Haber–Bosch process became redundant especially due to requirements of large energy inputs. Although many new methods have been reported recently where the requirement of high pressure and temperature has been reduced considerably, none of them has been good enough to be upgraded for industrial production. A few processes using ruthenium-based catalysts, such as the Kellogg advanced ammonia process, show some promise. However ruthenium catalysts are strongly inhibited by hydrogen adsorption at low reaction temperatures thereby exhibiting low activity below 350°C.

Details of the Haber–Bosch process

In this process, a mixture of hydrogen and nitrogen gases react over an iron or ruthenium catalyst at elevated temperatures and pressures, to produce ammonia. The unconverted part of the gaseous mixture is recycled and an overall maximum conversion of 97% is achieved. Due to the industrial importance of ammonia as the precursor for almost all nitrogen-based compounds, most of the metals present in the periodic table were tested for catalytic activity and the best results were obtained from the iron group of metals.[13]

Some interesting facts about the Haber–Bosch process

Approximately 3 out of 7 billion people on earth would have perished of hunger if sufficient amounts of ammonia was not generated by the Haber–Bosch process for making fertilisers. One estimate suggests that about 50% of the nitrogen present in the human tissue has had its origins in the Haber–Bosch process!

More than 50% of the hydrogen produced in the world is used for the synthesis of ammonia by this process. The Haber–Bosch process consumes around 1% of the world's energy sources. For every ton of ammonia produced, 1.87 tons of carbon dioxide is generated which is about 3% of the world's CO_2 emission. For the records, in 2010, 245 million tons of CO_2 was released to the atmosphere by the process.

Production of hydrogen for the reaction by steam reforming of methane is still the best method for hydrogen generation as electrolysis of water is 12 times costlier.

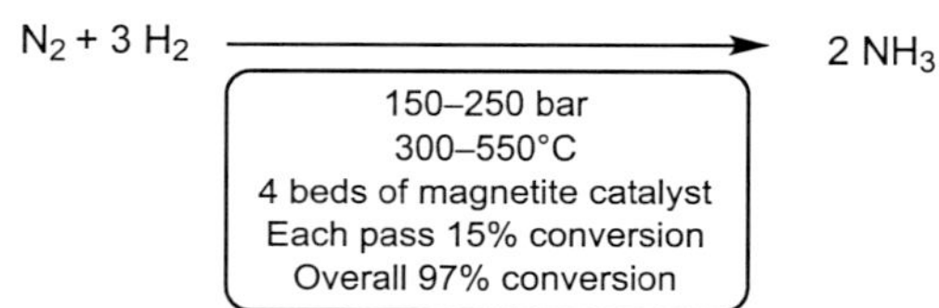

Scheme 8.3 General reaction conditions of the Haber–Bosch process

The nitrogen needed for this process is obtained by the cryogenic method from air or by separation techniques using carbon molecular sieves. However, preparing pure hydrogen for the process is more cumbersome as CO must be completely removed (as it can poison the catalyst). There are basically six steps in the industrial Haber–Bosch process. Out of this, five are methods to prepare and purify hydrogen gas which is produced from methane. These are:

(i) Sulfur removal from methane

$$H_2 + RSH \xrightarrow[Al_2O_3,\ 700\ K]{\text{Co and Mo Oxides}} RH + H_2S \tag{8.10}$$

$$ZnO + H_2S \xrightarrow[230\text{–}430°C]{} ZnS + H_2O \tag{8.11}$$

ZnO can be regenerated from ZnS

(ii) Steam reforming

$$CH_4 + H_2O \xrightarrow[700\text{–}1110°C]{\text{Ni, 40 atm}} CO + 3 H_2 \tag{8.12}$$

(iii) Shift conversion (water gas shift reaction)

$$CO + H_2O \xrightarrow[\substack{Cr/Fe_3O_4\ 360°C \\ Cu/ZnO/Cr\ 210°C}]{10\ atm} CO_2 + H_2 \quad (8.13)$$

(iv) Carbon dioxide removal: This is carried out by pressure swing adsorption method using 13x molecular sieves and activated charcoal. Two other chemical methods used by the industry are the reversible reactions with (a) K_2CO_3 forming $KHCO_3$ or (b) monoethanolamine $HOCH_2CH_2NH_2$ forming the carbonate (Chapter 5; Eqs 5.18, 5.19).

(v) Methanation: Removal of small amounts of CO and CO_2 remaining in the hydrogen gas is done by the methanation process.

$$CO + 3\ H_2 \xrightarrow[600°C]{Ni/Al_2O_3} CH_4 + H_2O \quad (8.14)$$

$$CO_2 + 4\ H_2 \xrightarrow[600°C]{Ni/Al_2O_3} CH_4 + 2\ H_2O \text{ (Sabatier reaction)} \quad (8.15)$$

(vi) Ammonia synthesis: In this step, purified H_2 and N_2 are mixed and reacted over a heterogeneous catalyst to make ammonia (Scheme 8.3).

8.5.4 The Catalysts for the Haber-Bosch Process

A catalyst and promoter are used in the process. The catalyst is partially reduced magnetite (Fe_3O_4) encased in a shell of Wüstite (FeO) covered by an iron shell. Promoters such as K_2O, CaO, SiO_2 and Al_2O_3 are also used with the catalyst to maintain a high surface area. An essential requirement of the promoter oxide is that it should not get reduced in the presence of hydrogen gas.[13]

Among other metals, only ruthenium and osmium showed catalytic activity superior to iron. A study comparing the ammonia synthesis activity of late transition metals supported on activated carbon along with potassium metal promoter gave the following relative activities. (mL of NH_3/h of catalyst (5.0 wt% of carbon) and potassium (25 wt% of carbon) kept at 300°C and 600 torr: flow rate 4.5 L/h).[14]

Table 8.3 Comparison on catalytic activity of different metals for Haber–Bosch process

Fe :0.72	Ru:22.20	Os:5.60
Co:0.40	Rh:0.52	Ir:0.68
Ni:0.04	Pd:0.00	Pt:0.008

Although osmium was the original catalyst used by Haber in 1909, it never became an industrial catalyst due to its limited availability, relative cost and poisonous nature. In 1979, British Petroleum (BP) came out with a new ruthenium-based catalyst and the firm Kellogg started commercialisation of this catalyst-based process (known as Kellogg advanced

Fritz Haber: A chemist most cherished and also most dreaded[24,25]

The life of Fritz Haber, a German chemist of Jewish origin and recipient of the Nobel Prize of 1918 for the conversion of nitrogen to ammonia, is little known to chemists but has been one of high drama and contradictions. Haber was indeed a truly outstanding chemist. In chemistry circles he is remembered for the Haber–Bosch Process, for the Born–Haber cycle and also for a process for extracting gold from sea water. In fact it has been remarked about the Haber–Bosch process that out of the 7 billion people now on earth, 3 billion would not have survived had this process of converting nitrogen to ammonia not been discovered to replenish the depleting nitrogen content of the soil. Another estimate suggests that roughly 50% of nitrogen present in the body of every human being can be traced back to ammonia made by the Haber–Bosch process. Born as a Jew in Breslau, he was eager to get accepted by Germany and since he thought his faith will be an obstacle, he converted to Christianity. Haber had an enviable education record as he was taught by brilliant chemists of those times—Robert Bunsen from the university of Heidelberg, Hoffmann from University of Berlin and Libermann from the Technische University, Berlin.

What many are unaware of is the dark and often horrifying side of the life and deeds of Haber. The BBC came out with a programme about his life entitled 'Bread from the air, gold from the sea, the chemist of life and death' which truly describes the life of this scientist. German TV in 2014 came out with a movie focussing on the life of his wife, Clara Immerwhar. Haber's motto which ultimately resulted in an ethical dilemma was 'a scientist should serve mankind in times of peace and the fatherland during war'. So during World War I, he joined the German army to fight against the Allies. The Allies had by then faultlessly implemented and executed 'trench warfare' where their armies dug trenches which were interconnected and the traditional warfare using gunfire and direct person to person shooting was found to be ineffective in killing the soldiers of the Allies who could effortlessly hide and move through the interconnected trenches. The use of metal helmets kept causalities to a minimum. Being a highly innovative chemist, Haber came up with a novel idea to stop the progress of the enemy soldiers through the trenches. He advised the German army that chlorine gas cylinders should be left open near the trenches and since it was a dense gas, it would seep down into the trenches and kill the soldiers hiding and moving through the trenches. On 22 April 1915, Haber went with the German army, taking 6000 cylinders of chlorine gas to the war front and opened them on 30,000 allied troops spread over 7 km. Within 15 minutes 5000 of them died and 10,000 were seriously wounded. The German command was delighted with the outcome and made him a captain. This was the first time chemical warfare was carried out and that too successfully.

However, his wife Clara Immerwahr, a virtuous lady, and trained chemist and also the first German woman to receive a Ph.D, stridently opposed her husband's idea of using chemistry for killing human beings. But her arguments fell on deaf ears. He ignored her protests and went again for another kill loaded with more chlorine cylinders on 2 May 1915. She, to register her strong dislike to her husband's unethical action, took the service revolver provided by the German army to Haber, came out into the garden of her house and shot herself in the chest. Her son Hermann (who also committed suicide later in 1946, while in the USA) found her bleeding with the gunshot wound and she died in his hands. The ruthless Dr Haber (also called Dr Evil) on hearing the news of his wife's suicide, did not even bother to give her a decent burial but went ahead to the eastern warfront loaded with more chlorine cylinders!

After World War I, Haber also experimented with the idea of enriching gold from sea water which did show promise but was found to be economically unviable. Later, during World War II, Haber had to flee from Germany to England as Hitler did not trust him due to his original

Jewish background. He was hated for his criminal background by upright scientists of his times and it is written that Lord Rutherford refused to shake hands with him when he was introduced to him during his stay in Cambridge. Haber died a lonely death in a hotel of a heart attack while he was on his way to Israel to take up an academic position at the newly formed Weizmann Institute.

Incidentally, learning from the Germans, the Allies also started trying out poison gas warfare and their contribution to this dastardly method of killing was mustard gas, a poisonous sulfur-based dense gas. Ultimately, both sides gave up this practice as it became evident that simply the change in direction of the wind could result in fatalities on both sides!

ammonia process, KAAP). The catalyst for this process was prepared by subliming $Ru_3(CO)_{12}$ on a carbon support impregnated with rubidium nitrate and the formed catalyst had 5 wt% Ru and 10 wt% Rb. This catalyst with high surface area had an activity that was 10–20 times more than the conventional Fe-based catalyst. The KAAP process also operates at a lower pressure compared to the iron-based processes (90 bar pressure versus 140–170 bar pressure; 375–400°C versus 400–450°C) thus reducing the cost of production. While ruthenium catalysts work under milder conditions at lower reaction temperatures, they exhibit lower activity. Ruthenium catalysts are strongly inhibited by hydrogen adsorption at lower reaction temperatures.

A third commercially viable catalyst was reported in 2000 by chemists working for the Haldor–Topsøe company of Denmark. This catalyst involved ternary nitrides such as Fe_3Mo_3N, Co_3Mo_3N and Ni_2Mo_3N. Along with cesium promoters, these catalysts showed catalytic activities 2–3 times greater than commercial iron-based catalysts under similar operating conditions. These are also less expensive compared to ruthenium based catalysts used in the KAAP process.

8.6 Chemistry of Ammonia

8.6.1 Reactions in Liquid Ammonia

Ammonia is a well known non-aqueous solvent and many reactions when carried out in liquid ammonia differ significantly when compared to reactions carried out in aqueous media. One major reason is the solubility difference of salts in ammonia compared to water and another reason is the difference in the solvation capability of ammonia when compared to other solvents. For example, in the reaction of $AgNO_3$ and $BaCl_2$, AgCl is precipitated in aqueous media while $BaCl_2$ gets precipitated in liquid ammonia medium.[13]

$$2\,AgNO_3 + BaCl_2 \underset{NH_3}{\overset{H_2O}{\rightleftharpoons}} 2\,AgCl + Ba(NO_3)_2 \tag{8.16}$$

Ammonia also shows the unique capability of solvating electrons when alkali metals and most of the alkaline earth metals are dissolved in liquid ammonia. The solution has a characteristic blue colour irrespective of the metal dissolved (λ_{max}= 1400 nm). It is also one of the ideal solvents used for realising Zintl phases which are naked clusters of heavy main group metals and metalloids.

8.6.2 Ammonia as a Precursor for Nitrogen-Based Compounds

About 85% of all ammonia produced in the world is used for making nitrogen-based fertilisers. Out of this, about 40% is used for making urea. The rest of the ammonia is currently used for making nitric acid (5%), polyamides, melamine, polyurethane polymers (5%) fine chemicals and wood-pulp treatment (5%). It is also used for many other compounds of academic interest and industrially useful products (given in Schemes 8.4, 8.5). The major

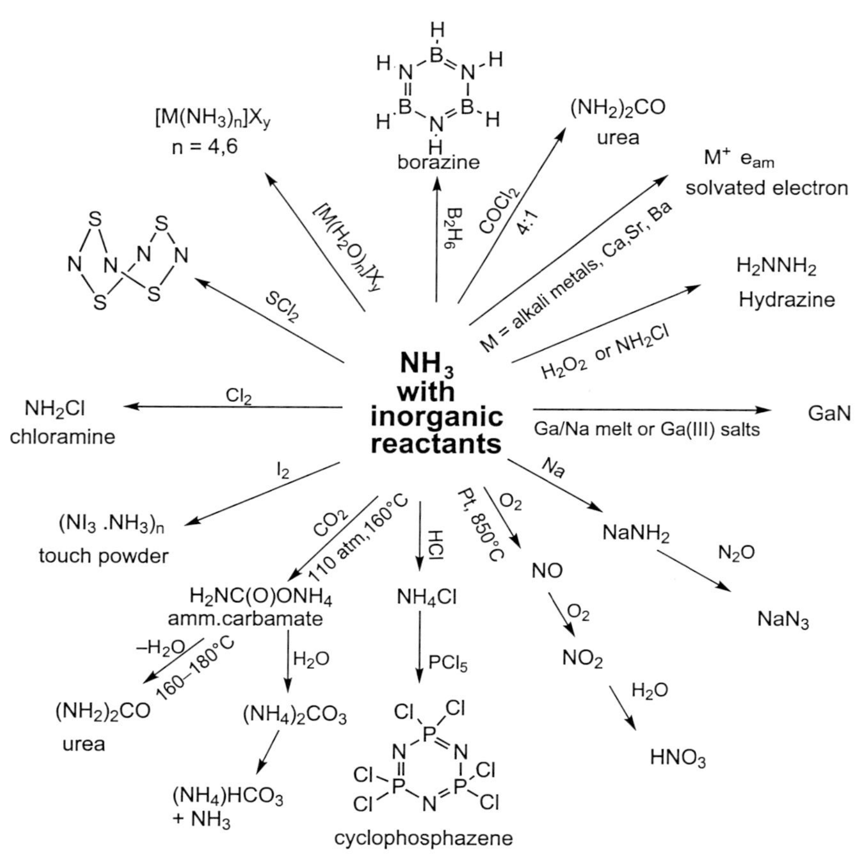

Scheme 8.4 Major reactions of ammonia with inorganic reagents

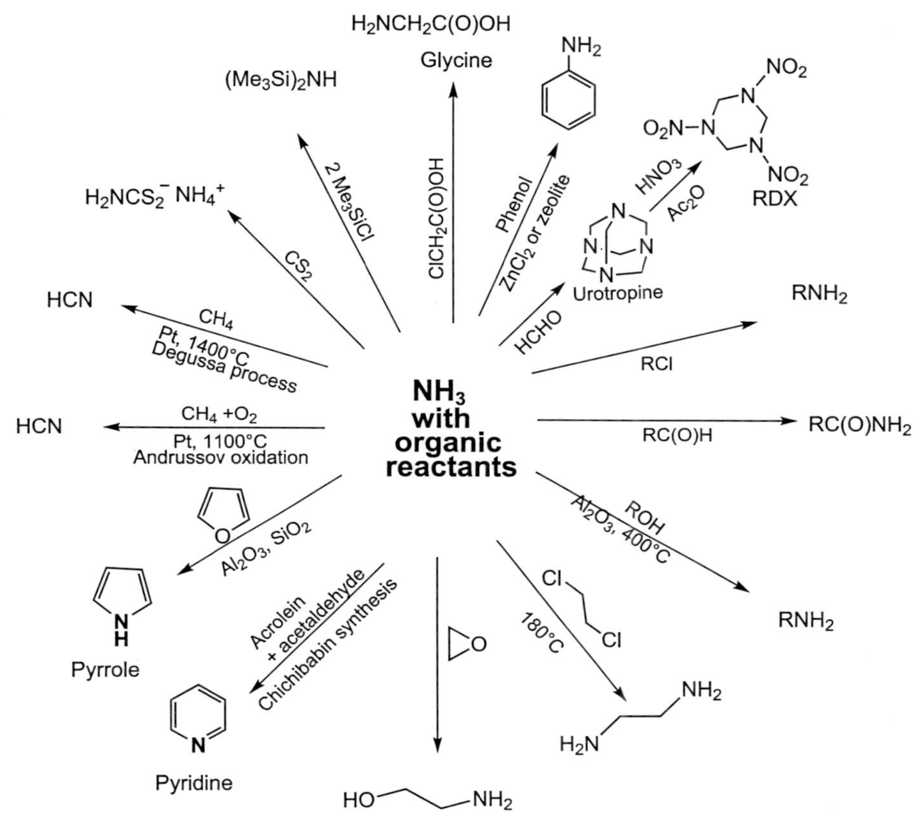

Scheme 8.5 Major reactions of ammonia with organic reagents

reactions of ammonia with inorganic substrates leading to useful products are given in Scheme 8.4. It is interesting to learn that ammonia is the source of nitrogen for some of the nitrogen-based inorganic heterocycles.[13, 15] Ammonia also reacts with all acids forming salts. The most industrially important among these are ammonium nitrate (fertiliser and explosive) and ammonium phosphates (fertiliser).

Ammonia is also directly or indirectly the source of nitrogen in the industrial synthesis of almost all organic nitrogen-based compounds. Nitric acid required for nitration and synthesis of explosives is also made from ammonia by the Ostwald process. Other major organic products of industrial importance obtained from ammonia include HCN, nitrogen-based heterocycles such as pyrrole and pyridine, amines, amides and amino acids. The ubiquitous complexing agent EDTA is synthesised by the Singer synthesis using ethylene diamine and HCN, both prepared from ammonia (Eq. 8.17). Well known reactions with organic substrates are given in Scheme 8.5.

$$H_2NCH_2CH_2NH_2 + 4\,HCHO + 4\,HCN \longrightarrow (NCCH_2)_2NCH_2CH_2N(CH_2CN)_2 \xrightarrow{\text{aq.NaOH}} (NaO(O)CCH_2)_2NCH_2CH_2N(CH_2C(O)ONa)_2 \qquad (8.17)$$

8.6.3 Urea

The major industrial product obtained from ammonia is urea and 90% of it is used as a nitrogen release fertiliser. Urea is the precursor for a few industrially useful compounds and formulations. The major use of urea other than as fertiliser, is in the synthesis of melamine and diesel exhaust fluid (DEF). Melamine is used in the manufacture of a few drugs and more importantly for the melamine formaldehyde resin which is a hard thermosetting plastic and one of the most sought after materials as an inexpensive alternative to ceramic crockery and kitchenware and for the manufacture of laminates such as formica and for wall panels and white boards.

$$6\,(NH_2)_2CO \xrightarrow[175°C]{} C_3N_3(NH_2)_3 + 6\,NH_3 + 3\,CO_2 \qquad (8.18)$$

$$C_3N_3(NH_2)_3 + 6\,HCHO \longrightarrow C_3N_3(N(CH_2OH)_2)_3 \longrightarrow \text{Melamine–formaldehyde resin} \qquad (8.19)$$

Melamine–formaldehyde resin

Diesel exhaust fluid is a 32.5% aqueous solution of urea in demineralised water which is used as a consumable for selective catalytic reduction of nitrogen oxides in the diesel exhaust emission of diesel engines especially of heavy duty vehicles. Diesel engines can be run with an excess of air so that soot is not produced and undergoes full combustion. But in this process, nitrogen which is sucked in gets oxidised to nitrogen oxides (NO_x), a well-known pollutant. Selective catalytic reduction converts NO_x into nitrogen and water. DEF is generally stored in a tank in diesel vehicles and is injected into the exhaust stream by a metering device. Since the solution freezes at –11°C it has to be kept warm below 60°C as DEF starts decomposing above this temperature.

$$(NH_2)_2CO \longrightarrow 6\,NH_3 + HNCO \qquad (8.20)$$

$$HNCO + H_2O \longrightarrow NH_3 + CO_2 \qquad (8.21)$$

$$2\,NH_3 + 2\,NO + 1/_2O_2 \longrightarrow 2\,N_2 + 3\,H_2O \qquad (8.22)$$

$$4\,NH_3 + 3\,NO_2 \longrightarrow 3\,1/2\,N_2 + 6\,H_2O \quad (8.23)$$

Urea also finds use in creams for rehydration of skin and a 40% formulation is used for many skin diseases such as psoriasis, eczema, and so on. It is also used in de-icing of roads as an alternative to rock salt, in animal feeds as a source of nitrogen and as a cloud seeding agent.

In 1828, Wöhler the father of synthetic organic chemistry, prepared urea by the reaction of silver cyanate with ammonium chloride debunking the *vitalism* theory prevalent during that period which assumed that chemicals produced by living organisms are fundamentally different from that of lifeless objects. Wöhler wrote to his contemporary Berzelius: '*I must tell you that I can make urea without the use of kidneys, either man or dog*'.

$$AgNCO + NH_4Cl \longrightarrow (NH_2)_2CO + AgCl \quad (8.24)$$

8.7 Oxides of Nitrogen

Nitrogen forms many oxides with a formal oxidation state varying from +1 to +5. In these, N_2O, NO and NO_2 are relatively more stable compared to the others. NO is paramagnetic in the gaseous phase and forms a weak dimer in liquid and solid phases mostly of the type ONNO which is diamagnetic.

Table 8.4 Stable oxides of nitrogen—structure and properties

Stable oxides of nitrogen

Name and Formula	Oxidation state of nitrogen	Resonance structures	Nature and properties	Melting and boiling points
Dinitrogen oxide (nitrous oxide) N_2O(laughing gas)	+1	$\bar{N}{=}\overset{+}{N}{=}O \longleftrightarrow N{\equiv}\overset{+}{N}{-}\bar{O}$	Colourless, linear, green house gas; anesthetic, analgesic, stable at room temperature. Above 577°C, it decomposes to O_2 and N_2	m.p. –91°C b.p. –88.5°C
Nitrogen monoxide (nitric oxide) NO	+2	N≡O (with dashed bond)	Colourless gas, free radical; signalling molecule; made by oxidation of NH_3, reacts with O_2	m.p. –164°C b.p. –152°C
Nitrogen dioxide NO_2	+4	$:\ddot{O}{-}\dot{N}{=}\ddot{O}: \longleftrightarrow :\ddot{O}{=}\dot{N}{-}\ddot{O}:$ (bent)	Reddish brown gas toxic, paramagnetic, bent (134.3°)	m.p. –11°C b.p. 21°C

Nitrous oxide, a colourless gas which is quite unreactive at room temperature was discovered by Joseph Priestley by the reduction of NO using Fe or Fe/S mixtures. The compound commonly known as *laughing gas* has many beneficial applications especially as an inhalation anesthetic in medicine and dentistry. In the presence of heated catalysts above 577°C, N_2O decomposes to give N_2 and O_2 accounting for its high temperature reactivity. The *barking dog reaction* is a classic example of its high temperature chemistry. The sound of a dog barking is produced when a mixture of CS_2 and N_2O is ignited in a long measuring vessel giving, N_2, S_8 and CO_2.

NO_2 gas is brown and paramagnetic and dimerises at low temperatures to the colourless N_2O_4. It dissolves in water giving nitric acid. The resonance structures, nature and physical properties of the oxides are given in Tables 8.4 and 8.5.[16]

8.7.1 Synthesis of Oxides of Nitrogen

Nitrous oxide

$$NH_4NO_3 \xrightarrow[\Delta]{250°C} N_2O + 2\,H_2O \tag{8.25}$$

$$2\,H_2NC(O)NH_2 + 2\,HNO_3 + H_2SO_4 \longrightarrow 2\,N_2O + 2\,CO_2 + (NH_4)_2SO_4 + 2\,H_2O \tag{8.26}$$

Nitric oxide

$$4\,NH_3 + 5\,O_2 \xrightarrow[Pt]{850°C} 4\,NO + 6H_2O \tag{8.27}$$

$$8\,HNO_3 + 3Cu \longrightarrow 2\,NO + 3\,Cu(NO_3)_2 + 4H_2O \tag{8.28}$$

Nitrogen dioxide

$$Pb(NO_3)_2 \longrightarrow NO_2 + PbO + O_2 \tag{8.29}$$

$$2\,NO + O_2 \longrightarrow 2NO_2 \tag{8.30}$$

$$\text{Conc } HNO_3 + Cu \longrightarrow 2NO_2 + Cu(NO_3)_2 + 2H_2O \tag{8.31}$$

Dinitrogen trioxide

$$NO + NO_2 \xrightarrow[\text{low temp}]{} N_2O_3 \tag{8.32}$$

Dinitrogen tetroxide

$$2\,NO_2 \xrightarrow[\substack{\text{low temp}\\ \text{(below 21°C)}}]{} N_2O_4 \tag{8.33}$$

Dinitrogen pentoxide

$$\text{Conc } 12\,HNO_3 + P_4O_{10} \longrightarrow 6N_2O_5 + 4H_3PO_4 \tag{8.34}$$

$$2\,NO_2 + 2O_3 \longrightarrow 2N_2O_5 \qquad (8.35)$$

Nitrosylazide N_4O[17]

$$NO^+BF_4^- + NaN_3 \longrightarrow NON_3 + Na\,BF_4 \qquad (8.36)$$

Trinitramide $N(NO_2)_3$[18]

$$NO_2^+BF_4^- + KN(NO_2)_2 \xrightarrow[-40°C]{} N(NO_2)_3 + KBF_4 \qquad (8.37)$$

Table 8.5 Unstable oxides of nitrogen—structures and properties

Name and formula	Oxidation state of nitrogen	Resonance structures	Nature and properties	Melting and boiling points
Dinitrogen trioxide N_2O_3	+3	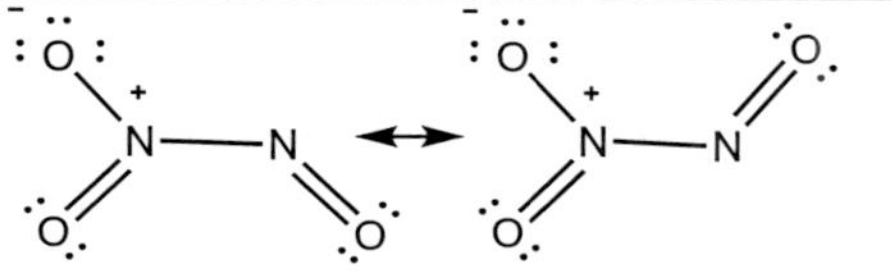 A symmetrical isomer, O=N–O–N=O is also obtained on near-infrared irradiation of N_2O_3	Isolable only at low temperatures as a deep blue solid or liquid, anhydride of HNO_2	m.p. −101°C b.p. 3.5°C
Dinitrogen tetroxide N_2O_4	+4	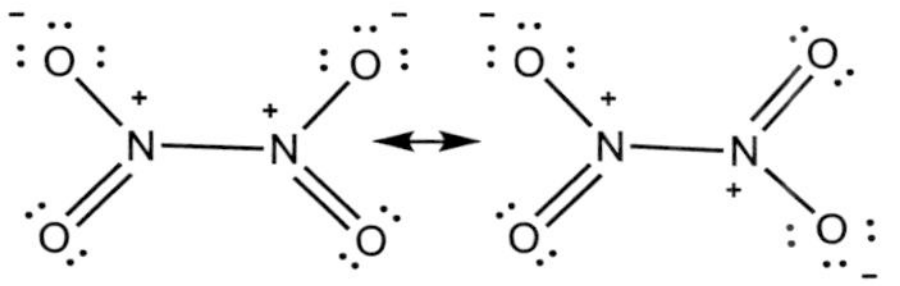	Colourless, dia-magnetic, converts to brown NO_2 at higher temperatures	m.p. −11°C b.p.22°C
Dinitrogen pentoxide N_2O_5	+5	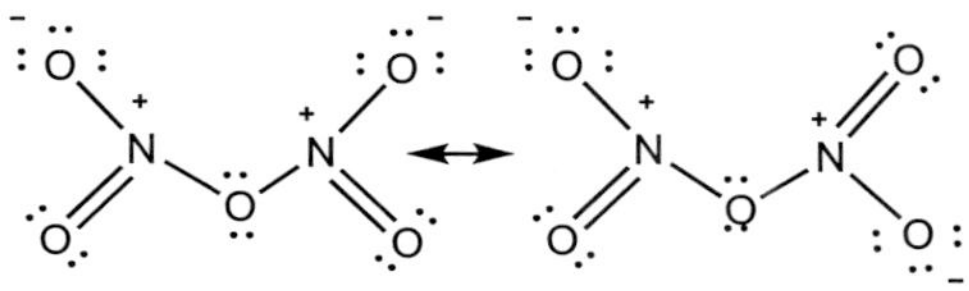	Colourless solid, anhydride of HNO_3, unstable, oxidiser	m.p. 41°C b.p. 47°C
Nitrosylazide N_4O	+1/2	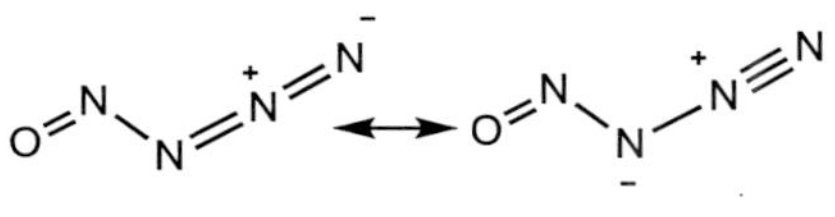	Pale yellow solid below −50°C, highly unstable	dec. −50°C
Trinitramide $N(NO_2)_3$	+3	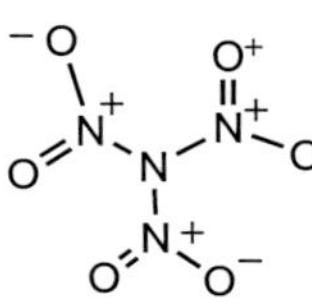 1 2 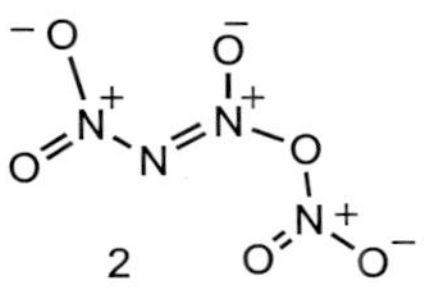	Stable below −10°C, identified by IR and NMR spectroscopy, largest nitrogen oxide known	dec. −10°C

8.7.2 The Importance of Nitric Oxide

Before 1986, the only importance ascribed to nitric oxide NO, was as a precursor for the industrial synthesis of nitric acid by the Ostwald process. NO was obtained by the oxidation of ammonia in the presence of platinum (catalyst) which was made to react further with O_2 to generate NO_2. In 1986, the pioneering work of Robert F Furchgott demonstrated that NO is indeed a long sought after molecule formed in the blood vessels with significant activity (till then termed as EDRF, endothelium derived relaxing factor).[19] It was observed that EDRF released from the endothelial cells which form a single cell inner lining of all blood vessels has unusual vasodilatory effects. The molecule with a half life of a few seconds in the blood was found to be the first gaseous molecule to act as a biological messenger and was acting as a cardiovascular signal and vasodilator (arterial vasodilation increases blood flow). In living systems, nitric oxide was biosynthesised by the oxidation of L-arginine in the presence of NADPH, O_2 and an enzyme called nitric oxide synthase (NOS). Functionally, NO has been found to help produce cyclic guanosine monophosphate (c-GMP), which seems to have a role in initiating the relaxation of blood vessels. Nitric oxide is a free radical and toxic, but one that is not as reactive as other biological free radicals, diffusing only micron distances before engaging in a reaction with another molecule. Although structurally simple, it has been found to initiate an astonishing complexity of reactions that results from its unique biochemistry. c-GMP production itself has biological responses such as vasodilation, platelet inhibition, cell adhesion, neurotransmission and penile erection.[20]

$$H_2N-C(=NH_2^+)-NH-(H_2C)_3-CH(NH_2)-C(O)OH \xrightarrow[O_2,\ NADPH]{NOS} H_2N-C(=O)-NH-(H_2C)_3-CH(NH_2)-C(O)OH + NO + H_2O + NADP^+ \quad (8.38)$$

L- Arginine L-Citrulline

NOS oxidises the guanidine group of L-arginine in a process that utilises five electrons and produces NO with stoichiometric formation of L-citrulline. The process (Eq. 8.38) involves the oxidation of NADPH and the reduction of molecular oxygen. The Nobel Prize in Medicine in 1998 was awarded to F Murad, Robert F Furchgott and L Ignarro for resolving the mystery of EDRF and determining the signaling properties of nitric oxide.[20]

8.7.3 Nitrosyl Complexes and NO Releasing Chemicals

Nitric oxide has received considerable attention due to its cardiovascular benefits. It has been found to act as a vasodilator, as it can expand blood vessels and thereby reduce blood pressure. It can also act as a platelet inhibitor preventing aggregation of platelets and inhibit blood clot formation. There are many compounds which are used as drugs for releasing NO in the blood stream. It is recorded that over 100 years ago Alfred Nobel was prescribed nitroglycerin by his doctors to help him with his heart-related problems. As the man who made dynamite and who used nitroglycerin extensively, he was skeptical. However, nitroglycerin helped him with his heart condition. Little did he know nitroglycerin acts by releasing nitric oxide in the blood stream which helps to relax narrowed blood vessels and increase the blood flow. Amyl nitrite, $(CH_3)_2CHCH_2CH_2ONO$ is another well known alkyl nitrite which has been used for release of NO and also as a cyanide antidote as it also has the ability to oxidise hemoglobin to methemoglobin which has a high propensity to bind free cyanide ions. Among inorganic compounds, sodium nitroprusside, $Na_2[Fe(CN)_5NO].2H_2O$ (a test reagent for thiol/sulfhydryl group) is a well known vasodilator which does not depend on intracellular mechanisms for the release of NO.[21] In vivo, the mechanism of release of NO from sodium nitroprusside is hypothesised to involve the sulfhydryl-containing compounds glutathione and cysteine, leading to the formation of the corresponding disulfides and S-nitrosothiols, NO and cyanide ions.[22] Many more drugs are now being tested and approved which have NO release function.

8.8 Nitric Acid

Nitric acid, one of the most well known oxidising mineral acids is industrially prepared by the Ostwald process starting with the oxidation of ammonia. Before this, it was made by the Birkeland process in which nitric oxide was produced in poor yields (4%) using the electric arc and the overall method was highly energy consuming. From ancient times, the use of HNO_3 has been related to the synthesis of nitrogen-based explosives, especially nitrates. In fact it is known that Haber himself synthesised NH_3 for making nitric acid and thereby explosives. Commercial grade nitric acid solutions are usually between 52% and 68% concentration. Production of nitric acid by the Ostwald process uses anhydrous ammonia which is oxidised to nitric oxide, in the presence of a Pt/Rh (9:1) gauze catalyst at about 500 K and 9 bar ($\Delta H = -905.2$ kJ/mol). Nitric oxide formed is then reacted with oxygen in the air to form nitrogen dioxide ($\Delta H = -114$ kJ/mol). NO_2 is absorbed in water to form nitric acid and nitric oxide ($\Delta H = -117$ kJ/mol). The nitric oxide is cycled back for re-oxidation. Alternatively, if the last step is carried out in air, NO formation can be avoided.[10]

$$4\,NH_3 + 5\,O_2 \xrightarrow[500\text{ K, }9\text{ bar}]{Pt/Rh} 4\,NO + 6\,H_2O \tag{8.39}$$

$$2\,NO + O_2 \longrightarrow 2\,NO_2 \tag{8.40}$$

$$3\,NO_2 + H_2O \longrightarrow 2\,HNO_3 + NO \tag{8.41}$$

$$4\,NO_2\,(g) + O_2\,(g) + 2\,H_2O\,(l) \longrightarrow 4\,HNO_3\,(aq) \tag{8.42}$$

The dilute HNO_3 obtained can be concentrated to about 68% by mass by distillation. Further concentration to 98% can be achieved by dehydration using concentrated H_2SO_4 or by using anhydrous magnesium nitrate.[23]

Anhydrous magnesium nitrate rapidly forms the hexahydrate with the formula $Mg(NO_3)_2{\cdot}6H_2O$. It can be synthesised by the reaction between nitric acid and magnesium metal or MgO.

$$2\,HNO_3 + Mg \longrightarrow Mg(NO_3)_2 + H_2 \tag{8.43}$$

$$2\,HNO_3 + MgO \longrightarrow Mg(NO_3)_2 + H_2O \tag{8.44}$$

Magnesium nitrate hexahydrate gives off water of hydration and further decomposes into magnesium oxide, oxygen, and nitrogen oxides.

$$2\,Mg(NO_3)_2 \longrightarrow 2\,MgO + 4\,NO_2 + O_2 \tag{8.45}$$

This MgO can be reconverted to $Mg(NO_3)_2$ and so, unlike concentrated sulfuric acid as a dehydrating agent, it can be easily recycled.

8.9 Polynitrogen Compounds: Azide, Pentazenium and Pentazolate

Polynitrogen compounds, which are also considered as allotropic modifications of nitrogen, are cyclic or chain-like, neutral, anionic or cationic compounds consisting only of nitrogen atoms. They are promising candidates as high energy density materials (HEDM). According to theoretical and experimental data, they are characterised by high enthalpy of formation and relatively high density in the condensed phase. Since the known examples are solids, it is expected that their use will permit solid rocket propellants to compete in energetic efficiency with liquid propellants.

A unique feature of polynitrogen compounds is the fact that their decomposition is accompanied by the release of a large amount of heat and voluminous amount of N_2. This is due to the fact that, in these compounds, N–N single or double bonds are present which are much weaker than the N≡N triple bond of N_2. The N≡N bond energy is 945 kJ/mol, whereas the N=N bond energy is 419 kJ/mol, and the N–N bond energy is only 159 kJ/mol. Consequently, the sum of the energies of three single bonds, a double bond and a single bond or one-and-a-half double bonds is much lower than the triple bond energy of N_2. See Table 8.6 for a similar comparison of a few triple bonds involving the first row elements of the *p*-block.[26]

Note that, for carbon–carbon bonds, the opposite ratio between the bond energies is observed, that is, the triple bond energy is lower than the sum of the energies of three single bonds, a single bond and a double bond or one-and-a-half double bonds. This in turn implies that carbon compounds easily form stable polymers, whereas nitrogen compounds prefer formation of stable triple bonded monomers (N_2). This comparison also explains the fact that although the C≡O bond enthalpy (the highest known for any compound) is higher than the N≡N bond enthalpy, the formation of the C≡O bond in a reaction is seldom explosive and does not involve the release of large quantities of energy when compared to formation of N_2 (3 × C–O bond enthalpy is slightly less than that of C≡O). But a comparison to the corresponding enthalpy of 3 × N–N indicates that the latter, 477 kJ/mol, is only about half of the N≡N bond enthalpy).

Table 8.6 Comparison of triple bond energies of N, C and O based compounds

kJ/mol	N and N	C and C	C and O	C and N
Single bond energy	159	356	326	304.6
Double bond energy	419	599	803	625
Triple bond energy	945	812	1075	889.5
3 × single bond energy	477	1068	978	913.8
Single + double bond energy	578	955	1129	919.6
Double + ½ double bond energy	628.5	898.5	1204	922.5

Polynitrogen chemistry has great scope for preparing high energy density materials (HEDM). Based on theoretical predictions, polynitrogen compounds can be used to make monopropellants with specific impulses (change in momentum delivered per unit of propellant consumed) in the 350–500 seconds range which is comparable to that of the liquid hydrogen/oxygen bipropellant systems. The large differences in bond dissociation energies between single, double and triple bonds makes the polynitrogens highly endothermic and thermodynamically unstable. In general, sensitivity to decomposition increases with endothermicity. They have only kinetic stability which is set by the activation energy barrier towards dissociation. Since the weakest bond in a polynitrogen compound determines this barrier, it is important to stabilise such weak bonds through resonance. An additional difficulty is the absence of suitable methods for preparing such compounds, suitable solvents that can absorb any liberated reaction energy and to prevent the product from thermal decomposition.

8.9.1 Azides

Azide, a pseudohalide which is highly heat and shock sensitive and highly poisonous, was first prepared by Curtius in 1890. One azide group contributes about 335kJ/mol (~80 kcal/mol) of endothermicity to a compound. Inorganic azides such as silver and lead azide are perhaps best known for their ability to explode and detonate when subjected to shock. An explosion can be initiated in the azides in a variety of ways. These include heat, shock, light and electric discharge.[27] Sodium azide has been used as a propellant in automobile air bags as it decomposes to give sufficient nitrogen gas to fill the air bag. Allen

K Breed, in 1967, demonstrated that sodium azide NaN_3, along with a collision sensor can initiate an automobile airbag within the 30 millisecond time period necessary for collision safety.

$$2\,NaN_3 \longrightarrow 2\,Na + 3N_2 \tag{8.46}$$

Heavy metal salts such as lead azide, $Pb(N_3)_2$, are shock-sensitive detonators used regularly by the military all over the world. These which decompose to the corresponding metal and nitrogen.

$$Pb(N_3)_2 \longrightarrow Pb + 3N_2 \tag{8.47}$$

2-Dimethylaminoethylazide (DMAZ) is an organic azide and liquid fuel being investigated for use as a spacecraft propellent to replace the toxic, carcinogenic monomethylhydrazine and unsymmetrical dimethyl hydrazine (UDMH). It can be prepared by the reaction of 2-dimethylaminoethylchloride and sodium azide at 50°C and concentrated by vacuum distillation below room temperature ($P = 40$ kPa and $T = 24–28$°C).

$$(CH_3)_2NCH_2CH_2Cl + NaN_3 \xrightarrow{50°C} (CH_3)_2NCH_2CH_2{-}N{=}\overset{+}{N}{=}\overset{-}{N} + NaCl \tag{8.48}$$

There are two well known methods for the synthesis of sodium azide, one starting from ammonia and the other from sodium nitrite. The Wislecenus method is used to prepare azide in large quantities while the Curtius method remains a laboratory synthesis route.

Wislecenus process

$$Na + NH_3 \longrightarrow NaNH_2 + H_2 \tag{8.49}$$

$$NaNH_2 + N_2O \longrightarrow NaN_3 + NaOH + NH_3 \tag{8.50}$$

Curtius method

$$2\,NaNO_2 + 2\,C_2H_5OH + H_2SO_4 \longrightarrow 2\,C_2H_5ONO + Na_2SO_4 + 2\,H_2O \tag{8.51}$$

$$C_2H_5ONO + N_2H_4 . 2H_2O + NaOH \longrightarrow NaN_3 + C_2H_5OH + 4\,H_2O \tag{8.52}$$

The click reaction

Organic azides are relatively less shock-sensitive than inorganic azides and in the recent past have been used for the preparation of triazoles by the azide–alkyne Huisgen 1,3-dipolar cycloaddition reaction, more commonly known as the *click reaction*. This reaction between an azide and a terminal or internal alkyne gives a 1,2,3-triazole. Rolf Huisgen was the first person to realise the scope and possibilities with this reaction. Barry Sharpless, Nobel Prize recipient, had referred to this cycloaddition as the cream of the crop of click chemistry and the premier example of a click reaction.[28, 29]

There are two variations of this reaction depending on the type of catalyst used for selectivity. The copper(I)-catalysed azide–alkyne cycloaddition (CuAAC) preferentially gives the 1,4-regio isomer while a ruthenium catalysed reaction (RuAAC) preferentially gives the 1,5-regioisomer as the product. Since Cu(I) is unstable in the aqueous medium, it is often generated *insitu* by using a Cu(II) salt and a reducing agent such as sodium ascorbate. $Cp^*RuCl(PPh_3)_2$, $Cp^*Ru(COD)$ and $Cp^*[RuCl_4]$ are common ruthenium catalysts used for the preparation of the 1,5-regioisomer.[30]

$$\text{PhOCH}_2\text{C}\equiv\text{CH} + \text{N}\equiv\overset{+}{\text{N}}-\overset{-}{\text{N}}-\text{CH}_2\text{Ph} \xrightarrow[\text{98°C, 18h}]{} \text{1-benzyl-5-(phenoxymethyl)-1,2,3-triazole} \quad (8.53)$$

8.9.2 Pentazenium

In 1999, 109 years after the discovery of the azide, K O Christe and coworkers described the unprecedented synthesis of the pentanitrogen cation N_5^+ which was trapped as a salt with AsF_6^-. The novel cation was a white powder, sparingly soluble in anhydrous HF, marginally stable at room temperature and was characterised by low-temperature vibrational NMR spectroscopy and theoretical calculations.[31] The N_5^+ ion has proven to be a powerful oxidiser and reacts explosively with water and organic materials, thus becoming the third stable member of the polynitrogen family. It was observed that explosion of a few milligrams of the shock-sensitive salt actually destroyed a sample chamber of a Raman spectrometer! Subsequently, the same group reported that the $(N_5^+SbF_6^-)$ and $(N_5^+Sb_2F_{11}^-)$ salts are stable and the latter was found to have a V-shaped structure for N_5^+ from X-ray diffraction studies. The synthetic route for N_5^+ starting from urea is shown in Scheme 8.6. The cation is V-shaped and is a very high energy material with a calculated enthalpy of formation (ΔH_f°) of 1471.8 kJ/mol.[32]

$$\text{N}\equiv\overset{\oplus}{\text{N}}-\overset{\ominus}{\text{N}}-\overset{\oplus}{\text{N}}\equiv\text{N} \quad (111°)$$

Fig. 8.3 Structure of pentazenium ion

$$(NH_2)_2CO \xrightarrow{F_2/N_2} H_2N-\overset{O}{\overset{\|}{C}}-NF_2 \xrightarrow{\text{Conc } H_2SO_4} NF_2H$$

$$2\,NF_2H + 2\,KF \longrightarrow N_2F_2 + 2\,KHF_2$$

Trans to *cis* isomerisation of N_2F_2 using AsF_5/ NaF if necessary

$$cis\ N_2F_2 + AsF_5 \longrightarrow N_2F^+\ AsF_6^-$$

$$NaN_3 + \text{dry HF} \longrightarrow HN_3 + NaF$$

$$N_2F^+\ AsF_6^- + HN_3 \xrightarrow[\text{dry HF, - 78°C}]{} N_5^+AsF_6^- + HF$$

Scheme 8.6 Synthetic route for N_5^+ starting from urea

This compound is resonance-stabilised (Fig. 8.4). The resonance structures D–F have one or more terminal nitrogens with only 6 valence electrons and these have been added to account for charge distributions calculated from theoretical studies and shielded nitrogen signals in nitrogen NMR studies.

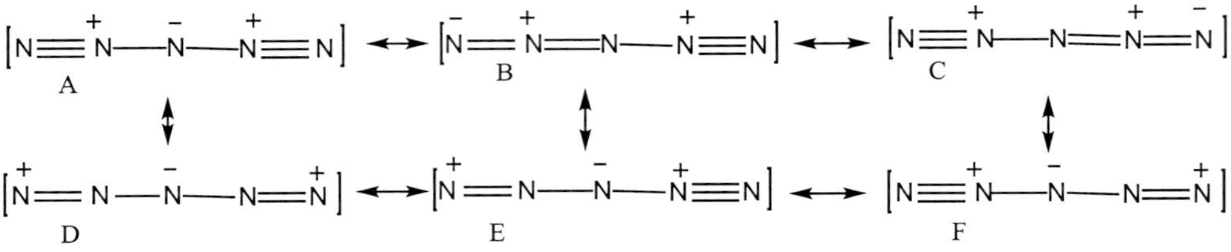

Fig 8.4 Resonance structures of the pentazenium cation

The N_5^+ ion is capable of oxidising water, NO, NO_2 and Br_2 but not Cl_2 or O_2. Its electron affinity is 10.44 eV (1018.4 kJ/mol).

8.9.3 Pentazolate (cyclo N_5^-)

Substituted arylpentazole ring compounds of the type Ar–N_5 have been known for quite some time and many examples with varying substituents on the aryl ring have been made and well characterised. Christe and coworkers obtained mass spectral evidence for the pentazolate anion from studies on aryl pentazolates in 2002 by using electrospray negative-ion mass spectrometry.[33] However, isolation of the parent HN_5 molecule or its anion N_5^- remained unsuccessful till 2017. Selective cleavage of the C–N bond in arylpentazoles while keeping the cyclo-N_5^- unit intact was a great challenge. The breakthrough of isolating and structurally characterising pentazolate anions was made possible by Zhang and coworkers of Nanjing University of Science and Technology, China in 2017. The introduction of electron donating groups on the arene ring additionally helped with selective cleavage.

Zhang and coworkers were able to selectively cleave the C–N bond of 3,5-dimethyl-4-hydroxyphenylpentazole and stabilise the resulting pentazolate anion using ferrous bisglycinate, $Fe(Gly)_2$.[34] The N_5^-anion was isolated as part of a salt, $(N_5)_6(H_3O)_3(NH_4)_4Cl$ in 19% yield. The ferrous glycinate stabiliser also served as a mediator for *m*-chloroperbenzoic acid, the oxidant used in the cleavage.

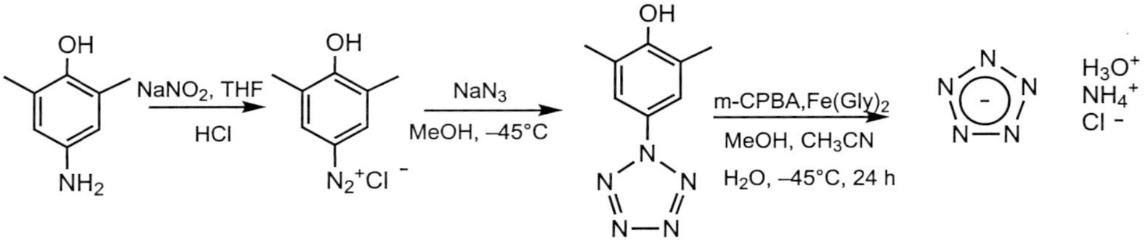

Scheme 8.7 Synthetic route for N_5^- starting from 3,5-dimethyl-4-hydroxyaniline

Upon purification using a silicagel column and crystallisation, $(N_5)_6(Cl)(H_3O)_3(NH_4)_4$ was obtained as an air-stable white solid. The pentagonal N_5^- ring comprises five nitrogen atoms in a perfectly planar arrangement, which is evident from the torsion angles which were 0°. The structure was further confirmed using NMR studies as well as infrared and Raman spectroscopy. Due to the hydrogen bonding present, the salt was found to have remarkable thermal stability up to 117°C. The cations in the salt apparently play a role in stabilising the pentazolate anion as attempts to replace them resulted in the decomposition of the salt.[34] A reaction of $(N_5)_6(Cl)(H_3O)_3(NH_4)_4$ with $Co(NO_3)_2.6H_2O$ at room temperature yielded $Co(N_5)_2(H_2O)_4.4\ H_2O$ as an air stable orange complex whose crystal structure indicated an octahedral complex with *trans*-oriented N_5^- units.[35] However, both these compounds having pentazolate units were quite stable and unlike azides and pentazenium, were not sensitive to explosion due to the presence of solvent units and non-energetic counterions. To overcome these issues, a solvent free N_5^- was prepared in 2018 by the reaction of $[Mg(H_2O)_6][(N_5)_2(H_2O)_4]$ with $AgNO_3$. The $Ag^+N_5^-$formed being light-sensitive and insoluble, was converted to a water-free soluble complex $[Ag(NH_3)_2]^+[Ag_3(N_5)_4]^-$ by reaction with ammonia.[36]

8.10 Pseudohalogens and Pseudohalides[37]

A pseudohalogen or halogenoid is a substance composed of two or more atoms, mostly electronegative which resembles the halogens in physical and chemical properties. It is interesting that the well known pseudohalides are nitrogen-based compounds. The most common pseudohalides are azide, cyanide, cyanate and thiocyanate.

8.10.1 Cyanide (CN^-)

$$2\ CH_4 + 2\ NH_3 + 3\ O_2 \xrightarrow{\text{Pt, 1100°C}} 2\ HCN + 6\ H_2O \tag{8.54}$$

Cyanide is derived from the pseudohalogen, cyanogen $(CN)_2$. The acid is hydrocyanic acid with a pKa of 9.21. The classical test for cyanide is the formation of Prussian blue $Fe_4[(Fe(CN)_6]_3$. This is obtained when a solution of cyanide is reacted with ferrous and ferric ions.

8.10.2 Cyanate (CNO^-)

$$CN^- + H_2O_2 \longrightarrow CNO^- + H_2O \tag{8.55}$$

$$CNO^- + 2H_2O \longrightarrow NH_4^+ + CO_3^{2-} \tag{8.56}$$

Cyanate is derived from the pseudohalogen, oxycyanogen $(OCN)_2$. The acid is isocyanic acid with a pKa of 3.7. Cyanate gives a transient violet colour for a spot test with 2% hydroxylamine followed 2% $FeCl_3$ solutions. The colour is transient as hydroxylamine can reduce the iron to the +2 oxidation state. In addition, if $Co(OAc)_2$ is added to the cyanate ions, a blue solution of $[Co(OCN)_4]^{2-}$is obtained.

8.10.3 Thiocyanate (SCN⁻)

$$8\ CN^- + S_8 \longrightarrow 8\ SCN^- \quad (8.57)$$

$$CN^- + S_2O_3^{2-} \longrightarrow SCN^- + SO_3^{2-} \quad (8.58)$$

Thiocyanate is derived from the pseudohalogen, thiocyanogen $(SCN)_2$. The acid is thiocyanic acid with a pKa of 1.1. The classic test for thiocynate is that it gives an intense blood red colour with dilute Fe^{3+} solution indicating the formation of $[Fe(NCS)(H_2O)_5]^{2+}$. In addition, if $Co(NO_3)_2$ is added to the thiocyanate ions, a blue solution of $[Co(SCN)_4]^{2-}$ is obtained.

8.10.4 Azide (N_3^-)

$$N_2O + 2\ NaNH_2 \longrightarrow NaN_3 + NaOH + NH_3 \quad (8.59)$$

Azide is in principle derived from the pseduohalogen, nitrine $(N_3)_2$ which does not exist due to very high instability. However, the corresponding acid, hydrazoic acid, exists which has a pKa of 4.72. The classic test for azide is that it gives a brisk effervescence of N_2 on reaction with I_3^- and decolourisation of I_3^- from dark yellow. In addition, azide solution acidified with HCl followed by reaction with $FeCl_3$ gives a red colour of ferric azide.

8.11 The Chemistry of Nitrogen-Based Explosives

Explosives are fundamentally of three types—mechanical (air pressurised containers), chemical and nuclear. A chemical explosive is a compound or mixture which, upon application of heat or shock, decomposes or rearranges with extreme rapidity, yielding large volumes of gases and heat. A chemical explosive essentially contains a fuel and an oxidiser which may be separate [for example, black powder having 75% saltpeter (KNO_3) as oxidiser, 11.5% sulfur, and 13.5% charcoal as fuel], or both fuel and oxidiser as part of the same molecule [for example, nitrocellulose (guncotton) or RDX (research department explosive)]. It is interesting to note that almost all of the chemical explosives have nitrogen as one of the elements bound to other elements such as oxygen (nitroglycerin), carbon (TNT) and nitrogen (RDX) with single or double bonds and the decomposition products of these explosives have dinitrogen N_2, as one of the components.

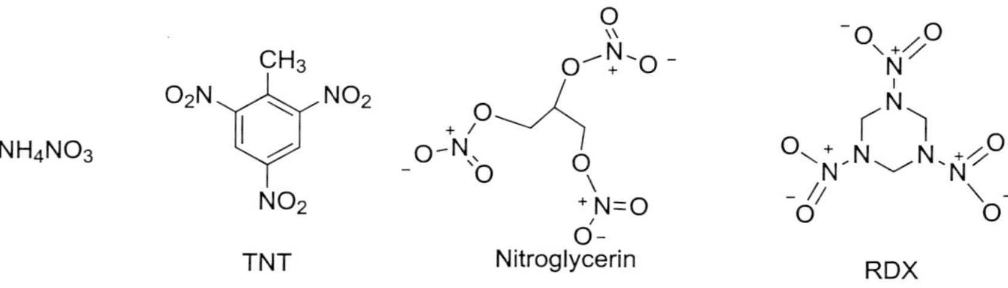

Fig. 8.6 Examples of common explosives

High energy materials are classified as explosives, propellants or pyrotechnics and the difference between the three is mostly based on whether a given substance undergoes only deflagration, or detonation or both. Deflagration is characterised by a subsonic (less than the speed of sound) flame propagation velocity, typically below 100 m/s, and relatively modest overpressures, below 0.5 bar. Although all combustion (fires) can be defined as a deflagration, these are typically associated with burning of natural gas or propane releases (*gas explosion*), or gasoline and hydrocarbon vapours (*vapour explosion*). A detonation on the other hand is characterised by supersonic flame propagation velocities, may be up to 2000 m/s, and substantial overpressures, up to 20 bars. Being very fast reactions, these explosions create a high-pressure shock wave that causes significant damage at large distances from the site of the blast. Detonations usually are associated with blasting agents or munitions (*high explosives*). An example of a classic compound which undergoes both deflagration and detonation is ammonium nitrate.

Explosives and propellants have applications ranging from blasting of ores, breaking down coal in coal mines, mining of salts, blasting holes in mountains, construction of roads and excavating earth, in military applications as ammunition for guns, grenades and bombs, propellants for missiles and torpedoes. Applications in space include powering rockets and maneuvering satellites and space stations. In addition to strength, explosives display a second characteristic, which is their shattering effect or brisance (from the French word meaning to 'break'), which is distinguished and separate from their total work capacity. This characteristic is of practical importance in determining the effectiveness of an explosion. The speed with which a given explosive reaches the peak pressure it is able to provide is a measure of its brisance.[38]

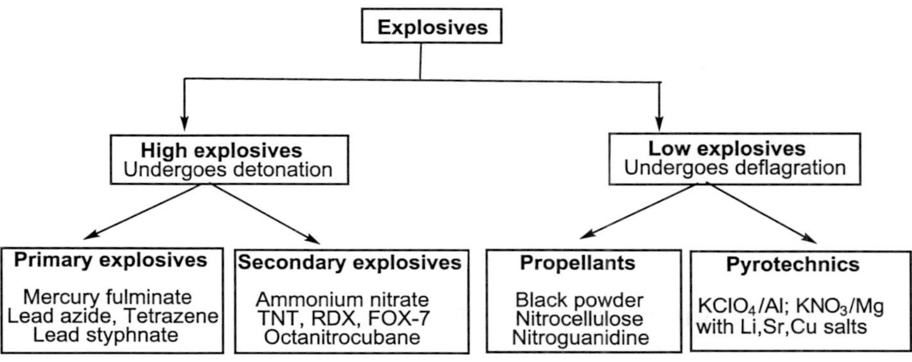

Fig. 8.7 Classification of explosives

Explosives are also classified as primary and secondary explosives. Primary explosives usually explode or detonate when they are heated or subjected to shock. They do not burn and often may not contain elements necessary for combustion and the explosion occurs even if they are not confined to a space. Primary explosives are extremely sensitive to stimuli such as friction, impact, static electricity, heat or electromagnetic radiation. A relatively small amount of energy is required for their initiation. As a practical measure, primary explosives are sufficiently sensitive to be reliably initiated with a blow from a hammer. Primary explosives are often used in detonators or to trigger larger charges of less sensitive secondary explosives and are commonly used in blasting caps and percussion caps to translate a physical shock signal. Some examples of primary explosives are shown in Scheme 8.8.

Hg + HNO_3 → (EtOH) → $\bar{O}$–$\overset{+}{N}$≡C-Hg-C≡$\overset{+}{N}$–$\bar{O}$
Mercury fulminate
VoD 4250 m/s

$Pb(NO_3)_2$ + NaN_3 → $\bar{N}$=$\overset{+}{N}$=N–Pb–N=$\overset{+}{N}$=$\bar{N}$
Lead azide
4630 m/s

Mg styphanate + Lead acetate → Pb^{2+} [O^-, O_2N, NO_2, O^-, NO_2 substituted benzene]
Lead styphanate
5200 m/s

$(CH_3)_2CO$ + H_2O_2 → (H^+) → Me, Me, O–O, O–O, Me, Me (cyclic peroxide)
Acetone peroxide
5300 m/s

Scheme 8.8 Synthesis of common primary explosives

Secondary explosives or high explosives, on the other hand detonate only under the influence of the shock of the explosion of a primary explosive. Not all of them are combustible and most of them can even be ignited by a flame and generally burn without vigour and can be extinguished also easily. They differ from primary explosives in not exploding by simple heat or shock, and require substantially more energy to be initiated and when exploded using a primary explosive are more brisant and powerful. Some secondary explosives have also other applications (as fertilisers) and are safe to handle and store. Consider the case of the most well-known secondary explosive, ammonium nitrate. If it is carefully heated, around 250°C, it will get converted to N_2O and H_2O and is a method for preparation of N_2O. There are many more ways by which ammonium nitrate can decompose (Scheme 8.9). When sufficient initial energy is given in the form of a detonation using a primary explosive, ammonium nitrate will explode releasing the highly stable molecules N_2, H_2O and O_2.[39]

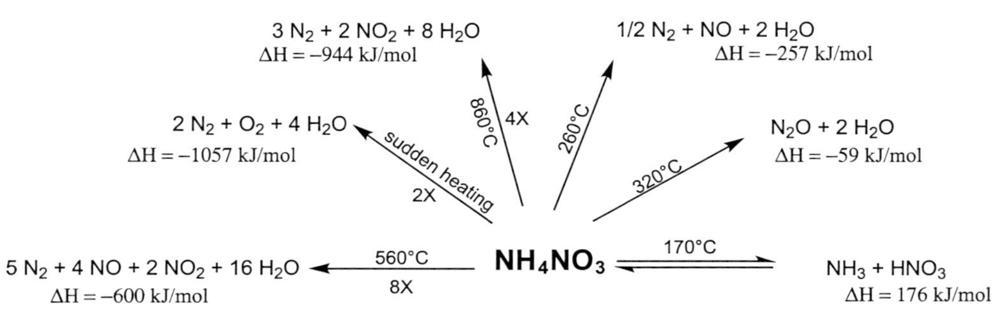

Scheme 8.9 Different ways by which ammonium nitrate has been found to decompose

History reveals that some of the greatest explosions reported involved ammonium nitrate which was intended to be used as fertiliser. On 16 April 1947, the cargo ship *Grandcamp* was being loaded with fertiliser grade ammonium nitrate at the port town of Texas city, USA when a fire was detected in the hold (storage sections below the deck). 2600 tonnes of ammonium nitrate packed in sacks were already taken aboard. The captain responded to the fire by closing the hold and pumping in pressurised steam to douse the fire. An hour later, the ship exploded and killed more than five hundred people and also set fire to another cargo ship moored nearby. This ship called *High Flyer* and moored 250 m away from *Grandcamp* already had on board 1050 tonnes of sulfur and 960 tons of ammonium nitrate. The explosion of *Grandcamp* created a powerful shockwave that broke windows of houses in Houston, Texas, which was 40 miles away and knocked down two small planes flying at 1,500 feet out of the sky. After burning for sixteen hours, the *High Flyer* also exploded the next day killing more people. Similar to ammonium nitrate, all secondary explosives which when simply ignited will only burn but if initiated by the blast of a primary explosive, will explode.

Oxygen balance

Oxygen balance provides valuable information on the oxygen requirement of an explosive and also on the types of gases liberated. A large and negative oxygen balance indicates that there is insufficient oxygen for carbon dioxide to be formed. As a result, carbon monoxide can get liberated. Oxygen balance is defined as the amount of oxygen expressed in weight per cent liberated when an explosive material is completely converted to CO_2, H_2O, N_2, SO_2, Al_2O_3, and so on. With regard to oxygen balance (OB%), zero per cent oxygen balance is the situation required for a very strong explosive which is achieved when all of the C gets converted to CO_2, all of the H gets converted to H_2O, N to N_2 and all metals to stable metal oxides with no excess of oxygen remaining or being deficient in oxygen.

The calculation of OB% for TNT and RDX are as follows

(TNT structure: benzene ring with CH_3, O_2N, NO_2, NO_2) $\longrightarrow$ $n\ CO_2 + n\ H_2O + n\ N_2$

$C_7H_5N_3O_6 \longrightarrow 7\ CO_2 + 2\ 1/2\ H_2O + 1\ 1/2\ N_2$

$6\ O \longrightarrow 14\ O + 2\ 1/2\ O = 16.5\ O$

Oxygen balance = –10.5 O

Mol. Weight of TNT = 227 Total oxygen in product = –10.5 x16 = –168

Percentage of oxygen liberated or taken = (–168 X 100)/ 227 = –74%

RDX $\longrightarrow$ $n\ CO_2 + n\ H_2O + n\ N_2$

$C_3H_6N_6O_6 \longrightarrow 3\ CO_2 + 3\ H_2O + 3\ N_2$

$6\ O \longrightarrow 6\ O + 3\ O = 9\ O$

Oxygen balance = - 3 O

Mol. Weight of RDX = 222 Total oxygen in product = - 3 x16 = –48

Percentage of oxygen liberated or taken = (–48 X 100)/ 222 = –21.62 %

Oxygen balance can also be determined by using the formula

$$OB = \frac{-1600}{\text{mol. wt.}} [2X + Y/2 + M - Z]$$

where X = number of carbon atoms; Y = number of hydrogen atoms; Z = number of oxygen atoms and M = number of atoms of metal (which gets converted to oxide).

Table 8.7 Percentage of oxygen balance (OB%) of selected explosives

Explosive	Formula	%OB	Explosive	Formula	%OB
Potassium chlorate	$KClO_3$	+39.2	RDX	$C_3H_6N_6O_6$	−21.6
Ammonium perchlorate	NH_4ClO_4	+34.0	Picric acid	$C_6H_3N_3O_7$	−45.4
Potassium nitrate	KNO_3	+47.5	Tetryl	$C_7H_5N_5O_8$	−47.4
Lithium perchlorate	$LiClO_4$	+60.0	TATB	$C_6H_6N_6O_6$	−55.8
PETN	$C_5H_8N_4O_{12}$	−10.1	EGDN	$C_2H_4N_2O_6$	0.0

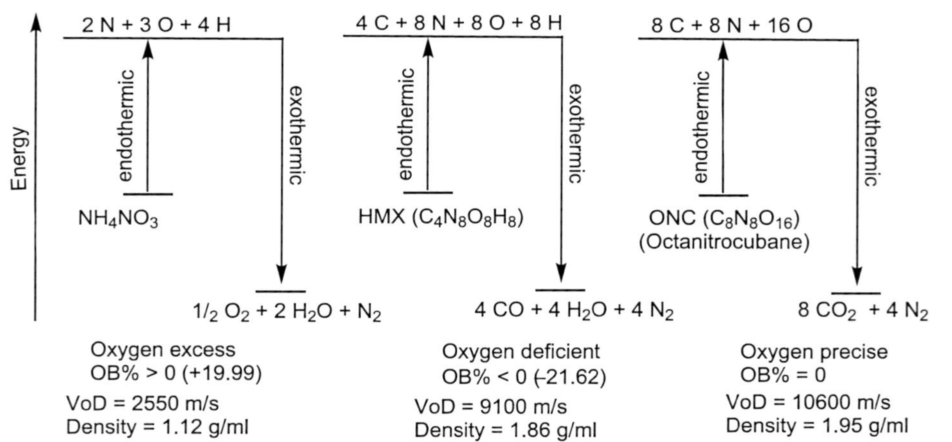

Scheme 8.10 Comparison of energetics of explosion of ammonium nitrate, HMX and ONC

Often the combination of an oxygen excess explosive with calculated amounts of fuel can result in better performance and lesser sensitivity and ease of handling [for example, ANFO (94% AN + 6% fuel oil) has better performance (VoD = 5270 m/s) than AN (VoD = 2550 m/s)].

Most of the nitrogen-based explosives are made using nitric acid or more precisely, a nitrating mixture. Scheme 8.11 shows how some of the well-known O, C and N-nitrated explosives have been synthesised.

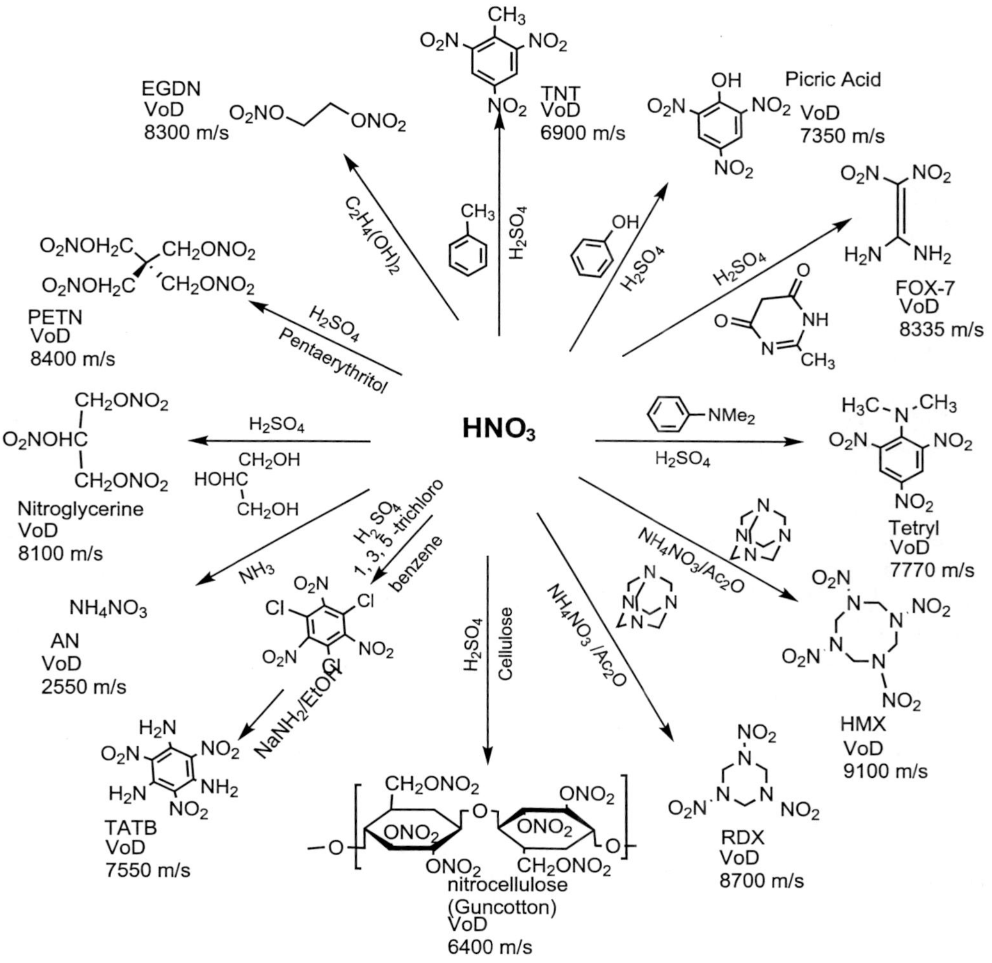

Scheme 8. 11 Synthesis of primary and secondary explosives using HNO_3

While nitration using nitrating mixture or combination of HNO_3/ammonium nitrate is the most common methodology, other nitrating reagents are also used especially when these systems are not quite effective. N_2O_4 and N_2O_5 are often suitable under such situations. Amines and isocynates have been oxidised to nitro groups by dimethyldioxirane (DMDO).

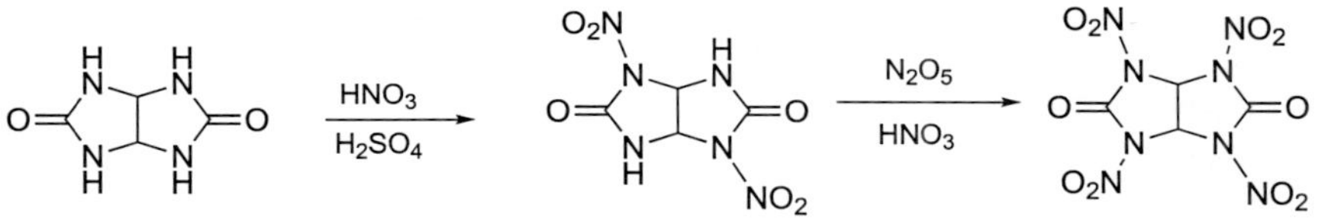

(8.60)

$$\text{ethylene oxide} \xrightarrow[CH_2Cl_2]{N_2O_4} ONO\text{-}CH_2CH_2\text{-}ONO_2 \xrightarrow{N_2O_5} O_2NO\text{-}CH_2CH_2\text{-}ONO_2 \ (\text{EGDN}) \quad (8.61)$$

$$\text{OCN-cubane-NCO} \xrightarrow[\text{Acetone } H_2O]{\text{DMDO (Dimethyldioxirane)}} O_2N\text{-cubane-}NO_2$$

$$H_2N\text{-}C_6H_4\text{-}COOH \xrightarrow[\text{RT 30 min}]{\text{DMDO}} O_2N\text{-}C_6H_4\text{-}COOH \ \ 95\% \quad (8.62)$$

8.11. 1 History

The very first explosive to be used was *black powder*, a combination of KNO_3, sulfur and charcoal which originated in china in the 9th century. According to some historians, alchemists added potassium nitrate and sulfur to the gold ore in the alchemists' furnace but in the first step of the reaction they forgot to add charcoal. To rectify this error they added charcoal in the final step which resulted in a tremendous explosion. The credit for the accidental discovery of gun cotton or nitrocellulose goes to Christian Friedrich Schönbein, a German–Swiss chemist of Basel, who in 1846, after cleaning up a spill of concentrated nitric acid and sulfuric acid on the kitchen table with a cotton apron hung the apron on the stove door to dry. As soon as it was dry, there was a flash as the apron ignited and burned without smoke. Nitrocellulose, after slight modification in processing, was used as smokeless powder for guns. Incidentally the very first gun was invented by the Arabs which was a bamboo tube reinforced with iron and a charge of black powder was used to fire an arrow! The initial development in modern explosives owes its credit to Immanuel Nobel and his two sons Emil and Alfred for experimenting with nitroglycerine for the first time. One explosion in their factory with nitroglycerine killed Emil Nobel. His brother, Alfred Nobel later managed to stabilise the extremely shock-sensitive nitroglycerine with diatomaceous earth, named it dynamite and the rest is the history of the Nobel prizes.

8.11.2 Comparison of Different Explosives

When the efficacy of an explosive is compared with another, there are many factors that need to be taken into consideration. A common parameter is velocity of detonation (VoD) which is defined as the velocity with which the shock wave front travels through a detonated explosive. It can be theoretically estimated and also be measured in some cases but with difficulty. A higher value of this indicates a stronger explosive. Also, a higher density indicates a stronger explosive.

In general, a stronger explosive have some or all of the following features:

- higher velocity of detonation (VoD),
- higher density,

- zero oxygen balance (OB%) indicating a good fuel to oxidiser ratio. This indicates the degree to which an explosive can be oxidised,
- high heat of explosion (Q_v), and
- higher volume of gases generated.

Examples of some of the strongest explosives known so far along with their velocity of detonation are shown in Fig. 8.7.

FOX-7 8335 m/s; CL-20 9380 m/s; DDF 10,000m/s; ONC 10,600m/s

Fig. 8.7 Examples of the strongest explosives

Heat of explosion (Q_v) and volume of gas generated V [dm^3g^{-1}] are used to compare the efficacy of different explosives. Heat of explosion is defined as the difference between the energies of formation of a material and the energies of formation of the products of the explosion. Experimentally (Q_v) can be measured as the heat released when a material is ignited and burned inside a bomb calorimeter in an inert atmosphere. The heat of explosion (Q_v) can be determined from heat of detonation (ΔH_d) as shown below.

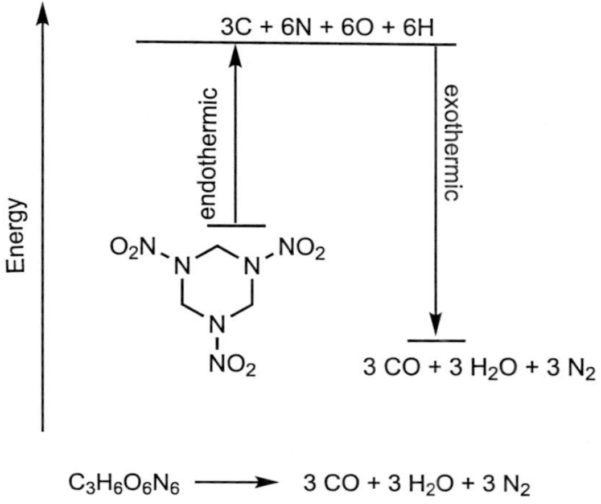

Heat of formation of RDX = +62.0 kJ/mol
Heat of formation of CO = –110.0 kJ/mol
Heat of formation of H_2O = –242.0 kJ/mol
ΔH_f (RDX)= +62.0 kJ/mol

$3\ \Delta H_f\,(CO) + 3\ \Delta H_f\,(H_2O) = -1056.0$ kJ/mol
ΔH_d (RDX)=−1056 − (+62.0)= −1118 kJ/mol
Therefore the heat of detonation of RDX is −111g kJ/mol
Heat of explosion $Q_v = [\Delta H_d\,(RDX) \times 1000]/M$
where M is the molar mass of RDX

$$Q_v = (-1118 \times 1000)/222 = -5036 \text{ kJ/kg}$$

The higher the absolute value of Q_v, the larger will be the heat generated from an explosion. The negative sign of Q_v is ignored as it denotes only an exothermic reaction.

Table 8.8 Heat of detonation and heat of explosion for some primary and secondary explosives

Explosive	Empirical formula	ΔH_d (kJ mol^{-1})	Q_v(kJ kg^{-1})
Mercury fulminate	$HgC_2N_2O_2$	−500	1755
Lead styphnate	$PbC_3H_3N_3O_9$	−868	1855
Lead azide	PbN_6	−469	1610
Nitroguanidine	$CH_4N_4O_2$	−257	2471
Picric acid	$C_6H_3N_3O_7$	−744	3249
TNT	$C_7H_5N_3O_6$	−964	4247
RDX	$C_3H_6N_6O_6$	−1118	5036
HMX	$C_4H_8N_8O_8$	−1483	5010
Nitroglycerine	$C_3H_5N_3O_9$	−1406	6194
EGDN	$C_2H_4N_2O_6$	−1012	6658

The volume of gas generated can be determined as per the following calculations. At STP (0°C and 1 atm) one mole of a gas will occupy 22.4 dm^3.

$$C_3H_6O_6N_6 \longrightarrow 3\ CO + 3\ H_2O + 3\ N_2$$

Hence, from 1 mol of RDX, 9 mol of gaseous products are produced. Therefore the total volume of gases released is 9 × 22.4 = 201.6 dm^3. Therefore 1 g of RDX will produce 201.6/222 = 0.908 dm^{-1}g^{-1}.

The volumes of gases produced from other common explosives are TNT = 0.740 dm^{-1}g^{-1}, picric acid=0.831 dm^{-1}g^{-1}, PETN = 0.780 dm^{-1}g^{-1}.

There are different methods used for comparing explosives with respect to their power and effectiveness. One of them is named power index where the product of heat of explosion and volume of gases released during explosion is compared to that of picric acid. Table 8.9 lists the power indices determined for some explosives.

$$\text{Power index} = \frac{Q.\,V}{Q_{pic}\,.\,V_{pic}} \times 100$$

Table 8.9 Power index of some common explosives

	Q(kJ g^{-1})	***V*(dm^{3}g^{-1})**	**Power index**
Primary explosives			
Mercury fulminate, $Hg(C_2N_2O_2)$	1755	0.215	14
Lead styphanate	1885	0.301	21
Lead azide	1610	0.218	13
Secondary explosives			
NG	6194	0.740	170
PETN	5794	0.780	167
RDX	5036	0.908	169
Picric acid	3249	0.831	100
TNT	4247	0.740	116

Questions for self study (multiple choice)

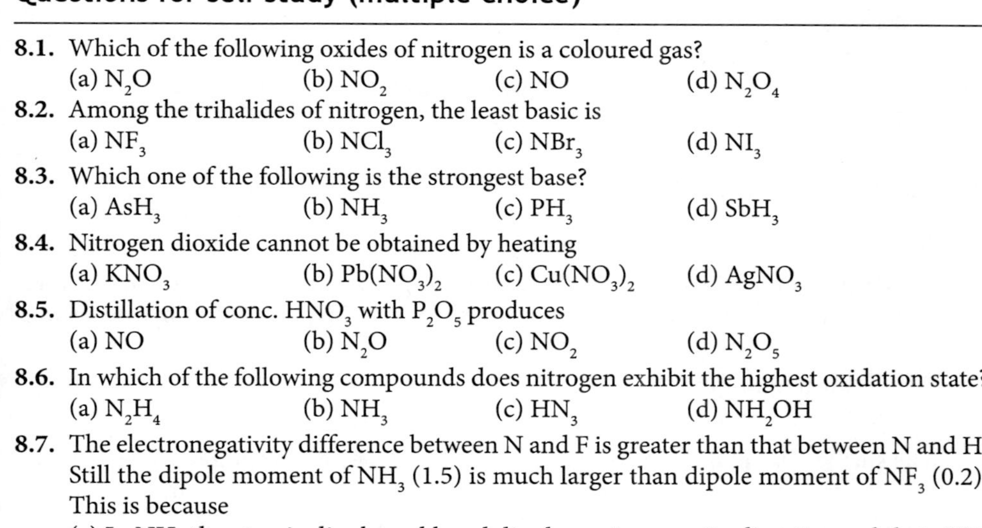

8.1. Which of the following oxides of nitrogen is a coloured gas?
(a) N_2O (b) NO_2 (c) NO (d) N_2O_4

8.2. Among the trihalides of nitrogen, the least basic is
(a) NF_3 (b) NCl_3 (c) NBr_3 (d) NI_3

8.3. Which one of the following is the strongest base?
(a) AsH_3 (b) NH_3 (c) PH_3 (d) SbH_3

8.4. Nitrogen dioxide cannot be obtained by heating
(a) KNO_3 (b) $Pb(NO_3)_2$ (c) $Cu(NO_3)_2$ (d) $AgNO_3$

8.5. Distillation of conc. HNO_3 with P_2O_5 produces
(a) NO (b) N_2O (c) NO_2 (d) N_2O_5

8.6. In which of the following compounds does nitrogen exhibit the highest oxidation state?
(a) N_2H_4 (b) NH_3 (c) HN_3 (d) NH_2OH

8.7. The electronegativity difference between N and F is greater than that between N and H. Still the dipole moment of NH_3 (1.5) is much larger than dipole moment of NF_3 (0.2). This is because

(a) In NH_3 the atomic dipole and bond dipole are in opposite directions while in NF_3 these are in the same direction.

(b) Both in NH_3 and NF_3, the atomic dipole and the bond dipole are in the same direction.

(c) In NH_3 the atomic dipole and bond dipole are in the same direction while in NF_3 these are in opposite directions.

(d) Both in NH_3 and NF_3 the atomic dipole and the bond dipole are in opposite directions.

8.8. Which of the following oxides of nitrogen is paramagnetic?
(a) NO_2 (b) N_2O_3 (c) N_2O (d) N_2O_5

8.9. Which of the following is a nitric acid anhydride?
(a) NO (b) NO_2 (c) N_2O_5 (d) N_2O_3

8.10. Pure nitrogen is prepared in the laboratory by heating a mixture of
(a) NH_4OH + NaCl (b) NH_4NO_3 + NaCl
(c) NH_4Cl + NaOH (d) $NH_4Cl + NaNO_2$

8.11. Concenetrated nitric acid upon standing for long turns yellow-brown in colour. This is due to the formation of
(a) NO (b) NO_2 (c) N_2O (d) N_2O_4

8.12. Very pure nitrogen gas can be obtained by heating
(a) $NH_3$with CuO (b) NH_4NO_3
(c) $(NH_4)_2Cr_2O_7$ (d) $Ba(N_3)_2$

8.13. Ammonia gas can be dried by
(a) Conc. H_2SO_4 (b) P_2O_5 (c) CaO (d) Anhydrous $CaCl_2$

8.14. Which of the following is not true for NO
(a) It is paramagnetic in the solid state
(b) It is a neutral oxide
(c) It reacts with O_2 forming NO_2
(d) Its bond order is 2.5

8.15. A pale blue liquid obtained by equimolar mixing of two gases at –30°C is
(a) N_2O (b) N_2O_3 (c) N_2O_4 (d) N_2O_5

Solved concept-based questions

8.16. NH_3 is a better electron donor than PH_3. Why?

In NH_3 the lone pair of electrons occupies an sp^3 hybrid orbital which has formed from the more compact 2*s* and 2*p* orbitals while in PH_3 it is an sp^3 hybrid orbital formed from the relatively less compact 3*s* and 3*p* orbitals. The former will have more directionality and hence stronger H bonds.

8.17. The N–N single bond is weaker than the P–P single bond. Why?

The smaller size and shorter bond length of the single N–N bond results in greater interelectronic repulsion between the non-bonded electrons of both nitrogen atoms thereby weakening the bond in comparison to an analogous P–P bond.

8.18. NO_2 dimerises while ClO_2 does not. Why?

The unpaired electron on NO_2 facilitates its dimerisation at low temperature. On bringing to higher temperature it reverts to the monomeric brown NO_2. Although ClO_2 has an unpaired electron, the presence of an extra lone pair of electrons on the Cl atom and its repulsion prevents ClO_2 from dimerisation.

8.19. NF_5 does not exist while PF_5 exists. Why?

N and P both have 5 valence electrons out of which two are paired in the 2*s* and 3*s* orbitals. For NF_5 to exist, its nitrogen has to promote one of its 2*s* electrons to a *d*-orbital. But there are no *d*-orbitals of low energy available for the first row elements. In PF_5 one of the 3*s* electrons is promoted to the empty 3*d* orbital and an sp^3d hybrid results with the orbitals being half-filled. These can pair with the electrons from the five fluorine atoms.

8.20. The bond dissociation energy of carbon monoxide is +1084 kJ mol^{-1} while that of nitrogen N_2, is +945 kJ mol^{-1}. CO has a stronger bond but it is more reactive than N_2. Why?

At the basic level one can see that unlike N_2 which is not polar, CO is polar due to a positive dipole moment. This makes the CO more reactive than N_2. At the advanced level one has to compare the MO diagrams of N_2 and CO which are isoelectronic. While the former is simple and symmetrical, the latter is complicated by the fact that it involves two atoms with an electronegativity difference and hence a complex MO diagram. The basic difference in the MO diagram of CO when compared to that of N_2 is, the HOMO which is a σ_{2p} orbital, is partly antibonding with respect to O while it is bonding in the case of N_2. This orbital, which in MO terminology is a carbon-based orbital, can donate its electron pair especially to metals thus acting as a good donor ligand and gaining energy since it is getting rid of antibonding electrons. The partly antibonding nature of the HOMO also indicates that the actual bond order should be less than 3 while it is 3 in case of N_2 where the HOMO is clearly bonding in nature.

8.21. NCl_3 is explosive while NF_3 is stable. Why?

Despite the similarities of the Pauling electronegativities of nitrogen and chlorine, the NCl_3 molecule is very polar with the negative charge residing on nitrogen. Theoretical studies indicate that the nitrogen in NCl_3 is in the –3 oxidation state and the chlorine atoms are in the +1 oxidation state. In contrast, due to the higher electronegativity of fluorine, the N atom of NF_3 has a +3 oxidation state and fluorines are in the –1 oxidation state. A halogen in the +1 oxidation state is a good oxidising agent and it can oxidise the nitrogen to the highly stable N_2 and get itself reduced to Cl_2, both stable gases with a release of energy. This does not happen in NF_3 as the fluorine in –1 state is not an oxidiser and F_2 is not as stable as F^-. The size of Cl also matters as three Cl on one N is an unstable situation.

8.22. The triple bond between N and N in N_2 is very strong while the quadruple bond between Re and Re in $Re_2Cl_8^{2-}$ is not as strong. Why?

In principle, a quadruple bond should be stronger than a triple bond. But unlike N_2 which has a triple bond resulting from $p\pi$–$p\pi$ interaction, the quadruple bond of $Re_2Cl_8^{2-}$ involves bonding overlap (1σ, 2π, 1δ) between *d*-orbitals. *d*-Orbitals are larger in size (especially 5*d*) and are more diffuse in nature compared to the *p*-orbitals (especially 2*p* of N_2). Hence the bonding overlap in $Re_2Cl_8^{2-}$ is not as strong as that of N_2.

8.23. The boiling point of NH_3 is 238.5 K while that of PH_3 is only 185.5 K. Explain.

The high electronegativity of N in the NH_3 molecule facilitates intermolecular hydrogen bonding resulting in a higher boiling point. In PH_3, the lower electronegativity and larger size results in a weaker, less directional hydrogen bond which is less significant and hence results in a lower boiling point.

Problems and exercises

8.24. How are carbon molecular sieves different from activated carbon?

8.25. CO bond energy is higher than that of N_2. However, reactions in which CO is formed are not explosive unlike those in which N_2 is formed. Why?

8.26. Starting from ammonia how is (a) melamine synthesised (b) nitroglycerin synthesised?

8.27. What is touch powder? How is it prepared?

8.28. Nitric acid produced by the Ostwald process is dilute. How is it concentrated efficiently?

8.29. Which transition metal shows the best catalytic activity for the Haber–Bosch process?

8.30. What are the chemical uses of N_2O?

8.31. You are provided with a gas (at room temp) whose chemical analysis indicates the presence of only nitrogen and oxygen. How will you find out its identity (without smelling the gas)?

8.32. What is the KAAP process? What advantage does it have over conventional the iron-based process?

8.33. You are given three white crystalline solids which are NaN_3, NaCN and NaNCS. Suggest chemical tests to distinguish between the three.

8.34. Suggest a method to make N_8. In your estimate, will it be stable?

8.35. What is the difference between CUAAC and RUAAC?

8.36. Starting from ammonia show how EDTA is synthesised. You are provided with formaldehyde, methane 1,2-dichloroethane and NaOH as main reactants.

8.37. Reaction of ethylene with a dihalogen A gave a compound B. Compound B on reaction with excess of ammonia gave compound C. Excess of compound C on reaction with blue coloured $[Ni(NH_3)_6]^{2+}$ gave a purple coloured compound D. Identify compounds A to D and draw the structure of compound D.

8.38. Reaction of NH_3 with an organic compound with the empirical formula H_2CO gave compound A. Compound A upon reaction with excess of HNO_3 gave B, while reaction of A with HNO_3, $NH_4NO_3$and acetic anhydride in acetic acid gave compound B as the minor product along with compound C as the major product. One mole of compound B upon detonation exploded giving 3 moles each of CO, H_2O and N_2, while one mole of compound C upon explosion gave 4 moles each of CO, H_2O and N_2. Identify compounds A to C.

8.39. Calculate %OB for picric acid, EGDN and FOX-7 and based on oxygen balance arrange them in the increasing order of strength as a secondary explosive.

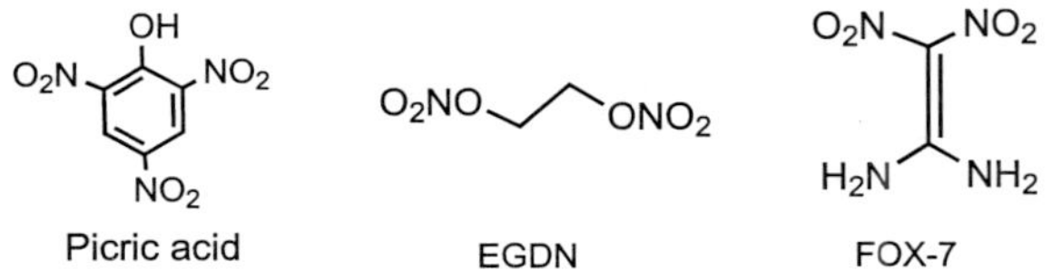

Picric acid EGDN FOX-7

Answer key for self study questions

8.1. (b) 8.2. (a) 8.3. (b) 8.4. (a) 8.5. (d) 8.6. (c) 8.7. (c) 8.8. (a) 8.9. (c) 8.10. (d) 8.11. (b) 8.12. (d) 8.13. (c) 8.14. (a) 8.15. (b)

Supplementary reading

1. Almqvist E, 2003, *History of Industrial Gases*, Springer, 163.
2. Mukhopadhyay M, Fundamentals of cryogenic engineering, *PH Learning Pvt Ltd*, 2010, 109.
3. Oliveira E C de, Pires C T G V M T and Pastore H O, Why are carbon molecular sieves interesting? *J. Braz. Chem. Soc.*, 2006. Vol. 17, 16.
4. Smith A R and Klosek J, A review of air separation technologies and their integration with energy conversion processes, *Fuel Processing Technology*, 2001, Vol. 70, 115.
5. Gambarotta S, Dinitrogen fixation and activation after 30 years: A puzzle still unsolved, *J. Organomet. Chem.*,1995,Vol.500, 117.
6. Fryzuk M D and Johnson S A, The continuing story of dinitrogen activation, *Coord. Chem. Rev.*, 2000, Vol. 200–202, 379.
7. Wagner S C, Biological nitrogen fixation, *Nature Education Knowledge*, 2011, Vol. 3, 15.
8. Seefeldt L C, Hoffman B M and Dean D R, Mechanism of Mo-dependant nitrogenase, *Ann. Rev. Biochem.*, 2009, Vol. 78, 701.
9. Postgate J R, *The fundamentals of nitrogen fixation*, 1982, Cambridge University Press, New York, NY.
10. Thiemann M, Scheibler E and Wiegand K W, Nitric acid, nitrous acid and nitrogen oxides. *Ulmann's encyclopedia of industrial chemistry*, Wiley VCH, 2012, Vol. 24, 177.
11. Appl M, Ammonia, 1 Introduction, *Ulmann's encyclopedia of industrial chemistry*, Wiley VCH, 2012, Vol. 3, 107.
12. Modak J M, Haber process for ammonia synthesis, *Resonance*, 2002, Vol. 7, 1159.
13. Appl M, Ammonia, 2, Production process, *Ulmann's encyclopedia of industrial chemistry*, Wiley VCH, 2012, Vol. 3, 139.
14. Ozaki A, Development of alkali-promoted ruthenium as a novel catalyst for ammonia synthesis, *Acc. Chem. Res.*, 1981, Vol. 14, 16.
15. Erisman J W, Sutton M A, Galloway J, Klimont Z and Winiwarter W, How a century of ammonia synthesis changed the world, *Nature Geosciences*, 2008, Vol. 1, 636.
16. Gruenhut N S, Goldfrank M, Cushing M L, Caesar G V, Caesar P D and Shoemaker C, Nitrogen(V) oxide, *Inorg. Synth.*, 1950, Vol. 3, 78.
17. Schulz A, Tornieporth-Oetting I C and Klapötke T M, Nitrosyl azide, an intrinsically unstable oxide of nitrogen, *Angew Chem. Int. Ed.*, 1993, Vol. 32, 1610.

18. Rahm M, Dvinskikh S V, Furo I and Brinck T, Experimental detection of trinitramide $N(NO_2)_3$, *Angew Chem. Int. Ed.*, 2011, Vol. 50, 1145.
19. Furchgott R F and Zawaszki J V, The odligatoryrol of the endothelium in the relaxation of arterial smooth muscle by acetylcholine, *Nature*, 1980, Vol. 288, 373.
20. Furchgott R F, Endothelium-derived relaxation factor: discovery, early studies and identification as nitric oxide, *Angew Chem.*, 1999, Vol. 38, 1870.
21. Butler A R and Megson I L, Non-heme iron nitrosyls in biology, *Chem. Rev.*, 2002, Vol. 102, 1155.
22. Grossi L and D'Angelo S, Sodium nitroprusside, mechanism of NO release mediated by sulfhydryl-containing molecules, *J. Med. Chem.*, 2005, Vol. 48, 2622.
23. Hunt L B, The ammonia oxidation process for nitric acid manufacture, *Platinum metals rev.*, 1958, Vol.2, 129.
24. Hager T, *The alchemy of air: a Jewish genius, a doomed tycoon, and the scientific discovery that fed the world but fueled the rise of Hitler*, 1st Ed., 2008, Harmony Books, New York.
25. Smil V, *Enriching the earth: Fritz Haber, Carl Bosch, and the transformation of world food production*, 2004, MIT Press, Cambridge, Mass.
26. Christe K O, Recent advances in the chemistry of N_5^+, N_5^-, and high oxygen compounds, *Propellants, Explosives, Pyrotechnics*, 2007, Vol. 32, 194.
27. Evans B L, Yoffe A D and Gray P, Physics and chemistry of the inorganic azides, *Chem. Rev.*, 1959, Vol. 59, 515.
28. Kolb H C, Finn M G and Sharpless K B, Click chemistry; Diverse chemical function from a few good reactions, *Angew Chem. Int. Ed.*, 2001, Vol. 40, 2004.
29. Spiteri C and Moses J E, Copper–catalysed azide-alkyne cycloaddition: Regioselective synthesis of 1,4,5- trisubstituted1,2,3 triazoles, *Angew Chem. Int. Ed.*, 2010, Vol. 49, 31.
30. Moses J E and Moorhouse A D, The growing applications of click chemistry, *Chem. Soc. Rev.*, 2007, Vol. 36, 1249.
31. Christe K O, Wilson W W, Sheehy J A and Boatz J A, N_5^+: A novel homoleptic polynitrogen ion as a high energy density material, *Angew Chem. Int. Ed.*, 1999, Vol. 38, 2004.
32. Vij A, Wilson W W, Vij V, Tham F S, Sheehy J A and Christe K O, Polynitrogen chemistry. Synthesis, characterisation and crystal structure of surprisingly stable fluoroantimonate salts of N_5^+, *J. Am. Chem. Soc.*, 2001, Vol. 123, 6308.
33. Vij A, Pavlovich J G, Wilson W W, Vij A and Christe K O, Experimental detection of the pentaazacyclopentadienide (pentazolate) anion, cyclo-N_5^-, *Angew. Chem. Int. Ed.*, 2002, Vol. 41, 3051.
34. Zhang C, Sun C, Hu B, Yu C and Lu M Synthesis and characterisation of the pentazolate anion cyclo-N_5^- in $(N_5)_6(H_3O)_3(NH_4)_4Cl$, *Science*, 2017, Vol. 355, 374.
35. Zhang C, Yang C, Hu B, Yu C, Zheng C and Sun C, A symmetric $Co(N_5)_2(H_2O)_4.4\ H_2O$ high nitrogen compound formed by cobalt (II) cation trapping of a cyclic N_5^- anion, *Angew. Chem. Int. Ed.*, 2017, Vol. 56, 4512.
36. Sun C, Zhang C, Jiang C, Yang C, Du Y, Zhao Y, Hu B, Zheng Z and Christe K O, Synthesis of AgN_5 and its extended 3D energetic framework, *Nature Commun.*, 2018, Vol. 9, 1269.
37. Kauffman G B, Faust G E and Tun P, Pseudohalogens, *J. Chem. Educ.*, 1968, Vol. 45, 141.
38. Akhavan J, *The chemistry of explosives*, 2nd Ed., RSC, 2004.
39. Chaturvedi S and Dave P N, Review on thermal decomposition of ammonium nitrate, *J. Energetic Mater.*, 2013, Vol. 31, 1.

THE CHEMISTRY OF PHOSPHORUS

CHAPTER

9

UNIQUE PROPERTIES OF PHOSPHORUS AND ITS COMPOUNDS

- *Phosphorus mostly forms tri- and penta-valent compounds and the latter in its tetracoordinate form is the most commonly observed oxidation state and geometry.*
- *P=O is one of the strongest double bonds of heavier p-block elements (P=O: 544 kJ/mol; P=S: 335 kJ/mol) and for many reactions involving phosphorus (III) and phosphorus (V) reagents, the driving force is the formation of the P=O bond (for example, the Wittig reaction).*
- *Trivalent phosphorus–carbon compounds are excellent ligands with varying steric and electronic properties and are used as spectator ligands in metal complexes for homogenous catalysis.*
- *Phosphates are essential for life processes and many molecules of the living system such as DNA, ATP and phosphacreatine contain phosphate groups. ATP is the universal currency for energy. The average adult's need for phosphorus is about 700 mg per day.*
- *Phosphorus forms many allotropes of which white phosphorus is the most reactive and ignites in air at ~30°C forming phosphorus pentoxide. This property is used in making phosphorus bombs and smoke screen bombs.*
- *Since it is one among the three essential plant nutrients (N, P and K), the major use of phosphates is as fertilisers.*
- *Diammonium phosphate along with ammonium sulfate and a red dye is used as an environmentally friendly fire retardant in large quantities to control forest fires by spraying from aeroplanes.*
- *Phosphorus shows formal oxidation states –3 (Ca_3P_2, PH_3), –2 (P_2H_4), 0 (P_4), +1 (H_3PO_2), + 2 ($H_4P_2O_4$), + 3 (PCl_3), + 4 ($H_4P_2O_6$) and +5 (PCl_5) in its compounds.*

- *Phosphazenes which are phosphorus–nitrogen double bonded compounds with pentavalent tetracoordinate phosphorus, find use as inorganic polymers and superbases.*
- *Organophosphates inhibit the functioning of the enzyme acetyl cholinesterase which is the reason for their use in the synthesis of pesticides and nerve gases. Unlike chlorine-based pesticides, phosphorus-based pesticides degrade rapidly by hydrolysis on exposure to sunlight, air and soil.*
- *Herbicides constitute the biggest part of the world pesticide use (40%) and glyphosate [N-(phosphonomethyl)glycine] is the most widely used herbicide in the world and works by inhibiting a plant enzyme involved in the synthesis of aromatic amino acids.*
- *Similar to petroleum, the available phosphates in the earth are getting depleted and this is a looming crisis as before the end of this century, readily available global supplies of phosphorus will start running out.*
- *Nearly 40% of the global reserves of phosphorus as phosphates for future are present in a single country—Morocco, which has been named the future 'Saudi Arabia of phosphorus'.*
- *Sterically hindered phosphines along with sterically hindered Lewis bases form frustrated Lewis pairs (FLPs).*

9.1 About the Element

Phosphorus was first isolated in 1669 by the alchemist Hennig Brandt of Hamburg, Germany while trying to make gold from a concentrate of urine. Urine is 95% water and has in addition to urea and some organic compounds, inorganic ions such as Na^+, K^+, Ca^{2+}, NH_4^+ Cl^-, HPO_4^-, SO_4^{2-} and HCO_3^-. After concentrating urine to a thick syrup, he distilled a red oily substance from the residue. This red oil, along with a black residue, when heated strongly for 16 hours gave white phosphorus which solidified under water. Brandt could make 120 g of white phosphorus from 5500 litres of urine! He called it cold fire (*kaltes Feuer*) as it emitted a faint glow on exposure to oxygen.

This process of making white phosphorus was modified by another alchemist Kunckel who added sand to the urine concentrate before distilling. An assistant of Robert Boyle, Ambrose Hankwitz (1660–1741) developed the process on a commercial scale but kept the process a secret. By 1743, a better method was published that gave good quality white phosphorus which involved the use of lead salts and potassium carbonate along with concentrated urine. It is believed that $Na(NH_4)HPO_4$ in the urine, on heating gets converted to sodium metaphosphate, $(NaPO_3)_6$. In the meantime K_2CO_3 and carbon reduces lead salts to lead. The carbon and lead reduces sodium metaphosphate to sodium pyrophosphate $Na_2H_2P_2O_7$ and finally to phosphorus.[1]

$$2\,(NaPO_3)_6 + 30\,C + 6\,SiO_2 \longrightarrow 6\,Na_2SiO_3 + 30\,CO + 3\,P_4 \tag{9.1}$$

Table 9.1 Physical properties of phosphorus

Phosphorus	*Atomic mass*	*Electronegativity (Pauling scale)*	*1st ionisation energy*	*Electronic configuration*	*van der Waals radius*
Symbol: P	30.973 g.mol^{-1}	2.19	1011.8 kJ.mol^{-1}	[Ne]$3s^2 3p^3$	0.180 nm
At. no. 15	*Atomic radius*	*Melting point*	*Boiling point*	*Electron affinity*	*Density*
	0.95 Å	44.15°C	280.5°C	72.037 kJ.mol^{-1}	1.82 g.cm^{-3} at m.p.

9.1.1 Isolation and Purification of the Element

Around 10 lakh tonnes of elemental phosphorus is produced per year, mostly from fluorapatite $Ca_5(PO_4)_3F$ (Eq. 9.2). The product is condensed into a white powder under water. Common reactions of white phosphorus are given in Scheme 9.1.[2]

$$4\,Ca_5(PO_4)_3F + 18\,SiO_2 + 30\,C \xrightarrow[1200\text{–}1500°C]{} 3\,P_4 + 30\,CO + 18\,CaSiO_3 + 2\,CaF_2 \quad (9.2)$$

P_4 is also prepared from calcium phosphate.

$$2\,Ca_3(PO_4)_2 + 6\,SiO_2 + 10\,C \longrightarrow P_4 + 10\,CO + 6\,CaSiO_3 \quad (9.3)$$

$$2\,Ca_3(PO_4)_2 + 8\,C \longrightarrow P_4 + 8\,CO_2 + 6\,Ca \quad (9.4)$$

White phosphorus, even on storage for long periods of time gets converted to the thermodynamically more stable red allotrope.

$$\underset{\text{white}}{P_4} \longrightarrow \underset{\text{red}}{(P_4)_n} \quad (9.5)$$

White phosphorus has been used for making smoke screen bombs and incendiary bombs. White phosphorus results in painful chemical burn injuries since it is highly lipid soluble and has rapid skin penetration. Because of its enhanced lipid solubility, injuries caused by white phosphorus result in delayed healing of wounds. P_4 is used for making PCl_3, PCl_5, P_4S_{10} and P_4O_{10} which are the precursors to a host of important useful products such as pesticides, phosphorus-based acids and organophosphorus compounds. Reactions of P_4 are summarised in Scheme 9.1.

Phosphorus-based compounds are used for making fertilisers to nerve gases and are also additives in many daily use products from cola drinks to toothpaste. White phosphorus is used by industry to produce pure phosphoric acid and phosphates as food additives, and as cleaning compounds.

In recent years, concentrated phosphoric acid, containing 70% to 75% of P_2O_5, has become very important in agriculture. Worldwide demand for fertilisers has caused record phosphate production. Calcium phosphate, also known as *bone ash* is used to make bone china crockery and to produce monocalcium phosphate, used in some baking powders.

Phosphorus is also important in the production of special steels and phosphor bronze [a corrosion and wear resistant alloy of copper with tin and phosphorus (0.01%–0.40%].

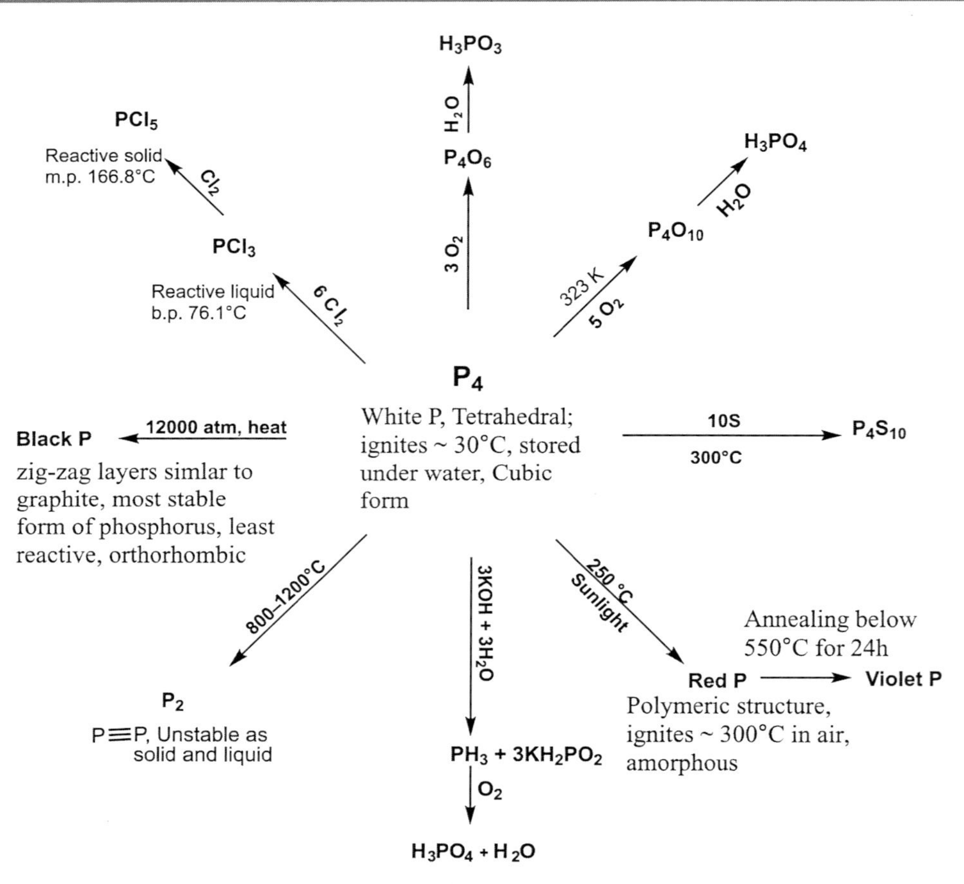

Scheme 9.1 Reactions of white phosphorus

9.1.2 Allotropy of Phosphorus

Although about twelve allotropic forms of phosphorus have been reported, the most well established among these are the white, red, violet (Hittorf's) and black allotropes. Among these, the most sensitive to oxygen is white phosphorus which has a tetrahedral P_4 arrangement of three-membered rings and the lower stability of this allotrope originates possibly from the strain associated with the bond angles of 60°. White phosphorus has to be stored under water to prevent auto-ignition and oxidation to P_2O_5. Heating white phosphorus in an inert atmosphere around 540 K converts it to red phosphorus which has lesser strain and is polymeric in nature. Violet phosphorus or Hittorf's phosphorus is a well characterised form of the red phosphorus having interlocking chains of five-membered rings. It is obtained by annealing red phosphorus under vacuum or in the presence of I_2 catalyst. Black phosphorus with six-membered rings is the most thermodynamically stable allotrope, whose structure and electrical conductivity resemble that of graphite. Black phosphorus is kinetically inert and does not ignite in air even at 670 K. Above 1070 K or at high pressures, P_4 is in equilibrium with P_2. Structures of white, red, violet and black phosphorus are shown in Fig. 9.1.[2, 3]

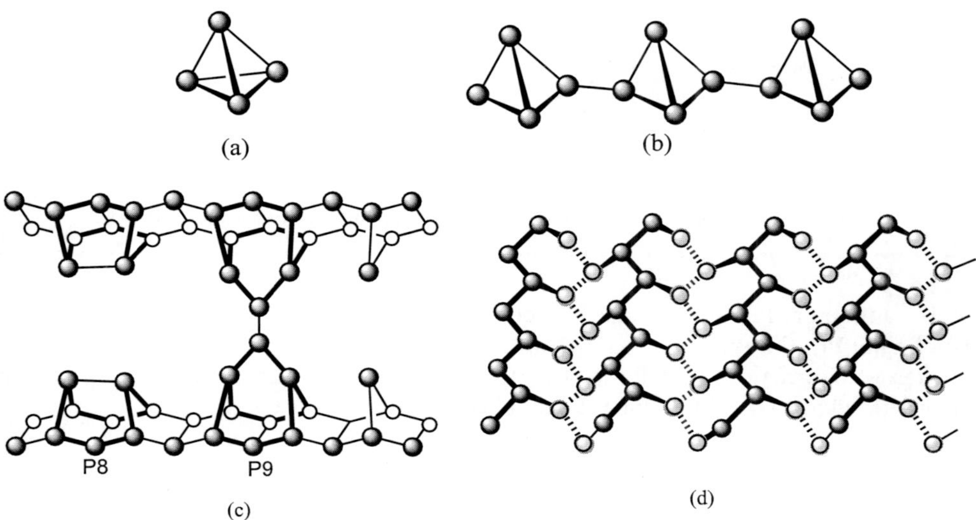

Fig. 9.1 Structures of (a) white phosphorus, (b) red phosphorus, (c) violet phosphorus, and (d) black phosphorus

9.2 Hydrides of Phosphorus; Phosphines

Phosphine is the name given to both phosphorus hydrides and phosphorus alkyl/aryl compounds. Although phosphorus forms many cyclic and acyclic hydrides, the one which is most well known is PH_3, first prepared by Gengembre in 1783 by heating white phosphorus in an aqueous solution of potassium carbonate.[4] The solubility of PH_3 in water is only slight; 0.22 mL of gas dissolves in 1 mL of water. Although all classic demonstration of synthesis of this gas shows it catching fire upon coming into contact with air, pure PH_3 ignites in air only at 150°C. However, minor amounts of diphosphine P_2H_4, if present along with PH_3 will make it catch fire spontaneously on coming in contact with air. PH_3 has the smell of rotten fish and this is one of the reasons for aluminium and zinc phosphides being used as rat bait as they slowly release PH_3 upon hydrolysis attracting the pest. Its toxicity is similar to that of cyanide as it inhibits the functioning of cytochrome c oxidase which is part of the electron transport chain in oxidative phosphorylation and ATP synthesis. There are many methods for the synthesis of PH_3 (Scheme 9.2). The toxicity and pyrophoricity of this gas is a deterrent for its usage. PH_3 is a trigonal pyramidal molecule similar to ammonia with a boiling point of −87.7°C. The H–P–H bond angles are 93.5° and the dipole moment is 0.58 D. For comparison, in ammonia the H–N–H bond angles are 106.7°, the dipole moment is 1.42 D and its boiling point is −33.3°C.[5, 6]

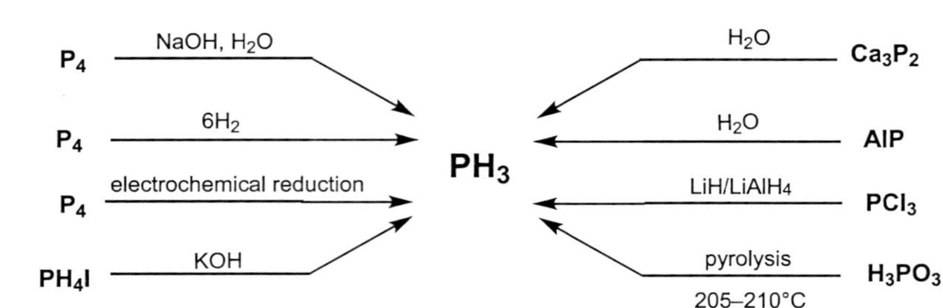

Scheme 9.2 Various methods for the preparation of phosphine, PH_3

Unlike ammonia, PH_3 has a lesser number of reactions reported due to the difficulty in handling of PH_3. The most well known among these from the industrial perspective is the reaction with formaldehyde followed by HCl to give tetrakis-hydroxymethylphosphonium chloride, which is a fire retardant added to phenol formaldehyde resins.[7]

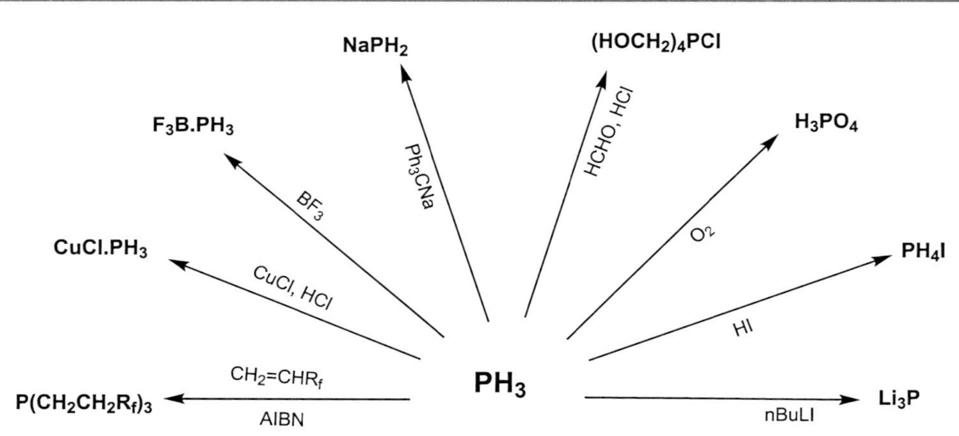

Scheme 9.3 Well known reactions of phosphine, PH_3

9.3 Alkyl and Aryl Phosphines

Alkyl and aryl phosphines are some of the most sought after phosphorus-based ligands in coordination and organometallic chemistry, especially as ligands for providing varying steric and electronic effects for the design and fine-tuning of transition metal catalysts. The most well known among the phosphines – Ph_3P – is omnipresent, both in inorganic and organic chemistry. Except a few such as Ph_3P, all phosphines are highly oxygen-sensitive and easily convert to the corresponding phosphine oxides, $R_3P{=}O$. Analogous to multidentate amine ligands, a variety of multidentate phosphine ligands also exist. The useful aspect of phosphines is that their electron donor and acceptor capabilities as well as steric bulkiness can be modified by choosing the right set of substituents. They are both σ-donors and π-acceptors. The π-acidity of phosphine ligands is due to the availability of low lying σ* anti-bonding orbitals. Alkyl phosphines are some of the best sigma donors, second only to N-heterocyclic carbenes. Aryl phosphines are lesser σ-donors but stronger π-acceptors than alkyl phosphines. The π-acidity of $P(CF_3)_3$ approaches that of the carbonyl ligand. Among alkyl phosphines, $P(t\text{-Bu})_3$ is the strongest σ-donor and also used as a benchmark for comparing electron donor capabilities of other phosphines.[8] Different synthetic routes have been adopted for the synthesis of tertiary alkyl and aryl phosphines (Scheme 9.4).

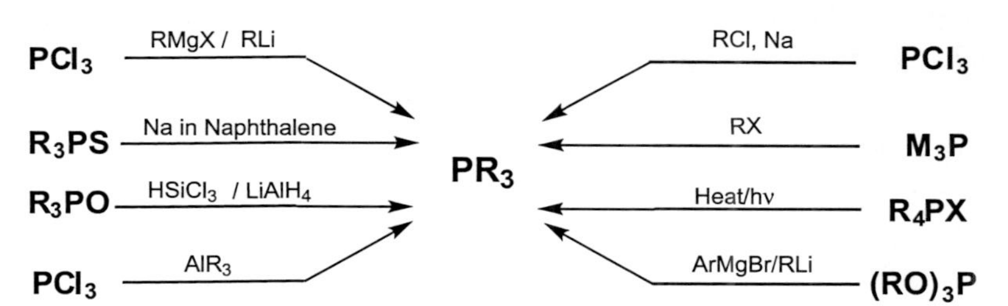

Scheme 9.4 Well known synthetic routes to tri alkyl and aryl phosphines, PR_3

The electronic parameter χ, which is the difference in the υ_{CO} carbonyl stretching frequencies of the $Ni(CO)_3$(phosphine) complex, with that of $Ni(CO)_3P(t\text{-Bu})_3$ was used to classify the σ-donor capability of phosphines. The more electron density the phosphine ligand donates to the metal centre, the more π-backbonding occurs to the carbonyl ligands, weakening the C≡O triple bond and thereby lowering the υ_{CO} carbonyl stretching frequency. Table 9.2 shows the relative electron donor capacities of some selected P–C bonded phosphines.

The steric bulkiness of phosphines was first compared by Tolman in 1977 using what is now well known as the cone angle determination.[9,10] Tolman defined cone angle as the solid angle formed with the metal at the vertex and the ligand substituents at the perimeter of a cone. The distance of 2.28 Å was taken as the average distance between the metal and phosphorus atom of the phosphines. Tertiary phosphine ligands are commonly classified using this parameter and these angles were originally determined by taking measurements from accurate physical models of the phosphines. Relatively recently, a new factor namely 'per cent buried volume' (%V_{bur}) has been introduced by S P Nolan as an equally good method to compare the steric bulkiness of

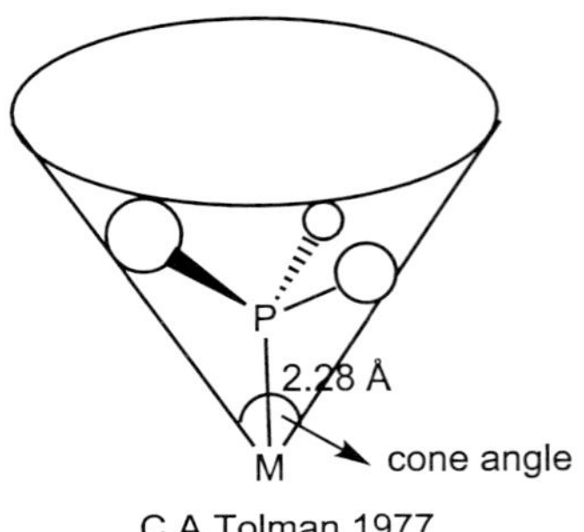

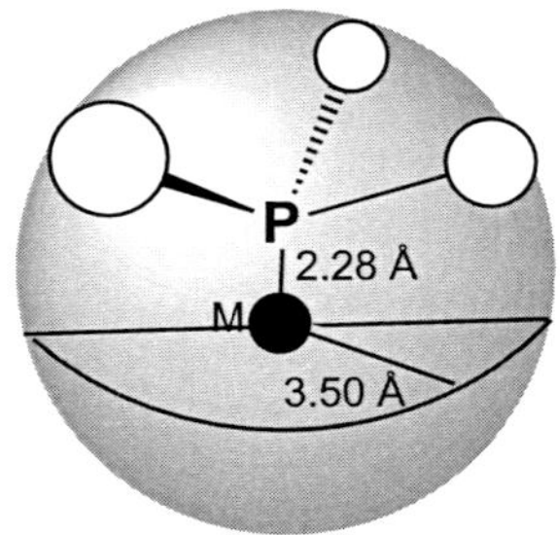

phosphines and also N-heterocyclic carbenes which are the best ligand competitors for electron donor capability, compared to phosphines. In 2010, Nolan made a comprehensive comparison using around 700 X-ray crystallographic data of phosphines and NHC complexes especially using AuCl as the metal unit and compared the steric properties. He defined $\%V_{bur}$ simply as the per cent of the total volume of a sphere occupied by a ligand when bound to a metal. The volume of this sphere is indicative of the potential coordination sphere space around the metal bound by a ligand or fragment of ligand. The sphere has a defined radius (which was fixed at 3.50 Å) and the metal to phosphine distance was fixed at 2.00 Å and 2.28 Å, the latter for a one to one comparison with cone angles. $\%V_{bur}$ is calculated using crystallographic data and using a dedicated software. Table 9.3 compares cone angles and $\%V_{bur}$ for selected phosphines.[11, 12]

Table 9.2 The electronic parameters (χ) of selected phosphines

PR_3	ν_{CO} (cm−1)	χ (cm−1)	PR_3	ν_{CO} (cm−1)	χ (cm−1)
$P(t\text{-}Bu)_3$	2056.1	0.0	$PPh_2(C_6F_5)$	2074.8	18.7
PCy_3	2056.4	0.3	$P(p\text{-}C_6H_4\text{-}CF_3)_3$	2076.6	20.5
$P(i\text{-}Pr)_3$	2059.2	3.1	$PPhH_2$	2077.0	20.9
PEt_3	2061.7	5.6	$P(OMe)_3$	2079.5	23.4
PMe_3	2064.1	8.0	$PPh(OPh)_2$	2079.8	23.7
PBz_3	2066.4	10.3	PPh_2Cl	2080.7	24.6
$P(o\text{-}Tol)_3$	2066.6	10.5	$P(C_6F_5)_3$	2090.9	34.8
PPh_3	2068.9	12.8	$P(CF_3)_3$	2115.0	58.9

It was observed that a simple correlation exists between the cone angle θ and the per cent buried volume especially for angles and % determined for gold(I) monochloride complexes. The only exceptions which did not fit into the correlation were phosphites such as $(PhO)_3P$, possibly due to their variable steric factors resulting from increased flexibility.[12]

Phosphines are part of ligands for a wide variety of synthetically and industrially important homogeneous catalysts. Multidentate phosphine ligands which are excellent

Table 9.3 Cone angles and $\%V_{bur}$ values for selected phosphines

S. no.	Phosphine	Cone angle (θ)	$\%V_{bur}$ for R_3PAuCl At 2.00 Å (at 2.28 Å)
1	Me_3P	118	27.3 (23.3)
2	Et_3P	132	31.7 (27.1)
3	Ph_3P	145	34.8 (29.9)
4	iPr_3P	160	39.1 (24.0)
5	Cy_3P	170	38.8 (33.4)
6	t-Bu_3P	182	43.9 (38.1)
7	$(C_6F_5)_3P$	184	42.6 (37.3)
8	$(o\text{-Tol})_3P$	194	44.8 (39.4)
9	Mes_3P	212	50.5 (45.0)
10	$(PhO)_3P$	128	36.5 (31.9)

chelators are also known as part of many catalysts. Recently, there was a renewed interest in the synthesis of electron rich and at the same time sterically hindered phosphines which find potential use for activation of less active aryl halides in palladium catalysed cross-coupling reactions. Some of the well known phosphine bound catalysts are listed in Fig. 9.2.[8]

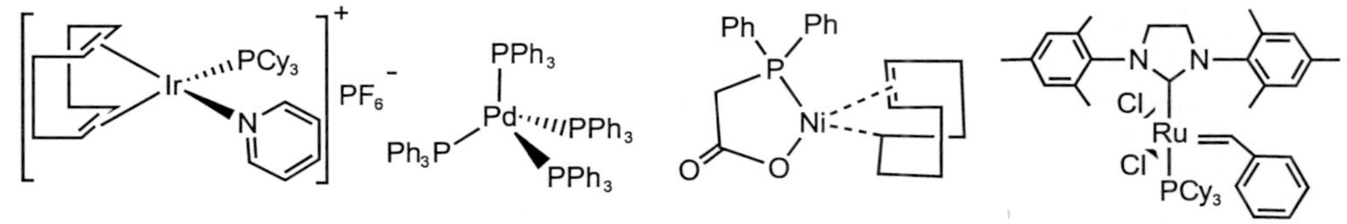

Crabtree's catalyst Hydrogenation of alkenes | Pd(0) catalyst for cross-coupling reactions | Shell's higher olefin Process (Ethylene oligomerization) | Grubbs 2nd gen. catalyst (Olefin metathesis)

Fig. 9.2 Well known phosphine bound homogenous catalysts

In addition, phosphines are used as reagents in organic reactions. The following reactions make use of phosphines (mostly PPh_3) as such (Mitsunobu, Appel and Staudinger) or a phosphine-derived reagent (phosphorus ylide for Wittig reaction). Mechanistically Mitsunobu, Staudinger, Appel and Wittig reactions are driven by the strong oxophilicity of tertiary phosphines and formation of phosphine oxide as a concomitant byproduct.

9.3.1 Mitsunobu Reaction

This reaction converts an alcohol into a variety of functional groups, such as an ester, using PPh_2 and an azodicarboxylate such as diethyl azodicarboxylate (DEAD) or diisopropyl azodicarboxylate (DIAD) with inversion of configuration.[13]

$$R^1CH(OH)R^2 + R^3C(O)OH \xrightarrow{PPh_3\ /\ Diethylazodicarboxylate} R^1CH(OC(O)R^3)R^2 + Ph_3P{=}O \quad (9.6)$$

9.3.2 Appel Reaction

This reaction converts an alcohol to an alkyl halide using a tetrahalomethane and PPh_3. The reaction begins with the halogenation of PPh_3 and is followed by alkoxide formation from the alcohol. The alkoxide then attacks the phosphorus, releasing the halide and forming phosphine oxide. Of late, this reaction is sparingly used due to the environmental issues related to CCl_4.[14]

$$R_1CH(OH)R_2 \xrightarrow[X=\ Cl,\ Br]{CX_4,\ PPh_3} R_1CH(X)R_2 + CHCl_3 + Ph_3P{=}O \quad (9.7)$$

Substitution and inversion of configuration

9.3.3 Staudinger Reaction

This reaction converts an organic azide to a primary amine using a PR_3 (preferably PPh_3) and water. The mechanism begins with the attack of PPh_3 on the far nitrogen of the organic azide to give a phosphazide intermediate. This intermediate rearranges and releases a molecule of N_2 and forms a phosphinimine. Water attacks the phosphorus atom and after a series of proton transfer steps results in the formation of the primary amine and $R_3P{=}O$ as a side product.[15, 16]

$$RN_3 + Ph_3P \xrightarrow[-N_2]{} Ph_3P{=}NR \xrightarrow{H_2O} RNH_2 + Ph_3P{=}O \quad (9.8)$$

9.3.4 Wittig Reaction

The Wittig reaction of an aldehyde or ketone uses a phosphonium ylide ($Ph_3P{=}CH_2$) (generated from the reaction of PPh_3 and MeI followed by *n*-BuLi). Phosphine oxide is the side product.[17-19]

$$\text{cyclohexanone} \xrightarrow{Ph_3\overset{\oplus}{P}-\overset{\ominus}{C}H_2} [\text{oxaphosphetane: } O-PPh_3,\ CH_2] \longrightarrow \text{methylenecyclohexane} + Ph_3P{=}O \quad (9.9)$$

9.3.5 The Arbuzov Rearrangement

The Arbuzov rearrangement also known as the Michaelis–Arbuzov rearrangement, is one of the versatile pathways for the formation of carbon–phosphorus bonds and involves the reaction of an ester of trivalent phosphorus (phosphites) with alkyl halides. The driving force for this reaction is the formation of a pentavalent tetracoordinate phosphorus with a phosphorus–oxygen double bond. The reaction, originally discovered by Michaelis in 1898, was explored in great detail by Arbuzov and several others. The rearrangement is one of the well-studied ones among organophosphorus reactions and is widely employed for the synthesis of phosphonates and phosphinic acid esters.[20, 21]

$$P(OR)_3 + R'X \xrightarrow{-RX} R'P(=O)(OR)_2 \tag{9.10}$$

R = alkyl, aryl... R' = alkyl, acyl... X = Cl, Br, I

Arbuzov reactions can also proceed without the addition of halides, probably catalysed by impurities during the distillation of the phosphonites. Uncatalysed Arbuzov rearrangements of ordinary pure phosphonites occur at 240°C or higher.

$$Ph{-}P(OCH_2CH_2Cl)_2 \xrightarrow[\Delta,\ 150°C]{decalin} Ph(CH_2CH_2Cl)P(=O)(OCH_2CH_2Cl) \tag{9.11}$$

9.4 Phosphides

It has been known for a long time that phosphorus combines with nearly all elements. A phosphide is a compound containing the phosphorus ion in the –3 oxidation state or its equivalent. The most commonly encountered ones are the binary phosphides, which consist of phosphorus and a less electronegative element, mostly a metal. There are also many polyphosphides, which are anionic chains, rings or clusters of phosphorus. Phosphides are known with many less-electronegative elements, the exceptions being with mercury and a few heavy main group metals. Some phosphides are molecular. Notable examples of phosphides include aluminium phosphide (AlP) and zinc phosphide (Zn_3P_2), which are used as baits and pest eradication agents (rat poison), exploiting their tendency to release toxic phosphine (PH_3) upon hydrolysis. Magnesium phosphide and calcium phosphide are moisture sensitive. Indium phosphide (InP) and gallium phosphide (GaP) are used as semiconductors and as solar cell materials, often in combination of related arsenides. Phosphides with electropositive metals hydrolyse, liberating PH_3.[22]

$$Ca_3P_2 + 6\ H_2O \longrightarrow 3\ Ca(OH)_2 + 2\ PH_3 \tag{9.12}$$

The general method by which phosphides are prepared is by the heating the metal with red phosphorus under an inert atmosphere or vacuum. In principle, all metal phosphides and polyphosphides can be synthesised from red phosphorus by this method. Holmes signal is an age old technique used by sailors for making emergency lighting on the sea surface during night. A can containing both calcium carbide and calcium phosphide is pierced and thrown into the sea. When they come into contact with water, carbide generates acetylene and phosphide generates a mixture of PH_3 and P_2H_4. The latter being pyrophoric, catches fire and also ignites the acetylene produced, providing a flame which will last till the gas production lasts.

9.5 Phosphorus Oxides and Sulfides

Phosphorus forms two major oxides namely P_4O_6 and P_4O_{10}, the former containing trivalent tricoordinate phosphorus units and the latter containing pentavalent tetracoordinate phosphorus units. P_4O_6 readily burns in air to give P_4O_{10}. Similar to other P(III) compounds, P_4O_6 can act as a ligand and forms complexes with metal carbonyls. It also reacts with ozone forming P_4O_{18} at lower temperatures which releases O_2 and forms P_4O_{10} on increasing the temperature. P_4O_6 gets converted to phosphorous acid, H_3PO_3, on reaction with water.[2]

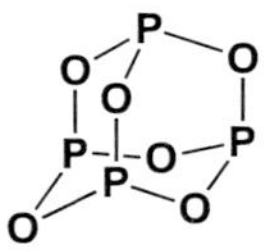

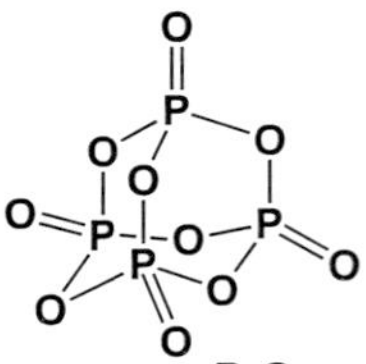

Phosphorus pentoxide P_4O_{10}, the anhydride of H_3PO_4, on the other hand acts as one of the strongest dehydrating agents known so far. Its strength as a dehydrating agent is displayed from the fact that it can dehydrate HNO_3 to N_2O_5, H_2SO_4 to SO_3 and $HClO_4$ to Cl_2O_7. P_4O_{10} gets converted to H_3PO_4 on reaction with water. It is also used for preparing C_3O_2, carbon suboxide by reaction with malonic acid. With anhydrous HF it gives HPF_6 and with CaF_2, PF_5 can be generated. Phosphorus pentoxide has been used to prepare concentrated pure phosphoric acid and pyrophosphoric acid. In the thermal process of preparing H_3PO_4, the phosphorus pentoxide obtained by burning white phosphorus is dissolved in dilute phosphoric acid to produce concentrated phosphoric acid.[2]

In contrast, the two most common phosphorus sulfides are P_4S_3 (phosphorus sesquisulfide) and P_4S_{10} (phosphorus pentasulfide). The former, containing P–P bonds and trivalent phosphorus, was used for the making of 'strike anywhere matches' while the latter is a very common thionation agent (also called Berzelius reagent). It is especially used in the industry for the synthesis of the lubrication additive zinc dithiophosphate and in the synthesis of many pesticides such as parathion, chlorpyrifos and malathion. Around 150,000 tons of P_4S_{10} are produced annually for the above mentioned uses.

While P_4S_6 is known to be present as six different structural isomers, none of these isomers is identical to the structure of P_4O_6. Also unlike tetraphosphorus oxides, tetraphosphorus sulfides varying in sulfur content from S3 to S10 are known, many of them having P–P bonds. This pattern indicates the relatively higher affinity of phosphorus for oxygen than sulfur.

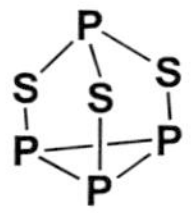

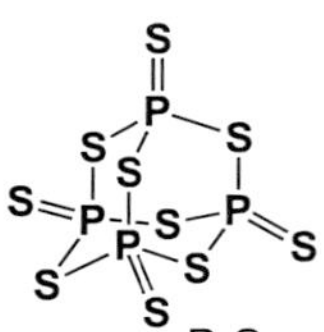

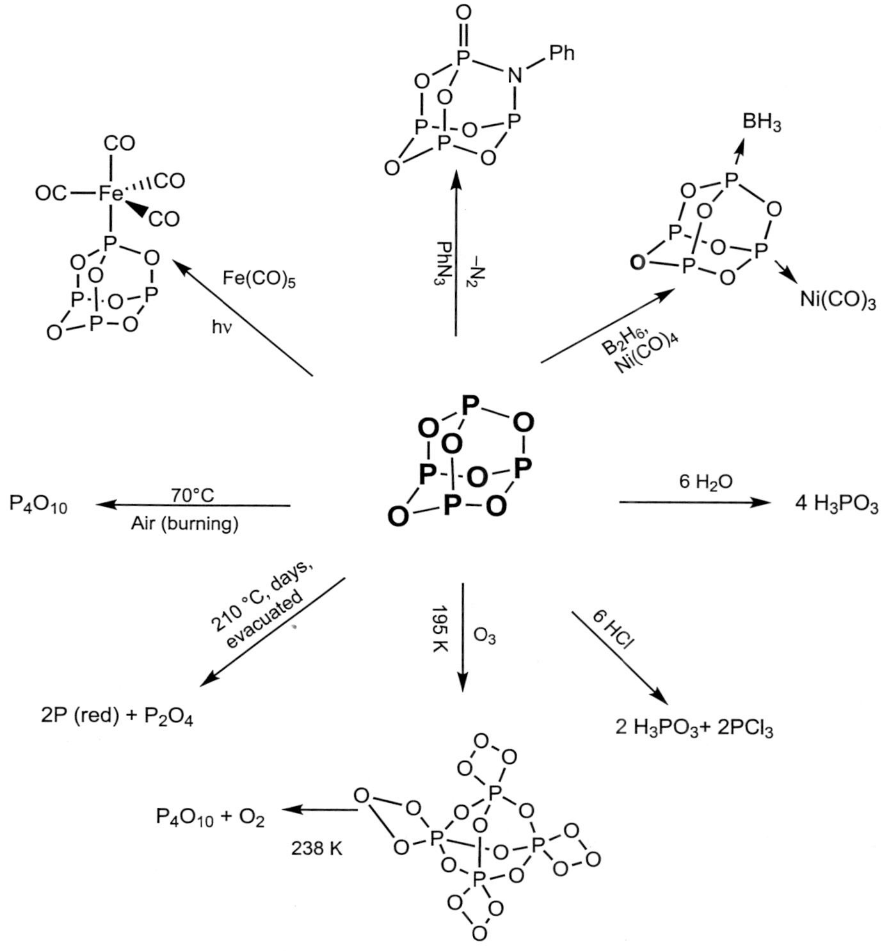

Scheme 9.5 Reactions of P_4O_6

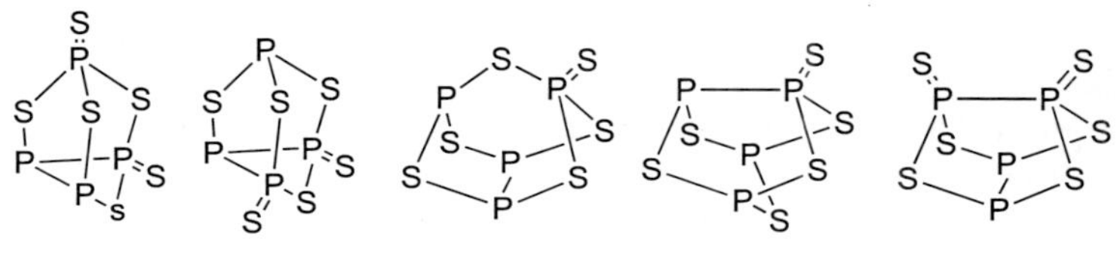

Fig. 9.3 Structural isomers of P_4S_6

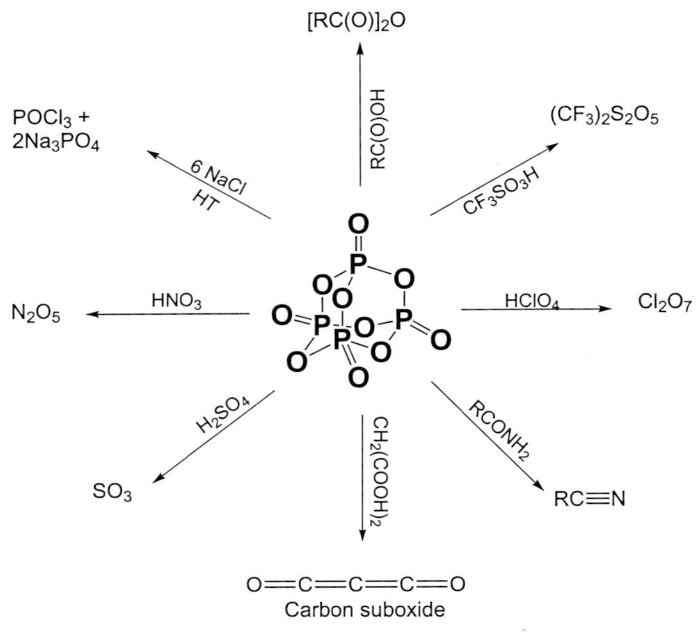

Scheme 9.6 Reactions of P_4O_{10}

Phosphorus pentasulfide has a structure which is almost identical to that of P_4O_{10}. It is a yellow high melting solid (m.p. 288°C). It is obtained by the reaction of liquid white phosphorus and elemental sulfur above 300°C or by the reaction of elemental sulfur or iron pyrites with ferrophosphorus (mostly Fe_3P_2), a byproduct of P_4 production from phosphate rocks.[23]

P_4S_{10} also known as Berzelius reagent, has a lot of industrial value as it is the source of phosphorus for making a very large number of pesticides as well as zinc dialkyldithiophosphate (ZDDP), a well known anti-wear agent added as lubricant to motor oils, greases and gear oil of vehicles. The starting materials for making both pesticides and ZDDP are the dithiophosphoric acids prepared by the reaction of P_4S_{10} with an alcohol. In addition, P_4S_{10} is a well-known thiolating agent which by itself, or along with hexamethyl disiloxane, converts the C=O groups of ketones, esters and amides to C=S groups (Scheme 9.7). Reaction with anisole converts P_4S_{10} to the well-known Lawesson's reagent having a P_2S_4 unit. Lawesson's reagent does the same chemistry as P_4S_{10} but with better yields and less side reactions, and ease in handling.

The synthesis of four well known phosphorus-based pesticides and ZDDP starting with the corresponding dithiophosphoric acid is shown in Scheme 9.8.

Parathion, also known as Folidol, is a potent insecticide and acaricide (compound which kills ticks and mites). It was originally developed by I G Farben in the 1940s. It is

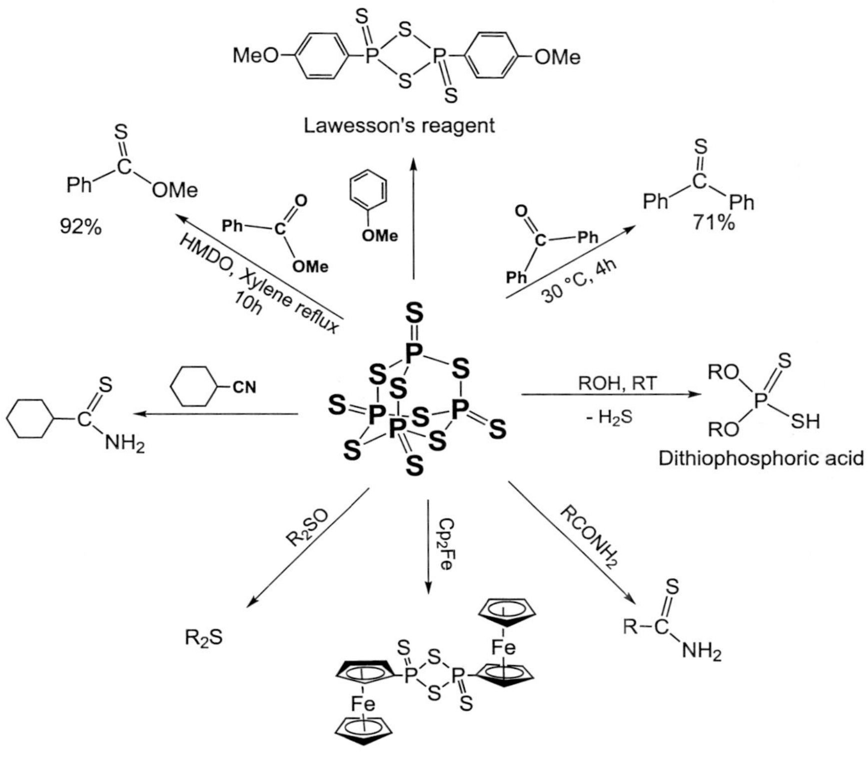

Scheme 9.7 Reactions of P_4S_{10}

highly toxic to non-target organisms, including humans. Malathion is an insecticide of relatively low human toxicity and is widely used in agriculture, residential landscaping, public recreation areas, and in pest control programmes such as mosquito eradication. Diazinon, developed by Ciba Geigy in 1952 is a contact insecticide which kills insects by altering normal neurotransmission within the nervous system of the insect. Chlorpyrifos introduced by Dow Corning in 1965 is moderately toxic to humans, and exposure has been linked to neurological effects, persistent developmental disorders, and autoimmune disorders. It remains one of the most widely used organophosphate insecticides. Acute poisoning with chlorpyrifos results mainly from interference with the acetylcholine neurotransmission pathway.

Although both phosphorus-based pesticides and nerve gases affect the activity of the enzyme acetyl cholinesterase, pesticides are relatively slower in activity and are considered less toxic. The reason for the difference is that the P=S moiety has to be first oxidised to a P=O moiety by biological oxidants such as cytochrome P-450 so that the potent form of the compound is generated.

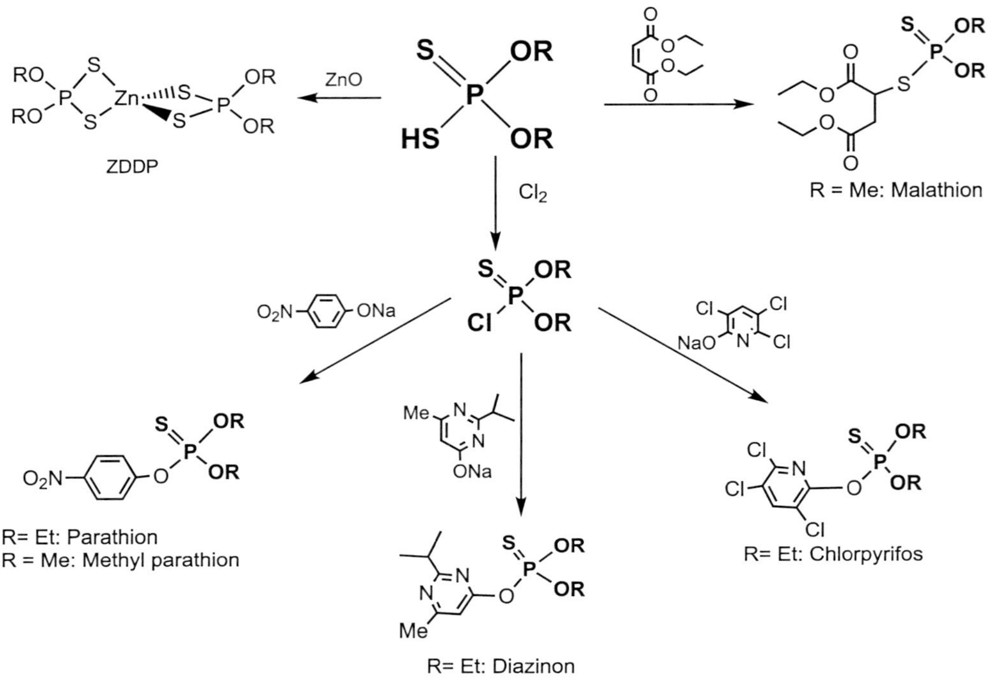

Scheme 9.8 Reactions of dithiophosphoric acid

9.6 Phosphorus Halides

9.6.1 Phosphorus Trichloride[2]

PCl_3 is a volatile liquid which fumes in air and rapidly gets hydrolysed to form HP(O)$(OH)_2$ and hydrochloric acid. It is prepared by the combustion of yellow phosphorus in a dry stream of chlorine. In the Hoechst continuous process, molten white phosphorus in PCl_3 is reacted with chlorine gas and the formation of PCl_5 is prevented by the presence of a slight excess of phosphorus throughout the synthesis. PCl_3 formed is distilled off around 76°C. Its main use is in the synthesis of other phosphorus reagents such as $POCl_3$, PCl_5, $PSCl_3$ and is one of the main starting materials for organophosphorus compounds. It is an important intermediate in the production of phosphate-ester-based insecticides. Phosphorus trichloride is used as a chemical intermediate to produce a variety of pesticides, surfactants, flame retardants, additives for lubricants and plastic stabilisers. PCl_3 can be used to produce chemical weapons especially nerve gases by multistage chemical synthesis which can be carried out with bare infrastructure and facilities. Because of this reason, its production, use and export are stringently controlled by the International Chemical Weapons Convention treaty.

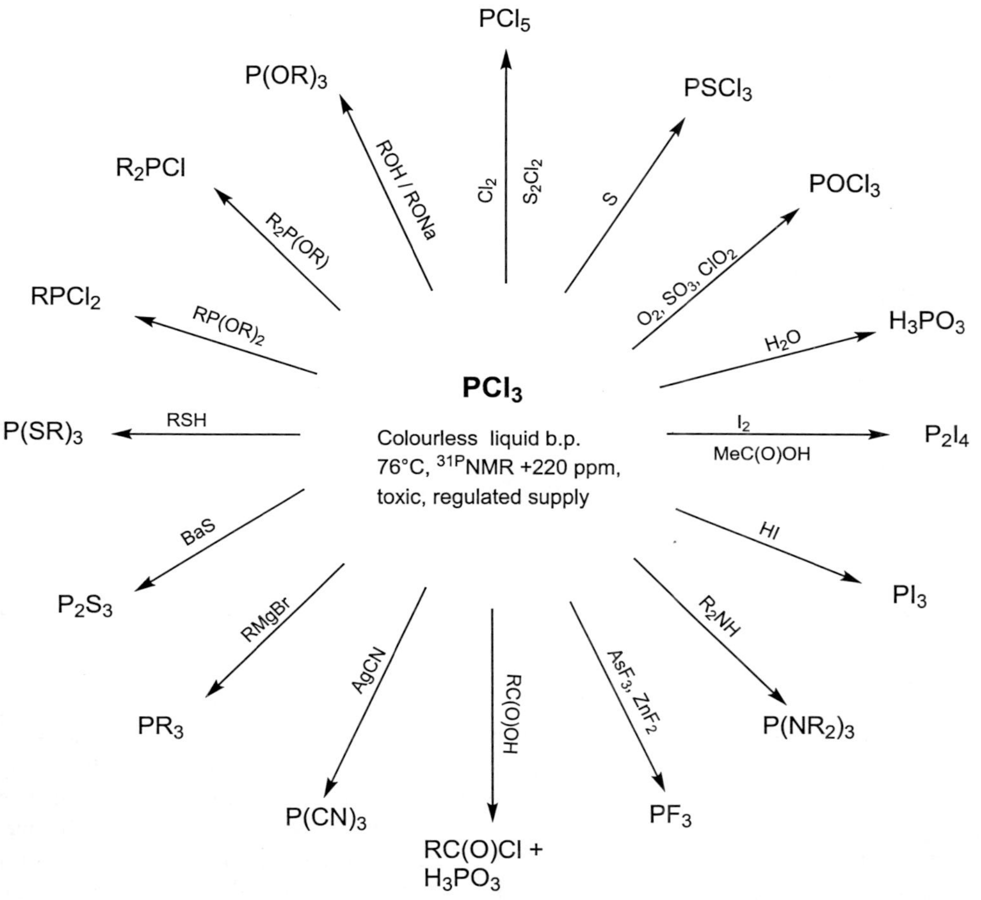

Scheme 9.9 Selected reactions of PCl_3

One of the major industrially important products prepared from PCl_3 is glyphosate (N-phosphonomethyl glycine), the world's largest selling single crop-protecting chemical and most used herbicide. In 2014, 826 million kg of glyphosate was made globally. It works by blocking an enzyme which many plants need for making aromatic amino acids such as phenylalanine, tyrosine and tryptophan. Glyphosate is not useful to control weeds in a crop which is already growing as the herbicide will kill both the crop plants and the weeds. Thus, crops have been genetically modified to resist glyphosate. If not, glyphosate can be used to kill weeds in a field before a crop is sown or after it has been harvested. It is also sprayed on crops before they are harvested to make them dry out and harvesting of crops such as cereals, oilseeds, maize and sunflowers can be made easy.

Two methods adopted for the synthesis of glyphosate are shown in Scheme 9.10.[25, 26]

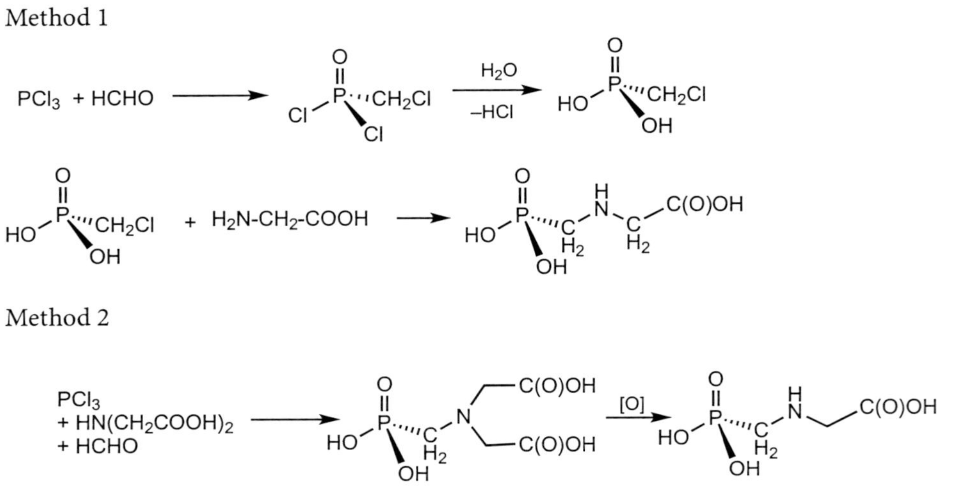

Method 1

Method 2

Scheme 9.10 Synthesis routes to glyphosate from PCl_3

9.6.2 Phosphorus Pentachloride

Phosphorus pentachloride, the hypervalent phosphorus(V) halide is a well-known, moderately strong, solid, chlorinating agent which is also the starting material for many organophosphates as well as cyclic and acyclic phosphorus nitrogen compounds known as phosphazenes. PCl_5 can bring about two types of chlorinations—oxidative and substitutive. Recently, for oxidative chlorination, its use has been reduced due to the availability of a better chlorinating agent sulfuryl chloride, SO_2Cl_2. The relative disadvantage of PCl_5 is the fact that the side product of substitutive chlorination is $POCl_3$ and oxidative chlorination is PCl_3, both reactive liquids with moderately high boiling points. With SO_2Cl_2, SO_2 gas is the side product and can be easily removed from the reaction medium. It is still a well-known reagent used for transforming organic acids and alcohols to alkyl chlorides and acyl chlorides. It is also a specific reagent for the chlorination of aliphatic alcohols and it gives $LiPF_6$ with LiF which is an electrolyte for lithium ion batteries.[2, 27, 28]

$$PCl_5 \rightleftharpoons [PCl_4]^+ Cl^- \quad (9.13)$$

$$2\,PCl_5 \rightleftharpoons [PCl_4]^+ [PCl_6]^- \quad (9.14)$$

In solutions of polar solvents, PCl_5 undergoes autoionisation. Dilute solutions dissociate to give $PCl_4^+Cl^-$. At higher concentrations, a second equilibrium involving PCl_6^- becomes more significant. This can be seen in the ^{31}P-NMR spectra as PCl_5 gives a peak at –80 ppm while PCl_4^+ and PCl_6^- give peaks at +86 and –295 ppm respectively with 85% H_3PO_4 as reference. Well known reactions of PCl_5 are shown in Scheme 9.11.[28]

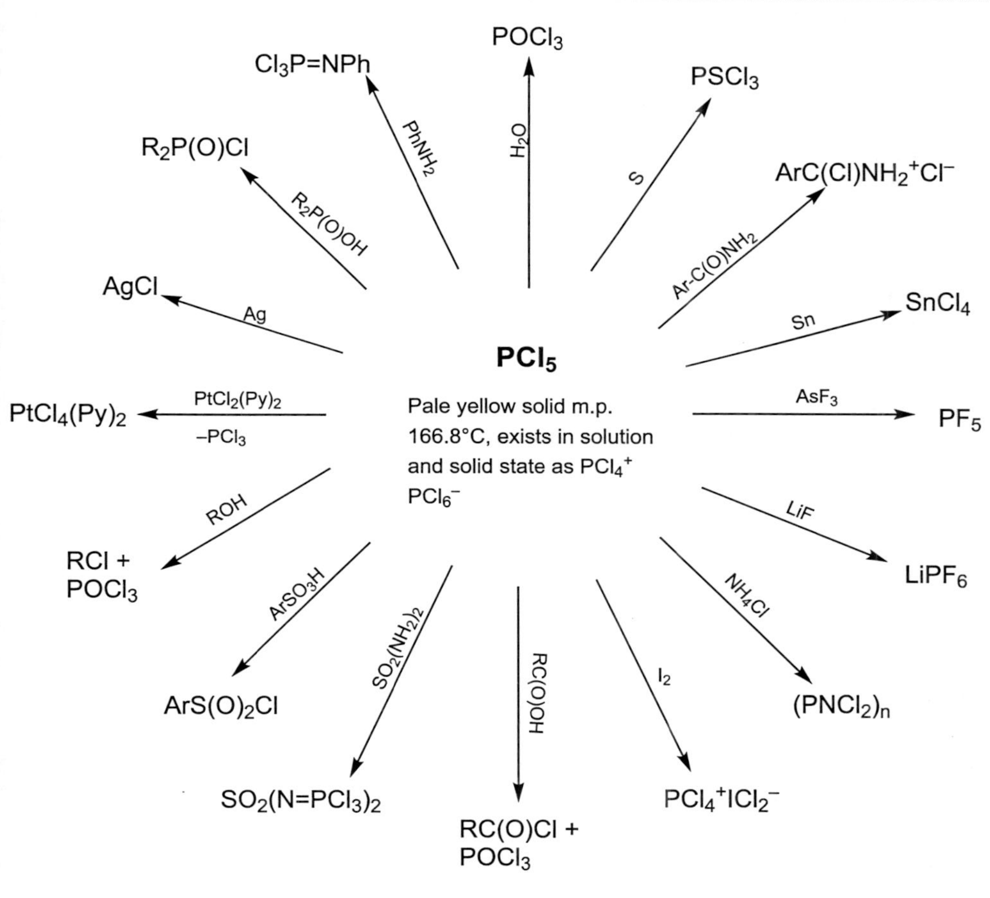

Scheme 9.11 Selected reactions of PCl_5

Is the world heading towards an impending phosphate scarcity related crisis?[29,30]

Phosphorus, traditionally a low-profile element, is essential to all living things. It is an indispensable component of DNA, an ingredient present in the universal energy currency (ATP) and a building block of all cell membranes and of bone (as hydroxyapatite). Being an essential element for plant growth, one can see phosphorus in the given nutrient content of all fertilisers indicated by the NPK ratio (ratio of nitrogen : phosphorus : potassium) which is often a multiple of 3:1:2. Without phosphorus in a fertiliser, the yields of crops can get considerably reduced. Rice yields, for example, can get reduced by over 15% if phosphorus is absent in the fertiliser; yields of wheat and corn crops could decline even more.

As the atmosphere has 78% nitrogen and humanity has discovered and applied the Haber–Bosch process, there is no future scarcity predicted for nitrogen at all. The same is the case for potassium, which makes up about 2.1% of the weight of the earth's crust and is the seventh most abundant element on earth. But there is definitely an impending scarcity for phosphorus. Phosphorus is substantially low in abundance (0.12% of the earth's crust) and adding to that, excessive use of phosphate fertilisers in farming over the decades has leached out the available phosphorus on land to water bodies such as ponds, lakes and the oceans. The huge growth of blue green algae (cyanobacteria) which thrive on dissolved phosphates in many large water bodies such as lakes and ocean river fronts, visible even from outer space and satellite photographs are clear evidence for the massive amounts of phosphates leached out into the water bodies from the landmass. With about 160 million tons of annual production and 16 billion tons of reserves, it looks like we have sufficient phosphorus for about 100 years. While that sounds like a decent margin, the global peak, after which production can no longer keep up with demand, may be even just 30 years away. The demand for phosphorus increased sharply in the mid-20th century with the success of the green revolution and is predicted to increase by 50%–100% by 2050. *The global population which was threatened by a lack of soil fertility in 1850 was then less than 1.5 billion but by 2050 the population is expected to grow to over 9 billion, all of whom will depend on a sustainable solution to replenish phosphates present in the soil.*

The fact that only very few countries have phosphorus deposits that can be mined, creates more concern. About 90% of the world's known reserves (phosphate rocks) are located in or are controlled by five countries—Morocco, Jordan, South Africa, the United States and China. Morocco (which sits on more than one third of proven worldwide phosphate reserves) has been named the 'Saudi Arabia of phosphorus' and has monopoly on western Saharan reserves. China, which recently raised its tariffs on phosphorus, is expected to keep more phosphate at home. India, whose population is expected to overtake China in the future is almost totally dependent on imports for phosphates. The declining quality, or purity, of known phosphate rock sources is only heightening the anxiety. So will future wars be fought for gaining phosphorus?

Unless humanity takes this situation seriously and develops methods to reuse and recirculate phosphates before it gets leached out to the sea or find more efficient ways of using it, the future of agriculture and food security will be in grave distress in about 100 years. Perhaps the best advice comes from Victor Hugo who in 1862 wrote in *Les Miserables*: *'Science, after having long groped about, now knows that the most fecundating and the most efficacious of fertilisers is human manure. The Chinese, let us confess it to our shame, knew it before us. Not a Chinese peasant goes to town without bringing back with him, at the two extremities of his bamboo pole, two full buckets of what we designate as filth. Thanks to human dung, the earth in China is still as young as in the days of Abraham. Chinese wheat yields a hundred fold of the seed.'* Capturing urine at the source (at the toilet) seems to have possibilities as a way of recovering phosphorus significantly.

In Sweden for example, some municipalities have imposed rules so that all new toilets made must be urine diverting. Sweden has also proposed that a major chunk of phosphorus in sewage should be returned to the land in the coming years.

9.7 Phosphorus Oxyacids and Their Salts

Phosphorus forms many oxyacids. The ones which are well known are mostly monomeric and dimeric in nature. The geometry around the phosphorus centre is tetracoordinate, while the formal oxidation state of phosphorus varies from 0–5. The well known examples of phosphorus oxyacids are given in Fig. 9.4.[31]

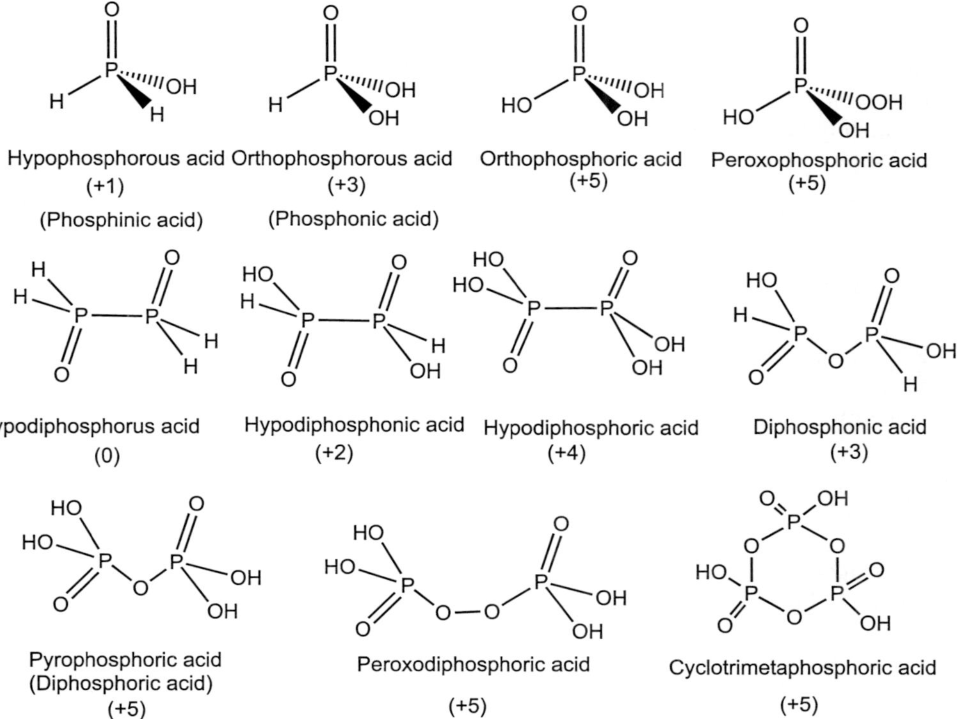

Fig. 9.4 Formal oxidation states of phosphorus in its oxyacids

9.7.1 Hypophosphorus Acid

Hypophosphorus acid is a powerful reducing agent with a molecular formula H_3PO_2. $HOP(O)H_2$ exists in equilibrium with the minor tautomer $HP(OH)_2$. The minor tautomer is referred to as hypophosphorus acid while the major tautomer is called phosphinic acid.

The industrial synthesis of the acid involves two steps. First, the hypophosphite salts of the alkali and alkaline earth metals are prepared from the reaction of white phosphorus with hot aqueous solution of the hydroxide—NaOH or $Ca(OH)_2$.

$$P_4 + 3\ NaOH + 3\ H_2O \longrightarrow PH_3 + 3\ NaH_2PO_2 \quad (9.15)$$

The free acid can be prepared by the action of a strong acid on these salts.

Electroless nickel plating

Electroless nickel plating is an auto-catalytic chemical technique used to deposit a layer of nickel-phosphorus alloy on a solid surface, such as metal or plastic. The process is carried out using a reducing agent which is often sodium hypophosphite, NaH_2PO_2 along with a Ni(II) solution. Observed by Wurtz in a chemical accident in 1844, electroless nickel plating has several advantages over traditional electroplating. It is free from flux density and requirement of power supply, it provides an even deposit regardless of the geometry of the material plated, and with the proper pre-plate catalyst, can deposit on even non-conductive surfaces. In addition, there is flexibility in plating volume and thickness and the method plates recesses and blind holes with stable thickness. One of the disadvantages is the limited lifespan of the chemicals utilised due to their reactivity. Hypophosphite baths are regularly used for this plating as it is inexpensive and provides better corrosion resistance.

$$Ni^{2+} + H_2PO_2^- + H_2O \longrightarrow Ni^0 + H_2PO_3^- + 2H^+ \quad (9.16)$$

9.7.2 Phosphorous Acid

H_3PO_3 is more clearly described with the structural formula $HPO(OH)_2$. This species exists in equilibrium with a minor tautomer $P(OH)_3$. $HPO(OH)_2$ is the product of the hydrolysis of its acid anhydride:

$$P_4O_6 + 6\ H_2O \longrightarrow 4\ HP(=O)(OH)_2 \quad (9.17)$$

(An analogous relationship connects H_3PO_4 and P_4O_{10}.)

On an industrial scale, the acid is prepared by hydrolysis of phosphorus trichloride with water or steam.

$$PCl_3 + 3\ H_2O \longrightarrow HPO(OH)_2 + 3\ HCl \quad (9.18)$$

H_3PO_3 on heating at 200°C undergoes disproportionation to give phosphoric acid and phosphine.

$$4\ H_3PO_3 \xrightarrow{200°C} 3\ H_3PO_4 + PH_3 \quad (9.19)$$

Phosphorous acid is a powerful reducing agent. When treated with a cold solution of mercuric chloride, a white precipitate of mercurous chloride is formed.

$$H_3PO_3 + 2\ HgCl_2 + H_2O \longrightarrow Hg_2Cl_2 + H_3PO_4 + 2\ HCl \quad (9.20)$$

9.7.3 Phosphoric Acid

Phosphoric acid also known as orthophosphoric acid, has a large number of applications such as precursor to fertilisers, rust inhibitor, food additive, dental etchant, electrolyte, flux, dispersing agent and component of home cleaning products. 80% of the industrially produced phosphoric acid is used for making fertilisers and 12% in synthetic detergents mostly as sodium tripolyphosphate. The most common source of phosphoric acid is an 85% aqueous solution which is also the standard for ^{31}P-NMR measurements. As a food additive, it provides a tangy or sour taste due to its acidity. Salts of phosphoric acid, such as monocalcium phosphate, are used as leavening agents.

Phosphoric acid also works as a rust remover by direct application to rusted iron, steel tools, or other surfaces. The acid changes the reddish-brown Fe_2O_3 to ferric phosphate, $FePO_4$ (Eq. 9.21).

$$2\ H_3PO_4 + Fe_2O_3 \longrightarrow 2\ FePO_4 + 3\ H_2O \quad (9.21)$$

In dentistry, phosphoric acid is used as an etching solution, and to clean and roughen the surfaces of teeth where dental appliances or fillings need to be introduced. It is also used as a solution for anodising, as a buffer agent, as an electrolyte in fuel cells, as a catalyst in many organic reactions, as a pH adjuster in skin-care products and as a sanitising agent in dairy and brewery industries.

It is industrially made by two processes. The *wet process* uses reaction of sulfuric acid on apatite. The lesser used method is the *thermal process* in which white phosphorus is oxidised to P_2O_5 and this on hydrolysis gives pure phosphoric acid.

$$Ca_5(PO_4)_3X + 5\ H_2SO_4 + 10\ H_2O \longrightarrow 3\ H_3PO_4 + 5\ CaSO_4.2H_2O + HX \quad (9.22)$$

X: = OH Hydroxyapatite; X = F: Fluorapatite

Phosphate in aqueous solutions exists primarily as $H_2PO_4^-$ or HPO_4^{2-} and it is an effective buffer at pH 7.0–7.5. The dissociation constants of phosphoric acid can be expressed as follows.

$$H_3PO_4 \underset{pKa_1\ 2.12}{\overset{-H^+}{\rightleftharpoons}} H_2PO_4^- \underset{pKa_2\ 7.21}{\overset{-H^+}{\rightleftharpoons}} HPO_4^{2-} \underset{pKa_3\ 12.35}{\overset{-H^+}{\rightleftharpoons}} PO_4^{3-} \quad (9.23)$$

$Ka_1 = 7.5 \times 10^{-3}$; $Ka_2 = 6.2 \times 10^{-8}$; $Ka_3 = 4.46 \times 10^{-13}$

Phosphoric acid on heating to 200–250°C converts to pyrophosphoric acid, $H_4P_2O_7$. On further heating to 300–600°C it converts to metaphosphoric acid $(HPO_3)_n$.

9.7.4 Sodium Tripolyphosphate

Sodium tripolyphosphate (STPP) is the most common phosphate additive in detergent powder. 20 lakh tonnes of this compound was made in 2000. Since the five sodium ions are exchangeable with calcium and magnesium ions, it efficiently reduces the hardness of water, facilitating the cleaning action of synthetic detergents. The use of STPP is being reduced as it has been found to have positive effects in eutrophication of lakes due to the growth of blue green algae. STPP is a preservative for seafood, meats, poultry and animal feed. In foods, STPP is used as an emulsifier and to retain moisture.[33]

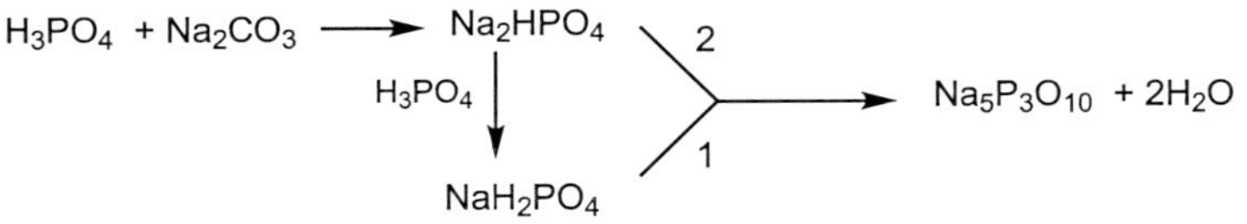

$$H_3PO_4 + Na_2CO_3 \longrightarrow Na_2HPO_4 \; (2); \quad Na_2HPO_4 \xrightarrow{H_3PO_4} NaH_2PO_4 \; (1); \quad 2\,Na_2HPO_4 + NaH_2PO_4 \longrightarrow Na_5P_3O_{10} + 2H_2O \qquad (9.24)$$

9.7.5 Adenosine Triphosphate

The three biomolecules, ATP, ADP and AMP are interrelated and form the most well known phosphorus-based molecules in living systems. The energy generated by the body by oxidation of food (oxidative phosphorylation) or the energy producing analogous process in plants (photo phosphorylation) result in converting ADP molecules to ATP as an extra phosphate unit is added by a P–O–P chemical bond to the ADP.

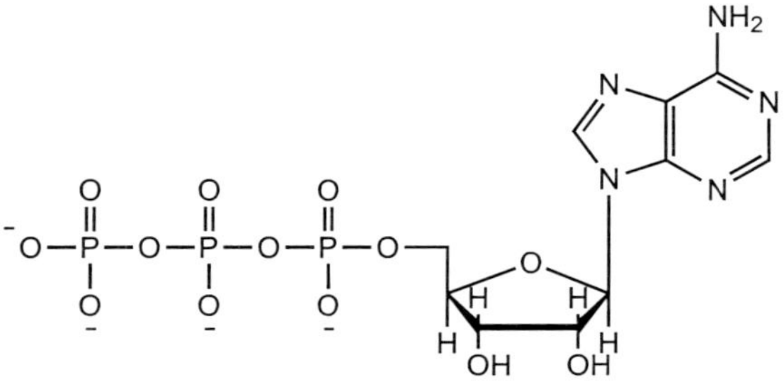

$$C_6H_{12}O_6 + 36\,ADP + 36\,Pi + 36\,H^+ + 6\,O_2 \longrightarrow 6\,CO_2 + 36\,ATP + 42\,H_2O \qquad (9.25)$$

$$2\,H_2O + 2\,NADP + 3\,ADP + 3\,Pi \xrightarrow[\text{Chlorophyll}]{\text{light}} O_2 + 2\,NADPH + 3\,ATP + 2H^+ \qquad (9.26)$$

The ATP molecule cleaves its P–O bond when energy is required and 30.5 kJ/mol of energy is released when the ATP to ADP conversion takes place. Plant cells use the ATP for converting CO_2 to sugar molecules while the major use of ATP (1/5th) in animals is running the sodium potassium pump. ATP is the most widely distributed high-energy compound within plants and animals and a macromolecule which is second in importance only to DNA. This ubiquitous molecule is used for building other complex molecules, to contract muscles, to generate electricity in nerves and a myriad other energy required activities. All fuel sources of nature produce ATP, which in turn powers virtually every activity of the cell and of an organism. The structure of ATP shows a purine base (adenine) connected by one of its nitrogen atoms to the carbon atom of a pentose sugar. Three phosphate groups as a chain are attached to this pentose sugar. ATP, ADP and AMP interconvert by the addition and removal of these phosphate groups. Each cell in the human body contains about a billion ATP molecules.

Hydrolysis of the phosphate groups in ATP is a reaction that releases energy, as the resulting orthophosphate group (P_i) is highly stabilised by resonance structures, making the products (ADP and P_i) much lower in energy than the reactant (ATP). The high electron density associated with the three adjacent phosphate units of ATP also destabilises the molecule, making its energy higher. The energy used by human cells requires the hydrolysis of 100 to 150 moles of ATP on a daily basis. There is only about 5 g of ATP in the entire human body. Up to 40 kg equivalent of ATP is used in 24 hours. In human beings, ATP molecules are recycled up to 750 times per day. Since these molecules cannot be stored, they are consumed immediately following synthesis.

$$ATP + H_2O \longrightarrow ADP + P_i \quad \Delta G^\circ = -30.5 \text{ kJ/mol } (-7.3 \text{ kcal/mol}) \tag{9.27}$$

$$ADP + H_2O \longrightarrow AMP + P_i \quad \Delta G^\circ = -30.5 \text{ kJ/mol } (-7.3 \text{ kcal/mol}) \tag{9.28}$$

9.7.6 Uses of Phosphates

Phosphates have a wide range of industrial and agricultural uses, the major one being as fertilisers followed by uses in detergents for water softening, additives in food, soft drinks and baked products as well as in medicines.

Phosphorus-based fertilisers

With regard to bulk and economy, the most important use of phosphorus-based compounds are as fertilisers. The two major minerals hydroxyapatite and fluorapatite are converted to soluble and neutral forms of phosphorus via phosphoric acid or directly to calcium salts. In fact more than 80% of all sulfuric acid produced in the world is used for making phosphoric acid or single superphosphate of lime which is a mixture of monocalcium phosphate and gypsum. Mono and diammonium phosphates which supply two of the main elements required for plant growth are products of the reaction of ammonia and phosphoric acid. The industrially important ammonium and calcium phosphate based compounds and their properties are shown in Table 9.4.[37]

ATP: The universal energy currency whose discovery missed the Nobel prize[34-36]

All organisms – from the simplest form of bacteria, plants and all animals to humans – use ATP as their primary energy currency. Every day an adult converts a quantity of ATP corresponding to at least half his/her body weight. Quite surprisingly, the discovery of such an important molecule missed receiving the Nobel Prize!

Two claims have been made for the discovery of ATP. The first by Karl Lohmann of the Kaiser Wilhelm institute of biology in Berlin, Germany and the other by C H Fiske and his coworker, Yellapragada Subbarow of Harvard medical school, USA. Both the groups made their discovery of ATP by separate methods and published their work in the same year 1929; one in the journal *Naturwissenschaften* and the other in the journal, *Science*. The correct structure of adenosine-5′-triphosphate was finally confirmed 20 years later by Alexander Todd by chemical synthesis. The beginning of the World War II in 1930 and the rise of Germany under Hitler and related war between the US led allies and Germany as well as the close competition between the US and German research groups might have contributed to the discovery of ATP being ignored by the Nobel Committee. Later in 1997, Paul D Boyer and John E Walker shared the Nobel prize for their work on how the enzyme ATP synthase acts as a catalyst to synthesise ATP. Boyer and his co-workers have proposed a mechanism for the formation of ATP from ADP and inorganic phosphate (P_i).

The discovery of ATP brings to limelight the contributions of a relatively unknown Indian born scientist Y Subbarow. Subbarow who is widely respected for his contributions in biochemistry was born in India in 1895 and as a medical student went from Madras Medical College to the USA for higher studies. While working with Fiske, his mentor at Harvard Medical School, he made many fundamental and path breaking discoveries other than that of ATP. These include synthesis of phosphocreatine which helps in replenishing ATP while resting, methoxytrexate an anticancer drug, folic acid (vitamin B_9), many antibioitcs such as aureomycin, the first tetracycline and Hetrazan, the most widely used drug against filariasis which leads to the deformity causing elephantiasis. The Fiske–Subbarow method of colorimetrically estimating phosphorus, is still one of the best methods and is in the top 100 cited scientific articles ever, with more than 20,000 citations. He worked at Harvard Medical School until 1940 when he went to Lederle Laboratories to direct its medical research. He died at the young age of 53 in the US without even becoming a US citizen. The New York Herald Tribune described Dr Yellapragada Subbarow as 'An eminent medical mind of the century'. He discovered cures for many killer diseases which brought great relief to millions of suffering people throughout the world. Writing in the April 1950 issue of *Argosy*, Doron K Antrim observed, 'You've probably never heard of Dr Yellapragada Subbarow. Yet because he lived, you may be alive and are well today. Because he lived, you may live longer'.

Diammonium phosphate along with ammonium sulfate and a few other chemicals in minor quantities such as ammonium polyphosphate and monoammonium phosphate, coloured with a red dye is used as a fire retardant in large quantities to control forest fires by spraying from aeroplanes. This is an environmentally friendly composition as they do not support combustion, are non-explosives (unlike ammonium nitrate) and they will only enrich the forest soil as nitrogen and phosphorus fertilisers after their intended use.

Table 9.4 Synthesis of common phosphates and their uses

Chemical name (Short form)	Industrial name	Synthesis	Major uses
Mono ammonium phosphate MAP	Mono ammonium phosphate	$NH_3 + H_3PO_4 \longrightarrow (NH_4)H_2PO_4$	Fertiliser
Diammonium phosphate DAP	Diammonium phosphate	$2\,NH_3 + H_3PO_4 \longrightarrow (NH_4)_2HPO_4$	Fertiliser
Monocalcium phosphate MCP (TSP)	Triple superphosphate	$Ca_5(PO_4)_3F + 7\,H_3PO_4 \longrightarrow 5\,Ca(H_2PO_4)_2 + HF$	Fertiliser
Monocalcium phosphate + gypsum SSP	Single superphosphate	$Ca_3(PO_4)_2 + 2\,H_2SO_4 \longrightarrow Ca(H_2PO_4)_2 + 2\,CaSO_4$	Fertiliser
	Superphosphate	$Ca(OH)_2 + 2\,H_3PO_4 \longrightarrow Ca(H_2PO_4)_2 + 2\,H_2O$	Leavening agent
Calcium mono-hydrogen phosphate	Dicalcium phosphate	$CaCl_2 + H_3PO_4 + 2\,NaOH \longrightarrow CaHPO_4 + 2\,NaCl + 2\,H_2O$ $H_3PO_4 + Ca(OH)_2 \rightarrow CaHPO_4$	Dietary tableting agent, poultry feed

9.8 Phosphorus Nitrogen Compounds

Phosphorus forms both cyclic and acyclic, as well as single- and multiple-bonded compounds with nitrogen. Among these, cyclic and acyclic phosphazenes having phosphorus in the pentavalent state are important compounds due to their diverse range of applications. In addition, saturated cyclophosphazanes also abound in the literature where phosphorus is in the trivalent state.[38]

9.8.1 Cyclophosphazanes

The most well-known and well-studied examples of saturated P–N rings are the four-membered *cyclodiphosphazanes*, also known as *diazadiphosphetidenes*. These are prepared by the reaction of PCl_3 and primary amines, especially sterically hindered amines such as *t*-butyl amine.

$$[ClP(\mu\text{-}NBu^t)_2PCl] \xleftarrow{1:3 \text{ ratio}} PCl_3 + t\text{-}BuNH_2 \xrightarrow{1:10 \text{ ratio}} [(Bu^tNH)P(\mu\text{-}NBu^t)_2P(NHBu^t)] \quad (9.29)$$

The reactivity of the P–Cl bond and the trivalency of the phosphorus site has been utilised to make a range of derivatives of these compounds which have been used as ligands in coordination chemistry.

Phosphazenes are compounds having pentavlent tetracoordinated phosphorus atoms bound to nitrogen through a double bond. In general, they can be prepared by two different synthetic methods—the *Staudinger method* and the *Kirsanov method*. In the Staudinger method, a trivalent phosphorus compound is converted to the phosphazene unit by using an azide. In the Kirsanov method, a pentavalent phosphorus halide is reacted with ammonia or a primary amine to realise the phosphazene unit.

$$R_3PCl_2 + R'NH_2 \xrightarrow[-HCl]{} R_3P{=}NR' \quad \text{Kirsanov method} \quad (9.30)$$

$$R_3P + R'N_3 \xrightarrow[-N_2]{} R_3P{=}NR' \quad \text{Staudinger method} \quad (9.31)$$

9.8.2 Acyclic Phosphazenes

Acyclic phosphazenes have been prepared for two basic purposes—as precursors for making phosphazene polymers and as peralkylated aminophosphazenes which have been found to act as very strong bases which are non-nucleophilic and air stable.

Phosphazene bases

Synthetic chemists require very strong bases for activation of organic substrates that have not-so-acidic C–H bonds. The strong bases often used are alkali metal stabilised anions such as lithium dialkylamides (LDA), alkyl lithium, and so on. These are highly sensitive to water and oxygen. Non-ionic and non-nucleophilic bases also have an important role in base-catalysed organic reactions. These should be bulky so that the proton can attach to the basic centre but no alkylation or complex formation should occur. Sodium and potassium bis-trimethylsilyl amides (NaHMDS, KHMDS) are examples of alkali metal based non-nucleophilic strong bases. There are also many examples of bulky amines which are non-nucleophilic bases such as DBU, DBN, Hünigs base (N,N-diisopropyl ethyl amine). However, the lesser basicity and the instability of some of these bases towards O_2, hydrolysis and electrophilic attack (e.g. carbonyls) limit their use.[39]

Reinhard Schwesinger in 1987 came out with a set of peralkylated polyamino acyclic phosphazenes, which are extremely strong, uncharged non-nucelophilic bases with a high level of steric hindrance. These compounds, now also known as Schwesinger bases or phosphazene superbases, can be fine-tuned to obtain a range of basicities and have been applied to many organic reactions with excellent success. Highly reactive enolates produced by Schwesinger bases are more stable towards side reactions than lithium enolates obtained from conventional lithium dialkylamides. In addition, the very low Lewis acidity of the BH^+ cation (compared to Li^+) helps in the alkylation of otherwise inaccessible 'naked' anions. The ability of these phosphazene bases to produce highly reactive anions together with their low nucleophilicity and inertness towards the electrophilic component has enabled reactions of very weak C–H acidic substrates via deprotonation in excellent yields. They are not only used for removing protons but are also used for anionic polymerisation where one can avoid the presence of metal ions and even cleavage of C–Si bonds leading to anion formation. Also, these bases can be easily isolated and regenerated due to their excellent chemical and thermal stability. Examples of these bases and their basicities are listed in Fig. 9.5. Also given is a reaction catalysed by these bases (Eq. 9.32). As one can see, due to conjugation, there is better charge delocalisation on the cation formed.

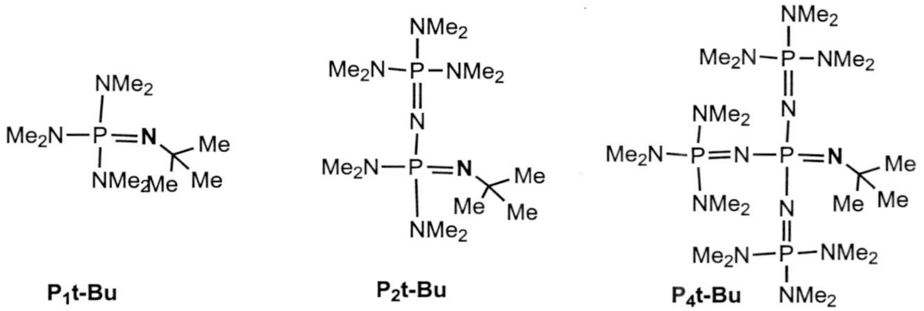

P_1t-Bu $^{MeCN}pK_{BH+} = 26.9$

P_2t-Bu $^{MeCN}pK_{BH+} = 33.5$

P_4t-Bu $^{MeCN}pK_{BH+} = 42.7$

$$B + H_3O^+ \rightleftharpoons BH^+ + H_2O$$

$$K_{BH^+} = \frac{[B]\,[H_2O]}{[BH^+]}$$

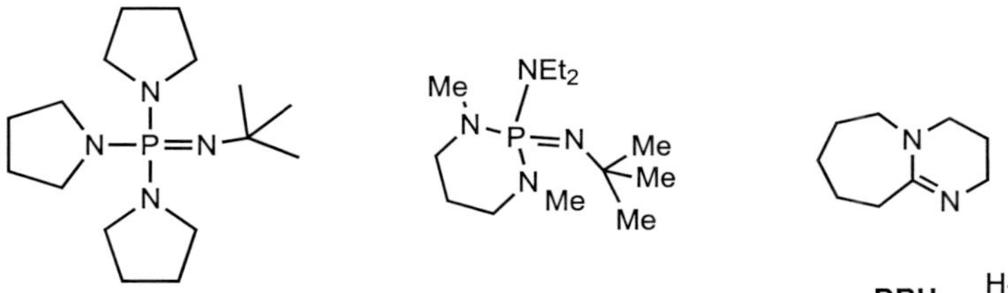

P_1tBu(Pyrr) $^{MeCN}pK_{BH+} = 28.4$

BEMP $^{MeCN}pK_{BH+} = 27.6$

DBU $H_2O_{pKa} = 13.5$ $^{MeCN}pK_{BH+} = 24.3$

Fig. 9.5 Selected examples of phosphazene superbases in comparison with DBU

Most of the phosphazene-based superbases have been synthesised by the reactions of $Bu^tN{=}PCl_3$ and $(Me_2N)_3P{=}NH$ and Me_2NH in varying stoichiometries. The utility of these superbases has been demonstrated in the following reaction.[40]

(9.32)

The above reaction does not proceed with *n*-BuLi, NaOBu^t or P_1-^{t}Bu bases. The advantages of phosphazene bases include high solubility in apolar to moderately polar solvents such as hexane, toluene and THF, stability towards air and moisture and electrophilic attack. Depending on their base strength, these compounds have been found to be slightly to extremely hygroscopic. Applications of phosphazene bases include in situ generation of highly reactive 'naked' anions, for example, for alkylation reactions and applicability in reactions where ionic bases cause solubility problems. These are useful in reactions where ionic bases are sensitive towards oxidation or acylation and also result in Lewis acid catalysed side reactions (as in aldol reactions, epoxide-opening, hydride shifts and anionic polymerisation, and so on). In addition, the work-up through cleaner reactions is easier and occurs often with reaction rate enhancement. Scheme 9.11 shows how anionic polymerisation of butyraldehyde has been made possible by using benzylic anion generated by P_4-*t*-Bu superbase. This polymerisation does not happen with DBU or P_1-*t*-Bu bases.[41]

Scheme 9.11 Anionic polymerisation of butyraldehyde using P_4-*t*-Bu superbase

P_4-*t*-Bu is highly sterically hindered, extremely hygroscopic and highly stable towards hydrolysis. It is 18 orders of magnitude more basic than DBU with a basicity range typical of organolithium bases. It is particularly suitable for deprotonation (activation) of pronucleophiles. It may be noted that unlike acyclic phosphazenes, amino-substituted cyclophosphazenes do not show superbasicity. The pKa reported for $(Me_2N)_6P_3N_3$ is only 7.8.

Verkade's superbases (Proazaphosphatranes)

Verkade's superbases which are ball shaped proazaphosphatrane molecules, are very strong bases due to the extraordinary stability of their protonated forms where the phosphorus centre is stabilised by the distal nitrogen as well. Verkade's bases are about 8 orders of magnitude stronger as a Lewis base than any amine known, including the well known diazabicyclo-5,4,0-undecane (DBU) and diazabicyclo-4,3,1-non-5-ene (DBN). These superbases have been successfully used in many organic reactions, especially C–C bond formation reactions such as alkylations, dehydrohalogenations and acylations. A second characteristic of proazaphosphatranes is their ability to act as catalysts for reactions such as the synthesis of α,β-unsaturated nitriles. They have also been used for synthesis of a type of FLPs which are termed as inverse frustrated Lewis pairs (Scheme 9.22).

$N(CH_2CH_2NHR)_3$ (R = i-Bu, i-Pr...) $\xrightarrow[-Me_2NH]{ClP(NMe_2)_2}$ Azaphosphatrane Cl^- $\xrightarrow{KOBu^t}$ Verkade's superbase (9.33)

$\xrightarrow[R = iBu, iPr]{H^+}$ (9.34)

MeCN pK_{BH+} (iPr) = 33.4

Equation 9.35 shows the utility of Verkade's superbase as a catalyst for the direct α-arylation of nitriles with aryl chloride.[42]

R_1R_2CHCN + C_6H_5Cl $\xrightarrow{Pd(OAc)_2, RT}$ $R_1R_2C(CN)C_6H_5$ (9.35)

R = i-Bu

82%

Cyclophosphazenes

Cyclophosphazenes, which are part of the broad family of phosphorus–nitrogen compounds, are among the oldest inorganic heterocycles to be made. These were first prepared by Liebig and Wöhler in 1834 and further studied in detail by Stokes in the 1890s. The most common cyclophosphazenes – although known with varying ring sizes from 4–24 – are the trimer and the tetramer. The rings are characterised by a valence unsaturated skeleton of repeating units of $[N{=}PR_2]_n$ in which phosphorus is pentavalent and tetracoordinated and the nitrogen is trivalent and dicoordinated (Fig. 9.6).

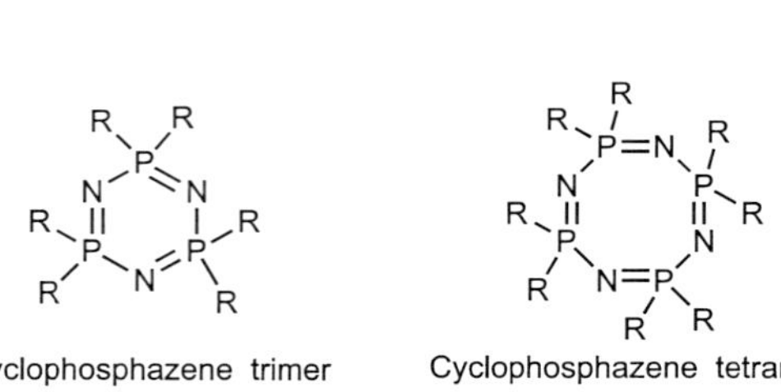

Cyclophosphazene trimer Cyclophosphazene tetramer

R = halogen, alkyl, aryl, alkoxide, aryloxide, amine...

Fig. 9.6 Cyclic trimeric and tetrameric phosphazenes

The most extensively studied among phosphorus–nitrogen heterocycles are the halogenated cyclophosphazenes. Their syntheses, reactions, properties and potential applications have been well documented. Some of the differences in the physical properties among halogenated cyclophosphazenes are summarised in Table 9.5.

Table 9.5 Physical properties of halogenated cyclophosphazenes

	$N_3P_3F_6$	$N_4P_4F_8$	$N_3P_3Cl_6$	$N_4P_4Cl_8$	$N_3P_3Br_6$	$N_4P_4Br_8$
m.p. (°C)	27.8	30.5	114	124	192	202
b.p. (°C)	50.9	89.7	256	dec.	dec.	dec
P–X [bond distance (Å)]	1.52	1.51	1.98	1.99	2.20	-
Ring conformation	Planar	Puckered (saddle)	Slightly puckered	Puckered (tub and chair)	Puckered (chair)	Puckered (saddle)

The chlorinated cyclophosphazenes which are the precursors to a whole lot of other phosphazenes, have been prepared traditionally by the reaction of PCl_5 and ammonium chloride in a high boiling chlorinated solvent such as 1,1,2,2-tetrachloroethane. The trimer and tetramer are the major products along with other cyclic phosphazenes as well as linear polymers. The cyclic compounds are soluble in petroleum ether and can be separated by extraction. Due to the relatively higher basicity of $N_3P_3Cl_6$, it can be selectively extracted from the mixture using concentrated H_2SO_4. The rest of the cyclic compounds can be separated by vacuum distillation or sublimation.[43]

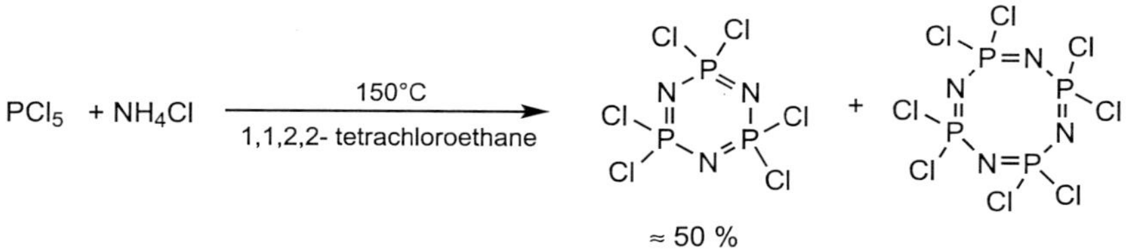

$$PCl_5 + NH_4Cl \xrightarrow[\text{1,1,2,2- tetrachloroethane}]{150°C} N_3P_3Cl_6\ (\approx 50\ \%) + N_4P_4Cl_8 \qquad (9.36)$$

The proposed mechanism for the formation of the six-membered heterocycle is as follows:

$$3PCl_5 + NH_4Cl \longrightarrow [Cl_3P{=}N{=}PCl_3]^+ PCl_6^- + 4\,HCl \tag{9.37}$$

$$NH_4^+ + PCl_6^- \longrightarrow [Cl_3P{=}NH] + 3HCl \tag{9.38}$$

Cl_3PNH has not been isolated but analogous compound

$Cl_3PNSiMe_3$ has been made and characterised

$$[Cl_3P{=}N{=}PCl_3]^+ + Cl_3PNH \longrightarrow [Cl_3P{=}N{-}PCl_2{=}N{=}PCl_3]^+ \xrightarrow{Cl_3PNH} [Cl_3P{=}N{-}PCl_2{=}N{-}PCl_2{=}N{=}PCl_3]^+ \xrightarrow{-PCl_4^+} [NPCl_2]_3 \tag{9.39}$$

Supporting evidence for this mechanism includes isolation of $Cl_3P{=}N{-}PCl_2{=}NPCl_3^+PCl_6^-$ and its reaction with NH_4Cl at 150°C which resulted in $[NPCl_2]_3$ in 80% yield.[44]

Unlike isoelectronic siloxanes, the phosphorus–halogen bonds of cyclophosphazenes can be made to undergo substitution reactions easily. The studies of substitution reactions of halocyclophosphazenes include reactions with mono and difunctional alcohols, amines, phenols, thiols as well as main group and transition metal organometallic reagents. These studies continue to attract considerable interest not only for their exploratory and fundamental value, but also for a variety of potential applications. The nucleophilic substitution reactions of the phosphorus–halogen bonds is the principal route for the synthesis of a variety of fully and partially substituted cyclophosphazenes containing organic and/or organometallic substituents.

These rings have shown the potential for a range of applications. The chlorinated trimer undergoes ring opening polymerisation to the well known polyphosphazene $(Cl_2P{=}N)_n$. Another class of polymers is pendant polymers where cyclophosphazene rings are made as pendants on a conventional organic polymer chain. The cyclophosphazene rings, especially the trimers, also have the unique orientation of the substituents projecting above and below the mean plane of the ring thus providing scope for acting as rigid cores for making starburst type dendrimers.[45]

Molecular structures and aromaticity of cyclophosphazenes

The fact that many six-membered cyclophosphazenes have a planar or almost planar ring with equidistant P–N bonds has made them candidates to be considered for aromaticity. The P–N bond distances in the six-membered N_3P_3 ring lie in the range of 1.55–1.61 Å which are shorter than a P–N single bond (1.70–1.77 Å). The P–N–P angles are close to 120° and the phosphorus shows tetrahedral geometry. While there are many examples of trimeric cyclophosphazenes which are planar having substituents such as F, OR, and so on, $N_4P_4F_8$ was also reported as planar. The now obsolete suggestion of contraction of *d*-orbitals by

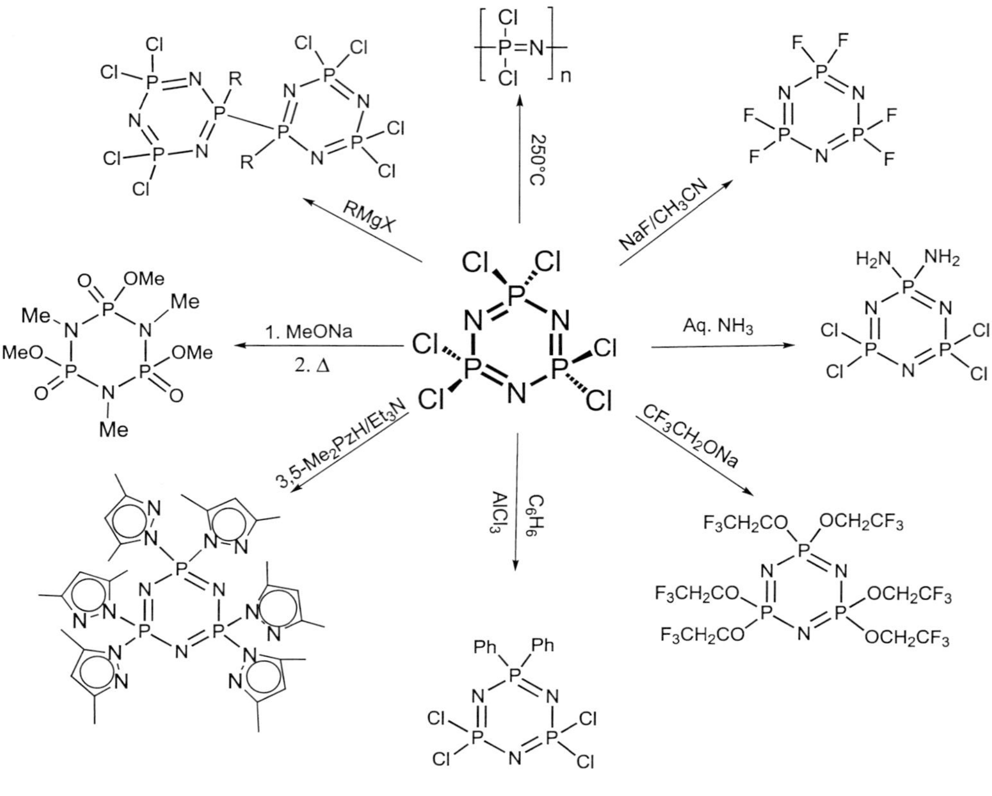

Scheme 9.12 Reactions of chlorocyclotriphosphazene

electronegative substituents for effective overlap for π-bond formation was often used to explain the planarity of fluorine bound phosphazenes. However, re-determination of the structure of $N_4P_4F_8$ by Elias and others in 2000 indicated that $N_4P_4F_8$ is not planar as it was considered to be but is also puckered with a saddle type configuration similar to other 8-membered phosphazene rings.[46]

The factors which support assigning aromaticity to cyclophosphazenes are planarity, presence of $4n+2$ π electrons, equal P–N bond distances which lie between classical single and double bond distances and the observation that cyclophosphazenes are hard to reduce electrochemically. However, there are also factors which do not support classifying these inorganic rings as aromatic. The spectral effects which are associated with organic π-systems are absent in cyclophosphazene rings. For example, introduction of substituents that can conjugate with the ring double bonds should generally cause a bathochromic shift in the UV spectra due to increased π delocalisation. This was found to be absent in cyclophosphazene derivatives. Secondly, aromatic rings normally have ring currents which will induce a magnetic field, these cause shifts in the peaks of the NMR spectra. To verify this by ^{1}H-NMR studies is not possible on cyclophosphazenes as P–H substituted phosphazenes are almost nonexistent. But the ^{19}F-NMR spectra of $N_3P_3F_6$, $N_4P_4F_8$, $N_5P_5F_{10}$ gave fluorine NMR peaks at 71.9, 71.9 and 69.0 ppm which was almost the same as that obtained for the acyclic polymer $(NPF_2)_n$ (70.5 ppm). Since the δ values for cyclic and acyclic phosphazenes are almost the same, one can clearly rule out the presence of ring currents in cyclophosphazenes. Therefore cyclophosphazenes are often referred to as quasi-aromatic compounds.

Structural isomerism in cyclophosphazenes

There has also been interest in the sequential substitution of phosphorus–halogen bonds in view of stereo- and regioselectivities that are involved in such reactions. The unique orientation of the substituents on the phosphazene ring also provides the scope for realising lots more types of structural isomers for substituted cyclophosphazenes, unlike in normal aromatic rings whose substituents lie in the plane of the ring. Figure 9.7 depicts the possible types of isomers for mono-functional and di-functional nucleophiles reacting with a chlorinated cyclophosphazene trimer.[47]

Hydrolysis products of cyclophosphazenes

The halophosphazenes and primary amino-substituted cyclophosphazenes have been found to undergo hydrolysis leading to harmless products. The steps involved in the hydrolysis of $N_3P_3Cl_6$ are shown in Scheme 9.13.

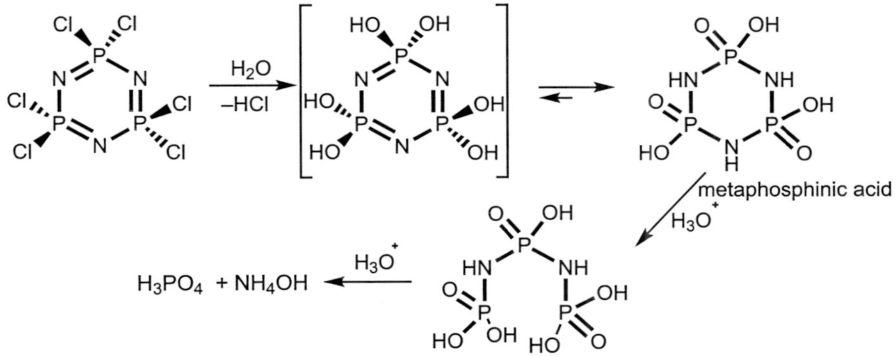

Scheme 9.13 Hydrolysis of $N_3P_3Cl_6$

Polyphosphazenes

Polyphosphazenes form the largest family of inorganic polymers with close to a thousand examples. Stokes, towards the end of the 19th century, showed that $N_3P_3Cl_6$ polymerises but what was obtained was an insoluble rubber. This was possibly due to extensive cross-linking. A workable soluble form of chlorinated polyphosphazene was first made by H R

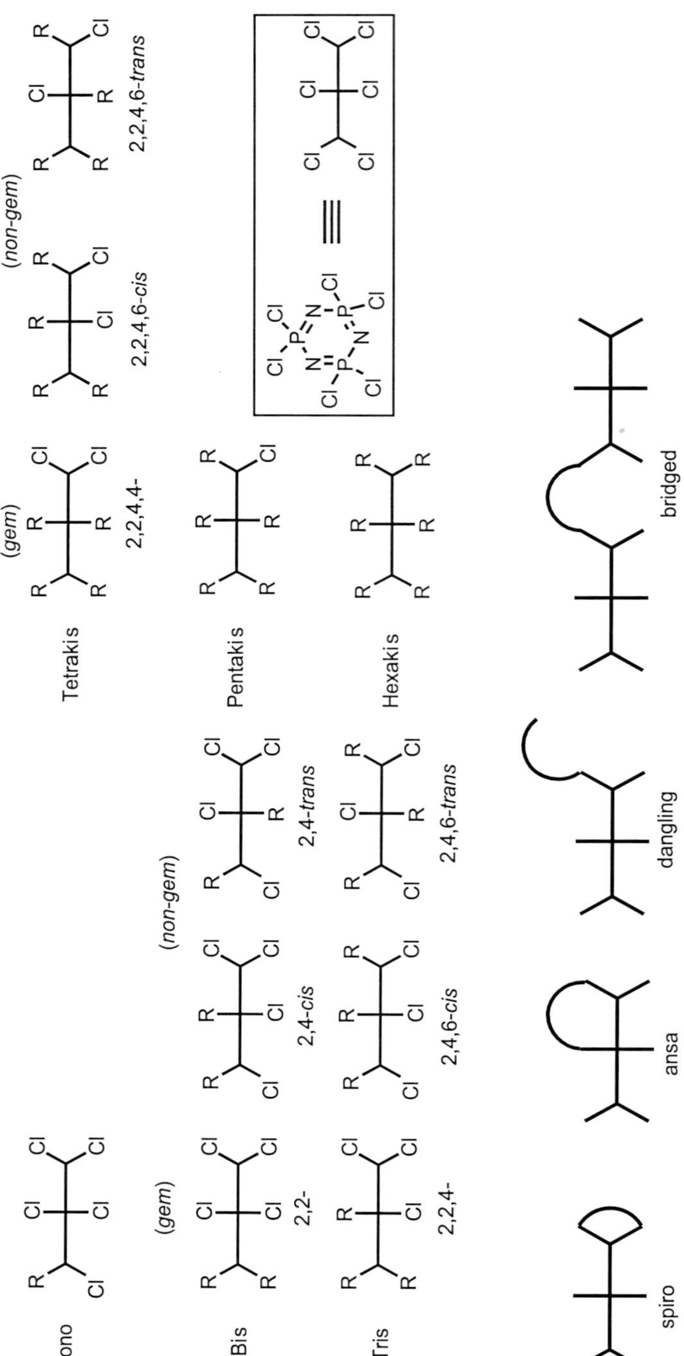

Fig. 9.7 Structural isomers possible for cyclotriphosphazenes

Allcock in 1960 when he heated $N_3P_3Cl_6$ in vacuum at 250°C and allowed the conversion to proceed only up to 70%. Linear polyphosphazene was thus isolated and later this type of ring opening polymerisation was extended to other hexa-substituted phosphazenes as well but at different temperatures (Eq. 9.40).[48]

$$N_3P_3X_6 \xrightarrow{\text{Vacuum}} -[N{=}PX_2]_n- \quad \text{(PDI >2)} \tag{9.40}$$

X= Cl; 250°C
X= F; 350°C
X= Br; 200–250°C
X=NCS; 145°C

The chlorinated polymer $-[N{=}PCl_2]_n-$ is soluble in many organic solvents such as benzene, toluene, chloroform, THF, cyclohexane and has a relatively low glass transition temperature ($T_g = -63$°C). It has an estimated molecular weight (M_w) of around 1.2×10^6. It has been shown that this ring opening polymerisation is initiated by the heterolytic cleavage of a P–Cl bond. The highly electrophilic phosphazenium cation generated can attack the skeletal nitrogen of another $N_3P_3Cl_6$ propagating chain growth.

In addition to the ring opening polymerisation method, there are many other procedures which involve condensation polymerisation. The best among these is the cationic polymerisation of $Cl_3P{=}NSiMe_3$ which is not only a living polymerisation which can be carried out at room temperature yielding molecular weights up to 10^6, but also provides polymer chains having molecular weights in a narrow range with a good polydispersity index in the range of 1.01–1.18.

$$Cl_3P{=}NSiMe_3 \xrightarrow[25^\circ C,\ CH_2Cl_2]{PCl_5} -[N{=}PCl_2]_n- \tag{9.41}$$

M.W 10^6
PDI 1.1–1.18

Although $-[N{=}PCl_2]_n-$ has been obtained with high molecular weight and linearity, it is not useful for practical applications due to very high hydrolytic instability of the P–Cl bonds eventually leading to HCl, NH_3 and H_3PO_4 based decomposition products. Hence to make stable phosphazene polymers, a macromolecular substitution method is utilised (Scheme 9.14).

By using bulky alkoxides, aryloxides and amines, one can also achieve partial substitution of the P–Cl bonds paving the way for the synthesis of hetero-substituted phosphazene polymers.

A wide range of polyphosphazene polymers have been prepared and many of them have been found to have excellent application potential as substitutes for existing polymers. These polymers can be made hydrophobic or hydrophilic, converted to

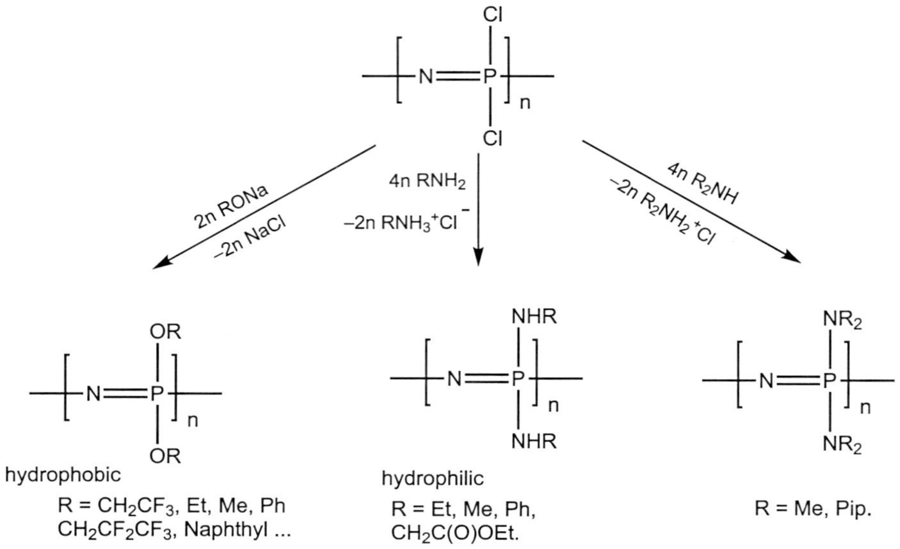
Cl
N=P
n
Cl
2n RONa
−2n NaCl
4n RNH2
−2n RNH3+Cl−
4n R2NH
−2n R2NH2+Cl
OR
N=P
n
OR
hydrophobic
R = CH2CF3, Et, Me, Ph
CH2CF2CF3, Naphthyl ...
NHR
N=P
n
NHR
hydrophilic
R = Et, Me, Ph,
CH2C(O)OEt.
NR2
N=P
n
NR2
R = Me, Pip.

Scheme 9.14 Substitution reactions on chlorophosphazene polymers

polymer electrolytes for ion transport, can be made bio-erodable or bio-inert, converted to anticoagulants for blood transfer and also made fire resistant. Some of the specific examples of such polyphosphazenes are shown in Fig. 9.8.[48]

Phosphorus-based nerve gases and their mode of action

The notorious nerve gases are an unexpected product of research on organophosphorus pesticides (Fig. 9.9). The first among these, Sarin, was made by the team led by Gerhard Schrader of IG Farben, Germany. Schrader found that the combination of P–F bonds and P=O was too toxic for human use as a pesticide. It may be noted that most of the well known phosphorus-based pesticides have P=S groups present. Chemically, this P=S unit will have to get oxidised to become P=O and the oxidation is usually performed by biological oxidants such as cytochrome P-450. The name Sarin comes from the first names of the members of Schrader's team namely Schrader, Ambros, Rudringer and van der Linde. Following the success of sarin, the group made Tabun with a P–CN bond that was even more toxic. Since it was easier to prepare, it was made on a large scale during 1942. It is reported that by the end of World War II, the Germans had made enough quantities of nerve gases to exterminate human life on earth! It was never deployed as they had the feeling that their enemies, the Allies, would also have some nerve gases in stock which would be deployed on them.

During 1952, the team lead by Ranajit Ghosh of Imperial chemical industry, UK, discovered even more potent nerve gases such as VX. The V series had P–S bonds which were found to cleave much more easily than P–F bonds. Table 9.6 compares the toxicity properties of some of these nerve gases.[49]

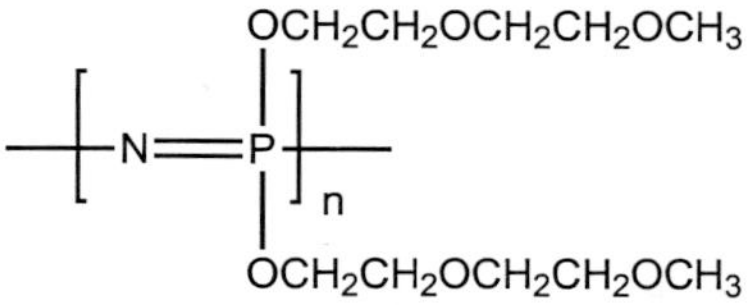

$$\left[\!-N{=}P(OCH_2CH_2OCH_2CH_2OCH_3)_2-\!\right]_n$$

Polybis(methoxy ethoxy ethoxy) phosphazene
MEEP

Polymer electrolyte for lithium batteries

Flexible at low temp ($T_g = -83°C$)
Conductivity = 3×10^{-5} S/cm at 25°C
Conductivity higher than polyethylene oxide
Unlike PEO it is fire resistant
Unlike PEO need not be heated for achieving higher conductivity

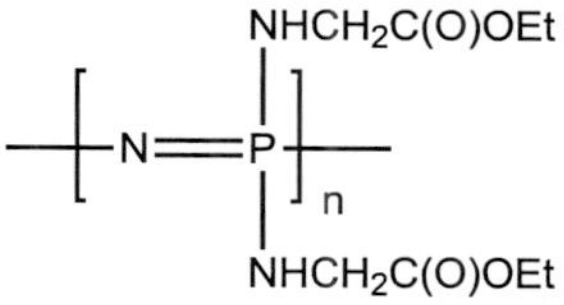

$$\left[\!-N{=}P(NHCH_2C(O)OEt)_2-\!\right]_n$$

Polybis(ethylglycinato) phosphazene

Bio-erodable polymer

The NH unit bound to P forms hydrogen bonds with water facilitating slow hydrolysis.
Products of hydrolysis are non-toxic namely
EtOH, $H_2NCH_2C(O)OH$, H_3PO_4 and NH_4OH

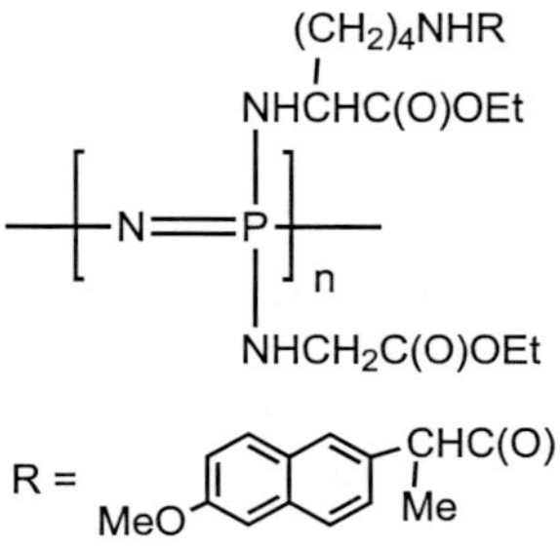

$$\left[\!-N{=}P(NHCH((CH_2)_4NHR)C(O)OEt)(NHCH_2C(O)OEt)-\!\right]_n$$

R = 6-MeO-naphthalen-2-yl–CH(Me)C(O)

Bio erodable slow drug releasing polymer

The Naproxen unit is linked through lycine ethyl ester. Ethyl glycinate units facilitate slow degradation
At pH 7.4 and 37°C 6% release of drug in 7 days

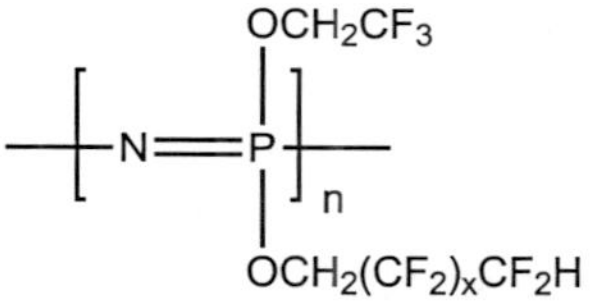

$$\left[\!-N{=}P(OCH_2CF_3)(OCH_2(CF_2)_xCF_2H)-\!\right]_n$$

Flexible at low temp ($T_g = -75$ °C)

Bio-inert polymer

Impenetrable barrier for silicone fluid gels in mammary impants, FDA approved for soft denture lining application. X-ray opaque and resistant to bacterial and fungal colonisaton.

Fig. 9.8 Polyphosphazenes with diverse application potential

Table 9.6 Comparison of toxicities of well known nerve gases

	Tabun	Sarin	Soman	Cyclosarin	VX
b.p. (°C)	247	147	167	239	300
Vapour pressure (mm Hg)	0.07	2.86	0.30	0.06	0.0007
LCt_{50} (inhalation, mg.min/m^3)	70	35	35	50	15
LD_{50} skin (mg/individual 70 kg)	4000	1700	300	–	10

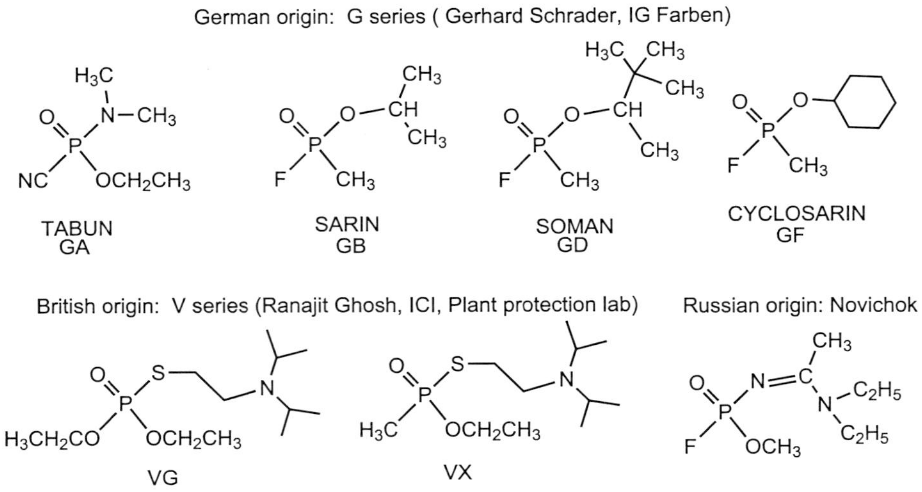

German origin: G series (Gerhard Schrader, IG Farben)

TABUN GA   SARIN GB   SOMAN GD   CYCLOSARIN GF

British origin: V series (Ranajit Ghosh, ICI, Plant protection lab)

VG   VX

Russian origin: Novichok

Fig. 9.9 Structures of well known nerve gases

The nerve gases act by disrupting the mechanism by which nerves transfer messages to organs. The disruption is caused by blocking acetylcholinesterase, an enzyme that normally breaks down acetylcholine, a neurotransmitter to the acetyl group and choline. In fact both organophosphorus pesticides and organophosphorus nerve agents work by the same mechanism of disrupting the activity of acetylcholinesterase. Incidentally acetylcholinesterase is the third most efficient enzyme in the living system coming after catalase and carbonic anhydrase in terms of turnover frequency. The exact mechanism by which the organophosphate residue disrupts the activity of acetyl cholinesterase is schematically shown in Fig. 9.10.

In the event of a normal nerve impulse activity, when a signal reaches a nerve ending or synapse it initiates the release of a chemical called acetylcholine into the synaptic cleft. The molecules of acetylcholine travel though this area and bind to the acetylcholine receptor protein units (which are part of closed sodium channels) present on the post synaptic membrane. The binding of acetylcholine opens up the sodium channels associated with it. Positive sodium ions flow into the cell and initiate a signal which is carried further down the neuron network thereby completing the signalling activity. The used acetylcholine molecule moves into the active site of the enzyme acetyl cholinesterase (AChE) where the serine unit (–CH_2OH) of the enzyme binds to the acetyl unit and cleaves the choline part. The binding of the acetate unit to the enzyme is also cleaved very soon in the presence of water molecules and the free choline units and acetate units move back to the synapse for reforming acetylcholine units and for repeating the cycle.[50]

When a nerve gas molecule or pesticide molecule enters the synaptic cleft, it binds to the active site of acetyl cholinesterase (AChE). The enzyme cleaves part of this pesticide unit but the phosphate unit binds to the serine unit through the CH_2OH group. The binding of the phosphate is much stronger than the binding of the acetyl unit. As a result, the phosphate residue is left bound to the active site of acetyl cholinesterase. When fresh

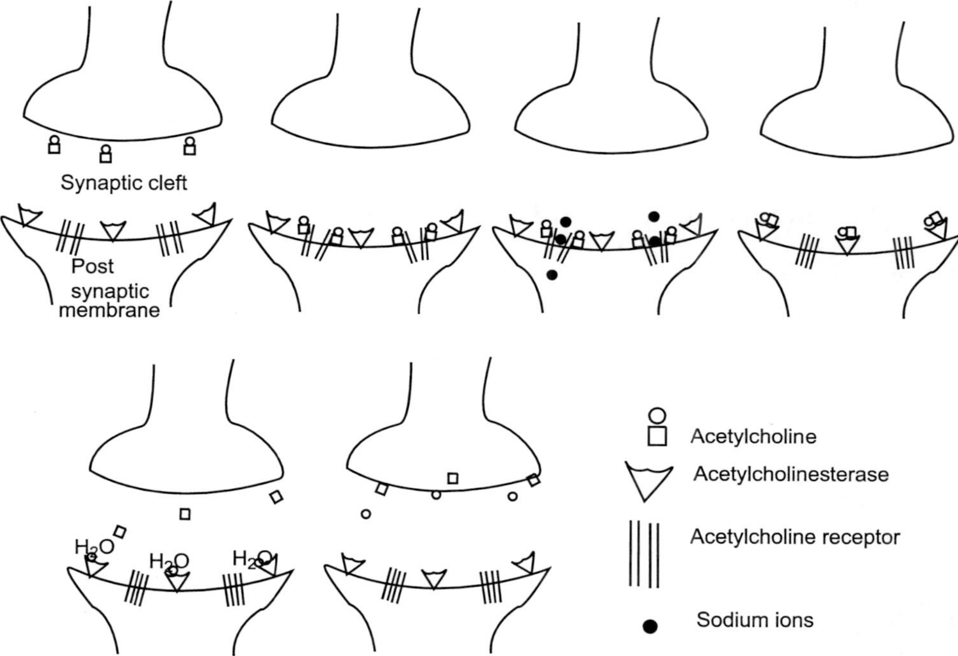

Fig. 9.10 Sequence of actions during the normal functioning of acetylcholinesterase

molecules of acetylcholine are released, they do the signalling part but do not get broken down to the acetyl and choline fragments due to the inactivity of the acetylcholinesterase. The acetylcholine molecules keep on signalling and at the end, the whole signalling process is disrupted and leads to no more electrical signals and ultimate death.

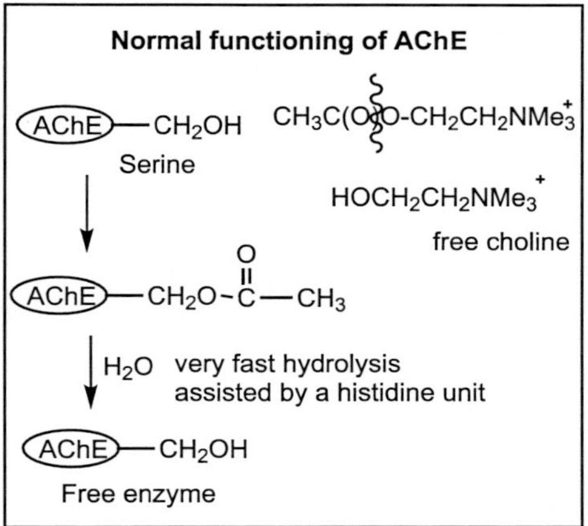

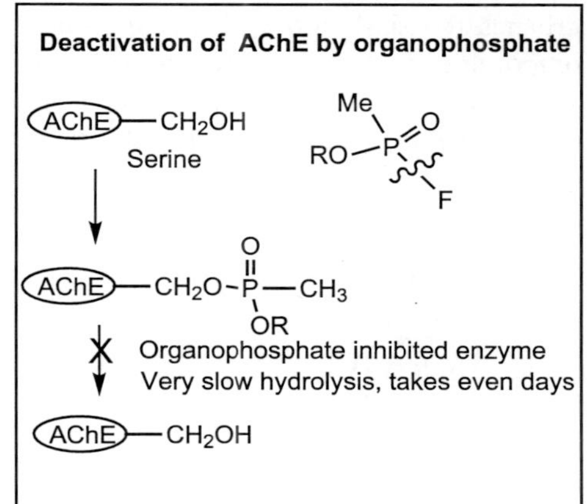

Fig. 9.11 Comparing normal functioning versus deactivation of AChE

Atropine and pralidoxime are antidotes used for nerve gas poisoning. Atropine is expected to block the acetylcholine receptors which are likely to be over-stimulated by excessive acetylcholine. Pralidoxime is expected to release the active-site-bound phosphates.

9.9 Recent Developments

9.9.1 Phosphorus Stabilised in Zero Oxidation State

Work carried out by G H Robinson and coworkers in 2008 and 2013 on N-heterocyclic carbene-stabilised PCl_3 showed some interesting results. A reduction of NHC bound PCl_3 with KC_8 resulted in a novel P–P bonded phosphorus compound where the P is in a zero formal oxidation state. Oxidation of this dimer with dry oxygen resulted in the formation of a pentavalent tetracoordinate phosphorus having a unique P_2O_4 core similar to that of N_2O_4.[51]

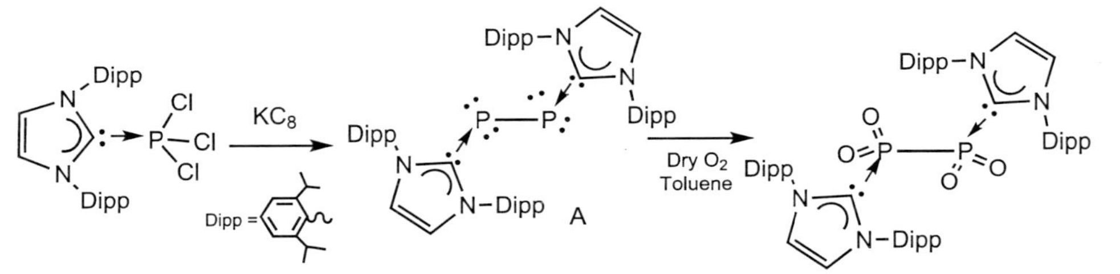

Scheme 9.15 Formation and oxygen reactivity of phosphorus(0) stabilised by NHC

Later G Bertrand and coworkers, by the reaction of sterically less hindered cyclic-alkylamino carbenes (cAACs) with white phosphorus (P_4) was able to make two novel cAAC stabilised compounds, one of them having a P_2 unit analogous to the P_2 compound prepared by Robinson (Eq. 9.42).[52] An analogous reaction with a sterically hindered cAAC resulted in a structure having a chain of 4 phosphorus atoms (Eq. 9.43).[53]

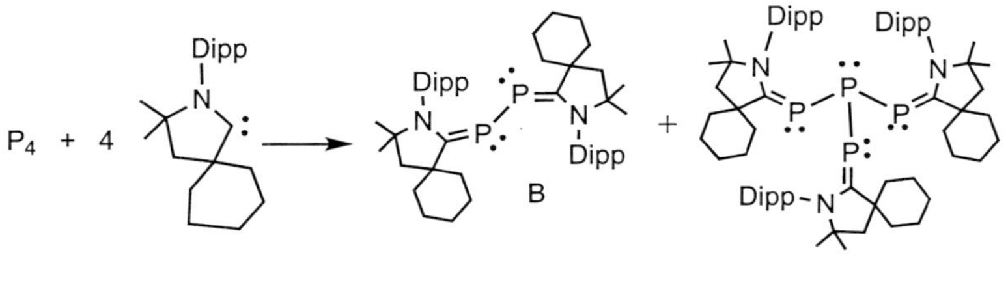

(9.42)

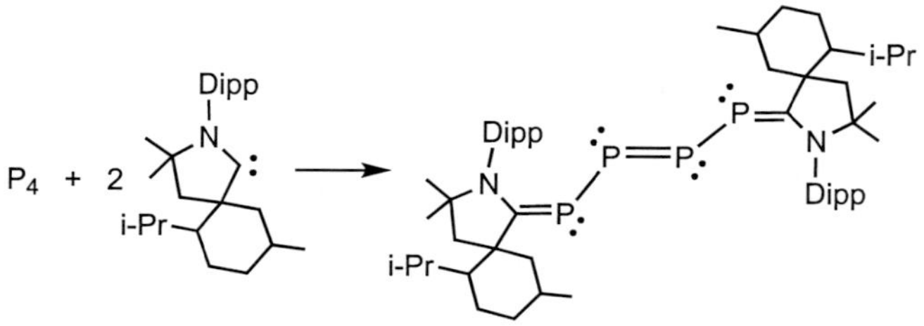

(9.43)

A comparison of structures A and B indicated that A should be considered as a diphosphinidine while B should be considered as a diphosphacumulene (2,3-diphosphabutadiene type). The P–C bond length in A is 1.750 Å while the same distance in B is 1.719 Å. This difference is expected due to the different electronic properties of cAACs and NHCs. cAACs are more electrophilic (π-acceptor) than NHCs and therefore they strengthen the P–C bonds of B.

By a careful synthetic strategy using both NHCs and cAACs, the first example of a carbene stabilised phosphorus mononitride (PN) was prepared. This compound was found to be air-stable at room temperature for indefinite periods of time. Both C–P (1.719 Å) and C–N (1.282 Å) distances are rather short and the central P–N distance (1.709 Å) is in the normal range for a P–N single bond.[54]

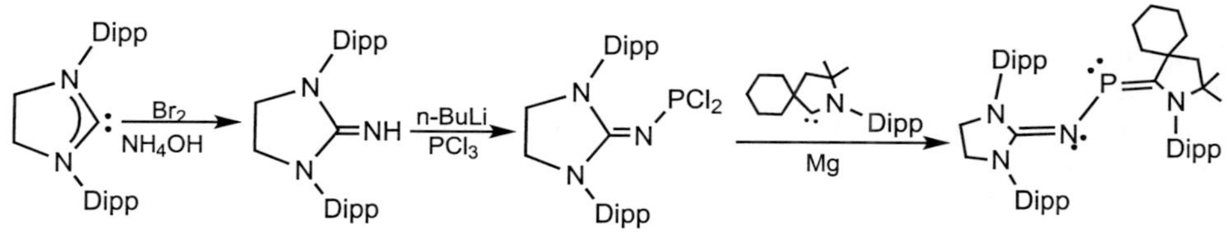

Scheme 9.16 Synthesis of the first carbene stabilised phosphorus mononitride

9.9.2 Frustrated Lewis Pairs (FLPs)

Frustrated Lewis pairs are defined as combinations of sterically hindered Lewis acid–Lewis base pairs with unquenched acidity and basicity. Such combinations or pairs have been found to be highly effective in activating small molecules such as H_2, CO_2, N_2O, SO_2, CO, NO, alkenes and alkynes.

The term, *Frustrated Lewis Pair* (FLP), was first coined by Douglas Stephan of University of Toronto in 2006 when he showed that combinations of sterically encumbered Lewis acid base pairs can be used for reversibly activating hydrogen gas. Among many uses, the area shows great promise in the development of the first examples of hydrogenation catalysts where no metal is involved.[55] Scheme 9.17 shows how Stephan and coworkers made the first example of a frustrated Lewis pair.

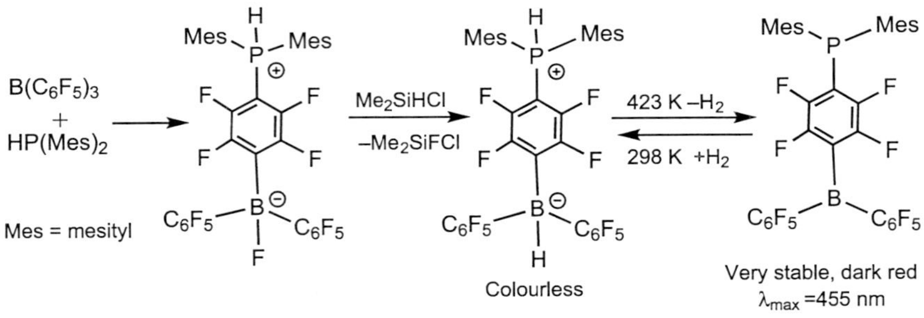

$$B(C_6F_5)_3 + HP(Mes)_2 \longrightarrow Mes_2PH^{+}\text{–}C_6F_4\text{–}BF^{-}(C_6F_5)_2 \xrightarrow[-Me_2SiFCl]{Me_2SiHCl} Mes_2PH^{+}\text{–}C_6F_4\text{–}BH^{-}(C_6F_5)_2 \underset{298\ K\ +H_2}{\overset{423\ K\ -H_2}{\rightleftharpoons}} Mes_2P\text{–}C_6F_4\text{–}B(C_6F_5)_2$$

Mes = mesityl; Colourless; Very stable, dark red, λ_{max} =455 nm

Scheme 9.17 Reactions which resulted in the discovery of the first FLP

The landmark discovery involved the reaction of $B(C_6F_5)_3$ with $HP(Mes)_2$ which resulted in a zwitterionic complex with a hydride unit on the phosphorus and fluoride on the boron. Replacement of the fluoride by hydride using Me_2SiHCl resulted in a colourless zwitterionic compound possessing a hydridic H on the phosphorus and a protic H on the boron. A dark red coloured aryl linked phosphine–borane species $Mes_2P–C_6F_4–B(C_6F_5)_2$ was obtained on heating this complex around 150°C with the release of H_2. Stephan showed that $Mes_2P–C_6F_4–B(C_6F_5)_2$ heterolytically cleaves H_2 giving back the colourless $Mes_2PH^+–C_6F_4–BH^-(C_6F_5)_2$. They had thus demonstrated the first reversible metal-free hydrogen activation. Later on, combinations of a sterically hindered phosphine base and strong Lewis acid or such a pair with a short alkyl chain connecting them, were found to be equally good in bringing about H_2 activation. Also, other molecules such as CO_2 and alkenes were activated by the FLPs.[56]

$$(t\text{-}Bu)_3P\ /\ Mes_3P + B(C_6F_5)_3 \xrightarrow{H_2,\ 25°C} [R_3PH]^+\ [HB(C_6F_5)_3]^- \quad R = t\text{-}Bu\ /\ Mes \tag{9.44}$$

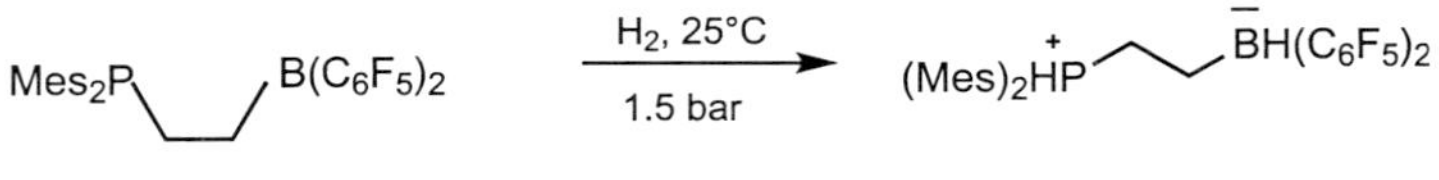

$$Mes_2P\text{–}CH_2CH_2\text{–}B(C_6F_5)_2 \xrightarrow[1.5\ bar]{H_2,\ 25°C} (Mes)_2\overset{+}{H}P\text{–}CH_2CH_2\text{–}\overset{-}{B}H(C_6F_5)_2 \tag{9.45}$$

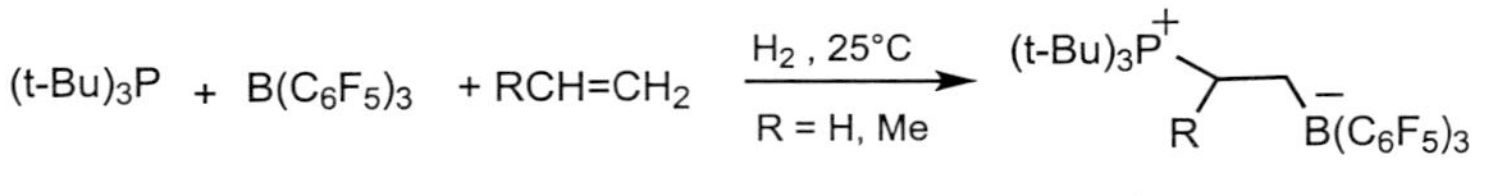

$$(t\text{-}Bu)_3P + B(C_6F_5)_3 + RCH{=}CH_2 \xrightarrow[R = H,\ Me]{H_2,\ 25°C} (t\text{-}Bu)_3\overset{+}{P}\text{–}CH_2\text{–}CH(R)\text{–}\overset{-}{B}(C_6F_5)_3 \tag{9.46}$$

Historically, the fact that certain Lewis acid base pairs will not form adducts was first observed by H C Brown in 1942, when he noticed that lutidine formed adducts with BF_3 but not with BMe_3 while pyridine did. The lutidine story was revisited in 2009 by Stephan who showed that the combination of lutidine and the strong Lewis acid $B(C_6F_5)_3$ can activate and cleave THF and the cleaved ether can form a link for the FLP (Eq. 9.47). Also Wittig in 1959, showed that 1,2-didehydrobenzene generated from 1,2-bromofluorobenzene can add PPh_3 and BPh_3 ortho to each other to give ortho-phenylene bridged phosphonium borate.[57]

N–BF3 ← (pyridine) BF3 → (2,6-lutidine) N–BF3 (9.47)

N–BMe3 ← (pyridine) BMe3 → (2,6-lutidine) N–BMe3 (9.48)

H C Brown 1942

F, Br → (Mg; PPh3, BPh3) +PPh3, –BPh3 (9.49)

Wittig, 1959

N + B(C6F5)3 → (THF) N+ ... O– B(C6F5)3 (9.50)

Stephan 2009

Fluoride ion affinity

For a detailed understanding of FLPs, one needs to know the relative Lewis acidities of various participants in such pair formation. A standard and reliable way of measuring Lewis acidity is by using the fluoride ion affinity (FIA) which indicates the strength of a Lewis acid and the energy released when it combines with a fluoride ion. FIA is defined as the negative of the enthalpy ΔH for the reaction of a Lewis acid (A) with fluoride ion (F^-) giving AF^-, all species being in the gaseous form. FIA values are obtained from quantum chemical calculations in isodesmic reactions. SbF_5 is the strongest conventional Lewis acid that is stable under normal conditions, easily accessible and used in many technical applications. Similar to Brønsted super acids, molecular Lewis acids which are stronger than monomeric SbF_5 in the gaseous phase are considered as Lewis superacids. As the FIA values suggest, $B(C_6F_5)_3$ is only a moderately strong Lewis acid but what makes it a useful candidate for FLP formation is its steric bulkiness, ease of preparation and air stability combined with this Lewis acidity.[58]

Table 9.7 Fluoride ion affinity of selected Lewis acids

Lewis acid/anion	FIA	Lewis acid/anion	FIA
$CB_{11}F_{11}$/ $CB_{11}F_{12}^-$	716	$Ga(C_6F_5)_3$ / $[FGa(C_6F_5)_3]^-$	447
$Sb(OTeF_5)_5$/ $[FSb(OTeF_5)_5]^-$	633	$B(C_6F_5)_3$ / $[FB(C_6F_5)_3]^-$	444
$B(CF_3)_3$ / $[FB(CF_3)_3]^-$	552	BBr_3 / $[FBBr_3]^-$	433
$Al(C_6F_5)_3$ / $[FAl(C_6F_5)_3]^-$	530	$GaCl_3$ / $[FGaCl_3]^-$	431
SbF_5 / $[SbF_6]^-$	489	BCl_3 / $[FBCl_3]^-$	405
$AlCl_3$ / $[FAlCl_3]^-$	457	PF_5 / $[PF_6]^-$	394
BI_3 / $[FBI_3]^-$	448	BF_3/ $[BF_4]^-$	338

Small molecule activation

Small molecules such as CO_2, SO_2, N_2O, NO, alkenes, alkynes and conjugated dienes have been activated in a manner similar to the activation of H_2 by a range of FLPs. Schemes 9.18 and 9.19 show some of the well known examples. While CO_2 activation is reversible, that of most of the other small molecules is irreversible.[59,60]

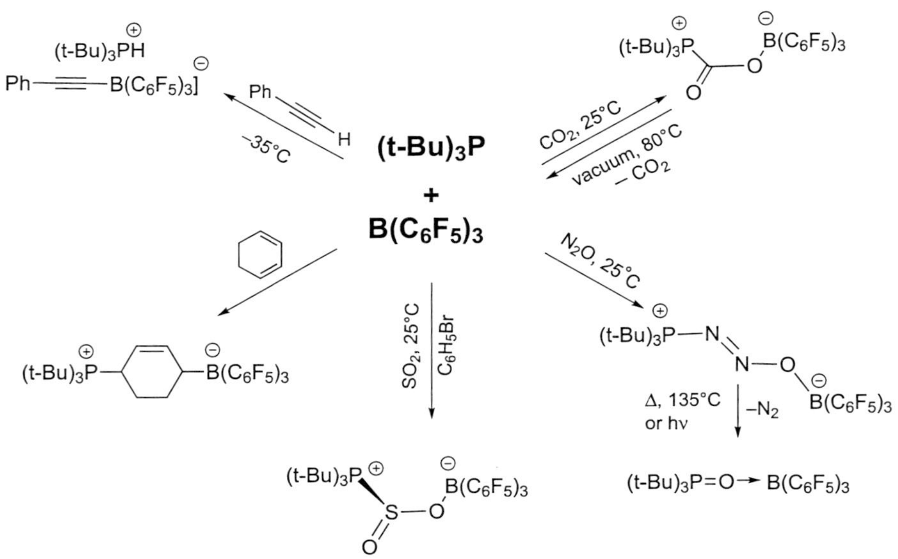

Scheme 9.18 Activation of small molecules by FLP formed by $(t\text{-Bu})_3P$ and $B(C_6F_5)_3$

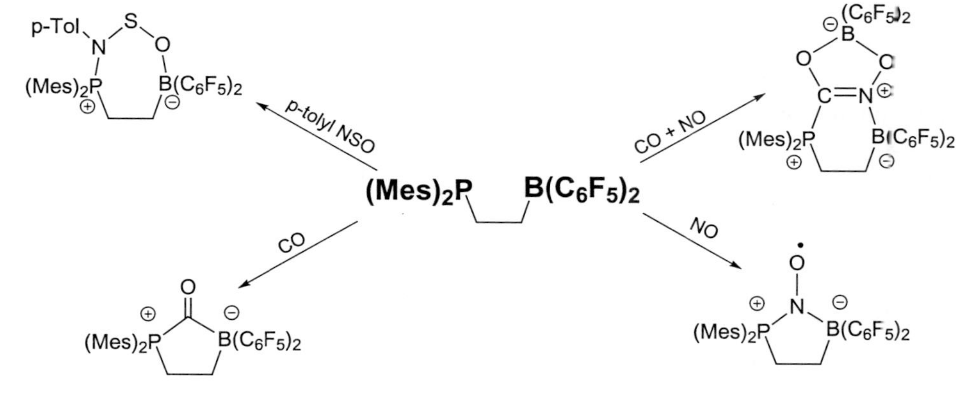

Scheme 9.19 Activation of small molecules by an ethylene bridged FLP

9.9.3 Hydrogenation: Reduction of Multiple Bonds and Aromatic Rings

The range of reducible substrates which can be reduced by FLPs include imines, enamines, olefins, alkynes, silylenol ethers, oximes, polyaromatics, ketones and aldehydes. With less basic Lewis bases, higher pressures of H_2 are required for the reduction to occur. Studies of FLP hydrogenation using calorimetry and using the first FLP $Mes_2P–C_6F_4–B(C_6F_5)_2$ gave a reaction enthalpy of −43(4) kJ/mol. Time resolved calorimetric studies using varying hydrogen pressures showed pseudo first order kinetics indicating that H_2 activation by FLP is rate limiting. Examples of hydrogenation of silyl enol ethers, ketones and aromatic hydrocarbons are given.[61]

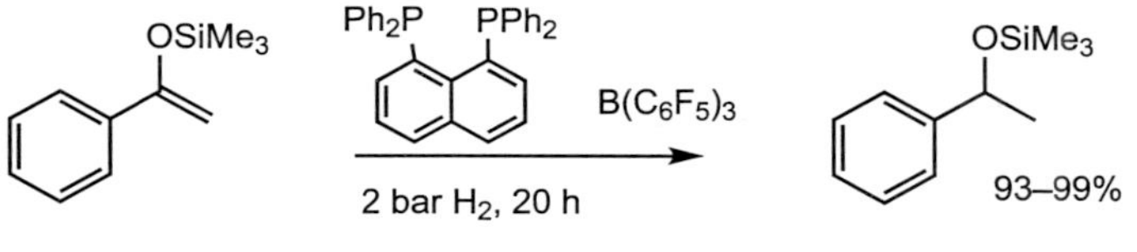

OSiMe3 Ph2P PPh2 B(C6F5)3 2 bar H2, 20 h OSiMe3 93–99% (9.51)

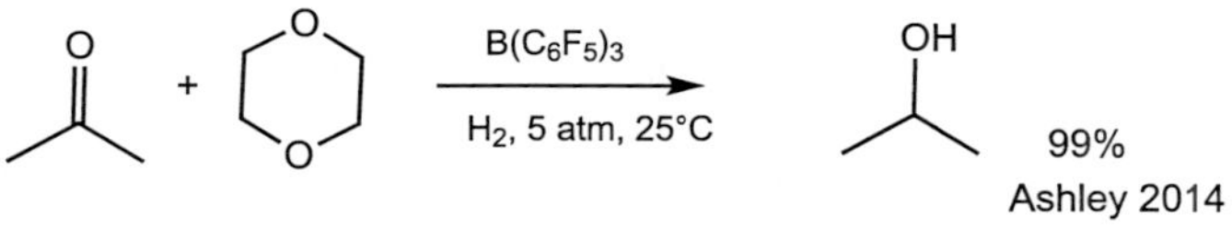

O + O O B(C6F5)3 H2, 5 atm, 25°C OH 99% Ashley 2014 (9.52)

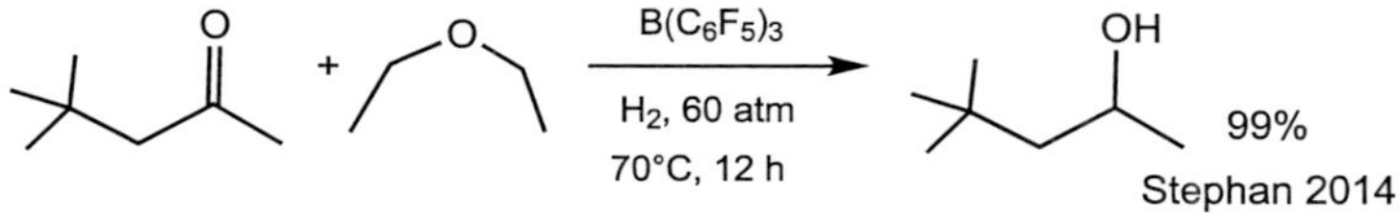

O + O B(C6F5)3 H2, 60 atm 70°C, 12 h OH 99% Stephan 2014 (9.53)

$Ph_2P(C_6F_5)$ / $B(C_6F_5)_3$

102 atm H_2, 80°C

97%

(9.54)

Reduction of aromatic rings

Aromatic secondary amines at 110°C and 4 atm hydrogen pressure have been found to undergo conversion of aryl rings bound to nitrogen to cyclohexyl rings.

R–NH, $B(C_6F_5)_3$, H_2, 4 atm, 25°C; R–$NH_2^{\oplus}$ $HB(C_6F_5)_3^{\ominus}$; 110°C, 96 h, H_2, Toluene; R–$NH_2^{\oplus}$ $HB(C_6F_5)_3^{\ominus}$

R = t-Bu, i-Pr

(9.55)

R–NH, $B(C_6F_5)_3$, H_2, 4 atm, 25°C; R–$NH_2^{\oplus}$ $HB(C_6F_5)_3^{\ominus}$; 110°C, 36–96 h, H_2, Toluene; $NH_2^{\oplus}$ $HB(C_6F_5)_3^{\ominus}$

R = Ph, Cy

(9.56)

CH_3, N; $B(C_6F_5)_3$, H_2, 4 atm, 110°C, 96 h, Toluene; CH_3, $NH_2^{\oplus}$ $HB(C_6F_5)_3^{\ominus}$

(9.57)

N, N; $B(C_6F_5)_3$, H_2, 4 atm, 110°C, 72 h, Toluene; $H_2N^{\oplus}$, $H_2N^{\oplus}$; 2 $HB(C_6F_5)_3^{\ominus}$

(9.58)

Efficiency as hydrogenation catalysts

As catalysts, FLPs have been found to be efficient for hydrogenation of imines. One of the well-studied systems involves the use of NHC borenium catalysts for metal-free hydrogenation of imines (Scheme 9.20). A maximum turn over frequency (TOF) of 940 h^{-1} has been observed for the archetypal substrate N-benzylidene-*t*-butylamine using the NHC borylenium catalyst having N-methyl and C-chloro substituents on the NHC ring. In comparison with traditional hydrogenation catalysts for the hydrogenation of *n*-hexene, Wilkinson catalyst gave a TOF of 650 h^{-1} and $RuCl(PPh_3)_3$ 9000 h^{-1}. So FLP catalysts for hydrogenation should only be viewed as a complementary hydrogenation methodology that may find applications where trace metals need to be avoided completely or metal toxicity is a critical issue and catalyst cost are very high (for example, pharmaceuticals and electronic grade materials).

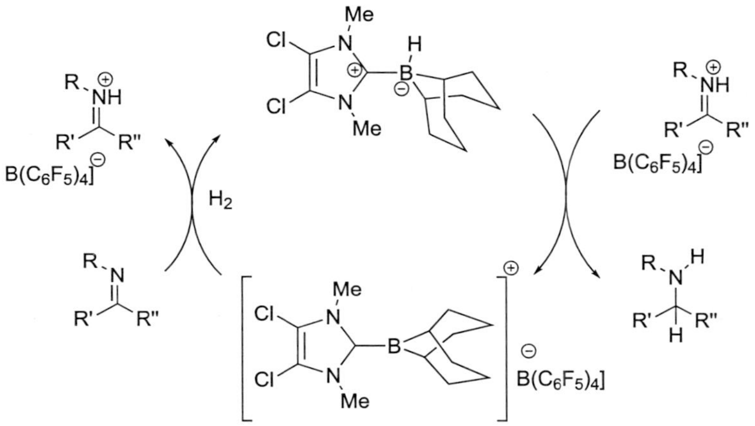
Me
Cl
N
H
B
Cl
N
Me
R
NH
R'
R''
$B(C_6F_5)_4$
H2
R
N
R'
R''
R
NH
R'
R''
$B(C_6F_5)_4$
R
N
H
R'
R''
H
Me
Cl
N
B
Cl
N
Me
$B(C_6F_5)_4$
NHC–Borenium ion catalyst

Scheme 9.20 Mechanism of activity of FLPs as hydrogenation catalysts

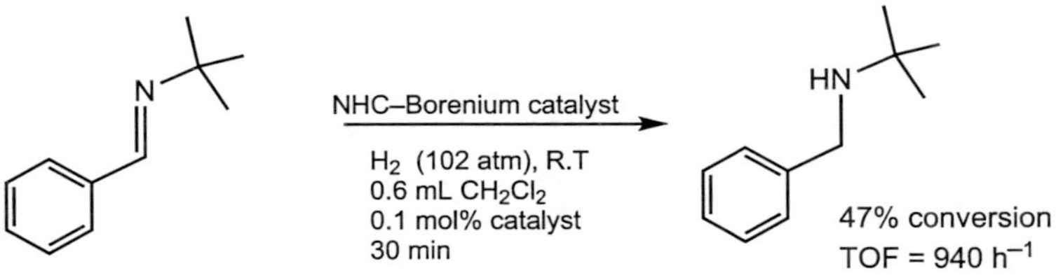
N
NHC–Borenium catalyst
H2 (102 atm), R.T
0.6 mL CH2Cl2
0.1 mol% catalyst
30 min
HN
47% conversion
TOF = 940 h−1

(9.59)

Catalytic alkyne hydrogenation to *cis* alkenes

A unique ansa-aminohydroborane was used as a FLP based catalyst for the catalytic reduction of alkynes to *cis*-alkenes. The mechanism proposed and substantiated by experimental and theoretical studies is shown in Scheme 9.2.[62]

Asymmetric hydrogenation using FLPs

Many chiral FLPs have been prepared and asymmetric hydrogenation has been attempted, however, most of them resulted in poor enantiomeric excess. Chiral amino borane FLP catalysts based on binapthyl units were prepared and asymmetric hydrogenation was attempted for imines and enamines. These gave good yields and good enantiomeric excess at room temperature and 2 bar hydrogen pressures in MTBE as solvent.

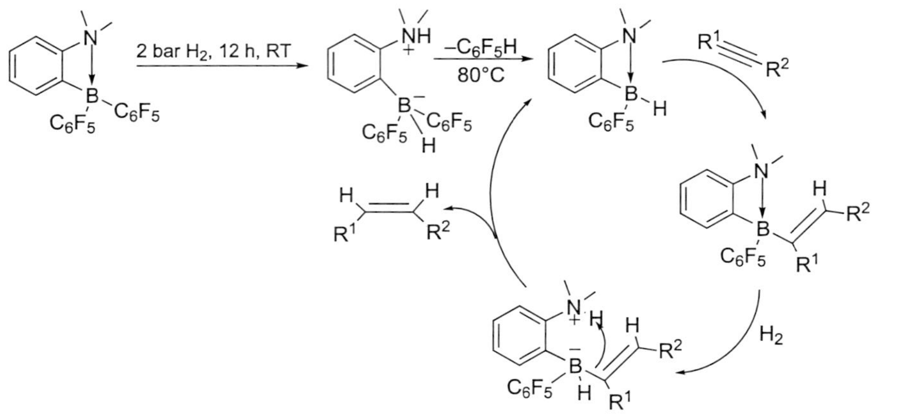

Scheme 9.21 Mechanism proposed for the catalytic alkyne hydrogenation to *cis*-alkenes

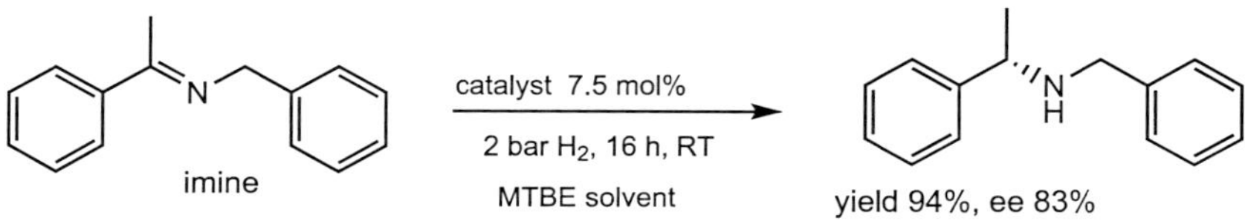

imine — catalyst 7.5 mol%, 2 bar H_2, 16 h, RT, MTBE solvent → yield 94%, ee 83% (9.60)

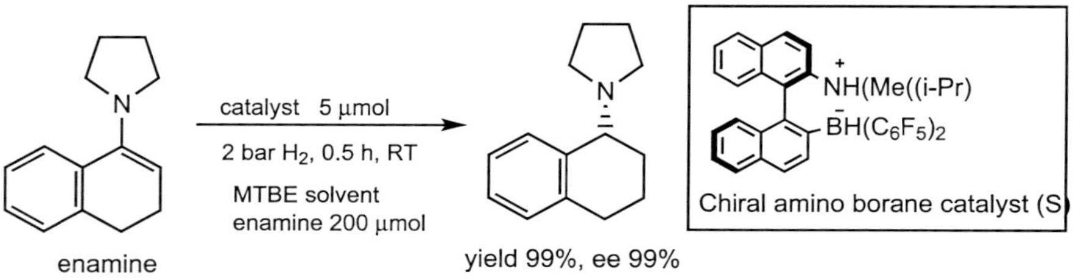

enamine — catalyst 5 µmol, 2 bar H_2, 0.5 h, RT, MTBE solvent, enamine 200 µmol → yield 99%, ee 99% (9.61)

Chiral amino borane catalyst (S): $\overset{+}{N}H(Me((i\text{-}Pr)$, $\bar{B}H(C_6F_5)_2$

9.9.4 FLPs Based on Organometallic Substrates

In the following reaction, an organometallic aryl ruthenium cationic complex having a hindered boron was reacted with the hindered base $(t\text{-Bu})_3P$. Two complexes were formed where the hydride partially saturates the η^6-bound aryl ring. The Lewis acidity of this phenyl ring is enhanced by the ancillary metal centre and therefore instead of going to the boron atom, the more Lewis acidic aryl ring is attacked by the hydride ion. The 18 electron rule of the complex also is satisfied by this partial saturation of the aryl ring.

(9.62)

Inverse FLPs

The Lewis acid components of FLPs have been relatively limited to a few strong Lewis acids such as $B(C_6F_5)_3$ while a number of Lewis base amines and phosphines have been tried out. For heterolytic activation of H_2, it appeared that the Lewis acid partner of the FLP must incorporate strongly electron withdrawing substituents. It was therefore proposed that moderate to weak Lewis acids may be utilised in FLP mediated H_2 cleavage in the presence of sufficiently strong hindered bases. This suggestion was verified by Krempner and coworkers in the design of FLPs with phosphorus-based superbases and moderate to weak Lewis acids.[63]

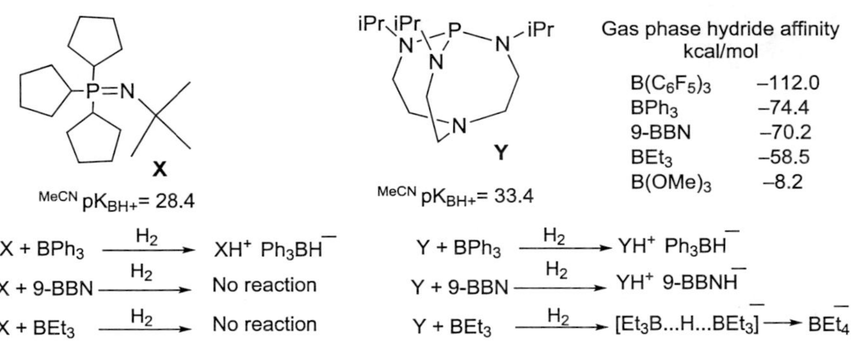

Scheme 9.22 Inverse FLPs made using superbases and their reactivity

Mechanism of FLP Action

Initial theoretical studies attempting to explain the mechanism of hydrogen activation by FLPs considered dimerisation of two FLP units in a head to tail fashion. However, studies carried out later, especially kinetic studies indicated that dimerisation of FLPs is not necessary to explain the hydrogen activation. On the basis of static density functional theory calculations, it has been shown that one mole of FLP, one mole of H_2 and one mole of hydrogenated product [PH^+BH^-] are related through only one transition state and the reaction follows a concerted reaction path for both intermolecular an intramolecular FLPs. The polarisation of H_2 occurs and after the heterolytic splitting of H_2, the proton (H^+) and the hydride (H^-) are simultaneously captured by the Lewis acid and the Lewis base.

Two different models have been proposed for explaining the polarisation of the H–H bond using FLP—the electron transfer (ET) model and the electric field (EF) model.[65]

The electron transfer model was proposed by Papai and coworkers. Since no intermediate was observed experimentally in the reaction of $(t\text{-Bu})_3P$ and $B(C_6F_5)_3$, a computation study was initiated to get a detailed atomistic input of the mechanism. A single early transition state was identified along the reaction pathway from the reactants to the products, that is, where the H_2 is closer to the P and B centres and interacts with them simultaneously. A synergistic electron transfer was proposed from the lone pair of the Lewis base to the σ^* orbital of H_2 and from the σ orbital of H_2 to the empty orbital on the Lewis acid. This tends to weaken and finally heterolytically cleave the H–H bond. Another feature of this weakening is the formation of a 'frustrated complex' or 'encounter complex' which are weak flexible adducts (as a consequence of non-covalent interactions between the bulky substituents) of the pair where the donor and acceptor centres are pre-organised for the reaction with the H_2 molecule. For intramolecular-linked FLPs, such a preorganisation of the active centres may be provided by the covalent back bone itself.

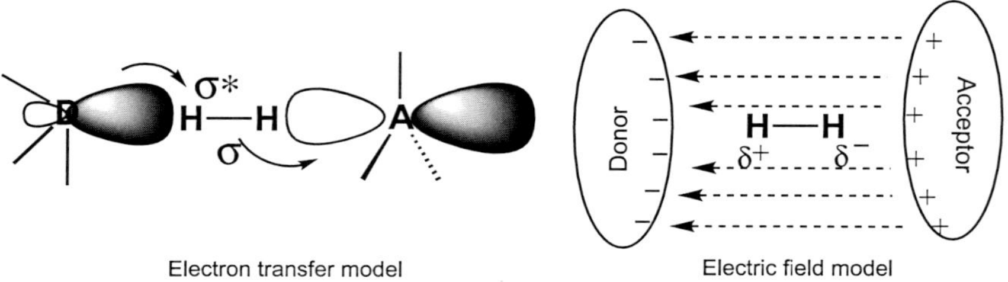

Scheme 9.23 Differences in the two models proposed for mechanism of H_2 activation by FLPs

The electric field model (EF) was proposed by Grimm and coworkers for the same $(t\text{-Bu})_3P/B(C_6F_5)_3$ system. This model is based on the statement that the frustrated complex possesses a significant electric field in its interaction. Computational studies also indicated that the H_2 molecule can be cleaved heterolytically in a barrier-less process when exposed to a strong and homogeneous electric field. From the calculations, it was found that the linear alignment of the reacting PHHB fragment inside the FLP undergoes barrier-less cleavage. As per this model, there is no requirement to consider specific FLP/H_2 orbitals or electron donation–acceptance of electrons for the activation process. According to this model, the transition state is the entry of H_2 into the cavity of the FLP and a preparation step where H_2 reaches the region of high electric field strength where its heterolytic cleavage can occur.

This simple model was however emphasised to be appropriate only for the initial part of the reaction. In short, the electric field model proposes that the heterolytic bond breaking of H_2 occurs as a result of the polarisation of the molecule by the strong electric field present in the cavity formed between the reactive Lewis acid–base centres.

Both models consider that the course of the reaction goes through intermediates consisting of pre-organised Lewis acid–base partners. However, studies by Papai and coworkers have indicated that the electric field model may be improper to understand the actual occurrences at the molecular level.[66]

Reversibility versus non reversibility of FLP hydrogenations

Vankova and coworkers considered the thermodynamic and kinetic effects of four selected reactions to understand the non-reversibility versus reversibility of hydrogenation using frustrated Lewis pairs.[67] These systems are given below.

System	Type of reaction	Overall solvent phase Gibbs energy T = 298.15 K, p = 1atm
1. t-Bu_3P + B(p-$C_6F_4H)_3$	Non reversible H_2 activation	–32.4 kJ/mol
2. Mes_3P + $B(C_6F_5)_3$	Non reversible H_2 activation	–22.1 kJ/mol
3. (o-$C_6H_4Me)_3P$ + B(p-$C_6F_4H)_3$	Reversible H_2 activation	–1.0 kJ/mol
4. $(C_6F_5)_3P$ + $B(C_6F_5)_3$	No reaction	+ 108.0 kJ/mol

The calculated solvent phase Gibbs free energies of these systems indicate that the process is exergonic (accompanied by release of energy) for systems 1 and 2, very slight exergonicity for system 3 suggesting that the competing reverse reaction (release of H_2) can also occur and high positive Gibbs energy for system 4 indicating a thermodynamically unfavourable reaction. For non-reversible H_2 activation, the kinetic barrier for the hydrogen uptake is predicted to be 20–30 kJ/mol less than the reverse reaction (H_2 release). The reversibility of system 3 can be rationalised by very similar kinetic profiles of the forward and backward reactions with H_2 at room temperature. The inertness of system 4 can be attributed to a combination of factors such as high activation barrier for H_2 uptake, the late character of the located transition state and the kinetic instability of the hydrogenated product.

Questions for self study (multiple choice)

9.1. When orthophosphoric acid is heated to 600°C, the product obtained is
(a) PH_3 (b) P_2O_5 (c) H_3PO_3 (d) $(HPO_3)_n$

9.2. Strong reducing behaviour of H_3PO_2 (hypophosphorous acid) is due to
(a) High electron gain enthalpy of phosphorus
(b) High formal oxidation state of phosphorus
(c) Presence of two OH groups and one P–H group
(d) Presence of one OH group and two P–H groups

9.3. Which of the following statements is not valid for oxoacids of phosphorus?
(a) Orthophosphoric acid is used in the manufacture of triple superphosphate fertiliser
(b) Hypophosphorus acid is a diprotic acid

(c) All oxoacids have tetrahedral four coordinate phosphorus
(d) All oxoacids contain at least one P=O unit and one P–OH group

9.4. Oxidation states of P in $H_4P_2O_5$, $H_4P_2O_6$ and $H_4P_2O_7$ are
(a) +3, +5, +4 (b) +5, +3, +4, (c) +5, +4, +3 (d) +3, +4, +5

9.5. How many bridging oxygen atoms are present in P_4O_{10}?
(a) 6 (b) 4 (c) 2 (d) 5

9.6. Among the following oxides, the most acidic oxide is
(a) As_2O_5 (b) P_2O_5 (c) N_2O_5 (d) Sb_2O_5

9.7. Among the following combinations of reactants the one which does not produce PH_3 as a product is
(a) P_4 and NaOH (b) Mg_3P_2 and H_2O
(c) PCl_3 and H_2O (d) PH_4I and KOH

9.8. The phosphorus compound present in human urine is
(a) NaH_2PO_4 (b) $NH_4H_2PO_4$ (c) KH_2PO_4 (d) $NaNH_4(H)PO_4$

9.9. The reaction which does not require PPh_3 as a reagent is
(a) Mitsunobu reaction (b) Appel reaction
(c) Wittig reaction (d) Arbusov rearrangement

9.10. Which one of the following hydrides has the lowest boiling point?
(a) NH_3 (b) PH_3 (c) AsH_3 (d) SbH_3

Solved concept-based questions

9.11. Between yellow, red and black phosphorus, which is the most stable allotrope of phosphorus and why?

Yellow phosphorus has four P units bonded in a tetrahedral arrangement while red phosphorus is a polymer of this tetrahedron where one of the P–P bonds of the tetrahedron is broken and acts as a bridge connecting the 4-atom P units. Therefore, red phosphorus has less angle strain in its P_4 unit. Yellow phosphorus fumes in air while getting converted to P_4O_{10}, a process which releases the angle strain. Black phosphorus is the most thermodynamically stable allotrope, whose structure and electrical conductivity resemble that of graphite. Black phosphorus is kinetically inert and does not ignite in air even at 670 K.

9.12. Boiling point of NH_3 is 238.5 K while that of PH_3 is only 185.5 K. Explain.

In NH_3, the lone pair of electrons occupies an sp^3 hybrid orbital which has formed from the more compact 2*s* and 2*p* orbitals, while in PH_3 it is an sp^3 hybrid orbital formed from the relatively less compact 3*s* and 3*p* orbitals. The former will have more directionality and hence stronger H bonds.

9.13. Between PCl_4^+ and PCl_4^- which is not likely to exist and why?

PCl_4^+ obeys the octet rule and is obtained by removing an electronegative chlorine from PCl_5. PCl_4^- will be a 10 electron species and its synthesis requires the removal of a Cl^+ from PCl_5, which is unlikely. PCl_3 is also has an octet, so addition of a Cl^- to it to realise PCl_4^- is going to destabilise PCl_3. Hence PCl_4^+ is more likely to exist (in fact PCl_5 exists as $PCl_4^+PCl_6^-$ in the solid state).

9.14. PCl_5 is ionic in the solid state and covalent in the gaseous state. Why?

PCl_5 exists as $PCl_4^+PCl_6^-$ in the solid state and hence it is ionic while in the gaseous phase it exists as PCl_5.

9.15. The formal oxidation state of H_3PO_2 is 1 while that of H_3PO_4 is 5. How?

Although the Pauling electronegativity scale gives very close electronegativity values to both P and H, other detailed calculations such as Mulliken–Jaffe electronegativity scale confers slightly more electronegativity to sp^3 hybridised P making it δ^- compared to H. H_3PO_2 has two direct P–H bonds while H_3PO_4 has none. Hence the H_3PO_2 has a formal +1 oxidation state.

9.16. Zinc phosphide Zn_3P_2 is a well known rat bait and poison. Why do rats get attracted to it and how does it kill them?

Tablets of zinc phosphide on slow hydrolysis liberate PH_3 gas. PH_3 has a garlic/rotten fish smell which attracts rats to the bait. Once consumed, the acidic digestive juices make the hydrolysis faster and the poisonous PH_3 released kills the rat from inside. PH_3 has been found to act in a manner similar to cyanide, affecting the functioning of cytochrome C oxidase.

9.17. Me_3P and Et_3P catch fire on exposure to air while Me_3N and Et_3N do not. Why?

Alkyl phosphines are electron rich, small and volatile liquids and unlike alkyl amines they can easily expand the phosphorus valence using 3*d* orbitals forming a very strong P=O bond in the presence of oxygen. The heat generated in this oxidation reaction is sufficient to make the molecule go beyond its auto ignition temperature, making it pyrophoric.

9.18. The toxic effect of organophosphorus pesticides is slower than organophosphorus nerve gases. Why?

The activity of acetylcholinesterase is disrupted by organophosphate units by binding in an irreversible manner. Most of the organophosphorus pesticides have P=S units which have to get converted to P=O units normally by biological oxidants such as cytochrome C oxidase. Hence there is a delay in activity.

9.19. Between PCl_5, Cl_2 and SO_2Cl_2 which is a better chlorinating agent. Why?

While all of them can carry out the same type of chlorination reactions, there are difficulties in handling and controlling stoichiometry in reactions involving Cl_2 gas. PCl_5 often leaves behind a side product, $POCl_3$, which is difficult to remove from sensitive reaction mixtures. However SO_2Cl_2 being a liquid is easy to measure and the side product is often SO_2 which is easily removed from the reaction mixture.

Problems and exercises

9.20. What is the unique structural feature shown by the four major allotropes of phosphorus?

9.21. Reaction of white phosphorus with 6 moles of Cl_2 resulted in an air sensitive liquid product A. Compound A on reaction with excess of PhMgBr resulted in a covalent solid compound B. Compound B on reaction with an equimolar amount of MeI resulted in an ionic compound C. Compound C on reaction with *n*-BuLi resulted in compound D. Compound D is also a well known reagent used in organic synthesis. Write the formulae of compounds A–D.

9.22. Between sulfuryl chloride and PCl_5, which is the more convenient oxidative chlorinating agent and why?

9.23. Given P_4, S_8, EtOH, Cl_2, NaOH and *p*-nitrophenol, suggest steps for the industrial synthesis of the pesticide parathion.

9.24. What products are obtained when $N_3P_3Cl_6$ is reacted with (a) water and (b) heated under vacuum at 250°C?

9.25. Phosphorus-based pesticides are relatively less toxic compared to phosphorus-based nerve gases. Why?

9.26. Cyclophosphazenes are excellent precursors for making dendrimers. Why?

9.27. What are the advantages of cyclophosphazenes over cyclosiloxanes?

9.28. Starting from P_4S_{10} and using the given pyridine, show the steps for the synthesis of the pesticide chlorpyriphos.

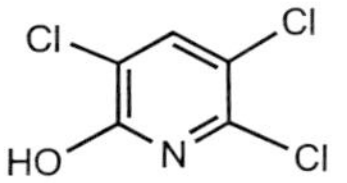

Cl, Cl, HO, N, Cl

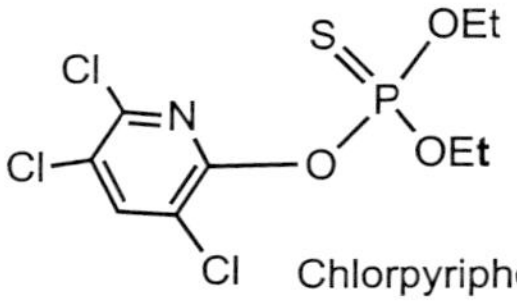

Cl, Cl, N, Cl, O, P, S, OEt, OEt

Chlorpyriphos

9.29. Define and represent $\%V_{bur}$ schematically. Which P–M distance, 2.00 or 2.28 Å, used in $\%V_{bur}$ determination has a better correlation with cone angles?

9.30. Match the phosphorus oxyacids with the property application given.

H_3PO_2	**+3 formal oxidation state of P**
H_3PO_3	Presence of P–P bond but not reducing
H_3PO_4	+2 formal oxidation state of P
$H_4P_2O_6$	Sodium tripolyphosphate synthesis
$H_4P_2O_7$	Contains peroxo bond
$H_4P_2O_4$	Electroless nickel plating
	Pyrophosphoric acid

9.31. Draw the structure of products A, B and C in the following reactions

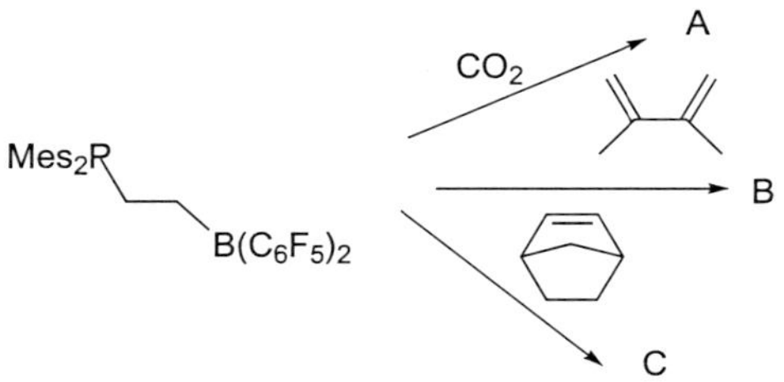

Mes_2P, $B(C_6F_5)_2$, CO_2, A, B, C

9.32. Suggest a method (with equation) to prepare carbon suboxide.

9.33. Show how the herbicide glyphosate [$(HO)_2P(=O)–CH_2–NH–CH_2–COOH)$] is prepared starting from PCl_3 and NH_3 as basic starting materials.

9.34. The reaction of white phosphorus with an excess of a group 16 element resulted in compound A. Compound A on reaction with excess of ethanol gave compound B with the evolution of a gas. Compound B on reaction with 1 mol of Cl_2 gave compound C. Compound C on reaction with the sodium salt of para-nitrophenol resulted in the formation of a well known pesticide D. Identify and draw the structures of compounds A–D.

9.35. Reaction of white phosphorus with an excess of a gaseous dihalogen resulted in compound A. Compound A on reaction with an ammonium halide resulted in two compounds B and C. B on heating at 250°C resulted in a polymer D. B and D on hydrolysis gave H_3PO_4 as one of the products. Write the structures of compounds A–D.

9.36. Arrange the following bases in the increasing order of base strength.

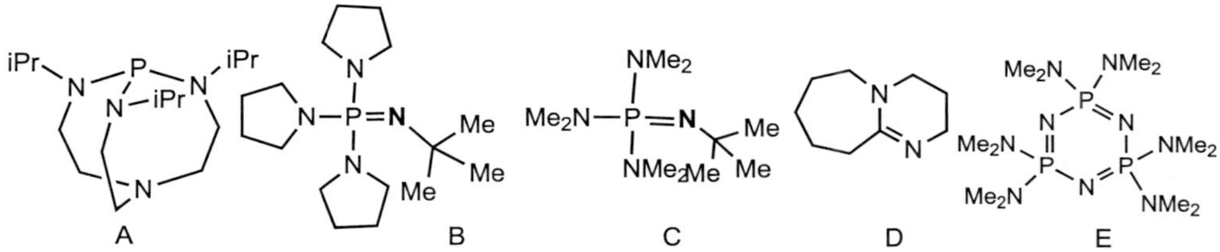

A B C D E

9.37. Suggest phosphorus-based reagents for the following conversions.

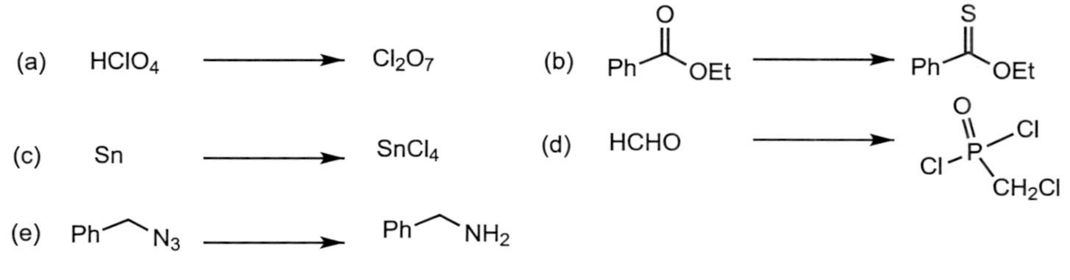

(a) $HClO_4 \longrightarrow Cl_2O_7$

(b) $PhC(=O)OEt \longrightarrow PhC(=S)OEt$

(c) $Sn \longrightarrow SnCl_4$

(d) $HCHO \longrightarrow Cl_2P(=O)CH_2Cl$

(e) $PhCH_2N_3 \longrightarrow PhCH_2NH_2$

9.38. $Al(C_6F_5)_3$ is a new compound which is used to make FLPs replacing $B(C_6F_5)_3$. Given $AlCl_3$, Cl_2, KF/HF, Br_2, Mg and Cp_2ZrH_2 show how you will make $Al(C_6F_5)_3$ starting from C_6H_6.

9.39. The reaction of white phosphorus with Cl_2 in 1:6 molar ratio resulted in a reactive liquid compound A. Compound A on Friedel–Crafts reaction with benzene (followed by removal of $AlCl_3$) gave a compound B. Compound B on reaction with SO_2Cl_2 gave compound C along with a gaseous side product. Compound C on reaction with excess of NaF gave compound D. The analysis of compounds A to C showed the presence of chlorine and their ^{31}P-NMR gave single peaks. The ^{19}F-NMR of compound D gave two well separated doublets of triplets at very low temperature. Identify compounds A to D.

9.40. Starting from hydroxyapatite, show the steps for obtaining sodium tripolyphosphate (STPP), the well known detergent additive.

Supplementary reading

1. Emsley J, *The shocking history of phosphorus. A biography of the devil's element*, London: MacMillan, 2000.
2. Betterman G, Krause W, Retired G R and Hofmann T, Phosphorus compounds, inorganic, *Ulmann's encyclopedia of industrial chemistry*, Wiley VCH, 2000, Vol. 27, 1.
3. Corbridge D E C, *Phosphorus: Chemistry, biochemistry and technology*, 6th Ed., CRC Press, 2013.
4. Gengembre, Mémoiresurun nouveau gas obtenu, par l'action des substances alkalines, sur le phosphore de Kunckel (Memoir on a new gas obtained by the action of alkaline substances on Kunckel's phosphorus), *Mémoires de mathématique et de physique*, 1783, Vol. 10, 651.
5. Thénard P, Mémoiresur les combinaisons du phosphore avec l'hydrogène (Memoir on the compounds of phosphorus with hydrogen), *Comptes rendus*, 1844, Vol. 18, 652.
6. Gokhale S D and Jolly W L, Phosphine, *Inorg. Synth.*, 1967, Vol. 9, 56.
7. Fluck E, The chemistry of phosphine, *Fortschritte der Chemischen Forschung*, 1973, Vol. 35, 1.
8. Gupta B D and Elias A J, *Basic organometallic chemistry: Concepts syntheses and applications*, CRC Press: Boca Raton, FL, 2010, Chapter 4.
9. Tolman C A, Steric effects of phosphorus ligands in organometallic chemistry and homogeneous catalysis, *Chem. Rev.*, 1977, Vol. 77, 313.
10. Tolman C, Steric effects of phosphorus ligands in organometallic chemistry and homogeneous catalysis, *Chem. Rev.*, 1977, Vol. 77, 313.
11. Bunten K A, Chen L, Fernandez A L and Poe A J, Cone angles: Tolman's and Plato's, *Coord. Chem. Rev.*, 2002, Vol. 41, 233.
12. Clavier H and Nolan S P, Per cent buried volume for phosphine and N-heterocyclic carbene ligands: steric properties in organometallic chemistry, *Chem. Commun.*, 2010, Vol. 46, 841.
13. Mitsunobu O and Yamada Y, Preparation of esters of carboxylic and phosphoric acid via quaternary phosphonium salts, *Bull. Chem. Soc., Jpn*, 1967, Vol. 40, 2380.
14. Appel R, Tertiary phosphane/tetrachloromethane, a versatile reagent for chlorination, dehydration, and P–N linkage, *Angew. Chem. Int. Ed.*, 1975, Vol. 14, 801.
15. Gololobov Y G, Sixty years of Staudinger reaction, *Tetrahedron*, 1981, Vol. 37, 437.
16. Staudinger H and Meyer J, Über neue organische phosphorverbindungen III. Phosphinmethylen derivate und phosphinimine, *Helv. Chim. Acta*, 1919, Vol. 2, 635.
17. Hoffmann R W, Wittig and his accomplishments: still relevant beyond his 100th birthday, *Angew. Chem. Int. Ed.*, 2001, Vol. 40, 1411.
18. Vedejs E and Marth C F, Mechanism of the Wittig reaction: the role of substituents at phosphorus, *J. Am. Chem. Soc.*, 1988, Vol. 110, 3948.
19. Wittig G and Schöllkopf U, Über triphenyl-phosphin-methylene also lefinbildendereagenzien I, *Chem. Ber.*, 1954, Vol. 87, 1318.
20. Bhattacharya A K and Thyagarajan G, Michaelis–Arbuzov rearrangement, *Chem. Rev.*, 1981, Vol. 81, 415.
21. Arbuzov B A, Michaelis–Arbusow- und Perkow-Reaktionen, *Pure Appl. Chem.*, 1964, Vol. 9, 307.
22. Pöttgen R, Hönle W and Schnering H G V, Phosphides: Solid state chemistry, *Encyclopedia of Inorganic and Bioinorganic Chemistry*, John Wiley and Sons, 2011.
23. Ozturk T, Ertas E and Mert O, A Berzelius reagent, phosphorus decasulfide (P_4S_{10}), in organic syntheses, *Chem. Rev.*, 2010, Vol. 110, 3419.
24. Fletcher J H, Hamilton J C, Hechenbleikner I, Hoegberg E I, Sertl BJ and Cassaday J T, The synthesis of parathion and some closely related compounds, *J. Am. Chem. Soc.*, 1950, Vol. 72, 2461.

25. Duke SO and Powles SB, Glyphosate: a once-in-a-century herbicide: Mini-review, *Pest Management Science*, 2008, Vol. 64, 319.
26. Zhou J, Li J, An R, Yuan H and Yu F, Study on a new synthesis approach of glyphosate, *J. Agric. Food Chem.*, 2012, Vol. 60, 6279.
27. Maxson R N, Phosphorus pentachloride, *Inorg. Synth.*, 1939, Vol. 1, 99.
28. Burks Jr. J E, Phosphorus (V) chloride, *Encyclopedia of Reagents for Organic Synthesis* (Ed: L Paquette), J. Wiley & Sons, New York, 2004.
29. Vaccari D A, Phosphorus, A looming crisis, *Scientific American*, 2009, Vol. 300, 54.
30. Cordell D, Drangerta J O and White S, The story of phosphorus, global food scarcity and food for thought, *Global Env. Change*, 2009, Vol. 19, 292.
31. Holleman A F and Wiberg E, *Inorganic chemistry*, San Diego: Academic Press, 2001, 715.
32. Davies D R and Corbridge D E C, The crystal structure of sodium triphosphate, $Na_5P_3O_{10}$, phase II, *Acta Cryst.*, 1958, Vol. 11, 315.
33. Schrödter K, Bettermann G, Staffel T, Wahl F, Klein T and Hofmann T, Phosphoric acid and phosphates, *Ullmann's Encyclopedia of Industrial Chemistry,* Wiley-VCH, Weinheim, 2008.
34. Lohmann K, Über die Pyrophosphat fraktionim Muskel (On the pyrophosphate fraction in muscle), *Naturwissenschaften*, 1929, Vol. 17, 624.
35. Fiske C H and Subbarow Y, Phosphorus compounds of muscle and liver, *Science*, 1929, Vol. 70, 381.
36. Langen P and Hucho F, Karl Lohmann and the discovery of ATP, *Angew. Chem. Int. Ed.*, 2008, Vol. 47, 1824.
37. Gowariker V, Krishnamurthy V N, Gowariker S, Dhanorkar M and Paranjape K, *The fertiliser encyclopedia*, John Wiley & Sons, 2009.
38. Allcock H R, *Phosphorus–nitrogen compounds: cyclic, linear, and high polymeric systems*, Academic Press, New York, 1972.
39. Schwesinger R and Schlemper H, Peralkylated polyamino phosphazenes—extremely strong, neutral nitrogen bases, *Angew. Chem. Int. Ed.*, 1987, Vol. 26, 1167.
40. Kanazawa C, Goto K and Terada M, Phosphazene base-catalysed intramolecular cyclisation for efficient synthesis of benzofurans via carbon–carbon bond formation, *Chem. Commun.*, 2009, 5248.
41. Wiedermann L, Pfukusa R and Klumperman B, Phosphazene base promoted anionic polymerisation of *n*- butyraldehyde, *Eur. Polym. J.*, 2017, 97.
42. You I and Verkade J.G, A general method for the α-arylation of nitriles with aryl chlorides, *Angew. Chem. Int. Ed.*, 2003, Vol. 42, 5051.
43. Chandrasekhar V, *Inorganic rings and polymers*, Springer, 2005.
44. Gleria M and Jaeger R D, *Phosphazenes-A worldwide insight*, Nova Science, New York, 2004.
45. Chandrasekhar V and Krishnan V, Advances in the chemistry of chlorocyclophosphazenes, *Adv. Inorg. Chem.*, 2002, Vol. 53, 159.
46. Elias A J and Shreeve J M, Perfluorinated cyclic phosphazenes, *Adv. Inorg. Chem.*, 2001, Vol. 52, 335.
47. Allen C W, Regio- and stereochemical control in substitution reactions of cyclophosphazenes, *Chem. Rev.*, 1991, Vol. 91, 119.
48. Mark J E, West R and Allcock H R, *Inorganic polymers*, Prentice hall, Englewood Cliffs, 1992.
49. Ellison D H, *Handbook of chemical and biological warfare agents*, New York: CRC Press, 2007.
50. Quinn DM, Acetylcholinesterase: enzyme structure, reaction dynamics, and virtual transition states, *Chem. Rev.*, 1987, Vol. 87, 955.
51. Wang Y Z, Xie Y M, Wei P R, King R B, Schaefer H F, Schleyer P V and Robinson G H, Carbene-stabilized diphosphorus, *J. Am. Chem. Soc.*, 2008, Vol. 130, 14970.

52. Back O, Kuchenbeiser G, Donnadieu, B and Bertrand G, Non-metal mediated fragmentation of P_4. Isolation of P_1 and P_2 bis-carbene adducts, *Angew. Chem. Int. Ed.*, 2009, Vol. 48, 5530.
53. Masuda J D, Schoeller, W W, Donnadieu B and Bertrand G, Carbene activation of P_4 and subsequent derivatization, *Angew. Chem. Int. Ed.*, 2007, Vol. 46, 7052.
54. Kino R, Donnadieu B and Bertrand G, Isolation of a carbene-stabilized phosphorus mononitride and its radical cation (PN+). *Angew. Chem. Int. Ed.* 2010, Vol. 49, 5930.
55. Stephan DW, Frustrated Lewis pairs, *J. Am. Chem. Soc.*, 2015, Vol. 137, 10018.
56. Weicker S and Stephan D W, Main group Lewis acids in frustrated Lewis pair chemistry: beyond electrophilic boranes, *Bull. Chem. Soc. Jap.*, 2015, Vol. 88, 1003.
57. Stephan D W and Erker G, Frustrated Lewis pair chemistry: Development and perspectives, *Angew. Chem. Int. Ed.*, 2015, Vol. 54, 6400.
58. Mueller L O, Himmel D, Stauffer J, Sein field G, Slattery J, Quinones G S, Brecht V and Crossing I, Simple access to the non-oxidising Lewis superacid PhF–Al(OR*)$_3$ (R* = C(CF$_3$)$_3$, *Angew Chem. Int. Ed.,* 2008, Vol. 47, 7659.
59. Stephan D W, Frustrated Lewis pairs: from concept to catalysis, *Acc. Chem. Res.*, 2015, Vol. 48, 306.
60. Stephan D W, The broadening reach of frustrated Lewis pair chemistry, *Science*, 2016, Vol. 354, 1248.
61. Stephan D W, Metal-free hydrogen activation, *Chem. Commun.*, 2010, Vol. 46, 8526.
62. Chernichenko K, Madarasz A, Papai I, Nieger M, Leskela M and Repo T, A frustrated Lewis pair approach to catalytic reduction of alkynes to *cis*-alkenes, *Nature Chem.*, 2013, Vol. 5, 718.
63. Mummadi S, Unruh D K, Zhao J, Li S and Krempner, C, Inverse frustrated Lewis pairs—activation of dihydrogen with organosuperbases and moderate to weak Lewis acids, *J. Am. Chem. Soc.*, 2016, Vol. 138, 3286.
64. Mukherjee S and Thilagar P, Frustrated Lewis pairs: enabling via inability, *Resonance,* 2014, 1017.
65. Mukherjee S and Thilagar P, Frustrated Lewis Pairs: Design and reactivity, *J. Chem. Sci.,* 2015, Vol. 125, 241.
66. Rokob T A, Bako I, Stirling A, Hamza A and Papai I, Reactivity Models of Hydrogen Activation by Frustrated Lewis Pairs: Synergistic Electron Transfers or Polarisation by Electric Field?, *J. Am. Chem. Soc.*, 2013, Vol.135, 4425.
67. Zeonjuk L L, Vankova N, Mavrandonakis A, Heine T, Roschenthaler G-V and Eicher J, On the mechanism of hydrogen activation by frustrated Lewis pairs, *Chem. Eur. J.*, 2013, Vol., 19, 17413.

THE CHEMISTRY OF ARSENIC, ANTIMONY AND BISMUTH

CHAPTER 10

UNIQUE PROPERTIES OF ARSENIC AND ITS COMPOUNDS

- *Arsenic compounds, especially trivalent compounds, are highly toxic. They bind with SH groups of proteins disturbing their function and activity. Arsenic shows oxidation states of +5, +3, 0 and –3.*
- *Arsenic trioxide As_2O_3 and arsenic acid H_3AsO_4 are structurally similar to their phosphorus analogues while the structure of As_2O_5 is different compared to P_2O_5.*
- *The main application of arsenic trioxide is in the making the wood preservative CCA (chromated copper arsenate), discovered and patented by the Indian scientist S Kamesam in 1933 which is still in use as no replacement chemical is available.*
- *High purity arsenic is used in the manufacture of semiconductors (such as gallium arsenide). $AsCl_3$ and AsH_3 are used as precursors in different methods for making gallium arsenide and indium arsenide which are used in the fabrication of solar cells, lasers, LEDs and infrared detectors.*
- *Salvarsan, discovered by Paul Ehrlich in 1909, was one of the very early medications for sleeping sickness (trypanosomiasis) and syphilis (spirochacte bacterium).*
- *Paris green [copper(II) acetate triarsenite or copper(II) acetoarsenite] is a highly toxic emerald-green crystalline powder that has been used as a rodenticide and insecticide and also as a pigment in paintings and green wall papers, despite its toxicity. Scheele's green and Schweinfurth's green, two well known pigments, also contain arsenates.*
- *Realgar, As_4S_4 is present in red glass from ancient times. On a large scale, red glass has been made from arsenic ores and sulfur ores.*
- *Arsenic pentafluoride is a strong Lewis acid.*

UNIQUE PROPERTIES OF ANTIMONY AND ITS COMPOUNDS

- *Antimony is used in the electronics industry to make semiconductor devices, such as infrared detectors and diodes.*

- *It is alloyed with lead or other metals to improve their hardness and strength. The grids of lead–acid batteries are usually made of lead–antimony alloys containing 5–11 wt% of antimony. During charging, this procedure reduces the production of unwanted hydrogen and improves charging characteristics. The necessary mechanical strength and castability are easily achieved with the addition of antimony. Other uses of antimony alloys include type metal (in printing presses), bullets and cable sheathing.*
- *Pure lead was used to cast hollow bullets during the American civil war along with black gun powder. Later development of higher velocity smokeless powder loadings of the gun caused the lead to melt and get torn off from soft bullets to get deposited in the barrel after firing. Adding antimony to lead solved the problem. Linotype metal is a eutectic alloy of 3% tin, 12% antimony, and 85% lead used for casting modern bullets.*
- *SbF_3 is one of the first fluorinating agents in organic chemistry (Swarts reagent). This application was discovered by the Belgian chemist Frédéric Jean Edmond Swarts in 1892, who demonstrated its usefulness. The method involved treatment with antimony trifluoride with chlorine or with antimony pentachloride to give the active species antimony trifluorodichloride ($SbCl_2F_3$) which converts organochlorine compounds to organofluorine compounds.*
- *Antimony compounds are used in the preparation of glass, enamels, paints and pottery.*
- *SbF_5 is a strong Lewis acid, and forms a very stable anion $[SbF_6]^-$ with the fluoride ion. Hexafluoroantimonate $[SbF_6]^-$, is also a weakly coordinating anion.*
- *Fluoroantimonic acid $H_2F[SbF_6]$ is prepared by reacting anhydrous HF with SbF_5 in a stoichiometric ratio of 2:1. It is the strongest known superacid, and even gives pentacoordinate carbocations by protonation of hydrocarbons.*
- *Antimony compounds, especially Sb_2O_3, are used for making flame retardant materials, in combination with halogenated flame retardants except for polymers having halogens. The flame retarding effect of antimony trioxide is a consequence of the formation of halogenated, especially brominated compounds, which have the ability to react with free radicals of oxygen, hydrogen and OH radicals, thus helping to prevent the spreading of fire.*

Unique Properties of Bismuth and its Compounds

- *Bismuth has unusually low toxicity for a heavy metal. For this reason, there is increasing use of bismuth-based alloys (almost a third of bismuth production) as an alternative for lead.*
- *Among metals, it has one of the lowest values of thermal conductivity and the highest Hall coefficient (Hall coefficient is the ratio of the induced electric field to the product of the current density and the applied magnetic field). Only liquid mercury has a lower thermal conductivity. It has a high electrical resistivity.*

- *Bismuth compensates for the contraction of the other alloying components and therefore is a component of low-melting typesetting alloys. Isostatic bismuth–lead eutectic alloy is an example.*
- *To meet the 'lead-free' mandates imposed in some countries (beginning 2014) bismuth serves as a substitute, especially for lead, in valves used for drinking water systems.*
- *Bismuth subsalicylate is a well known component of some antidiarrheal drugs. It is the active ingredient of Pepto-Bismol and Kaopectate (for treating heartburn, upset stomach, indigestion, nausea, diarrhea, and so on). However, its use in the treatment of gastrointestinal diseases is decreasing due to increased awareness of toxicity when used above a limit.*
- *Bismuth subsalicylate along with bismuth subcitrate is used in the treatment against the bacteria causing peptic ulcers.*
- *The difference in density between lead (density 11.32 g·cm^{-3}) and bismuth (density 9.78 g·cm^{-3}) is relatively small and therefore bismuth can substitute for lead in many ballistics and weighting applications. Bismuth is replacing lead as a dense material in bullets and shots and gun ammunition.*
- *It was discovered in 2003 that bismuth-209, its only naturally occurring isotope, is very weakly radioactive. It decays by α particle emission and its half life is more than a billion times the approximate age of our universe.*
- *Bismuth telluride shows Peltier effect, a thermoelectric effect resulting in substantial lowering of temperature. Bi_2Te_3 has therefore been used for the manufacture of thermoelectric modules present in mobile refrigerators and air conditioners, cooling units for computer CPUs, and infrared spectrophotometer detectors.*
- *Similar to germanium and gallium, on solidification, the volume of bismuth increases. This increase is by about 3.3%.*
- *Bismuth is diamagnetic with a specific susceptibility of 16109 m^3/kg at 293 K.*
- *Bismuth has one of the lowest absorption cross-sections for thermal neutrons (with the exception of highly toxic beryllium).*
- *BiF_5 is a strong fluorinating agent.*

10.1 About the Elements

10.1.1 Arsenic

Arsenic is a brittle silver-grey metal whose surface gets tarnished when exposed to moist air. The naturally occurring arsenic sulfides, orpiment (As_2S_3) and realgar (As_4S_4) were known very early and were mentioned by Aristotle (384–322 BCE). The Greek name 'arsenikon' means fearless or brave which was assigned to the arsenic sulfides because of their reactivity with metals.

Arsenic oxide like sulfur, was considered as a basic constituent of metals. Albertus Magnus, a catholic Bishop (Saint Albert the great) apparently made free arsenic in 1250. The alchemists knew the poisonous nature of arsenic oxide (As_2O_3, called white arsenic). During the renaissance period, Paracelsus introduced arsenic compounds in medicine. Scheele in 1775 discovered arsenic acid and arsine, AsH_3. Arsenic burns with a bluish-white flame when heated in air, and forms dense fumes of arsenic trioxide. It also gives off a characteristic garlic-like odour when heated.[1]

Due to the high toxicity, the demand for arsenic and its compounds have decreased drastically over the years. Currently its main use is in wood preservatives such as chromated copper arsenate [where copper arsenate $Cu_3(AsO_4)_2$ is used] and for semiconductor and solar cell applications involving group III–group V binary compounds such as gallium arsenide or indium arsenide. For the latter use, pure As_4, $AsCl_3$ or AsH_3 has been used. Arsenic oxide (As_2O_3) is obtained in the pure form and it is reduced by carbon to get pure arsenic.

Arsenic occurs in three common allotropes—yellow, grey and black forms. Yellow arsenic As_4 has a structure similar to that of white phosphorus, P_4. It is a volatile, waxy solid and is the most toxic. This thermally and photochemically unstable allotrope is produced by rapid cooling of arsenic vapour. Black arsenic has a structure similar to that of black phosphorus and is obtained by cooling arsenic vapour at around 100–200°C. Grey arsenic, which is the most stable form of arsenic, has a double layered structure made of six-membered rings which are interlocked. It is brittle and has low hardness.

Table 10.1 Physical properties of arsenic

Arsenic	*Atomic mass*	*Electronegativity (Pauling scale)*	*1st ionisation energy*	*Electronic configuration*	*van der Waals radius*
Symbol: As	74.921 g.mol^{-1}	2.18	947 kJ.mol^{-1}	[Ar]$3d^{10}4s^24p^3$	0.185 nm
At. no. 33	*Atomic radius*	*Melting point*	*Boiling point*	*Electron affinity*	*Density*
	1.19 Å	816.8°C	613°C	77.65 kJ.mol^{-1}	5.22 g.cm^{-3} at m.p.

Table 10.2 Physical properties of antimony

Antimony	*Atomic mass*	*Electronegativity (Pauling scale)*	*1st ionisation energy*	*Electronic configuration*	*van der Waals radius*
Symbol: Sb	121.760g.mol^{-1}	2.05	834 kJ.mol^{-1}	[Kr]$4d^{10}5s^25p^3$	0.206 nm
At. no. 51	*Atomic radius*	*Melting point*	*Boiling point*	*Electron affinity*	*Density*
	1.40 Å	630.6°C	1635°C	101.06 kJ.mol^{-1}	6.53 g.cm^{-3} at m.p.

Table 10.3 Physical properties of bismuth

Bismuth Symbol: Bi At. no. 83	*Atomic mass*	*Electronegativity (Pauling scale)*	*1st ionisation energy*	*Electronic configuration*	*van der Waals radius*
	208.980 g.mol^{-1}	2.02	703 kJ.mol^{-1}	[Xe]$4f^{14}5d^{10}6s^{2}6p^{3}$	0.207 nm
	Atomic radius	*Melting point*	*Boiling point*	*Electron affinity*	*Density*
	1.56 Å	271.5°C	1546°C	90.93 kJ.mol^{-1}	10.05 g.cm^{-3} at m.p.

10.1.2 Antimony

Antimony is a white lustrous metal and is brittle. When it solidifies slowly, it has a foliated structure and when it solidifies rapidly, it has a granular structure. Compounds of antimony were used in cosmetics and medicine from ancient times and the Arabic name for it is *kohl* (referring to its use in beautifying eyes). Powdered Sb_2S_3 was also a cosmetic and ophthalmic ointment. The Italian metallurgist Vannoccio Biringuccio discovered this element in 1540 although there was some confusion about it probably being thought of as lead. The only stable allotrope is metallic antimony. Pure antimony is not affected by air, moisture or pure water. When heated to redness in air, it ignites. As_2O_3 is formed when liquid antimony is oxidised by steam above 750°C, accompanied by hydrogen evolution.[2]

The most well known application for metallic antimony is in lead–acid batteries where an alloy of it with lead and tin is used as the lead antimony plates. The presence of antimony in such alloys of lead and tin has improved properties for making bullets, solders and bearings. It is used as an additive in brominated fire retardants.

10.1.3 Bismuth

Bismuth, known since ancient times, was often confused with lead and tin. While no single person is credited with its discovery, it wasn't until 1753 that Claude Geoffroy, a French chemist discovered that it was a separate element and proved it to be distinct from lead. Naturally occurring forms of bismuth include the sulfide (Bi_2S_3), oxide (Bi_2O_3) and sometimes carbonate [$(BiO)_2CO_3$]. It is industrially also obtained as a byproduct in the ore processing and purification of lead, zinc and copper. Between +3 and +5 oxidation states, the +3 state is the most common and the most stable form, a consequence of the inert pair effect.

The largest deposits of bismuth in the world are found in South America, mainly Bolivia. It is usually obtained as a byproduct of mining and refining lead and coinage metals. Bismuth metal is brittle and is 86% as dense as lead and it is usually mixed with other metals to make it useful. It is naturally diamagnetic and possesses one of the lowest thermal conductivity values among metals. Bismuth alloys with tin or cadmium have low melting points and are used for temperature sensing in electric fuses and solders, automatic water sprinklers which should work when temperature increases, electric fuses and safety devices in fire detection systems. For example, the eutectic alloy In(19.1)–Cd(5.3)–Pb(22.6)–Sn(8.3)–Bi(44.7) melts at 47°C, a convenient temperature to detect excessive heating in equipment. Bi–Cd–Pb–Sn is another low melting alloy (m.p. 70°C) and used in automotive and aviation parts. Bismuth's unusual propensity to expand upon freezing makes it useful in applications such as in casting of printing type. Since it has unusually low toxicity for a heavy metal, there is an increasing use of bismuth alloys (presently about a third of bismuth production) as a replacement for lead.[3]

Bismuth oxide and bismuth(III) chlorideoxide (BiClO) are used in cosmetics since the times of ancient Egypt. BiClO is part of the pearly pigment found in eye shadows, nail polishes and hair sprays. Bismuth subcitrate and salicylate are used in medicines prescribed for stomach related problems. Experiments carried out in 2002 showed that bismuth-209 is not stable as thought to be and undergoes radioactive decay to thallium-205 through alpha particle decay. The half-life of bismuth-209 has been roughly estimated as 1.9×10^{19} years.

10.2 Isolating the Pure Element

10.2.1 Arsenic

The major ores of arsenic are:

Arsenopyrite (FeAsS) (Arsenic 46%)
Lollingite ($FeAs_2$) (Arsenic 73%)
Orpiment (As_2S_3) (Arsenic 61%)
Realgar (As_4S_4) (Arsenic 70%)

During roasting or smelting of the ores, the volatility of arsenic sulfides and oxides make them remain in the gas phase. After roasting, As_2O_3 with a melting point of 312°C and boiling point of 465°C is obtained.

The oxide is reduced with carbon to get commercial grade arsenic.

$$2\,As_2O_3 + 3C \longrightarrow As_4 + 3\,CO_2 \qquad (10.1)$$

As_4 produced by the above reaction from As_2O_3 contains >99% arsenic. This is useful for the preparation of alloys. High-purity arsenic required for semiconductor and solar cell applications (often as GaAs) is produced from this arsenic by the following method. As_4, which is commercial arsenic is sublimed and reacted with O_2 to make pure As_2O_3. After recrystallisation and sublimation in hydrogen stream, the oxide is reduced to arsenic metal with hydrogen gas at 800–850°C. The arsenic made by this procedure has a purity > 99.999%.

Another process which gives arsenic of >99.99% purity involves the reaction of As_2O_3 with HCl to give arsenic trichloride, $AsCl_3$. Purification of $AsCl_3$ is carried out by fractional distillation (b.p. 130.2°C). Reduction with hydrogen and condensation of the metal vapour gives pure As_4.

$$As_2O_3 + 6\,HCl \longrightarrow 2\,AsCl_3 + 3\,H_2O \qquad (10.2)$$

$$4\ AsCl_3 + 6\ H_2 \xrightarrow{800\text{–}850°C} As_4 + 6\ HCl \qquad (10.3)$$

10.2.2 Antimony

The most important ore of antimony is Sb_2S_3, stibnite (also known by the names antimony glance or antimonite). Sb_2S_3 contains 71.7% antimony and occurs as black crystals. Other ores include the antimony oxides – valentinite (Sb_2O_3), senarmontite (Sb_2O_3), cervantite, (Sb_2O_4), antimony hydroxides (stibiconite) – and double sulfide jamesonite (Sb_2S_3).

Stibnite has a relatively low melting point (546–548°C), which allows it to be separated from the gangue by melting. Antimony metal is extracted from its ore by pyrometallurgical techniques. Either antimony(III) sulfide is converted to its oxide, followed by reduction, or the ore is partially roasted and then allowed to react with Sb_2S_3 to form the metal and sulfur dioxide (self reduction method).

$$2\ Sb_2S_3 + 9\ O_2 \xrightarrow{290\text{–}340°C} 2\ Sb_2O_3 + 6\ SO_2 \qquad (10.4)$$

If excess of oxygen is present or is made available, antimony(IV) oxide is also formed.

$$Sb_2S_3 + 5\ O_2 \longrightarrow Sb_2O_4 + 3\ SO_2 \qquad (10.5)$$

However, during roasting, antimony(IV) oxide can react with antimony(III) sulfide to give antimony(III) oxide.

$$Sb_2S_3 + 9\ Sb_2O_4 \longrightarrow 10\ Sb_2O_3 + 3\ SO_2 \qquad (10.6)$$

Reduction to the metal is carried out by the reaction with carbon whose requirement depends on the composition of the oxide. Pulverised charcoal or anthracite dust (a hard, compact variety of coal) can also be used for reduction.

Self reduction method has also been utilised for making pure antimony. Antimony(III) sulfide, by itself or mixed with antimony oxide, is smelted in the blast furnace. The sulfide gets oxidised and the oxide formed reacts with unconverted sulfide to yield metallic antimony with the liberation of sulfur dioxide.

$$2\ Sb_2S_3 + 9\ O_2 \longrightarrow 2\ Sb_2O_3 + 6\ SO_2 \qquad (10.7)$$

$$Sb_2S_3 + 2\ Sb_2O_3 \longrightarrow 6\ Sb + 3\ SO_2 \qquad (10.8)$$

The overall equation for self reduction method can be given as:

$$3\ Sb_2S_3 + 9\ O_2 \longrightarrow 6\ Sb + 9\ SO_2 \qquad (10.9)$$

Pure electronic grade antimony needed for semiconductor and LED applications is obtained by further purification. Antimony of 99.8% purity is dissolved in hydrochloric acid, and the antimony trichloride formed (b.p. 220.3°C) is repeatedly distilled to remove arsenic impurities. The purified antimony trichloride is reduced with $Fe(CO)_5$. Zone refining gives a product of 99.999% purity. Another process uses hydrogen gas as the reducing agent to reduce antimony pentachloride (b.p. 140°C or 92°C at 30 mm Hg) that has been purified by repeated distillation in an atmosphere of argon.[2]

10.2.3 Bismuth

In the earth's crust, bismuth is about twice as abundant as gold. Earlier, bismuth was extracted principally from native ores of elemental bismuth by the process of liquation. The ore was heated in inclined cylindrical furnaces and the molten bismuth allowed to flow away from the gangue and afterwards refined.

The ores from Bolivia which form the major ore of bismuth (basically Bi_2S_3) is calcined in reverberatory furnaces and subsequently smelted with the addition of about 3% of coal, lime, sodium carbonate and fluorspar. The lowest layer of the molten mass is principally metallic bismuth which is separated. Sulfur-based ores are smelted with metallic iron. Oxide and carbonate ores are treated by smelting with carbon. Carbon reduces any oxide, either present originally in the ore or produced in the calcination, and the iron combining with the sulfur of the bismutite.

$$2\,Bi_2S_3 + 9\,O_2 \longrightarrow 2\,Bi_2O_3 + 6\,SO_2 \qquad (10.10)$$

$$Bi_2O_3 + 3\,C \longrightarrow 2\,Bi + 3\,CO \qquad (10.11)$$

In the wet process, the ores are dissolved in hydrochloric acid, the solution of metallic chlorides so obtained is precipitated by iron, the metallic bismuth filtered, washed with water, and finally fused in graphite crucibles, the surface layer is protected by a layer of charcoal. Another process consists in adding water to the solution and so precipitating the bismuth as its oxychloride, BiOCl, which is then reduced with hydrogen to get the metal.[3]

$$Bi_2S_3 + 6\,HCl \longrightarrow 2\,BiCl_3 + 3\,H_2S \qquad (10.12)$$

$$2\,BiCl_3 + 3\,Fe \longrightarrow 2\,Bi + 3\,FeCl_2 \qquad (10.13)$$

$$BiCl_3 + H_2O \longrightarrow BiOCl + 2\,HCl \qquad (10.14)$$

10.3 Chemical Reactions of Arsenic

Chlorine combines readily with arsenic to form arsenic trichloride, $AsCl_3$, while HCl does not react. When arsenic is heated with sulfur, sulfides such as AsS, As_2S_3 or As_2S_5 are obtained, depending on the ratio of the elements used. Arsenic combines with metals to form arsenides. With oxygen, arsenic forms three oxides, which are arsenic trioxide, As_2O_3; arsenic pentoxide, As_2O_5; and As_2O_4. As_2O_4 has both +3 and +5 arsenic units. Arsenous acid $As(OH)_3$, is obtained from arsenic trioxide by slow hydrolysis and is stable only as the aqueous solution. Arsenic acid (H_3AsO_4) is obtained from arsenic pentoxide. Arsenates are similar to phosphates in their stoichiometry but solid arsenic acids similar to phosphorus-based acids are not known; instead, $H_3AsO_4.\frac{1}{2}H_2O$ or $As_2O_5.4H_2O$ crystallises from solutions of arsenic acid. The arsenic hydride AsH_3, is formed from arsenic compounds in acidic solution in the presence of strong reducing agents such as

Zn or from some metal arsenides. Arsine is used extensively in metal–organic chemical vapour deposition of GaAs and is prepared by the reaction of $NaBH_4$ with $AsCl_3$.[4]

$$4\ AsCl_3 + 3\ NaBH_4 \longrightarrow 4\ AsH_3 + 3\ BCl_3 + 3\ NaCl \qquad (10.15)$$

$$Ga(CH_3)_3 + AsH_3 \longrightarrow GaAs + 3\ CH_4 \qquad (10.16)$$

10.4 Important Compounds of Arsenic

10.4.1 Arsenic Trioxide

Arsenic trioxide has the same structure as that of phosphorus trioxide and antimony trioxide. However, the structure of the pentoxides of As and Sb differ from that of P_4O_{10}. It is called white arsenic and occurs in two crystalline and one amorphous form. The octahedral (or cubic) modification, arsenolite, is stable at room temperature. It is formed by the condensation of As_4O_6 vapour. The poisonous nature of white arsenic has been known for centuries and in the 17th and 18th century it was often used as a poison. Arsenic trioxide is the precursor for a lot of other arsenic compounds and is therefore produced in large quantities.

In an open vessel, arsenic trioxide sublimes above 135°C. It is moderately soluble in water. The toxicity is so high that even 0.1 g As_2O_3 can lead to death if it happens to enter the stomach. Thiol-based metal chelating agents such as dimercaprol are effective antidotes for arsenic trioxide poisoning.[1, 4]

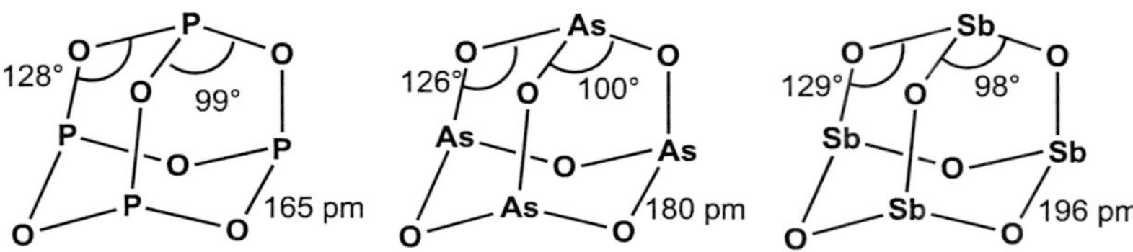

Fig 10.1 Comparison of structures of P_4O_6, As_4O_6 and Sb_4O_6

10.4.2 Arsenic Pentoxide

Arsenic pentoxide, As_2O_5 is prepared by the reaction of O_2 and As_2O_3 or by dehydrating H_3AsO_4 above 200°C. It is a glassy, white solid, relatively unstable compared to As_4O_6. The structure consists of tetrahedral AsO_4 and octahedral AsO_6 units, linked by sharing of corners, and is therefore completely different from the structure of P_4O_{10}.

In the crystal structure of As_2O_5, arsenic(V) is found in equal numbers in both octahedral and tetrahedral coordination. In the structure, each AsO_6 unit shares a corner with two other octahedra in a *cis* fashion, forming chains. Two such chains are connected to each other by AsO_4 tetrahedra through vertex sharing. Each tetrahedron shares its corners with four octahedra and each octahedron shares its corners with two octahedra and four tetrahedra. The overall composition of the solid is $[\{AsO_{6/2}\}\{AsO_{4/2}\}] \equiv As_2O_5$. All the polyhedra are

Arsenic: The king of poisons and the poison of kings[5]

From the time of Hippocrates, the father of modern medicine who lived in 370 BC, the toxic nature of arsenic was known. He described abdominal colic in a miner as due to arsenic, and similar properties of arsenic were also described by Pliny the Elder in the first century BC. The Greek physician in the court of the great Roman emperor Nero, Dioscorides, identified the poisonous nature of arsenic. It is believed that Nero poisoned his step-brother Tiberius Britannicus with arsenic to make his position safe as Roman Emperor.

The tasteless and odourless properties of arsenic trioxide (white arsenic) made it an ideal poison. As_2O_3, readily made by heating an arsenic ore, is a white crystalline powder which is moderately soluble in water and is almost undetectable in food or drink preparations. Arsenic poisoning symptoms initially are that of food poisoning and difficult to detect. A single dose of As_2O_3 could result in severe diarrhea and vomiting, paralysis and even death with cholera-like symptoms. Because of these reasons, it was the favourite poison of the rulers to kill their rivals and adversaries and therefore came to be known as the 'Poison of kings and the king of poisons'.

In Europe, during the Renaissance period, poisoning was very common with arsenic compounds. Arsenic trioxide was notoriously named *'poudre de succession'*, meaning 'inheritance powder'. However, from the 18th century, the use of arsenic compounds as poison decreased as chemical tests were being developed to detect arsenic during postmortems. Interestingly, one of the earlier tests which is used even today for testing groundwater for arsenic was developed by Dr Samuel Hahnemann, who is known as the father of homeopathic system of medicine. Hahnemann's test involved treating the suspect solution with H_2S and HCl which will result in a yellow precipitate of As_2S_3 if arsenic is present. However this test failed in detecting very small amounts of arsenic and often the yellow precipitate deteriorated with time making it difficult to present before a jury. A new and more reliable test was developed by James Marsh, an English chemist, in 1836. Since Berzelius modified this test it was known as the Marsh–Berzelius test. This test is highly sensitive for the detection of arsenic, and was found to be useful in forensic toxicology.

The test involves treating the sample with sulfuric acid and arsenic-free zinc. Zinc reduces the arsenic present to As^{3+} and the acid liberates AsH_3 gas.

$$As_2O_3 + 6\,Zn + 6\,H_2SO_4 \longrightarrow 2\,AsH_3 + 6\,ZnSO_4 + 3\,H_2O$$

Arsine gas (along with hydrogen) is formed and is led through a drying tube to a hard glass tube in which it is heated. The arsenic is deposited as a mirror in the vicinity of the heated area and on any cold surface held in the burning gas emanating from the jet. Antimony also gives a similar test, but the mirror formed by it is insoluble in sodium hypochlorite, whereas arsenic mirror dissolves. The mirror can also be formed by burning the gas and allowing the vapours to condense on a ceramic or glass surface.

linked by corners. Along the [001] plane, there are strands of octahedra which are linked by tetrahedra in such a way that tubular cavities are formed in the direction of [001].[6]

Strong oxidising agents such as O_3, H_2O_2 and HNO_3 can convert arsenic trioxide to the pentoxide. Roasting of orpiment, a well known naturally occurring arsenic sulfide also produces the pentoxide.

$$As_2O_3 + O_2 \longrightarrow As_2O_5 \qquad (10.17)$$

$$2\,As_2S_3 + 11\,O_2 \longrightarrow 2\,As_2O_5 + 6\,SO_2 \qquad (10.18)$$

10.4.3 Arsenic Acid and Arsenates

Arsenic acid, H_3AsO_4, has a structure similar to that of orthophosphoric acid. It is known in the solid state only as the hemihydrate $H_3AsO_4{\cdot}½H_2O$ in the form of rhombic, deliquescent crystals. Solutions in water behave as a tribasic acid. Arsenate and phosphate salts behave very similarly. Arsenic acid has not been isolated, but only found in solution where it is largely ionised. Its hemihydrate form ($H_3AsO_4{\cdot}½H_2O$) does form stable crystals. The dissociation constants of arsenic acid are determined as follows.

H_3AsO_4 $K_1 = 5.6 \times 10^{-3}$; $K_2 = 1.7 \times 10^{-7}$; $K_3 = 3 \times 10^{-12}$

In comparison, the dissociation constants for phosphoric acid are

H_3PO_4 $K_1 = 7.5 \times 10^{-3}$; $K_2 = 6.3 \times 10^{-8}$; $K_3 = 2.2 \times 10^{-13}$

Structurally, arsenic acid is almost identical to phosphoric acid.

Arsenate can replace phosphate in many biochemical reactions because they have similar structure and properties. For example, arsenate reacts in vitro with glucose and gluconate to form glucose-6-arsenate and 6-arsenogluconate, respectively. These compounds resemble glucose-6-phosphate and 6-phosphogluconate, respectively. Inorganic As(V), like P(V), is negatively charged over a range of physiological pH, as $H_2AsO_4^-$, $HAsO_4^{2-}$ or AsO_4^{3-} (an inorganic arsenate or As_i similar to P_i). As shown above, the dissociation constants for H_3AsO_4 are similar to those of H_3PO_4 and as a consequence As_i and P_i follow very similar speciation patterns. As_i like its phosphorus analogue, is capable of retaining a negative charge even when it is bound to two other molecules. As a result of these striking similarities, living systems cannot easily distinguish between inorganic arsenate and inorganic phosphate. Thus, arsenate can be taken up by cells via phosphate transporters and can substitute for P_i in the early steps in many of the phosphate-based metabolic pathways. Arsenic is generally thought to decouple oxidative phosphorylation, the crucial process in the production of ATP and as a result no ATP is produced. Polyarsenates hydrolyse much faster than polyphosphates which is the key step in the making and interconversion of ATP and ADP. These properties account for the biological toxicity of inorganic arsenate. A second difference when phosphorus is compared to arsenic is that while it is quite difficult to reduce P(V) in the tetracoordinate state to tricoordinate P(III), As(V) easily gets reduced to As (III).[7]

10.4.4 Realgar

Realgar is a monoclinic arsenic sulfide mineral with a brilliant red colour having a chemical composition of As_4S_4. Well-formed realgar crystals resemble the red gemstone ruby and therefore this mineral was often called 'ruby sulfur' or 'ruby arsenic'. However, realgar cannot be used as a gemstone as it is very soft, with a Mohs hardness between 1 to 2, easily ground to a red powder, and also toxic. The fascinating colour caused it to become a favourite pigment in many parts of the ancient world until people realised its toxicity. Realgar has a cradle shaped structure which is isostructural to As_4Se_4 with arsenic oriented to the four corners of a tetrahedra and sulfur forming a square. The As–As distances are 2.57 Å while the As–S distance is 2.23 Å.

Once its toxicity due to the presence of arsenic was realised, the mineral was used as a poison to kill rodents, insects and weeds. Realgar was also used in leather processing to remove the hair from raw hides of animals. Today it is hardly used for any of these purposes because more effective and less toxic alternate substitutes have been developed. Today the main use of realgar is as an ore of arsenic metal.

10.4.5 Orpiment

Orpiment As_4S_6, similar to realgar, is a historically important compound and occurs as a lemon-yellow mineral which is formed by reactions of trivalent arsenic compounds with H_2S. Arsenic trisulfide sublimes before melting and melts at 310°C to form a red liquid which boils at 707°C. It was used in cosmetics earlier but is now used in the semiconductor industry. The structure of orpiment consists of layers of covalently bound S and As atoms lying normal to the [010] direction.

10.4.6 Lewisite

Lewisite is an organoarsenic compound. During World War II, it was manufactured for use as a chemical weapon, causing blisters and irritating lungs. Impure samples of Lewisite are often amber coloured oily liquids with a unique odour similar to the smell of geranium flowers. The compound is prepared by the addition of arsenic trichloride to acetylene in the presence of a suitable catalyst.[8]

$$AsCl_3 + HC\equiv CH \longrightarrow ClCH{=}CH{-}AsCl_2 \tag{10.19}$$

Lewisite, like other arsenous chlorides, hydrolyses to form chlorovinyl arsenous oxide which is a less powerful blister-causing agent. This hydrolysis is accelerated in alkaline solutions and forms acetylene and trisodium arsenate.

$$ClCH{=}CH{-}AsCl_2 \xrightarrow[-HCl]{H_2O} ClCH{=}CH{-}As(OH)_2 \tag{10.20}$$

Lewisite will also react with metals to form hydrogen gas. It is combustible, but is not easy to ignite. Dimercaprol also known as British antilewisite, is the antidote for Lewisite. Chemically, dimercaprol binds to the arsenic present in Lewisite.

10.4.7 Salvarsan

Arsphenamine, also known as Salvarsan or compound 606, was a drug that was introduced at the beginning of the 1910s as the first effective treatment for the sexually transmitted disease Syphilis (a disease caused by the spirochaete bacterium *Treponema pallidum*), and was also used to treat trypanosomiasis (sleeping sickness). The identity of this compound was also thought by many as a cyclic organo-arsenic hexamer $[(ArAs)_6]$ (Ar = 3-amino,4-hydroxybenzene) for a long time. In 2005, an extensive mass spectral analysis showed Salvarsan to have As–As single bonds and gave evidence for it to be a mixture of cyclic trimer and pentamer of the same basic unit.[9,10]

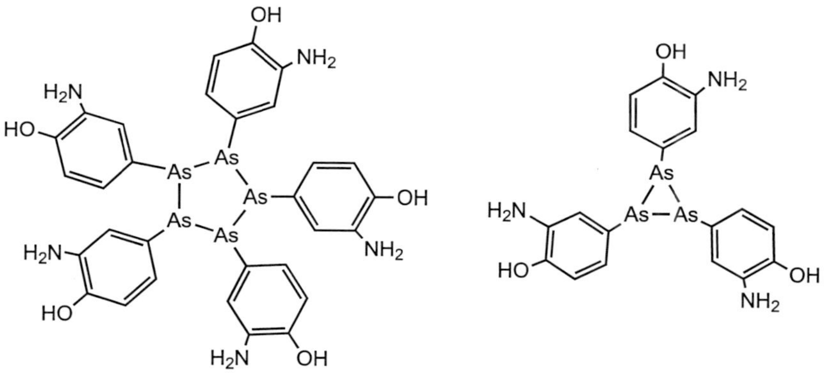

Fig. 10.2 Structures of compounds present in Salvarsan

Paul Ehrlich synthesised Salvarsan in 1909 by the reaction of 3-nitro-4-hydroxyphenylarsonic acid with dithionite and, with the chemist Bertheim's help, synthesised hundreds of related organoarsenic compounds (nowadays called a *library of compounds*). Salvarsan proved to be quite effective, particularly when compared to the conventional therapy of mercury salts used at that time. Salvarsan quickly became the most widely prescribed drug in the world and it was the most effective drug for syphilis till penicillin took over in the 1940s.

10.4.8 Cacodyls

Cacodyl also known as tetramethyldiarsine, is an organo-arsenic compound that constitutes a minor part of 'Cadet's fuming liquid' (after the French chemist Louis Claude Cadet de Gassicourt). $(CH_3)_2As–As(CH_3)_2$ is a poisonous oily liquid with a garlic-like odour. Cacodyl undergoes spontaneous combustion in dry air. It is considered as the earliest organometallic compound to be discovered and was investigated by Edward Frankland and Robert Bunsen. In Bunsen's own words, '*the smell of this body produces instantaneous tingling of the hands and feet, and even giddiness and insensibility...It is remarkable that when one is exposed to the smell of these compounds the tongue becomes covered with a black coating, even when no further evil effects are noticeable*'. It was originally made from arsenic distilled with potassium acetate.[11]

10.4.9 Arsine

Arsenic hydride, AsH_3, arsine, b.p. −62.5°C, m.p. −111.2°C, is a colourless extremely toxic gas with an unpleasant, garlic-like odour. It burns with a greyish blue flame converting to As_2O_3. While pure arsine is stable at room temperature, on passing the gas through a red hot glass tube, it decomposes to H_2 and metallic arsenic. The best and most convenient method to prepare arsine is by the reaction of zinc arsenide and dilute HCl. Arsine is used in chemical vapour deposition of GaAs.

$$As_2Zn_3 + 6\ HCl \longrightarrow 2\ AsH_3 + 3\ ZnCl \qquad (10.21)$$

10.4.10 Trimethyl Arsine, Me_3As

In 1893, the Italian physician Bartolomeo Gosio published his results on the poisonous '*Gosio gas*' that was subsequently found to be trimethylarsine. Under wet conditions, the mold *Scopulariopsis brevicaulis* produces significant amounts of methylarsines via methylation of arsenic-containing inorganic pigments, especially the green coloured pigments, which were once used in the making the indoor wallpapers. Copper–arsenic dyes for wall papers such as Paris Green, Bremer Green and Kaiser Green when attacked by *Scopulariopsis brevicaulis* resulted in the formation of Me_3As. Reaction of arsenic oxide with trimethylaluminium results in the formation of trimethylarsine.

$$As_2O_3 + 1.5\ (AlMe_3)_2 \longrightarrow 2\ AsMe_3 + 3/n\ (MeAlO)_n \qquad (10.22)$$

10.5 Toxicity of Arsenic Compounds

Almost all arsenic compounds are highly toxic, especially the inorganic ions. Toxicity of Lewisite, used during World War II is well documented. The trivalent arsenic compounds, such as As_2O_3, AsH_3, $AsCl_3$ and AsF_3 are more toxic than the pentavalent compounds. The reason for same has been attributed to the higher solubility of the As(III) compounds compared to the As(V) compounds. Elemental arsenic is not very toxic, but it gets converted readily to toxic compounds when ingested. After absorption in the body, arsenic compounds are distributed through the bloodstream to tissues, organs, bone marrow and skin, where they bind mainly to the sulfhydryl groups of proteins. Arsenic also gets accumulated in the hair and nails, and analysis of hair and nails has been used for the detection of arsenic poisoning in the dead. Arsenic replaces phosphorus in bones and can remain there for many years. Some of the antidotes for arsenic poisoning such as 2,3-dimercapto-1-propanol (British anti-lewisite) and 2,3-dimercapto-1-propanesulfonic acid utilise the affinity of oxoarsenic moieties to the vicinal mercapto groups of these compounds.

The metabolism of arsenic plays a significant role in its toxicity. It involves reduction to a trivalent state and oxidative methylation to a pentavalent state. The exact mechanism of the action of arsenic is not known, but several hypotheses have been proposed. At a biochemical level, inorganic arsenic in the pentavalent state has been found to replace phosphate in several reactions. In the trivalent state, inorganic and organic (methylated) arsenic may react with the thiol groups of proteins resulting in their inhibition and loss of activity.[1]

10.6 Chemical Reactions of Antimony

Antimony does not get ignited in hydrogen atmosphere. Halogens react vigorously with antimony, forming antimony trihalides. Chlorine yields either $SbCl_3$ or a mixture of $SbCl_3$ and $SbCl_5$. Sulfur, H_2S or dry SO_2 gives antimony(III) sulfide. Molten antimony forms binary compounds with P, As, Se and Te. Antimony is readily soluble in a mixture of nitric and tartaric acids and in aqua regia but concentrated HF, dilute HCl and dilute HNO_3 have no action. Some of the organic acids excluding acetic acid also dissolve the metal. It reacts with concentrated H_2SO_4 at 90–95°C although at room temperature there is no reaction.

10.7 Important Compounds of Antimony

10.7.1 Antimony Trioxide

Antimony trioxide, Sb_2O_3, is a white crystalline powder with a melting point of 656°C. On heating, it becomes yellow but becomes white on cooling. Antimony trioxide dissolves in acids and alkalies. Methods for preparing antimony trioxide include roasting antimony trisulfide, burning antimony in oxygen, alkaline hydrolysis of halides except fluoride, and hydrolysis of antimony trisulfide in superheated steam.

Antimony trioxide in combination with a halogen compound imparts flame-retardant properties to various materials. As antimony trioxide it is ineffective and for flame retardant activity the antimony must be intimately associated with the halogen, which is usually present as a chlorinated alkylated resin, PVC, or a chlorinated plasticiser. During a fire, antimony trihalide and/or oxide halide is generated which promotes reactions that result in the formation of carbonaceous char and not volatile gases. The char formed acts as a heat shield which slows down the decomposition of the burning plastics, thereby preventing the release of flammable gases.

Another use of antimony trioxide as an additive in plastics is as an opacifier (substance added to make a material opaque). The major non-flame-retardant uses of antimony trioxide include pigments, as an opacifier for ceramic glazes, in the refining of lead crystal glass, and glass used for television tubes.

10.7.2 Antimony Pentachloride

Antimony pentachloride, $SbCl_5$, m.p. 4°C, b.p. 140°C, is a pale yellow fuming liquid with an unpleasant odour. It can be produced by reacting molten $SbCl_3$ with Cl_2, followed by vacuum distillation. Antimony trichloride is prepared by the reaction of chlorine with antimony, antimony tribromide, antimony trioxide, or antimony trisulfide. It also may be made by treating antimony trioxide with conc. HCl. Other methods to make

$SbCl_5$ include direct chlorination of the metal, and chlorination of its oxides or sulfides. It is used as a polymerisation catalyst. $SbCl_5$ is readily converted to SbF_5 which is a fluorinating agent and stronger Lewis acid and precursor for superacids.

10.7.3 Antimony Trifluoride

Antimony trifluoride, SbF_3, m.p. 292°C, b.p. 319°C, is a white solid compound that crystallises readily. This compound was first prepared by Berzelius in 1824 by evaporation of a solution of Sb_2O_3 in HF. It has also been prepared by fluorination of $SbCl_3$ with HF in an organic solvent. SbF_3 decomposes in the presence of moisture, releasing HF. It has lesser solubility than $SbCl_3$ in organic solvents, but aqueous solutions are stable. It is a traditional fluorinating agent, well known as Swarts reagent (along with $SbCl_5$ or Cl_2) and replaces chlorine or bromine by fluorine atoms in both inorganic and organic compounds.

$$C_6H_5CCl_3 \xrightarrow{SbF_3} C_6H_5CF_3 \tag{10.23}$$

$$CCl_4 \xrightarrow{SbF_3\,/\,Br_2} CFCl_3 \tag{10.24}$$

$$SO_2Cl_2 \xrightarrow{SbF_3} SO_2F_2 \tag{10.25}$$

It is very useful for converting acid chlorides to the fluorides but fluorination can sometimes be slow. Addition of antimony pentachloride or bromine can increase the rate of reaction considerably. Antimony trifluoride is also used in the production of porcelain.

10.7.4 Antimony Pentafluoride

Antimony pentafluoride SbF_5, is a colourless oily liquid at room temperature. It is prepared by the reaction of $SbCl_5$ with anhydrous HF. It solidifies at 7°C and boils at 150°C. SbF_5 carbonises and fluorinates organic substances. The chlorine in chlorinated hydrocarbons is replaced by fluorine. Many inorganic compounds form complexes with it and both SbF_6^- and $Sb_2F_{11}^-$ are excellent counter ions to stabilise highly unstable cations (for example, N_5^+).[12]

$$C_4Cl_6 \xrightarrow{SbF_5/\,SbCl_5} C_4F_6Cl_2 \tag{10.26}$$

$$POCl_3 \xrightarrow{SbF_5/\,SbCl_5} POCl_2F \tag{10.27}$$

$$SOCl_2 \xrightarrow{SbF_5/\,SbCl_5} SOF_2 \tag{10.28}$$

Being a strong Lewis acid, SbF_5 has a special affinity to sources of F^- resulting in pulling a fluoride out from it and forming the highly stable $[SbF_6]^-$ anion. This is a weakly coordinating anion and is weakly basic. $[SbF_6]^-$ reacts with excess SbF_5 giving a centrosymmetric adduct $Sb_2F_{11}^-$ which is formed in many of the reactions of SbF_5. For example, the first gold xenon complex $AuXe_4^{2+}$ was stabilised by $Sb_2F_{11}^-$ anions.

$$SbF_5 + SbF_6^- \longrightarrow Sb_2F_{11}^- \qquad (10.29)$$

10.7.5 Fluoroantimonic Acid

Earlier, fluoroantimonic acid was known as $HSbF_6$, obtained from various reactions. Its superacidic properties were first observed by Olah in 1969 when he was studying the effect of HF on the magic acid mixture HSO_3F/SbF_5, which he had accidentally discovered. Olah reported the formation of $H_2F^+SbF_6^-$.[13,14] It is the most reactive superacid and can be stored only in teflon containers. Hammett acidity function measurements carried out later indicated it to be the strongest superacid known so far with a H_0 value of –28 estimated on the basis of isomerisation kinetics data while similar calculations indicated a H_0 of –25 for magic acid. The free acid $HSbF_6$ remained elusive and X-ray structure analysis revealed the acid existing as $[H_2F]^+[Sb_2F_{11}]^-$ and $[H_3F_2]^+[Sb_2F_{11}]^-$. In 2015, work by Gandon and coworkers on the reaction of the NHC adduct IPr·$InBr_3$ with $AgSbF_6$ indicated the presence of two molecules of the free fluoroantimonic acid ($HSbF_6$) to be present in the crystal structure of the product.[15]

The superacidity of fluoroantimonic acid is indicated in the following reaction with neopentane.

$$C(CH_3)_4 + H_2F^+\,SbF_6^- \xrightarrow[\text{HF–SbF}_5]{} (CH_3)_3C^+\,SbF_6^- + CH_4 \qquad (10.30)$$

10.8 Toxicity of Antimony Compounds

Antimony, a non-essential element, is comparable to arsenic and bismuth in its toxicological behaviour. Similar to arsenic, trivalent antimony compounds are more toxic. Exposure to antimony-based particles which are dispersed in the air can lead to poisoning in the workplace.

10.9 Chemical Reactions of Bismuth

According to hard–soft acid–base (HSAB) theory, Bi(III) is a borderline metal ion. Bi(III) ions bind with both oxygen- and nitrogen-based ligands in aqueous solutions. The coordination number of Bi in metal complexes has been found to vary between 3 and 10. There is an interesting structural similarity between $[Bi(H_2O)_9]^{3+}$ and $[Ln(H_2O)_9](SO_3CF_3)_3]$. In this compound, bismuth is bound by nine molecules of water in a tricapped trigonal prismatic structure.

Unlike nitrogen, phosphorus and arsenic, bismuth does not form a stable hydride. Bismuth hydride, bismuthine (BiH_3), is endothermic in nature, is stable only below −60°C and decomposes spontaneously to bismuth and hydrogen at room temperature.[16]

Table 10.4 Comparison of properties of AsH_3, SbH_3 and BiH_3

Property	AsH_3 (Arsine)	SbH_3 (Stibine)	BiH_3 (Bismuthine)
Stability	Pyrophoric, highly toxic	Decomposes at RT slowly and rapidly at 200°C	Decomposes below 0°C
b.p.	−62.5°C	−17.0°C	Extrapolated value 16.8°C
Density	1.64 g/mL	5.48 g/cc (gas)	8.66 kg/m^3
$\Delta_f H°$ (298 K) kJ/mol	66.4	145.1	277
E–H bond pm	151.1	170.7	177.6
∠ H–E–H /deg	92.1	91.7	90.5

Bismuth is stable both in dry and moist air at room temperature. When it is red-hot, it reacts with water to form Bi_2O_3 with the evolution of H_2.

The oxide Bi_2O_3 is purely basic unlike As_2O_3 and Sb_2O_3 which are amphoteric in nature. With sulfuric acid, Bi_2O_3 gives bismuth sulfate and SO_2 while with nitric acid it gives bismuth nitrate and NO_2. It also reacts with HCl but only in the presence of O_2 forming $BiCl_3$ and water. $Bi(OH)_3$ is not a fully characterised chemical compound and is produced as white flakes when alkali is added to a solution of a bismuth salt and is usually described as bismuth oxide–hydrate. In several Bi(III) thiolate complexes, bismuth exhibits coordination numbers varying from 3 to 7 with thiolate sulfur atoms. Bismuth(III) sulfide can be prepared by the reaction of elemental bismuth and elemental sulfur in an evacuated silica tube at 500°C for 96 hours. It is also prepared by the reaction of bismuth trihalides with H_2S.[17]

Many of the Bi(V) complexes known so far are five-coordinate and unstable in aqueous media, although Bi(V) in $Ar_3Bi(HCO_2)_2$ (triarylbismuthate complexes Ar = Ph, *p*-tol), is six-coordinate. Bismuth trifluoride reacts with F_2 forming BiF_5 at 500°C. ClF_3 also reacts with BiF_3 oxidising it to the pentafluoride. BiF_5 is a powerful oxidising and fluorinating agent and reacts with XeF_4 forming $XeF_3^+BiF_6^-$. It also reacts with water converting it to OF_2 and O_3, UF_4 to UF_6, Br_2 to BrF_3 and Cl_2 to ClF. Other halogens are incapable of oxidising it to the +5 state and it forms only Bi(III) halides. The trihalides are corrosive, react with moisture, forming oxyhalides with the formula Bi(O)X.[14]

10.10 Important Compounds of Bismuth

10.10.1 Medicinal Compounds: Bismuth Subcitrate and Bismuth Salicylate

Bismuth has a long association with medical applications with mentions of bismuth subnitrate appearing even as early as the middle ages. During the 1800s various bismuth compounds such as subnitrate, subgallate, subcitrate and subsalicylate) have been administered for the treatment of infections, syphillis, blood pressure and stomach disorders.[18]

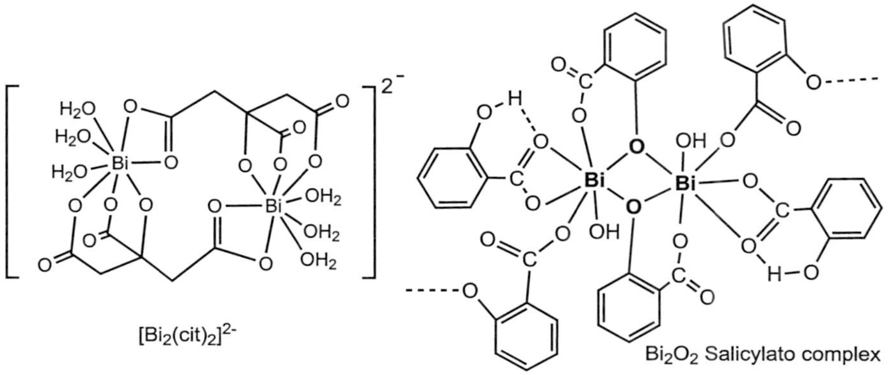

$[Bi_2(cit)_2]^{2-}$

Bi_2O_2 Salicylato complex

Fig. 10.3 Bismuth subcitrate and bismuth subsalicylate

Starting from the 1970s, two bismuth compounds have been commonly used worldwide—bismuth subsalicylate (BSS, present in Pepto-Bismol made by Procter & Gamble) for the prevention and treatment of diarrhea and dyspepsia, and colloidal bismuth subcitrate (CBS, brand name Pylera and De-Nol tab launched in 1976) for the treatment of peptic ulcers. The latter (CBS) has been successfully used in the treatment of gastric and duodenal ulcers. Compared to bismuth subnitrate and subsalicylate, CBS is highly water soluble and gives a colloidal solution with water. (The non conventional 'sub' terminology is common to many bismuth compounds indicating +3 oxidation state and multidentate ligands binding to bismuth with O-donor sites.) The efficacy of these bismuth complexes have been attributed to their bactericidal action against *Helicobacter pylori.* Many bismuth compounds as stand-alone compounds or in combination with antibiotics have been found to be effective against *Helicobacter pylori.* Clinical studies with CBS and BSS show that due to bactericidal action of these two complexes, patients treated with bismuth alone experienced a slower relapse than patients treated with other ulcer-healing agents. Attempts to determine the crystal structure of the active component of colloidal bismuth subcitrate indicated dimers having octacoordinate Bi^{+3} complex units (Fig. 10.3). The salicylate forms polymeric sheets based on Bi_2O_2 rings. In addition, irregular Bi_6O_7 octahedra have been observed. Complexes of Bi(III) with carboxylate ligands form insoluble coatings on the stomach wall from which Bi^{3+} can be released slowly and is taken up by the bacterial cells. Once inside the bacterial cell, Bi(III) possibly interferes with the enzyme activities that are vital for the bacteria's survival in the gastric environment, thus killing them. The Bi(III) compounds are also taken up by gastric mucous to form a protective coating on the crater of the ulcer, thereby limiting the adherence of *Helicobacter pylori.*[19]

How Pepto-Bismol became a household medicine for stomach complaints

The history of the use of bismuth subsalicylate (BSS) and to some extent bismuth subcitrate is actually the history of Pepto-Bismol. During the early 1900s, approximately two out of every ten children died before the age of four, and most often the death resulted from an illness called *cholera infantum*, a disease characterised by diarrhea and upset stomach. Although a variety of treatments were tried, none proved successful. On the basis of reports from Europe of the use of bismuth salts for the treatment of various ailments of the gastrointestinal tract, a physician in New York developed a liquid preparation that contained bismuth subsalicylate [a colloidal substance obtained by the hydrolysis of bismuth salicylate $[Bi\{C_6H_4(OH)CO_2\}_3]$ with the empirical formula $C_7H_5BiO_4$ along with zinc salts (for astringency), pepsin, oil of wintergreen and phenyl salicylate (for its antiseptic ability)]. A pinkish red dye was also added to make the product pink coloured in order to appeal to children. Initially this preparation was called *mixture cholera infantum* and, unlike previous preparations was successful for the treatment of cholrea infantum. When demand increased, the physician contracted the newly formed Norwich Pharmacia Company for its bulk production. With increased usage, doctors discovered that this mixture was effective against other types of stomach upsets especially during summer months with a range of symptoms including acute gastroenteritis, diarrhea, stomach cramps and vomiting. In 1918 Norwich Pharmacia began to market this product as Bismosal. However, the name was changed to Pepto-Bismol when they found that the Bismosal name was already given to another product. Over a century after its initial launch, Pepto-Bismol has become a household name in the USA and Europe.

In BSS, each molecule contains 58% of bismuth and 42% salicylate by weight. BSS is believed to be completely hydrolysed in the stomach. In the colon, non-dissociated BSS and other bismuth salts are believed to react with H_2S generated by bacteria to produce bismuth sulfide, a highly insoluble black salt. This black salt makes the stool of the patient darker during the use of this medicine which is temporary and of no harm. While salicylate is absorbed by the body, very little bismuth is absorbed from the stomach and intestines as these bismuth salts are insoluble in biological fluids of the stomach and intestines. More than 99% of the bismuth taken by the oral route of drugs like Pepto-Bismol get excreted in the feces. Black stool and black tongue are the often reported side effects of this drug.

Pepto-Bismol came to the Procter & Gamble company as part of the company's acquisition of Norwich Eaton Pharmaceuticals in 1982. It's now sold in several countries around the world.

10.10.2 Bismuth Telluride

Peltier discovered in 1834 an effect quite different from resistance heating, a cooling or heating effect when electric current flows through a junction formed by two different metals. Many attempts were made since then to utilise this property to devise thermoelectric refrigerators using the Peltier cooling effect which would not require volatile freons which could otherwise harm the ozone layer or cause a greenhouse effect. All initial attempts were unsuccessful since no materials were available which had the required properties for use as thermocouple elements. Another problem was the fact that the Peltier cooling is generally much less than the Joule heating. The challenge was to find a material with a high thermoelectric power and a high ratio of electrical conductivity to thermal conductivity.

The ratio of electrical to thermal conductivity is almost constant for most metals at a given temperature. The best metals suitable for thermoelectric refrigeration are therefore those having the highest thermoelectric powers. Antimony, bismuth and some of their alloys were found to be good candidates.[20]

In 1954, while working with a thermocouple made of bismuth telluride, Goldsmid and Douglas noticed an observable temperature decrease as a result of the Peltier effect. Direct fusion of bismuth and tellurium gave Bi_2Te_3. It was observed that *p*-type samples of Bi_2Te_3 had excellent properties that were required to get a good Peltier effect (thermoelectric power = 220 $\mu V.°C^{-1}$; electrical conductivity = 4.0×10^2 $\Omega^{-1}cm^{-1}$; thermal conductivity = 2.1×10^{-2} $Wcm^{-1}.°C^{-1}$). A thermocouple was constructed using the *p*-type Bi_2Te_3, in conjunction with pure bismuth. The couple was found to have a calculated figure of merit (a numerical expression representing the performance or efficiency of a given material) of 0.475 and a predicted maximum temperature difference of 28°C. In actual practice, a temperature difference of 26°C was obtained when a current of about 6.5 A was passed through the couple. In order to demonstrate the refrigerating effect, Goldsmid and Douglas kept a small copper vessel with 4 mL of water on top of the copper connecting strip. Within 10 minutes, this water was found to be frozen with frost deposit observed on the outer side of the vessel as well.

Over the years, the performance of these thermoelectric modules has improved and the figure of merit, ZT, has increased from the initial value of 0.475 to values greater than 1. Bi_2Te_3 has been used in the fabrication of thermoelectric modules and refrigeration systems have been marketed. Currently Bi_2Te_3 is one of the best materials available for freon-free refrigeration near room temperature.[21]

10. 11 Recent Developments

10.11.1 Arsenic, Antimony and Bismuth in Low Oxidation States

Similar to the low valent chemistry of phosphorus, synthesis and characterisation of low valent compounds of arsenic, antimony and bismuth has also been achieved by the use of ligands such as N-heterocyclic carbenes and cyclic alkylamino carbenes. G H Robinson and coworkers prepared the first examples of As(0) compounds stabilised by Dipp substituted NHCs in 2010. The *trans* bent structure of this compound had an As–As distance of 2.442 Å and C–As–As angle of 101.11° (Eq. 10.31).[22]

Dipp N N Dipp $\xrightarrow[\text{Hexane}]{AsCl_3}$ NHC:→$AsCl_3$ $\xrightarrow[\text{THF}]{3\ KC_8}$ (NHC)As–As(NHC), Dipp = 2,6-diisopropylphenyl (10.31)

However, attempted synthesis of NHC–$SbCl_3$ and NHC–$BiCl_3$ was not successful. Recent studies have indicated that because of the smaller HOMO–LUMO gap cyclic-alkylamino

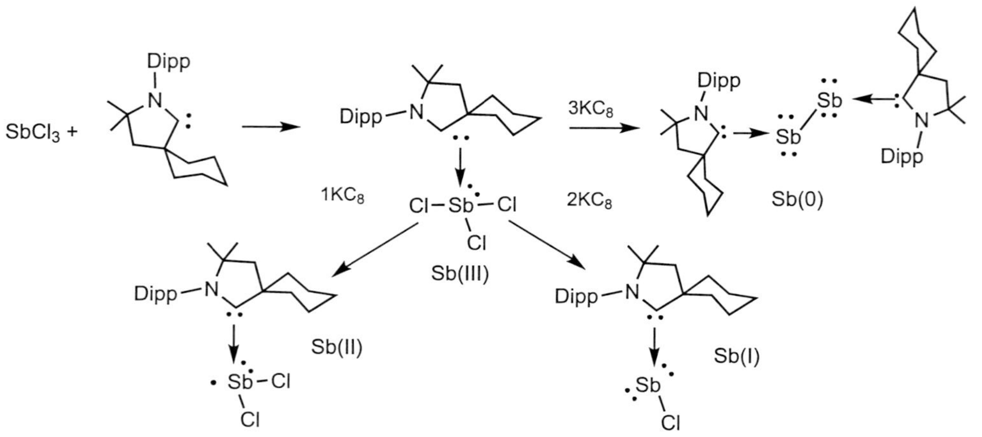

Scheme 10.1 Stabilisation of four different oxidation states of antimony by cAACs

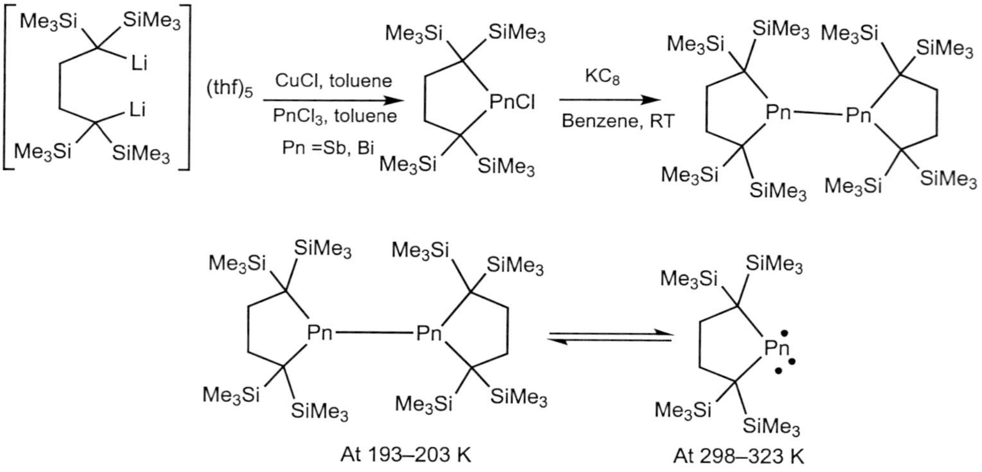

Scheme 10.2 Synthesis of radicals of low valent antimony and bismuth

carbenes can stabilise and help in the isolation of reactive species which NHCs cannot. Bertrand and coworkers in 2014 prepared the cyclic alkylamino carbene adduct of $SbCl_3$ and by using varying amounts of KC_8 carried out one, two and three electron reduction of the same giving antimony(II), antimony(I) and antimony(0) compounds (Scheme 10.1). One-electron reduction gave an example of a neutral antimony(II) centred radical, which has been characterised by EPR spectroscopy in solution.[23]

In 2014 persistent antimony(II) and bismuth(II) centred radicals were prepared by Ishida, Iwamoto and coworkers by facile dissociation of the corresponding dimers having bulky substituents (Scheme 10.2). The radicals were characterised by NMR and UV-Vis spectroscopy and by reaction with a stable nitroxyl radical, TEMPO. The extraordinarily long Sb–Sb (3.030 Å) and Bi–Bi (3.182 Å) bonds have been found to dissociate reversibly into the radicals in solution.[24]

Questions for self study (multiple choice)

10.1. The allotrope(s) of arsenic which are structurally very similar to that of phosphorus is (are)
(a) Yellow arsenic (b) Yellow and grey arsenic
(c) Black arsenic (d) Yellow and black arsenic

10.2. The structurally similar oxides/sulfides are
(a) P_4O_6 and As_2O_3 (b)P_4O_{10} and As_2O_5
(c) P_4S_3 and As_2S_3 (d) As_2S_3 and As_2O_3

10.3. Ruby sulfur is the name given to
(a) As_2S_3 (b) As_4S_4
(c) Sb_2S_3 (d) Sb_2S_5

10.4. Which of the following statements is not true for Salvarsan
(a) First effective drug for treatment of syphilis
(b) The structure of this compound is $(ArAs)_6$
(c) This drug was discovered by Paul Ehrlich
(d) This drug was replaced by penicillin in the 1940s

10.5. Swarts reagent for conversion of C–Cl bond to C–F bond is
(a) SbF_3 (b) $SbF_3 + SbCl_5$
(c) $SbCl_5$ (d) $SbCl_3$

10.6. Fluoroantimonic acid, the strongest superacid known so far is
(a) $HSbF_6$ (b) HSO_3F/SbF_5
(c) $H_2F^+SbF_6^-$ (d) CF_3SO_3H/SbF_5

10.7. Bismuth(III) aqua complex has the formula
(a) $Bi(H_2O)_6^{3+}$ (b) $Bi(H_2O)_9^{3+}$
(c) $Bi(H_2O)_4^{3+}$ (d) $Bi(H_2O)_5^{3+}$

10.8. Marsh–Berzelius test is used for identifying presence of traces of the element
(a) Arsenic (b) Antimony
(c) Bismuth (d) Thallium

10.9. The element which is a non-toxic replacement for lead in many of its applications is
(a) Tin (b) Antimony
(c) Tellurium (d) Bismuth

Solved concept-based questions

10.10. Why is arsenic toxic?

The toxicity of arsenic is due to many factors. Living systems not being able to differentiate between inorganic phosphate and inorganic arsenate is one of the factors.

Thus arsenate can be taken up by cells and can substitute phosphate in many phosphate-based metabolic pathways. Arsenic is thought to affect oxidative phosphorylation which is responsible for making ATP. Polyarsenates hydrolyse orders of magnitude higher than polyphosphates which is a key step in ATP–ADP interconversion. Another reason is it is easy to reduce As(V) to As(III) unlike P(V) to P(III).

10.11. What makes SbF_5 one of the strongest Lewis acids?

The high stability of SbF_6^- is one of the major reasons for SbF_5 to act as a strong Lewis acid especially in reactions leading to abstraction of an F^-.

10.12. Why are bismuth compounds useful for the treatment of stomach ulcers?

One of the views is that complexes of Bi(III) with carboxylate ligands form insoluble coatings on the stomach wall from which Bi^{3+} gets released slowly and is taken up by the bacterial cells. Once inside the bacterial cell, Bi(III) interferes with the enzyme activities required for the survival of the bacteria killing them. The Bi(III) compounds are also taken up by the gastric mucus to form a protective coating on the crater of the ulcer, thereby limiting the adherence of the bacteria *Helicobacter pylori.*

Problems and exercises

10.13. A main group metal oxide of the type A_2O_3 on treatment of 6 moles of HF gave the compound B and water. Compound B on reaction with F_2 gas at 500°C gave compound C. Compound C on reaction with water was found to convert it to O_3 and F_2O and converted UF_4 to UF_6. Identify element A and compounds B and C.

10.14. A main group metal oxide of the type A_2O_3 on treatment with conc. HCl was found to give compound B. B on reaction with Cl_2 gave compound C. C on reaction with anhydrous HF have compound D. A combination of compounds C and D was found to convert C–Cl bonds to C–F bonds of many organic substrates. Identify element A and compounds B to D.

10.15. How does BiH_3 compare with AsH_3 and SbH_3 with respect to (a) stability (b) heat of formation?

10.16. Which are the major applications of arsenic, antimony and bismuth compounds?

10.17. How does arsenic acid compare with phosphoric acid? How is the toxicity of the former explained?

10.18. Compare the structures of the following pairs of compounds (a) P_4O_6 and As_4O_6 (b) P_4O_{10} and As_4O_{10}.

10.19. Suggest two tests for identifying the presence of arsenic in a solution.

10.20. What is Peliter effect? Which main group compound shows Peltier effect?

Answer key for self study questions

10.1. (d) 10.2. (a) 10.3. (b) 10.4. (b) 10.5. (b) 10.6. (c) 10.7. (b) 10.8. (a) 10.9. (d)

Supplementary reading

1. Grund S C, Hanusch K and Wolf H U, Arsenic and arsenic compounds in *Ullmann's encyclopedia of industrial chemistry*, 2005, Weinheim, Wiley-VCH.
2. Grund S C, Hanusch K, Breunig H J and Wolf H U, Antimony and antimony compounds in *Ullmann's encyclopedia of industrial chemistry*, 2006, Wiley-VCH, Weinheim.
3. Kruger J, Winkler P, Luderitz E, Lueck M and Wolf H U, Bismuth, bismuth alloys and bismuth compounds in *Ullmann's encyclopedia of industrial chemistry*, 2003, Weinheim, Wiley-VCH.
4. Mandal B K and Suzuki K T, Arsenic round the world: A review, *Talanta*, 2002, Vol. 58, 201.
5. Emsley J, *Arsenic, The element of murder: a history of poison*, Oxford University Press, 2006, 93.
6. Jansen M, The Crystal structure of As_2O_5, *Angew. Chem. Int. Ed.*, 1977, Vol. 16, 314.
7. Tawfik D S and Viola R E, Arsenate replacing phosphate: Alternate life chemistries and ion promiscuity, *Biochemistry*, 2011, Vol. 50, 1128.
8. Vilensky J A, *Dew of Death—The Story of Lewisite, America's World War I Weapon of Mass Destruction*, 2005, Indiana University Press, 4.
9. Lloyd N C, Morgan H W, Nicholson B K and Ronimus R S, The composition of Ehrlich's Salvarsan: resolution of a century old debate, *Angew. Chem. Int. Ed.*, 1977, Vol. 16, 314.
10. Gibaud S and Jaouen G, Arsenic based drugs: From Fowler's solution to modern anticancer chemotherapy, *Topics in organometallic Chemistry*, 2010, Vol. 32, 1.
11. Seyferth D, Cadet's fuming arsenical liquid and cacodyls compounds of Bunsen, *Organometallics*, 2001, Vol. 20, 1488.
12. Olah G A, Prakash, G K S, Wang Q and Li X-Y, *Antimony(V) fluoride in encyclopedia of reagents for organic synthesis*, Paquete L, Ed., 2004, Wiley & Sons, NY.
13. Olah G A, A life of magic chemistry: autobiographical reflections of a Nobel prize winner, John Wiley & sons, 2001.
14. Olah G A, Crossing conventional boundaries in half a century of research, *J. Org. Chem.*, 2005, Vol. 70, 2413.
15. Bour C, Guillot R and Gandon V, First evidence for the existence of hexafluoroantimonic(V) acid, *Chem. Eur. J.*, 2015, Vol. 21, 6066.
16. Jerzembeck, W, Buerger H, Constantin, L, Margules L, Demaison J, Breidung J and Thiel W, Bismuthine BiH_3: Fact or fiction? High resolution infrared, millimeter-wave and Ab Initio studies, *Angew. Chem. Int. Ed.*, 2002, Vol. 41, 2550.
17. Hermann W A, Herdtwek E and Pajdla L, Colloidal bismuth subcitrate (CBS): Isolation and structural characterisation of the active substance against Helicobacter pylori, a causal factor of gastric diseases, *Inorg. Chem.*, 1991, Vol. 30, 2579.
18. Lee S P, The mode of action of colloidal bismuth subcitrate, *Scand J. Gastroenterol. Suppl.*, 1991, Vol. 185, 1.
19. Sadler P J, Li H and Sun H, Coordination chemistry of metals in medicine: Target sites for bismuth, *Coord. Chem. Rev.*, 1999, Vol. 185–186, 689.
20. Goldsmid H J and Douglas R W, The use of semiconductors in thermoelectric refrigeration, *Br. J. Appl. Phys.*, 1954, Vol. 5, 386.
21. Goldsmid H J, Bismuth telluride and its alloys as materials for thermoelectric generation, *Materials*, 2014, Vol. 7, 2577.
22. Abraham M Y, Wang Y, Xie Y, Wei P, Schaefer H F, Schleyer P R and Robinson G H, Carbene stabilisation of diarsenic: From hypervalency to allotropy, *Chem. Eur. J*, 2010, Vol. 16, 432.
23. Ishida S, Hirakawa F, Furukawa K, Yoza K and Iwamoto T, Persistent antimony and bismuth centred radicals in solution, *Angew. Chem.*, 2014, Vol. 126, 11354.
24. Kretschmer R, Ruis D A, Moore C E, Rheingold A L and Bertrand G, One-, two- and three-electron reduction of a cyclic alkyl(amino)carbene–$SbCl_3$ adduct, *Angew. Chem. Int. Ed.*, 2014, Vol. 53, 8176.

THE CHEMISTRY OF OXYGEN AND SULFUR

CHAPTER

11

UNIQUE PROPERTIES OF OXYGEN, SULFUR AND THEIR COMPOUNDS

OXYGEN

- *While oxygen constitutes 20.94% of the earth's atmosphere, its only allotrope, ozone, constitutes 0.6 ppm of the atmosphere.*
- *More than 50% of the commercially produced oxygen is used in the smelting of iron ore to steel. Oxygen injected into molten iron removes sulfur as SO_2 and carbon as CO.*
- *25% of the commercially produced oxygen is used by the chemical industry for the synthesis of ethylene oxide which is further converted mainly to ethylene glycol.*
- *Other industrial uses of oxygen include, medical use, for making oxyacetylene flames for welding and as liquid rocket fuel.*
- *The importance of breathing oxygen is reflected in the fact that glucose upon oxidation in the presence of O_2 produces 18 times more energy than in the absence of O_2.*
- *Plants produce oxygen during photosynthesis by splitting of water molecules.*
- *Oxygen molecule is reversibly bound to the iron centres of hemoglobin which are carried by the red blood cells from lungs to the muscles where it is transferred to the myoglobin.*
- *Ozone strongly absorbs ultraviolet radiation (UV-B) and the stratospheric ozone layer helps to protect the earth from harmful UV radiation.*
- *Ozone with an oxidation potential of 2.07 V, is a stronger oxidising agent than H_2O_2, O_2 and $KMnO_4$.*
- *Robert H Goddard developed a rocket engine that used a liquid fuel with liquid oxygen as the oxidiser. He successfully flew a small rocket to a height of 56 m in 1926 with this fuel.*

- *Oxygen is composed of three stable isotopes, ^{16}O, ^{17}O and ^{18}O, with ^{16}O being the most abundant (99.762% natural abundance).*
- *Oxygen constitutes 49.2% of the earth's crust by mass as part of oxide compounds such as iron oxides, silicates and alumino silicates. It is also the major element of the world's oceans (88.8% by mass).*
- *65% of the human body consists of oxygen where it is the largest constituent. Sulfur is present in 0.3%.*
- *O_2 accumulation in the earth's atmosphere is believed to have begun 2.5 billion years ago during the Great Oxygenation Event (GOE). Oceanic cyanobacteria, which evolved into multicellular forms more than 2.3 billion years ago are believed to be the first microbes to produce oxygen by photosynthesis.*
- *In acidic solutions, H_2O_2 (1.78 V) is one of the most powerful oxidisers known—stronger than $KMnO_4$ and gets reduced to water. In basic solutions, H_2O_2 (−0.68 V) acts as a reducing agent, while it gets oxidised to O_2 gas.*
- *Fenton's reagent, a solution of hydrogen peroxide with ferrous iron as a catalyst is used to oxidise many unreactive organic compounds. For example, hydroxylation of arenes.*

SULFUR

- *Sulfur is known from very ancient times, is mentioned in the Bible and Greek mythology as brimstone and was used by alchemists.*
- *Sulfur and mercury were called philosophical elements (along with the Aristotelian elements air, water, earth and fire) during the time of alchemy and it represented the principle of combustibility.*
- *Sulfur mostly forms divalent, tetravalent and hexavalent compounds and the latter in its tetracoordinate form is the most stable geometry and oxidation state of sulfur.*
- *Sulfur is stable in its elemental state and was mined directly from the earth for a long time.*
- *The most industrially important compound of sulfur is sulfuric acid which is used for making sulfate- and phosphate-based fertilisers. 93% of sulfur produced is used for making sulfuric acid.*
- *Sulfur forms the maximum number of allotropes (over 30) and many of them are cyclic compounds with ring sizes varying from 6–20.*
- *Sulfur was extracted earlier from sulfur deposits, but this method has been obsolete since the late 20th century. Today, almost all of the elemental sulfur (~97%) is produced from natural gas and petroleum as a byproduct from its*

sulfur-containing contaminants. Another 2% is obtained from coal gassification processes.

- *Sulfur is the fourth major plant nutrient after N, P and K. About 12.5 million tons of sulfur was applied as fertilisers worldwide in 2015.*
- *Sulfur is a component of almost all proteins, present in the amino acids, cysteine (having SH group) and methionine (having S–CH_3 group). Disulfide bonds are very important for building the tertiary structure of many proteins and are largely responsible for the mechanical strength and insolubility of proteins such as keratin, found in the hair.*
- *Sulfur in the organic form is present in the vitamins biotin (B_7) and thiamine (B_1). Sulfur is also an important part of many enzymes and in antioxidant molecules such as glutathione peroxidase and thioredoxin.*
- *Sulfonamides which are structural analogues of para-aminobenzoic acid are a class of bacteriostatic drugs (which stops bacteria multiplication).*
- *Thionyl chloride is a chlorinating and dehydrating agent while sulfuryl chloride is a chlorinating agent similar in reactivity to Cl_2.*
- *Sulfur tetrafluoride and related compounds are useful as reagents for selective conversion of C=O to CF_2 groups.*
- *In the early literature the spelling 'sulphur' was dominant, especially in Europe and Asia over the spelling 'sulfur'. The IUPAC and the nomenclature committee of the Royal Society of Chemistry, UK adopted the spelling 'sulfur' in 1990 and 1992 respectively.*

11.1 About the Element: Oxygen

The discovery of oxygen was mired in controversies and is strongly related to the phlogiston theory. Joseph Priestly in his book, *Experiments and Observations on Different Kinds of Air*, described it as dephlogisticated air in 1774. Claims for its discovery have also been made by his contemporaries Lavoiser and Scheele. The discovery of oxygen is also credited by some historians to a Polish alchemist called Michael Sendivogius who found that heated saltpeter (KNO_3) produced 'the elixir of life' and Sendivogius, in 1604, described his experiments in an authoritative book. Priestly's classic experiment showed that by focusing the sun's rays on a sample of mercuric oxide, a gas was released. He first tested this gas on mice which survived for quite a while, and then on himself, writing that it was 'five or six times better than common air for the purpose of respiration, inflammation, and, I believe, every other use of common atmospherical air'. Priestley did not name it oxygen, Scheele called it *fire air* and it was Lavosier who named it oxygen.[1]

Table 11.1 Physical properties of oxygen

Oxygen	*Atomic mass*	*Electronegativity (Pauling scale)*	*1st ionisation energy*	*Electronic configuration*	*van der Waals radius*
Symbol: O	15.999 g.mol^{-1}	3.44	1313.9 kJ.mol^{-1}	[He]$2s^2 2p^4$	0.152 nm
At. no. 8	*Atomic radius*	*Melting point*	*Boiling point*	*Electron affinity (O^{16})*	*Density*
	0.90 Å	–218.8°C	–182.9°C	140.976 kJ.mol^{-1}	1.141 g.cm^{-3} at b.p.

Table 11.2 Physical properties of sulfur

Sulfur	*Atomic mass*	*Electronegativity (Pauling scale)*	*1st ionisation energy*	*Electronic configuration*	*van der Waals radius*
Symbol: S	32.066 g.mol^{-1}	2.58	999.6 kJ.mol^{-1}	[Ne]$3s^2 3p^4$	0.180 nm
At. no. 16	*Atomic radius*	*Melting point*	*Boiling point*	*Electron affinity*	*Density*
	1.11 Å	115.2°C	444.6°C	200.410 kJ.mol^{-1}	1.819 g.cm^{-3} at m.p.

Oxygen is a colourless gas, without smell or taste, is slightly heavier than air, is sparingly soluble in water, has a boiling point of –183°C, and the liquid is pale blue in colour showing appreciable paramagnetism. At still lower temperatures, light-blue solid oxygen is obtained, which has a melting point of –218.4°C.

On a commercial scale, oxygen can be obtained from the atmosphere by the liquefaction and fractional distillation of air. Liquid air is a mixture of liquid nitrogen (b.p. –196°C), and liquid oxygen (b.p. –183°C). The nitrogen boils off first during evaporation of the mixture. Separation of the two gases can be brought about by careful fractionation.[2]

11.2 The Great Oxygenation Event

The early atmosphere of earth, formed after its formation about 4.5 billion years ago, was devoid of oxygen and had mostly nitrogen, carbon dioxide and water vapour. The photosynthesis started by early forms of bacteria especially cyanobacteria (old name *blue green algae*) is responsible for the build up of oxygen in the atmosphere. While cyanobacteria, formed in the oceans about 3.5 billion years ago were producing oxygen, it was immediately taken up by the dissolved iron in the oceans and ocean floor to form iron oxides. Later, the oxygen converted the land based iron, silicon, aluminum, and so on, to their oxides and only after these reactions were over, the atmospheric build up of oxygen started and that is estimated to have begun 2.3–2.4 billion years ago. This event of building up oxygen in the atmosphere has been termed as the great oxygenation/great oxidation event and resulted in the atmosphere having its present 20.94% dioxygen. It also resulted in the death of almost all anaerobic organisms present on the earth at that time.[3–5]

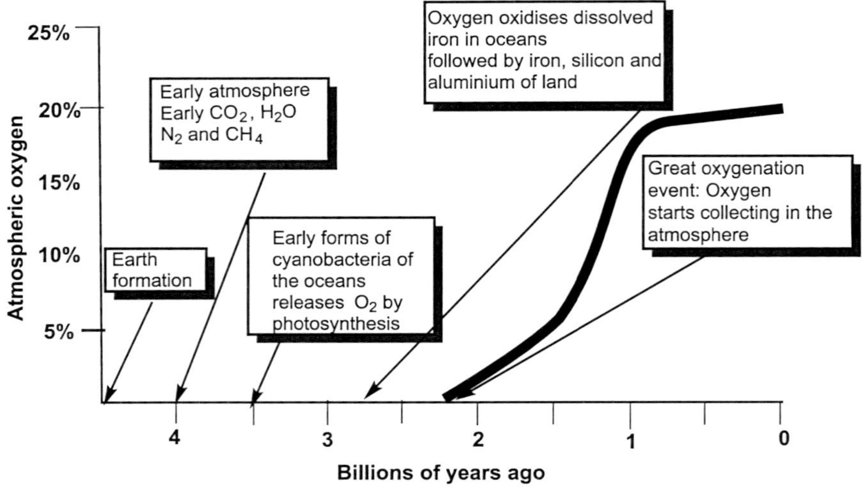

Fig. 11.1 Milestones in the collection of oxygen in the atmosphere

$$6\,CO_2 + 6\,H_2O \xrightarrow[\text{Chlorophyll}]{\text{Sunlight}} C_6H_{12}O_6 + 6\,O_2 \tag{11.1}$$

11.3 Preparation of Oxygen

Being a major component of air, oxygen has been studied extensively over the centuries and there are many methods for its preparation. The most common methods involve heating of a range of metal and nonmetal oxides such as $KClO_3$, KNO_3, PbO_2, $KMnO_4$ and CrO_3. Industrially, it is obtained by the liquefaction of air after separating nitrogen and other components of air.[6]

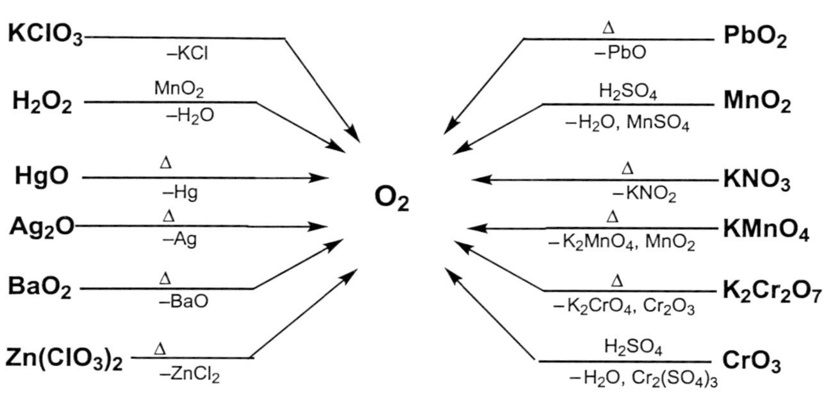

Scheme 11.1 Different methods for preparing oxygen

11.4 Reactive Oxygen Species and Reactive Oxygen Intermediates

When the concentration of O_2 in the atmosphere increased, leading to the evolution of aerobic respiration, complex eukaryotic organisms emerged (organisms with a complex cell or cells, in which the genetic material is organised into a membrane-bound nucleus or nuclei). For all living aerobic species, O_2 is the molecule crucial for cellular respiration. However, some oxygen-based species are highly toxic and the terms ROS (reactive oxygen species) and ROI (reactive oxygen intermediates) have been coined to address this class of highly reactive, oxygen-based species some of which are radicals and radical anions too. According to a more specific classification, the term ROI depicts the chemical species formed upon incomplete reduction of molecular oxygen, namely superoxide radical anion ($O_2^{\bullet -}$), hydrogen peroxide (H_2O_2) and hydroxyl radicals ($OH^{\bullet}$), while ROS includes all of the ROI, singlet oxygen ($^1O_2$) and ozone (O_3).[7, 8]

$$O_2 \xrightarrow[-0.16V]{e^-} O_2^{\bullet -} \xrightarrow[+0.94V]{e^- + 2H^+} H_2O_2 \xrightarrow[+0.38V]{e^- + H^+} OH^{\bullet} \xrightarrow[+2.33V]{e^- + H^+} H_2O \qquad (11.2)$$

The response of O_2 and reactive oxygen species to biomolecules can be compared as follows. O_2, with a redox potential of −0.16 V (for oxygen concentration of 1 M, pH 7 as the standard state, −0.33 V for 1 bar as the standard state), is a relatively poor monovalent electron acceptor. As a result, molecular oxygen is not very reactive on biomolecules. The unpaired electrons of oxygen can, in principle, interact with unpaired electrons of transition metal ions or organic radicals. One-electron reduction of molecular oxygen to superoxide radical anion is thermodynamically less favoured than the two-electron reduction to hydrogen peroxide. But the simultaneous $2e^-$ requirement of the reduction to hydrogen peroxide is also unfavourable. Therefore, for processes which are auto-oxidative in nature, formation of superoxide radical anion is characteristic but not H_2O_2 formation.

With electron-rich centres, the superoxide radical anion has limited reactivity due to its anionic charge. Despite its high reduction potential of +0.94 V, O_2^- can oxidise very few biological compounds. One-electron reduction of O_2^- leads to the formation of hydrogen peroxide.

Under physiological conditions H_2O_2 is relatively stable, although it has a positive one-electron reduction potential of 0.38 V. In contrast, the hydroxyl radical, with a one-electron reduction potential of +2.33 V, is a very strong oxidant reacting with organic matter.

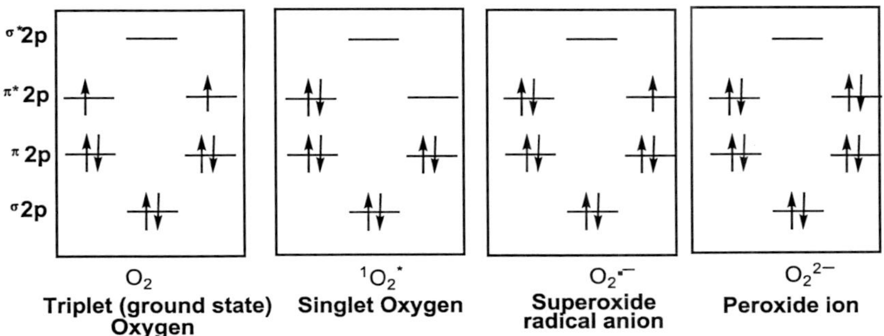

Fig. 11.2 MO pictures of reactive dioxygen species

Why is dioxygen a stable molecule in spite of being a diradical?

Dioxygen O_2, with two unpaired electrons in its stable form is the only paramagnetic molecule found in abundance in our environment that also has a triplet diradical ground state. In contrast, sulfur and selenium form rings and chains in their stable forms. Generally unpaired electrons of radicals and diradicals undergo dimerisation or oligomerisation leading to stabilisation by forming bonds with each other or with other radicals. Therefore, an interesting question is, what makes O_2 so stable?

Oxygen is thermodynamically stable by itself. However, in the presence of other elements (with the exception of gold) it can be made to react, often exothermically. It is energy rich and therefore is used as a liquid propellant in rockets. While radicals such as Cl• or OH• are extremely reactive abstracting hydrogens from C–H bonds of organic compounds, O_2 under normal circumstances does not abstract hydrogens from organic molecules. So it is clear that O_2 must have a high barrier to reactions with other molecules.

This problem was studied in detail by Roald Hoffmann and coworkers in 2017.[9] Experimental heats of formation and enthalpies obtained from theoretical calculations (Gaussian 4) indicated that resonance stabilisation of the two unpaired electrons in triplet O_2 is around 100 kcal/mol which is around 37–38 kcal/mol greater than the resonance stabilisation of unpaired electrons in two hydroxyl radicals. Although O_2 is a triplet biradical, its thermodynamic unfavourability towards ring/chain formation and hydrogen atom abstraction can be attributed to this very large resonance stabilisation energy of 100 kcal/mol. In spite of the resonance stabilisation resulting from the π-system of the triplet O_2, it is the relative weakness of the O–O σ bond which makes reactions of O_2 possible leading eventually to the cleavage of this bond.

The strong π-bonding in •OO• is the reason for this diradical to become persistent. The weak σ bond, on the other hand, permits the formation of two water molecules from •OO• delivering large amounts of energy. This unusual combination of properties is what gives O_2 the ability to support aerobic life on the earth.

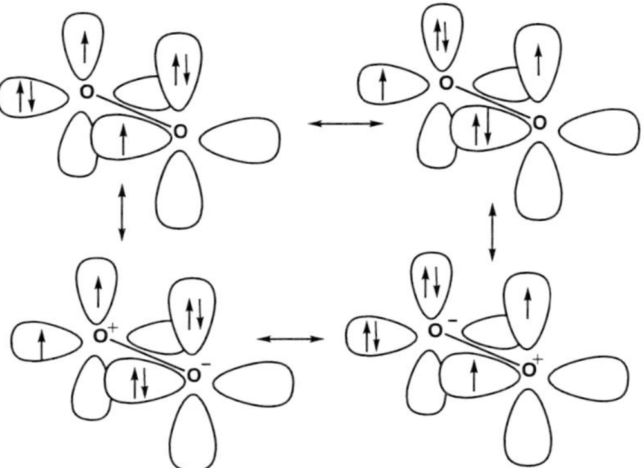

Gaussian 4 calculations have also indicated that the oligomerisation of •SS• to S_6 or S_8 is thermodynamically favourable. The resonance energy calculated for triplet •SS• is only 50.9 kcal/mol. The enthalpy changes calculated are exothermic (–70.8 kcal/mol) for trimerisation to S_6 and for tetramerisation to S_8 (–103.0 kcal/mol). In contrast, the enthalpy change for the trimerisation of •OO• to O_6 is endothermic by 79.4 kcal/mol.

Reactive oxygen species have for quite some time been responsible for the oxidative damage inflicted on fatty acids, DNA, proteins as well as many other cellular components. Diseases such as cancer, cardiovascular, inflammation, Parkinson's and Alzheimer's diseases have been related to oxidative stress caused by the imbalance between excessive formation of reactive oxygen species and limited antioxidant defenses. According to the 'free radical theory of ageing' suggested by D Harman in 1956, the toxic effects of ROS, generated during cellular respiration at the mitochondrial level are directly involved in the processes of ageing.

Triplet molecular oxygen in its ground state is a paramagnetic diradical and its HOMO has two electrons occupying two π^* orbitals with parallel spins. In contrast, organic molecules are mostly diamagnetic, having paired electrons. Therefore, a spin restriction comes into force for O_2 to participate in redox reactions with such molecules. O_2 is therefore not able to efficiently oxidise biomolecules through addition reactions which are 2-electron processes.

11.5 Industrial Uses of Oxygen

A major industrial use of oxygen gas (~50%) is in the smelting of iron ore to steel. It enriches and increases combustion temperature in blast furnaces. The carbon monoxide formed by the reaction of C and O_2 brings about the reduction of Fe_2O_3 to Fe. Oxygen gas when introduced into molten iron removes sulfur as SO_2 and converts carbon to CO. Around 25% of the commercially produced oxygen is used by the chemical industry for the synthesis of ethylene oxide which is a very important industrial precursor. This low boiling liquid is industrially prepared by the oxidation of ethylene over a heterogeneous catalyst involving Ag embedded in a matrix such as silicagel, alumina, aluminosilicates, and so on. Temperature (220–280°C) and stoichiometry are crucial for this reaction as over-oxidation can lead to complete conversion to carbon dioxide.

$$7\ H_2C{=}CH_2 + O_2 \xrightarrow[\text{Al}_2\text{O}_3 \text{ or silica gel matrix, } 220\text{–}280°\text{C}]{\text{Ag}} 6\ \underset{\text{b.p. } 10.4°\text{C}}{H_2C\overset{O}{—}CH_2} + 2\ CO_2 + 2\ H_2O \qquad (11.3)$$

Some of the important and industrially useful products which can be prepared from ethylene oxide are shown in Scheme 11.2. It is significant that more than 65% of ethylene oxide is used for the synthesis of ethylene glycol which is the most widely used antifreeze in vehicles throughout the world [20 mmt (million metric tons) in 2010 while world production of ethylene was 150 mmt].

Industrial uses of oxygen also include its use as a gas for cutting, welding and melting of metals. Oxygen gas is capable of generating temperatures of 2800–3000°C. Acetylene/oxygen

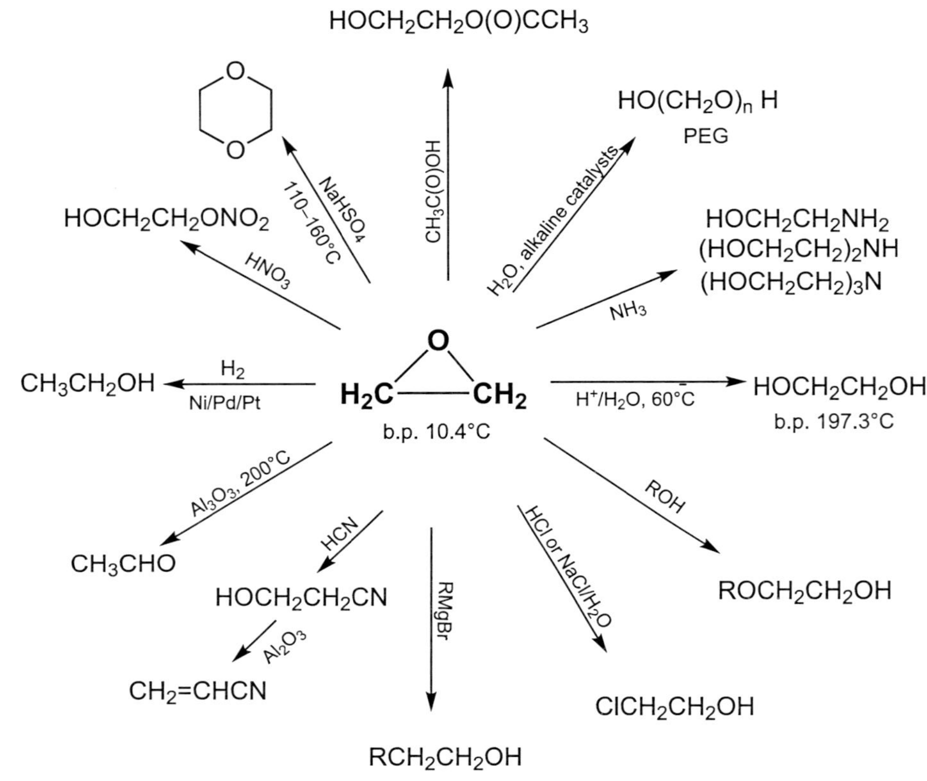

Scheme 11.2 Some reactions of ethylene oxide

combination gives temperatures of up to 3500°C. Liquid oxygen is used as an oxidant along with spacecraft fuel; oxygen cylinders are essential for life support in hospitals.

11.6 Dioxygen as a Ligand

Although dioxygen is a comparatively stable molecule, its reaction with most of the organic molecules is spin forbidden. Therefore, a great deal of interest lies in the activation of the dioxygen molecule by coordination to a metal ion. This interest is further increased by the fact that the respiration process involves reversible binding of dioxygen to metals. These metal-coordinated species come under the category of superoxo ($M–O_2^-$) or peroxo ($M–O_2^{2-}$). Since the reduction of the O_2 involves electron donation to the $2p\pi^*$ antibonding orbitals of dioxygen, the superoxo and peroxo species can be distinguished by the differences in their O–O bond strengths. With respect to the bond order of the O–O bond, molecular dioxygen (O_2), superoxide (O_2^-), and peroxide (O_2^{2-}) have predicted bond orders 2, 1.5

and 1, respectively. The O–O IR stretching frequencies vary continually from 1,300 cm^{-1} to 700 cm^{-1} and dioxygen (O_2), superoxide (O_2^-), and peroxide (O_2^{2-}) can have O–O stretching frequencies of 1555 cm^{-1}, 1200–1070 cm^{-1}, and 930–750 cm^{-1}, respectively.[10]

The dioxygen metal complexes are mostly classified into four categories.

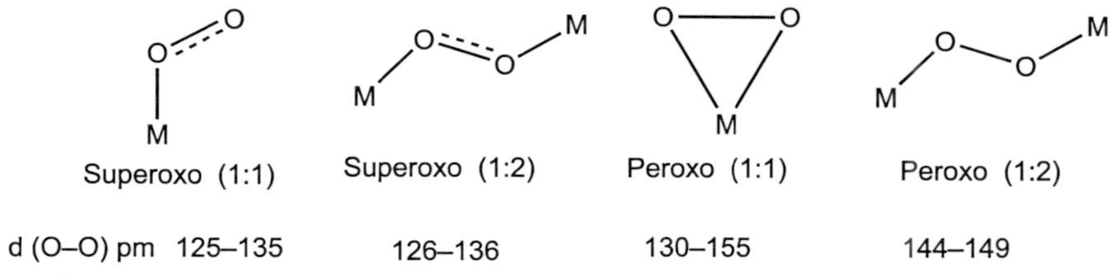

	Superoxo (1:1)	Superoxo (1:2)	Peroxo (1:1)	Peroxo (1:2)
d (O–O) pm	125–135	126–136	130–155	144–149

Fig. 11.3 Different modes of binding of dioxygen to metal centres

In 1963, L Vaska discovered that the dioxygen molecule can reversibly bind to the yellow square planar complex $IrCl(CO)(PPh_3)_2$ which has thereafter been famously called *Vaska's complex*. The dioxygen ligand is bound to iridium with both oxygen atoms by a side-on bonding. The resulting peroxo-dioxygen adduct of Vaska's complex reverts to the parent complex upon heating or purging the solution with an inert gas, signalled by a colour change from orange back to yellow.[11, 12]

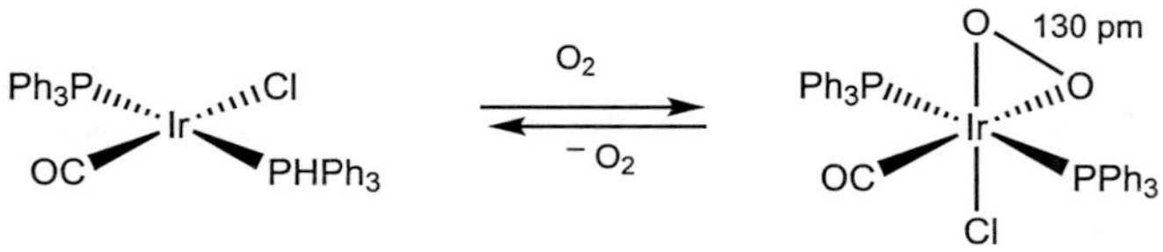

$$IrCl(CO)(PPh_3)_2 \underset{-O_2}{\overset{O_2}{\rightleftharpoons}} IrCl(CO)(PPh_3)_2(O_2)\ (\text{O–O } 130\ pm) \quad (11.4)$$

Well known examples of singly bridging peroxo complexes are the cobalt-based dinuclear complex μ-peroxo-bis[(1,9-bis(2-pyridyl)-2,5,8-triazanonane)cobalt(III)] tetraiodide (PYDIEN) (O–O distance of 1.489 Å), and its closely related compound μ-peroxo-bis[(ethylenediamine) (diethylenetriamine)cobalt(III)] perchlorate (O–O = 1.488 Å).[13,14]

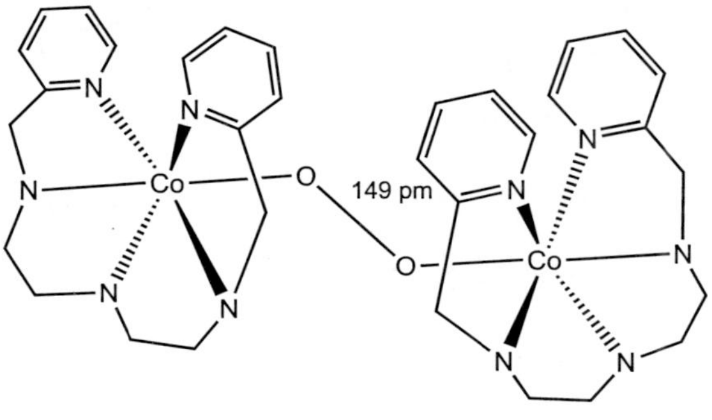

Fig. 11.4 μ-peroxo-bis[(1,9-bis(2-pyridyl)-2,5,8-triazanonane)cobalt(III)]

The binding of O_2 to the iron of hemoglobin has been envisaged to be as a superoxide ion. The first step in the activation of O_2 by the iron–porphyrin complex involves binding to the Fe(II) ion. Initial attempts to prepare such 1:1 Fe–O_2 adducts were met with difficulty due to the propensity of these early models to undergo auto-oxidation via the (μ-η^1:η^1-peroxo) diiron intermediates. The presence of superoxide has been found to be a better model and was confirmed by various techniques, including infrared, X-ray absorption and Mössbauer spectroscopy. Since superoxide in the free state is highly toxic, nearly all organisms living in the presence of oxygen have the superoxide-scavenging enzyme called superoxide dismutase, or SOD.[15]

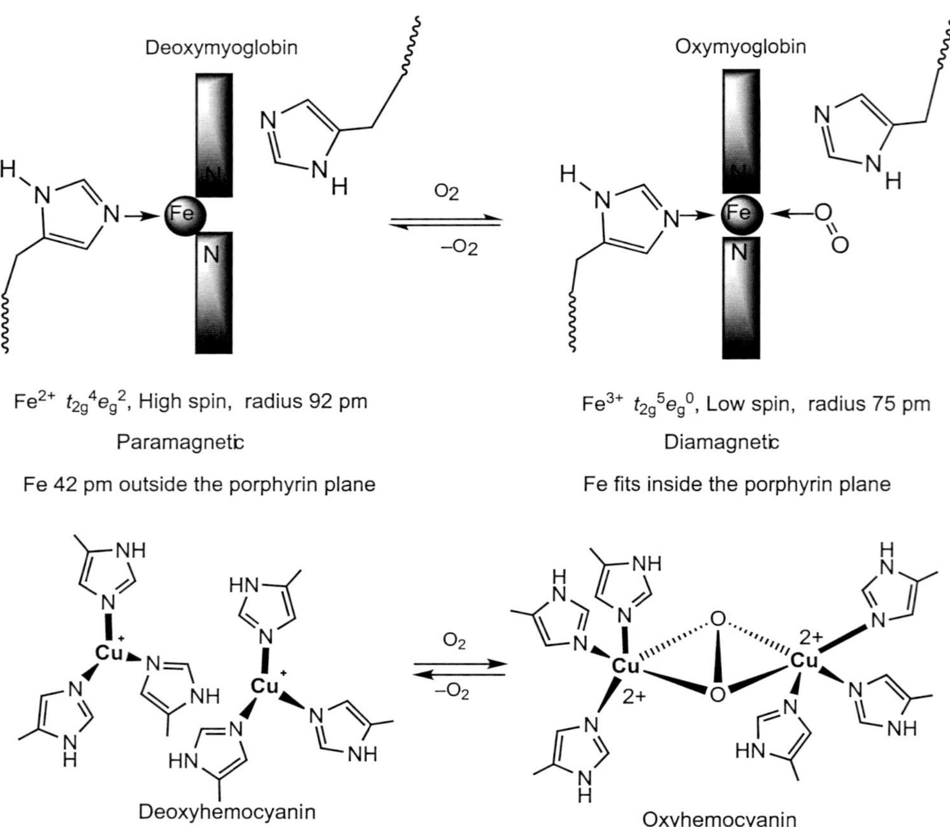

Fig. 11. 5 Deoxygenated and oxygenated forms of hemoglobin and hemocyanin

There are four major biological oxygen carriers which bind in different modes to metal centres such as iron and copper. These four oxygen carriers are hemoglobin, hemocyanin, hemerythrin and chlorocruorin. Hemocyanin, found in spiders, crustaceans, some molluscs, octopuses, squids, and so on, is copper-based and O_2 binds by the side on mode. When deoxygenated, it is colourless and when oxygenated it is blue coloured. Hemerythrin is found in some types of marine worms such as peanut worms, brachiopods, and so on. It is iron-based and oxygen binds in the end-on mode but one end is bound to the H of a bridging OH unit. It is colourless in the deoxygenated state and violet-pink coloured in the oxygenated form.

Hemoglobin found in blood is made up of four subunits, each having a heme unit made of a porphyrin ligand. The deoxygenated form is deep red in colour while the oxygenated form is red in colour. Chlorocruorin is very similar to the heme unit except that one of the vinyl substituents ($CH{=}CH_2$) of the porphyrin unit is replaced by an aldehyde (CHO) unit. Chlorocruorin is light green in the deoxygenated form and the oxygenated form is green in colour.

Deoxyhemerythrin
colourless

O_2 / $-O_2$

Oxyhemerythrin
purple red

Fig. 11. 6 Deoxygenated and oxygenated forms of hemerythrin

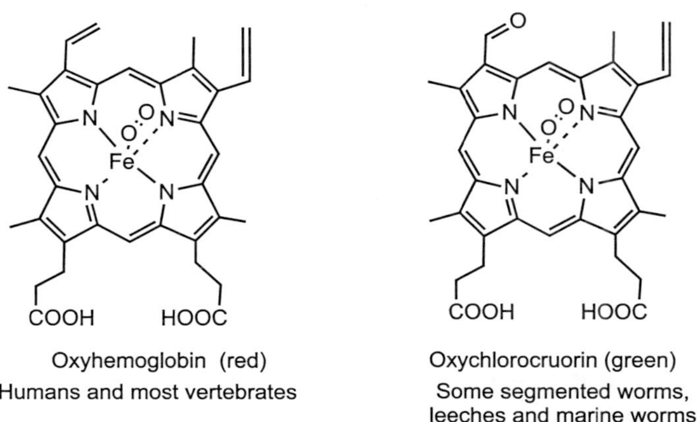

Fig. 11.7 Comparison of structures of oxyhemoglobin and oxy-chlorocruorin

11.7 Ozone: Properties

Ozone is the only allotrope of oxygen. This pale blue gas with a distinctive pungent smell can be detected by human beings especially after a thunderstorm. Electric motors using brushes have the ability to generate ozone as a result of the repeated sparking inside the unit. As a result, even traces of O_3 near rubber tyres have been found to lead to cracks in the tyres reflecting its high reactivity towards alkene bonds. Ozone is diamagnetic and on cooling, converts to a dark blue liquid at −112°C and further to a violet-black solid on cooling to −192.2°C. Experimental evidence obtained from microwave spectroscopy indicates that ozone is a bent molecule and the O–O distance is 1.272 Å and the O–O–O angle 116.78°.[16]

The discovery of ozone is attributed to two scientists. In 1785, Martinus van Marum from the Netherlands, while conducting experiments involving electrical sparking above water noticed a rather unusual smell. He attributed this smell the electrical reactions, and did not realise that he had created a new allotrope of oxygen. Half a century later, C F Schönbein detected the same smell associated with lightning bolts. In 1839, Schönbein was successful in isolating the gaseous chemical and named it 'ozone', as *ozein* in Greek means 'to smell'.

11.7.1 Synthesis of Ozone

Two basic methods of producing ozone, namely, ultraviolet and corona discharge methods, have been developed. Corona discharge makes ozone when a high voltage is applied to a metallic grid sandwiched between two dielectrics. The high voltage travels through the dielectric material to a grounded screen and during the process, creates ozone from the oxygen present around the grid. Formation of a corona, seen as a bluish glow, is observed when the electric field strength around a conductor is high enough to form a conductive region, but not enough to cause electrical arcing to nearby objects.

Ultraviolet (UV) light creates ozone when an oxygen molecule is irradiated with a wavelength of 254 nm, which also happens in the atmosphere. The molecule (O_2) splits into two atoms (O), one of which combines with another oxygen molecule (O_2) to form ozone (O_3). Standard UV ozone generators are not expensive, but they usually produce ozone with a low concentration (0.5% or lower). Also it requires the oxygen to be exposed to the UV for a longer period.

The relative stability indicated by half life for ozone in distilled water at 25°C is about 15 minutes. The half life for ozone gas in a glass container at 20°C is 3 days, however, humidity of 50% will cut that down to 11 hours.

11.7.2 Reactions of Ozone

Molecular ozone with an oxidation potential of 2.07 V is among the most powerful oxidising agents known, more powerful than H_2O_2 (1.78 V), permanganate (1.70 V) and O_2 (1.23 V). It is also unstable at high concentrations, decaying to ordinary diatomic oxygen. It is also a useful reagent to deactivate toxic species such as cyanide converting it to non-toxic cyanate.

The most utilised reaction of ozone is ozonolysis. Alkenes can be oxidatively cleaved by ozone (ozonolysis), giving alcohols, aldehydes, ketones and carboxylic acids, depending on the second step of the workup. Ozone oxidises H_2S to sulfurous acid and sodium chloride to sodium hypochlorite.[15]

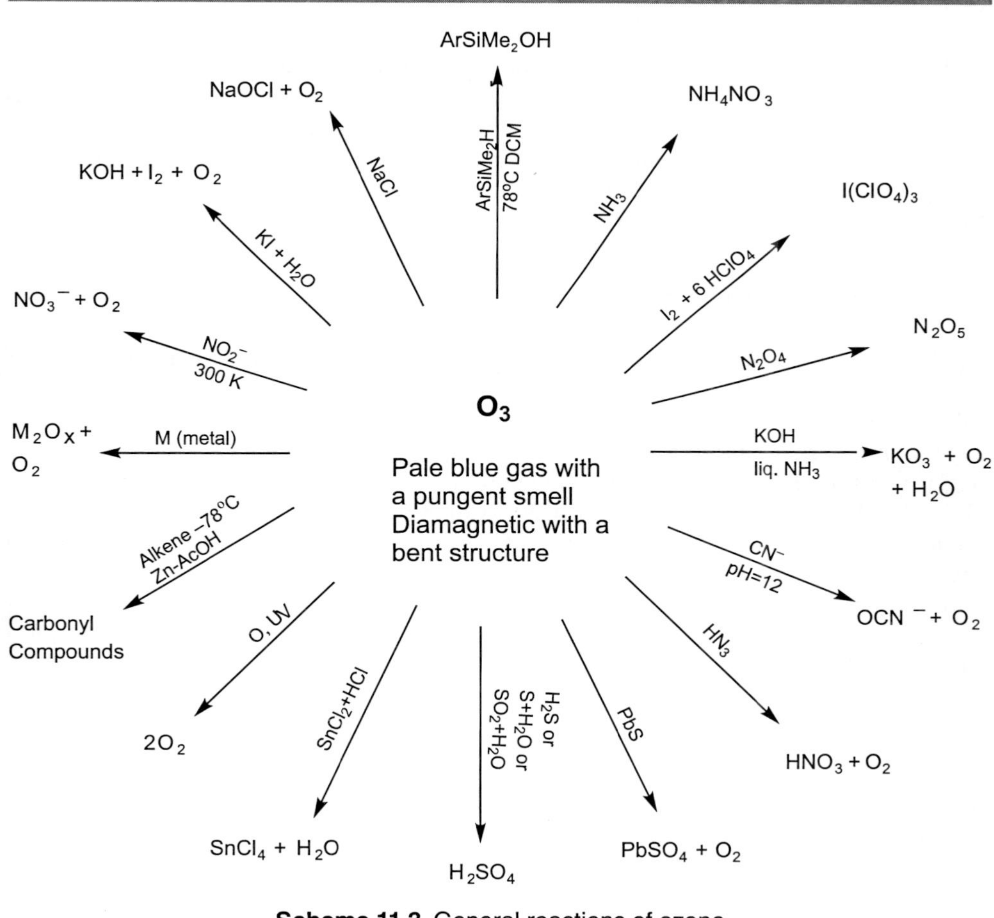

Scheme 11.3 General reactions of ozone

11.7.3 Uses of Ozone

Ozone is very powerful and an effective eliminator of different types of smoke and odours such as tobacco smoke, phenol vapours, cooking odours, smell of pets, mould and other household odours. Turning on an ozone generator in a room with a bad odour, removes all odours and converts it into a fresh odour-free room often in less than 20–60 minutes. Ozone is an extremely useful tool in the area of fire and flood restoration of buildings. In such a damaged building, ozone will eliminate smoke odours in the carpets, curtains, wood, cupboards, clothing, and so on. In a flood-damaged building, ozone can be used once all the moisture has been removed. The ozone will eliminate musty odours and also prevent growth of fungus and other microorganisms. Odours and bacteria are significantly reduced by ozone in public toilets and washrooms and also kills the germs than can lurk on surfaces of bathroom and toilet fittings.

The presence of sulfur in materials can lead to the formation of sulfur-based compounds such as hydrogen sulfide, carbonyl sulfide, sulfur dioxide, carbon disulfide, and thiols during a fire. Thiols tend to get adsorbed on surfaces and produce a lingering odour which gets removed once it is oxidised. A systematic analysis of compounds generated from a home on fire indicates the presence of aldehydes, ketones, alcohols, carboxylic acids, acetophenone, benzyl alcohol, methoxyphenol, hydroxybenzaldehyde, methylphenols and naphthalene. Most of these compounds upon oxidation with ozone lose their characteristic smell.

11.8 Hydrogen Peroxide: Properties

In the chemical industry, hydrogen peroxide (H_2O_2) is widely used as an oxidant principally as its degradation products are only water and oxygen. It has therefore played a significant role in environmentally friendly methods in the chemical industry. Although there are 4–5 methods for its synthesis, it was produced on an industrial scale by the anthraquinone oxidation (AO) process. It was first prepared by Thenard in 1818 by the reaction of barium peroxide with nitric acid. It is an excellent oxidant which is effective over the whole pH range with high oxidation potential ($E^0 = 1.763$ V at pH 0, $E^0 = 0.878$ V at pH 14). An added advantage of using it is that water is the only byproduct of the oxidation reaction. The impure H_2O_2 is purified and concentrated and is usually marketed with an added stabiliser as aqueous solutions at concentrations of 30, 35, 50 and rarely 70% by weight. The most common stabiliser of H_2O_2 is acetanilide.

11.8.1 Synthesis of H_2O_2

Initially hydrogen peroxide was prepared by the reaction of barium peroxide (prepared by high temperature oxidation of barium oxide) with sulfuric acid. Thenard who prepared this for the first time had used nitric acid instead of sulfuric acid.[17]

$$2\,BaO + O_2 \underset{820^\circ C}{\overset{500^\circ C}{\rightleftharpoons}} 2\,BaO_2 \xrightarrow{H_2SO_4} BaSO_4 + H_2O_2 \qquad (11.5)$$

Another early method involves the hydrolysis of ammonium peroxydisulfate (persulfate).

$$(NH_4)_2SO_4 \xrightarrow[-NH_3]{\Delta} (NH_4)HSO_4 \xrightarrow{\text{electrolysis}} (NH_4)_2\,[{}^{-}O-S(=O)_2-O-O-S(=O)_2-O^{-}] \qquad (11.6)$$

$$(NH_4)_2S_2O_8 \xrightarrow[-H_2O_2]{H_2O} (NH_4)HSO_4$$

Today, more than 95% of the world's hydrogen peroxide is manufactured by the anthraquinone process, which was developed by BASF in 1936 and patented in 1939. Many derivatives of anthraquinone can be used but 2-ethylanthraquinone is chosen because of its high selectivity. The hydrogenation can reach 90% selectivity by using 2-ethylanthraquinone. It begins with the reduction of 2-ethylanthraquinone to the corresponding anthrahydroquinone, typically by hydrogenation on a supported palladium catalyst. The anthrahydroquinone then undergoes oxidation by the oxygen in air to regenerate the starting anthraquinone, with hydrogen peroxide being produced as a byproduct. One reason for the success of the anthraquinone method is that H_2O_2 is produced continuously at mild temperatures and without direct contact of O_2 and H_2. However, this process does have some inefficiencies caused by mass-transport limitations, as well as organic contamination of H_2O_2 during production.

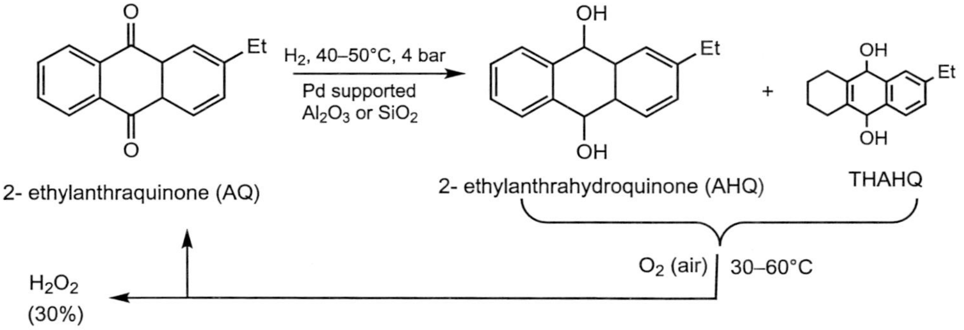

2- ethylanthraquinone (AQ) $\xrightarrow[\text{Pd supported } Al_2O_3 \text{ or } SiO_2]{H_2,\ 40–50°C,\ 4\ bar}$ 2- ethylanthrahydroquinone (AHQ) + THAHQ

AHQ + THAHQ $\xrightarrow{O_2 \text{ (air)},\ 30–60°C}$ AQ + H_2O_2 (30%)

Scheme 11.4 Industrial synthesis of H_2O_2

Other methods of lesser importance are, aerial oxidation of secondary alcohols at high temperatures and pressures, and electrochemical methods involving aqueous NaOH solutions.

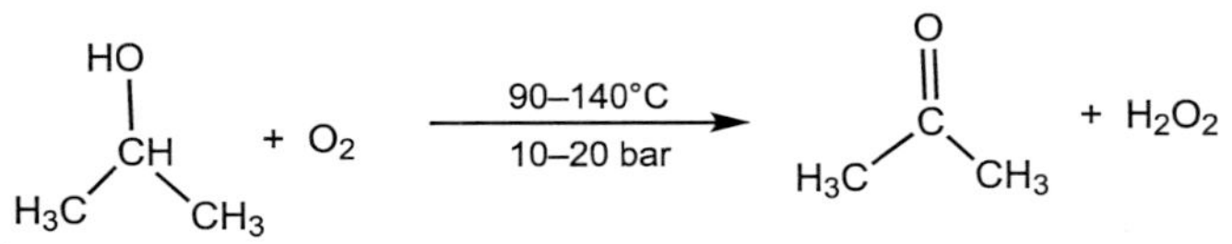

$$(CH_3)_2CH\text{–}OH + O_2 \xrightarrow[10–20\ bar]{90–140°C} (CH_3)_2C{=}O + H_2O_2 \tag{11.7}$$

11.8.2 Reactions of H_2O_2

Hydrogen peroxide can act as both an oxidising agent and a reducing agent in its reactions.

$$\underset{0}{O_2} \xleftarrow{\text{As reducing agent: gets oxidised to oxygen}} \underset{-1}{H_2O_2} \xrightarrow{\text{As oxidising agent: gets reduced to water}} \underset{-2}{H_2O} \tag{11.8}$$

Oxygen valence **0** **–1** **–2**

$$H_2O_2 + 2H^+ + 2e \longrightarrow 2\,H_2O \quad E^0 = 1.776\text{ V} \tag{11.9}$$

$$H_2O_2 \longrightarrow O_2 + 2H^+ + 2e \quad E^0 = -0.682\text{ V} \tag{11.10}$$

Examples of oxidation and reduction reactions of H_2O_2 in acidic, basic and neutral medium are listed below. Reduction reactions are always accompanied by the release of dioxygen gas.

Oxidising action in acidic solution

$$PbS + 4\,H_2O_2 \longrightarrow PbSO_4 + 4\,H_2O \tag{11.11}$$

$$2\,Fe^{2+} + 2\,H^+ + H_2O_2 \longrightarrow 2\,Fe^{3+} + 2\,H_2O \tag{11.12}$$

$$I^- + H_2O_2 + 2\,H^+ \longrightarrow I_2 + 2\,H_2O \tag{11.13}$$

$$SO_3^- + H_2O_2 \longrightarrow SO_4^{2-} + H_2O \tag{11.14}$$

$$NO_2^- + H_2O_2 \longrightarrow NO_3^- + H_2O \tag{11.15}$$

$$SO_2 + H_2O_2 \longrightarrow H_2SO_4 \tag{11.16}$$

$$AsO_3^{3-} + H_2O_2 \longrightarrow AsO_4^{3-} + H_2O \tag{11.17}$$

Reducing action in acidic solution

$$2\,ClO_3^- + H_2O_2 + 2\,H^+ \longrightarrow O_2 + 2\,ClO_2 + 2\,H_2O \tag{11.18}$$

$$2\,MnO_4^- + 6\,H^+ + 5\,H_2O_2 \longrightarrow 2\,Mn^{2+} + 8\,H_2O + 5\,O_2 \tag{11.19}$$

$$HOCl + H_2O_2 \longrightarrow H_3O^+ + Cl^- + O_2 \tag{11.20}$$

$$2\,Ce^{4+} + H_2O_2 \longrightarrow 2\,Ce^{3+} + 2\,H^+ + O_2 \tag{11.21}$$

$$2\,CrO_5 + 7H_2O_2 + 6\,H^+ \longrightarrow 2Cr^{3+} + 7O_2 + 10\,H_2O \tag{11.22}$$

Reducing action in alkaline solution

$$I_2 + H_2O_2 + 2\,OH^- \longrightarrow 2\,I^- + 2\,H_2O + O_2 \tag{11.23}$$

$$2\,MnO_4^- + 3\,H_2O_2 \longrightarrow 2\,MnO_2 + 3\,O_2 + 2\,H_2O + 2\,OH^- \tag{11.24}$$

$$2\,Fe^{3+} + H_2O_2 + OH^- \longrightarrow 2\,Fe^{2+} + 2\,H_2O + O_2 \tag{11.25}$$

$$ClO_2 + NaOH + H_2O_2 \longrightarrow NaClO_2 + O_2 + 2\,H_2O \tag{11.26}$$

Reducing action in neutral medium

$$H_2O_2 + Cl_2 \longrightarrow 2\,HCl + O_2 \qquad (11.27)$$

$$H_2O_2 + O_3 \longrightarrow H_2O + 2\,O_2 \qquad (11.28)$$

$$H_2O_2 + Ag_2O \longrightarrow H_2O + O_2 + 2\,Ag \qquad (11.29)$$

$$3\,H_2O_2 + Au_2O_3 \longrightarrow 2\,Au + 3\,H_2O + 3\,O_2 \qquad (11.30)$$

In addition to these reactions, H_2O_2 also brings about addition or adduct formation with species such as urea and acetone and also brings about oxidation and peroxide formation of organic compounds.

Addition

$$O_3 + H_2O_2 \longrightarrow H_2O_5 \qquad (11.31)$$

$$NH_2CONH_2 + H_2O_2 \longrightarrow H_2NCONH_2.HOOH \qquad (11.32)$$

$$CH_3COCH_3 + 3\,H_2O_2 \longrightarrow \qquad (11.33)$$

H3C CH3
O O
O O
H3C CH3
H3C O—O CH3

Oxidation and peroxide formation of organic compounds

$$RCOOH + H_2O_2 \longrightarrow RCOO(O)H + H_2O \qquad (11.34)$$

$$RCOR + H_2O_2 \longrightarrow RC(O)OR + H_2O \qquad (11.35)$$

$$RSR + H_2O_2 \longrightarrow RS(O)R + H_2O \qquad (11.36)$$

$$ArNH_2 + KOH + H_2O_2 \longrightarrow ArNO \qquad (11.37)$$

$$R_2CHOH + H_2O_2 \longrightarrow RCOR \qquad (11.38)$$

Disproportionation

On storing for long periods, it can also undergo disproportionation leading to the formation of O_2 and H_2O.

$$2\,H_2O_2 \longrightarrow 2\,H_2O + O_2 \qquad (11.39)$$

11.8.3 Fenton's Reaction

H J H Fenton, in 1893, published a paper in which he described the reaction of tartaric acid with H_2O_2 and Fe^{2+}. He showed that an intense violet colour was produced and the tartaric acid gets converted to 2,3-dihydroxymaleic acid.

$$\text{(2R,3R) tartaric acid} \xrightarrow{Fe^{2+},\ H_2O_2} \text{2,3-dihydroxy maleic acid} \quad (11.40)$$

(2R,3R) tartaric acid

2,3-dihydroxy maleic acid

He later showed that some other hydroxy acids are also oxidised in the presence of this reagent and that Fe^{3+} does not bring about this reaction. Haber and Weiss suggested in 1934 that in the decomposition of H_2O_2 catalysed by iron salts, the OH radical is formed as an active intermediate via the oxidation of ferrous ion by hydrogen peroxide.[18]

$$Fe^{2+} + H_2O_2 + H^+ \longrightarrow Fe^{3+} + H_2O + OH^{\bullet} \quad (11.41)$$

However since the requirement of iron is only in catalytic amounts, detailed studies showed that the ferrous iron is regenerated in the reaction.

The modern view of this reaction is that iron(II) is oxidised by hydrogen peroxide to iron(III), forming an $OH\bullet$ and an OH^- in the process. Iron(III) is then reduced back to iron(II) by another molecule of hydrogen peroxide, forming a hydroperoxyl radical ($HOO\bullet$) and a proton. The overall effect is a disproportionation of hydrogen peroxide to create two different oxygen-radical species, with water ($H^+ + OH^-$) as a byproduct.

$$Fe^{2+} + H_2O_2 \longrightarrow Fe^{3+} + OH^- + OH^{\bullet} \quad (11.42)$$

$$Fe^{3+} + H_2O_2 \longrightarrow Fe^{2+} + HOO\bullet + H^+ \quad (11.43)$$

Fenton's reaction was initially a preparative tool to synthesise hydroxylated organic compounds. A century later, this reaction plays a very important role in free radical chemistry. It is now used to oxidise many organic contaminants and destroy chlorinated unsaturated compounds such as tri- and tetrachloroethylenes. Fenton's reaction is very useful for the hydroxylation of arenes, such as, in the conversion of benzene to phenol.

$$Fe^{2+} + H_2O_2 + C_6H_6 \longrightarrow C_6H_5OH + Fe^{3+} + H_2O \quad (11.44)$$

11.8.4 Uses of H_2O_2

The major industrial uses of H_2O_2 are shown in Scheme 11.5. Almost 50% of it is used in the manufacture of paper where it is used for bleaching. It is used for making perborates and percarbonates which are additives used in detergents especially where bleaching is required to remove stains and a chlorine-free milder bleach is required.

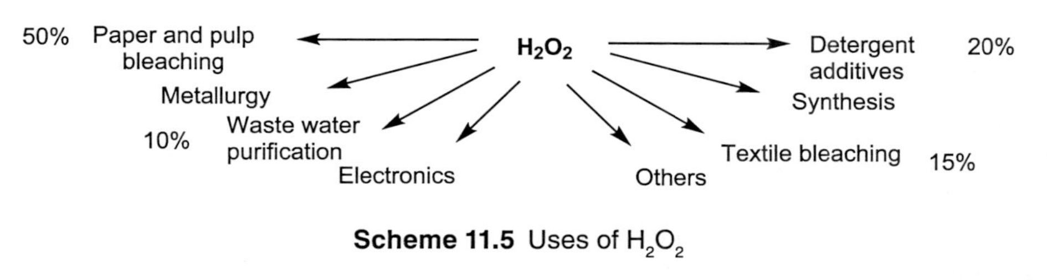

Scheme 11.5 Uses of H_2O_2

11.9 About the Element: Sulfur

Sulfur which makes up 3% of the earth's mass, is an odourless, brittle, air-stable, pale yellow solid and has been known to humanity from prehistoric times. It was referred to as 'brimstone' as it was often found at the brim of natural hot springs and geysers and is mentioned in the Bible also. In the classic litreature 'Odyssey' written by Homer around 800 BC, Odysseus says '*Bring sulfur, old nurse, that cleanses all pollution, and bring me fire, that I may purify the house with sulfur*'. In 424 BC, the tribe of Bootier destroyed a city's walls using a burning mixture of coal, sulfur, and tar. A Chinese litreature in the year AD 808 provides the first method for making gunpowder using saltpeter (KNO_3), carbon and sulfur. When the Shakespearian character Othello asks for punishment, a suggestion he makes is to '*roast him in sulfur*'. Sulfur was recognised as a chemical element by 1789 when Antoine Lavoiser included it in his famous list of elements. In 1823, German chemist Eilhard Mitscherlich discovered sulfur's allotropy—he showed that the crystal shapes of sulfur obtained from cooling molten sulfur (monoclinic) were different from those obtained when the element crystallised from a solution (orthorhombic). It is interesting to note that Berzelius introduced the term allotropy only in 1841 and used monoclinic and rhombic forms of sulfur to explain the concept. However, in the modern definition of allotropy and polymorphism these are classified as polymorphs rather than allotropes.

> *"It is no exaggeration to say, we may fairly judge the commercial prosperity of a country from the amount of sulfuric acid it consumes"*
> *Justus von Liebig 1843*

Most of the sulfur that is produced in the world is used in the manufacture of sulfuric acid (H_2SO_4). Large amounts of sulfuric acid are used each year to make fertilisers, lead–acid batteries, and in many industrial processes. Smaller amounts of sulfur are used to vulcanise natural rubbers and as an insecticide. Sulfur is a component of many common minerals, such as galena (PbS), gypsum ($CaSO_4 \cdot 2H_2O$), pyrites (FeS_2), sphalerite (ZnS or FeS), cinnabar (HgS), stibnite (Sb_2S_3), epsomite [$MgSO_4 \cdot 7(H_2O)$], celestite ($SrSO_4$) and barite ($BaSO_4$). In addition to sulfuric acid, sulfur forms other interesting compounds. Hydrogen sulfide (H_2S) is a gas that smells like rotten eggs. Sulfur dioxide (SO_2), formed by burning sulfur in air, is used as a bleaching agent, solvent, disinfectant and as a refrigerant. When combined with water (H_2O), sulfur dioxide forms sulfurous acid (H_2SO_3), a weak acid which is found to be stable in the gas phase.

11.10 Isolation and Purification of the Element

The traditional method of isolation of elemental sulfur was by the Frasch process which made use of superheated water. In this method three concentric tubes are introduced into the sulfur deposit. Superheated water (165°C, 2.5–3.0 MPa) is introduced into the deposit through the outermost tube. Sulfur (m.p. 115°C) melts and flows into the middle tube. As the molten sulfur has greater density, water pressure alone is unable to force the sulfur to the surface and therefore hot air is introduced via the innermost tube to froth the sulfur thus making it less dense and pushing it to the surface. However, the Frasch process has almost become redundant nowadays and only a couple of such sulfur mines exist in the world. Currently a major percentage of sulfur which is industrially produced is from hydrodesulfurisation processes of petroleum products.

The feedstocks of petroleum such as naptha, kerosene, diesel and heavier oils contain a range of organic sulfur compounds in varying concentrations. These include thiols, thiophenes, sulfides and disulfides which are basically products of degradation of sulfur-containing biological products formed under pressure during the formation of fossil fuels such as crude oil. These compounds if present in fuels will result in SO_2 formation during combustion which will lead to air pollution and therefore have to be removed. In the first step of hydrodesulfurisation, these sulfur-based compounds are catalytically converted to H_2S and hydrocarbons. The normal catalyst used for this is MOS_2 although RuS_2 or binary combination of Co and Mo (called cobalt-modified MoS_2 catalysts) have been found to be even more active.

$$C_2H_5SH + H_2 \xrightarrow[MoS_2]{} C_2H_6 + H_2S \tag{11.45}$$

$$C_4H_4S + 4\,H_2 \xrightarrow[[CoMo]]{} C_4H_{10} + H_2S \tag{11.46}$$

Scheme 11.6 Proposed mechanism for hydrodesulfurisation of sulfur containing compounds of petroleum

The gaseous products containing H_2S called *sour gas* is made to undergo a 'sweetening process' where the H_2S is absorbed in amines mainly aliphatic alkanolamines such as diethanolamine (DEA), monoethanolamine (MEA) and methyldiethanolamine (MDEA).

The chemistry involved in the amine treatment of such gases varies to some extent with the particular amine being used. Often it is a reversible ammonium salt formation. In the absorber, the downflowing amine solution absorbs H_2S and CO_2 from the upflowing sour gas to produce a gas free of hydrogen sulfide and carbon dioxide as a product. The resultant amine-rich side product is then routed to a regenerator (a stripper with a reboiler) to produce regenerated amine which is recycled for reuse in the absorber. The stripped overhead gas from the regenerator contains H_2S and CO_2.

Hydrogen sulfide thus produced is converted to elemental sulfur by the Claus process designed by the German chemist Carl Frederich Claus and patented in 1883. The Claus process has two steps thermal and catalytic. The first thermal step of the Claus process involves burning one-third of the H_2S with air in a reactor furnace to form sulfur dioxide (SO_2) according to the following reaction:

$$2\,H_2S + 3\,O_2 \xrightarrow[980\text{–}1540^\circ C]{} 2\,SO_2 + 2\,H_2O + \text{heat} \quad (11.47)$$

In the second catalytic step of the Claus process, the remaining unreacted two-thirds of the hydrogen sulfide undergoes Claus reaction (reacts with SO_2) to form elemental sulfur by the self reduction method as follows:

$$2\,H_2S + SO_2 \underset{\substack{\text{Alumina} \\ 200\text{–}310^\circ C}}{\rightleftharpoons} 3\,S + 2\,H_2O + \text{heat} \quad (11.48)$$

In fact, the majority of the 70 million metric tons of sulfur produced worldwide in 2015 was byproduct sulfur from refineries and other hydrocarbon processing plants.

One can write the overall reaction as

$$10\,H_2S + 5\,O_2 \longrightarrow 2\,H_2S + SO_2 + 7/2S_2 + 8\,H_2O \quad (11.49)$$

This equation shows that in the thermal step alone, two-thirds of the hydrogen sulfide has been converted to sulfur.

11.11 Allotropy and Polymorphism of Sulfur

Among the *p*-block elements, sulfur along with carbon form the maximum number of allotropes. There are about 30 well-characterised allotropes of sulfur and they are broadly classified under ambient pressure allotropes and high pressure allotropes and some of them are only polymorphs. Ambient pressure allotropes have been grouped by their molecular structure as cyclic or acyclic. The 20 crystalline phases, most of them characterised by single crystal X-ray structure determination are:[19]

S_6; S_7 (4 phases); S_8 (3 phases); S_9 (2 phases); S_{10}; $S_6.S_{10}$; S_{11}; S_{12}; S_{13}; S_{14}; S_{15}; S_{18} (exo and endo); S_{20}.

While interest in these allotropes is mostly academic in nature, more than one synthetic route is available for their preparation. The traditional methods involve reaction of sodium thiosulfate with hydrochloric acid (S_6: 69%) or reactions of S_2Cl_2 with NaI or KI. In the latter method, I_2 along with S_6, S_8, S_{12}, S_{18} and S_{20} are formed. A more recent and versatile method involves the use of titanocene pentasulfide which is easily prepared from Cp_2TiCl_2 and aqueous sodium or ammonium polysulfide. The air-stable dark red solid Cp_2TiS_5 reacts with many S–Cl containing compounds at 0–20°C as a sulfur transfer agent and the driving force for this reaction is the formation of Cp_2TiCl_2. Some of the rare polysulfur dichlorides are prepared by reaction of the corresponding cyclic sulfur compound with equimolar amounts of chlorine.

$$S_n + Cl_2 \longrightarrow Cl{-}S_n{-}Cl \qquad n = 2, 6, 8 \tag{11.50}$$

$$Cl{-}S_6{-}Cl + Cl_2 \longrightarrow Cl{-}S_4{-}Cl + S_2Cl_2 \tag{11.51}$$

$$2\,NH_3 + H_2S + 1/2\,S_8 \longrightarrow (NH_4)_2S_5 \tag{11.52}$$

$$(NH_4)_2S_5 + Cp_2TiCl_2 \xrightarrow{CHCl_3} \underset{\text{88\% dark red, air stable solid}}{Cp_2TiS_5} + NH_4Cl \tag{11.53}$$

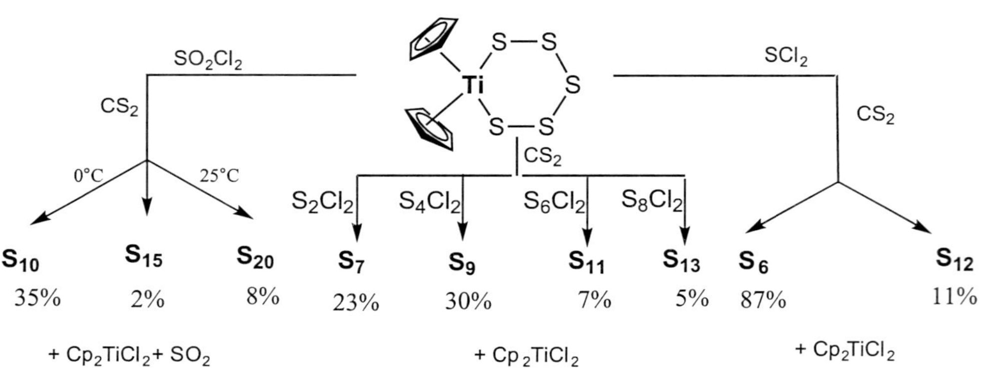

Scheme 11.7 Synthesis of different allotropes of sulfur using Cp_2TiS_5

11.12 Reactions of Sulfur

Reactions of S_8 where the ring structure is retained are not known due to the very weak nature of the S–S bonds, but the closest is the formation of ClS_8Cl at low temperatures. With fluorine, sulfur is oxidised readily to SF_6, an unreactive greenhouse gas while with chlorine it initially forms sulfur monochloride which can be further chlorinated with $FeCl_3$ to sulfur dichloride. On burning in air it forms SO_2, a well known atmospheric pollutant but also the precursor to two well known sulfur-based reagents namely thionyl chloride and sulfuryl chloride. Thionyl chloride is better prepared by the reaction of SO_3 with SCl_2. With NaOH, sulfur gives sodium thiosulfate and sodium sulfide.

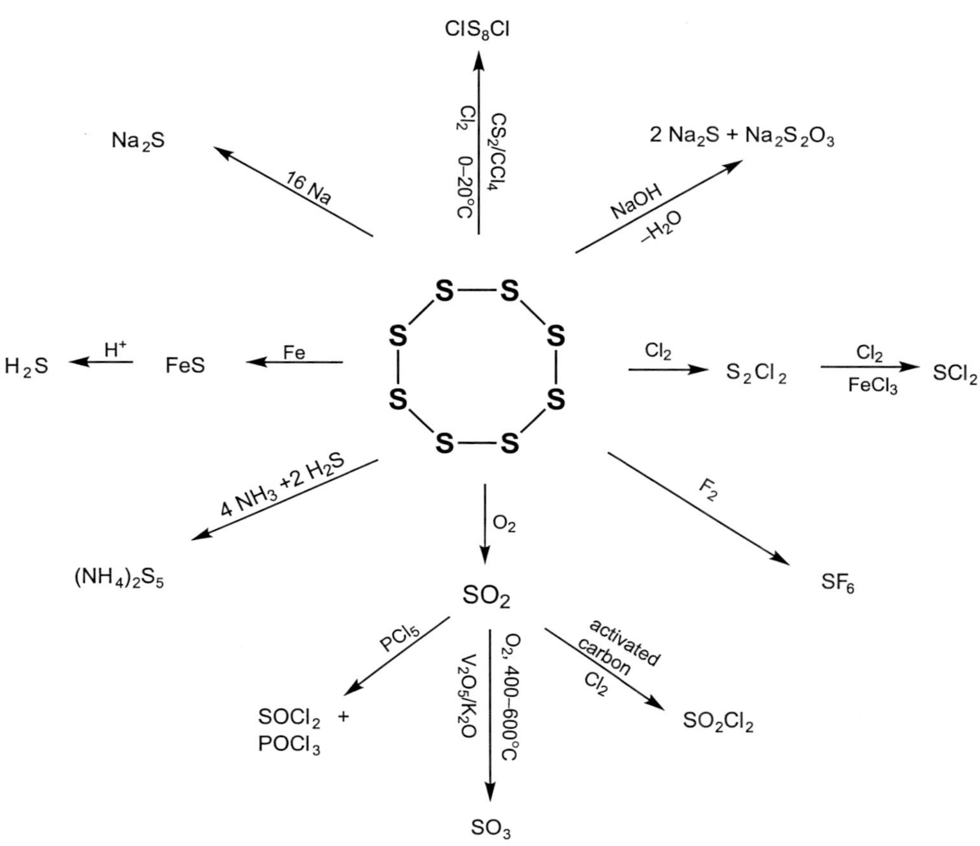

Scheme 11.8 Reactions of elemental sulfur to produce useful reagents

Sulfur dichloride is also the precursor for SF_4, a well known fluorinating agent, S_4N_4 the first sulfur–nitrogen heterocycle and also the notorious mustard gas, $(ClCH_2CH_2)_2S$.

11.13 Hydrogen Sulfide

Hydrogen sulfide is well known as an extremely toxic gas with the smell of rotten eggs which is colourless and also flammable. It is produced by the breakdown of organic matter and human/animal wastes and sewage and also found in natural gas. Being heavier than air it can collect in low-lying and enclosed, poorly ventilated areas. Unlike water, H_2S does not readily form metal complexes. It oxidatively adds to low-valent metal ions as SH and H units. It is slightly soluble in water and acts as a weak acid, giving the ion HS^- ($pK_a = 6.9$ in 0.01–0.1 mol/litre solutions at 18°C).[20]

SCl2
(b.p. 59°C)
1. Cl2 2.NaF
NH3
SO3
H2C=CH2
SF4
(b.p. −38°C)
S4N4
tetasulfur tetranitride
S(O)Cl2
(b.p. 74.6°C)
(ClCH2CH2)2S
mustard gas

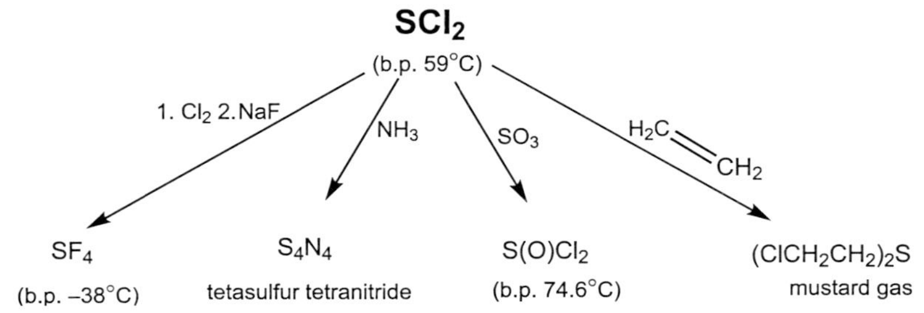

Scheme 11.9 Conversion of SCl_2 to useful reagents and products

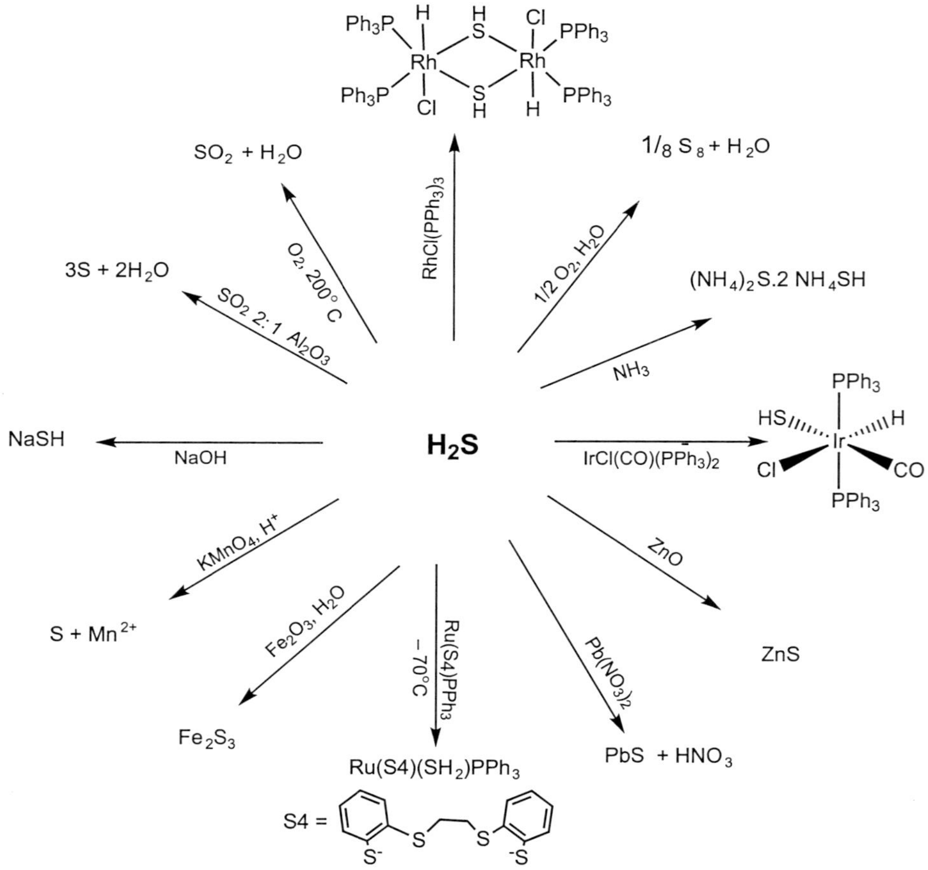

H2S
RhCl(PPh3)3
SO2 + H2O
O2, 200° C
1/8 S8 + H2O
1/2 O2, H2O
3S + 2H2O
SO2 2:1 Al2O3
(NH4)2S.2 NH4SH
NH3
NaSH
NaOH
IrCl(CO)(PPh3)2
KMnO4, H+
S + Mn2+
ZnO
ZnS
Fe2O3, H2O
Fe2S3
Ru(S4)PPh3
− 70°C
Pb(NO3)2
PbS + HNO3
Ru(S4)(SH2)PPh3
S4 =

Scheme 11.10 General reactions of hydrogen sulfide

Compressed H_2S shows superconductivity at –70°C

The phenomenon of superconductivity of certain materials has attracted scientists from the very early days. Conventional superconductivity was first observed in solid mercury at 4.5 K by Kamerlingh Onnes in 1911 while he was exploring how materials behave near liquid helium temperatures. The first non-metallic superconductor was polythiazyl $(SN)_x$, which showed superconductivity at 0.26 K. Superconductivity is demonstrated by two main features. The first being zero electrical resistance and the second, Meissner effect. Meissner effect occurs when a material makes a changeover from the normal to the superconducting phase upon cooling and excludes magnetic flux fields from its interior. The phenomenon of magnetic levitation is caused by the Meissner effect. Among conventional superconductors the highest critical temperature (T_c) was determined for magnesium boride, MgB_2 (39 K). In 1986, a class of cuprate compounds (yttrium barium copper oxide, $YBa_2Cu_3O_7$; T_c = 93 K, –180°C) was found to show unconventional superconductivity which was above liquid nitrogen temperatures. Subsequent research led to more such cuprates and the best among the lot was $HgBa_2Ca_2Cu_3O_y$ which showed superconductivity at 134 K (–139°C) at ambient pressure and 153 K (–120°C) at high pressure (15 GPa). This was the highest T_c achieved till 2015. Post 2000, another class of superconducting compounds based on iron, namely ferropnictides, were also discovered. However, the highest T_c achieved for such compounds was only 55 K. The superconductivity mechanism of both cuprates and ferropnictides were not fully understood but is generally believed to be distinct from the BCS theory of superconductivity which considers superconductivity as a microscopic effect caused by condensation of Cooper pairs into a boson-like state.

In 2015, compressed hydrogen sulfide was found to show superconductivity with the highest T_c observed for any material so far, 203 K (–70°C). Drozodov, Eremets and coworkers of the Max Planck institute for chemistry at Mainz, Germany, subjected a tiny sample of hydrogen sulfide to a pressure of 153 GPa, observed Meissner effect and zero electrical resistance at a record high T_c of 203 K.[21, 22] Interestingly, this temperature actually exists on the earth's surface at places such as Alaska and Antarctica (–74 to –89°C). Note that this particular discovery was not accidental and was motivated by theoretical studies which predicted such a phenomenon. Although the high transition temperature is very encouraging, the fact that the pressures needed (153 GPa), which is almost half of the pressure inside earth's core prevents finding an immediate practical application of H_2S superconductivity. Nevertheless, it has encouraged scientists to look for new superconductors by applying pressure on insulators and to look beyond traditional cuprates and ferropnictides for high temperature superconductivity.

Preliminary studies trying to determine the species responsible for this superconductivity have indicated that H_2S on application of pressure undergoes decomposition to higher hydrides. Although initially it was predicted that H_3S ($3H_2S \rightarrow 2H_3S + S$) is the species responsible, later studies based on DFT calculations indicated that the perovskite compound $(SH)^-(H_3S)^+$ is thermodynamically more favourable than H_3S. The decomposition of H_2S can also take the route ($2H_2S \rightarrow H_3S^+ + SH^-$).[23]

New studies have predicted that it may be even possible to raise the T_c to 7°C by substituting a small fraction of sulfur in H_2S by phosphorus and applying high pressures. It is highly encouraging that indications are there for room temperature superconductivity to be eventually possible with polyhydrides but at high pressures. This breakthrough, apart from being of fundamental significance, opens up a new dimension to explore the diversity of chemical bonding and structures of compounds at high pressures.

The reason for the difference in the chemical behaviour of H_2S compared to water is that equilibrium with respect to the formation of solid sulfides is established much more rapidly with H_2S than the corresponding aqua systems. Hydrogen sulfide gas has been used as a reagent for the identification of metal ions in inorganic qualitative analysis (by generating it using a Kipps apparatus). The following general colour variations of metal sulfides were utilised to segregate and identify some of the metal ions.

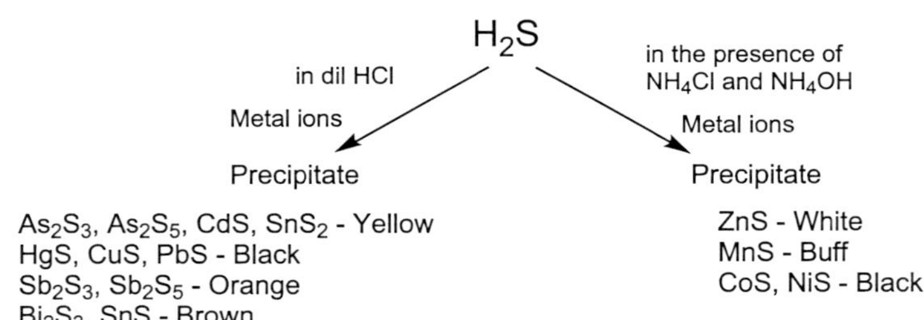

Scheme 11.11 Identification of metal ions by using H_2S as a test reagent

Due to the high toxicity of H_2S gas, and the possible chance of its leakage, these tests have been recently discarded from use in schools and undergraduate laboratories and have been restricted to research labs.

Recent studies have demonstrated that hydrogen sulfide (H_2S) protects against multiple cardiovascular disease states in a similar manner as nitric oxide (NO). H_2S therapy also has been shown to augment NO bioavailability and signalling. The cardio-protective role of garlic is possibly related to the conversion of the polysulfide group in allicin to H_2S.[20]

11.14 Sulfur Oxides

While sulfur is known to form many oxides such as SO, SO_2, SO_3, S_7O_2, S_6O_2, S_2O_2, the two most prominent and useful oxides of sulfur are sulfur dioxide and sulfur trioxide. Scheme 11.12 lists some of the methods of preparing SO_2.

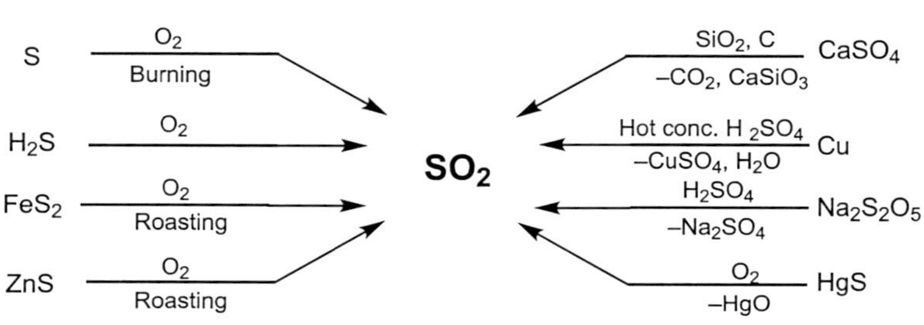

Scheme 11.12 Synthetic methods for SO_2

The synthesis and uses of sulfur trioxide are given along with the contact process for the industrial preparation of sulfuric acid.

11.15 Oxoacids of Sulfur

There are at least 10 documented examples of sulfur oxoacids with the oxidation state of sulfur varying from +2 to +6. While some of these acids are unstable, their salts are quite stable. Figure 11.8 lists these oxoacids along with their structures. The common and well known among these are the sulfurous and sulfuric acids whose chemistry is discussed in more detail.

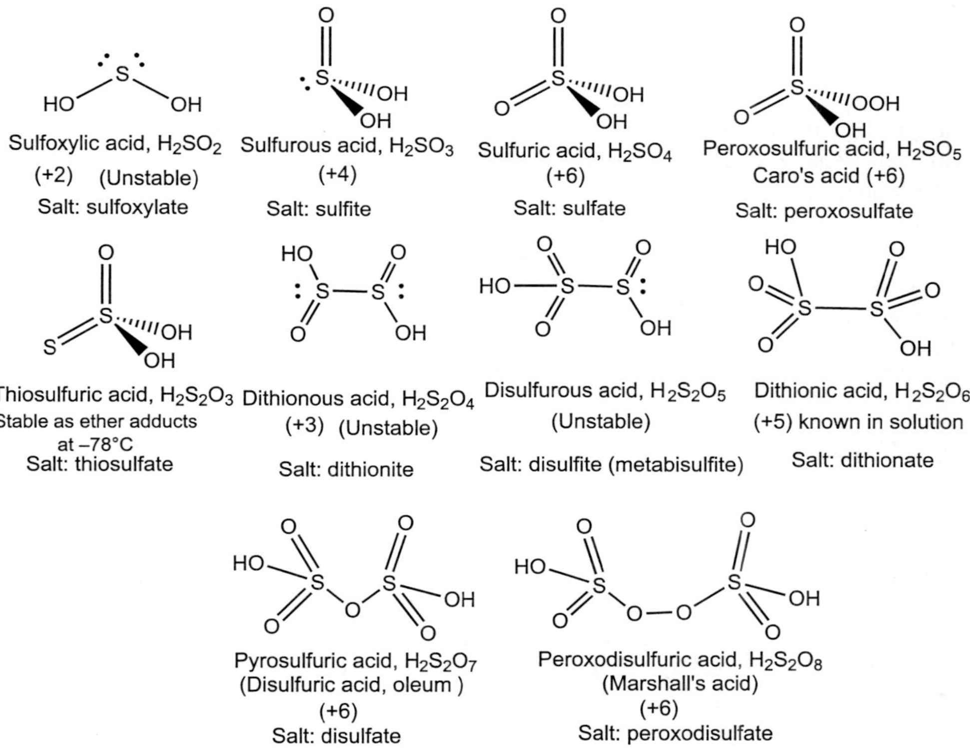

Fig. 11.8 Oxoacids of sulfur

11.16 Sulfurous Acid

The acid solution obtained when sulfur dioxide is dissolved in water is called sulfurous acid, H_2SO_3. Similar to carbonic acid, H_2CO_3 and carbamic acid $H_2NC(O)OH$, sulfurous acid is unstable and is nonexistent in the free form in the solution phase and is reported to be stable only in the gas phase. The nonexistence in solution is not an intrinsic instability but a result of bimolecular reactions. Salts containing the sulfite ion, SO_3^{2-}, such as Na_2SO_3 are well known and are also good reducing agents. It is used in food preservation.

Sodium metabisulfite $Na_2S_2O_5$, and potassium metabisulfite $K_2S_2O_5$, are well known antioxidants and food preservatives. Protonation of SO_3^{2-} leading to HSO_3^- and saturating the solution with SO_2 yields metabisulfite.

$$2[HSO_3]^- \rightleftharpoons H_2O + [S_2O_5]^{2-} \qquad (11.54)$$

Sulfurous acid is also a mild bleach, and is used for bleaching materials which may be damaged by chlorine-based bleaches.

11.17 Sulfuric Acid

Sulfuric acid, one of the most well known and widely used chemicals known to humanity was little known before the 16th century. Historical documents from Asia indicate that it was first prepared in 940 AD by the physician, alchemist and writer from Baghdad, Abu Bekr Mohammad Ibn Zakariya Razi (Rhazes), possibly by roasting green vitriol ($FeSO_4$). The European scientific community credits its discovery to Johann Van Helmont (c.1600) by destructive distillation of the same compound, green vitriol, and by burning sulfur. Initial short-lived methods for its synthesis include roasting $FeSO_4$ and burning sulfur with potassium nitrate. The two methods which became popular later on were the lead chamber process invented by John Roebuck in 1746, and the contact process originally developed (ca.1830) by Peregrine Phillips in England.[24]

11.18 The Lead Chamber Process

The lead chamber process, the older of the two processes, produces a relatively dilute acid (62%–78% H_2SO_4). Many people initially take for granted that lead is a catalyst for this reaction but the fact is that the catalysts are oxides of nitrogen. The idea of introducing lead in this process came from the need to find a material which is not very reactive with sulfuric acid and at the same time can absorb sudden heat fluctuations which occur during the formation of H_2SO_4 from SO_3 which used to result in the breaking of low quality sodaglass containers of the times. Obviously modern day borosilicte glass which has the ability to withstand sudden temperature variations was not available then. (Borosilicate glass was first developed by the German glassmaker Otto Schott in the late 19th century.)

In the lead chamber process, hot sulfur dioxide gas enters the bottom of a reactor (Glover tower) where it is washed with nitrous vitriol (sulfuric acid with nitric oxide and nitrogen dioxide dissolved in it). Part of the sulfur dioxide is oxidised to sulfur trioxide and dissolved in the acid wash to form Glover acid. From the Glover tower, a mixture of gases gets transferred to a chamber lined with lead blocks where it is reacted with more water. The chamber used to be a large, box-like room. Sulfuric acid formed condenses on the walls and collects on the floor of the lead chamber. Many lead chambers were connected in a series and the gases pass through each in succession. The acid produced, also called fertiliser acid, contains 62%–68% H_2SO_4. After the gases have passed through the chambers they are passed into a reactor called the Gay-Lussac tower where they are washed with cooled concentrated acid and the nitrogen oxides and unreacted sulfur dioxide dissolve in the acid to form the nitrous vitriol, which is recirculated to the Glover tower.

In 1910, about 80% of the sulfuric acid made in Europe and North America came from the lead chamber process. However by 1960, the total output of acid from the lead chamber process came down to 15%. Currently, contact process has completely taken over the sulfuric acid manufacture.[24]

11.19 The Contact Process

In the contact process, purified sulfur dioxide and air are mixed, heated to about 450°C, and passed over a catalyst. The catalyst is usually platinum on a silica or asbestos carrier or vanadium pentoxide on a silica carrier. The sulfur dioxide is oxidised to sulfur trioxide. The sulfur trioxide is cooled and passed through two towers. In the first tower it is washed with oleum (fuming sulfuric acid, 100% sulfuric acid with sulfur trioxide dissolved in it). In the second tower it is washed with 97% sulfuric acid; 98% sulfuric acid is usually produced in this tower. Waste gases are usually discharged into the atmosphere. Acid of any desired concentration may be produced by mixing or diluting the products of this process.

$$S + O_2 \xrightarrow{\text{burning}} SO_2 \tag{11.55}$$

$$2SO_2 + O_2 \xrightarrow[\text{Pt/SiO}_2\text{ or V}_2\text{O}_5\text{/SiO}_2]{450°C} SO_3 \tag{11.56}$$

$$SO_3 + H_2SO_4\ (98\%) \xrightarrow[\text{400 K}]{} H_2S_2O_7\ (\text{oleum}) \tag{11.57}$$

$$H_2S_2O_7\ (\text{oleum}) \xrightarrow[H_2O]{} 2\,H_2SO_4 \tag{11.58}$$

Note that directly dissolving SO_3 in water is impractical due to the highly exothermic nature of the reaction. Sulfuric acid vapour or mists are formed instead of a liquid which is difficult to condense when prepared in large scale.

The role of the catalyst in this reaction is possibly as shown below.

$$SO_2 + V_2O_5 \longrightarrow SO_3 + 2\,VO_2 \tag{11.59}$$

$$2\,VO_2 + \tfrac{1}{2}\,O_2 \longrightarrow V_2O_5 \tag{11.60}$$

Concentrated sulfuric acid is a very strong dehydrating agent. It is often used as a drying agent and can be used to dehydrate many compounds such as carbohydrates. The acid reacts similarly with plant and animal matter such as skin and cellulose. When the concentrated acid mixes with water, a large quantity of heat is released.

Blue jeans, indigo carmine and the demand for sulfuric acid

Sulfuric acid, earlier called *oil of vitriol* was produced for centuries in small quantities in the laboratories of alchemists and might have remained a laboratory curiosity but for an interesting development in the textile industry. One of the most sought after dyes during the middle ages was a special type of blue called *indigo*. Blue dye was available in Europe from a plant named woad which grows only in cool climates. But indigo, a tropical plant, produced a much more intense blue. *Indigofera tinctoria* is a shrub commonly found in tropical India with the Hindi name *neel*. Marco Polo (13th century) was the first to report to the Europeans on the availability and preparation of indigo in India. Indigo was quite often used in European easel painting since the middle ages. It was also referred to as *blue gold* and was an ideal, high value trading commodity during that time. During the British rule, peasants in Bengal were forced to cultivate indigo for exporting to England. These peasants obtained only a meagre price for their indigo, the fertility of their land was lost and they ended up in debts. This situation led to an uprising in 1859 by them against the British indigo planters, referred to in history as the '*indigo revolt*'.

The primary use for indigo is as a dye for cotton yarn, which nowadays is mainly for the production of denim cloth for blue jeans. On average, a pair of blue jeans requires 3–12 g of indigo. Indigo was soluble in its reduced, yellow form and insoluble in its oxidised, blue form. If clothes were dyed in tropical climates, it was easy to use indigo. During the dying process, the leaves of the plant are kept in water and allowed to ferment anaerobically, and clothes are dipped in the yellow coloured dye bath. After exposing the clothes to air, the dye became blue and insoluble in water. Importing indigo to Europe presented a problem. Once the dye was oxidised, it was insoluble in water and could not be used as dye. For a while this fundamental chemical problem was the cause for indigo to be expensive, and consequently there was great demand for it. In 1743, Johann Christian Barth, a German lawyer discovered that indigo dissolves in conc. sulfuric acid, forming a water-soluble blue dye which was named Indigo carmine.[25]

It was realised that if conc. sulfuric acid can be produced economically, indigo dyed blue cloth can be made available at very low prices. It had its disadvantages also in the initial stages as water solubility resulted in wash-fastness. About twenty thousand tons of indigo carmine is now produced annually, mainly for the colouring of blue jeans. This discovery initiated a search to find more efficient and cheaper ways to produce conc. sulfuric acid. As the price dropped, more and more uses for sulfuric acid were discovered. The demand for sulfuric acid has now shifted from dyes to fertilisers and more than 80% of it is used for making phosphoric acid and phosphate fertilisers.

Indigo carmine is also used for estimation of dissolved ozone as ozone attacks the double bond between the two heterocyclic rings and produces chemiluminescence.

The largest amount of sulfuric acid (more than 60%) is used to make phosphoric acid, which is in turn used to make phosphate fertilisers, calcium dihydrogen phosphate and the ammonium phosphates. Aluminium sulfate, which is used in paper manufacture and water purification, is another useful product. It is also used to make ammonium sulfate, which is a particularly important fertiliser in sulfur-deficient soils. More of sulfuric acid is made each year than any other industrially produced chemical and more than 200 million tons of it was produced in the world in 2011.

In addition to fertilisers, H_2SO_4 is widely used in the manufacture of fine chemicals. For example, in making hydrochloric acid, nitric acid, sulfate salts, dyes and pigments, detergents, and pharmaceuticals. Sulfuric acid is used in processing metals, for example, in pickling (cleaning) iron and steel before plating them with tin or zinc. It is the electrolyte in lead acid batteries, commonly used in household power inverters and automobiles (often called battery acid, which contains about 33% H_2SO_4).[24]

A carefully made mixture of 30% H_2O_2 and conc. H_2SO_4 in a 1:3 ratio with the nickname 'Piranha solution' is used extensively in the electronic industry to remove photoresist residues from silicon wafers. This solution is used for oxidatively removing organic residues from electronic surfaces and also to make silicon surfaces hydrophilic by introducing hydroxyl groups. The formation of the strong oxidant peroxymonosulfuric acid (Caro's acid; H_2SO_5) ($E^0 = 2.51$ V) and nascent oxygen are two reasons for the very high reactivity of this solution. The vigour with which this reaction eats away organic residues is similar to the way the flesh-eating piranha do so. Thus, the name. The potassium salt of Caro's acid is available commercially as a stable triple salt with the trade name Oxone ($2KHSO_5.KHSO_4.K_2SO_4$) and is useful for many oxidations in organic chemistry.

$$H_2SO_4 + H_2O_2 \longrightarrow H_2SO_5 + H_2O \quad (11.61)$$

$$H_2SO_4 + H_2O_2 \longrightarrow H_3O + HSO_4^- + [O] \quad (11.62)$$

11.20 Sodium Thiosulfate

Sodium thiosulfate is one of the most common sulfur-based inorganic compounds and has a range of important applications based on its reactivity. It is industrially produced by the following reaction.

$$S_8 + 8\,SO_3^{2-} \xrightarrow[\text{water, 100°C}]{} 8\,S_2O_3^{2-} \quad (11.63)$$

It is also prepared by the reaction of SO_2 gas with NaOH giving Na_2SO_3 which further reacts with sulfur or sulfide/carbonate liquor (which is a byproduct of paper manufacturing).

Traditionally it has been used as a primary standard in iodometric titrations as $Na_2S_2O_3.5H_2O$. It has an excellent shelf life and is not hygroscopic, thus facilitating its easy and accurate weighing. In an iodometric titration it reduces I_2, or more specifically its soluble form in water, I_3^- to I^- and gets itself oxidised to tetrathionate.[26]

$$[S_2O_3]^{2-} + I_3^- \longrightarrow 3\,I^- + {}^-O_3S{-}S{-}S{-}SO_3^- \quad (11.64)$$

The same chemistry when extended to Cl_2 helps as a method to dechlorinate tap water and lower chlorine level is swimming pools. When further extended to Br_2 in water, it helps in the safe disposal of bromine water. Thiosulfate also reduces hypochlorite rendering commercial bleach inactive.

$$4\,NaClO + Na_2S_2O_3 + 2\,NaOH \longrightarrow 4\,NaCl + 2\,Na_2SO_4 + H_2O \quad (11.65)$$

Another very important utility of thiosulfate is in deactivation of cyanide as it converts cyanide to thiocyanate which has only minimal toxicity compared to cyanide.

$$CN^- + Na_2S_2O_3 \longrightarrow SCN^- + 2\,Na_2SO_3 \quad (11.66)$$

This reaction also manifests in a similar way in the biological detoxification of cyanide by the enzyme *rhodanase* or more precisely known as *thiosulfate thiotransferase*.

$$\text{R–SH (Cystene (247))} \xrightarrow[-SO_3^{2-}]{S_2O_3^{2-}} \text{R–S–SH} \xrightarrow{CN^-} \text{R–SH} + SCN^- \quad (11.67)$$

It can be prepared in the laboratory by heating an aqueous solution of sodium sulfite with sulfur or by boiling aqueous sodium hydroxide and sulfur (Eq. 11.68).

$$6\,NaOH + 4\,S \longrightarrow 2\,Na_2S + Na_2S_2O_3 + 3\,H_2O \quad (11.68)$$

Thiosulfate also has the capability of forming water soluble complexes with AgBr and this property has been utilised in converting undeveloped AgBr grains to soluble silver thiosulfate complexes and removing it from black and white photographic films and X-ray sheets. This application as a photographic fixer was discovered by John Herschel. Sodium thiosulfate is known as a photographic fixer, and is often referred to as 'hypo'.

$$AgBr + 2\,S_2O_3^{2-} \longrightarrow Ag(S_2O_3)_2^{3-} + Br^- \quad (11.69)$$

A similar complex also forms with gold(I) which provides a non-toxic alternative to the traditional cyanide route in the extraction of gold.

$$4\,Au + 8\,S_2O_3^{2-} + O_2 + 2\,H_2O \longrightarrow 4\,Au(S_2O_3)_2^{3-} + 4\,OH^- \quad (11.70)$$

Sodium thiosulfate solution is used for tackling spills of Cl_2, Br_2 and I_2 and ClO^- as it reduces these to the corresponding halides, getting itself oxidised to Na_2SO_4.

11.21 Sulfur-Based Fluorinating Agents

One of the unique applications of sulfur–fluorine compounds is based on the chemical reactivity of sulfur tetrafluoride with carbonyl compounds. SF_4, a highly reactive gas, has the ability to convert carbonyl (C=O) groups to CF_2 groups. It also has the unique ability to convert a carboxylic acid group to a CF_3 group. Because of the difficulty in handling of SF_4 due to its extreme reactivity and instablity, a few SF_3 and SF_2 based reagents have been developed which are liquids or solids and which are easier to store and handle. The reactivity of these SF_4 derivatives is less than that of SF_4. The reactions with aldehydes, ketones and alcohols are same as that of SF_4 while with carboxylic acid groups, conversion to the CF_3 group is possible only with a few of these reagents (Table 11.3).[27–32]

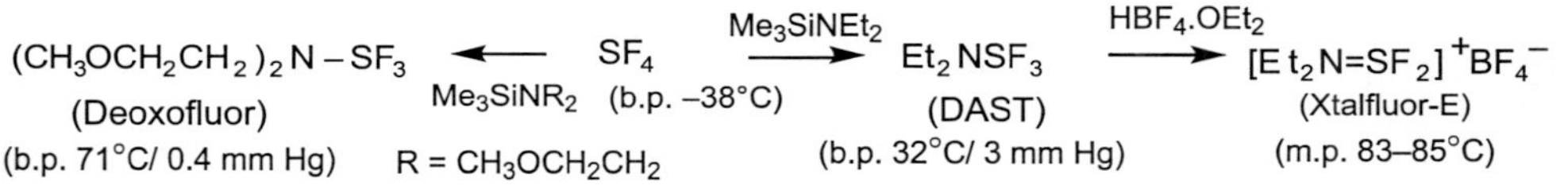

$(CH_3OCH_2CH_2)_2N-SF_3$ (Deoxofluor) (b.p. 71°C/ 0.4 mm Hg) $\xleftarrow{Me_3SiNR_2}$ SF_4 (b.p. −38°C) $\xrightarrow{Me_3SiNEt_2}$ Et_2NSF_3 (DAST) (b.p. 32°C/ 3 mm Hg) $\xrightarrow{HBF_4.OEt_2}$ $[Et_2N{=}SF_2]^+BF_4^-$ (Xtalfluor-E) (m.p. 83–85°C)

R = $CH_3OCH_2CH_2$

Scheme 11.13 Synthesis and properties of sulfur-based fluorinating agents

Table 11.3 Products of reaction of acids, ketones, alcohols and aldehydes with sulfur-based fluorinating agents

Fluorinating agent / Substrate	SF_4	$(CH_3OCH_2CH_2)_2N-SF_3$ (Deoxofluor)	Et_2NSF_3 (DAST)	$[Et_2N{=}SF_2]^+BF_4^-$ (Xtalfluor-E)
RC(O)OR' or RC(O)OH	$R-CF_3$	$R-CF_3$ (above 85°C); RC(O)F (at 0°C)	RC(O)F	RC(O)F
RC(O)R	R_2CF_2	R_2CF_2	R_2CF_2 With traces of HF	R_2CF_2 With $Et_3N.HF$
RCH(OH)R	R_2CHF	R_2CHF	R_2CHF	R_2CHF
RC(O)H	$RCHF_2$	$RCHF_2$	$RCHF_2$ With traces of HF	$RCHF_2$ With $Et_3N.HF$

11.22 Chemistry of Thionyl Chloride and Sulfuryl Chloride

Thionyl chloride is one of the very common reagents used by organic chemists especially for the conversion of OH groups to Cl groups. It is also an excellent dehydrating agent and is often used for removing the water of crystallisation/hydration of metal salts and convert them to anhydrous salts. The advantage of this reaction is that the side products of the reaction are highly volatile HCl and SO_2 which can be easily driven out from the reaction flask by an inert gas. Another important reaction is the conversion of amide to nitriles. Its boiling point (74.6°C) and volatility helps to remove excess thionyl chloride after a reaction by application of vacuum. There are three main ways by which thionyl chloride can be prepared.[33]

$$SO_3 + SCl_2 \longrightarrow SOCl_2 + SO_2 \quad (11.71)$$

$$SO_2 + COCl_2 \longrightarrow SOCl_2 + CO_2 \quad (11.72)$$

$$SO_2 + PCl_5 \longrightarrow SOCl_2 + POCl_3 \quad (11.73)$$

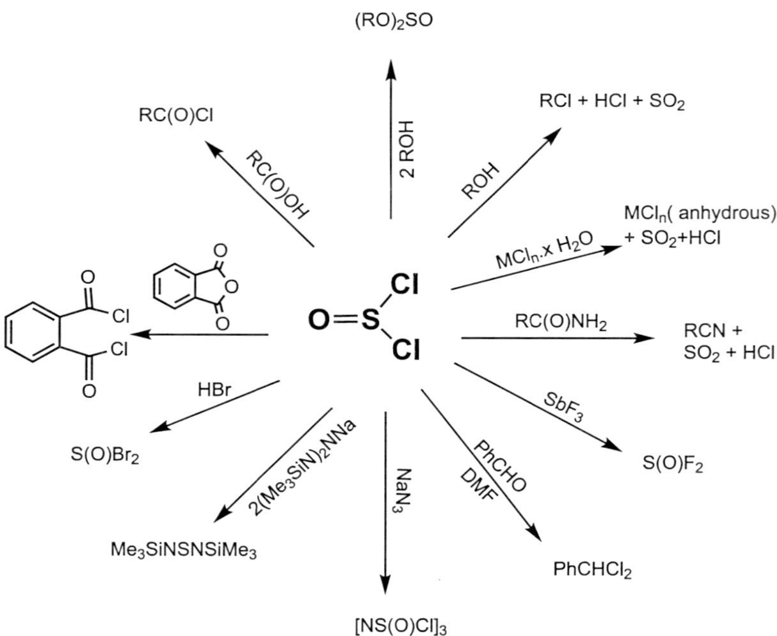

Scheme 11.14 General reactions of thionyl chloride

Sulfuryl chloride (b.p. 69.4°C) is well known as a very useful chlorinating agent. It brings about almost all reactions which chlorine gas does. Unlike Cl_2 gas, whose reaction in stoiochiometric ratio with a substrate is often difficult, SO_2Cl_2 can be used for controlled chlorination, especially when an organic compound has multiple double bonds. Similar to Cl_2, sulfuryl chloride also undergoes substitutive chlorination and oxidative chlorination. Unlike thionyl chloride, sulfuryl chloride when reacted with water results in the formation of sulfuric acid. The advantage of using sulfuryl chloride in oxidative chlorination is that the side product formed is SO_2 gas which can be easily driven out from the system.[34]

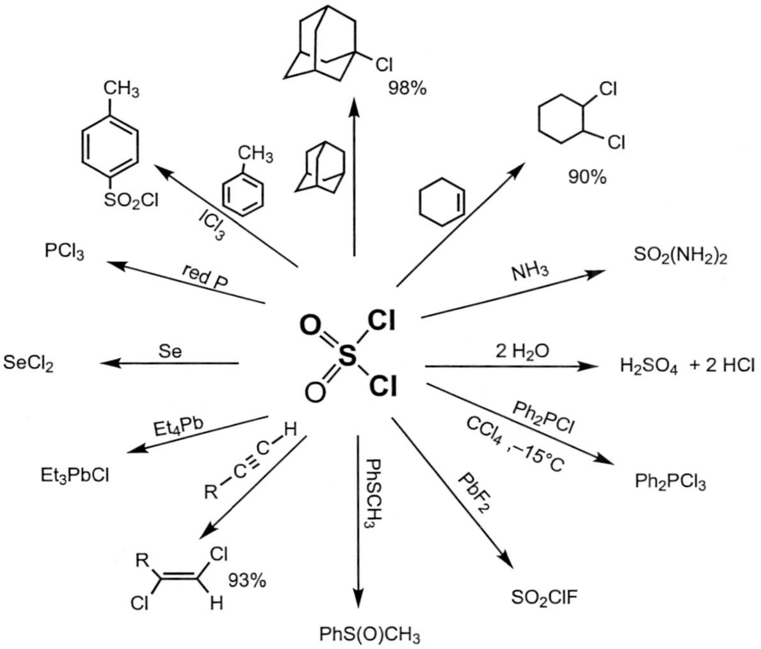

Scheme 11.15 General reactions of sulfuryl chloride

Chlorination of cyclohexene, with well known chlorinating reagents, indicates the very high chlorination efficiency of sulfuryl chloride.[35]

(11.74)

Reactant/Solvent	Product yield (%)
SO_2Cl_2 /CCl_4	90
$PhICl_2$ / CH_2Cl_2	80
PCl_5	75
Cl_2 / CCl_4	70
$MoCl_5$ / CH_2Cl_2	70

11.23 Sulfur Hexafluoride

Sulfur hexafluoride is a compound, first prepared in 1900 by Henri Moissan by the reaction of elemental sulfur and fluorine gas. It is a colourless, tasteless, non-flammable, non-toxic and insoluble gas at room temperature. SF_6 is an octahedral-shaped molecule, with a bond length of 1.561 Å (gas phase). It is poorly reactive due to kinetic reasons. The SF_6 gas is one of the heaviest gases. The density of this gas at 20°C and 1 atm, is 6.139 kg/m^3 which is approximately 5 times higher than air under the same conditions. In 1937, General Electrical Company first realised that SF_6 gas can be used as an insulating material. By 1960, using SF_6 gas as a dielectric in high voltage switchgear and related power equipments became common. With the increase in demand, manufacturers in Europe and North America started making SF_6 gas on a large scale. Initially it was only used for insulating in electrical systems but later it was observed that SF_6 has excellent arc quenching properties as well. Therefore, it was also used in circuit breakers as an arc quenching medium. The thermal conductivity of this gas is lower than air. During dissociation of SF_6 molecules around an electric arc, the molecules absorb large amounts of heat and this heat is released when the molecules re-form at the periphery of the arc.

SF_6 when heated to 500°C does not decompose. It does not react with glass, even up to its softening temperature. It does not react with metals such as magnesium and copper, even at red heat. It also does not react with molten sodium, and only with boiling sodium to form Na_2S and NaF. Phosphorus and arsenic can be distilled in SF_6 vapour without decomposition.

The relatively poor reactivity of SF_6 is probably due to a combination of factors. Since fluorine forms strong covalent bonds, the S–F bonds have high bond energies and are therefore difficult to break. Since there are six fluorine atoms surrounding the relatively small sulfur atom in an octahedral geometry, it is shielded quite well from attacking molecules too. It is likely that its reactivity is extremely slow due to high activation energies, as hydrolysis is energetically not favourable. SF_6 is also one of the most potent greenhouse gases with a greenhouse effect 22,000 times that of the reference compound, CO_2. It has a predicted lifetime in the atmosphere of 3200 years. There has been an added interest in discovering the reactions of SF_6, especially with low valent organometallic substrates so that ways can be developed to chemically degrade this otherwise highly potent greenhouse gas. Some of the reactions which SF_6 undergoes at room temperature are given below.[36–40]

$$V(C_5H_5)_2 \xrightarrow{SF_6} (C_5H_5)_2V{-}F \quad (11.75)$$

$$Cr(C_5Me_5)_2 \xrightarrow[DMAP]{SF_6} (C_5Me_5)CrF_2(DMAP) \quad (11.76)$$

$$RhH(PEt_3)_3 \xrightarrow[-SPEt_3,\ -F_2PEt_3,\ -FSiEt_3]{SF_6,\ PEt_3,\ HSiEt_3} [Rh(H)_2(PEt_3)_4]^+\ HF_2^- \qquad (11.77)$$

$$R\text{–CH=CH–CH(OH)–}R' \xrightarrow[\text{Ru(bipy)}_3(PF_6)_2,\ \text{DIPEA, 470 nm}]{SF_6} R\text{–CH=CH–CHF–}R' + R\text{–CHF–CH=CH–}R' \qquad (11.78)$$

In addition, a few low valent titanium, zirconium and nickel complexes have also been reported to react with SF_6. Three of the high temperature reactions of SF_6 are given in the following equations.

$$SF_6 + 2\ SO_3 \xrightarrow[250°C]{} 3\ SO_2F_2 \qquad (11.79)$$

$$SF_6 + AlCl_3 \xrightarrow[200°C]{} AlF_3 + Cl_2 + SCl_n \qquad (11.80)$$

$$SF_6 \xrightarrow[\text{boiling sodium}]{} Na_2S + 6\ NaF \qquad (11.81)$$

SF_6 reacts with sodium in liquid ammonia also. The important uses of SF_6 are, as dielectric material for high voltage transformers, as a blanket for molten melt of magnesium and as a filler for multipane windows for noise reduction. It was also used in the soles of 'Nike Air' shoes for its flexibility. Due to concerns raised on the greenhouse effect, Nike has now stopped using SF_6.

11.24 Tetrasulfur Tetranitride

Tetrasulfur tetranitride S_4N_4, is the first sulfur–nitrogen heterocycle to be made and is also the precursor for many sulfur–nitrogen compounds. It is prepared by the reaction of sulfur dichloride or sulfur monochloride or a mixture of both with ammonia and is a bright orange, air-stable crystalline compound.[41–44]

$$6S_2Cl_2 + 16\ NH_3 \xrightarrow[CCl_4]{50°C} S_4N_4 + 2\ NH_4Cl + 8S \qquad (11.82)$$

$$6SCl_2 + 16\ NH_3 \xrightarrow[CCl_4]{50°C} S_4N_4 + 12\ NH_4Cl + 2S \qquad (11.83)$$

The formation of S_4N_4 is endothermic and $\Delta H_f^0 = 460 \pm 8$ kJ/mol. The compound when struck detonates forming sulfur and N_2.

Although discovered just 10 years after the discovery of benzene, the unusual structure of S_4N_4, like that of benzene, was not elucidated for over 100 years. The application of diffraction techniques revealed the unusual cage arrangement with two weak cross-ring

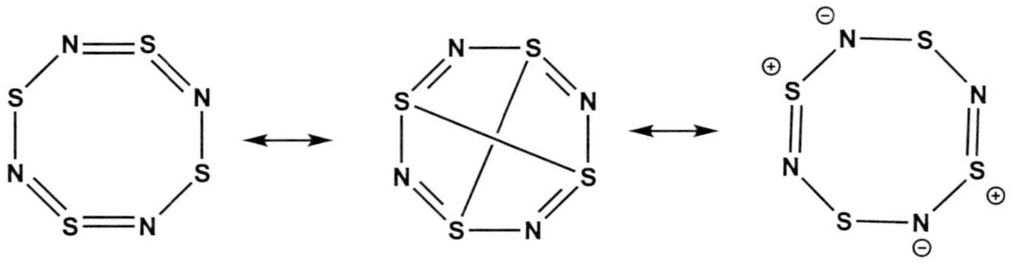

Fig. 11.8 Resonance structures possible for S_4N_4

sulfur–sulfur interactions. The structure of S_4N_4 is described as a bisphenoid and all the S–S bond distances are very close to each other (162 pm). The structure also shows two weak S...S interactions (258 pm which is higher than the S–S van der Waals distances and lower than the S–S singe bond distances 208 pm) resulting in the bisphenoid.

The details of the electronic structure of this molecule are still a matter for debate. The polymer, $(SN)_x$, was first obtained in 1910 and its metallic character was noted. The discovery in 1973 that $(SN)_x$, a polymer comprising only nonmetallic elements behaves as a superconductor at 0.26 K, sparked new interest in sulfur–nitrogen (S–N) chemistry. It was proposed that planar S–N heterocycles belong to a class of 'electron-rich aromatics' that conform to the well-known Hückel $(4n + 2)$ π-electron rule of organic chemistry. This suggestion, which was based on simple electron-counting concepts, provided an additional interest for theoretical investigations of S–N systems. The combination of structural studies and molecular orbital calculations has provided reasonable rationalisations of the structure–reactivity relationships of these fascinating compounds. Among potential applications, the compounds of the type S-nitrosothiols (RSNO) have been shown to be important species in the storage and transport of nitric oxide. As NO donors, these S–N compounds have potential medical applications. Some of the well known reactions of tetrasulfur tetranitride are shown in Scheme 11.16.

11.25 Sulfonamides

Sulfonamide drugs are some of the very first antibiotics to be used systemically, and paved the way for the revolution of antibiotics in medicine. The first sulfonamide (trade name, Prontosil), a pro-drug (meaning it becomes effective only in the presence of body fluids) made by Bayer laboratories, was the first medicine ever discovered that could effectively treat a range of gram positive bacterial infections of the body. It had a strong protective action against infections caused by *streptococci,* including blood infections. However, it had no effect at all in the testtube, exerting its antibacterial action only in live animals. Later, it was discovered by a French research team led by Ernest Fourneau at the Pasteur Institute, that the drug was metabolised into two pieces inside the body, releasing from the inactive portion a smaller, colourless, active compound called sulfanilamide.[45]

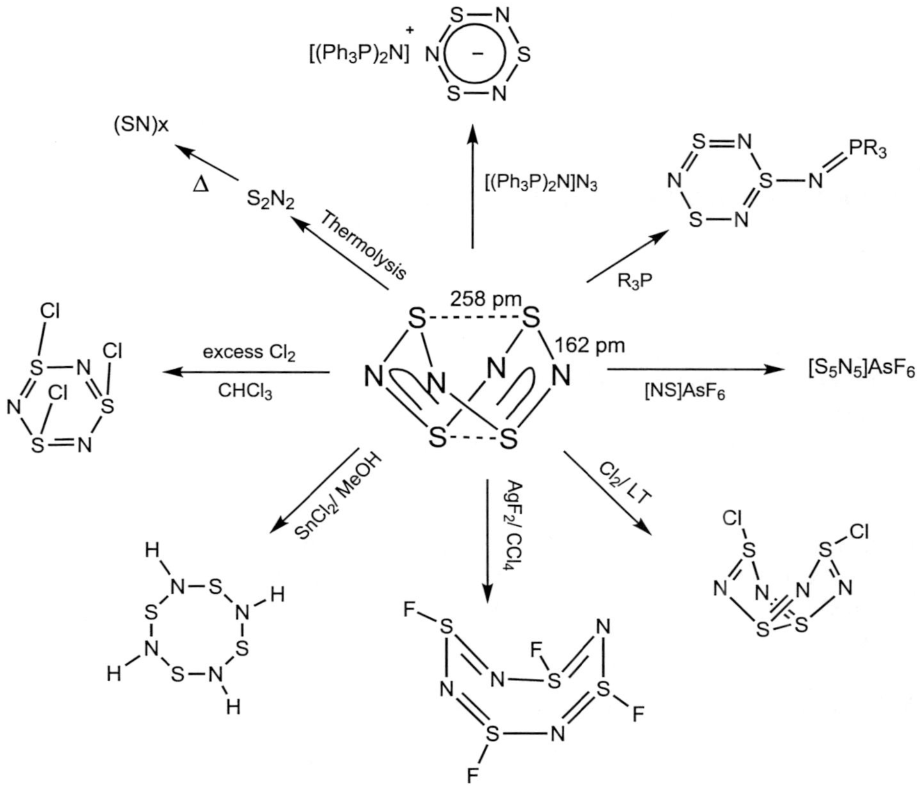

Scheme 11.16 General reactions of tetrasulfur tetranitride, S_4N_4

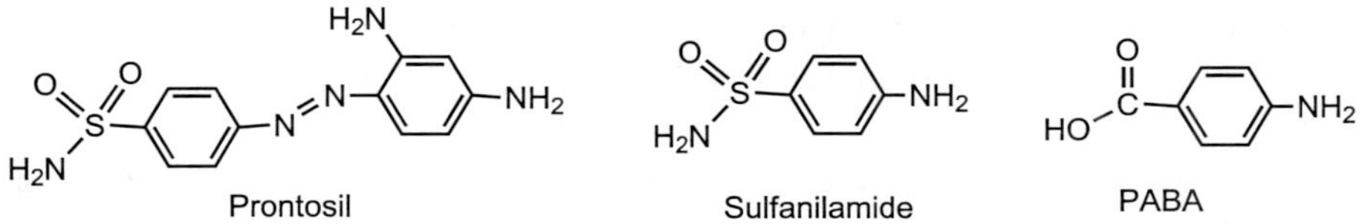

Prontosil

Sulfanilamide

PABA

Human beings get their folic acid from food while bacteria have to make folic acid from the medium and the organic compound it requires is para-aminobenzoic acid (PABA). Sulfonamides are structural analogues of PABA. Sulfonamides can competitively inhibit the folate synthesis. Therefore sulfonamides are bacteriostatic (stopping multiplication) and not bactericidal (killing bacteria).

Sulfonamides can be synthesised from chlorosulfonic acid, which is obtained from the reaction of SO_3 and HCl.

$$HCl + SO_3 \longrightarrow ClSO_3H \xrightarrow[-HCl]{C_6H_6} C_6H_5SO_3H \xrightarrow{ClSO_3H} C_6H_5SO_2Cl \xrightarrow{R_2NH} C_6H_5SO_2NR_2 \quad (11.84)$$

Questions for self study (multiple choice)

11.1. Which of the following compounds does not give oxygen upon heating?
(a) $K_2Cr_2O_7$ (b) $(NH_4)_2Cr_2O_7$ (c) $KClO_3$ (d) $Zn(ClO_3)_2$

11.2. Which of the following statements is incorrect?
(a) Ozone molecule is bent
(b) ONF is isoelectronic with O_2N^-
(c) Ozone is a colourless liquid upon cooling
(d) Ozone is a stronger oxidising agent than H_2O_2

11.3. Oxygen will directly react with each of the following elements except
(a) P (b) Cl_2 (c) Na (d) S

11.4. It is possible to obtain oxygen from air by fractional distillation because
(a) Oxygen is in a different group of the periodic table than nitrogen
(b) Oxygen is more reactive than nitrogen
(c) O_2 has a higher boiling point than N_2
(d) O_2 has a lower density than nitrogen

11.5. Which among the following is the correct order of oxidising capability?
(a) $O_3 > H_2O_2 > KMnO_4$ (b) $O_3 > KMnO_4 > H_2O_2$
(c) $H_2O_2 > O_3 > KMnO_4$ (d) $H_2O_2 > KMnO_4 > O_3$

11.6. Which of the following statements is not true for H_2O_2?
(a) It releases H_2 gas when acting as an oxidising agent
(b) H_2O_2 can act as an oxidising and reducing agent
(c) It can be prepared by the reaction of BaO_2 and H_2SO_4
(d) In acidic medium it is a stronger oxidising agent than $KMnO_4$

11.7. The fluorinating agent that can convert COOH group to CF_3 group is
(a) DAST (b) SF_4 (c) Xtalflour-E (d) Me_3SnF

11.8. The sulfur-based compound that is ideal for removing the water of crystallisation from metal salts is
(a) SO_2Cl_2 (b) $SOCl_2$ (c) SO_2 (d) S_2Cl_2

11.9. The catalyst for the production of sulfuric acid using lead chamber process is
(a) Lead (b) V_2O_5 (c) Platinum (d) NO_x

11.10. SO_3 can be produced from heating of which of the following compounds/ mixture?
(a) $CaSO_4 + C$ (b) $Fe_2(SO_4)_3$ (c) $S + H_2SO_4$ (d) $H_2SO_4 + PCl_5$

Solved concept-based questions

11.11. H_2O is a liquid whereas H_2S is a gas at room temperature. Why?

The electronegativity of oxygen (3.5) is much higher than that of sulfur (2.6). Hence strong hydrogen bonds are formed in water than in H_2S where only very weak hydrogen bonding is present making it a gas at room temperature (b.p. –60°C).

11.12. H_2S is more acidic than H_2O. Why?

The bond dissociation energy of O–H in H_2O is 463 kJ/mol while that of S–H in H_2S is 347 kJ/mol. Hence H_2S is more acidic than H_2O.

11.13. SF_6 is known while OF_6 is unknown. Why?

Oxygen cannot expand its valency since it is a first row main group element as it does not have *d*-orbitals of lower energy.

11.14. Why is H_2O_2 always stored in dark plastic bottles?

Degradation of hydrogen peroxide is catalysed by UV light. As the peroxide breaks down, it forms oxygen gas. This gas can build up and expand, therefore H_2O_2 also has to be stored in a dark preferably plastic bottle.

11.15. Sulfur has a greater tendency to form catenation compared to oxygen. Why ?

The ability of an element to catenate is primarily based on the bond energy of the element binding to itself, which decreases with more diffuse orbitals overlapping to form the bond. However in the case of oxygen, due to small size, there will be inter-electronic repulsion between the lone pairs of electrons present on each oxygen atom which will make the single O–O bond weaker than the S–S bond. Hence sulfur has a greater tendency to form catenation than oxygen.

11.16. For the industrial synthesis of H_2O_2 why is 2-ethylhydroquinone or a similar 2-alkyl hydroquinone used, although in principle hydroquinone can bring about the reaction?

The process involves the use of a solvent mixture having a nonpolar (benzene) and polar solvent (higher secondary alcohols). 2-ethyl anthraquinone has better solubility in this solvent mixture than simple anthraquinone.

11.17. Sulfur is diamagnetic in the solid state and paramagnetic in the vapour phase. Why?

In the vapour phase sulfur exists as S_2 whose highest occupied molecular orbital is similar to that of O_2 having two unpaired electrons. Hence in the vapour phase it is paramagnetic like O_2. In the solid phase it is present in the catenated or ring form.

11.18. The bleaching action of SO_2 on colours is reversible while that with O_3 or bleaching powder is irreversible. Why?

SO_2 in presence of moisture forms H_2SO_3 which bleaches by a reduction reaction, while ozone and bleaching power bleaches by an oxidation reaction. The reversibility of the former reaction is due to the fact that oxygen in the atmosphere can reoxidise the reduced colours back to their normal colour.

11.19. Rubber tyres kept near electric spark generating equipments (such as motors) for long periods of time develop cracks on the surface. Give reasons.

An electric spark in an oxygen-based atmosphere generates ozone and ozone has been found to react with double bonds of organic compounds (in this case the double bonds present in natural rubber), breaking the polymer chain which in turn become visible as cracks.

11.20. A yellow solid A forms a volatile foul smelling hydride B which gives a black precipitate with lead nitrate solution. On heating in air, A forms a colourless pungent gas C. The gas C in the presence of moisture was found to temporarily bleach flowers. C in the presence of a vanadium catalyst was found to get oxidised to a low boiling liquid D. This on reaction with water was found to result in a dibasic acid E. Predict the compounds A to E.

A = S; B = H_2S; C = SO_2; D = SO_3; E = H_2SO_4

Problems and exercises

11.21. Hydrogen peroxide can act as both an oxidising agent and as a reducing agent in both acidic and alkaline media. How does one identify from a chemical equation involving H_2O_2 as a reactant if it had acted as an oxidising or reducing agent?

11.22. Starting from elemental sulfur (S_8) show steps with reagents for the synthesis of (a) S_6 and (b) phenyl sulfonic acid ($C_6H_5SO_3H$) in good yields.

11.23. Reaction of a group 16 element with excess of dihalogen A in the presence of an iron halide resulted in the formation of a liquid compound B. Compound B on reaction with A and sodium salt of halide C, gave a gaseous compound D. Compound D on reaction with benzoic acid converted it to trifluoromethyl benzene. Identify compounds A–D.

11.24. How are S_6 and S_{10} prepared starting with $(NH_4)_2S_5$?

11.25. The reaction of SCl_2 with the hydride of a group 15 element resulted in a crystalline cyclic compound A. Compound A on reaction with AgF_2 in CCl_4 gave another cyclic compound B and with $SnCl_2$/MeOH gave a third cyclic compound C. The number of sulfur atoms present in compounds A, B and C are the same. Draw the structures of A, B and C indicating type of bonds and weak interactions, if any.

11.26. Suggest the most convenient and high yielding chlorinating reagent for the following transformations; justify your answer in 1–2 lines.

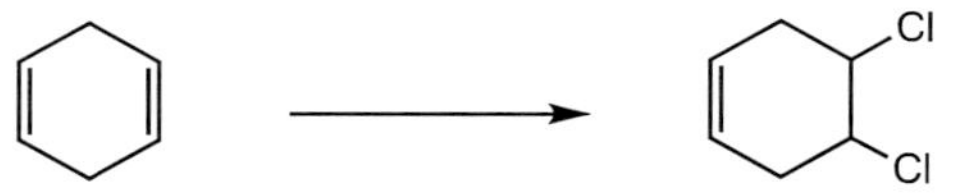

Cl, Cl

$Ph_2PCl \longrightarrow Ph_2PCl_3$

11.27. SF_6 in general is very poorly reactive while TeF_6 readily reacts with water getting converted to $Te(OH)_6$. Give reasons.

11.28. Reaction of SO_3 with SCl_2 was found to give a reactive liquid A and a gas B. Gas B in presence of moisture was found to bleach colours in a reversible manner. Liquid A on reaction with yellow phosphorus gave another liquid compound C as the major product. C on hydrolysis gave an acid D which on heating released a colourless gas E with smell of rotten fish. Identify compounds A–E.

11.29. Reaction of a group 16 element with excess of dihalogen P in the presence of an iron halide resulted in the formation of a liquid compound Q. Compound Q on reaction with NH_3 gave a crystalline yellow cyclic compound R which was found to be endothermic in nature. Compound R on reaction with PPh_3 gave a red coloured stable compound S in which three atoms of the group 16 element were present in two different oxidation states. Identify compounds P to S.

11.30. Suggest sulfur-based reagents for the following reactions. Some may involve more than one step.

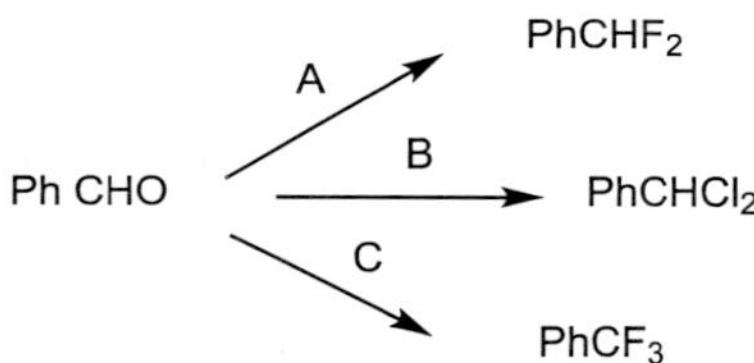

Answer key for self study questions

11.1. (b) 11.2. (c) 11.3. (b) 11.4. (c) 11.5. (a) 11.6. (a) 11.7. (b) 11.8. (b) 11.9. (d) 11.10. (b)

Supplementary reading

1. Priestly J, An account of further discoveries in air, *Philosophical Trans.*, 1775, Vol. 65, 285.
2. Weiss H M, Appreciating oxygen, *J. Chem. Educ.*, 2008, Vol. 85, 1218.
3. Holland H D, The oxygenation of the atmosphere and oceans, *Phil. Trans. R Soc. B*, 2006, Vol. 361, 903.
4. Allen J F and Martin W, Evolutionary biology: out of thin air, *Nature*, 2007, Vol. 445, 610.
5. Schirrmeister B E, de Vos J M, Antonelli A and Baghari H C, Evolution of multicellularity coincided with the increased diversification of cyanobacteria and the great oxidation event, *Proc. Nat. Acad. Sci.*, 2013, Vol. 110, 1791.
6. Greenwood N N and Earnshaw A, *Chemistry of the elements*, Pergamon Press, 1984.
7. Krumova K and Cosa G, *Overview of reactive oxygen species in singlet oxygen: Applications in biosciences and nanosciences*, RSC 2016, Vol. 1, 1.
8. Hayyan M, Hashim MA and Al Nashef I M, Superoxide ion, Generation and chemical implications, *Chem. Rev.*, 2016, Vol. 116, 3029.
9. Borden W T, Hoffmann R, Stuyver T and Chen B, Dioxygen: What makes this triplet diradical kinetically persistent?, *J. Am. Chem. Soc.*, 2017, Vol. 139, 9010.
10. Berry R E, Reactivity and structure of complexes of small molecules: dioxygen, *Comprehensive Coordination chemistry*, 2004, Vol. II, 625.

11. Vaska L and Diluzio K W, Carbonyl and hydrido carbonyl complexes of iridium by reaction with alcohols. Hydrido complexes by reaction with acid, *J. Am .Chem. Soc.*, 1961, Vol. 83, 2784.
12. Vaska L, Oxygen carrying properties of a simple synthetic system, *Science*, 1963, Vol. 140. 809.
13. Marteill A E, Formation and stabilities of cobalt dioxygen complexes in aqueous solution, *Acc. Chem. Res*, 1982, Vol. 15, 55.
14. Timmons J M, Niswander R N, Clearfield A and Martell A E, Crystal and molecular structure of μ-peroxo-bis[(1,9-bis(2-pyridyl) 2,5,8-triazanonane) cobalt(III)] tetraiodide. Effect of chelate ring size on structures and stabilities of dioxygen complexes, *Inorg. Chem.*, 1979, Vol. 18, 2977.
15. Lippard S J and Berg J M, *Principles of bioinorganic chemistry*, University science books, 1994.
16. Streng A G, Table of ozone properties, *J. Chem. Educ.*, 1961, Vol. 6, 431.
17. Campos-Martin J M, Blanco-Brieva G and Fierro J L G, Hydrogen peroxide synthesis: an outlook beyond the anthraquinone process, *Angew Chem., Int. Ed.*, 2006, Vol. 45, 6962.
18. Fenton H J H, Oxidation of tartaric acid in the presence of iron, *J. Chem. Soc. Trans.*, 1894, Vol. 65, 899.
19. Steudel R, Homocyclic sulfur compounds, *Top. Curr. Chem.*, 1982, 149.
20. Lefer D J, A new gaseous signaling molecule emerges: Cardioprotective role of H_2S, *Proc. Natl. Acad. Sci.*, 2007, Vol. 104, 17907.
21. Drozdov A P, Eremets M I, Troyan I A, Ksenofontov V and Shylin S I, Conventional superconductivity at 203 kelvin at high pressures in the sulfur hydride system, *Nature*, 2015, Vol. 525, 73.
22. Troyan I, Gavriliuk A, Ruffer R, Chumakov A, Mironovich A, Lyubutin I, Perkalin D, Drozdov A P and Ermets M I, Observation of superconductivity in hydrogen sulfide from nuclear resonant scattering, *Science,* 2016, Vol. 351, 1303.
23. Yao Y and Tse J S, Superconducting hydrogen sulfide, *Chem. Eur. J.*, 2018, Vol. 24, 1769.
24. Mueller H, Sulfuric acid and sulfur trioxide in *Ullmann's encyclopedia of industrial chemistry*, Wiley-VCH, 2000.
25. Steingruber E, Indigo and Indigo Colourants in *Ullmann's Encyclopedia of Industrial Chemistry*, Wiley-VCH, 2004.
26. Dhawale S W, Thiosulfate, An interesting oxoanion that is useful in both medicine and industry but is implicated in corrosion, *J. Chem. Educ.*, 1993, Vol. 70, 12.
27. Wang C-L J, *Sulfur tetrafluoride, Encyclopedia of reagents for organic synthesis*, Paquette L, Ed., 2001.
28. Fawcett F S and Tullock C W, Sulfur(IV) fluoride, *Inorg. Synth.*, 1963, Vol. 7, 119.
29. Middleton W J and Bingham E M, Diethylaminosulfur trifluoride, *Org. Synth.*, Coll. Vol. 6, 440.
30. Middleton W J, New fluorinating agents, Dialkylamino sulfur fluorides, *J. Org. Chem.*, 1975, Vol. 40, 574.
31. Lal G S, Pez G P, Pesaresi R J, Prozonic F M and Cheng H, Bis(2-methoxyethyl) aminosulfur trifluoride: a new broad spectrum deoxofluorinating agent with enhanced thermal stability, *J. Org. Chem.*, 1999, Vol. 64, 7048.
32. Heureux A L, Bealieu F, Bennett C, Bill D R, Clayton S, La Flamme F, Mirmehrabi M, Tadayon S, Tovell D and Couturier M, Aminodifluorosulfinium salts: Selective fluorinating agents with enhanced thermal stability and ease of handling, *J. Org. Chem.,* 2010, Vol. 75, 3401.
33. Wirth D D, Thionyl Chloride, *Encyclopedia of reagents for organic synthesis*, Paquette L, Ed., 2001.
34. Maynard G D, Sulfuryl chloride, *Encyclopedia of reagents for organic synthes*is, Paquette L, Ed., 2001.

35. Moussa V N, Sulfuryl chloride, A versatile alternative to chlorine, *Aust. J. Chem.*, 2012, Vol. 65, 95.
36. Holze P, Horn B, Limberg C, Matlachowski C and Mebs S, The activity of SF_6 at highly reduced low coordinate nickel-dinitrogen complexes, *Angew Chem. Int. Ed.*, 2014, Vol. 53, 2750.
37. Harvey B G, Arif A M, Glocuox A and Ernst R D, SF_6 as a selective and reactive fluorinating agent for low valent transition metal complexes, *Organometallics*, 2007, Vol. 26, 2872.
38. Basta R, Harvey B J, Arif A M and Ernst R D, Reactions of SF_6 with organotitanium and organozirconium complexes, *J. Am. Chem. Soc.*, 2005, Vol. 127, 11924.
39. Zamostna L and Braun T, Catalytic degradation of SF_6 by rhodium complexes, *Angew Chem. Int. Ed.*, 2015, Vol. 54, 10652.
40. Mc Teage T A and Jamison T F, Photoredox activity of SF_6 for fluorination, *Angew Chem. Int. Ed.*, 2016, Vol. 55, 15072.
41. Chivers T, *A guide to chalogen nitrogen chemistry*, World Scientific, Singapore, 2005.
42. Heal H G, *The inorganic heterocyclic chemistry of sulfur, nitrogen and phosphorus*, Academic Press 1980.
43. Chivers T and Manners I, *Inorganic rings and polymers of the p-block elements*, RSC publishing, 2009.
44. Chivers T, Sulfur nitrogen compounds, *Encyclopedia of inorganic and bioinorganic chemistry*, 2011.
45. Actor P, Chow A W, Dutko F J and McKinlay M A, 2005, Chemotherapeutics in *Ullmann's Encyclopedia of Industrial Chemistry*, Weinheim: Wiley-VCH.

The Chemistry of Selenium and Tellurium

CHAPTER 12

Unique Properties of Selenium and its Compounds

- *Selenium is a photoconductor, a material that changes light energy into electrical energy. Furthermore, it becomes better at this conversion as the intensity of light increases. Once an essential material present in photocopiers, it has largely been replaced by organic photoconductors.*
- *Although selenic acid (H_2SeO_4) resembles sulfuric acid, it is a more powerful oxidant, capable of releasing Cl_2 from concentrated HCl and of dissolving gold to form gold(III) selenite.*
- *Selenium is incorporated in proteins through the amino acids selenocysteine and selenomethionine in which it replaces the sulfur atom of cysteine and methionine, respectively.*
- *Selenocystiene which is the 21st amino acid found in living systems, is present in at least 54 human proteins. Selenoproteins are essential components of metabolic pathways, including thyroid hormone metabolism, antioxidant defense systems or immune functions.*
- *Red selenium is a Se_8 macrocycle similar to the sulfur allotrope S_8.*
- *Grey selenium, the most stable form, is a semiconductor that conducts electricity better in the light than in the dark, and is used in photovoltaic cells.*
- *Most of the selenium produced worldwide is isolated from the mud that forms at the anode during the electrolytic refining of copper.*
- *Combined with bismuth, it has replaced lead in plumbing brasses since the 1990s to meet lead-free environmental standards. EnviroBrass, a type of modern brass for making brass drinking water faucets and shower heads is made of a lead free alloy (SeBiLOY) which contains 2% Bi and 1% Se along with Cu (86%–88%), tin (4%–6%) and zinc (4%–6%). A combination of bismuth and selenium provides*

the same beneficial effect on machinability as does lead in addition to safety from lead poisoning.

- *The grey form of selenium also converts electric current from AC to DC, because of which it is a component in rectifiers.*
- *The largest worldwide use of selenium, especially as selenium dioxide is in manufacturing of red and pink coloured glass, where it is used as a dopant. The ubiquitous red traffic light owes its colour to selenium additives, sometimes along with CdS.*
- *SeO_2 is a well known oxidising agent by itself or along with other oxidants as catalyst. During oxidation, SeO_2 gets reduced to red elemental selenium and gets precipitated due to the poor solubility of selenium.*
- *Selenium is a well established essential trace element with a recommended dietary intake of 55 μg per day for adults (in both USA and Europe). This amount can be supplied by a single dried Brazil-nut. The upper limit for the intake of selenium has been fixed by FDA as 400 μg per day.*
- *Although selenium is not toxic and is considered a micronutrient, many of its compounds, such as hydrogen selenide (H_2Se) are quite toxic.*
- *Woolins' reagent, $[PhPSe_2]_2$, is a useful selenating agent that converts certain carbonyl groups (C=O) to the selenocarbonyl group (C=Se).*

Unique Properties of Tellurium and its Compounds

- *Unlike sulfur and selenium, tellurium is not an essential element in living systems.*
- *Tellurium compounds are stable at higher oxidation states and readily form hexavalent hexacoordinate species such as $Te(OH)_6$ and $TePh_6$.*
- *Dihalides of tellurium TeX_2 are unstable. Unlike sulfur and selenium, direct chlorination of tellurium gives $TeCl_4$.*
- *The formation of π bonds with tellurium is difficult and tellurium oxide prefers single bonds.*
- *Adding tellurium to steel and copper during metallurgy makes them more machineable and reduces metal fatigue.*
- *Thermoelectric effect refers to a phenomenon by which either a temperature difference creates an electric potential or an electric potential creates a temperature difference in certain materials. A substantial lowering of temperature by means of the Peltier effect has been observed in a thermocouple incorporating bismuth telluride, Bi_2Te_3. This compound has been extensively used in the construction of thermoelectric modules such as Freon-free portable food coolers and refrigerators.*

- *Tellurium is used in the manufacture of rewritable CDs, DVDs and blue ray discs, and also for phase change memory chips. The telluride used is a silver–indium–antimony–tellurium (AgInSbTe) alloy.*
- *CdTe is used to make thin film solar cells, accounting for about 8% of all solar cells installed worldwide in 2011. These are among the relatively inexpensive types of solar cells and the second-most widely used solar cells after silicon. Cadmium telluride absorbs sunlight at close to the ideal wavelength with a band gap of 1.5 eV at 300 K, capturing energy at shorter wavelengths than is possible with silicon panels.*
- *CdTe can be alloyed with mercury to make a versatile infrared detector material HgCdTe. CdTe alloyed with a small amount of zinc makes an excellent solid-state X-ray and gamma ray detector, CdZnTe.*

12.1 About the Elements

Selenium was discovered in 1817 by the Swedish scientist Jöns Jacob Berzelius (1779–1848). The word selenium comes from the Greek word for moon, seléné. This element, exists only in trace amounts in nature and therefore did not receive much attention until the 1930s, when it was identified as a potent toxic substance that was dangerous for cattle and other livestock. Investigation into the origin of 'alkali disease' or 'blind staggers' disease of cattle indicated selenosis due to the ingestion of high dose of selenium present in certain plants, especially cereal crops. While selenium was earlier known for its toxicity, research done in the 1950s reported rather contradictory findings. For example, they showed that dietary intake of selenium helped to prevent liver necrosis in rats and prevented a type of muscular dystrophy known as white muscle disease, which was particularly lethal in young animals.

Table 12.1 Physical properties of selenium

Selenium **Symbol: Se** **At. no. 34**	*Atomic mass*	*Electronegativity (Pauling scale)*	*1st ionisation energy*	*Electronic configuration*	*van der Waals radius*
	78.971 g.mol^{-1}	2.55	941 kJ.mol^{-1}	[Ar]$3d^{10}4s^2 4p^4$	0.190 nm
	Atomic radius	*Melting point*	*Boiling point*	*Electron affinity*	*Density*
	1.20 Å	221°C	685°C	194.959 kJ.mol^{-1}	3.99 g.cm^{-3} at m.p.

Table 12.2 Physical properties of tellurium

Tellurium **Symbol: Te** **At. no. 52**	*Atomic mass*	*Electronegativity (Pauling scale)*	*1st ionisation energy*	*Electronic configuration*	*van der Waals radius*
	127.60 g.mol^{-1}	2.1	869.3 kJ.mol^{-1}	[Kr]$4d^{10}5s^2 5p^4$	0.206 nm
	Atomic radius	*Melting point*	*Boiling point*	*Electron affinity*	*Density*
	1.40 Å	449.5°C	988°C	190.161 kJ.mol^{-1}	5.70 g.cm^{-3} at m.p.

Selenium is obtained commercially from the anode mud deposited during the electrolytic refining of copper which also contains sulfides of other coinage metals. Selenium forms several allotropes that interconvert with variation in temperature changes. When formed in chemical reactions, especially after oxidation using SeO_2, selenium is usually a brick-red amorphous powder. When rapidly melted, it forms the black, vitreous form. The structure of black selenium is complex and consists of polymeric rings with up to 1000 atoms per ring. It is slightly soluble in CS_2. Upon heating, it softens and gets converted to grey selenium at 180°C. The most stable and dense form of selenium is the grey from which has a hexagonal crystal lattice consisting of helical polymeric chains, where the Se–Se distance is 237.3 pm and Se–Se–Se angle is 130.1°. The grey form has interesting properties such as photoconductivity and ability to convert AC current to DC.

Tellurium was discovered by in 1782 by Franz Josef Mueller von Reichenstein from a gold mine in Hungary as gold tellurides are the most common gold containing minerals.[3] The word tellurium comes from the Latin word for earth namely 'tellus'. Tellurium is present at around 1 ppb in the earth's crust and is less abundant than gold and platinum. Gold tellurides are the most well-known minerals for tellurides. It is obtained commercially from the anode mud produced during electrolytic refining of copper. This contains up to 8% tellurium. Unlike selenium, tellurium has no known biological role. It is toxic and teratogenic (disturbs the development of an embryo or fetus).[1]

12.2 Isolation and Purification of the Elements

The first step in the purification process involves roasting the copper chalcogenide in air with soda ash which results in the formation of alkali selenite and tellurite. Selenium dioxide can also be obtained by heating the copper selenide in air at 650°C.

$$Cu_2Z + Na_2CO_3 + 2\,O_2 \longrightarrow 2\,CuO + Na_2ZO_3 + CO_2 \qquad Z = Se, Te \tag{12.1}$$

$$Cu_2Se + 3/_2O_2 \xrightarrow[650°C]{} 2\,CuO + SeO_2 \tag{12.2}$$

The reaction of a mixture of alkali selenite and tellurite with sulfuric acid makes the tellurite precipitate as a hydrous dioxide while the selenite remains in solution as selenous acid, H_2SeO_3. Selenous acid on reduction with NH_3, N_2H_4 or SO_2 gives selenium. Dehydration of selenous acid gives selenium dioxide. Sodium tellurite on electrolytic reduction in water gives tellurium metal.

$$H_2SeO_3 + H_2O + 2\,SO_2 \longrightarrow Se + 2\,H_2SO_4 \tag{12.3}$$

$$Na_2TeO_3 + H_2O \xrightarrow[\text{reduction}]{\text{electrolytic}} Te + 2\,NaOH + O_2 \tag{12.4}$$

12.3 Compounds of Selenium and Tellurium

12.3.1 Hydrides

Hydrides of selenium and tellurium have similarities to H_2S; they are also gases and are toxic. Hydrogen selenide has been prepared industrially by heating hydrogen and selenium together above 350°C, but hydrogen telluride is not attainable by this method due to its thermal instability as it decomposes in moisture, light and at temperatures above 0°C.

Table 12.3 Comparison of properties of H_2S, H_2Se and H_2Te

Property	H_2S	H_2Se	H_2Te
m.p. (°C)	−85.6	−65.7	−51
b.p. (°C)	−60.3	−41.3	−4.0
pKa	7.0	3.9	2.6
X–H bond length (pm)	133.6	146	169
H–X–H bond angle	92.1°	91°	90°
Dissociation constant (HX^-, K_1)	1.3×10^{-7}	1.3×10^{-4}	2.3×10^{-3}
Toxicity	Toxic	Toxic	Toxic

H_2Se is a colourless, flammable gas under standard conditions. It is considered as the most toxic selenium compound and has an exposure limit of 0.05 ppm over a period of 8 hours. Even at very low concentrations, H_2Se has an irritating smell resembling that of decayed horseradish or of rotten eggs at higher concentrations. H_2Se is usually prepared by the action of water on Al_2Se_3.[2]

$$Al_2Se_3 + 6H_2O \rightleftharpoons 2\,Al(OH)_3 + 3\,H_2Se \tag{12.5}$$

$$FeSe + 2\,HCl \longrightarrow H_2Se + FeCl_2 \tag{12.6}$$

Jöns Jacob Berzelius who discovered selenium also described the poisonous effect of hydrogen selenide; *'In order to get acquainted with the smell of this gas, I allowed a bubble not larger than a pea to pass into my nostril; in consequence of its smell I so completely loss my sense of smell for several hours that I could not distinguish the odour of strong ammonia even when held under my nose. My sense of smell returned after five or six hours, but severe irritation of the mucous membrane set in and persisted for a fortnight'.*

H_2Te is rarely encountered because it is unstable in air above 0°C and decomposes rapidly to tellurium and water. However, the gas can exist at very low concentrations long enough to be readily detected by the odour of rotting garlic at extremely low concentrations or by the revolting odour of rotting onions/leeks at higher concentrations. Hydrogen telluride is prepared by the acidification of salts of Te^{2-}, such as Al_2Te_3 and Na_2Te. Na_2Te in turn can be prepared by the reaction of Na and Te in liquid ammonia. Both H_2Se and H_2Te burn in air with a blue flame giving the dioxide. Volatile tellurium compounds often have unpleasant odours, reminiscent of decayed leeks or garlic.

12.3.2 Oxides

Unlike sulfur dioxide which is a gas, the oxides of selenium and tellurium are solids with different structures. Selenium dioxide is polymeric in nature with interconnected SeO_3 pyramidal units carrying pendant terminal oxygen atoms while TeO_2 exists in α- and β-forms. α-TeO_2 is monomeric with a see-saw like structure and shares all vertices to form a rutile like structure. It has a distorted octahedral core with two Te–O close contacts as secondary bonding interactions. β-TeO_2 has a layer structure constituting dimeric units with two TBP units connected to each other using two oxygen atoms. Quite interestingly, tellurium shows a reluctance to form double bonds with oxygen. At around 350°C, SeO_2 sublimes giving a monomeric form with structure similar to that of SO_2. TeO_2 is nonvolatile and β-TeO_2 occurs naturally as the mineral tellurite.[6]

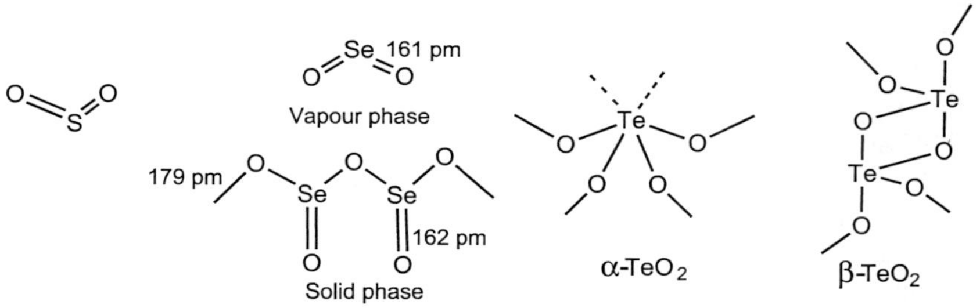

Fig 12.1 Structures of the oxides of sulfur, selenium and tellurium

12.3.3 Selenium Dioxide

Selenium dioxide is usually prepared by adding commercial selenium powder in small amounts to an excess of concentrated nitric acid. The initial reaction is highly exothermic with the evolution of NO and NO_2. The clear solution of selenous acid, H_2SeO_3, obtained is fumed to remove excess nitric acid and then evaporated to dryness to dehydrate selenous acid to selenium dioxide. The crude selenium dioxide can be purified by sublimation at 317°C to form shining white, monoclinic crystals of SeO_2.

$$3\,Se + 4\,HNO_3 + H_2O \xrightarrow[-4\,NO]{} 3\,H_2SeO_3 \xrightarrow[-3\,H_2O]{\Delta} 3\,SeO_2 \qquad (12.7)$$

Crystals of selenium dioxide melt at 340°C in a sealed tube. The compound has a density of 3.95 at 15°C and a heat of vapourisation of 21,600 cal/mol. In the vapour state, selenium dioxide shows a typical greenish yellow colour.

Selenium dioxide is highly soluble in water and to some extent in alcohols. It absorbs moisture from the air to form solid selenous acid, but even at ordinary temperatures this reaction is reversible.[2]

$$SeO_2 + H_2O \rightleftharpoons H_2SeO_3 \qquad (12.8)$$

While SO_2 is considered as a mild reducing agent, SeO_2 functions as a mild oxidising agent over a wide temperature range. In such oxidations, selenium dioxide undergoes reduction to free selenium. This selenium can be recovered and reoxidised back to selenium dioxide. In hot concentrated sulfuric acid, however, it is possible for selenium and selenium dioxide to function as catalytic agents.

Uses of selenium dioxide

(i) For making red coloured glass: When dissolved in water, SeO_2 forms selenous acid, which is how the oxide is often used in reactions. Industrially selenium dioxide is used mainly in glass making. Although the colour of SeO_2 is between white and yellowish pink, it imparts a rich, red colour to glass. Selenium is one of the most important colouring agents for making pink and red glass. Selenium in higher concentrations imparts a red colour, caused by selenium nanoparticles dispersed in the glass. One can observe that if the amounts of selenium are increased and added to a cadmium sulfide glass, the yellow colour changes to orange and finally to a brilliant red known as 'selenium ruby'.[3]

(ii) As an oxidising agent: As early as 1947, selenium dioxide was reported in *Nature* as being used to oxidise more than three hundred different organic compounds. Selenium (IV) oxide-mediated oxidation of substituted olefins is regarded as one of the most reliable and predictable methods for introducing a hydroxyl group into the allylic position or conversion of alkenes to allylic alcohols. Riley oxidation shows regio and stereoselectivity when applied to trisubstituted olefins, resulting preferentially in (*E*)-allylic alcohols.[4]

R–CH=CH–CH(H)–R' $\xrightarrow{SeO_2}$ R–CH=CH–CH(OH)–R' (12.9)

SeO_2 is used to oxidise the α-methylene group adjacent to a carbonyl group to give a 1,2-dicarbonyl compound and oxidation of a similar α-methyl group leads to an aldehyde. A number of organic transformations come under this category, such as the conversion of paraldehyde – the trimer of acetaldehyde – to glyoxal and cyclohexanone to cyclohexane 1,2-dione. The glyoxal reaction is useful as it is a good building block for organic synthesis. During these reactions, SeO_2 gets reduced to elemental selenium as its red allotrope which gets precipitated and is easily separated.[5]

$[H_3C{-}CHO]_3$ (paraldehyde) $\xrightarrow[\Delta,\ -Se]{SeO_2,\ H_2O}$ H–CO–CO–H (glyoxal) (12.10)

4-acetyl benzoic acid $\xrightarrow[95°C,\ -Se]{SeO_2,\ H_2O}$ 4-glyoxal benzoic acid (12.11)

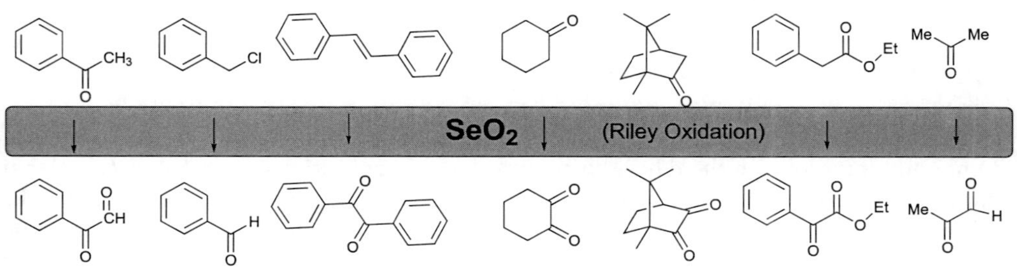

Scheme 12.1 Oxidation of organic substrates using SeO_2

(iii) As a testing reagent: When a selenium dioxide solution is mixed with concentrated sulfuric acid, it produces a fluid known as the Mecke reagent [100 mL of concentrated (95%) sulfuric acid with 1 g of selenous acid], which is used by a number of testing agencies to provide spot chemical tests for a range of illegal substances from LSD to opium as it undergoes a colour change that indicates the presence of alkaloids as well as other compounds. For example LSD gives greenish black colour while morphine gives dark bluish green colour with the Mecke reagent.

$$SeO_2 \underset{-H_2O}{\overset{H_2O}{\rightleftharpoons}} H_2SeO_3 \xrightarrow{O_3 \text{ or } H_2O_2} H_2SeO_4 \quad (12.12)$$

Selenous acid and selenic acid

Selenous acid [also called selenious acid, $(OH)_2SeO$] is readily prepared by dissolving selenium dioxide in water (Eq. 12.8). Unlike sulfurous acid which is unstable and non-existent in the free form, selenous acid is stable and is a solid with a m.p. of 70°C. Selenic acid H_2SeO_4, on the other hand, is obtained by oxidation of selenous acid by a host of reagents such as H_2O_2, O_3, $KMnO_4$ and $HClO_3$. It is a much stronger acid than sulfuric acid and is fully dissociated in aqueous solution with respect to loss of the first proton. For the second proton, the pK_a is 1.92. Oxidation of selenium metal with chlorine or bromine water also gives the same acid. Its higher oxidation strength is exemplified from the observation that it liberates Cl_2 from conc. HCl and converts metallic gold to Au(III). Sulfuric acid only dehydrates hydrochloric acid and does not react with gold. Selenic acid on heating to 200°C converts to selenous acid liberating O_2. Selenic acid shows many reactions similar to H_2SO_4, such as losing water to give SeO_3 which combines with H_2SeO_4 to give pyroselenic acid, $H_2Se_2O_7$. Unlike sulfuric acid, selenic acid cannot be prepared by hydration of SeO_3 since SeO_3 itself is unstable.

$$H_2SeO_4 + 2\,H^+ + 2\,Cl^- \longrightarrow H_2SeO_3 + Cl_2 + H_2O \quad (12.13)$$

$$2\,Au + 6\,H_2SeO_4 \longrightarrow Au_2(SeO_4)_3 + 3\,H_2SeO_3 + 3\,H_2O \quad (12.14)$$

12.4 Organoselenium Compounds

Organoselenium compounds which contain Se–C bonds form a large class of selenium compounds and some of them have shown very useful applications especially as functional mimics of essential selenium containing enzymes such as glutathione peroxidase. Like sulfur, selenium can exist in –2, 0, +2, +4 and +6 oxidation states but the dominant state is Se(II). Comparing bonds with carbon, the C–S bond strength is 272 kJ/mol while C–Se bond is only 234 kJ/mol. The C–Se bond is also longer (198 pm) compared to the C–S bond (181 pm).

Selenium compounds are more nucleophilic in nature than the corresponding sulfur compounds and are also more acidic. The pK_a values are 7 for H_2S and 3.8 for H_2Se. While sulfoxides are stable compounds, the corresponding selenoxides are unstable in the presence of β-protons and this property is utilised in many reactions especially in selenoxide oxidations and in selenoxide eliminations.[6]

Some of the well known organoselenium compounds and their general syntheses are detailed below.

12.4.1 Selenols

Selenols, R–SH, which are the selenium equivalents of thiols have even more unpleasant odours and are more acidic. The pK_a of methyl selenol, MeSeH is 5.2 while that of methyl mercaptan, MeSH in 8.3. Similarly, pK_a of phenylselenol PhSeH, is 5.9 while that of thiophenol is 6.5.

Since the Se–H bond is weaker than the S–H bond, selenols get easily oxidised to the diselenides and serve as H-atom donors. Selenols are about 1000 times stronger acids than thiols. Selenols are important in many biological processes. There are three enzymes that contain selenol units at their active sites. These are glutathione peroxidase, iodothyronine deiodinase, and thioredoxin reductase. The selenols in these proteins belong to the amino acid selenocysteine.

$$\text{PhSeSePh} \xrightarrow{H_3PO_2 \text{ or } Bu_3SnH} \text{PhSeH} \xleftarrow{\text{1. Se/ether; 2. HBr}} \text{PhMgBr} \quad (12.15)$$

$$\text{PhSeSePh} \xrightarrow{LiEt_3BH} \text{PhSeLi} \xrightarrow[HCl]{} \text{PhSeH} \quad (12.16)$$

12.4.2 Diselenides

Diselenides (R–Se–Se–R) are the equivalents of peroxides and disulfides in selenium chemistry. They are stable starting materials for the preparation of more reactive organoselenium compounds such as selenenyl halides and selenols. The most widely studied among these is diphenyldiselenide, prepared from the reaction of a phenyl Grignard reagent with selenium and oxidation of the product, PhSeMgBr, with bromine. Many other diselenides have been prepared using alkali metal diselenides such as Na_2Se_2 and Li_2Se_2. Some of the substituted diselenides have been found to have a good capability to act as hydrogen peroxide reductants mimicking the enzyme, glutathione peroxidase. Well known reactions of diphenyldiselenide are given in Scheme 12.2.[7–15]

$$\text{PhBr} \xrightarrow{\text{Mg, ether}} \text{PhMgBr} \xrightarrow[\Delta]{\text{Se Powder}} \text{PhSeMgBr} \xrightarrow{Br_2} \text{PhSeSePh} \quad (12.17)$$

Benzene selenols are also easily oxidised by air to the respective diselenides.

12.4.3 Selenyl Halides

Selenyl halides (R–Se–Cl, R–Se–Br) are prepared by the halogenation of diselenides. Controlled bromination and chlorination of diphenyldiselenide gives phenylselenyl bromide and phenylselenyl chloride. Excess of bromine or sulfuryl chloride will result in the Se(IV) trihalogenated compounds.

$$\text{PhSeSePh} \xrightarrow[\text{hexane}]{Cl_2} \text{PhSeCl} \quad (12.18)$$

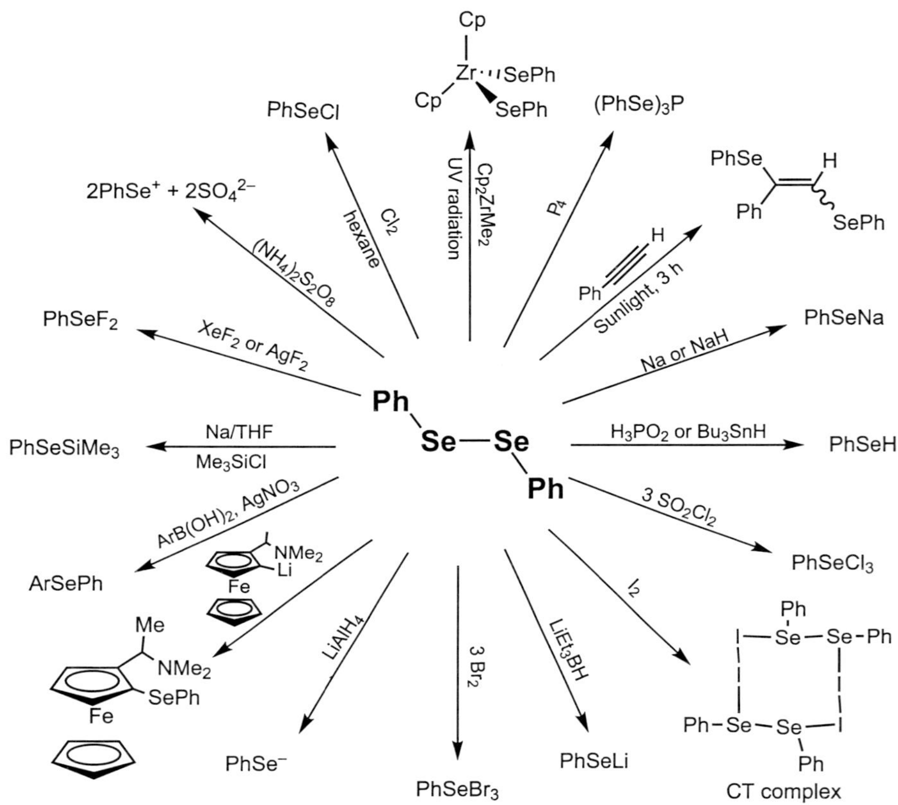

Scheme 12.2 Selected reactions of PhSe–SePh

12.4.4 Selenides

Selenides (R–Se–R), also known as selenoethers, are the selenium analogues of ethers or thioethers. These are regular organoselenium compounds and are usually prepared by alkylation of salts such as Na_2Se. Unsymmetrical selenides can be prepared by the alkylation of selenoates. Divalent selenium in these compounds can also interact with soft heteroatoms to form hypervalent selenium centres.

12.4.5 Selenoxides and Selenones

These are the selenium equivalents of sulfoxides with Se in the +4 oxidation state. R–Se(O)R can be further oxidised to the +6 oxidation state R–Se$(O)_2$R and such compounds are called selenones. Selenides are easily oxidised to selenoxides with reagents such as H_2O_2 or HIO_4.

After the development of sulfoxide elimination as an effective method for generating carbon–carbon double bonds, it was discovered that selenoxides undergo a similar process, even more rapidly. Most selenoxides decompose to the corresponding alkenes at temperatures between −50 and 40°C whereas sulfoxides require higher temperatures.

12.4.6 Selenoxide Elimination Reactions

One of the most important uses of selenium compounds in organic synthesis is for the selenoxide elimination reaction, developed by Sharpless, especially useful for preparation of α,β-unsaturated carbonyl compounds from the corresponding saturated molecules. Compared to the analogous sulfoxide eliminations, these reactions occur at much lower temperatures (ca. 100°C milder) and by reaction rates around 100,000 times faster making it a very convenient method for preparing alkenes.[16–18]

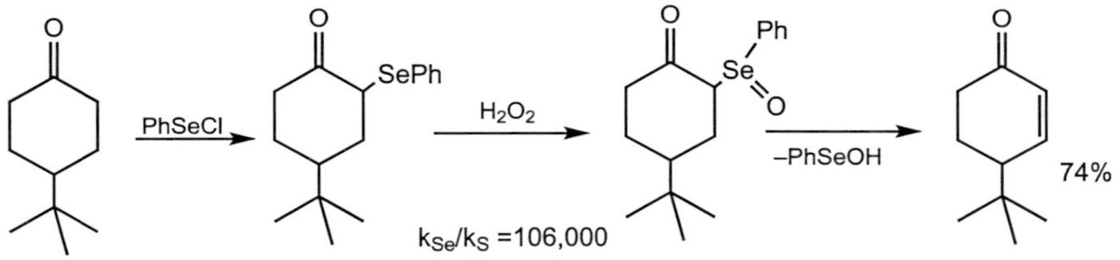

PhSeCl → SePh → H_2O_2 → Ph Se O → −PhSeOH → 74% (12.19)

k_{Se}/k_S =106,000

The same methodology has also been extended for the conversion of epoxides to allylic alcohols.

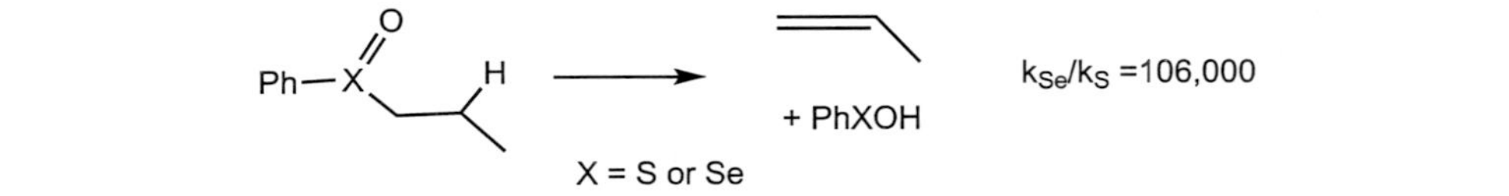

PhSeNa EtOH → HO SePh → H_2O_2 → HO Ph Se H O- → HO + PhSeOH

Ph—X O H → + PhXOH k_{Se}/k_S =106,000

X = S or Se

Scheme 12.3 Selenoxide elimination reactions

12.4.7 Selenenic Acids, Seleninic Acids and Selenonic Acids

Selenenic acids (RSeOH) with Se in the +2 state occur in selenoenzymes such as glutathione peroxidase. Selenenic acids derived from selenoenzymes are thought to be responsible for the antioxidant activity of these enzymes. Seleninic acids ($RSeO_2H$) similar to sulfinic acids, are generally obtained by the oxidation of selenenic acids and have Se in the +4 oxidation state. Selenenic acids are unstable due to self-condensation reaction, forming the corresponding selenoseleninates. They can also disproportionate to seleninic acids and diselenides. Seleninic acids (+4) are stable and are similar to alkyl sulfinic acids. Selenonic acids (+6) are sulfonic acid analogues. For example, polyalkylated benzenes can be readily converted to the corresponding selenonic acids by using a mixture of H_2SeO_4 and acetic anhydride.

$$\underset{\text{Selenoseleninate}}{RSe(O)SeR + H_2O} \xleftarrow{\text{self condensation}} \underset{\text{Selenenic acid}}{2\ RSeOH} \xrightarrow{\text{disproportionation}} \underset{\text{Seleninic acid}}{RSeO_2H} + 1/2\ RSeSeR \quad (12.20)$$

12.4.8 Alkali Metal Diselenides

These are nucleophilic reagents useful for the synthesis of organic diselenides. A suspension of Na_2Se_2 (or K_2Se_2) can be produced in high-boiling polar aprotic solvents, such as DMF, HMPA, NMP or liq. NH_3 by the direct reaction of selenium powder with sodium or potassium. However a convenient method is the reduction of selenium by the alkali metals lithium/sodium/potassium with catalytic amounts of naphthalene in a solvent such as THF. The reduction of selenium with lithium in THF has also been achieved by using diphenylacetylene as a catalyst to afford lithium diselenide. These alkali metal diselenides react readily with electrophiles giving alkyl or aryl diselenides.

$$2Se + 2Na \xrightarrow[\text{polar aprotic solvents}]{} Na_2Se_2 \xleftarrow[\text{THF}]{\text{naphthalene}} Se + Na \quad (12.21)$$

$$2Se + 2Li \xrightarrow[\text{THF, 20°C}]{\text{PhCCPh}} Li_2Se_2 \longleftarrow Se + LiHBEt_3 \quad (12.22)$$

$$\text{Bu}^t\text{-C}_6\text{H}_2(\text{CHO})_2\text{-Br} \xrightarrow[\text{THF}]{Na_2Se_2} (\text{Bu}^t\text{-C}_6\text{H}_2(\text{CHO})_2\text{-Se})_2 \quad (12.23)$$

Chiral diselenides have also been prepared by the ring opening of aziridines using dilithium diselenides

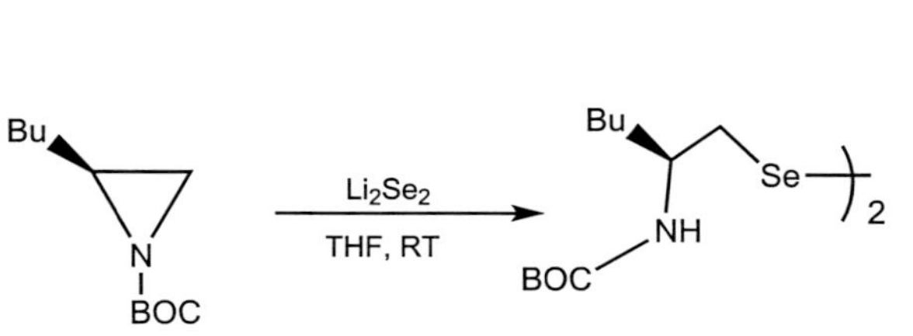

(12.24)

12.4.9 Woolins' Reagent

Similar to Lawesson's reagent (obtained from the reaction of P_4S_{10} and anisole), extensively used for converting carbonyl groups to thiocarbonyl groups, an analogous reagent made by J D Woolins shows selenation of selected carbonyl groups.[19–20]

Lawesson's reagent   Woolins' reagent

There are two methods for making Woolins' reagent and the one involving reaction of sodium selenide with phenyl dichlorophosphine is easier.

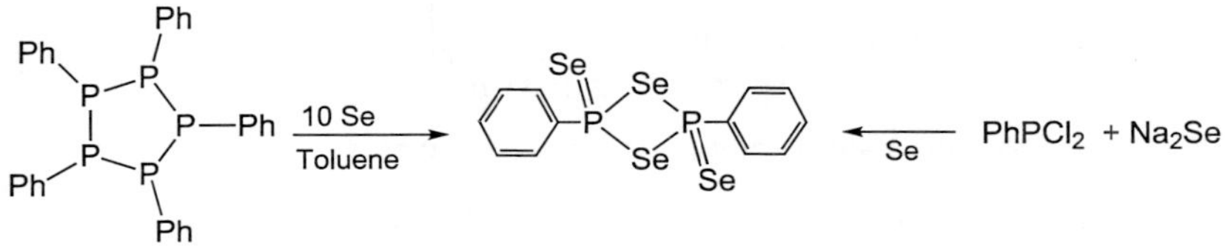

(12.25)

Woolins' reagent has been found to convert cyclic and acyclic amides to selenoamides and also nitriles to selenoamides.

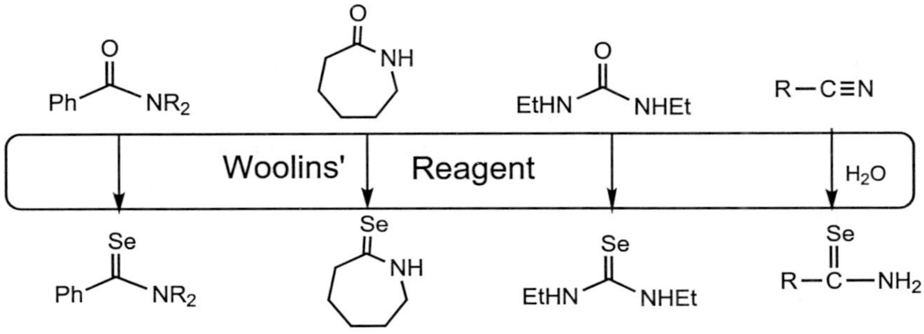

Scheme 12.4 Reactions of Woolins' reagent

12.5 Selenium Biochemistry

The importance of selenium in biochemistry became evident with the discovery of the 21st amino acid, selenocysteine (Sec) by Thressa Stadtman in 1972.[21] She found that glycine reductase, an enzyme that catalyses the reductive degradation of glycine contains selenium. Almost simultaneously, it was discovered that glutathione peroxidase, an enzyme present in the red blood cells also contains selenocysteine. As of 2016, fifty-four human proteins have been found to contain selenocysteine. In addition, selenium is also found in nucleic acids of certain bacteria as 5-methylaminomethyl-2-selenouridine. It is also found in selenomethionine (naturally occurring amino acid found in Brazil nuts, soya beans and so on), monoselenophosphate (a selenium donor which inserts selenium into selenium-based enzymes) and selenoneine (involved in demethylation of mercury in some fish). Selenocysteine units of proteins are also ligands in some of the Mo, Ni and Fe sulfur-based metalloenzymes.

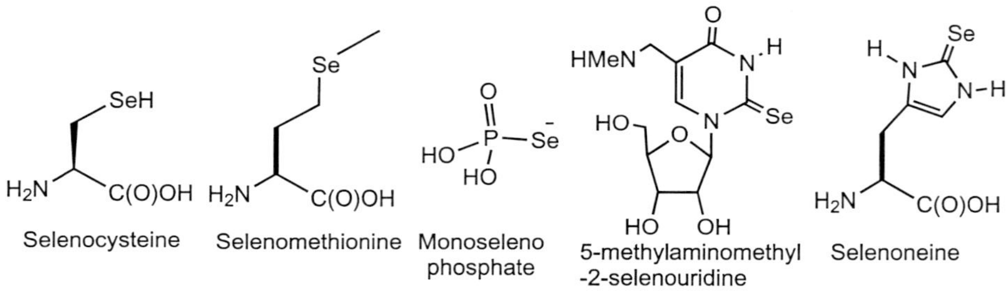

Selenocysteine Selenomethionine Monoseleno phosphate 5-methylaminomethyl -2-selenouridine Selenoneine

Fig 12.2 Selenium based compounds found in living systems

Selenocysteine is a true proteinogenic amino acid (amino acids which are biosynthetically inserted into proteins) and it satisfies the criteria met by the other 20 common amino acids. It is encoded by DNA, it has its own unique codon (UGA), it has an unique *t*-RNA which brings the residue to the ribosome before it is inserted into the polypeptide chains of selenoproteins.[22]

The best studied selenoenzymes that contain selenocysteine in their active site are the following with their functions.[23]

(i) Glutathione peroxidase (Gpx): Scavenges H_2O_2 and hydroperoxides thereby reducing oxidative stress and preventing oxidative damage.

(ii) Iodothyronine deiodinase (DIO): responsible for the activation/inactivation of the thyroid hormones. Reduces the Ar–I bonds present in thyroid hormones thyroxine (T4) and triiodothyronine (T3) to Ar–H.

(iii) Thioredoxin reductase (TrxR): Maintains the thiol–disulfide redox equilibrium. Plays an active role in DNA synthesis.

The enzyme glutathione peroxidase has been found to be quite effective in protecting cells from the oxidative breakdown brought about by low concentrations of ascorbic acid.

Selenium present in glutathione peroxidase is responsible for destroying lipid-damaging peroxides. Glutathione peroxidise is found within white blood cells and blood platelets, and has a significant role to play in the immune system and clotting of blood. Vitamin E and selenium tend to enhance the effect of one another, in that vitamin E works to prevent the formation of peroxides, and glutathione peroxide acts to destroy peroxides. Excess of selenium absorbed in the body is removed through the urine, and another molecule dimethyl selenide is excreted during respiration giving a garlic-like odour to breath which is characteristic of volatile selenium compounds.

The main reaction that glutathione peroxidise catalyses is:

$$2\ GSH + H_2O_2 \longrightarrow GSSG + H_2O \qquad (12.26)$$

GSH represents reduced monomeric glutathione, and GS–SG represents glutathione disulfide. The mechanism involves three steps. The selenol of a selenocysteine residue is oxidised by hydrogen peroxide. This results in a selenenic acid (RSeOH) group. The selenenic acid is then converted back to selenol by a two-step process.

$$RSeH + H_2O_2 \longrightarrow RSeOH + H_2O \qquad (12.27)$$

$$RSeOH + GSH \longrightarrow GS\text{–}SeR + H_2O \qquad (12.28)$$

$$GS\text{–}SeR + GSH \longrightarrow GS\text{–}SG + RSeH \qquad (12.29)$$

Chemistry at the active site of glutathione peroxidase has been studied in detail by the help of synthetic mimics. The compound Ebselen, first reported in 1984, was one of the early synthetic mimics of glutathione peroxidase.[24, 25] This compound also has additional properties such as anti-inflammatory, antioxidant and cytoprotective activity. It reacts with H_2O_2 and peroxynitrite. It is used as a drug for possible treatment of stroke, hearing loss and tinnitus (constant ringing sound in the ear). Following this in 1989, Spector and coworkers discovered that aminomethyl diselenides show much better glutathione peroxidise activity.[26] In 2015, Mugesh, D'Silva and coworkers prepared a different set of isoselenazoles having an OMe group on the aryl ring and found that these are 2–3 times more active than Ebselen and at the same time their toxicity level is several times lower than that of Ebselen.[27, 28]

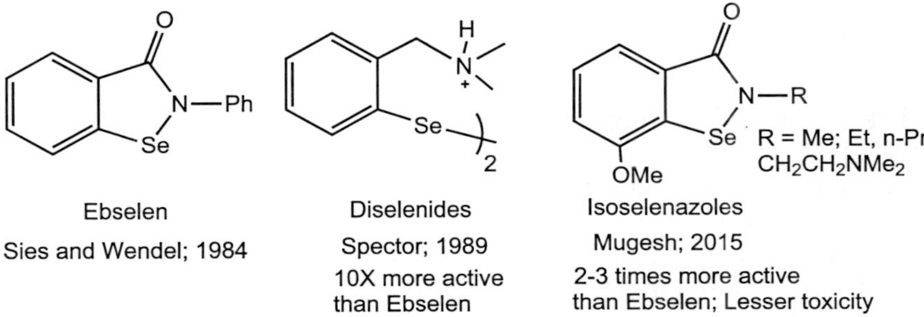

Fig 12.3 Synthetic mimics of glutathione peroxidase

An efficient copper catalysed method for synthesising Ebselen and related compounds was developed by Sangit Kumar and coworkers. The copper catalysed method uses selenium powder as source and potassium carbonate as base and is a one-step catalytic synthesis unlike previous methods reported for Ebselen.[29]

O, HN—Ph, I

CuI (20 mol%), 1,10-Phen

Se powder, K_2CO_3

DMF, 110°C, 8 h

O, N—Ph, Se

84%

(12.30)

12.6 Selenium Versus Sulfur: Why Nature Chose Selenium

While both sulfur and selenium are present in biological systems, the natural question that arises is, why did nature choose selenium if its chemistry is very similar to that of sulfur. To understand this, one has to see the basic differences between the properties of sulfur, selenium and their compounds.[22] Some of the interesting differences are:

- SO_2 is a mild reducing agent while SeO_2 is a mild oxidising agent.
- Selenium is more polarisable than sulfur, leading to rapid electrophilic and nucleophilic substitution at selenium.
- Bond breaking reactions are faster on selenium compared to sulfur due to weaker bonds originating from selenium.
- Higher oxidation states are relatively less stable for selenium and it is more tolerant to hypervalent bond formation.
- π-bonds are weaker in selenium compounds compared to sulfur due to larger size of selenium and longer bonds originating from it.
- Selenoesters are less stabilised than thioesters by resonance with carbonyl groups.
- Seleninic acids (RSe(O)OH) are weaker oxidising agents than sulfinic acids (pK_a of $PhSeO_2H$ is 4.79; pK_a of $PhSO_2H$ is 2.76) which is a consequence of better π acceptor properties and higher electronegativity of sulfur.

General differences in reactivity between some of the sulfur and selenium-based compounds with organic substrates are shown in Scheme 12.5.

12.7 The Selenium Paradox

Why nature chose selenium for certain biochemical processes is explained in part by the *selenium paradox*. The unusual ability of selenium to rapidly get oxidised and then rapidly get reduced back to the original oxidation state is termed as the selenium paradox and stands in stark contrast to the chemistry of sulfur and tellurium which are stable in their higher oxidation states and do not easily get reduced back. Unlike sulfur, selenium provides resistance to inactivation of its enzymes by oxidation.

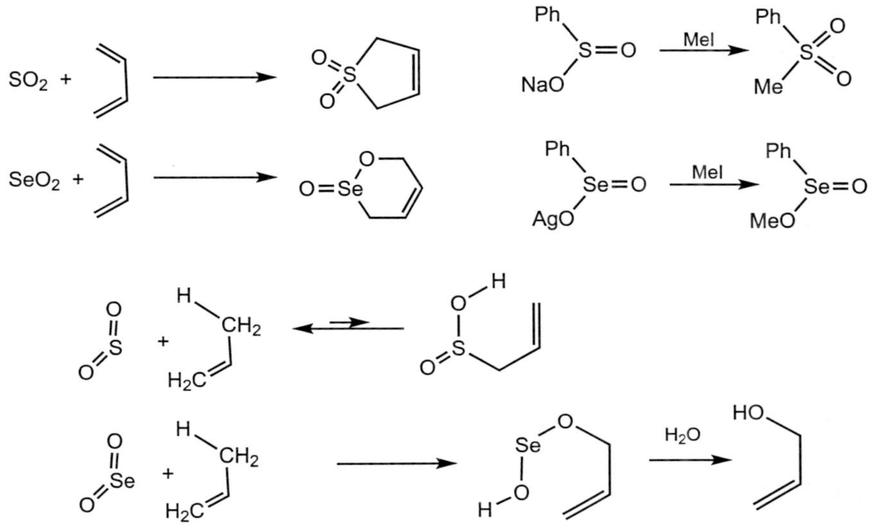

Scheme 12.5 Differences in reactivity of sulfur and selenium compounds

Both sulfur and selenium are good nucleophiles that react with reactive oxygenated species (ROS) in the living system by two electron oxidation events and as a result, get oxidised. However, the difference is that selenium has the unique ability to react with oxygen and ROS in a readily reversible manner. Also, since chemical reactions with selenium are faster than with sulfur, enhanced chemical activity is observed for selenoenzymes with accelerated rates compared to sulfur-based enzymes.

Selenium containing proteins are also able to withstand one-electron oxidation reactions better than sulfur based proteins—a consequence of the better stability of the selenyl radical compared to the thiyl radical.

12.8 Compounds of Tellurium

With regard to structure, properties and reactivities, tellurium compounds differ drastically when compared to compounds of sulfur and selenium. Like sulfur and selenium, tellurium can also have oxidation states from –2 to +6. However, due to lower ionisation energies, higher oxidation states are more stable for selenium and tellurium when compared to sulfur. For example $SeCl_4$ and $TeCl_4$ are found as stable solid compounds while SCl_4 does not form as it is thermally unstable. On the other hand, $SeCl_2$ and $TeCl_2$ are thermodynamically unstable towards disproportionation unlike SCl_2. There are also many examples of organotellurium compounds such as Ph_6Te and Me_6Te which are air-stable and for which there are no lighter chalcogen analogues.[30]

The single bond energies, both homonuclear and heteronuclear, involving tellurium are significantly lower than those involving the sulfur and selenium atoms as shown.

Te–Te	149 kJ/mol	Te–H	238 kJ/mol	Te–C	200 kJ/mol
Se–Se	192 kJ/mol	Se–H	276 kJ/mol	Se–C	234 kJ/mol
S–S	266 kJ/mol	S–H	366 kJ/mol	S–C	272 kJ/mol

The Pauling electronegativity of tellurium (2.1) is lower than that of selenium (2.6) and sulfur (2.6). So E–X bonds where X is more electronegative than the chalcogen E are more polar for tellurium which results in stronger secondary bonding interactions emanating from tellurium. The π bond energies for Te–C, Te–O and Te–N are significantly lower than those of sulfur and selenium and so tellurium shows less propensity to form multiple bonds with these elements thus showing vast structural differences in its compounds when compared to similar compounds of S and Se.

12.9 Tellurium Oxides and Acids

Tellurium dioxide is obtained by burning tellurium in air or by adding finely divided tellurium powder to conc. nitric acid followed by dilution with water when TeO_2 is produced as a precipitate. One basic difference with SeO_2 is that TeO_2 is very poorly soluble in water (1 part in 150,000 parts of water at room temperature). TeO_2 can be reduced to tellurium by heating with carbon or cyanide or at high temperatures with H_2 gas. Aqueous solutions of TeO_2 in acids also undergo reduction to elemental tellurium with phosphorus acid, sulfurous acid, stannous chloride, and so on.[31]

Tellurium dioxide is known in two forms—α-TeO_2 which is colourless and tetragonal and β-TeO_2 which is yellow and orthorhombic in nature. α-TeO_2 converts to the β-TeO_2 form at high pressures. α-TeO_2 is prepared by the reactions of Te with O_2, or by dehydrating tellurous acid, H_2TeO_3. Thermal decomposition of basic tellurium nitrate $[Te_2O_4H]^+(NO_3)^-$ above 400°C also gives α-TeO_2. The polymeric structures of α-TeO_2 and β-TeO_2 without any double bonds are shown in Fig. 12.1.

H_2TeO_4 is unknown and H_2TeO_3 is not as stable as H_2SeO_3 and is a weak acid [pK_a (1) = 2.48]. Telluric acid, $Te(OH)_6$ is a unique solid octahedral acid of tellurium, prepared when TeO_2 is oxidised with hydrogen peroxide or chromium trioxide. It behaves as a weak acid [pK_a (1) = 7.68]. It is however a fairly strong oxidant and gets reduced to elemental tellurium oxidising SO_2 to H_2SO_4.

$$TeO_2 \underset{-H_2O}{\overset{H_2O}{\rightleftharpoons}} H_2TeO_3 \tag{12.31}$$

$$TeO_2 + H_2O_2 + 2\,H_2O \longrightarrow Te(OH)_6 \tag{12.32}$$

$$Te(OH)_6 + 3\,SO_2 \longrightarrow Te + 3\,H_2SO_4 \tag{12.33}$$

12.10 Halides

Tellurium powder, on reaction with chlorine gas gives $TeCl_4$ as a pale yellow solid having a melting point of 224°C. Its structure is monomeric in the gas phase similar to that of SF_4 and in the solid state it is observed as a tetramer with a cubic core. The geometry around each tellurium is a distorted octahedron.

$$Te + 2\,Cl_2 \xrightarrow[\Delta]{} TeCl_4 \tag{12.34}$$

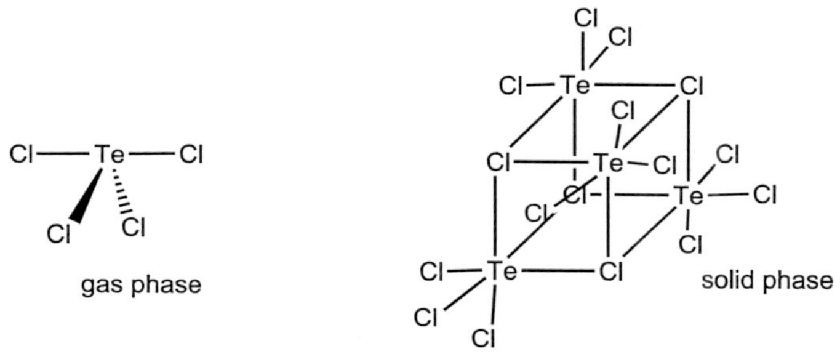

Fig. 12.4 Structure of $TeCl_4$ in gas and solid phases

It is used in making organic derivatives of tellurium with tellurium in the +4 oxidation state. Electron-rich arenes such as anisole react directly with $TeCl_4$ giving diaryldihalides which are useful Te(IV) compounds for further reactions. Tellurium also forms TeF_4 as a colourless solid and TeF_6 as a colourless gas. $TeBr_4$ and TeI_4 also exist in the tetrameric form as high melting solids.

12.11 Tellurides

There are many tellurides with interesting and diverse applications (Table 12.4).

Table 12.4 Different tellurides and their properties

Telluride	Unique property	Applications
Bismuth telluride	Thermoelectric material and topological insulator	Freon-free refrigerators, portable coolers
Cadmium telluride	Most suitable band gap (1.44 eV) for solar cells	Thin film solar cells
Silver–indium–antimony telluride	Reversible gain and loss of reflectivity and crystallinity upon heating	Rewritable CDs, DVDs and blue ray discs, phase change memory chips
Mercury cadmium telluride	Infrared detector	IR spectrometers
Zinc cadmium telluride	X-ray and γ-ray detector	Nuclear installations

12.12 Cadmium Telluride

CdTe solar cells (band gap 1.44 eV) are the second most widely used photovoltaic technology in the world after crystalline silicon (about 7% of the world market). CdTe solar cells can be manufactured with less cost, and are therefore a lower-cost alternative to normal silicon-based solar cells. The maximum conversion efficiency of a CdTe solar cell has reached 21.5% under laboratory conditions. Commercial module efficiency is still around 14.7%.

12.13 Silver-Indium-Antimony Telluride

A layer of silver–indium–antimony–tellurium (AgInSbTe) alloy, which in its original state, has a polycrystalline structure and reflective properties. When writing a CD, the material is heated to 50–70°C. This makes the alloy change its polycrystalline structure, assume an amorphous state and lose its reflectivity. The lost reflectivity serves the same function as bumps on manufactured CDs. To erase the writing on the disc, the write beam heats the amorphous regions with a low power of about 200°C. The alloy is not melted, but returns to the polycrystalline state and thus again becomes reflective.

12.14 Bismuth Telluride

Bismuth telluride, Bi_2Te_3 is a semiconductor which, when alloyed with antimony or selenium is an efficient thermoelectric material for refrigeration or portable power generation. Bi_2Te_3 has been extensively used in the construction of thermoelectric modules such as freon-free refrigerators using the Peltier effect. It is also a topological insulator, and thus exhibits thickness-dependent physical properties. Bismuth telluride is prepared by sealing a sample of bismuth and tellurium metal in a quartz tube under vacuum and heating it to 800°C in a muffle furnace. It is a grey powder and is also known as bismuth(III) telluride. More details of this compound are discussed in Chapter 10, Section 10.10.2.

12.15 Organotellurium Compounds

Unlike sulfur and selenium which have significant roles to play in many biochemical processes, tellurium is not present in the living systems and does not have any known biological role to play. However, many organotellurium compounds similar to their selenium analogues have been synthesised and their chemical properties explored and compared with that of selenium. Also some of them act as precursors for making semiconducting metal tellurides. The first organotellurium compound, diethyltelluride, was reported by Wöhler in 1840. Organotellurium compounds with tellurium in I (ditellurides), II, IV and VI oxidation states have been synthesised and characterised.[32]

An interesting conversation on tellurium with Linus Pauling

Human beings exposed to as little as 0.01 mg/m^3 or less of tellurium metal in air, give off a foul garlic-like odour known as 'tellurium breath'. Most organisms metabolise tellurium to the organotellurium compound dimethyl telluride, the source of this smell. Dimethyl telluride (b.p. 82°C) as a product of microbial metabolism was first discovered in 1939. Dimethyl telluride is produced by some fungi and bacteria (*Penicillium brevicaule*, *P. Chrysogenum*, and the bacterium *Pseudomonas fluorescen*).

Mathew S Meselson, (known for his discovery of *m*-RNA) who did Ph.D. with Nobel laureate Linus Pauling, recollects an interesting conversation he had with Pauling before he joined him for doing research when Pauling tried to make him take up research in tellurium chemistry.

After keeping a piece of rock on his table, Linus Pauling starts the conversation.

Linus Pauling: Well, Matt, this is a mineral of the element tellurium. Do you know about tellurium, the group VI element below selenium in the periodic chart of the elements?
Matthew Meselson: Uh, yes. Sulfur, selenium, tellurium . . .
Linus Pauling: I know that you know how bad hydrogen sulfide smells. Have you ever smelled hydrogen selenide?
Matthew Meselson: No, I never have.
Linus Pauling: Well, it smells much worse than hydrogen sulfide.
Matthew Meselson: I see.
Linus Pauling: Now, Matt, hydrogen telluride smells as much worse than hydrogen selenide as hydrogen selenide does compared to hydrogen sulfide.
Matthew Meselson: Ah.
Linus Pauling: In fact, Matt, some chemists were not careful when working with tellurium compounds, and they acquired a condition known as 'tellurium breath'. As a result, they have become isolated from society. Some have even committed suicide.
Matthew Meselson: Oh.
Linus Pauling: But Matt, I'm sure that you would be careful. Why don't you think it over and let me know if you would like to work on the structure of some tellurium compounds?

The story goes that the young Meselson went home and did not have to think twice after this frightening conversation. He came back and told Prof. Pauling that he wanted to work on something related to biology and only with elements such as C, H, N and O!

Adapted from "The Life and Work of Linus Pauling (1901-1994): A discourse on the art of biography." Matthew Meselson, Oral History Collection. © CSHL Digital archives, USA.

12.15.1 Alkali-Metal Ditelluride

One of the commonly used precursors for making organoditellurides are the alkali metal ditellurides. Disodium ditelluride has been prepared by the reduction of tellurium powder with hydrazine hydrate in alkaline medium in DMF. Dilithium ditelluride is obtained as a deep purple solution by refluxing both the elements in ethylenediamine.

$$4\,Te + N_2H_4.H_2O + 4\,NaOH \xrightarrow[DMF]{} 2\,Na_2Te_2 + 5\,H_2O + N_2 \quad (12.35)$$

$$2Te + 2\,Li \xrightarrow[\text{reflux}]{\text{ethylenediamine}} Li_2Te_2 \quad (12.36)$$

12.15.2 Dialkyl Ditelluride

Dipentylditelluride is obtained from the reaction of pentylbromide and dilithium ditelluride.

$$Li_2Te_2 + 2\ C_5H_{11}Br \longrightarrow C_5H_{11}Te{-}TeC_5H_{11} \quad (12.37)$$

12.15.3 Diphenyl Ditelluride

Diphenylditelluride is a well known organotellurium compound whose chemistry has been explored in detail. Its synthesis is shown below.

$$PhMgBr + Te \longrightarrow PhTeMgBr \quad (12.38)$$

$$2\ PhTeMgBr + 0.5\ O_2 + H_2O \longrightarrow PhTeTePh + 2\ MgBrOH \quad (12.39)$$

Benzene tellurol, PhTeH, is relatively unstable and diphenyl ditelluride is in principle its oxidised derivative. Oxidation of telluorphenolate by O_2 in water is one of the easy ways of preparing diphenyl ditelluride which is obtained as an orange coloured solid. The chemistry of this compound shows similarities as well as stark differences to that of PhSeSePh (Scheme 12.6).[33–36]

Other methods of making diorganoditelluirdes include the lithiation method, where an aryl bromide is lithiated using *n*-BuLi, followed by tellurium insertion and oxidation using a suitable oxidant. Grignard reagents can also be inserted with tellurium and further oxidised to prepare diorgano ditellurides.

$$Z{-}C_6H_4{-}Br \xrightarrow[K_3Fe(CN)_6]{2\ n\text{-BuLi},\ 2\ Te} Z{-}C_6H_4{-}Te{-}Te{-}C_6H_4{-}Z \quad (12.40)$$

$$\text{benzothiophene} \xrightarrow[\text{THF},\ -78^\circ C]{\text{LDA, Te}} \text{benzothienyl}{-}TeLi \xrightarrow{Br(CH_2)_2Br} (\text{benzothienyl}{-}Te)_2 \quad (12.41)$$

12.15.4 Tellurides

Tellurides are compounds of the type RTeR′ with tellurium in the +2 oxidation state. A common pathway to make tellurides is by detelluration of diorganoditellurides using excess of copper powder in refluxing dioxane solvent (Scheme 12.6). Raney nickel, Pd–C and palladium metal also were found to be good reagents for such detelluration reactions.

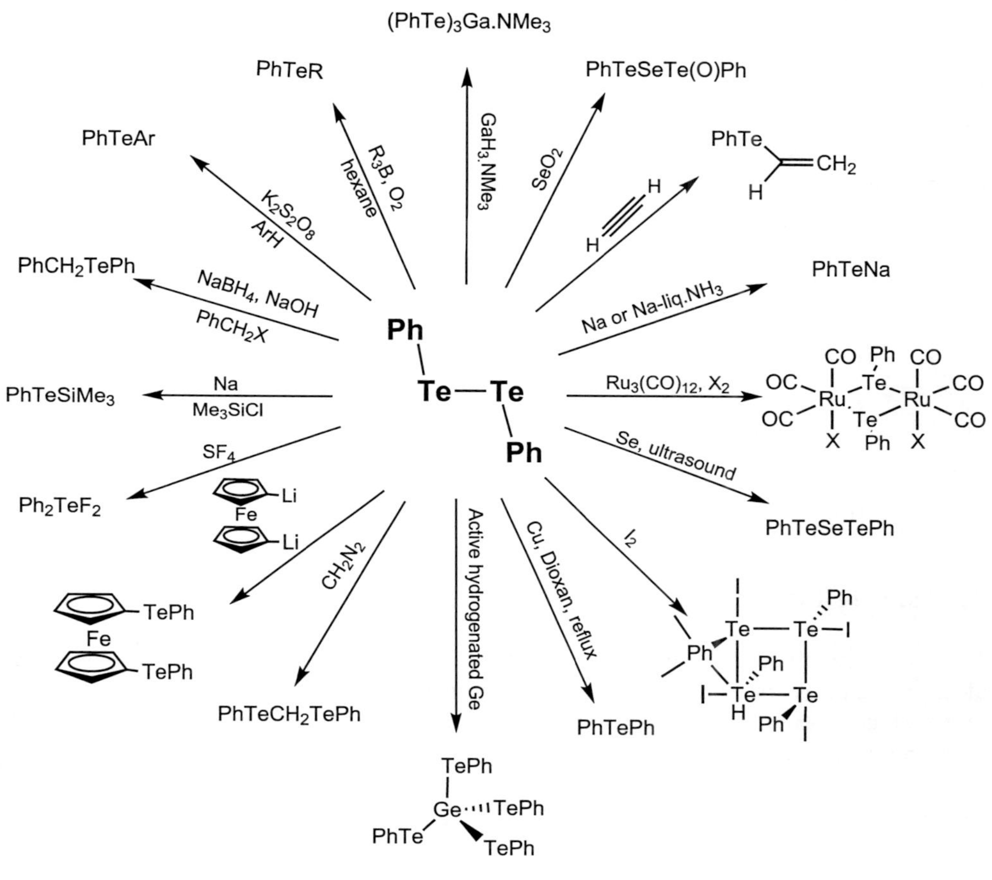

Scheme 12.6 Reactions of PhTeTePh

Tellurides having one tellurium have been made by ring opening of epoxides and aziridines.

$$\text{PhCH}_2\text{-aziridine(N-BOC)} \xrightarrow{\text{n-BuLi, Te}} \text{PhCH}_2\text{CH(NHBOC)CH}_2\text{TeBu} \quad (12.42)$$

$$\text{R'},\text{R-epoxide} \xrightarrow[\text{EtOH}]{\text{Ar}_2\text{Te}_2,\ \text{NaBH}_4} \text{R'CH(TeAr)CH(OH)R} \quad (12.43)$$

12.15.5 Tellurols

Among thiols, selenols and tellurols, RTeH are the least stable and due to their instability they are less studied. Since the Te–H bonds are weak, they also decompose to give ditellurides. Due to the absence of hydrogen bonding, they are low boiling liquids. Tellurols are easily oxidised and decompose thermally. Ultraviolet light also induces decomposition. For example, methanetellurol decomposes to dimethyl ditelluride and hydrogen. The first tellurol to be prepared was ethanetellurol which was prepared by the Grignard method. Reduction of diorganyl ditelluride especially using $(n\text{-Bu})_3SnH$ as the reducing agent is a method used for its synthesis. A few examples of tellurols synthesised and characterised are shown in Eq. 12.44.

$$R_2C{=}C(R')\text{–}Te\text{—}Te\text{–}C(R'){=}CR_2 \xrightarrow{(n\text{-}Bu)_3SnH} R_2C{=}C(R')\text{–}Te\text{—}H \qquad (12.44)$$

$$CH_2{=}CH\text{–}TeH \qquad c\text{-}C_3H_5\text{–}TeH \qquad c\text{-}C_3H_5\text{–}CH_2\text{–}TeH$$

12.15.6 Tellurol Esters

Tellurol esters are compounds of the type RC(O)TeR′ and they have been prepared from aldehydes and esters using diisobutylaluminium tellurolate (*i*-BuAlTeR) as the reagent. Diisobutylaluminium tellurolate itself is prepared by the reaction of DIBAL-H with diorganoditellurides.

$$(Bu^i)_2AlH + (n\text{-}Bu)TeTe(n\text{-}Bu) \longrightarrow n\text{-}BuTeAl(i\text{-}Bu)_2 \qquad (12.45)$$

$$PhCHO \xrightarrow[Et_2AlCl]{n\text{-}BuTeAl(i\text{-}Bu)_2} PhC(O)Te(n\text{-}Bu) \qquad (12.46)$$

$$PhC(O)OEt \xrightarrow[hexane]{n\text{-}BuTeAl(i\text{-}Bu)_2} PhC(O)Te(n\text{-}Bu) \qquad (12.47)$$

12.15.7 Organotellurium Compounds Having a Te(IV) Core

Many organotellurium compounds having tellurium in the +4 oxidation state have been prepared starting from $TeCl_4$ which is easily obtained by the chlorination of tellurium. Diaryl tellurium dichlorides are also easily prepared from $TeCl_4$. Oxidation of tellurides using dihalogens is also a method to make organotellurium (IV) compounds.

$$TeCl_4 + MeOC_6H_5 \longrightarrow (4\text{-}MeOC_6H_4)_2TeCl_2 \xrightarrow[\text{Fc = Ferrocenyl}]{FcCOOH} (4\text{-}MeOC_6H_4)_2Te(OC(O)Fc)_2 \quad (12.48)$$

12.15.8 Organotellurium Compounds Having a Te(VI) Core

Hexamethyl tellurium and a few examples of hexaaryl tellurium have been prepared. Structural studies on hexaaryl tellurium compounds indicate a regular octahedral environment around tellurium and $TePh_6$ structure indicates that it belongs to the T_h point group. Two methods for the synthesis of hexaryl tellurium compounds are shown. Unlike tetraaryl tellurium compounds, these are highly stable thermally and are not light sensitive.[37]

$$TeCl_4 + 4\ PhLi \longrightarrow TePh_4 \xrightarrow{XeF_2} TePh_4F_2 \xrightarrow{PhLi} TePh_6 \quad (12.49)$$

$$TeCl_4 + 5\ ArLi \longrightarrow [Ar_5Te]^- Li^+ \xrightarrow{SO_2Cl_2} Ar_5TeCl \xrightarrow{AgClO_4} [Ar_5Te]^+ ClO_4^- \quad (12.50)$$

$$[Ar_5Te]^+ ClO_4^- \xrightarrow{ArLi} Ar_6Te \quad (12.51)$$

$$Ar = Ph,\ 4\text{-}CF_3C_6H_4$$

Questions for self study (multiple choice)

12.1. Which one of the following statements is not true about selenium?
(a) It is an essential element for living systems
(b) It rarely forms hexavalent hexacoordinate species
(c) The most stable among red, black and grey forms of selenium is the red form
(d) It is prepared by the reduction of selenous acid, H_2SeO_3

12.2. The statement that is not true about SeO_2 is
(a) It is used for making red coloured glass
(b) It reacts with water forming selenous acid
(c) It is a mild oxidising agent
(d) It is a liquid at room temperature

12.3. The enzyme which does not have selenium is
(a) glutathione peroxidase (b) thioredoxin reductase
(c) ferredoxin hydrogenase (d) iodothyronindeiodinase

12.4. The order of pK_a values of MeSH, PhSH and MeSeH is
(a) MeSH > PhSH > MeSeH (b) MeSH > MeSeH > PhSH
(c) MeSeH > MeSH >PhSH (d) MeSeH > PhSH > MeSH

12.5. Woolins reagent converts
(a) P=O to P=Se (b) C=O to C=Se
(c) P=S to P=Se (d) C=S to C=Se

12.6. Which of the following is true for Ebselen?
(a) It is an enzyme
(b) It mimics glutathione peroxidase
(c) It contains selenocysteine units
(d) It is a diselenide

12.7. Which among the following type of compounds is only formed by tellurium and not by selenium?
(a) $Te(OH)_6$
(b) H_2TeO_3
(c) TeO_2
(d) ArTeTeAr

12.8. Bi_2Te_3 is used for
(a) making freon-free refrigerators
(b) thin film solar cells
(c) phase change memory chips
(d) infrared detectors

Solved concept-based questions

12.9. The π bonds formed by selenium are weaker than the π bonds formed by sulfur. Why?

The larger size of selenium compared to sulfur (atomic radius of 115 pm compared to 100 pm) results in larger hybridised orbitals and this along with longer bond length leads to weaker π overlap.

12.10. Why did nature choose selenium for biochemical functions although its chemistry is very close to that of sulfur?

Almost all biochemical reactions involving selenium-based enzymes are faster compared to sulfur; therefore, in one way, having selenium helps to accelerate enzymatic activity. More importantly, selenium unlike sulfur has the ability to react with reactive oxygenated species in a reversible manner. Due to the weak π bonds of Se=O it has a much better ability to get reduced back to the original state while it is difficult to reduce S=O bonds.

12.11. SeO_2 is used as a mild oxidising agent in organic chemistry but TeO_2 is not used as an oxidising agent for similar reactions. Why?

TeO_2 in acidic medium can act as an oxidising agent. But the main problem is the very poor solubility of TeO_2 in water, unlike SeO_2. TeO_2 is soluble only in strong acids which makes its applications in organic chemistry difficult.

Problems and exercises

12.12. Compare SO_2, SeO_2 and TeO_2 in terms of (a) structure and (b) reducing/oxidising properties.

12.13. Predict the products in the following reactions using SeO_2 as the reagent.

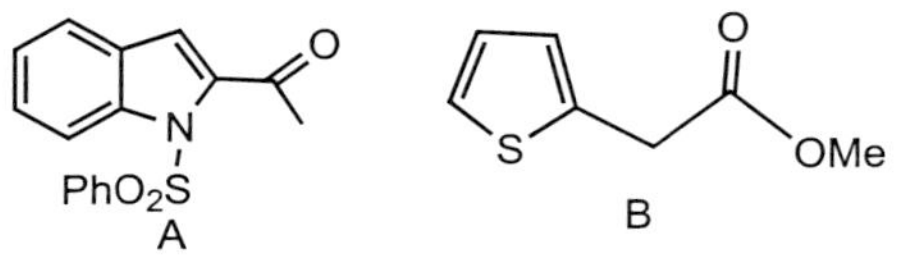

N
PhO_2S
A

S
OMe
B

12.14. Compare thiols, selenols and tellurols with reference to stability and hydrogen bonding.

12.15. Using a selenium-based reagent, suggest methods for the following conversions

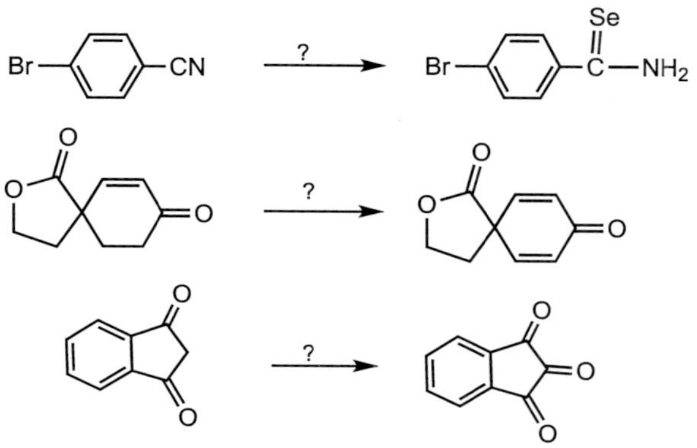

Br–C6H4–CN —?→ Br–C6H4–C(=Se)–NH2

12.16. Why is selenocysteine called the 21st amino acid?

12.17. What is the biochemical role of the enzyme glutathione peroxidase and how is selenium related to it?

12.18. Draw the structure of Ebselen and indicate its chemical utility.

12.19. What is meant by selenium paradox? Explain.

12.20. What properties of selenium make it an essential element in living systems?

12.21. Show two chemical reactions which indicate the differences between sulfur and selenium compounds.

12.22. Compare the chlorination reaction of tellurium with that of S and Se.

12.23. Despite the toxicity of tellurium, CdTe is one of the most sought after materials for solar cell manufacture. Why?

12.24. What is the compound used in rewritable CDs and what property of the material is used?

12.25. Compare reactions of I_2 with PhSeSePh and PhTeTePh.

12.26. Burning a chalcogen in its elemental form resulted in compound A. Dissolving compound A in water resulted in compound B. B on reaction with aqueous hydrogen peroxide resulted in compound C which is an octahedral shaped compound. Compound C on reaction with SO_2 gave an acid D along with the chalcogen element used in the first step. Identify compounds A to D.

Answer key for self study questions

12.1. (c) 12.2. (d) 12.3. (c) 12.4. (a) 12.5. (b) 12.6. (b) 12.7. (a) 12.8. (a)

Supplementary reading

1. Devillanova F A, *Handbook of chalcogen chemistry: New Perspectives in sulfur, selenium and tellurium*, RSC, 2007.
2. Langner B E, Selenium and selenium compounds, *Ullmans encyclopedia of industrial chemistry*, 2000.
3. Boyd R, Selenium Stories, *Nat. Chem.*, 2011, Vol. 3, 570.
4. Waitkins G R and Clark C W, *Chem. Rev.*, 1945, Vol. 36, 235.
5. Riley H L, Oxidation activity of selenium dioxide, *Nature*, 1947, Vol. 159, 571.
6. Zade S S and Singh H B, *Synthesis of organoselenium compounds in Patai's Chemistry of Functional groups*, 2012, John Wiley & Sons Ltd.
7. Masao S and Tadao Y, Selenium and tellurium derivatives of biscyclopentadienyl titanium IV and zirconium IV, *J. Organomet.Chem.*, 1974, Vol. 67, 395.
8. Goldani B, Ricordi V G, Seus N, Lenardao E J, Schumacher R F and Alves D, Silver-catalysed synthesis of diaryl selenides by reaction of diaryl diselenides with aryl boronic acids, *J. Org. Chem.*, 2016, Vol. 81, 11472.
9. Maxwell W M and Wynne K J, Aryl selenium trifluorides, *Inorg.Chem.*, 1981, Vol. 20, 1707.
10. Liotta D, Markiewicz W and Santiesteban H, The generation of uncomplexed phenyl selenide anion, *Tetrahedron Lett.*, 1977, Vol. 50, 4305.
11. Tiecco M, Testaferri L, Tingoli M, Bartoli D and Balducci R, Ring-closure reactions initiated by the peroxydisulfate ion oxidation of diphenyldiselenide, *J. Org. Chem.*, 1990, Vol. 55, 429.
12. Honeychuck R V, Okorafor M O, Shen L M and Brubaker C H, New ferrocenylthio and selenoether ligands. Preparation and characterisation of their palladium complexes as catalysts for selective hydrogenation, *Organometallics*, 1986, Vol. 5, 482.
13. Miyoshi N, Ishii H, Kondo K, Murai S and Sonoda N, Convenient synthesis of phenyl trimethylsilylselenide and benzene selenol, *Synthesis*, 1979, Vol. 4, 300.
14. Maier L, The direct synthesis of trialkyl and aryl selenophosphites and of tris Cp-anisyl tellurol phosphate, *Helv. Chim. Acta*, 1976, Vol. 59, 252.
15. Kubiniok S, du Mont W W, Pohl S and Saak W, The reagent diphenyldiselane/iodine: No phenylselenyliodine but a charge transfer complex with cyclic moieties, *Angew. Chem. Int. Ed.*, 1988, Vol. 27, 431.
16. Sharpless K B, Young M W and Lauer R F, Reactions of selenoxides: Thermal syn elimination and $H_2{}^{18}O$ exchange, *Tetrahedron Lett.*, 1973, Vol. 22, 1979.
17. Sharpless K B and Lauer R F, A mild procedure for the conversion of epoxides to allylic alcohols. The first organoselenium reagent, *J. Am. Chem. Soc.*, 1973, Vol. 95, 2697.
18. Sharpless K B, Lauer R F and Teranishi A Y, Electrophilic and nucleophilicorganoselenium reagents. New routes to α, β-unsaturated carbonyl compounds, *J. Am. Chem. Soc.*, 1973, Vol. 95, 6137.
19. Bhattacharya P and Woolins J D, Selenocarbonyl synthesis using Wollins' reagent, *Tetrahedron Lett.*, 2001, Vol. 42, 5949.
20. Pilkington M J, Slawin A M Z and Woolins J D, The preparation and characterisation of binary phosphorus-selenium rings, *Heteroatom Chem.*, 1990, Vol. 1, 351.
21. Stadtman T, Selenium Biochemistry, *Science*, 1974, Vol. 183, 915.
22. Reich H H and Hondal R J, Why nature chose selenium, *ACS Chem. Biol.*, 2016, Vol. 11, 821.
23. Roy G, Sarma B K, Phadnis P P and Mugesh G, Selenium containing enzymes in mammals, Chemical perspectives, *J. Chem. Sci.*, 2005, Vol. 117, 287.
24. Mueller A, Cadenas A, Graf P and Sies H, A novel biologically active seleno organic compound-1. Glutathione peroxidase like activity in vitro and antioxidant capacity of PZ 51 Ebselen, *Biochem. Pharmacol.*, 1984, Vol. 33, 3235.

25. Wendel A, Fausel M, Safayhi H and Tiegs G, A novel biologically active seleno organic compound-2. Activity of PZ 51 in relation to glutathione peroxidase, *Biochem. Pharmacol.*, 1984, Vol. 33, 3241.
26. Wilson S R, Zucker P A, Huang R R C and Spector A, Development of synthetic compounds with glutathione peroxidase activity, *J. Am. Chem. Soc.*, 1989, Vol. 111, 5936.
27. Bhowmick D, Srivastava S, D'Silva P and Mugesh G, Highly efficient glutathione peroxidase and peroxiredoxinmimetics protect mammalian cells against oxidative damage, *Angew. Chem. Int. Ed.*, 2015, Vol. 54, 8449.
28. Wirth T, Small organoselenium compounds: More than just glutathione peroxidase mimics, *Angew. Chem. Int. Ed.*, 2015, Vol. 54, 10074.
29. Balkrishna S J, Bhakuni B S, Chopra D and Kumar S, Cu catalysed efficient synthetic methodology for ebselen and related Se–N heterocycles, *Org. Lett.*, 2010, Vol. 12, 5394.
30. Chivers T and Laitinen R S, Tellurium: a maverick among chalcogens, *Chem. Soc. Rev.*, 2015, Vol. 44, 1725.
31. Ibers J, Tellurium in a twist, *Nat. Chem.*, 2009, Vol. 1, 508.
32. Zade S S and Singh H B, *Synthesis of organotellurium compounds in Patai's Chemistry of Functional groups*, 2012, John Wiley & Sons Ltd.
33. Boyle P D, Cross W I, Godfrey S M, Stephen M, McAuliffe C A, Pritchard RG, Sarwar S and Sheffied J M, Synthesis and characterisation of $Ph_4Te_4I_4$ containing a Te_4 square and $Ph_3PTePhI$, *Angew Chem. Int. Ed.*, 2000, Vol. 39, 1796.
34. Singh H B, Regini V A, Jaisinski J P, Paight E S and Butcher R J, Synthesis and characterisation of some ferrocene containing telluroether ligands. Crystal structure of 1,1'(bisphenyltelluro) ferrocene, *J. Organomet.Chem.*, 1994, Vol. 464, 87.
35. Dereu N L M, Zingaro R A, Meyers E A and Renson M, Reactions of selenium dioxide witharomaticditellurides, *Organometallics*, 1982, Vol. 1, 111.
36. Sadekov, I D, Bushkov A Ya, Markovskii L N and Minkia V I, Synthesis of diaryl tellurium difluorides, *Zh. Obsch. Khim.*, 1976, Vol. 46, 1660.
37. Minoura M, Sagami T, Akiba K, Modrakowski, C, Sudau A and Seppelt K, Hexaaryltellurium, the first neutral compound comprising hexaarylated elements, *Angew. Chem. Int. Ed.*, 1996, Vol. 35, 2660.

CHAPTER

13

The Chemistry of Fluorine

Unique Properties of Fluorine and its Compounds

- *Fluorine is the most electronegative element and the second smallest substituent (covalent radius 0.72 Å) after hydrogen (0.37 Å). The F–F bond strength (38 kcal/mol) is the second lowest among dihalogen molecules.*
- *Fluorine gas (F_2), well known as the 'tiger of chemistry' due to its wild reactivity is the most powerful oxidising element known ($E^0 = 2.87$ V), reacting with practically all organic and inorganic substances.*
- *^{19}F has almost 100% natural abundance and so fluorine is easily studied using ^{19}F-NMR spectroscopy.*
- *Fluoride ions present in drinking water in the range of 0.7 to 1.5 ppm prevent dental decay by conversion of hydroxyapatite of the teeth to fluorapatite, the latter being resistant to attack by acids produced in the mouth. Cavity fighting toothpastes contain fluoride sources such as sodium fluoride, stannous fluoride and sodium monofluorophosphate.*
- *Fluoride ions, if present above 1.5 ppm in drinking water can lead to disfiguration of bones leading to fluorosis and arthritis.*
- *Fluorine-based inorganic and organic covalent compounds are much more volatile compared to other halogenated compounds; a consequence of small size and high electronegativity resulting in very little secondary interactions. The very low polarisabiliy of organofluorine substituents affects secondary intermolecular interactions.*
- *Small size and high electronegativity help fluorine to form high oxidation state compounds of many main group, transition metal and inner transition metals which other halogens do not form. For example, SF_6, WF_6 and UF_6.*
- *Among carbon–halogen bonds, C–F bond has the highest bond dissociation energy making it stable for many uses. C–F bonds have very low polarisability, are resistant to oxidative metabolism and have increased fat solubility.*

- *Introducing a fluorine atom or the CF_3 group into pharmaceuticals is often accompanied by higher binding affinity, enhanced metabolic stability, improved bioavailability and in many cases enhanced selective activity of the drug.*
- *About 20% of all FDA approved drugs have fluorine present and some of the best selling drugs (Lipitor, Advair and Crestor) contain fluorine.*
- *About 28% of all agrochemicals contain fluorine and among them 54% are herbicides.*
- *The most commonly used radiotracer in clinical positron emission tomography (PET) is fluorodeoxyglucose (also called FDG), an analogue of glucose that is labelled with ^{18}F which has a half life of 110 min. PET produces a three dimensional image of functional processes in the body and the system detects pairs of gamma rays emitted indirectly by a positron-emitting radionuclide.^{18}F-PET is extremely sensitive and can trace drugs at very low concentrations (even 10^{-9} to 10^{-12} mol).*
- *^{19}F-MRI is a valuable tool for imaging guided drug delivery without background interference and without ionising radiation.*
- *Fluorous chemistry/technology utilises perfluorinated long chains (called ponytails) on molecules and catalysts to effectively separate them from reaction mixtures during extraction with perfluorinated solvents.*
- *Almost all of the inhalation anesthetics (except N_2O) used nowadays have fluorines or CF_3 groups present. For example, Sevoflurane is a highly fluorinated ether, having seven fluorine atoms, used for maintenance of general anesthesia. Fluorinated anesthetics show rapid onset and offset of consciousness due to the volatility and to low solubility in blood.*
- *Fluorine based hydrofluorocarbons (HFCs) and hydrofluoroolefins (HFOs) are used as refrigerants replacing chlorofluorocarbons (CFCs). Perfluorinated polyethylene, teflon, is a very useful 'non-stick' polymer which is also thermostable and waterproof.*
- *The largest application of fluorine gas, consuming up to 7,000 metric tons annually (about 75% of the world production), is in the preparation of uranium hexafluoride (m.p. 57°C) as a nuclear fuel. Fluorine gas is used to fluorinate UF_4, to convert it to UF_6. UF_4 is formed from uranium dioxide and hydrofluoric acid.*
- *HF, although a weak acid in aqueous solution, penetrates skin and forms an insoluble salt with Ca. Burns with concentrated HF involving even 2.5% of the body surface area are fatal and are also extremely painful.*
- *Anhydrous HF is a very good reagent and solvent for preparing not only conventional organofluorine compounds but also novel main group compounds such as N_5^+ (pentazenium), $[AuXe_4]^{2+}$, and so on. A combination of anhydrous HF with Lewis acids such as AsF_5 and SbF_5 gives very strong acids, which are stronger than 100% sulfuric acid and are classified as superacids.*

13.1 About the Element

Fluorine stands out as a main group element with unique properties which have been made use of by organic and inorganic chemists with equal interest. The name fluorine was coined by the French chemist Amperé, from '*le fluor*' after its ore fluorspar. It is the most active non-metallic element, reacting with virtually every other element, including some noble gases. The unusual properties of fluorine have been utilised in many ways by chemists for realising compounds and materials with exceptional properties. Till 2012, it was believed that F_2 gas does not occur free in nature (see box on page 489). The most common fluorine minerals are cryolite (Na_3AlF_6), fluorspar (also called fluorite, CaF_2), and fluorapatite [$Ca_5(PO_4)_3F$]. Fluorine is the 13th most common element on the earth's crust.[1]

Fluorine is a highly toxic, pale yellow gas about 1.3 times as heavy as air at atmospheric temperature and pressure. Fluorine gas is the second most powerful oxidising agent known, reacting with practically all organic and inorganic substances. Fluorine gas is corrosive to exposed tissues and to the upper and lower respiratory tract. It can penetrate deeply into body tissues unless neutralised. Fluorine reacts violently and decomposes to hydrofluoric acid on contact with moisture. Fluorine ignites on contact with ammonia, phosphorus, sulfur, copper wire, acetone and many unsaturated organic and inorganic compounds. It reacts with most compounds and often, violently. It has a higher oxidation potential than ozone. The high electronegativity of the fluorine and its small size indicates an electron cloud which is compact and unavailable for secondary interactions with other elements. The consequence of this is the high volatility of small organofluorine compounds as well as many small fluorinated compounds of *p*-block elements.

In 1529, Georgius Agricola, a German scholar, found a mineral known as fluorspar which came from the Latin word *fluere* meaning 'to flow'. Agricola found that fluorspar made molten metallic ores easier to work with, but did not realise that fluorspar was a mineral of fluorine. Fluorspar became more and more popular with the early chemists and in 1670 Heinrich Schwanhard, a German glass cutter, discovered that a certain mixture of fluorspar and acid formed a substance that could etch glass. The glass etching led to further research in hydrofluoric acid. In 1771, Swedish chemist, Carl Wilhelm Scheele, started an intense study of the acid to find ways to break hydrofluoric acid into its elements. The new experiments did end up harming many chemists because of the toxicity of hydrogen fluoride gas that they accidentally inhaled.[1]

In 1892 the Belgian chemist, Swarts, discovered the use of antimony trifluoride as an efficient reagent for converting organochlorine compounds to organofluorine compounds. The Schiemann reaction discovered in 1927 converted anilines via diazonium salts to aryl fluorine compounds, and helped the initial growth of industrial organofluorine chemistry. Many fluorochemicals such as the refrigerant gases (freons) and polytetrafluoroethylene (Teflon), were commercialised by the DuPont company. In the USA, fluoride is added to city water in the proportion of about one part per million to help prevent tooth decay. Tin(II) fluoride (SnF_2), sodium monofluorophosphate (Na_2PO_3F) and sodium fluoride (NaF), are fluoride compounds added to toothpaste, for the prevention of tooth decay. Crystals of calcium fluoride (CaF_2) are used to make lenses to focus infrared light.[2]

Table 13.1 Physical properties of fluorine

Fluorine	*Atomic mass*	*Electronegativity (Pauling scale)*	*Ist ionisation energy*	*Electronic configuration*	*van der Waals radius*
Symbol: F	18.99 g.mol^{-1}	3.98	1681 kJ.mol^{-1}	[He] $2s^2 2p^5$	0.146 nm
At. no. 7	*Atomic radius*	*Melting point*	*Boiling point*	*Electron affinity*	*Density*
	42 pm	–219.7°C	–188.1°C	–328 kJ mol^{-1}	1.696 g.cm^{-3}

13.2 Isolation and Purification of the Element

Ferdinand Frederic Henri Moissan, a French chemist, was the first to successfully isolate fluorine in 1886. His teacher, Fremy, had earlier tried electrolysis of pure HF and also of molten CaF_2 without success. As pure HF does not conduct electricity, Moissan made a solution of hydrofluoric acid in potassium hydrogen fluoride (KHF_2) and then cooled the solution to –23°C. Afterwards, he passed electric current through it upon which pale yellow F_2 gas appeared. Moissan also constructed special corrosion-resistant equipments such as containers crafted from a mixture of platinum and iridium (more chemically resistant than pure platinum) with fluorite (CaF_2) stoppers. It is worth mentioning that many well known chemists of that time tried but failed to isolate F_2 and some of them died and some were seriously injured due to HF poisoning. For his path-breaking work, Moissan received the Nobel Prize for chemistry in 1906. Currently, F_2 is still made by the electrolysis of potassium fluoride and hydrofluoric acid as well as through the electrolysis of molten potassium acid fluoride (KHF_2).[3] A schematic of the modern electrolytic cell for the industrial preparation of fluorine gas is shown in Fig. 13.1.

> ***Will fluorine ever have practical applications?***
> *It is very difficult to answer this question. I may, however, say in all sincerity that I gave this subject little thought when I undertook my researches, and I believe that all the chemists whose attempts preceded mine gave it no more consideration.*
>
> *A scientific research is a search after truth, and it is only after discovery that the question of applicability can be usefully considered.*
> *– Henri Moissan, 1897*

$$KHF_2 \longrightarrow 2\,KF + HF \xrightarrow{\text{electrolysis}} \begin{cases} K^+ + e \longrightarrow K \text{ (cathode)} \\ F^- \longrightarrow F + e \text{ (anode)} \end{cases} \quad (13.1)$$

$$K + HF \longrightarrow KF + H;\quad 2H \longrightarrow H_2 \quad \text{(cathode)} \quad (13.2)$$

$$2F \longrightarrow F_2 \quad \text{(anode)} \quad (13.3)$$

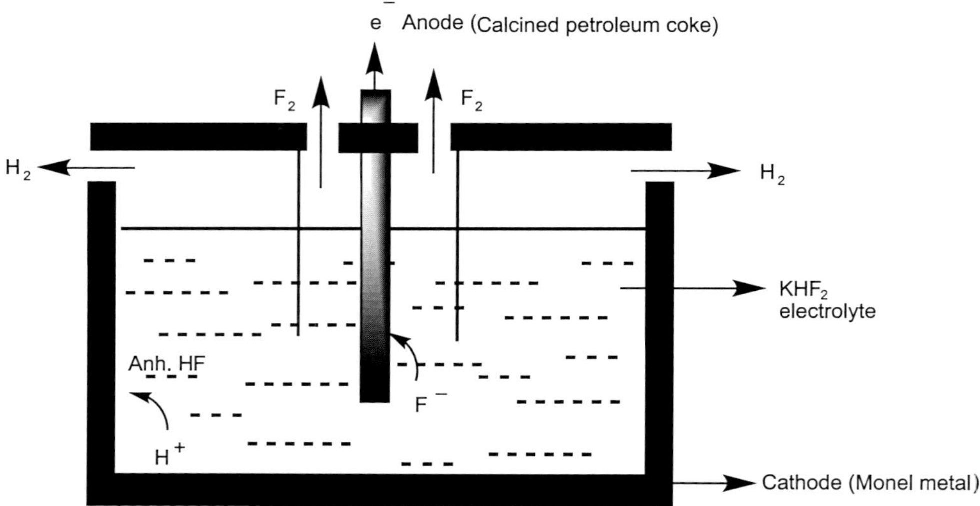

Fig. 13.1 Electrolytic cell for the industrial synthesis of fluorine gas

Since F_2 reacts with almost all the elements except a few rare gases, storage and transport of F_2 gas was also a challenge. Teflon is the preferred gasket material when working with fluorine gas. Equipments have to be kept dry as F_2 oxidises water giving a mixture of O_2, O_3 and HF. The reaction between metals and fluorine is relatively slow at room temperature, but becomes vigorous and self-sustaining at elevated temperatures. Fluorine can be stored in steel cylinders that have passivated interiors, or nickel cylinders at temperatures below 200°C.

In 1986, 100 years after the discovery of the Moissan's procedure, Karl O Christe came out with a simple laboratory method for the synthesis of small amounts of F_2 gas starting with $KMnO_4$ and HF. The synthesis is based on the fact that thermodynamically unstable higher oxidation state of transition metal fluorides can be stabilised by anion formation. Unstable NiF_4, MnF_4 and CuF_4 can be stabilised in the form of MF_6^{2-} anions. Also, weaker Lewis acids such as MnF_4 can be displaced from their salts by stronger Lewis acids such as SbF_5. MnF_4 is thermodynamically unstable and decomposes to a lower fluoride MnF_3 with the elimination of fluorine gas. The formed F_2 gas has been ascertained by the formation of the white solid HgF_2 (m.p. 645°C) on reaction with liquid mercury and by its characteristic pungent and unique odour.[4]

$$2\,KMnO_4 + 2\,KF + 10\,HF + 3\,H_2O_2 \longrightarrow 2\,K_2MnF_6 + 8\,H_2O + 3\,O_2 \qquad (13.4)$$

$$2\,K_2MnF_6 + 4\,SbF_5 \longrightarrow 2\,KSbF_6 + 2\,MnF_3 + F_2 \qquad (13.5)$$

Does elemental fluorine occur as such in nature?[5,6]

Being the most reactive of all elements, the above question on the reactivity of fluorine gas does attract considerable attention. One naturally would think that it is next to impossible, knowing the ferocity with which fluorine gas tears down anything which comes in its way. However, in 2012, scientists from the Technische Universität München (TUM) and the Ludwig-Maximilians-University, Munich (LMU) made a very interesting observation. A well known fluorite (CaF_2) mineral called antozonite that is normally dark violet or black in colour (also called as 'fetid fluorite' meaning stinking CaF_2), gives off a characteristic unpleasant smell when it is powdered. For over a century, there were suggestions that fluorine gas is responsible for this unpleasant smell since chemists, who regularly work with fluorine gas, could easily detect the offensive smell of F_2. The exact cause and identity of the odour however, remained a mystery.

Professor Florian Kraus who was a faculty member at the Technical University of Munich, read the story of fetid fluorite in 2012, and decided to investigate the origin of this smell by himself. He drove from Munich to a disused mine in Wölsendorf, 180 km away and picked up some rocks containing antozonite from a forest near the highway.

In his laboratory, he smashed the rock samples with a hammer which released the characteristic smell. Prof. Kraus, a fluorine chemist himself, could identify the smell as quite distinctive and similar to that of fluorine gas. Fluorine has a strong and characteristic odour that can be detected in very small amounts, as low as 20 parts per billion. He compared this smell by smelling pure samples of F_2, O_3, HF, OF_2 and XeF_2 gases very briefly and at very low concentrations. Since the scientific community still did not fully accept smell as proper scientific evidence, he tried to determine the molecular weight of the gas by using the well known mass spectrometric method. This was not successful as the extremely high reactivity of the gas possibly made it react with some parts of the spectrometer probe! He then sought the help of solid state NMR experts of the Ludwig-Maximilians-University, Munich and carried out ^{19}F-solid-state NMR on pea-sized samples of antozonite. They detected a peak at 425 ppm in the ^{19}F-NMR spectra that corresponded to the range predicted for molecular F_2. Clearly, F_2 does not react with CaF_2 and thus they were able to identify the fluorine gas non-destructively trapped in its natural environment.

How does one explain the presence of F_2 in the crystalline clusters of CaF_2? The mineral antozonite also contains tiny amounts of radioactive uranium-238, which decays into β-emitting daughter nuclides. The rocks have been lying around for over 100 million years and separate experiments carried out earlier on directly exposing samples of CaF_2 to β and γ radiation and high energy electron beams turned the mineral to violet colour indicating formation of calcium clusters. Tiny bubbles of a gas were also found to form during such irradiations which were never analysed earlier. The discovery thus resolved a nearly 200-year-old debate about why the mineral, antozonite or fetid fluorite, smells so bad when it is crushed.

13.3 Reactivity of Fluorine Gas

Fluorine is notoriously wild in its reactivity and has been found to react, often explosively, with many elements and compounds. Often the compounds it does not react with, are already fluorinated ones. It reacts with any element or compound, except for the lighter noble gases He and Ne, and highly oxidised fluorides such as CaF_2 and UF_6. Among elements, nitrogen requires an electric discharge at elevated temperatures for reaction, due to the strong triple bond. Oxygen does not react with fluorine under ambient conditions, but can be made to react using electric discharge at low temperatures and pressures. The best material to store fluorine gas is nickel or nickel-based alloys such as monel metal [Ni(~76%):Cu] or inconel [Ni(~70%):Cr:Fe]. Although nickel reacts with fluorine, a fine coating of nickel difluoride forms on the surface which prevents further attack by F_2. In a related experiment, three different metal wires of diameter 0.012–0.015 inches were kept in fluorine gas at atmospheric pressure and heated. Copper wire ignited between 689 and 701°C, iron wire between 618 and 644°C, and nickel wire between 1253 and 1266°C.[7] So nickel or monel metal cylinders are often used to store F_2 gas and sometimes stainless steel cylinders with passivated insides are also used if the temperature can be kept low. Teflon gaskets and reaction vessels are used but fluorine gas gets absorbed into the polymeric resin of Teflon and can become sensitive to a source of ignition such as sudden impact.

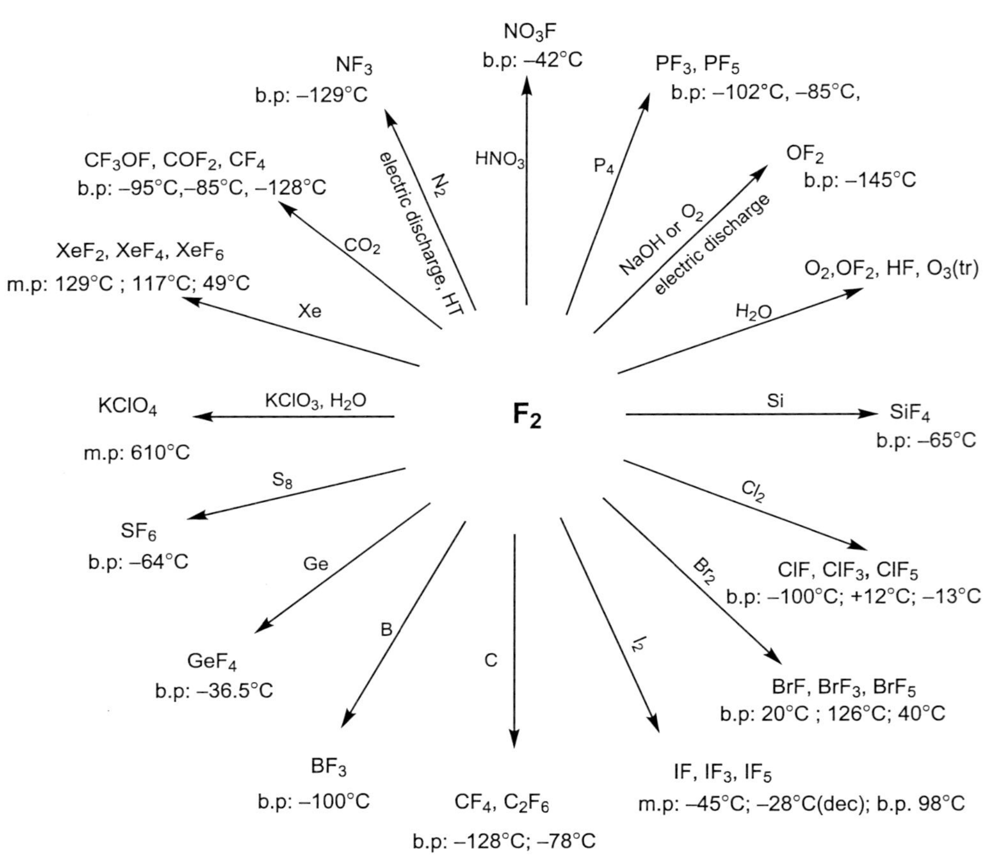

Scheme 13.1 General reactions of fluorine gas

Fluorine reacts explosively with hydrogen even in the dark and at low temperatures. It reacts violently with water forming hydrogen fluoride, and liberates oxygen and ozone when in excess, it combines with chlorine only on heating, forming the gaseous products ClF and ClF_3. Glass laboratory instruments cannot resist F_2 gas and even bricks burn when exposed to fluorine gas. Yet elemental fluorine has numerous industrial applications. It is used for the production of uranium hexafluoride in uranium enrichment and sulfur hexafluoride, which serves as insulating material in high voltage switches and transformers.

13.4 Ability of Fluorine to Stabilise Elements in High Oxidation States[8]

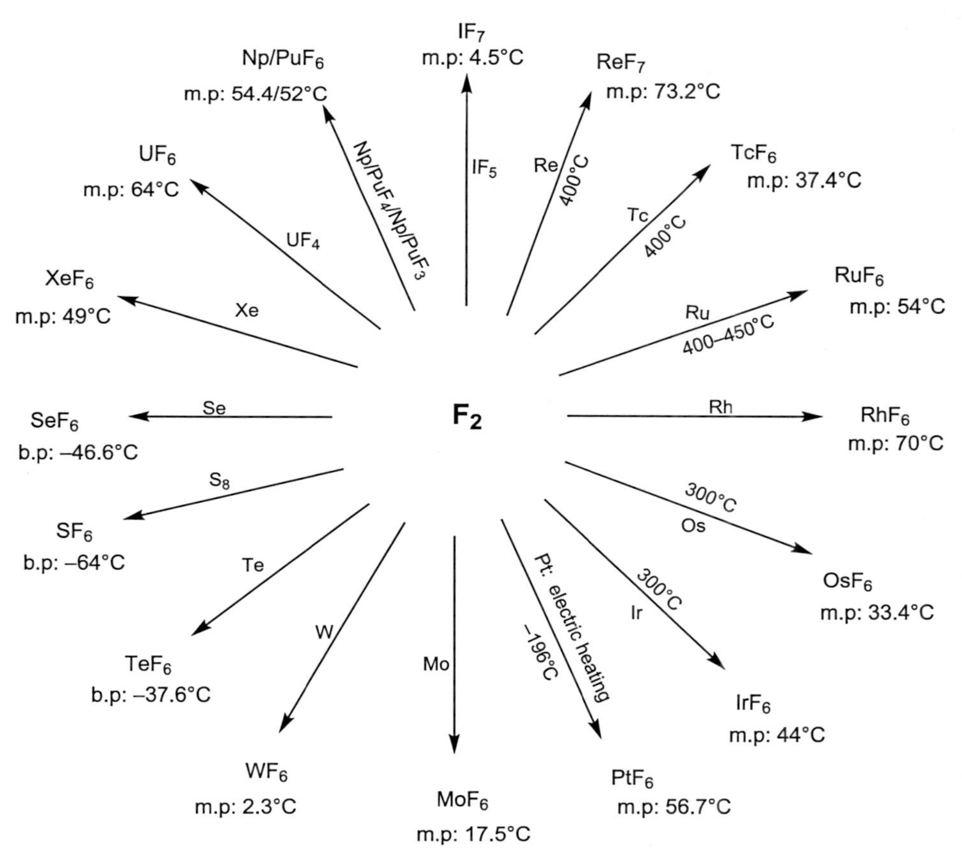

Scheme 13.2 Stable binary, neutral, hexa and hepta fluorides of elements

Although hydrogen is the smallest element, it does not have the ability to stabilise elements in high oxidation states due to its inherent low electronegativity. In contrast, fluorine which is the second smallest element, the most electronegative element and a strong oxidising agent has the ability to stabilise metals and nonmetals in high oxidation states. Binary neutral fluorides with oxidation states as high as 6–7 are known (Scheme 13.2). No electrically neutral octafluorides are currently known to exist but many anionic octafluorides are known, such as the octafluoroiodate (IF_8^-), octafluoroxenate (XeF_8^{2-}), octafluorozirconate (ZrF_8^{4-}), octafluorotantalate (TaF_8^{3-}), octafluoromolybdate (MoF_8^{2-}), octafluorotungstate (WF_8^{2-}) and octafluororhenate (ReF_8^-). The low melting and boiling points of these compounds are unique to the fluorides unlike other halides or oxides of the same elements. The effectiveness of UF_6 in the purification and enrichment of uranium-235 is the best example of utilisation of the low melting and boiling points of such metal fluorides.[6]

13.5 Fluorine in Inorganic Chemistry

13.5.1 Fluorine in Uranium Purification for Radioactive Use[1]

About 75% of F_2 produced in the world is used in uranium purification for nuclear reactions. The major isotopes of uranium are ^{235}U (0.71% natural abundance) and ^{238}U (99.28% natural abundance). Out of the two, ^{235}U is the only naturally occurring fissile isotope of uranium and undergoes the well known nuclear fission reaction when bombarded with thermal neutrons.

$$^{235}_{92}U \xrightarrow{^{1}_{0}n} {}^{146}_{56}Ba + {}^{92}_{36}Kr + 3\,{}^{1}_{0}n \qquad t_{1/2} = 704 \text{ million years} \tag{13.6}$$

In the enrichment process, the ore is first converted to sodium diuranate or ammonium diuranate. Extraction from partially spent fuel rods by nitric acid also gives uranyl nitrate. These are first converted to uranium trioxide and then reduced to uranium dioxide (Eq. 13.7). (UO_2 is sensitive and can get converted to U_3O_8 on contact with oxygen at 700°C.) UO_3 (m.p. 200–650°C) can also be converted to UF_6. Large scale systems reduce UO_3 to UO_2 (m.p. 2865°C) prior to fluorination to minimise production of oxyfluorides, to improve fluorine use efficiency, and to increase UF_4 (m.p. 1036°C) purity.

$$\text{Uranium ore} \xrightarrow{\text{Extraction}} Na_2U_2O_7 \text{ or } (NH_4)_2U_2O_7 \xrightarrow{\text{air, 500°C}} UO_3 \xrightarrow[700°C]{H_2} UO_2 \tag{13.7}$$

$$Na_2U_2O_7 \text{ or } (NH_4)_2U_2O_7 \xrightarrow{HNO_3} UO_2(NO_3)_2.6H_2O; \quad \text{Spent fuel rods} \xrightarrow{HNO_3} UO_2(NO_3)_2.6H_2O \xrightarrow[400–600°C]{O_2} UO_3$$

Uranium dioxide is converted to uranium tetrafluoride by treatment with anhydrous HF. UF_4 is oxidised to UF_6 by fluorine gas.

$$UO_2 \xrightarrow{\text{anhydrous HF}} UF_4 \xrightarrow{F_2 \text{ gas}} UF_6 \text{ (mixture of } ^{235}U \text{ and } ^{238}U) \tag{13.8}$$

UF_6 is a white crystalline solid below 57°C and a liquid at 64°C and 1.5 atm pressure. It readily sublimes at atmospheric pressure above 57°C. It is stable to dry air, O_2, N_2 and CO_2. UF_6 was originally reported by Ruff and Heinzelmann in 1911 by the direct reaction of uranium and F_2. However, its use increased exponentially during and after the World War II.

Since fluorine has only one stable isotope, the separation is made simpler. The enriched UF_6 containing ^{235}U is separated from ^{238}U by diffusion, gas centrifugation and laser isotopic separation. While the diffusion method is on the wane, the gas centrifugation method is increasingly being used and to a very small extent, the laser isotopic separation method.

The gas centrifugation process uses UF_6 as its feed and makes use of the slight difference in mass between ^{235}U and ^{238}U for centrifugal separation. The gas is fed into a series of vacuum tubes, each containing a rotor 3 to 5 m tall and 20 cm in diameter. When the rotors are spun in the speed range of 50,000–70,000 rpm, the heavier molecules with ^{238}U increase in concentration towards the outer edge of the cylinder. There is a corresponding increase in concentration of ^{235}U molecules near the centre. Using a countercurrent flow, the enriched product is drawn off axially, heavier molecules at one end and lighter ones at the other. The purchase and use of these sophisticated centrifuges are being monitored globally to observe how nations build up their nuclear arsenal.[1]

The laser separation process (also known as SILEX process) works on the principle of photoionisation, whereby a powerful laser is used to ionise particular atoms present in a vapour of uranium metal. The laser used is a CO_2 laser operating at a wavelength of 10.8 μm (micrometres) and optically amplified to 16 μm, which is in the infrared spectrum. Photodissociation of UF_6 to solid UF_5^+ is carried out using tuned laser radiation targeted to break the molecular bond holding one of the six fluorine atoms to a ^{235}U atom. (An electron can be ejected from an atom by light of a certain frequency.) The laser techniques for uranium use frequencies which are tuned to ionise a ^{235}U atom but not a ^{238}U atom. The positively-charged ^{235}U ions are then attracted to a negatively-charged plate and collected, thus achieving a separation of isotopes.

To obtain enriched ^{235}U metal, the enriched UF_6 is first converted to UF_4, a green coloured salt, and UF_4 on treatment with Ca or Mg give pure enriched metallic uranium.

$$^{235}UF_6 + H_2 \longrightarrow {}^{235}UF_4 + HF \tag{13.9}$$

$$^{235}UF_4 \xrightarrow{Ca/Mg} {}^{235}U \tag{13.10}$$

However, for use as nuclear fuel, enriched uranium hexafluoride is converted into UO_2 powder through uranyl fluoride (UO_2F_2) which is then processed into a pellet form. The melting points of various uranium oxides, fluorides and oxyfluorides (Table 13.2) indicate the reason for this conversion.

Table 13.2 Melting points of uranium oxides, fluorides and oxyfluorides

Compound	UO_3	UO_2	U_3O_8	UO_2F_2	UF_4	UF_6
m.p. (°C)	200–650	2865	1150	300 (dec)	1036	57

After sintering, these pellets are stacked into tubes made of corrosion-resistant alloys such as zirconium alloy. The tubes are sealed to contain the fuel pellets and these are called fuel rods. The finished fuel rods are grouped in special fuel assemblies that are then used to build up the nuclear fuel core of a power reactor.

$$UF_6\,(g) + H_2O\,(steam) \longrightarrow UO_2F_2\,(s) + HF(g) \tag{13.11}$$

$$UO_2F_2\,(s) + H_2(g) \longrightarrow UO_2(s) + HF(g) \tag{13.12}$$

Which is the most reactive fluorinated compound? [9,10]

Normally fluorination results in highly stable compounds such as the CFCs or SF_6. However, one compound does stand out among the fluorinated compounds for its extreme reactivity and ability to react and burn down even things which are normally fire fighters such as asbestos, water and concrete. This compound, a liquid with a boiling point of 12°C, is chlorine trifluoride (ClF_3) which is classified as one of the deadliest compounds ever made by man. To mention an example of the kind of devastation chlorine trifluoride can generate, a ton of ClF_3 was accidentally spilled inside a warehouse in the 1950s. The chemical burned straight through one foot of concrete and three feet of gravel of the warehouse flooring while releasing a deadly cloud of gas containing a cocktail of vapours that corroded every surface it came into contact with. The Nazis who discovered this compound were interested in its military applications. They were possibly fascinated by its property of melting concrete and reaction with water. It is also well known that a fire made by ClF_3 cannot be put out as it does not require atmospheric oxygen and it burns down all fire fighting equipments and chemicals.

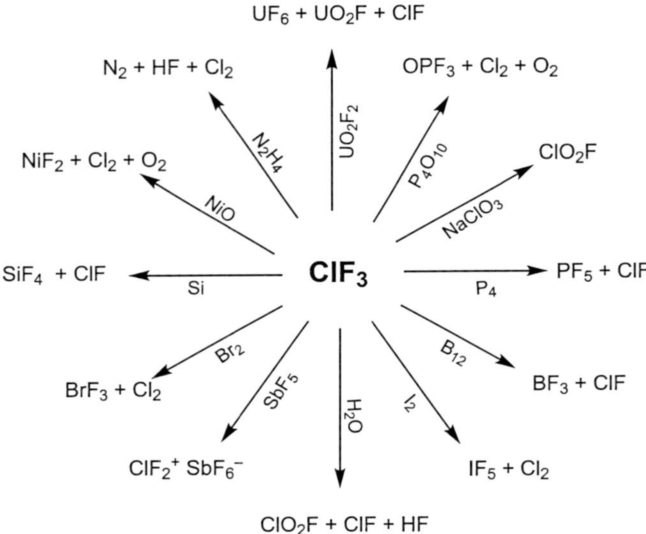

ClF_3 (b.p. 11.75°C) is prepared by the reaction of Cl_2 and F_2 at 250–280°C. ClF_5 is equally potent but requires reaction of ClF_3 with more F_2 and higher temperature and pressure and has a much lower boiling point (–13°C). Although chlorine trifluoride was never used during a war, the Nazis did succeed in making about 30 tons of ClF_3 in a secret facility known as the Falkenhagen Bunker before it was captured by the Russians in 1944. The Nazis referred to chlorine trifluoride simply as 'Substance N' or 'N-Stoff'. According to reports released by the Soviets following World War II, Nazi trials involving substance N were promising. But, of course, the weapon was incredibly dangerous for those carrying it around too. Some of the more unusual things chlorine trifluoride is known to set fire to on contact are glass, sand, asbestos, rust, concrete, people, pyrex, cloth, and so on. The substance is so highly reactive that elements like platinum, osmium and iridium will begin to corrode when they come into contact with it. Notably tough elements like titanium and tungsten are also regarded as being wholly unsuitable to storing the chemical because they are set on fire as soon as they come into contact with it.

The only known way to store chlorine trifluoride is to put it inside sealed containers made of steel, iron, nickel or copper which are passivated with flourine gas. This is because doing so will coat the metal in a thin fluoride layer, with which the chemical won't react. However, if this layer gets removed in anyway, or the metal isn't completely dry, chlorine trifluoride will begin to react violently and cause the vessel to explode.

A few of the other things known to be unreactive with chlorine trifluoride include nitrogen, the inert gases and polychlorotrifluoroethylene. Interestingly, it doesn't react with air unless it happens to contain a larger than average amount of water vapour. Due to its oxidising nature, there have been several attempts to use it as a low-cost, lightweight rocket and torpedo propulsion fuel. Since safe storage is extremely difficult, it was generally considered not to be worth the risk for this use. For instance, after studying and experimenting with this chemical for rocket use, the famous American rocket fuel developer John D Clark said about the best way to deal with potential chlorine trifluoride accidents—*'I have always recommended a good pair of running shoes'.*

There are also a couple of other useful applications. ClF_3 is great for plasma-less cleaning of certain surfaces used in semiconductor manufacturing and it also works well at cleaning radioactive metal residues off of the walls of nuclear power plants and removing built up oxides.

13.5.2 Anhydrous Hydrogen Fluoride and Hydrofluoric Acid

Anhydrous hydrogen fluoride is a colourless gas or liquid and is the principal industrial source of fluorine. Hydrogen fluoride boils near room temperature, while other hydrogen halides evaporate at much lower temperatures. Unlike other hydrogen halides, HF is lighter than air and diffuses relatively quickly through porous substances. HF is a highly dangerous gas, forming corrosive and penetrating hydrofluoric acid upon contact with living tissue where it will seek out calcium ions to form CaF_2. The French chemist, Edmond Frémy, is credited with discovering anhydrous hydrogen fluoride while trying to isolate fluorine. Unlike other hydrohalic acids, hydrogen fluoride is only a weak acid in dilute aqueous solution. This is due to its low dissociation constant.[11]

K_a	**Acid**	**Formula**
3.2×10^9	Hydroiodic acid	HI
1.0×10^9	Hydrobromic acid	HBr
1.3×10^6	Hydrochloric acid	HCl
1.0×10^3	Sulfuric acid	H_2SO_4
6.6×10^{-4}	Hydrofluoric acid	HF

It does not fully ionise in dilute aqueous solutions. However 100% HF is a strong acid as it undergoes homoassociation and formation of FHF^-. HF is widely used in the petrochemical industry and is a component of many superacids.

$$3\ HF \rightleftharpoons H_2F^+ + FHF^- \qquad (13.13)$$

HF is a colourless gas around room temperature. On freezing (melting point −83.6°C), HF forms orthorhombic crystals, consisting of zig-zag chains of HF molecules. The HF molecules, with a short H–F bond of 95 pm, are linked to neighbouring molecules by intermolecular H–F distances of 155 pm.

Hydrofluoric acid is a solution of HF in water and is highly corrosive especially with skin. Fluoride binds to calcium ions and potassium ions and enters the extra cellular space compensating for the loss of calcium. This irritates the nerve endings causing severe pain and deep tissue damage. Hydrofluoric acid exposure is often treated with calcium gluconate, a soluble source of Ca^{2+} that sequesters the fluoride ions. The solubility of CaF_2 is 0.0016 g/100 mL of water at 20°C while that of NaF is 40.4 g/L at 20°C.[12]

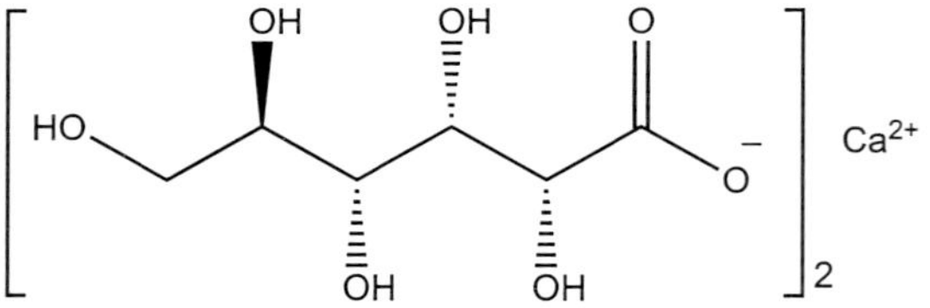

HO OH OH OH OH O O⁻]2 Ca2+

Fig. 13.2 Calcium gluconate

Due to its reaction with glass (forming H_2SiF_6) and many metals, it is stored in plastic containers. The conversion of fluorapatite to phosphoric acid by reaction with dilute sulfuric acid is the standard method for preparing dilute hydrofluoric acid.

$$Ca_5(PO_4)_3F + 5\ H_2SO_4 + 10\ H_2O \longrightarrow 3\ H_3PO_4 + 5\ CaSO_4.\ 2\ H_2O + HF \qquad (13.14)$$

Anhydrous HF is prepared by the reaction of fluorite with conc. H_2SO_4

$$CaF_2 + H_2SO_4 \xrightarrow[600°C]{} 2\ HF + CaSO_4 \qquad (13.15)$$

Many inorganic[1] and organic[2] fluorine compounds are prepared using anhydrous HF as the fluorine source; Schemes 13.3 and 13.4 consolidate these reactions. The industrial uses of HF (as of 2014) fall broadly under the following categories—fluorocarbon (CFCs, HFC, polymers) manufacture: 152,000 tons (63%), alkylation catalyst for gasoline: 16,700 tons (7%), nuclear fuel making: 13,000 tons (5%), aluminum fluoride production: 8,000 tons (3%), other uses (such as stainless steel pickling, electronics, specialty metal production, fine chemicals, and so on): 52,000 tons (22%).

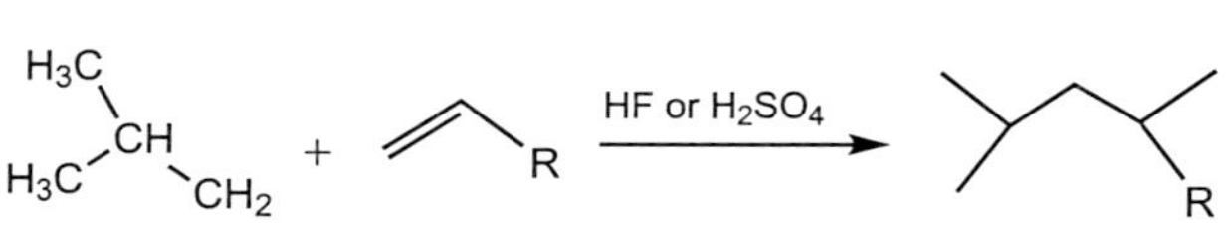

H_3C CH H_3C CH_2 + R $\xrightarrow{HF\ or\ H_2SO_4}$ R (13.16)

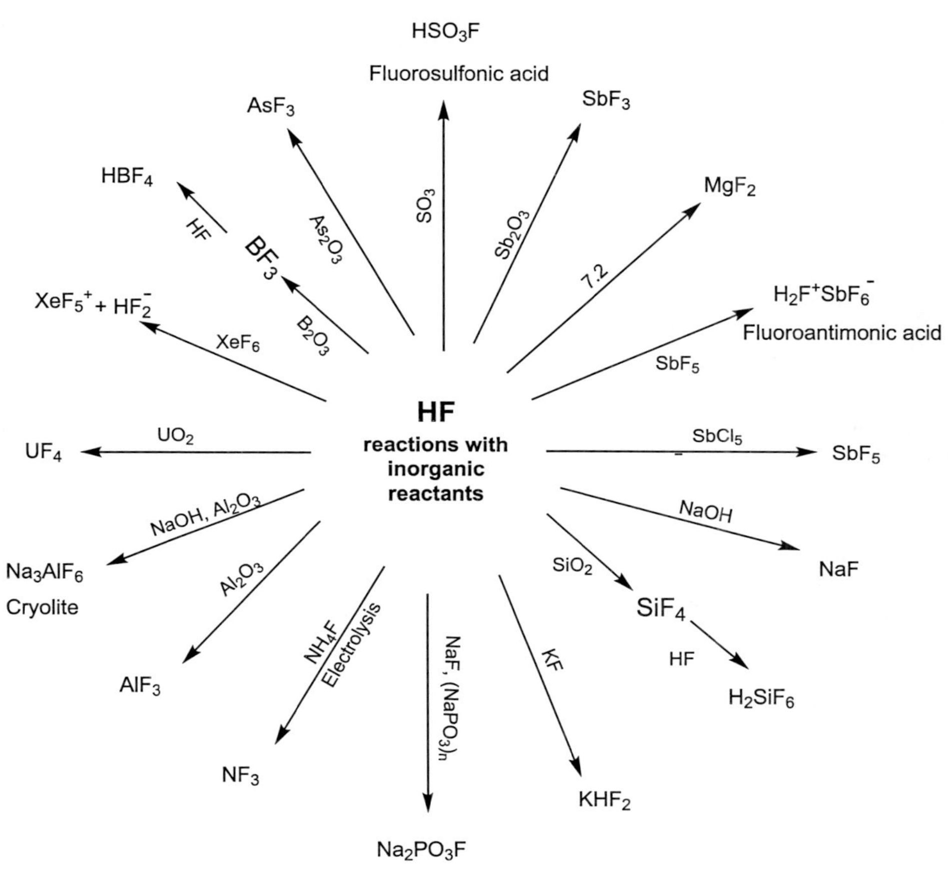

Scheme 13.3 Selected reactions of anhydrous HF with inorganic compounds[1]

13.5.3 Additives Having Fluoride for Use Against Tooth Decay [13, 14]

Fluoride-containing compounds, mainly sodium fluoride (NaF), tin(II) fluoride (SnF_2) and sodium monofluorophosphate (Na_2PO_3F) are added to toothpaste for preventing tooth decay by conversion of hydroxyapatite, $Ca_5(PO_4)_3OH$, of the teeth to fluorapatite, $Ca_5(PO_4)_3F$, the latter being resistant to attack by lactic acid produced in the mouth. Originally, sodium fluoride was used to fluoridate water but later hexafluorosilicic acid (H_2SiF_6) and its salt sodium hexafluorosilicate (Na_2SiF_6) were the more commonly used additives. The fluoridation of water is known to prevent tooth decay and is considered by the US centres for disease control and prevention as 'one of 10 great public health

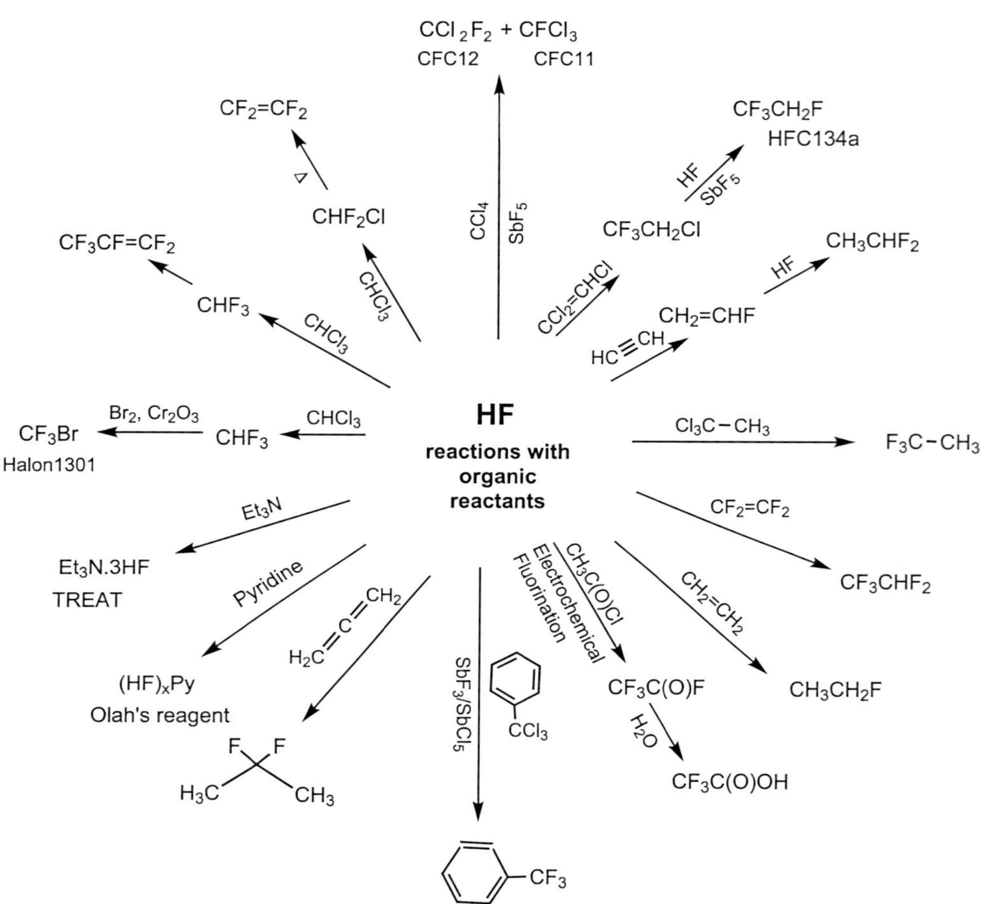

Scheme 13.4 Selected reactions of anhydrous HF with organic compounds[2]

achievements of the 20th century'. A 1994 World Health Organisation expert committee suggested a level of fluoride of 0.5–1.0 mg/L in drinking water. From 1945 onwards, the United States has been fluoridating drinking water; however in 2011, the US lowered its recommended level of fluoride to 0.7 mg/L. Fluoridation is not carried out in many countries as their ground water has substantially high concentration of fluoride. The upper limit of optimum fluoride level in drinking water for a tropical country like India is 0.5 mg/L. Endemic fluorosis has been recognised as a major public health problem in 18 states of India and in certain regions of India, water contains fluoride up to 38 mg/L, which is extremely high compared to the maximum permissible limit. 17 out of 21 districts in the state of Andhra Pradesh of India are affected by fluorosis and the fluoride levels of

Nalgonda district in this state range from 2 to 7 mg/L. Deformities of limb bones, which are notably seen in weight bearing lower limbs are observed in endemic areas of fluorosis. Quite interestingly, such deformities occur only in poorly nourished people whose diet is low in calcium intake. In regions of north India such as Haryana and Punjab, where intake of dairy products containing soluble calcium is quite high, deformities are hardly observed (average amount of calcium in the daily diet of a Nalgonda villager is 300 mg while it is 900 mg in Punjab). Calcium binds with fluoride in the gastrointestinal tract and the compound formed is eliminated through the feces. A diet poor in calcium increases the body's retention of fluoride. Fluoride increases bone metabolism and the diets deficient in calcium intake provokes parathyroid hyperactivity. This in turn mobilises calcium from bone to keep the serum levels, which causes weakening of bones by causing osteoporosis. Weight bearing lower limb bones suffer the maximum, leading to deformities.

13.5.4 Fluorinated Compounds as Superacids[15]

The term superacid was defined by R J Gillespie, also the father of VSEPR rules. According to Gillespie, protic acids stronger than 100% sulfuric acid can be considered superacids. Thus, perchloric acid ($HClO_4$), anhydrous hydrofluoric acid (HF), trifluoromethanesulfonic acid (CF_3SO_3H) and fluorosulfonic acid (HSO_3F) are all considered as superacids.

The name incidentally was given by J B Conant of Harvard for unusually strong acids, which are capable of protonating weak bases such as carbonyl compounds. In 1927, Conant observed that weak organic bases such as ketones and aldehydes can form salts with perchloric acid in non-aqueous solvents. Due to this ability of perchloric acid to protonate weak bases, he called it a superacid. Relative strengths of superacids were compared using the Hammet acidity function. The utility of superacids was first shown by George Olah and coworkers, of the University of Southern California, from an accidental laboratory incident. During a Christmas party in his lab, one of his post doctoral associates dropped a candle accidentally in a solution kept in the lab. The candle dissolved but the solution was not a hydrocarbon but a mixture of fluorosulfonic acid and antimony pentafluoride (was later given a nickname, *magic acid*). Analysis of the solution by ^{1}H-NMR indicated a single peak corresponding to the *t*-butyl cation. This surprising discovery indicated that the paraffin of the candle had not only undergone cleavage but had also undergone isomerisation in the mixture of acids used. This combination of a strong Brønsted acid and a strong Lewis acid was found to ideally form many examples of superacids.[15]

The Hammett acidity function (H_0) is a measure of acidity that is used for very concentrated solutions of strong acids and also superacids. It was proposed by the physical organic chemist Louis P Hammett. While the pH scale is useful for dilute aqueous solutions, the H_0 scale is used to extend the measure of Brønsted–Lowry acidity beyond the dilute aqueous solutions. The Hammett acidity function, H_0, is defined using the following equation:

$$H_0 = pK_{BH^+} + \log \frac{[B]}{[BH^+]} \qquad (13.17)$$

where $pK_{BH}{}^{+}$ is $-\log(K)$ for the dissociation of BH^{+}, which is the conjugate acid of a very weak base B, with a very negative $pK_{BH}{}^{+}$. One can observe that it is as if the pH scale has been extended to very negative values. Hammett originally used a series of anilines with electron-withdrawing groups for the bases.

On this scale, pure H_2SO_4 (18.4 M) has a value of −12. It may be noted that the Hammett acidity function avoids water in its equation. Fluoroantimonic acid $H_2F[SbF_6]$ is 10^{19} times stronger than 100% sulfuric acid. It will react with glass and plastic and explosively with water. It can protonate almost any organic molecule. For a container to hold fluoroantimonic acid, Teflon vessels are needed. Table 13.3 gives the H_0 values for the most well known superacids. There are also solid superacids such as carborane superacid and Nafion-H for which the H_0 value cannot be measured. The large complex fluoroanions facilitate dispersion of the negative charge and thus contribute to increased acidity of the solvated proton. Table 13.3 gives selected superacids and their Hammett acidity functions.

Table 13.3 Examples of superacids with their H_0 values

Acid	Formula	Hammett Acidity function (H_0)[11]
Sulfuric acid	H_2SO_4	−12
Fuming sulfuric acid (50 mol% SO_3)	$H_2S_2O_7$	−14.5
Chlorosulfonic acid	$ClSO_3H$	−13.8
Perchloric acid	$HClO_4$	−13.0
Trifluoromethylsulfonic acid (triflic acid)	CF_3SO_3H	−14.1
Fluorosulfonic acid	FSO_3H	−15.1
Hydrogen fluoride (anhydrous)	HF	−15.1
Magic acid	(FSO_3H–SbF_5) (90 mol% SbF_5)	−23
Fluoroantimonic acid	$H_2F[SbF_6]$	−28 to −31.3

Olah and coworkers have found out extensive uses for superacids, especially in the generation of highly stable carbocations and onium ions. Some of the well known reactions for two very strong superacids are given in Scheme 13.5. In addition, many novel inorganic and organic compounds which are difficult to synthesise have been prepared by using superacids as the medium.

13.6 Fluorine in Organic Chemistry

13.6.1 Volatile Organofluorine Compounds in Industry—CFCs, HFCs and HFOs

High volatility, low toxicity, low reactivity and low flammability of a series of mostly C1–C3 aliphatic organofluorine compounds have been utilised by the industry especially for

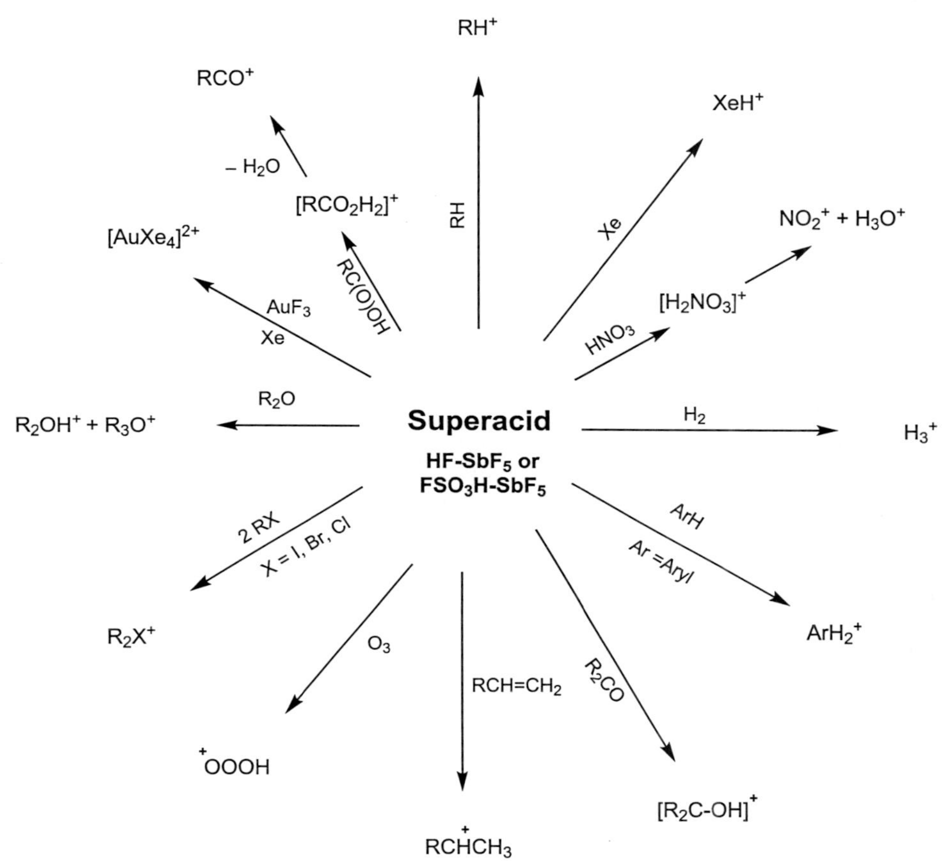

Scheme 13.5 Selected reactions of fluoroantimonic acid and magic acid[15]

applications such as refrigerants for refrigerators and air-conditioners, foaming agents for insulating materials and packing cushions, propellants in medicinal and perfume aerosol sprays, and degreasing solvents and cleaning agents for semiconductors and precision parts in electronic circuits. The first among these, known from 1928 and massively produced in the 1960s, are the CFCs, having only chlorine and fluorine substituents on the carbons. Halons, having fluorine and bromine (for example, CF_3Br) were also developed and used during this period especially for making buildings and aeroplanes fire safe. However in 1974 the roles of CFCs in depleting the ozone layer was discovered and for proof an ozone hole was discovered above Antarctica in 1985. The culprit in this scenario were the chlorine free radicals formed by the cleavage of the C–Cl bonds of the CFCs under UV irradiation as these compounds, due to high volatility and stability reach the higher atmosphere.[16, 17]

UV-A (315–400 nm) not too harmful, does not affect O_3
UV-B (280–315 nm) harmful, affects O_3
UV-C (100–280 nm) lethal, absorbed by O_3

$$O_2 \xrightarrow[240\text{–}160\ \text{nm}]{\text{UV}} O^{\bullet} + O^{\bullet} \qquad (13.18)$$

$$O_2 + O \longrightarrow O_3 \qquad (13.19)$$

$$O_3 \xrightarrow[290\text{–}230\ \text{nm}]{\text{UV}} O_2 + O \qquad (13.20)$$

The formation and degradation of ozone by UV radiation alone had stabilised over the years analogous to a constantly filled but also leaking bathtub in which the water level remains steady (bathtub analogy). However the entry of CFCs into the arena provided an additional pathway for the degradation of ozone (depleting the amount of ozone in the atmosphere is analogous to further reduction in water level of the bathtub). Since the chlorine free radical is regenerated after degrading a molecule of ozone, one Cl·, in principle, can destroy hundreds of thousands of ozone molecules if they are present in the vicinity.[18]

$$CCl_2F_2 + h\nu \longrightarrow {}^{\bullet}CClF_2 + Cl^{\bullet} \qquad (13.21)$$

$$\dot{C}l + O_3 \longrightarrow ClO + O_2 \qquad (13.22)$$

$$ClO + O \longrightarrow {}^{\bullet}Cl + O_2 \qquad (13.23)$$

CFCs were banned from 1987 after the Montreal protocol and analogous compounds having no C–Cl bonds known as hydrofluorocarbons (HFCs) having similar volatility and stability were introduced to replace the CFCs. However, these along with perfluorocarbons (PFCs) and SF_6 were found to have global warming properties. In 1997, the Kyoto protocol was adopted as a concrete step to control greenhouse gas emissions. Meanwhile, for comparison of different volatile compounds in terms of potential for ozone depletion and as greenhouse gas, two relative scales were defined. These are ozone depletion potential and global warming potential.

13.6.2 Ozone Depletion Potential (ODP)[19]

Ozone depletion potential is defined as the relative amount of degradation a compound causes to the ozone layer with trichlorofluoromethane, Cl_3CF (CFC-11) being fixed at an ODP of 1. ODP of some common volatile compounds are given in Table 13.4.

Table 13.4 Ozone depletion potential of some common fluorocarbons

Compound	ODP	Compound	ODP	Compound	ODP
Cl_3CF (CFC 11)	1	HFC 134a (CF_3CFH_2)	0.055	HCFC 113	0.8
CCl_2F_2 (CFC 12)	1 (0.82)	HCFC 225	0.033	CF_3I	0.013
1,1,2-Trichlorotrifluoroethane	0.8	HCFC 123	0.02	CH_3I	0.016
HCFC22 (CF_2HCl)	0.055	CH_3Br	0.7	CF_3Br (Halon 1301)	10

CFC and halon numbering scheme: *The industry-adopted CFC and halon numbering is obtained as follows: For CFCs, consider the number having three digits a, b and c. The digit a will be number of carbon atoms minus 1, b will be the number of hydrogen atoms plus 1 and c will be the number of fluorine atoms. If a is equal to zero it will not be displayed. For example for CCl_2F_2: a = 0, b = 1 and c = 2 and it is termed CFC 12. To find the chemical formula if the CFC number is provided, add 90 to the abc number to obtain a three digit xyz number. Now x will be the number of carbon atoms, y will be the number of hydrogen atoms and z will be the number of fluorine atoms. Also 2x+2–y–z will be the number of chlorine atoms for the CFC. For example, for CFC 11: 90 + 11 = 101; C = 1, H = 0, F = 1 and Cl = 2+2–0–1 = 3. Halons are fluorocarbons with at least one bromine atom and no hydrogen. Each halon has an abcd number where a = number of carbon atoms, b = number of fluorine atoms, c = number of chlorine atoms and d is the number of bromine atoms and if one of them is zero it is still mentioned. For example, Halon 1301 is CF_3Br.*

Global warming potential

Global warming potential is the relative measure of how much heat a greenhouse gas traps in the atmosphere. It compares the amount of heat trapped by a certain mass of a gas over a specific period of time interval (say 20, 100 or 500 years) to the amount of heat trapped by a similar mass of CO_2 in the same time interval.[20]

The GWP depends on the absorption of infrared radiation by a given species, the spectral location of its absorbing wavelengths and the atmospheric lifetime of the species. GWP (100 years) and atmospheric lifetime of few compounds are given in Table 13.5.

Table 13.5 Global warming potential of some common gaseous compounds

Compound	GWP (100 year)	Lifetime* (years)	Compound	GWP (100 year)	Lifetime (years)
CO_2	1	Not specified; taken as 35–200 years	N_2O	298	114
SF_6	22,800	3200	CH_4	25	12
HFC-23 (CHF_3)	14,800	270	CCl_4	1400	26
CFC-11 ($CFCl_3$)	5350	45	NH_3	0	< 0.019

**The atmospheric lifetime of a species measures the time required to restore equilibrium following a sudden increase or decrease in its concentration in the atmosphere.*

Currently, since CFCs are banned, HFCs have replaced most of the CFCs in routine applications and are the main source of fluorinated gas emissions. From the perspective of application, the maximum use of HFCs is as refrigerants. World use in 2011 for HFCs was 79% for refrigeration and air-conditioning, 11% for foam making, 5% for aerosols and 5% for other uses such as solvents, fire extinguishers and so on. However, the European Union has banned the use of all HFCs having GWP > 150 in all new vehicles made after 2011 and any transport equipment after 2017. So replacements having low GWPs for the main HFCs currently in use have already been proposed and some of them are HFOs (hydrofluoro olefins). A comparison of the CFC which was used for refrigeration and air-conditioning, its ideal replacement HFC currently in use (till 2017) and its proposed future replacement, an HFO, are given in Table 13.6.[18]

Table 13.6 Comparison of properties of CFC-12, HFC-13a and HFO-1234yf

Code	CFC-12	HFC-134a	HFO-1234yf
Name	Dichlorodifluoro methane	1,1,1,2-Tetrafluoroethane	2,3,3,3-Tetrafluoropropene
Formula	CCl_2F_2	$CF_3–CFH_2$	$CH_2=CF–CF_3$
Boiling Point	–29.8°C	–26.3°C	–30 °C
ODP	0.82	0.055	0.0
GWP (100 yrs)	10,900	1430	4

Synthesis of CFC-12

$$CCl_4 + 2HF \longrightarrow CCl_2F_2 + 2HCl \tag{13.24}$$

Synthesis of HFC 134a

$$H_2C{=}CH_2 + Cl_2 \longrightarrow ClCH_2CH_2Cl \xrightarrow[400°C,\ Cl_2]{KCl,\ AlCl_3} Cl_2C{=}CHCl \tag{13.25}$$

$$Cl_2C{=}CHCl \xrightarrow{3\ HF} CF_3CH_2Cl \xrightarrow[SbF_5]{HF} CF_3CH_2F \tag{13.26}$$

Synthesis of HFO 1234yf

$$CHCl_3 + HF \longrightarrow HCF_2Cl \xrightarrow[pyrolysis]{} F_2C{=}CF_2 \tag{13.27}$$

$$F_2C{=}CF_2 + CH_3F \xrightarrow[SbF_5]{} H_2C{=}CFCF_3 \tag{13.28}$$

13.6.3 Organofluorine Compounds in Pharmaceuticals and Agrochemicals

The chemical properties of the C–F bond[21]

Although hydrogen is much less electronegative (2.20) compared to fluorine (4.00), it has been substituted by fluorine in many bioactive molecules, pharmaceuticals and herbicides. One of the obvious consequences of the high electronegativity of fluorine is the polarisation resulting in less covalent and more electrostatic character to the C–F bond. The C–F bond is highly polarised with significant electrostatic character and the consequences of the same reflect in the stability, reactivity and structure of organofluorine compounds. Although it is polarised, it does not result in fluorine acting as a good electron donor as due to the high electronegativity, the three lone pairs on fluorine are held tightly by the atom making it reluctant to enter into resonance or interact as a hydrogen bond acceptor. So when a fluorine bond interacts with its immediate environment, it does through electrostatic dipole interactions. The following unique properties of the C–F bond qualify its introduction into pharmaceuticals.[14,15]

- Exchanging of H by F exerts a large electronic effect on neighbouring carbon centres altering both dipole moment and pK_a. In addition, the stability and reactivity of functional groups on the fluorinated molecule are also affected.
- Size wise, fluorine and hydrogen mimic each other and the van der Waals radii of fluorine (1.35 Å) is close to that of hydrogen (1.10 Å) which translates to the fact that only limited extra steric demands are present on the receptor sites by an exchange of H by F.
- The C–F bond distance (1.26–1.41Å) is marginally higher than that of the C–H bond (1.08–1.10Å). Therefore, there will be only a marginal change in bond distances when replacing H by F.
- The presence of C–F bonds aids hydrophobic interactions.
- Covalently bound fluorine also increases the lipophilicity thus increasing its bioavailability (property of a drug or substance which enters the blood circulation when introduced into the body and is able to have an active effect). CF_3 group is one of the most lipophilic of all substituents.
- C–F bonds have high oxidation and thermal stability. In the body it has been found to be not susceptible to oxidation by cytochrome P-450.
- The NMR activity of fluorine's sole natural isotope ^{19}F is convenient for characterisation and also to obtain valuable information about active site interactions.
- The high electronegativity and low polarisability of the C–F bond translates into very low secondary interactions resulting in high volatility for small molecules having mostly fluorine substituents.

13.6.4 Fluorine in Medicinal Chemistry[22]

Organofluorine compounds are virtually absent as natural products. However, currently more than 20% of the drugs available in the market and new pharmaceuticals entering the market has at least one fluorine atom present. In agrochemicals it is even higher, up to 30%. From 1957 to 2007 more than 150 fluorinated drugs have come to the market and the number is increasing every year. While Atrovastatin, an aryl fluorine containing drug belonging to the statin family of anticholesterols, held the record for the best selling drug, reports of 2015 indicate other fluorinated pharmaceuticals such as Fluticasone (Advair, a respiratory drug) and Sofosbuvir (Sovaldi, an antiviral drug) are overtaking the record in annual sales rapidly. Figure 13.3 shows examples of the well known fluorinated drugs having aryl fluorine bonds as well as alkyl fluorine bonds.[23]

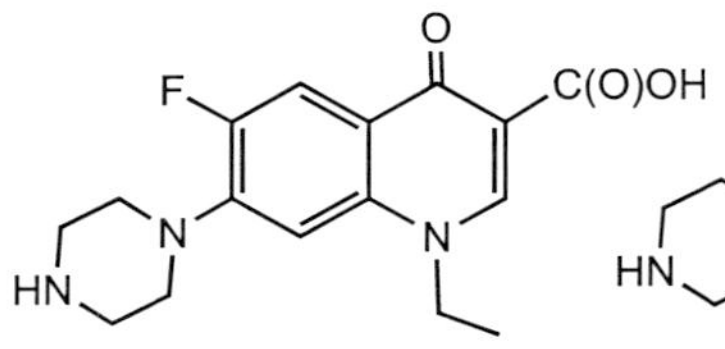

Norfloxacin
Antibacterial

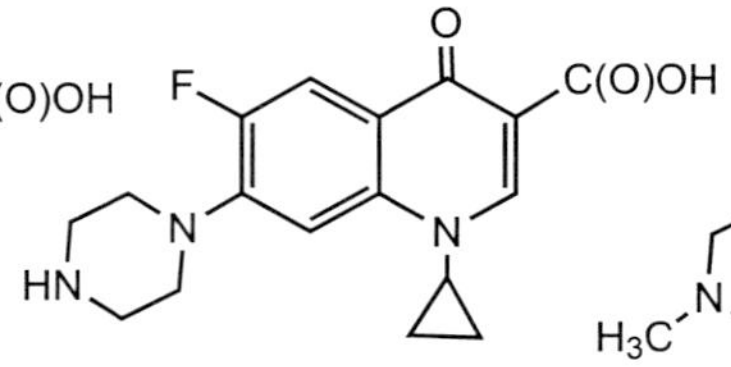

Ciploxacin
Broad spectrum antimicrobial

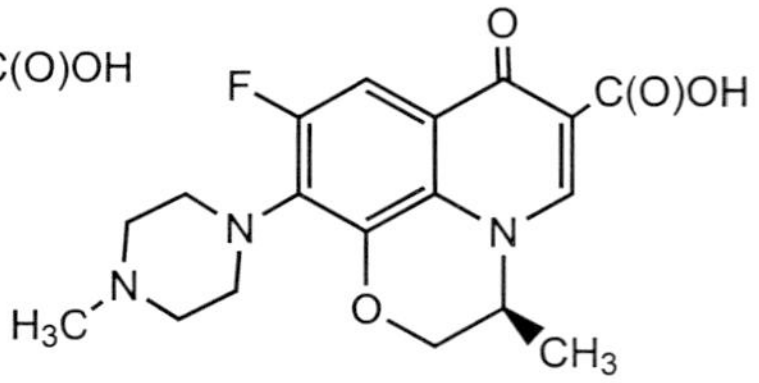

Levofloxacin
Broad spectrum antibiotic

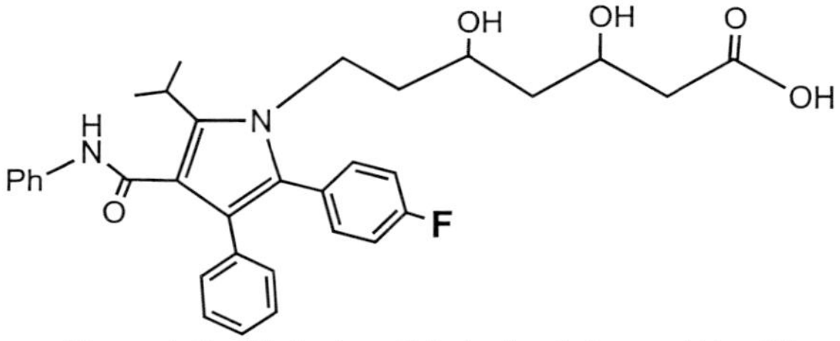

Atrovastatin (Lipitor) -anticholesterol drug sold by Pfizer
Worlds best selling drug of all times; US$ 125 billion in sales (1996–2012)

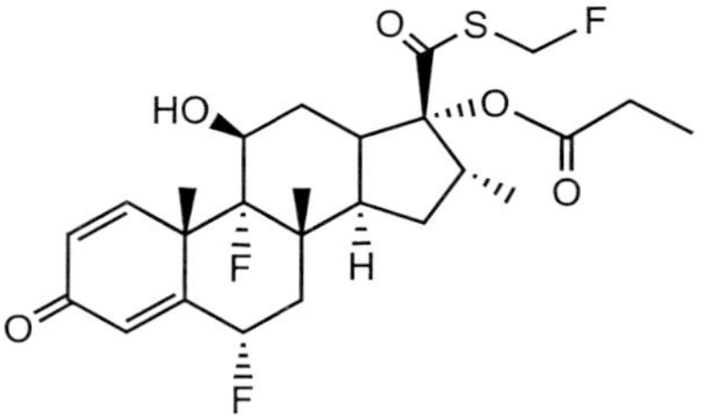

Fluticasone propionate (Advair)
A glucocorticoid; antiinflammatory for nasal allergy and so on

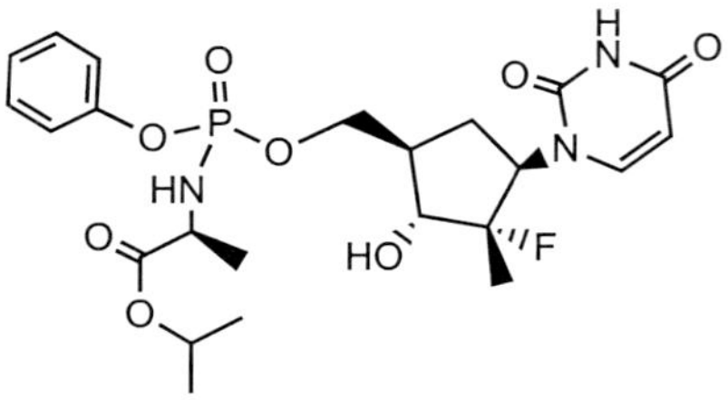

Sofosbuvir (Sovaldi)
For hepatitis C virus infection

Fig. 13.3 Common fluorine containing drugs

It is therefore interesting to see how fluorine affects or contributes to the efficiency of a drug. The unique nature of fluorine imparts a variety of properties to fluorinated drug molecules such as metabolic stability, enhanced binding interactions, changes in physical properties and selective reactivity. The first organofluorine compound used as a drug was 5-fluorouracil in 1971. It was used for malignancies such as breast, gastrointestinal, head and neck cancers but later on displayed neuro- and cardiotoxic effects. The proposed mechanism related to activity indicated that the cleavage of the C-5 proton is essential for the conversion of uracil to thymidine and blocking the C-5 position with a strong C–F bond disrupts the formation of thymidine.

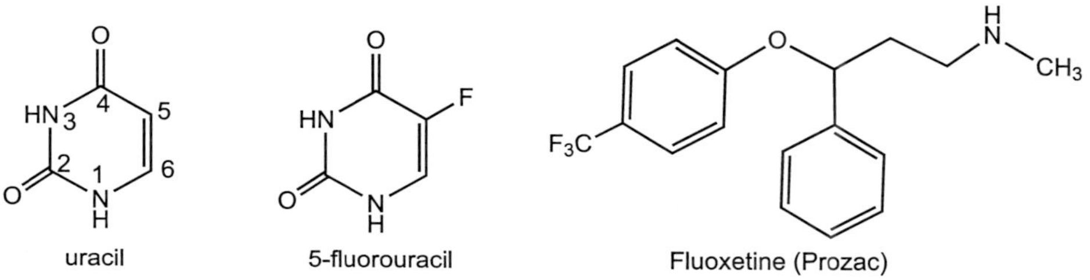

uracil

5-fluorouracil

Fluoxetine (Prozac)

Fig. 13.4 Uracil, 5-fluorouracil and fluoxetine (Prozac)

Prozac (Fluoxetine) is one of the most prescribed anti-depressant drugs worldwide, with sales of more than a billion dollars. Studies indicate that depression is linked to low levels of serotonin, a neurotransmitter (5-hydroxy tryptamine). Prozac acts by selectively preventing the re-uptake of serotonin by the seretonin transporter thus allowing the increase in quantity of serotonin to activate its specific receptor. The para-CF_3 group of the phenolic residue of Prozac increased the potency for inhibiting the serotonin uptake by six-fold, compared to the non-fluorinated compound. The CF_3 group allows the OPh ring to adopt a conformation which favours its binding to the serotonin transporter. The aryl-F substituents of Atrovastatin and Ciprofloxacin enhance potency and in the case of Atrovastain this has been explained with structural evidence indicating a favourable polar interaction of the C–F bond with the protein side chain. In the case of an orally taken drug it should be able to withstand the physiological pH in the stomach and be transported to the required site of action through the blood stream. It should also finally result in a non-toxic material after metabolism.

The following are the main effects of fluorine substituents on physicochemical and conformational properties of a drug molecule.[23]

(i) Effect on pK_a: Variation of the pK_a caused by introducing fluorine can strongly modify the binding affinity and pharmacokinetic properties of a drug. The absorption process of the drug can get affected thereby reflecting on the bioavailability of the drug (bioavailability indicates percentage of drug reaching the circulatory system such as blood). Unlike intravenously injected drugs, orally administered drugs can have lesser bioavailability due to poor absorption or getting metabolised. The high electronegativity of fluorine has a significant effect on the acidity and basicity of functional groups in the vicinity sometimes even up to γ and δ substituents (Table 13.7).

Table 13.7 Effect of fluorine on pK_a values of acids and bases

Carboxylic acid	pK_a	Amine	pK_b
CH_3COOH	4.76	$CH_3CH_2NH_2$	10.7
CH_2FCOOH	2.66	$FCH_2CH_2NH_2$	9.0
CHF_2COOH	1.24	$F_2CHCH_2NH_2$	7.3
CF_3COOH	0.23	$F_3CCH_2NH_2$	5.7
CH_3CH_2COOH	4.87	$CH_3CH_2CH_2NH_2$	10.7
CF_3CH_2COOH	3.06	$F_3CCH_2CH_2NH_2$	8.7

(ii) Modulation of lipophilicity: For a drug molecule to pass through the cell membrane, its lipophilicity should be suitable for it to pass into the lipid layer but not get trapped in the layer. Lipophilicity is normally expressed as a partition coefficient (log P) between octanol and water. The Lipinksi rule of 5 which is used as a guide to predicting good drug candidates indicates that a log $P > 5$ will indicate poor absorption of the drug molecule. It is also not true that fluorination always increases lipophilcity as monofluorination and trifluoromethylation of saturated alkyl groups quite often decreases lipophilicity due to fluorine being electron-withdrawing.[24, 25]

(iii) Conformational changes: Simple fluorine substitution for hydrogen does not significantly change the steric demand on a receptor site. In contrast, groups such as trifluoromethyl do influence the steric demand on a receptor site. The N,N-disubstituted, trifluoro-3-amino-2-propanols (Fig. 13.5) are potent inhibitors of cholesteryl ester transfer proteins. On changing the R group on the aryl ring from tetrafluoroethoxy to simple ethoxy, an eight-fold loss in potency of the inhibitor was observed. Molecular modelling experiments indicated that the tetrafluoroethoxy group prefers an out-of-plane orientation unlike the simple ethoxy group thus indicating importance of conformation of groups in the efficacy of the drug.

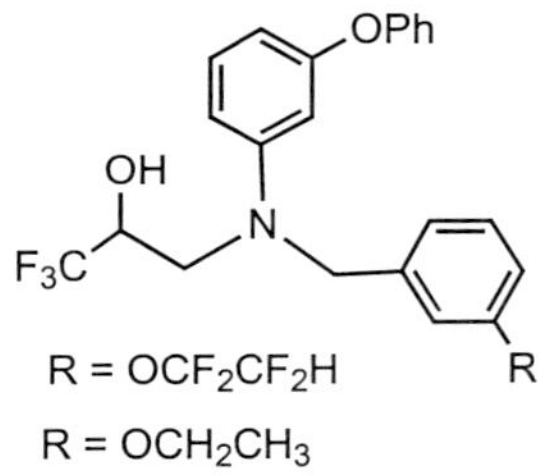

OPh
OH
N
F_3C
R
R = OCF_2CF_2H
R = OCH_2CH_3

Fig. 13.5 N,N-disubstituted trifluoro-3-amino-2-propanols

(iv) Metabolic stability: When a drug is given to a patient, the physiological response is to detoxify or eliminate the drug. Drugs are often metabolised before elimination from the system. Cytochrome P-450 monooxygenases, a family of the heme thiolate complexes, are the most important group of enzymes metabolising drugs. Upon oxidation, the lipophilicity of the drug is often decreased permitting faster clearance from the system. Low metabolic stability due to cytochrome P-450 mediated oxidation is a well known problem in drug discovery. This problem is overcome by blocking the metabolically labile sites by fluorination as C–F bonds are not affected by cytochrome P-450.

(v) Mechanism-based inhibition: This refers to the type of inhibition given by 5-fluorouracil against uracil where the C-5 proton cleavage is an essential step in the mechanism and which is inhibited by blocking the C-5 site with fluorine substitution.

13.6.5 Fluorine in the Design of Ingredients for Crop Protection[26]

A survey of halogenated crop protection products from 1940 onwards indicated that fluorine is present in more than 28% of products that are available. About 54% of the compounds are used as herbicides, 27% as insecticides and 19% as fungicides. Increased degradation stability, increased lipophilicity, increased metabolic stability are some of the reasons for fluorine being incorporated in crop protection agents. The synthesis of synthetic pyrethroids, analogues of pyrethin derived from the flowers of the chrysanthemum plant, is a classical example.

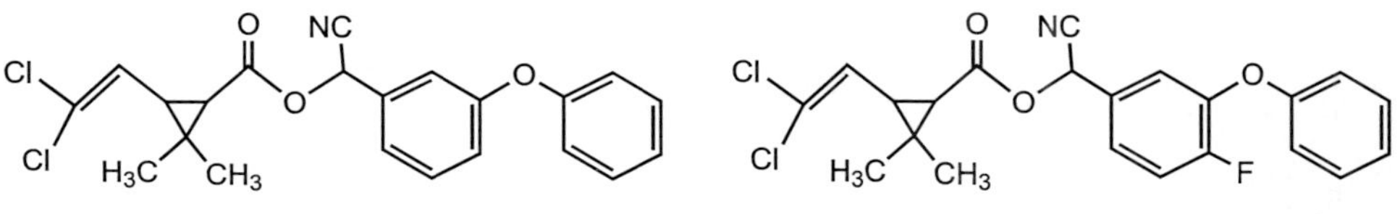

Cypermethrin　　　　Cyfluthrin

Fig. 13.6 Non-fluorinated and fluorinated crop protection agents

When compared to cypermethrin, a synthetic pyrethroid, cyfluthrin having an aryl C–F bond was found to show more than a three-fold reduction in the usage rate for the control of cotton pests. Due to its long lasting residual action and ingestion effect, cyfluthrin is also being used as a household insecticide for control of houseflies, mosquitoes and cockroaches. Similarly, among herbicides which destroy weeds by interfering with carotenoid biosynthesis and inhibit the synthesis of the carotinoids, a common substituent is the 3-trifluoromethylphenyl moiety.

13.6.6 Role of Fluorine in Inhalation Anesthetics[27, 28]

Anesthesia plays a vital role in surgery and surgical procedures and inhalation anesthetic compounds which are used for inducing anesthesia by smelling were known as early as 1846 when diethylether was used by W T G Morton as an anesthetic. This was soon followed by the use of nitrous oxide and chloroform. Inherent difficulties and problems with these compounds led to the discovery of a host of new compounds. Early 1900s saw the use of small organic molecules such as alkenes, cyclopropane and divinyl ether which all turned out to have some kind of complication, especially flammability. From 1950, the high volatility, excellent stability and non-flammability of organofluorine compounds, led to the use of fluorinated compounds as inhalation anesthetics. Fluorine-based compounds took over the inhalation anesthetic scenario with halothane leading the way. Later, a host of partially fluorinated ethers took over and currently these fluorinated ethers are the industry and medical field standards as inhalation anesthetics with Sevoflurane and Desflurane revolutionising and leading the way. Table 13.8 provides some historical landmarks in the development of inhalation anesthetics, their structures and properties and the role played by fluorinated organic compounds in finding the best molecule suitable for an easy anesthesia.

Among currently available inhalation anesthetics which are fluorinated ethers, Sevoflurane is the safest. It is useful for both adults and children and for both induction and maintenance of anesthesia during surgery. The physical, pharmacodynamic and pharmacokinetic properties of Sevoflurane are close to that of an ideal anesthetic. These properties include stability, low flammability, lack of irritation to airway passages with non-pungent odour, low blood:gas solubility which allows rapid onset and emergence from anesthesia, minimal side effects, minimal end-organ effects, low reactivity with other drugs, and a convenient vapour pressure and boiling point for delivery using standard methods of vaporisation. Consequently, Sevoflurane has become one of the most widely used inhalation anesthetics.

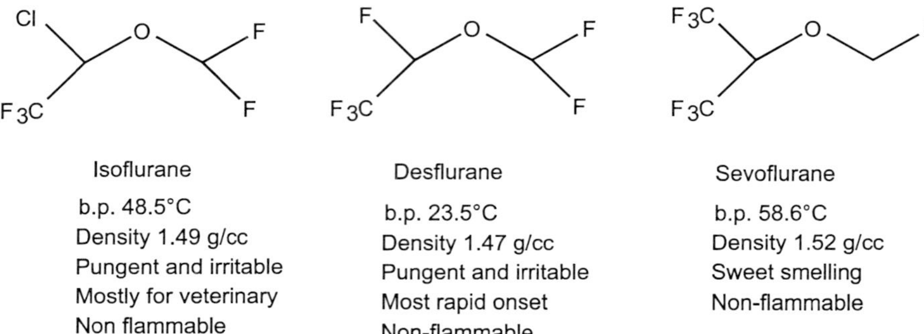

Fig. 13.7 Comparison of structure and properties of Isoflurane, Desfluorane and Sevo fluorane

Table 13.8 Historical landmarks in the development of inhalation anesthetics

Year of clinical use	Formula of inhalation anesthetic	Properties
1846	Diethyl ether Et-O-Et	Flammable; have undesirable side effects, such as post-anesthetic nausea and vomiting.
1847	Chloroform $CHCl_3$	Severe cardiovascular depression, hepatotoxicity, many reports of sudden deaths; was in use till 1976.
1774	Nitrous oxide N_2O	Non-ether, weak inhalation anesthetic, commonly used in combination with other more potent inhalational anesthetics, suitable for dental applications. Can produce headaches. Less potent and long acting compared to others. Remains in use even today.
1923	Ethylene $CH_2=CH_2$	High concentrations required, is an explosive with O_2, unpleasant smell, flammable.
1930	Divinylether $CH_2=CHOCH=CH_2$	Also called Vinethene; flammable and explosive.

Cont'd

Table 13.8 Cont'd

Year of clinical use	Formula of inhalation anesthetic	Properties
1930	Trichloroethylene $Cl_2C{=}CHCl$	Non-flammable alternative to ethylene. Decomposes to give phosgene when warmed in the presence of sodalime; human carcinogen and a non-carcinogenic health hazard.
1929	Cyclopropane △	Relatively potent, non-irritating and sweet smelling. Flammable, high cost, extreme reactivity under normal conditions: when the gas is mixed with oxygen, risk of explosion, patients could suffer a sudden decrease in blood pressure potentially leading to cardiac dysrhythmia, a reaction known as *cyclopropane shock*.
1954	Fluroxene $(CF_3)_2CH_2{-}O{-}CH{=}CH_2$	First fluorinated anesthetic, flammable, forms explosive mixtures with other gases, could lead to post operative liver and renal failure, withdrawn from the market in 1974.
1956	Halothane $CF_3CHClBr$	Non-ether compound, was in use for a long time, can cause nausea, flammable at high concentrations, rare cases of hepatitis following anesthesia linked to halothane.
1960	Methoxyflurane $(CHCl_2)CF_2{-}O{-}CH_3$	Biodegradation produces inorganic fluoride and dichloroacetic acid which are toxic for major organs of the human body, also produces nephrotoxicity.
1963	Enflurane $(CHFCl)CF_2{-}O{-}CHF_2$	Non-flammable, non-irritant, low toxic, high potency, rapid onset, cardiovascular depressant and convulsing properties shows hepatic dysfunction, hypotension.
1971	Isoflurane $CF_3CHCl{-}O{-}CHF_2$	Its pungency can irritate the respiratory system.
1992	Desflurane $CF_3CHF{-}O{-}CHF_2$	Pungent and irritable, has the most rapid onset and offset for any inhalation anesthetics, low blood solubility, high cost, low potency, it may cause tachycardia (increased heart beat rate).
1994	Sevoflurane $(CF_3)_2CH{-}O{-}CH_2F$	After desflurane, it has the fastest onset and offset of anesthesia, inherently stable, low flammability, sweet smelling, lack of irritation to airway passages, low blood:gas solubility, minimal cardiovascular and respiratory side effects, minimal end-organ effects, low reactivity with other drugs.

Synthesis of isoflurane

$$CF_3CH_2OH \xrightarrow[KOH,\ H_2O]{Me_2SO_4} CF_3CH_2OCH_3 \xrightarrow[h\nu]{Cl_2} CF_3CH_2OCHCl_2 + CF_3CHClOCHCl_2 \tag{13.29}$$

$$CF_3CHClOCHCl_2 \xrightarrow[SbCl_5]{HF} CF_3CHClOCHF_2 \tag{13.30}$$

$$CF_3CH_2OCHCl_2 \xrightarrow[SbCl_5]{HF} CF_3CH_2OCHF_2 \xrightarrow{Cl_2} CF_3CHClOCHF_2 \quad (13.31)$$

Synthesis of Sevoflurane[28]

$$S + CF_3CF{=}CF_2 \xrightarrow[DMF]{KF} \text{[(F}_3\text{C)}_2\text{C(}\mu\text{-S)}_2\text{C(CF}_3)_2] \xrightarrow[KIO_3]{KF} (F_3C)_2C{=}O \xrightarrow[catalyst]{H_2} (F_3C)_2CH{-}OH \xrightarrow[H_2SO_4,\ 65^\circ C]{(CH_2O)_n,\ HF} (F_3C)_2CH{-}O{-}CH_2F \quad (13.32)$$

13.6.7 Fluorinating Agents and Selective Fluorination of Organic Compounds

Since elemental fluorine is extremely reactive, controlled fluorination on organic substrates which is essential for preparing partially fluorinated compounds, especially pharmaceuticals and agrochemicals directly using F_2 is impossible. A host of fluorinating agents have been developed with excellent selectivity for specific types of carbon atoms and functional groups. The early methods by which fluorine was introduced into organic substrates were using the Balz–Scheimann reaction[29] via the aryl diazonium salt or by the use of Swarts reaction which uses antimony trifluoride along with chlorine or antimony pentachloride. These methods are still in use, but for more selectivity, one has to either resort to the use of electrophilic fluorinating agents or nucleophilic fluorinating agents.

Balz–Schiemann reaction

$$ArN_2^+BF_4^- \xrightarrow{\Delta} Ar{-}F + N_2 + BF_3 \quad (13.33)$$

Swarts reaction

$$C_6H_5CCl_3 \xrightarrow[\text{or } HF + (SbCl_3 + SbCl_5)]{SbF_3 + SbCl_5 \text{ or } Cl_2} C_6H_5CF_3 \quad (13.34)$$

Halex process

$$\text{EWG-}C_6H_4\text{-X} \xrightarrow{F^-} \text{EWG-}C_6H_4\text{-F} + X^- \quad (13.35)$$

Simons process (Electrochemical fluorination)

$$R_3CH + HF\ (\text{anhydrous}) \xrightarrow[5\text{–}6\ V]{\text{electrolysis}} R_3CF + H_2 \quad (13.36)$$

13.6.8 Electrophilic Fluorination

It is the reaction of a carbon-centred nucleophile with an electrophilic source of fluorine to prepare organofluorine compounds. Initially, elemental fluorine diluted with inert gases and reagents incorporating an oxygen–fluorine bond such as CF_3OF were used for this purpose. However, due to difficulty in preparation and handling they have largely been replaced by reagents containing nitrogen–fluorine bonds. XeF_2 was also developed as a stable electrophilic fluorinating agent, but its high oxidation potential to some extent limits its functional group tolerance. Compounds containing N–F bonds have proved to be the most economical, safe and stable, electrophilic fluorinating agents. Electrophilic N–F reagents can be neutral or cationic in nature and may possess either sp^2- or sp^3-hybridised nitrogen.[30, 31]

The development of crystalline, benchtop-stable fluorinating reagents with excellent selectivity such as N-fluorobis(phenyl)sulfonimide (NFSI), N-fluoropyridinium salts, N-fluoro-*o*-benzenedisulfonimide (NFOBS), Selectfluor I and Selectfluor II[25] have made electrophilic fluorination quite easy to perform. Although fluorinations employing N–F reagents do not use molecular fluorine directly, they are almost universally prepared from F_2 gas which requires great care and special apparatus for handling and usage. Even though N-fluoro reagents can formally behave as a source of fluoronium cations (F^+), the N–F bonds are polarised towards fluorine, with a partial negative charge on fluorine. Reactions may occur via S_N2 displacement with nucleophilic attack at the fluorine.[32]

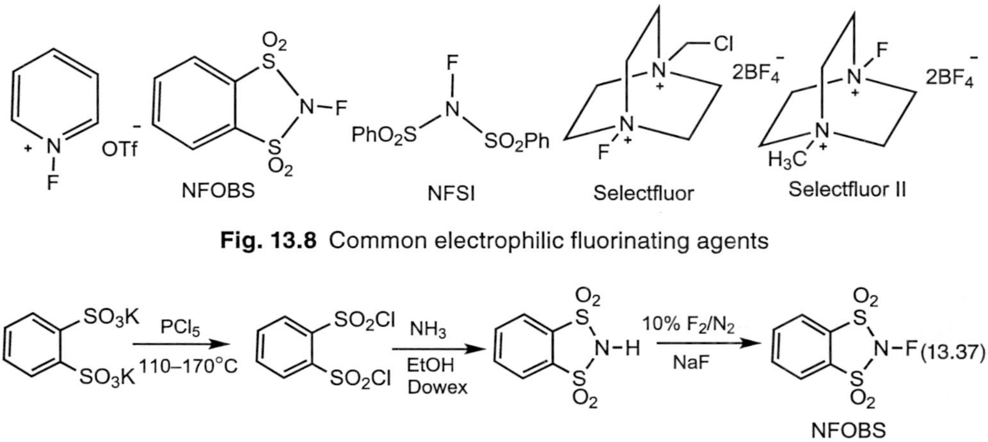

Fig. 13.8 Common electrophilic fluorinating agents

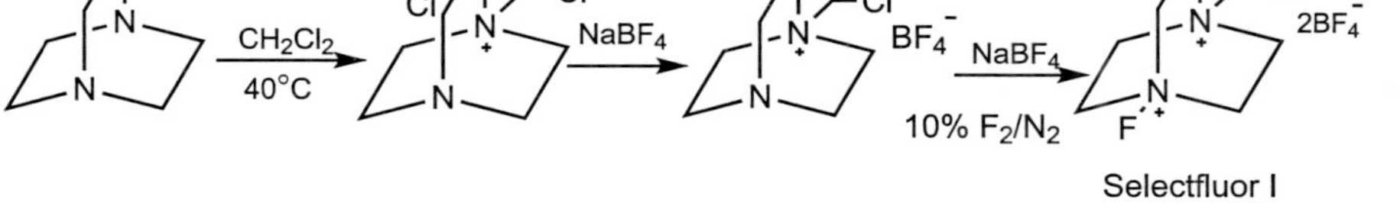

(13.37)

(13.38)

OTMS (silyl enol ether of cyclohexanone) $\xrightarrow{\text{NFOBS}}$ 2-fluorocyclohexanone 79% (13.39)

1,4-dimethoxybenzene $\xrightarrow{\text{NFOBS}}$ 2-fluoro-1,4-dimethoxybenzene 48% (13.40)

$C_6H_5MgBr \xrightarrow{\text{NFOBS}} C_6H_5F$ 80% (13.41)

Fluorobenzene $\xrightarrow[\text{TfOH}]{\text{Selectfluor}}$ 1,4-difluorobenzene 77% + 1,2-difluorobenzene 23% (13.42)

Ph–CH=CH–Me $\xrightarrow[\text{H}_2\text{O, Surfactant}]{\text{Selectfluor}}$ Ph–CH(OH)–CH(F)–Me (13.43)

13.6.9 Nucleophilic Fluorination

The main challenges of using fluoride ions for nucleophilic fluorination are poor solubility and dual activity as a nucleophile and as a base. Although thermodynamics favours the formation of C–F bonds which are the strongest carbon–heteroatom single bonds known so far, the high electronegativity of fluorine contributes to high kinetic barriers for forming the C–F bond. Exclusion of hydrogen bond donors are also required during such reactions as they can vary the nucleophilicity of the fluoride ion. Alkali metal fluorides such as NaF, KF, AgF and CsF are the primary nucleophilic fluorinating agents and their poor solubility can be overcome to some extent by using crown ethers or by replacing the cation by tetraalkylammonium cations.[33] However, the presence of water even in minute amounts will reduce the nucleophilicty of the fluoride as it readily forms hydrogen bonds with water and if HF is generated, it will form HF_2^- as well.

KF	CsF	AgF	$R_4N^+ F^-$
Potassium fluoride	Cesium fluoride	Silver fluoride	Tetraalkyl ammonium fluoride R = Me ; TMAF R = n-Bu: TBAF

SF_4 was found to be an excellent reagent for nucleophilic deoxofluorination. Carbonyl compounds were converted to geminal difluoromethylene derivatives by SF_4. It is also highly reactive and reaction with SF_4 was found to convert carboxylic acid groups to trifluoromethyl groups as well. The volatility and toxicity of SF_4 and difficulty in handling since it is a gas, led to the development of a series of less volatile SF_3 and SF_2 based reagents. Diethylamino sulfur trifluoride (DAST) has been a very commonly used reagent for the fluorination of carbonyl, hydroxyl, thiocarbonyl and sulfide-based substrates.[34] However, Deoxofluor is a more commonly and frequently used reagent these days as it more thermally stable.[28] XtalFluor E is non-explosive, crystalline and less moisture sensitive than both DAST and Deoxofluor but often requires a co-reagent such as amine hydrogen fluoride as a fluoride source.[35, 36]

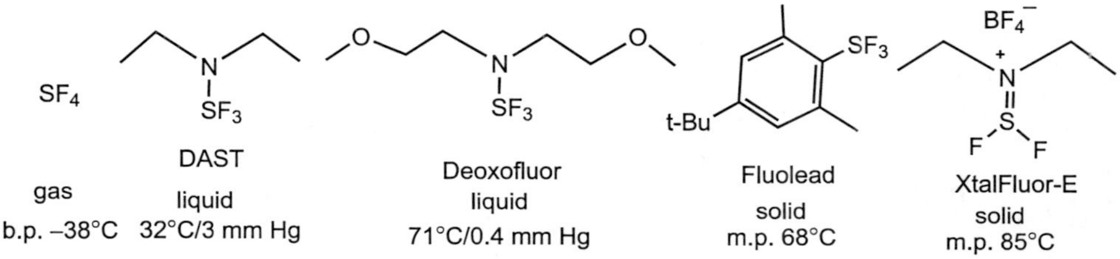

SF_4 — gas, b.p. –38°C

DAST — liquid, 32°C/3 mm Hg

Deoxofluor — liquid, 71°C/0.4 mm Hg

Fluolead — solid, m.p. 68°C

XtalFluor-E — solid, m.p. 85°C

Fig. 13.9 Common sulfur–fluorine based nucleophilic fluorinating agents

$$R\text{–}C(O)OH \longrightarrow RCF_3 \text{ or } RC(O)F \quad (13.44)$$

$$R\text{–}C(O)H \longrightarrow RCF_2H \quad (13.45)$$

$$RC(O)R \longrightarrow RCF_2R \quad (13.46)$$

$$ROH \longrightarrow RF \quad (13.47)$$

In addition to these sulfur-based reagents, deoxofluorination can be accomplished by several reagents such as pyridinium polyhydrogen fluoride (Olah's reagent), triethylamine trihydrogen fluoride (TREAT.HF), Yarovenko's reagent, Ishikawa's reagent, and so on. While Py.nHF is toxic and corrosive due to excessive HF presence, TREAT.HF is less hazardous and mild enough to be handled in normal borosilicate glassware without the problem of etching.[30]

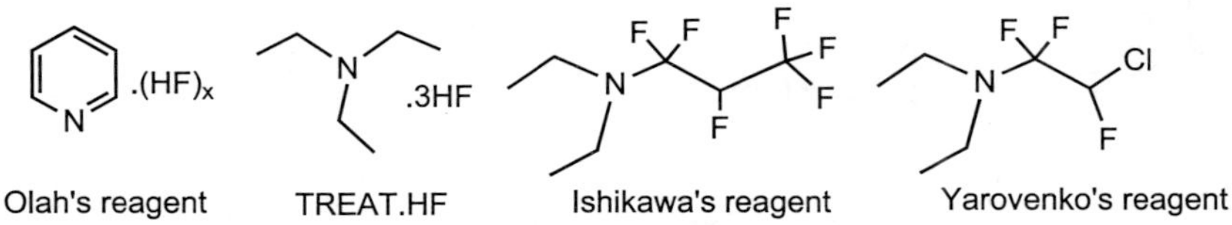

Olah's reagent: .(HF)x — TREAT.HF: .3HF — Ishikawa's reagent — Yarovenko's reagent

13.6.10 Chemoselective Deoxyfluorination of Phenols

Ritter and coworkers came out in 2011 with PhenoFluor[37] and in 2015 PhenoFluorMix,[38] reagents which resemble fluorinated N-heterocyclic carbenes and efficiently carry out deoxyfluorination of phenolic units. Aryl fluorides with varying electronic substituents in addition to heteroaromatic fluorides can be synthesised from the corresponding phenol precursors in a single step by using these reagents. Previously there was no efficient method for deoxofluorination of phenols. For example, catechol was monodeoxyfluorinated by the oxidation of the catechol to *o*-quinone followed by nucleophilic deoxofluorination with Deoxo-Fluor. Another method was conversion of a phenolic OH group to $SnBu_3$ by triflation /stannylation and further reacting with Selectfluor. Phenofluor is moisture sensitive but PhenoFluorMix is highly stable in the presence of moisture.

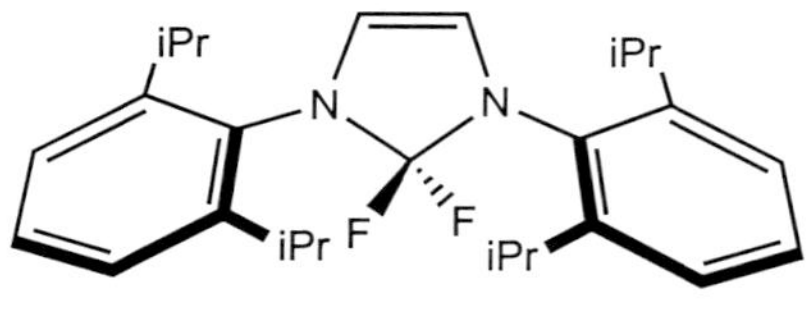

PhenoFluor

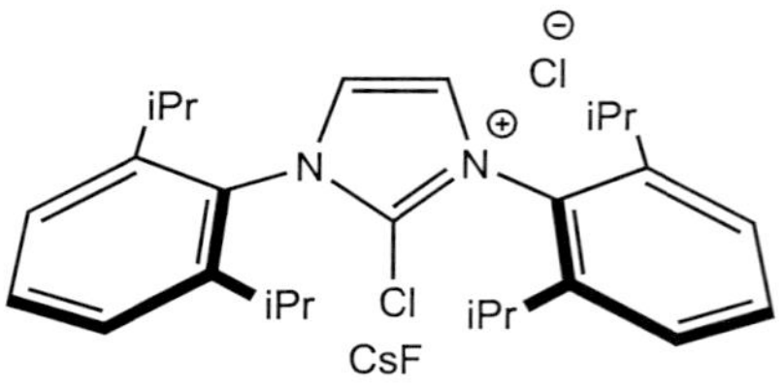

PhenoFluorMix

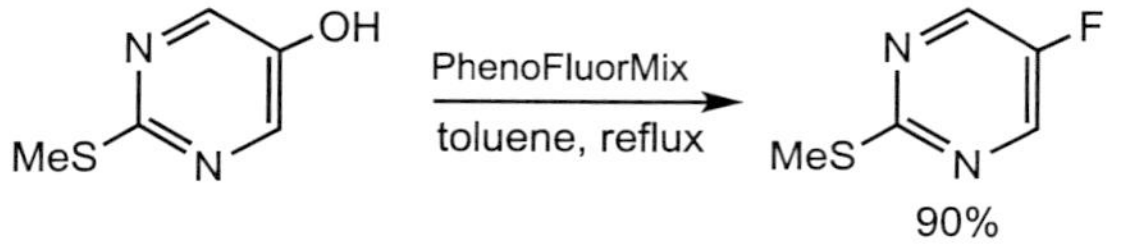

PhenoFluorMix, toluene, reflux; 90% (13.48)

Scheme 13.5 shows how both selective nucleophilic and electrophilic fluorinating agents have been used at the final step of synthesis of a well known drug molecule, fluticasone propionate.[39]

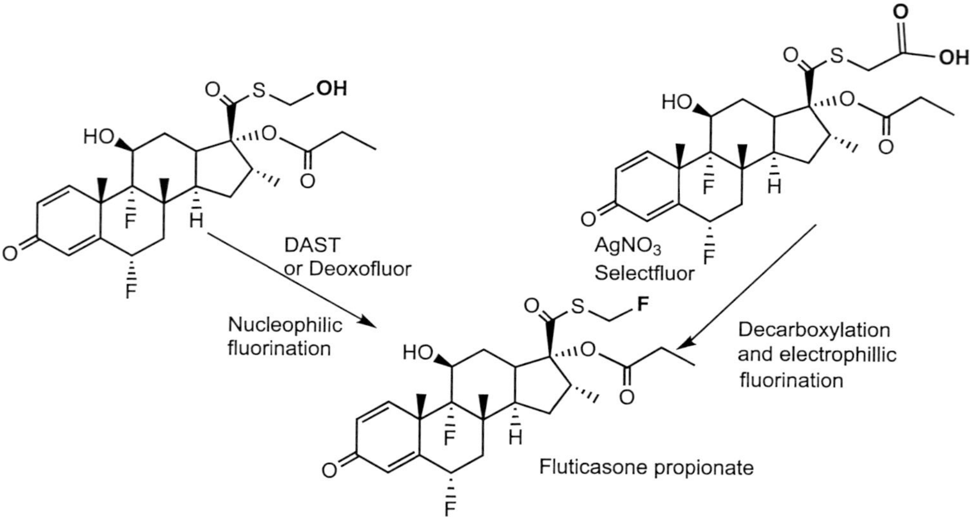

DAST or Deoxofluor — Nucleophilic fluorination

$AgNO_3$ Selectfluor — Decarboxylation and electrophillic fluorination

Fluticasone propionate

Scheme 13.5 Different synthetic routes to Fluticasone propionate

13.6.11 Trifluoromethylation[40]

Similar to selective fluorination, one often has to introduce trifluoromethyl groups on a substrate by both nulceophilic and electrophilic substitution methods. One of the most widely used reagents for nucleophilic trifluoromethylation is trimethyl(trifluoromethyl) silane, $(CH_3)_3SiCF_3$ ($TMSCF_3$). The reagent has become widely referred to as the Ruppert–Prakash reagent, after Ingo Ruppert who introduced the material in 1984 and Surya Prakash who popularised its use and extended its use for carbonyl compounds in 1989. Ruppert prepared this compound by the dehalogenation of a mixture of Me_3SiCl and CF_3Br using $P(NEt_2)_3$. $TMSCF_3$ was further modified for use with phenols and aniline. The broad applicability of $TMSCF_3$ makes it a popular reagent in the syntheses of various medicinal targets. An example is the trifluoromethylation of cyclohexanone in THF using tetrabutylammonium fluoride.[41]

R O O O H3C CH3 —TMSCF3 / F⁻, THF, 0°C→ HO CF3 R O O H3C CH3 (13.49)

Most reagents for electrophilic C- and S-trifluoromethylation are considerably less developed compared to nucleophilic trifluoromethylation. Antonio Togni and coworkers, in 2006, reported an electrophilic reagent based on hypervalent iodine, 3,3-dimethyl-1-(trifluoromethyl)-1,2-benziodoxole (Togni's reagent), which nicely complements the nucleophilic CF_3SiMe_3. This reagent is also easy to handle, and has air and moisture stability for short periods of time. β-Ketoesters react with the Togni reagent under phase-transfer catalysis conditions to yield the α-trifluoromethylated derivatives (Eq. 13.50).

O O OEt —Me Me O I CF3 / n-Bu4NI, K2CO3 MeCN, 28 h, rt→ O CF3 O EtO (13.50)

13.6.12 Hydrodefluorination Reactions

Similar to the conversion of C–H bonds to C–F bonds, the reverse reaction, namely selective conversion of C–F bonds to C–H bonds, commonly referred to as hydrodefluorination (HDF) is one of the most sought after reactions in the chemistry of organofluorine compounds. Fluorine is known to form the strongest σ bond to carbon. The C–F bond is significantly polar, is short, has a low polarisability and the σ* antibonding orbital is a low lying orbital. These features along with the fact that fluoride is a poor leaving group results in high thermodynamic stability and kinetic inertness for the C–F bond when compared to other carbon–halogen bonds. Partially fluorinated aromatics such as C_6F_5H, essential for the synthesis of pentafluorophenyl substituted compounds are difficult to prepare by direct fluorination and HDF is the best method to make them.[42] Both stoichiometric and catalytic reactions as well as reactions involving transition metal and main group element reagents are known for this reaction. Different methods for the conversion of C_6F_6 to C_6F_5H are given in Eq. 13.51.[43–46] Some of the well utilised reagents such as silyl hydrides, R_3SiH, exchange H for F as Si–F bonds are stronger than Si–H bonds. Reactions 13.52–13.55 indicate some of the well known HDF reactions.

$$C_6F_6 \xrightarrow[\text{conditions}]{\text{Reagent / Catalyst}} C_6F_5H \quad (13.51)$$

a. Cp_2ZrCl_2, Mg, $HgCl_2$ or PMe_3, THF, RT, 2 h — 93% yield
b. CuCl, Me_2SiHPh, t-BuOK, 12 h reflux — 100% yield
c. Zn, H_2O, NH_3 — 71% yield
d. Ph_3SiH, TBAT, THF, 60°C, 40 h — 83% yield

$$C_6F_5CF_3 \xrightarrow[\text{40°C 6 h, THF}]{Ph_3SiH,\ TBAT} 4\text{-}H\text{-}C_6F_4CF_3 \ (98\%) \quad (13.52)$$

TBAT = $Bu_4N^+[Ph_3SiF_2]^-$

$$C_6F_5CF_3 \xrightarrow[\text{25°C 24 h, neat or } o\text{-}C_6H_4Cl_2]{Et_3SiH,\ Ph_3C[HCB_{11}H_5Cl_6]} C_6F_5CH_3 \ (97\%) \quad (13.53)$$

$$C_6F_6 \xrightarrow[\text{5 equiv. } Ph_3SiH]{Ni_2(NHC)_4COD} 1,4\text{-}H_2C_6F_4 \ (60°C,\ 2\ days) \text{ OR } 1,2,4,5\text{-}H_4C_6F_2 \ (80°C,\ 4\ days) \quad (13.54)$$

NHC = 1,3-diisopropylimidazol-2-ylidene

$$Cp_2Zr(H)Cl + CH_2{=}CHF \xrightarrow{\text{insertion}} [Cp_2Zr(Cl)CH_2CH_2F] \xrightarrow{\beta\text{-F elimination}} Cp_2Zr(F)Cl + CH_2{=}CH_2 \quad (13.55)$$

13.6.13 Fluorinated Polymers[47]

Although fluoropolymers are less than 0.1% of all polymers produced in terms of weight, their consumption is growing at a higher rate. As of about 2006–2007, estimates of the global fluoropolymer production varied from over 100,000 to 180,000 metric tons per year. Polytetrafluoroethylene (PTFE, also well known as Teflon) is 60%–80% of the world's fluoropolymer production on a weight basis. Following this are FEP (fluorinated ethylene propylene) and PVDF, polyvinylidenedifluoride. The major application of Teflon is as an electrical insulation material due to its excellent dielectric properties and high chemical stability. In the chemical process industry, it is used where there is a requirement for corrosion resistance such as coating for pipes and as gaskets. Teflon coated fiberglass cloth is used for stadium roofs and similar architectural requirements. The major consumer application of telfon is in non-stick cookware. Other fluoropolymers also have high chemical resistance and good dielectric properties. While they are more easy to fabricate and work with, their upper use temperature is never close to that of PTFE. They are also more expensive than PTFE and have lower thermal stability. A study in 2012 indicated that three fluoropolymers constitute 85% of all fluoropolymers consumed across the world. These are PTFE (60%), PVDF (15%) and FEP (10%). The properties of these three polymers are compared in Table 13.8.

PTFE
Polytetrafluoroethylene

FEP
Fluorinated ethylene propylene

PVDF
Polyvinylidene difluoride

ETFE
Ethylene tetrafluoroethylene

PFA
Perfluoroalkoxy polymer

PCTFE
Polychlorotrifluoroethylene

Fig. 13.8 Common fluoropolymers

Table 13.8 Comparison of properties of PTFE, FEP and PVDF

Property	PTFE	FEP	PVDF
Melting point	327°C	265°C	177°C
Melt processability	Not melt processable	Melt processable	Melt processable
Melt viscosity	10^{11} poise	10^5 poise	0.2–17.0 × 10^3 poise
Upper use temperature	260°C	200°C	150°C
Tensile strength	5000	3000	4500
Glass transition temperature	115°C	60°C	–35°C
Flexural modulus (MPa)	340–620	655	1140–2240
Opacity	opaque	translucent	almost transparent

The syntheses of PTFE, FEP and PVDF from simple organic precursors are shown in Scheme 13.6.

$$CHCl_3 \xrightarrow{2HF} CHF_2Cl \xrightarrow[\Delta]{\text{pyrolysis}} F_2C{=}CF_2 \xrightarrow[\text{initiator}]{\text{peroxide}} \left[CF_2{-}CF_2 \right]_n \ (\textbf{PTFE})$$

$$CHCl_3 \xrightarrow{3HF} CHF_3 \xrightarrow[\Delta]{\text{pyrolysis}} F_3CFC{=}CF_2 \xrightarrow[\text{peroxide initiator}]{F_2C{=}CF_2} \left[CF_2{-}CF_2 \right]_x \left[CF_2{-}CF(CF_3) \right]_y \ (\textbf{FEP})$$

$$HC{\equiv}CH \xrightarrow{HF} CH_3CHF_2 \xrightarrow{Cl_2} CH_3CF_2Cl \xrightarrow[-HCl]{} F_2C{=}CH_2 \xrightarrow[\text{initiator}]{\text{peroxide}} \left[CF_2{-}CH_2 \right]_n \ (\textbf{PVDF})$$

Scheme 13.6 Synthesis of PTFE, FEP and PVDF

Discovery and properties of Teflon

The discovery of Teflon is also associated with the worldwide recognition of a company called Dupont. In the 1930s, Dupont, in a joint venture with General Electric Company started a small firm called Kinetic Chemical Company with a view to develop freons for replacing ammonia and sulfur dioxide in air conditioning. A young chemist by the name Roy J Plunkett joined this new venture and was assigned the job of making the hydrofluorocarbon, $CClF_2–CHF_2$ by the reaction of $CF_2=CF_2$ with HCl. The tetrafluoroethylene (b.p. –76°C) usually came from the supplier in small gas cylinders with a valve. On the morning of 6 April 1938, when Plunkett opened the valve of a new tetrafluoroethylene cylinder, no gas came out and there was no pressure. Since he was certain that it was a new cylinder full of gas, he checked the weight of the cylinder and found that there was no weight loss! He, together with his technician removed the valve and cut open the cylinder with a cutter. They saw a white slippery powder coated inside the cylinder. They realised that $CF_2=CF_2$ had polymerised inside the cylinder and made the polymer again by using a polymerisation initiator. However, preliminary analysis of this powder turned out to be quite disappointing. This new powder polymer was not soluble in any solvent, did not react with acids or bases, had very high melting point (327°C) and was not melt processable which was essential to make something useful out of the polymer. Although such properties would normally be discouraging to chemists, Dupont sensed the possible utility of this polymeric compound especially in handling of highly reactive fluorine-based reagents and compounds. Plunkett was rewarded with a patent (1941) and a promotion and moved to another division and the herculean task of taking this un-processable polymer to the fabrication level was assigned to some of the best polymer chemists of Dupont. Within a few years, new patents were filed on the processing and fabrication of this polymer. These include patents for making granular PTFE (M H Brubker, 1946), dispersing PTFE (M M Renfrew, 1950), paste extrusion of PTFE (J F Lontz, 1952) and high strength porous sheets of PTFE (R L Gore, 1976).

The first non-stick pan using teflon was called *Tefal* and was made by a French engineer by the name of Marc Grégoire, in 1954. Grégoire had begun coating his fishing gear with Teflon to prevent tangles. His wife Colette, asked him to coat one of her frying pans with Teflon which, to her delight she found to be non-sticking. Marc obtained a patent and in 1956, he started a company called Tefal to produce such pans in large numbers. Colette even set up a sales counter of the coated pans on the streets of Paris. Later on, Americans also followed suit. The first US based Teflon coated pan was marketed in 1961 under the name 'The Happy Pan'. Thus the discovery of Teflon which has been described by Plunkett as an example of serendipity, a lucky accident and a flash of genius became a very useful product to humanity.

Due to the fluorine atoms surrounding the carbon and its tight structure, teflon is chemically inert, as almost nothing can react with the carbon and with the tight bonding of C–F bonds, its molecular weight exceeds 30,000 daltons. Teflon is attacked by very few chemicals and those that react are among the most aggressive oxidisers and reducing agents known. Alkali metals such as sodium and potassium in intimate contact with fluorocarbons remove fluorine from the polymer chains. This reaction is widely used to etch the surface of PTFE so that the resins can be adhesively bonded. Mixtures of finely divided metal powders such as aluminum or magnesium with powdered fluorocarbon resins can react violently when ignited at high temperatures. There are also reports of F_2 and ClF_3 reacting or getting absorbed making them unsuitable for use with Teflon. In addition, near its maximum service temperate of 260°C, many chemicals such as NaOH, boranes and metal hydrides have been reported to react with Teflon.

Questions for self study (multiple choice)

13.1. Which among the following is the strongest oxidising agent?
(a) Br_2 (b) I_2 (c) Cl_2 (d) F_2

13.2. In which one of the following molecules are all the bonds not equal?
(a) NF_3 (b) ClF_3 (c) BF_3 (d) AlF_3

13.3. Which of the following will have a permanent dipole moment?
(a) SiF_4 (b) SF_4 (c) XeF_4 (d) BF_3

13.4. Which of the following fluorides does not exist?
(a) NF_5 (b) PF_5 (c) AsF_5 (d) SbF_5

13.5. Which of the following is the strongest Lewis acid?
(a) NF_5 (b) PF_5 (c) AsF_5 (d) SbF_5

13.6. KF combines with HF forming KHF_2. The compound contains the species
(a) K^+, F^- and H^+ (b) K^+, F^- and HF
(c) K^+ and $[HF_2]^-$ (d) $[KHF]^+$ and F^-

13.7. The most important industrial use of F_2 gas is
(a) Conversion of UO_2 to UF_4 (b) Conversion of UF_4 to UF_6
(c) Preparation of Teflon (c) Preparation of CFCs and HFCs

13.8. HF is not a very strong acid, yet burns resulting from exposure to HF can be very painful and even fatal. The compound HF forms in the living tissue is
(a) NaF (b) CaF_2 (c) MgF_2 (d) KF

13.9. The strongest acid known so far is
(a) Perchloric acid (b) Sulfuric acid
(c) Fluoroantimonic acid (d) Magic acid

13.10. The species originating from CFCs under UV irradiation responsible for ozone depletion is
(a) fluorine radical (b) chlorine radical
(c) hydrogen radical (d) HF

Solved concept-based questions

13.11. Fluorine does not show multiple oxidation states while other halogens do. Why?

Fluorine does not have available *d*-orbitals of lower energy in its valence shell and therefore cannot expand its octet, unlike the other halogens. Being the most electronegative element it exhibits only –1 oxidation state.

13.12. The negative election gain enthalpy of fluorine is less than that of chlorine. Why?

The negative electron gain enthalpy of fluorine should have been more than that of chlorine. This is due to the smaller size of fluorine atom. As a consequence of the small size, there are strong inter-electronic repulsions in the relatively small sized 2*p* orbital. As a result, the incoming electron does not experience much attraction.

13.13. Chlorofluorocarbons are highly volatile. Why?

The high volatility of the CFCs is due to the fact that there are very little intermolecular attractive forces or weak interactions. Electron density on the F atoms of the C–F bond is not polarisable due to the high electonegativity of fluorine. H atoms are absent, thus, there is no hydrogen bonding. In addition, their molecular weights are also low.

13.14. Only one hydrogen halide is liquid at near room temperature. Which one and why?

HF. Due to strong hydrogen bonding between FHF^- and H^+ units the compound remains a liquid.

13.15. Hydrofluoric acid is always stored in plastic bottles and reacted in plastic/teflon vessels. Why?

Hydrofluoric acid reacts with glass forming SiF_4. This is also called etching of glass.

13.16. Teflon is rigid even at room temperature while silicone is highly flexible. Why?

Adjacent carbon atoms of the teflon backbone have two fluorines attached to them. Fluorine being larger in size than H, packs this rigidly and hence teflon is rigid. In contrast, silicone has a Si–O–Si–O back bone and only the silicon atoms have substituents. Hence there is more space between the substituents and more flexibility.

13.17. Teflon coating provides a non-stick surface. Why?

The fluorine atoms on the surface of teflon which are covalently bound to the carbon chain due to their high electronegativity do not take part in secondary interactions.

As a result PTFE is hydrophobic, neither water nor water-containing substances wet PTFE. Fluorocarbons demonstrate no hydrogen bonds but only moderate London dispersion forces due to the high electronegativity of fluorine. PTFE has one of the lowest coefficients of friction against any solid.

13.18. Hydrofluorocarbons (HFCs) were projected and shown as good alternatives to ozone depleting chlorofluorocarbons (CFCs). Why?

HFCs do not have C–Cl or C–Br bonds. They do not harm the ozone layer since, while in the atmosphere it is the C–Cl and C–Br bonds which undergo homolytic cleavage resulting in Cl and Br free radicals. The presence of H reduces volatility to some extent as C–H bonds can also have weak intermolecular interactions.

13.19. Common reagent grade conc. HF (49%) is a weak acid but 100% HF (anhydrous HF) is a very strong acid. Why?

Unlike other hydrohalic acids such as hydrochloric acid, hydrogen fluoride is only a weak acid in dilute aqueous solution, with an acid dissociation constant (pK_a) of 3.19. This is partly a result of the strength of the hydrogen–fluorine bond which is higher than that of other hydrohalic acids. In contrast at 100% concentration, HF molecules undergo homoassociation to form polyatomic ions (such as bifluoride, FHF^-) and protons, thus greatly increasing the acidity.

13.20. Why is ClF_3 considered as one of the most dangerous chemicals made by man?

ClF_3 is a low boiling liquid (11.8°C) and a very strong oxidant. It is also a source of F_2 gas as ClF_3 degrades to give ClF and F_2. In the presence of moisture, ClF_3 decomposes to give HF, HCl and O_2, sometimes OF_2 as well. The combination of all these very reactive and highly corrosive chemicals released from its decomposition makes it one of the most dangerous chemicals made by man.

Problems and exercises

13.21. Although there are no organofluorine compounds present in the human body, more than 20% of the drugs have C–F bonds present in them. Give six reasons why C–F bonds are introduced in pharmaceutical products.

13.22. Given HF, F_2, H_2 and Mg, show with steps how bomb grade ^{235}U is prepared from UO_2.

13.23. Show steps for the laboratory synthesis of F_2 gas. What is the principle behind this methodology?

13.24. Starting from acetylene how is PVDF prepared?

13.25. Between PTFE, FEP and PVDF which

(a) Has the highest use temperature

(b) Has the highest transparency

(c) Is melt processable

(d) Lowest glass transition temperature

13.26. For the following trifluoromethylation reactions, provide the most suitable reagent.

A: COOH-benzene → CF_3-benzene

B: CCl_3-benzene → CF_3-benzene

C: cyclohexanone → 1-(CF_3)cyclohexan-1-ol (F_3C, OH)

D: CH_3-benzene → CF_3-benzene

13.27. Between the given two nitrazepam drugs A and B, one (notoriously known as the 'date rape drug') is more potent than the other. Identify this drug and justify in a couple of lines why it is more potent than the other.

A (N–H, phenyl) B (N–Me, 2-fluorophenyl)

13.28. Starting from methane and chlorine gases, show the steps for preparing 2,3,3,3-tetrafluoro propene (HFO-1234yf).

Answer key for self study questions

13.1. (d) 13.2. (b) 13.3. (c) 13.4. (a) 13.5. (d) 13.6. (c) 13.7. (b) 13.8. (b) 13.9. (c) 13.10. (b)

Supplementary reading

1. Aigueperse J, Mollard P, Devilliers D, Chemla M, Faron R, Romano R and Cuer J P, Fluorine compounds, *Inorganic Ulmann's encyclopedia of industrial chemistry*, Wiley VCH, 2012, Vol. 15, 398.
2. Siegemund G, Schwertfeger W, Feiring A, Smart B, Beher F, Vogel H and McKusik B, Fluorine compounds, *Organic Ulmann's encyclopedia of industrial chemistry*, Wiley VCH, 2012, Vol. 15, 443.
3. Tressaud A and Moissan H, Winner of the Nobel prize for chemistry 1906, *Angew. Chem. Int. Ed.*, 2006, Vol. 45, 6792.
4. Christe K O, Chemical synthesis of elemental fluorine, *Inorg. Chem.*, 1986, Vol. 25, 3722.
5. Schmedt auf der Gunne J, Mangstl M and Kraus F, Occurrence of difluorine F_2 in nature—In situ proof and quantification by NMR spectroscopy, *Angew. Chem. Int. Ed.*, 2012, Vol. 51, 7847.
6. Lockwood D, First fluorine gas found in nature, *C&E News*, 2012, Vol. 90.
7. Steindler M J and Vogel R C, Corrosion of materials in the presence of fluorine at elevated temperature, *Argonne Natl. Lab.*, ANL- 5662, 1957, 21.
8. Seppelt K, Molecular hexafluorides, *Chem. Rev.*, 2015, Vol. 115, 1296.

9. Ruff O and Krug H, Über einneues Chlorfluorid-ClF_3, *Z. Anorg. Allg. Chem.*, 1930, Vol. 190, 270.
10. Shishkov Y D and Opalovskii A A, Physicochemical properties of chlorine trifluoride, *Russ. Chem. Rev.*, 1960, Vol. 29, 357.
11. Ayotte P, Hébert M and Marchand P, Why is hydrofluoric acid a weak acid? *J. Chem. Phys.*, 2005, Vol. 123, 184501.
12. Roblin I, Urban M, Flicoteau D, Martin C and Pradeau D, Topical treatment of experimental hydrofluoric acid skin burns by 2.5% calcium gluconate, *J. Burn Care Res.*, 2006, Vol. 27, 889.
13. ten Cate J M, Contemporary perspective on the use of fluoride products in caries prevention, *Br. Dent. J.*, 2013, Vol. 214, 161.
14. Ismail A I and Hasson H, Fluoride supplements, dental caries and fluorosis: a systematic review, *J. Am. Dent. Assoc.*, 2008, Vol. 139, 1457.
15. Olah G A, Prakash G K S, Sommer J and Molnar A, *Superacid chemistry*, John Wiley and Sons, 2009.
16. Toon O B and Turco R P, Polar stratospheric clouds and ozone depletion, *Scientific American*, 1991, Vol. 264, 68.
17. Rowland F S, Stratospheric ozone depletion, *Phil. Trans. R. Soc. B.*, 2006, Vol. 361, 769.
18. WMO (World Meteorological Organization), Scientific Assessment of Ozone Depletion: *Global Ozone Research and Monitoring Project,* Report No. 52, Geneva, Switzerland, 2010.
19. Laube J C, Newland M J, Hogan C, Brenninkmeijer C A M, Fraser P J, Martinerie P, Oram D E, Reeves C E, Röckmann T, Schwander J, Witrant E and Sturges W T, Newly detected ozone-depleting substances in the atmosphere, *Nature Geoscience*, 2014, Vol. 7, 266.
20. Elrod M J, Greenhouse warming potentials from the infrared spectroscopy of atmospheric gases, *J. Chem. Educ.*, 1999, Vol. 76, 1702.
21. O'Hagen D, Understanding organofluorine chemistry, an introduction to the C–F bond, *Chem. Soc. Rev.*, 2008, Vol. 37, 308.
22. Mueller K, Faeh C and Diederich F, Fluorine in pharmaceuticals: Looking beyond intuition, *Science*, 2007, Vol. 317, 1881.
23. Purser S, Moore P R, Swallow S and Gouverneur V, Fluorine in medicinal chemistry, *Chem. Soc. Rev.*, 2008, Vol. 37, 320.
24. Hagmann W K, The many roles for fluorine in medicinal chemistry, *J. Med. Chem.*, 2008, Vol. 51, 4359.
25. Harsanyi A and Sandford G, Organofluorine chemistry: applications, sources and sustainability, *Green Chem.*, 2015, Vol. 17, 2081.
26. Jeschke P, The unique role of fluorine in the design of active ingredients for modern crop protection, *Chembiochem*, 2004, Vol. 5, 570.
27. Halpern D F, *Fluorinated Inhalation anesthetics, In organofluorine chemistry: Principles and commercial applications*, Banks R E, Smarts B E and Tatlow J C, Eds., Springer, 1994, 543.
28. Ramakrishna K, Behme C, Schure R M and Bieniarz C, A safe and efficient process for the synthesis of the inhalation anesthetic sevoflurane, *Org. Process Res. Dev.*, 2000, Vol. 4, 581.
29. Suschitzky H, The Balz–Schiemann reaction, *Adv. Fluorine Chem.*, 1965, Vol. 4, 1.
30. Lal G S, Pez G P and Syvret R G, Electrophilic N-fluorinating agents, *Chem. Rev.*, 1996, Vol. 96, 1737.
31. Singh R P and Shreeve J M, Recent highlights in electrophilic fluorination with 1-chloromethyl-4-fluoro-1,4-diazoniabicyclo[2.2.2]octane bis(tetrafluoroborate), *Acc. Chem. Res.*, 2004, Vol. 37, 31.
32. Nyffeler P T, Durón S G, Burkart M D, Vincent S P and Wong C H, Selectfluor: Mechanistic insight and applications, *Angew. Chem. Int. Ed.*, 2005, Vol. 44, 192.

33. Jacobson G G and Akhmetova N E, Alkali metal fluorides in organic synthesis, *Synthesis*, 1983, 169.
34. Markovskij L N, Pashinnik V E and Kirsanov A V, Application of dialkylamino sulfurtrifluorides in the synthesis of fluoroorganic compounds, *Synthesis*, 1973, 787.
35. Redlich M, Bis(2-methoxyethyl)amino sulfurtrifluoride (Deoxo-Fluor®), *Aldrich Chem Files*, 2008, Vol. 8.3, 3.
36. L'Heureux A, Beaulieu F, Bennett C, Bill D R, Clayton S, Laflamme F O, Mirmehrabi M,Tadayon S, Tovell D and Couturier M, Aminodifluorosulfinium salts: Selective fluorination reagents with enhanced thermal stability and ease of handling, *J. Org. Chem.*, 2010, Vol. 75, 3401.
37. Liang T, Newman C N and Ritter T, Introduction of fluorine and fluorine containing functional groups, *Angew. Chem. Int. Ed.*, 2013, Vol. 52, 8214.
38. Campbell M G and Ritter T, Modern carbon–fluorine bond forming reactions for aryl fluoride synthesis, *Chem. Rev.*, 2015, Vol. 115, 612.
39. Fujimoto T and Ritter T, PhenoFluorMix: Practical chemoselective deoxyfluorination of phenols, *Org. Lett.*, 2015, Vol. 17, 544.
40. Zhou J, Jin C and Su W, Improved synthesis of fluticasone propionate, *Org. Process Res. Dev.*, 2014, Vol. 18, 928.
41. Alonso C, Martínez de Marigorta E, Rubiales G and Palacios F, Carbon trifluoromethylation reactions of hydrocarbon derivatives and heteroarenes, *Chem. Rev.*, 2015, Vol. 115, 1847.
42. Kuehnel M F, Lentz D and Braun T, Synthesis of fluorinated building blocks by transition metal mediated hydrodefluorination reactions, *Angew. Chem. Int. Ed.*, 2013, Vol. 52, 3328.
43. Douvris C and Ozerov O V, Hydrodefluorination of perfluoroalkyl groups using silylium carborane catalysts, *Science,* 2008, Vol. 321, 1188.
44. Kikushima K, Grellier M, Ohashi M and Ogoshi S, Transition metal-free catalytic hydrodefluorination of polyfluoroarenes by concerted nucleophilic aromatic substitution with a hydrosilicate, *Angew. Chem. Int. Ed.*, 2017, Vol. 56, 16191.
45. Lv Hongbin, Cai Y-B and Zhang J-L, Copper catalysed hydrodefluorination of fluoroarenes by copper hydride intermediates, *Angew. Chem. Int. Ed.*, 2013, Vol. 52, 3203.
46. Fischer P, Götz K, Eichhorn A and Radius U, Decisive steps of the hydrodefluorination of fluoroaromatics using $Ni(NHC)_2$. *Organometallics*, 2012, Vol. 31, 1374.
47. Ebnesajjad S, *Introduction to fluoropolymers: Materials, technology and applications*, Elsevier, 2013.

The Chemistry of Chlorine, Bromine and Iodine

CHAPTER 14

Unique Properties of Chlorine and its Compounds

- *Chlorine has the highest electron affinity (348.6 kJ/mol) and the third highest electronegativity (3.16) of all the reactive elements. Its covalent radius is 1.02 Å and the Cl–Cl bond dissociation energy (58 kcal/mol) is the highest among dihalogen molecules. For comparison (F–F = 38 kcal/mol, Br–Br = 46 kcal/mol and I–I = 36 kcal/mol)*
- *Chlorine shows oxidation states 7, 6, 5, 4, 3, 2, 1, 0 and −1.* Cl_2 *and higher oxides are good oxidising agents.*
- *Chlorine is not only plentiful in seawater, but also in the earth's crust being the sixth most abundant element.*
- *Chlorine gas and many chlorine-based compounds are useful in water purification.* Cl_2 *reacts with water giving a mixture of HCl and HClO and the latter is responsible for breaching the bacterial cell wall and killing bacteria.*
- *Earth's oceans contain about* 2.6×10^{16} *metric tons of chlorine, mostly as sodium chloride. Among dissolved salts in sea water,* Cl^- *is 18.1 g/L while the amounts of the rest of the halides are 1.4 mg* F^-, *68 mg* Br^- *and 0.06 mg* I^- *per litre.*
- *Chlorine is not found free in nature while HCl is often found in volcanic gases. Chlorine is partly soluble in water. One litre of water dissolves 0.0921 moles of chlorine gas at 25°C.*
- *The gastric juices of stomach of human beings have 0.3% to 0.4% hydrochloric acid. White blood cells kill bacteria by producing HClO inside their cells.*
- *Being heavier than air if released in large quantities, it forms a poisonous blanket that can drift along with the wind. During World War I, chlorine was used as a chemical weapon, first by the Germans in 1915 followed by the Western Allies.*
- *Chlorine naturally occurs as a mixture of its two stable isotopes* ^{35}Cl *and* ^{37}Cl *with natural abundances being 75.8% and 24.2% respectively.*

- *It is used in the manufacture of household bleaches and disinfectants (such as chlorinated phenols for killing bacteria) and is used for bleaching paper pulp. It is an important substituent in organic chemistry, forming industrially important compounds such as carbon tetrachloride, chloroform, phosgene, PVC, PVDC and neoprene.*
- *The first chain reaction discovered was not a nuclear reaction but a chemical chain reaction. It was discovered in 1913 by Max Bodenstein, who saw a mixture of chlorine and hydrogen gases explode when triggered by ultraviolet light of 470 nm or stronger. The mechanism of this reaction was explained in 1918 by Walther Nernst.*
- *Chlorine is essential for the manufacture of about 93% of the top-selling drugs. Out of these, about 26% of the drugs manufactured contain chlorine as part of the molecule while the rest use chlorine chemistry in the synthesis mostly as chlorine-containing starting materials and intermediates.*
- *Polychlorinated dibenzodioxins, known as PCDDs or simply dioxins (especially 2,3,7,8-tetrachlorodibenzo-p-dioxin) are some of the most toxic compounds known to humans. These are known to form during burning of organochlorine compounds.*
- *From its first use as a germicide, as suggested by the Hungarian doctor Semmelweis, to prevent the spread of 'child bed fever' in the maternity wards of Vienna general hospital in Austria in 1847, chlorine-based compound have been one of society's most potent weapons against a range of life-threatening infections, viruses and bacteria for over 150 years.*
- *Chlorine is also used widely in the manufacture of many products and items directly or indirectly, that is, in paper production, antiseptics, dyestuffs, insecticides, paints, petroleum products, plastics, medicines, textiles, solvents, and many other consumer products.*

Unique Properties of Bromine and its Compounds

- *Bromine is the only halogen which is a liquid at room temperature (b.p. 58.8°C).*
- *Bromine naturally occurs as a mixture of its two stable isotopes ^{79}Br and ^{81}Br with natural abundances being 50.7% and 49.3% respectively.*
- *The mass of bromine in the oceans is about 1/300th of that of chlorine.*
- *At high temperatures, organobromine compounds easily get converted to free bromine radicals, a process that helps to slow down and arrest free-radical chemical chain reactions. Since fire progression is essentially a free radical process, this property makes organobromine compounds useful as fire retardants.*
- *Volatile organobromine compounds, similar to organochlorine compounds are ozone depleting agents as the C–Br bond is homolytically cleaved in the presence of UV radiation.*
- *The bromide ion has an inhibitory effect on the central nervous system and KBr was one of the very early anti-epileptic drugs.*

- *For many applications, organobromides represent a compromise of reactivity and cost as their reactivity is intermediate between the reactivity of organochlorine and organoiodine compounds.*
- *Ethylene bromide was used as an additive in gasolines containing tetraethyllead. It scavenges lead by forming volatile lead bromide.*
- *Organobromides are extensively used in palladium catalysed cross coupling reactions as they are less expensive than iodo compounds and more reactive than chloro compounds.*
- *The essential oil extracted from the Hawaiian algae Asperagopsis taxiformis consists of 80% bromoform. Most of the organobromine compounds found in the oceans are made by the action of the unique enzyme, vanadium bromoperoxidase present in seaweeds and marine microalgae.*

Unique Properties of Iodine and its Compounds

- *Iodine is the least abundant of the halogens, being even lesser in natural abundance than rare earth elements.*
- *The bond in I_2 (36 kcal/mol) is the weakest among the dihalogens.*
- *It is the heaviest among essential elements required for life and is present in the thyroid hormones.*
- *Iodine deficiency disorders affect about two billion people worldwide due to which iodised common salt has been made obligatory in many countries.*
- *Due to its high atomic number and ease of attachment to organic compounds, it has also been found useful as a non-toxic radiocontrast material.*
- *Among the stable halides, the iodide anion, I^-, is the strongest reducing agent, being the most easily oxidised back to diatomic I_2.*
- *I_2 is only sparingly soluble in water (1 g in 3450 mL of water at 20°C). To increase solubility, it is often converted to I_3^- by adding I^-.*
- *A deficiency of iodine in food results in enlargement of the thyroid gland (a disease known as goitre). Two thyroid hormones triiodothyronine (T_3) and thyroxine (T_4) are responsible for regulation of metabolism.*
- *The amylose chains of starch form a charge transfer complex when treated with I_2 and I^- (amylose I_5^- complex). This property is not only used in iodometric titrations but also utilised in the detection of counterfeit currency as genuine currency notes of many countries (for example, the US \$, Euro, Swiss franc, and so on) are printed on starch-free papers made of cotton fibres.*
- *I_2 forms deep brown charge transfer complexes when dissolved in polar solvents such as alcohols and amines. It is violet in non-polar solvents.*
- *Iodination of metals results in metal iodides of lower oxidation state compared to other halogens; for example, ReI_4 vs $ReBr_5$ vs $ReCl_6$ vs ReF_7.*

- *Iodine forms stable hypervalent organic compounds in +3 and +5 oxidation states and these compounds are selective and environmentally benign oxidising agents.*
- *Iodine (I_2) is readily displaced from alkali metal iodides by the displacement reactions with Cl_2 and Br_2.*
- *Iodine has only one stable isotope, ^{127}I and 36 other unstable isotopes among which ^{131}I is a common radioactive fission product formed during nuclear fission.*
- *Iodine-131 has a half life of eight days and undergoes β-decay to the excited state Xe-131, which emits γ-radiation while going to its ground state. This has been a problem associated with nuclear fission as radioactive iodine dispersed in the air can enter the body and get concentrated in the thyroid glands and emit γ-radiation, leading to cancer.*

14.1 About the Elements

14.1.4 Chlorine

The main natural source of chlorine is brine or sea water and rock-salt deposits found under the earth. Chlorine, as a free element, is obtained as a greenish-yellow gas and it is 2.5 times heavier than air. It liquefies at –34°C, and crystallises as a yellowish crystalline solid at –103°C. Liquid chlorine has been found to burn skin while gaseous chlorine is an irritant to mucous membranes. Breathing chlorine gas in large amounts can be fatal. Unlike other halogens, chlorine is produced by the chemical industry in very large quantities (2014: 56,000,000 tons). The major use of chlorine is for making polyvinyl chloride which is the third polymer (quantity wise) made across the world. The side product of PVC synthesis is HCl and currently this is one of the major industrial routes to HCl. Chlorine and many chlorinated compounds are used in the wood and paper industry (as a bleach), in the processing of metals, in the purification of water and the production of materials such as titanium dioxide. Chlorocarbon compounds range from the essential drugs (chloroquine, an antimalarial; vancomycin, an antibiotic) to some which have been banned (DDT and chlorofluorocarbons) to the downright toxic (polychlorinated biphenyls and dioxins).[1]

14.1.2 Bromine

Bromide is found dispersed throughout earth's crust and is also present in sea water (0.067 g/ kg). The water of the Dead sea with a high density and lesser NaCl content has a relatively high percentage of bromide also (4.2 g/kg). Natural salt deposits and brines are the main sources of bromine and its compounds. The most important application of bromine compounds as of now is as fire retardants. Polybrominated organic compounds are added to polymer compositions to make them fire safe. Till 2014, bromine was not thought to be an essential trace element. Ethylene bromide, obtained by addition of bromine

Is bromine an essential trace element? [3,4]

Bromine as bromide is universally present in human and animal diet as it is present in small quantities along with iodide in natural common salt. Yet unlike iodide, which is essential for the functioning of the thyroid, or chloride present in the digestive juices, or fluoride which is essential for healthy teeth, there seems to be no necessity for bromide. The proposed and widely accepted broader definition for essential trace elements says that 'an element is essential when a deficient intake consistently results in an impairment of a function from optimal to suboptimal level and when supplementation with physiological levels of this element, but not others, prevents or cures this impairment'. Based on this definition, Walter Mertz in 1981 proposed that out of the 92 naturally occurring elements of the universe, twenty-seven elements are essential for animal life. These 27 elements, which includes even elements such as arsenic and vanadium which are toxic to humans but are required for the health and reproduction of animals such as goats does not list bromine. Work by Hudson and coworkers in 2014 published in the journal *Cell* has proved beyond doubt that bromine is the 28th essential trace element indispensable for development and tissue architecture in all animals—from primitive sea creatures to humans.

Hudson and coworkers showed that bromide is a required cofactor for the formation of sulfimine crosslink formation within collagen fibres, an event critical for tissue and muscle development. Bromide is converted to hypobromous acid which in turn forms a bromosulfonium ion intermediate that energetically selects for sulfimine formation. They also showed that fruit flies (Drosophila) died when bromide was removed from their diet but survived when bromide was restored. They compared the effect of different halides (F^-, Cl^-, Br^- and I^-) and a pseudo halide (SCN^-) on sulfimine crosslink formation in cell cultures. SCN^- and I^- inhibited the reaction, F^- was found to be toxic, Cl^- was inconclusive while Br^- enhanced crosslink formation. Separate detailed studies comparing chloride and bromide indicated that chloride also forms the crosslinks but only very poorly along with an un-cross-linkable product namely methionine sulfoxide while bromide does it efficiently in a dose-dependant manner.

The study indicates that among many other useful applications, bromine supplementation may improve the health of patients on dialysis and related health issues.

to ethylene, was once of commercial significance as a component of leaded gasoline. Since leaded gasoline has been discontinued, its use has become insignificant. Another problem associated with volatile organobromine compounds is that UV radiation can homolytically cleave a C–Br bond releasing bromine free radicals which are equally bad for ozone depletion as chlorine free radicals. CF_3Br, a well-known freon (halon, used in closed spaces such as airplanes and high rise buildings to prevent fire) is now a banned chemical as its production and use are controlled by the Montreal Protocol.[2]

14.1.3 Iodine

The main sources of iodine are iodate minerals, especially Chile saltpetre which contains $NaIO_3$, natural brine deposits left by the evaporation of ancient seas and waters from oil and salt wells. Commercial production of I_2 is from the iodate obtained from processing of nitrate ores or by extracting iodine vapour from brine. Although iodine is present in seawater as iodide, it is present only in trace amounts (0.05 parts per million) and is assimilated and concentrated by seaweeds. Iodine is a dark purple brown shining solid element which readily sublimes and if kept in a closed bottle fills it with purple vapours.

Iodine's commercial uses include pharmaceuticals, printing inks and dyes, catalysts, animal feed supplements and photographic chemicals. It is also used to make polarising filters for LCD displays. Iodide deficiency disorders (IDD) affect populations which have less than the required amount of iodine in their diet. Therefore iodide or iodate is added in small amounts to table salt, to prevent iodine deficiency which affects the thyroid gland. Ethylene diammonium diiodide is a widely used iodine supplement in animal feeds. Iodide based rhodium and iridium complexes are used for the methanol to acetic acid catalytic processes namely Monsanto process and Cativa process respectively. The radioactive isotope iodine-131 is sometimes used to treat cancerous thyroid glands. This isotope is also in the radioactive fallout from nuclear explosions and disasters involving uranium-235 which the human body can accumulate from natural sources near a nuclear disaster site. Tincture of iodine and iodoform are used in disinfecting wounds.[5]

Table 14.1 Physical properties of chlorine

Chlorine	*Atomic mass*	*Electronegativity (Pauling scale)*	*1st ionisation energy*	*Electronic configuration*	*van der Waals radius*
Symbol: Cl	35.455 g.mol^{-1}	3.16	1251.2 kJ.mol^{-1}	[Ne]3$s^2$3p^5	0.175 nm
At. no. 17	*Atomic radius*	*Melting point*	*Boiling point*	*Electron affinity*	*Density*
	1.75 Å	–101°C	–34.04°C	348.575 kJ.mol^{-1}	1.56 g.cm^{-3} at b.p.

Table 14.2 Physical properties of bromine

Bromine	*Atomic mass*	*Electronegativity (Pauling scale)*	*1st ionisation energy*	*Electronic configuration*	*van der Waals radius*
Symbol: Br	79.904 g.mol^{-1}	2.96	1139.9 kJ.mol^{-1}	[Ar]3d^{10}4$s^2$4p^5	0.185 nm
At. no. 35	*Atomic radius*	*Melting point*	*Boiling point*	*Electron affinity*	*Density*
	1.85 Å	–7.2°C	58.8°C	324.537 kJ.mol^{-1}	3.10 g.cm^{-3} at r.t.

Table 14.3 Physical properties of iodine

Iodine	*Atomic mass*	*Electronegativity (Pauling scale)*	*1st ionisation energy*	*Electronic configuration*	*van der Waals radius*
Symbol: I	126.904 g.mol^{-1}	2.66	1008.4 kJ.mol^{-1}	[Kr]4d^{10}5$s^2$5p^5	0.198 nm
At. no. 53	*Atomic radius*	*Melting point*	*Boiling point*	*Electron affinity*	*Density*
	1.98 Å	113.7°C	184.3°C	295.153 kJ.mol^{-1}	4.93 g.cm^{-3} at r.t.

14.2 Isolation and Purification of the Element

14.2.1 Chlorine

Although he was unaware that it was an element, Carl Wilhelm Scheele is credited with the first preparation of chlorine in 1774 when he released a few drops of hydrochloric acid onto some manganese dioxide resulting in the evolution of a greenish-yellow gas. However, it was identified as an element by Sir Humphrey Davy only in 1810. Till then, it was presumed to be a compound of oxygen. Davy gave the name chlorine based on the greek word '*khloros*' meaning greenish yellow.

Chlorine was by this time already in use. In the small town of Javelle, France, chlorine added to alkaline water created bleach (hypochlorite) that was used in the fabric industry in the late-18th century. Bleach began to be used as a disinfectant in hospitals, and chlorination of the water supply in London during a cholera outbreak in 1850 saved many lives. Labarraque and Semmelweis are the pioneers who advocated the use of chlorine as a disinfectant (see box). Chlorine and chlorine-based disinfectants kill or deactivate a wide range of dangerous germs in hospitals, homes and public places. Chlorine's powerful disinfectant qualities primarily come from its ability to break open the cell walls of bacteria and viruses and leak-out its contents.

14.2.2 Bromine

The discovery of bromine dates back to 1826 when the French chemist A J Balard passed chlorine gas through the aqueous solution of the residues left over from the manufacture of sea salt which contained magnesium bromide. He liberated the element as dark brown-red vapours—a reaction now well known as the halogen displacement reaction. Distillation of the material with MnO_2 and H_2SO_4 produced more of the red vapours, which condensed to a dark liquid. Incidentally Liebig appears to have made bromine before Balard but he wrongly assigned it as iodine monochloride, another dark brown liquid compound. The French Academy of Sciences suggested the name bromine, from the Greek word *bromos*, meaning 'bad smell'.

14.2.3 Iodine

In 1811, Bernard Courtois of Paris prepared saltpetre (potassium nitrate, KNO_3) and heated ashes of seaweeds as his source of potassium. He added sulfuric acid to the residue and saw the evolution of purple fumes which condensed to form crystals with a metallic lustre. Courtois deduced that this was a new element and called it substance X. He gave some of it to fellow French scientists who carried out a systematic investigation and confirmed that it was a new element. In November 1813, they exhibited iodine at the Imperial Institute in Paris. In 1813, Sir Humphrey Davy who was passing through Paris on his way to Italy saw this substance and felt its properties are similar to that of chlorine. Davy sent a report to the Royal Institution in London and even suggested the name iodine from the Greek word *ioeides* meaning violet coloured. Therefore, in the early days it was mistakenly assumed that Sir Humphrey was the discoverer of iodine.

14.3 LABORATORY SYNTHESES

14.3.1 Laboratory Synthesis of Chlorine

There are many methods of synthesising chlorine gas in the laboratory and almost all of them use hydrochloric acid. A reaction of HCl with $KMnO_4$, MnO_2, PbO_2, Pb_3O_4, O_2, NaClO(*aq*) readily gives Cl_2 but often with water vapour. One of the efficient methods is by the reaction of trichloroisocyanuric acid (TCCA or TCICA) with hydrochloric acid, the former being a highly stable, free flowing and high melting white powder.[6, 7]

$$\text{TCCA} \xrightarrow{3\ HCl} \text{cyanuric acid} + 3\ Cl_2 \qquad (14.1)$$

14.3.2 Laboratory and Industrial Synthesis of Bromine

Bromine can be prepared in the laboratory by heating potassium bromide, KBr, with dilute sulphuric acid and manganese dioxide, MnO_2 according to Eq. 14.2.[8]

$$2\ KBr + MnO_2 + 3\ H_2SO_4 \longrightarrow Br_2 + 2\ KHSO_4 + MnSO_4 + 2\ H_2O \qquad (14.2)$$

It is industrially manufactured from sea water by allowing a solution of KBr to flow down a tower against a stream of chlorine gas going upstream in the presence of sulfuric acid.

$$2\ KBr + Cl_2 \longrightarrow 2\ KCl + Br_2 \qquad (14.3)$$

What results from this reaction is a solution of bromine which is dilute, from which Br_2 is removed by blowing air. The free bromine is then mixed with SO_2 gas, and the mixture is passed up a tower, in which water is trickling down. The following reaction takes place.

$$Br_2 + SO_2 + 2\ H_2O \longrightarrow 2\ HBr + H_2SO_4 \qquad (14.4)$$

The resulting mixture of acids is much richer in bromide ion than seawater. This mixture is treated with chlorine again which liberates Br_2, which is freed from chlorine and purified by passing it over moist filings of iron and dried by conc. H_2SO_4.

14.3.3 Laboratory Synthesis of Iodine

Although I_2 can be prepared by a procedure similar to that of bromine using sodium iodide, MnO_2 and H_2SO_4, or by halogen displacement reaction, a better method is the modification of the industrial process which employs the reduction of sodium iodate with sodium acid sulfite. Taking a mixture of sodium iodate with sodium sulfite and adding dilute sulfuric acid results in I_2 collecting at the bottom of the flask. It is necessary to use proper quantities in this preparation, since an excess of sodium sulfite will dissolve the iodine formed. These synthesis steps also form part of the well known experiment, *iodine clock reaction*.[9]

$$Na_2S_2O_5 + H_2O \longrightarrow 2\,NaHSO_3 \quad (14.5)$$

$$IO_3^- + 3\,HSO_3^- \longrightarrow I^- + 3\,HSO_4^- \quad (14.6)$$

$$IO_3^- + 5\,I^- + 6H^+ \longrightarrow 3\,I_2 + 3\,H_2O \quad (14.7)$$

14.4 Industrial Synthesis of Chlorine

14.4.1 The Castner-Kellner Process

The industrial synthesis of chlorine utilises the chloralkali process which is basically electrolysis of aqueous NaCl or brine. Although electrolysis of brine was done first in 1851 by Charles Watt there were problems in the separation of the products of electrolysis. Hamilton Castner, an American, invented the mercury cell and was granted a US patent in 1892. Almost simultaneously an Austrian chemist, Karl Kellner discovered the same process. In

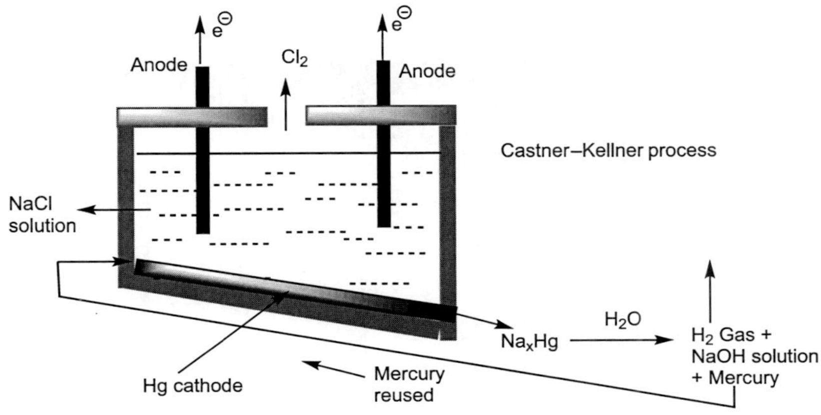

Fig. 14.1 Schematic diagram for Castner–Kellner process

order to avoid a legal fight, they became partners in 1895, founding the Castner–Kellner alkali company, which built plants employing the process throughout Europe. The mercury cell process continues to be in use in some parts of the world. Current-day mercury cell plant operations are disapproved due to the release of mercury into the environment, leading in some cases to high level of mercury poisoning like the Minamata bay incident in Japan in 1956. Due to these concerns, mercury cell plants are being phased out, and a sustained effort is being made to reduce mercury emissions from existing plants.

The electrolytic cell which produces chlorine and sodium hydroxide consists of graphite as anode and is filled with brine solution (28% NaCl). The iron cell employs liquid mercury as the cathode. The sodium liberated there forms an amalgam with the mercury, which is made to run off due to the slanting construction of the cell to another chamber and reacted with water to give sodium hydroxide (and hydrogen). The mercury is then re-used. Chlorine gas is produced at the anode. The process has high energy consumption. This process gives equivalent amounts of chlorine and sodium hydroxide. Also, for every mole of chlorine gas produced, one mole of hydrogen gas is also produced and this hydrogen is used to manufacture NH_3 and hydrochloric acid.[1]

14.4.2 The Membrane Process

The ion-permeable ion exchange membrane at the centre of the cell allows the sodium ions (Na^+) to pass to the second chamber where they react with the hydroxide ions to produce NaOH. A membrane cell is used to prevent the reaction between the chlorine and hydroxide ions. If this reaction were to occur, the chlorine would disproportionate to form chloride and hypochorite ions. Above 60°C, from the same reaction chlorate, $NaClO_3$, can also be formed.

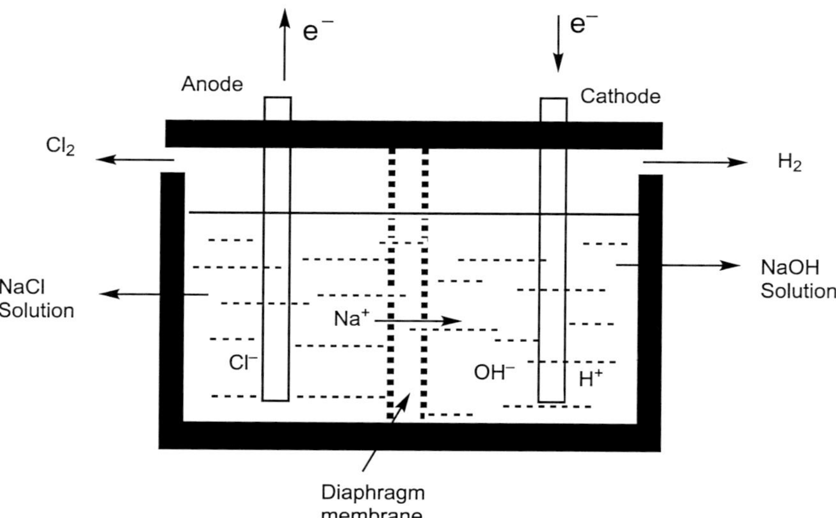

Fig. 14.2 Schematic diagram for membrane process of chlorine synthesis

14.5 Displacement Reactions of Halogens

One of the unique reactions of halogens is the displacement reaction in which a more reactive halogen will displace a less reactive halogen from its salt dissolved in water. In practice these reactions are carried out for chlorine, bromine and iodine as fluorine gas is too reactive to handle and reacts with water explosively. Selective evolution of Br_2 from a mixture of NaCl and NaBr by passing chlorine is one of the industrial uses of this reaction.

$$Cl_2 + 2\,NaBr \xrightarrow[H_2O]{} 2\,NaCl + Br_2 \qquad (14.8)$$

$$Cl_2 + 2\,NaI \xrightarrow[H_2O]{} 2\,NaCl + I_2 \qquad (14.9)$$

$$Br_2 + 2\,NaI \xrightarrow[H_2O]{} 2\,NaBr + I_2 \qquad (14.10)$$

14.6 Reactions of Chlorine

Chlorine gas undergoes a range of reactions such as oxidative chlorination, free radical reactions, addition reactions across double bonds and halogen exchange. Under a UV radiation of 470 nm or lower (calculated: 492 nm) it undergoes homolytic cleavage resulting in chlorine free radicals (Scheme 14.1). This property has been utilised in the synthesis of chloromethanes, HCl and a range of other organochlorine compounds. Its reaction with water leads to HOCl and HCl and the former also acts as an oxidant and bleach. Formation of sulfuryl chloride and phosgene are examples of oxidative addition. Scheme 14.2 shows a range of different types of reactions shown by Cl_2 gas.[1]

$$Cl_2 \xrightarrow[\text{470 nm or shorter}]{\text{UV}} 2\,Cl\cdot \qquad \text{Initiation}$$

$$CH_4 + Cl\cdot \longrightarrow CH_3\cdot + HCl$$
$$CH_3\cdot + Cl_2 \longrightarrow CH_3Cl + Cl\cdot \qquad \text{Propagation}$$

$$CH_3\cdot + Cl\cdot \longrightarrow CH_3Cl$$
$$Cl\cdot + Cl\cdot \longrightarrow Cl_2 \qquad \text{Termination}$$
$$CH_3\cdot + CH_3\cdot \longrightarrow C_2H_6$$

Scheme 14.1 Free radical reactions of chlorine leading to chloromethane

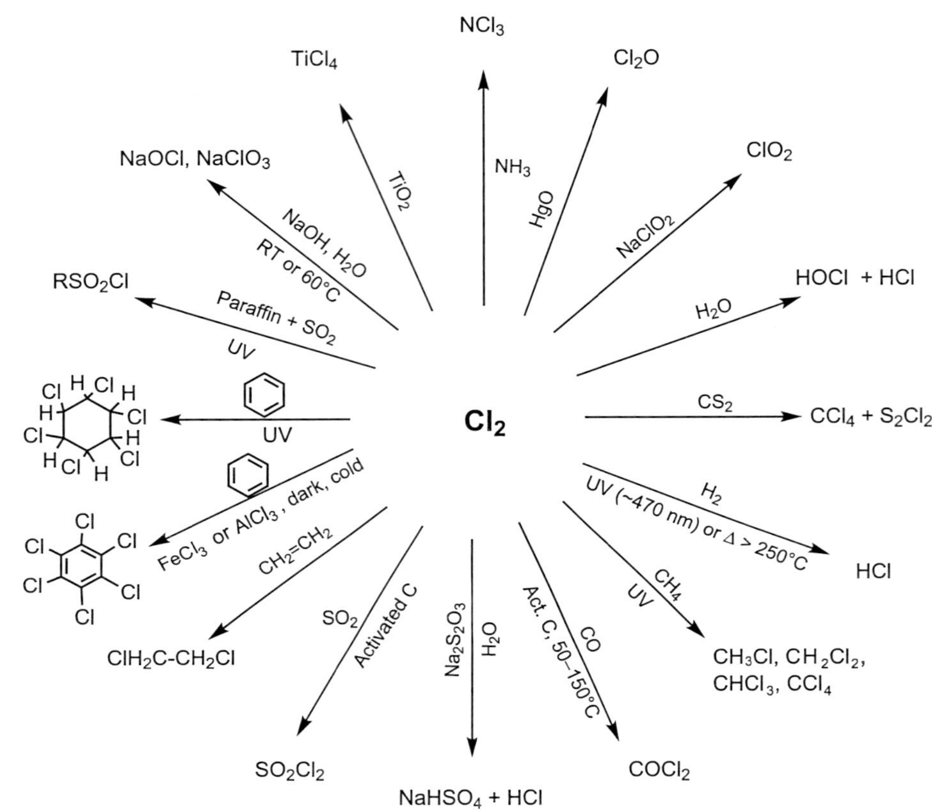

Scheme 14.2 General reactions of chlorine

14.6.1 Industrially Important Reactions of Chlorine

Chlorine is the most industrially important halogen and it is utilised in the processing and formation of a range of products currently useful to humanity. Chlorine and some of its products are also well known water purification agents and disinfectants. The processing of many industrially important polymers such as silicones, epoxy resins, alkyl sulfonic acids and polyurethanes utilise chlorinated precursors while polyvinyl chloride (PVC), polyvinylidene chloride (PVDC) and neoprene (synthetic rubber) have chlorine as one of the substituents on the polymer chain. Scheme 14.3 indicates in detail the role of chlorine in industrially significant chemical processes.[1] Scheme 14.4 shows how partial chlorination of ethylene is carried out leading to a range of industrially useful products.

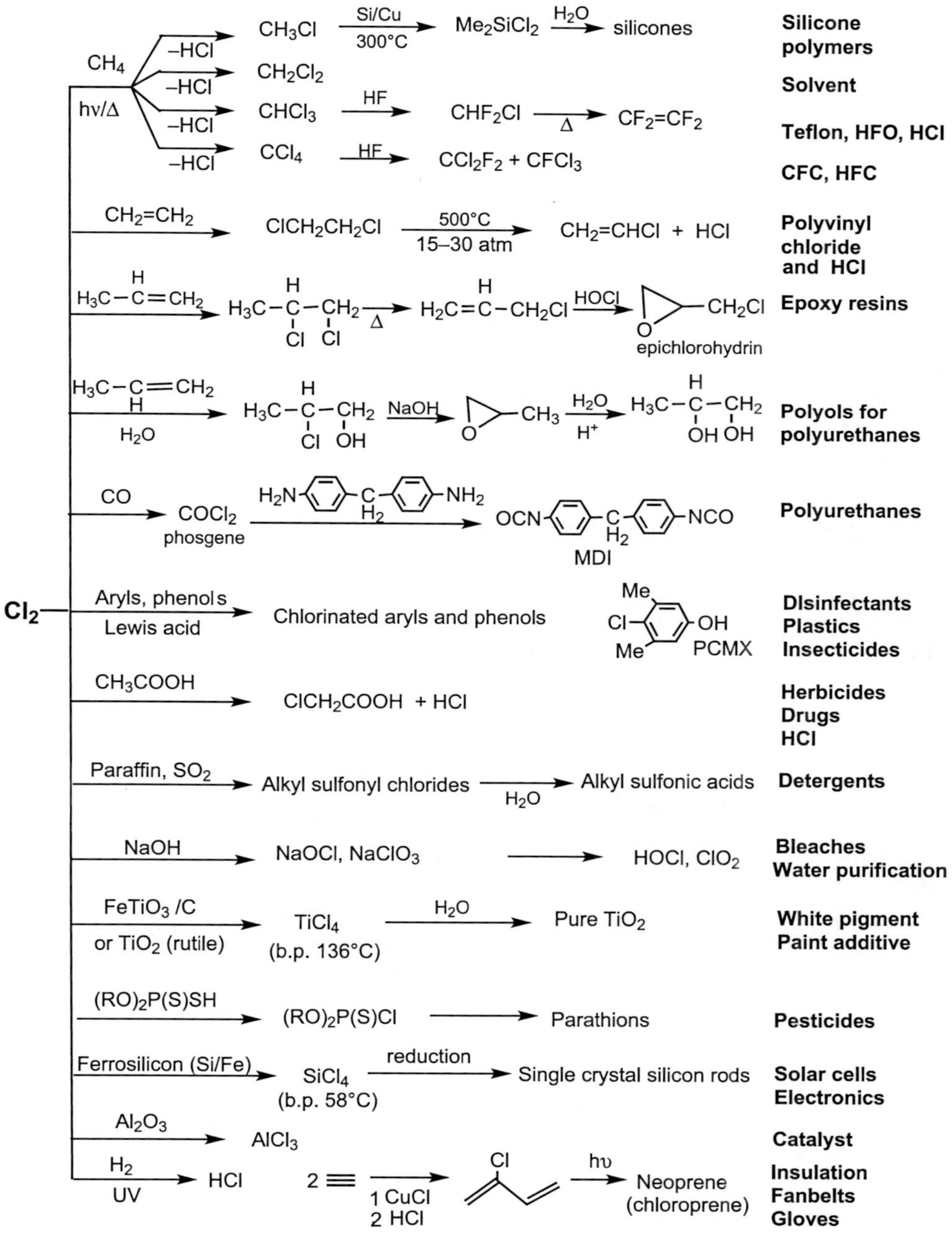

Cl_2 —

- CH_4, hν/Δ → CH_3Cl (−HCl) $\xrightarrow[300°C]{Si/Cu}$ Me_2SiCl_2 $\xrightarrow{H_2O}$ silicones — **Silicone polymers**
- → CH_2Cl_2 (−HCl) — **Solvent**
- → $CHCl_3$ (−HCl) $\xrightarrow{HF}$ CHF_2Cl $\xrightarrow{\Delta}$ $CF_2{=}CF_2$ — **Teflon, HFO, HCl**
- → CCl_4 (−HCl) $\xrightarrow{HF}$ $CCl_2F_2 + CFCl_3$ — **CFC, HFC**
- $CH_2{=}CH_2$ → $ClCH_2CH_2Cl$ $\xrightarrow[15–30\ atm]{500°C}$ $CH_2{=}CHCl + HCl$ — **Polyvinyl chloride and HCl**
- $H_3C{-}CH{=}CH_2$ → $H_3C{-}CH(Cl){-}CH_2Cl$ $\xrightarrow{\Delta}$ $H_2C{=}CH{-}CH_2Cl$ $\xrightarrow{HOCl}$ (epoxide)$-CH_2Cl$ epichlorohydrin — **Epoxy resins**
- $H_3C{-}CH{=}CH_2$, H_2O → $H_3C{-}CH(Cl){-}CH_2OH$ $\xrightarrow{NaOH}$ (epoxide)$-CH_3$ $\xrightarrow[H^+]{H_2O}$ $H_3C{-}CH(OH){-}CH_2OH$ — **Polyols for polyurethanes**
- CO → $COCl_2$ phosgene $\xrightarrow{H_2N-C_6H_4-CH_2-C_6H_4-NH_2}$ $OCN-C_6H_4-CH_2-C_6H_4-NCO$ MDI — **Polyurethanes**
- Aryls, phenols, Lewis acid → Chlorinated aryls and phenols; Me, Cl, OH, Me-substituted phenol PCMX — **Disinfectants, Plastics, Insecticides**
- CH_3COOH → $ClCH_2COOH$ + HCl — **Herbicides, Drugs, HCl**
- Paraffin, SO_2 → Alkyl sulfonyl chlorides $\xrightarrow{H_2O}$ Alkyl sulfonic acids — **Detergents**
- NaOH → NaOCl, $NaClO_3$ → HOCl, ClO_2 — **Bleaches, Water purification**
- $FeTiO_3$ /C or TiO_2 (rutile) → $TiCl_4$ (b.p. 136°C) $\xrightarrow{H_2O}$ Pure TiO_2 — **White pigment, Paint additive**
- $(RO)_2P(S)SH$ → $(RO)_2P(S)Cl$ → Parathions — **Pesticides**
- Ferrosilicon (Si/Fe) → $SiCl_4$ (b.p. 58°C) $\xrightarrow{reduction}$ Single crystal silicon rods — **Solar cells, Electronics**
- Al_2O_3 → $AlCl_3$ — **Catalyst**
- H_2, UV → HCl; 2 HC≡CH $\xrightarrow[2\ HCl]{1\ CuCl}$ $CH_2{=}C(Cl){-}CH{=}CH_2$ $\xrightarrow{h\nu}$ Neoprene (chloroprene) — **Insulation, Fanbelts, Gloves**

Scheme 14.3 Industrially important reactions of chlorine

$H_2C{=}CH_2$ —(1. Cl_2; 2. Δ, –HCl)→ $H_2C{=}CHCl$ (→ PVC) —(1. Cl_2; 2. Δ, –HCl)→ $H_2C{=}CCl_2$ (→ PVDC) —(1. Cl_2; 2. Δ, –HCl)→ $ClHC{=}CCl_2$ —(1. Cl_2; 2. Δ, –HCl)→ $Cl_2C{=}CCl_2$

$ClHC{=}CCl_2$ —(3 HF, –3 HCl)→ CF_3CH_2Cl —(HF, SbF_5)→ CF_3CH_2F (HFC 134a refrigerant)

$Cl_2C{=}CCl_2$ —(3 HF, –3 HCl)→ CF_3CHCl_2 —(O_2, Cl_2, hν)→ $CF_3C(O)Cl$

$CF_3C(O)Cl$ —(H_2O)→ $CF_3C(O)OH$

$CF_3C(O)Cl$ —(Reduction)→ CF_3CH_2OH

Scheme 14.4 Synthesis of partially chlorinated ethylenes and their industrially useful products

14.7 Oxides and Oxyacids of Halogens

Among halogens, chlorine forms the maximum number of oxides. Fluorine forms only two oxides where its oxidation state is –1 while iodine shows the most stable oxide, namely I_2O_5, which is stable till 300°C.[10]

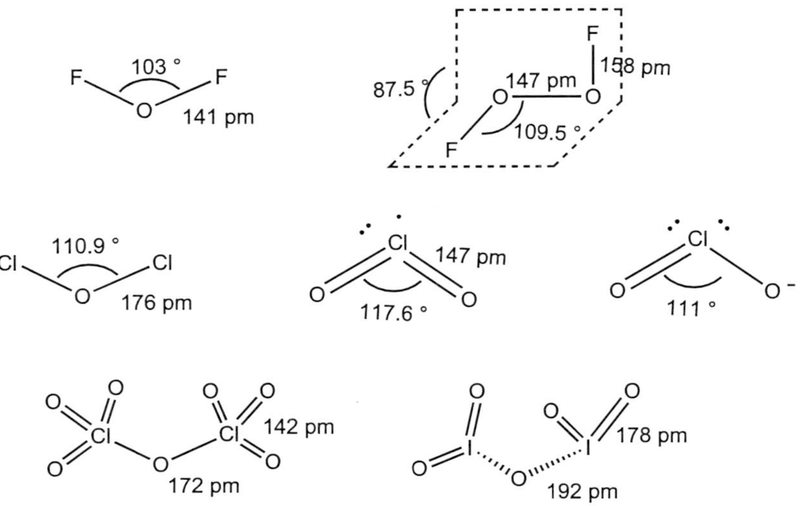

Fig. 14.3 Shapes of well-known halogen oxides

Table 14.4 Oxides of halogens, their colours and decomposition temperatures

Oxide type	Fluorine	Chlorine	Bromine	Iodine
E_2O	F_2O Colourless gas (200°C)	Cl_2O Yellow brown gas (60°C)	Br_2O Brown liquid (–40°C)	–
E_2O_2	F_2O_2 Orange red (–95°C)	–	–	–
E_2O_3	–	Cl_2O_3 Dark brown solid (0°C)	Br_2O_3 Orange (–40°C)	–
EO_2	–	ClO_2 Red-yellow liquid (45°C)	BrO_2 Yellow (~0°C)	I_2O_4 Yellow (85°C)
E_2O_5	–	–	–	I_2O_5 Colourless solid (300°C)
E_2O_6	–	Cl_2O_6 $(ClO_2^+ClO_4^-)$ Red liquid	–	–
E_2O_7	–	Cl_2O_7 Colourless liquid (85°C)	–	–

(decomposition temperatures in parenthesis)

14.7.1 Chlorine Dioxide

Chlorine dioxide (ClO_2), a compound of chlorine having an unpaired electron and bent shape is fast replacing chlorine as a selective oxidising agent for many industrial and household uses. ClO_2 is a low boiling yellow to red coloured liquid (b.p. 11°C) and is obtained by the reaction of $NaClO_2$ with chlorine or the electrolysis of brine at higher temperatures. Since ClO_2 is difficult to handle and is also an explosive, for all practical purposes it is generated from sodium chlorite, $NaClO_2$. Sodium chlorite worth over 18 million USD is produced per year. Sodium chlorite is indirectly obtained from sodium chlorate, $NaClO_3$. $NaClO_3$ can be produced by the disproportionation of sodium hypochlorite NaClO or by the high temperature electrolysis of brine. ClO_2 is prepared by oxidation of $NaClO_2$.[10]

$$Cl_2 + NaOH \longrightarrow NaClO + NaCl + H_2O \quad (14.11)$$

$$3\,NaClO \xrightarrow{-2\,NaCl} NaClO_3 \;(+5);\quad NaCl + H_2O \text{ (hot solution)} \xrightarrow[-H_2]{\text{Electrolysis}} NaClO_3$$
$$NaClO_3 \;(+5) \xrightarrow[\text{strong acid soln}]{\text{reduction}} ClO_2 \;(+4) \xrightarrow[\text{reduction}]{NaOH} NaClO_2 \;(+3) \qquad (14.12)$$

$$2\,NaClO_2 + Cl_2 \longrightarrow ClO_2 + 2\,NaCl \qquad (14.13)$$

For obtaining chlorine-free ClO_2 from $NaClO_3$, methanol or H_2O_2 can also be used as the reducing agent.

$$6\,NaClO_3 + CH_3OH + 4\,H_2SO_4 \longrightarrow 6\,ClO_2 + CO_2 + 2\,Na_3H(SO_4)_2 + 5\,H_2O \qquad (14.14)$$

Chlorine dioxide has been termed as a chlorine-free bleaching agent. It is quite soluble in water (8 g/L) which is ten times more than the solubility of Cl_2 in water. Unlike Cl_2, it does not react with water. For bleaching wood pulp it has been preferred over Cl_2 as it is a more selective oxidiser and less harmful side products are formed when it reacts. It does not chlorinate but only oxidises and is more selective than ozone.

Table 14.5 compares Cl_2 and ClO_2 in reactivity with common compounds encountered while purifying water. The standard reduction potential of ClO_2 is 0.95 V while that of Cl_2 is 1.36 V, HOCl is 1.49 V and ozone 2.07 V.

Table 14.5 Comparison of chemical reactions of Cl_2 with ClO_2

Compound	Reaction with Cl_2	Reaction with ClO_2
Water	HClO and HCl	No reaction
CH_3CHO	Cl_3CCHO	$H_3CC(O)OH$
NH_3	NH_2Cl, $NHCl_2$, NCl_3	No reaction
R_3N	R_3NCl^+	Oxidative dealkylation $R_2NH_2^+$, RCHO, $HClO_2$
Cyanide	$(ClCN)_3$, Cyanuric chloride	CNO^- (below pH 10) or $CO_2 + N_2$ (above pH 10)
H_2S	Colloidal sulfur	SO_4^{2-}
RSH	RSO_2Cl or RSSR	Sulfonic acid and sulfonates
Phenol	Trichlorophenol	Benzoquinone

14.7.2 Dichlorine Heptoxide

Dichlorine heptoxide is obtained when perchloric acid or magnesium perchlorate is dehydrated in the presence of P_2O_5. It is also reported to be formed when perchloric acid is treated with oleum, $H_2S_2O_7$.[11]

14.7.3 Halogen Oxyacids

Table 14.6 Examples of halogen oxyacids

Oxidation state of the halogen	Chlorine	Bromine	Iodine	Name of the acid
+1	HClO	HBrO	HIO	Hypohalous
+3	$HClO_2$	–	–	Halous
+5	$HClO_3$	$HBrO_3$	HIO_3	Halic
+7	$HClO_4$	$HBrO_4$	HIO_4, H_5IO_6	Perhalic

The strength of the oxyacids is related to the relative stability of the counter-ion and the one with more possible resonance structures will be the more stable oxyacid. Resonance structures possible for oxyacids of chlorine are given in Fig. 14.4.

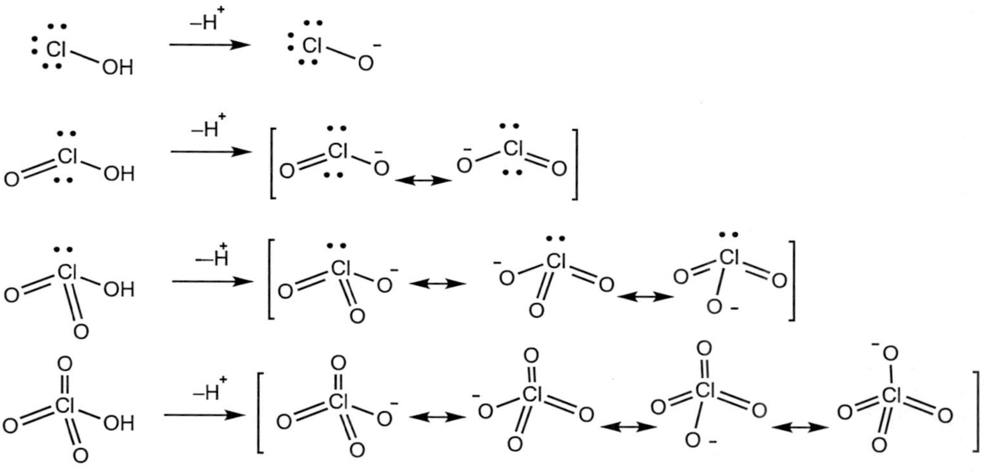

Fig. 14.4 Resonance structures of different oxyacids of chlorine

			E^0
$Cl_2(g) + 2e$	$\longrightarrow$	$2Cl^-$	1.36 V
$Br_2(l) + 2e$	$\longrightarrow$	$2Br^-$	1.09 V
$I_2(s) + 2e$	$\longrightarrow$	$2I^-$	0.54 V
$2ClO_3^- + 12H^+ + 10e$	$\longrightarrow$	$Cl_2(g) + 6H_2O$	1.49 V
$2BrO_3^- + 12H^+ + 10e$	$\longrightarrow$	$Br_2(l) + 6H_2O$	1.48 V
$2IO_3^- + 12H^+ + 10e$	$\longrightarrow$	$I_2(s) + 6H_2O$	1.20 V

Fig. 14.5 Standard reduction potentials of halogens and halates

14.8 Electrophilic Chlorinating Agents

Replacing hydrogen from an aryl or alkyl unit by chlorine requires the use of electrophilic chlorinating reagents. Such halogenations are useful in modifying the electronic structure as well as acting as points for diversification and derivatisation by reactions such as cross coupling. The presence of such C–Cl units provide molecules with more reactivity and the ability functionalise.

Traditional C–H chlorination by using chlorinating reagents such as Cl_2 or SO_2Cl_2 is the oldest method to prepare aromatic chlorides and can be considered as reactive and practical types of methods. Reactive methods generally have safety liabilities that limit their application. Mild methods use significantly less reactive reagents such as [N-chlorosuccinimide (NCS) and 1,3-dichloro-5,5-dimethylhydantoin (DCDMH)]. While these reagents are inexpensive and easy to handle, their reduced reactivity often results in limited application. While *t*-BuOCl is a valuable chlorinating reagent operating with a radical-based mechanism, its reactivity is variable on arenes and it is dangerously sensitive to light, heat and moisture. Recently a new chlorinating agent has been made, chloro(bismethoxycarbonyl)guanidine (CBMG), which is comparable in both cost and molecular weight to the classic reagent NCS yet exhibits a reactivity profile that is equivalent to or better than those of more aggressive reagents such as SO_2Cl_2, Cl_2 and *t*-BuOCl. It also shows functional group tolerance. Table 14.7 compares some of the well known electrophilic chlorinating agents.[12]

The synthesis and few typical reactions of these electrophilic chlorinating agents are given below.

$$H_3C\text{-}C_6H_4\text{-}SO_2NH_2 \xrightarrow{NaOCl} Na^{+}[H_3C\text{-}C_6H_4\text{-}SO_2\text{-}\overset{-}{N}Cl] \quad \text{(Chloramine-T)} \qquad (14.15)$$

$$ClCH_2CH_2SCH_2CH_2Cl \xrightarrow{\text{Chloramine-T}} ClCH_2CH_2S(=NTs)CH_2CH_2Cl \qquad (14.16)$$

$$Se \xrightarrow{\text{Chloramine-T}} TsN{=}Se{=}NTs \qquad (14.17)$$

Although chloramine-T has electrophilic chlorine present, it rarely carries out chlorination and often it oxidises the substrate resulting in a tosylimine.

Table 14.7 Different examples of electrophilic chlorinating agents

Name/ abbreviation	Structure	Physical constants (ΥC)	Advantages/ disadvantages	Relative reactivity
Sulfuryl chloride	SO_2Cl_2	b.p. 69.4°C	Highly reactive, highly air sensitive, source of chlorine, oxidiser	High
Chlorine gas	Cl_2	b.p. – 34°C	Highly reactive, poor selectivity	High
Chloramine-T	$\overset{+}{Na}$ [H3C–C6H4–S(=O)(=O)–$\overset{-}{N}$Cl]	m.p. 169°C	Good as oxidant, although electrophilic, products rarely have chlorine	Low
Trichloro isocyanuric acid [TCCA or TCICA]	Cl–N, N–Cl, N–Cl triazinetrione (O, O, O)	m.p. 234°C	Inexpensive, stable, can generate explosive NCl_3	Low
Dichloro dimethyl hydantoin [DCDMH or NDDH]	1,3-dichloro-5,5-dimethylhydantoin (Cl, N, O, N, Cl)	m.p. 132°C	Inexpensive, easily degraded after chlorination, reacts violently with xylene	Low
N-Chloro succinimide [NCS]	O, N–Cl, O (succinimide ring)	m.p. 151°C	Expensive, high toxicology, reacts violently with alcohols	Low
Tert butyl hypochlorite [*t*-BuOCl]	(CH3)3C–OCl	b.p. 79.6°C /750 mm Hg	Decomposes during transportation; light, heat and moisture sensitive	High
Chloro (bismethoxycarbonyl) guanidine [Palau' chlor, CBMG]	MeO–C(=O)–NH–C(=N–Cl)–NH–C(=O)–OMe	m.p. 110°C	Air and bench stable, free-flowing powder, functional group tolerant	Low

14.8.1 Trichloroisocyanuric Acid[12]

$$OC(NH_2)_2 \xrightarrow[\Delta,\ -NH_3]{175°C} \text{Cyanuric acid} \rightleftharpoons \text{Isocyanuric acid} \xrightarrow{Cl_2,\ NaOH} \text{Trichloroisocyanuric acid (TCCA or TCICA)} \quad (14.18)$$

Cyanuric acid   Isocyanuric acid   Trichloroisocyanuric acid (TCCA or TCICA)

$$\text{2-Methylquinoxaline} \xrightarrow[CHCl_3,\ reflux]{TCICA/DMF} \text{2-}(CH_2Cl)\text{quinoxaline} \quad 75\% \quad (14.19)$$

$$\text{Benzene} \xrightarrow[65\text{–}80°C]{TCICA/H_2SO_4} \text{Chlorobenzene} \quad 80\% \quad (14.20)$$

14.8.2 Dichlorodimethylhydantoin (DCDMH or DDH)

$$HCN + (CH_3)_2C{=}O + (NH_4)_2CO_3 \xrightarrow{68°C} \text{Dimethyl hydantoin} \xrightarrow{Cl_2/Na_2CO_3} \text{DCDMH or DDH} \quad (14.21)$$

Dimethyl hydantoin   DCDMH or DDH

$$C_6H_5COCH_3 \xrightarrow[SiO_2]{Reagent} C_6H_5COCH_2Cl \quad 14.22)$$

Reagent		
	DCDMH	95%
	NCS	70%

14.8.3 N-Chlorosuccinamide

$$\text{Succinimide (N–H)} \xrightarrow{Cl_2/NaOH/H_2O} \text{N–Cl (NCS)} \quad (14.23)$$

NCS

CN + N–Cl (N-chlorosuccinimide) —Pd(OAc)$_2$, PTSA, DCE→ 2-chlorobenzonitrile (CN, Cl) 90 % (14.24)

14.8.4 Tert-butyl Hypochlorite

tBuOH —NaOCl, AcOH→ tBuOCl (14.25)

liquid b.p. 79.6°C/750 mm Hg

Toluene —tBuOCl, SiO$_2$→ 2-chlorotoluene (Cl) + 4-chlorotoluene (Cl) 65:35 (14.26)

Oxirane —tBuOCl→ chlorooxirane (Cl) 95% (14.27)

14.8.5 Chloro(bismethoxycarbonyl) Guanidine (Palau'chlor)[13]

MeO(C=O)NH–C(=N–H)–NH(C=O)OMe —t-BuOCl→ MeO(C=O)NH–C(=N–Cl)–NH(C=O)OMe (14.28)

(CBMG)
Chloro(bismethoxycarbonyl) guanidine

1-[(2-chlorophenyl)diphenylmethyl]imidazole (N, N–C–Ph, Ph, Cl) —Reagent→ 4-chloro derivative (Cl, N, N–C–Ph, Ph, Cl) (14.29)

Reagent	Yield
NCS	0%
$CuCl_2$	0%
SO_2Cl_2	31%
Cl_2	41%
TCICA	43%
DCDMH	51%
tBuOCl	80%
CBMG	87%

N-methylindole-2-carbaldehyde (C(=O)H, N, Me) → 3-chloro-N-methylindole-2-carbaldehyde (Cl, C(=O)H, N, Me) (14.30)

CBMG 70%
NCS 15%

$$\text{1-phenylpyrazole} \xrightarrow[\text{NCS} \quad 0\%]{\text{CBMG} \quad 95\%} \text{4-chloro-1-phenylpyrazole} \quad (14.31)$$

14.9 Chlorine-Containing Disinfectants and Antibacterials

14.9.1 Chlorine and Hypochlorite

The disinfectant property of chlorine water and calcium hypochlorite (as part of bleaching powder) was known from the time of Labarraque who developed (in 1828) a method to prevent the putrefaction and degradation of catgut (used as strings for musical instruments and also in surgery as surgical sutures) prepared from sheep intestines. The success of Semmeleweis in arresting 'child bed fever', an unknown infection which was killing young mothers immediately after childbirth in the 'Vienna Lying-in Hospital', by making the doctors and medical students wash their hands in chlorine water, is recorded in history as a very significant development in the use of chlorine as a disinfectant.

Chlorine gas when dissolved in water gives HCl and HClO. HClO cannot be isolated as it is in equilibrium with Cl_2. However, sodium hypochlorite (m.p. 18°C; household bleach) and calcium hypochlorite (m.p. 100 °C; component of bleaching powder) are stable. The disinfectant action has been attributed to the oxidising ability of HClO. The half-cell oxidation–reduction potential, E_0, for HOCl at 25°C is 1.49.

$$Cl_2 + H_2O \rightleftharpoons HOCl + HCl \quad (14.32)$$

$$HOCl + H^+ + 2e \longrightarrow Cl^- + H_2O \qquad E_0 = 1.49\ V \quad (14.33)$$

During the period 1918 to 1944, the prevalent theory and belief was that HClO decomposes to give HCl and a monoatomic radical species termed nascent oxygen. It was believed that the nascent oxygen [O] was responsible for the inactivation of microbes and not chlorine or HClO. This theory was not supported by any experimental evidence and was widely accepted among scientists of that time. A 1944 publication by S L Chang finally put the controversy to rest. Chang attempted inactivation of *Entamoeba histolytica* cysts with high concentrations of H_2O_2 (3000 mg/l) which is expected to have 1500 times more nascent oxygen than HOCl. He observed that there was no inactivation or destruction of the cysts with H_2O_2 while chlorine water inactivated the cysts. Chang reported in his paper 'The action of chlorine and chloramine compounds on cysts was attributed to the active chlorine which may oxidise or chlorinate the proteins in the protoplasm. The possibility of action by nascent oxygen liberated by HOCl was indirectly studied, and the evidence strongly indicated that this was unlikely to occur'. Since Chang's publication, nascent oxygen has not been mentioned in professional publications and more detailed studies have indicated the direct role of chlorine in the inactivation of microbes.[14]

Labarraque and Semmelweiss: The pioneers who promoted use of chlorine for disinfection

Much before Louis Pasteur came out with the germ theory of diseases, there was a pharmacist and also a medical practitioner who advocated the use of chlorine (as chlorinated lime and chlorine water) to prevent putrefaction of flesh and spreading of diseases. The first was a French chemist and pharmacist, Antoine-Germain Labarraque who in 1828 developed a method to separate the peretonial membrane of animal intestines without putrefaction. Animal intestines were in great demand those days for making strings called 'catgut' of musical instruments such as violin and harp (the steel musical strings came only after 1900) and also for making goldbeaters skin with which very thin (1 μm) gold leaves were made. Catgut also finds use as surgical sutures. The secret of Labarraque was the use of sodium and calcium hypochlorite solutions which killed the germs responsible for putrefaction although at that time humanity was not even aware of germs. Labarraque also advised doctors to use his solutions to prevent putrefaction of wounds, breathe chlorine gas and also wash their hands with bleaching powder solution, which was grossly ignored by the doctors of that time.

The second person was a Hungarian born medical practitioner Ignaz Semmelweis who is now known as the pioneer of antiseptic procedures and described as the 'saviour of mothers'. Semmelweis, in 1847, was working in the maternity department of the Vienna Lying-in Hospital, Austria. There were two obstetrical clinics dealing with childbirth in this hospital. One was manned by midwives and student midwives and the other was managed by medical students and doctors. The surprising observation Semmelweis made was that death of mothers due to puerperal fever ('childbed fever', a common reason for death of mothers during childbirth) was three times more in the clinic manned by doctors and medical students than the clinic manned by midwives! He even noticed that even women who gave birth at homes or streets were less affected by childbed fever. So the medical students and doctors were the culprits! His keen observation led to the finding that the medical students were also practicing their surgical skills on cadavers/dead bodies in the autopsy room while attending to women giving childbirth. He concluded that the medical students carried 'cadaverous particles' on their hands from the autopsy room to the patients they examined in the first obstetrical clinic. This explained why the student midwives in the second clinic, who were not engaged in autopsies and had no contact with corpses, saw a much lower mortality rate.

Semmelweis was aware of the work of Labarraque on chlorinated lime. He therefore enforced a practice of using a solution of chlorinated lime (calcium hypochlorite solution) for washing hands between autopsy work and the examination of patients. He tried this because with his keen observation he noticed that the chlorinated solution worked best to remove the decaying smell of infected autopsy tissue, and hoped that perhaps it can destroy the causal contaminating agent being transmitted from the cadavers. This practice immediately reduced the incidence of fatal puerperal fever in the first clinic manned by medical students from about 10% to about 1–2% and finally to 0%. Semmelweis, confident with the success of his method started advocating the use of chlorinated lime or chlorine water to prevent childbed fever and death.

The medical community was unkind to Semmelweis. Diseases were attributed to many different causes at that time and his hypothesis, that cleanliness can simply get rid of diseases, was considered extreme at that period of time, and was ignored, rejected and ridiculed by doctors. He was dismissed from the hospital and was harassed by the medical community in Vienna, and was therefore forced to move to Budapest, Hungary. Semmelweis was angry and upset by the indifference shown by the medical community and even wrote open angry letters to well known European obstetricians, at times even calling them irresponsible murderers. His contemporaries thought he was losing his mind, and in 1865, he was put in a mental asylum by his own colleagues. After two weeks, he died there of blood poisoning at the age of 47, possibly due to being severely beaten by guards as he tried to escape.

Semmelweis' beliefs earned widespread acceptance only years after his death, when Robert Koch and Louis Pasteur developed the germ theory of disease, offering a theoretical explanation for his findings. He is now considered the pioneer of antiseptic procedures. Semmelweis reflex, a metaphor used to indicate a human behaviour type characterised by reflex-like rejection of new ideas and knowledge if it contradicts well-established norms and beliefs, is named after Ignaz Semmelweis, whose ideas were sadly rejected and ridiculed by his contemporary medical practitioners.

HOCl has been found to readily react with DNA and RNA, unsaturated bonds of lipids and sulfhydryl and amino group side chains of proteins. Based on experimental evidence, the mode of inactivation and destruction of microbes by chlorine and hypochlorite can be attributed to multiple factors. These are (i) interference of cell assisted membrane functions: permeability of the bacterial cell membrane getting affected by chlorination of the membrane resulting in leakage of critical cellular components, (ii) modification of the ionic environment of the cell leading to enzymes losing their activity, and (iii) irreversibly affecting the enzyme and protein functions of the microbes especially by irreversible oxidation of sulfhydryl (SH) groups of the proteins.[15]

14.9.2 Dettol

Many of the widely used disinfectants and antibacterial soaps and detergents owe their germ killing properties to organic compounds containing chlorine. Unlike in bleaching powder or hypochlorous acid, the role of the chlorine is different in these compounds and often is used to enhance the germ killing properties of a phenol or a similar compound. The most well known among such compounds is para-chloro-meta-xylenol (PCMX) which is the active germ killing ingredient in the disinfectant Dettol. PCMX is present in 4.8% along with pine oil (α-terpineol), isoprene and ricinoleic acid in Dettol. The relative germ killing property of PCMX with phenol and trichlorophenol is shown below.[16]

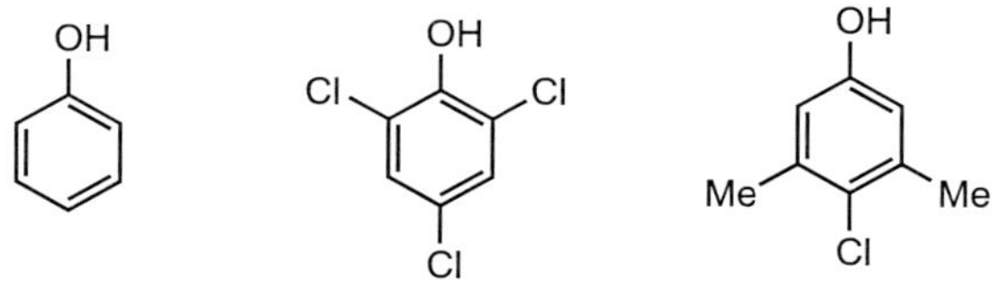

OH | Cl, OH, Cl, Cl | OH, Me, Me, Cl

1 | 23 | 280

The relative acidity of phenol compared to PCMX and ethanol are shown below.

PCMX	$pK_a = 9.7$
Phenol	$pK_a = 9.9$
Ethanol	$pK_a = 15.9$

The increased acidity results in increased activity with a need for a smaller amount of the compound. Increased methyl substitution on the aryl ring makes PCMX less water soluble than phenol. The water solubility of PCMX is 0.33 g/L while that of phenol is 84.2 g/L. Therefore unlike phenol, PCMX does not burn skin when applied and is slowly brought into water with the help of ricinoleic acid as a soap.

Another well known disinfectant triclosan is present in many antibacterial soaps (for example, Dial), toothpastes (for example, Colgate antibacterial), body wash, mouthwash, toys and surgical clean equipments. As of 2014, triclosan was used in more than 2,000 consumer products especially many liquid soaps and bar soaps. Triclosan has been found to interfere with lipid synthesis by inhibiting any enzyme that makes fatty acids. This results in weaker cell membranes in the bacterial cells resulting in their rupture and death. Human beings do not have this enzyme and hence are not affected.

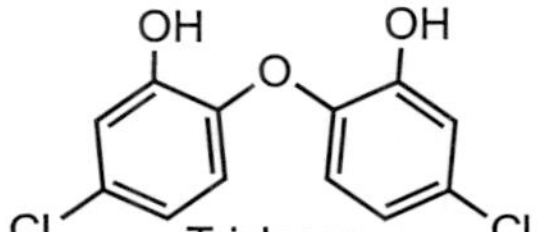

Triclosan

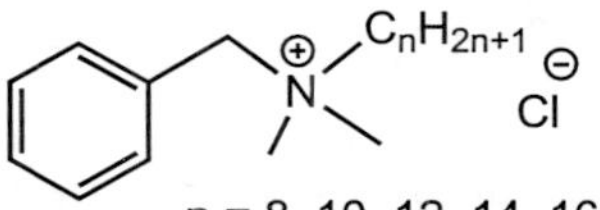

n = 8, 10, 12, 14, 16, 18
Benzalkonium chloride (Lysol)

The disinfectant Lysol (benzalkonium chloride) has chloride as counter ion to the alkyl ammonium unit.

14.10 Chlorine-Containing Drugs

Chlorine chemistry is highly instrumental in the manufacture of more than 93% of the top-selling drugs. About 26% of the drugs produced in the world contain chlorine as part of the molecule while more than two-thirds of them use chlorine-based reagents in the manufacturing process. There are many well known drugs in which chlorine is present as part of the molecule especially as an aryl–chlorine bond (Fig. 14.5). Some well known examples are the powerful antibiotic Vancomycin which kills the bacterium *Staphylococcus*. Chloroquine is a well known drug for malaria. Losartan and Amlodipin are very common drugs used for reducing blood pressure. Clopidogrel (brand name Plavix) is a medication that is used to reduce the risk of stroke and heart disease to those at high risk. It is on the WHO's list of essential medicines, the most important medications needed in a basic health system. Dichlofenac [2-(2,6-dichloranilino) phenylacetic acid] is taken or applied to reduce inflammation and as an analgesic to reduce pain in certain conditions.

14.11 Properties, Reactions and Applications of Bromine-Based Compounds

Bromine is a reddish-brown liquid which is volatile and dense and gives brown toxic vapours with a disagreeable and lung-irritating smell. It resembles chlorine more than iodine in many of its properties and reactions. However, it is less reactive than chlorine. Liquid Br_2 dispersed in water (bromine water), is used as a reagent to test for unsaturation in organic compounds and for phenols. Similar to chlorine, Br_2 reacts with water to form a mixture of HBr and hypobromous acid, HBrO. Br_2 also forms bromine free radicals with visible light of wavelength around 578 nm. Well known reactions of bromine are given in Scheme 14.4.

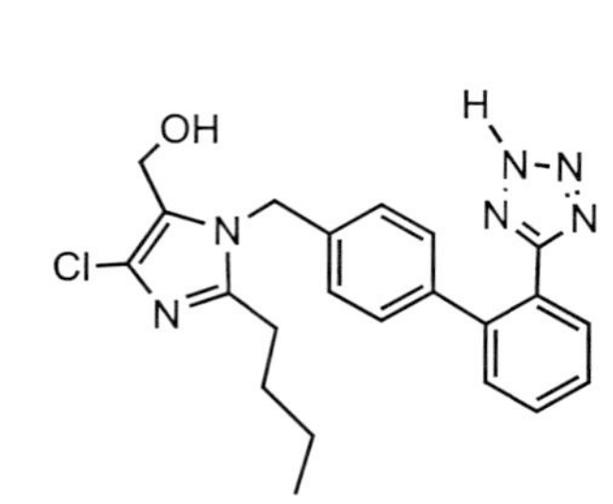

Losartan; angiotension II receptor antagonist

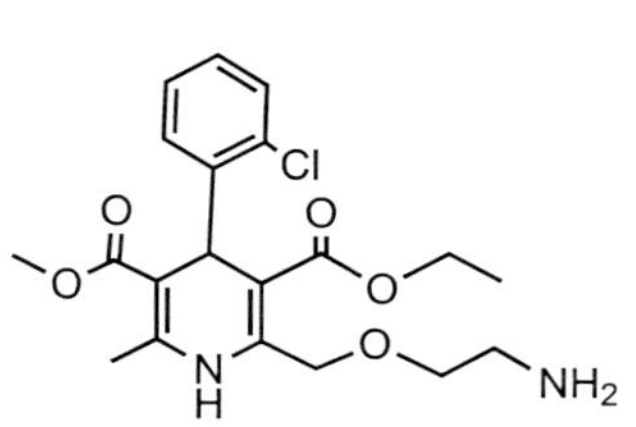

Amlodipin: b.p.medicine, Ca channel blocker

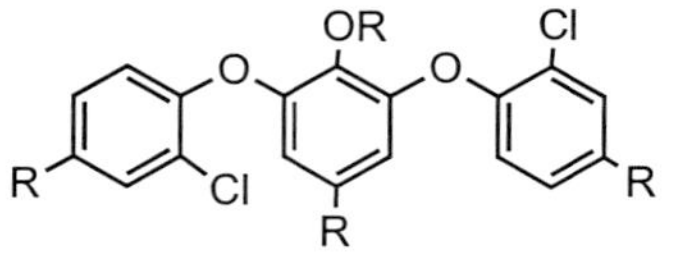

Vancomycin: antibiotic for bacterial infection

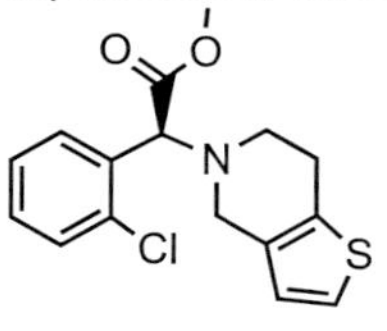

Clopidogrel: anti platelet agent for heart disease

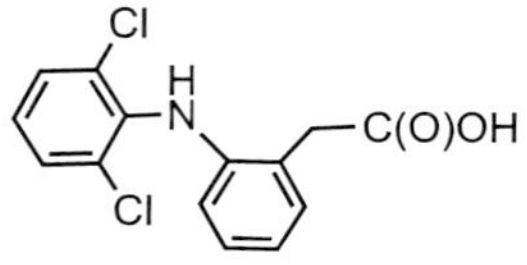

Dichlofenac: NSAID

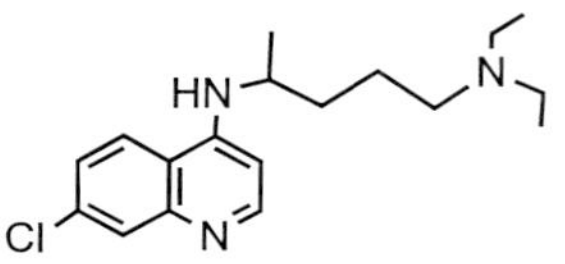

Chloroquin: anti malarial

Fig. 14.5 Examples of chlorine-containing drugs

14.11.1 Orgnobromine Compounds as Fire Retardants

Bromine in the gas phase helps to stop the chemical chain reaction that leads to flame formation and propagation. Brominated flame retardants (BFRs) either prevent a fire from starting, or slow down the progress of a fire significantly. Brominated compounds are added to materials such as polymers without altering their individual properties. Brominated flame retardants which are highly effective are used in a variety of materials, such as textiles, polymers, building materials, and electronics. Brominated fire retardants are often used in combination with antimony trioxide to enhance their activity.

Fire is considered as a chemical reaction between a fuel and oxygen as oxidiser in the gas phase. The mechanism by which flame retardants work are not simple, and is often specific to the type of the material to which the fire retardants are added. However, one common feature of flame retardants containing bromine compounds is their ability to release active bromine free radicals into the gas phase when the material decomposes by burning in the fire. These bromine atoms quench the free radical chemical reactions occuring in the burning flame, reduces the heat generated and slows down or even stops the burning process. There are four main classes of BFRs (Fig. 14.6).

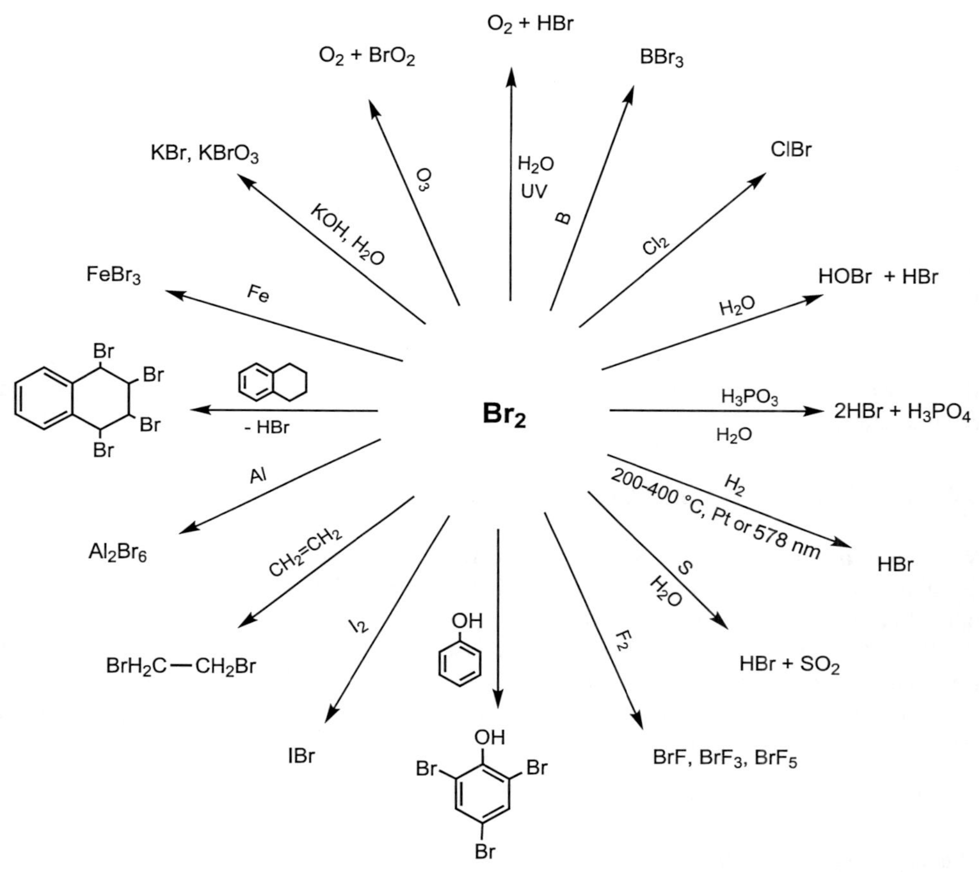

Scheme 14.4 General reactions of bromine

Br Br PBB

Br Br PBDE

Br Br Br Br Br Br HBCD

Br Br HO OH Br Br TBBPA

Fig. 14.6 Different types of brominated fire retardants

- Polybrominated diphenyl ethers (PBDEs)—for protecting plastics, textiles, electronic castings, circuitry.
- Hexabromocyclododecanes (HBCDs)—used as thermal insulation in the building industry.
- Tetrabromobisphenol A (TBBPA) and other phenols—used for printed circuit boards, thermoplastics (mainly in televisions).
- Polybrominated biphenyls (PBBs)—used in consumer appliances, textiles, plastic foams.

Some, such as the polybrominated biphenyls (PBBs), are no longer being produced. The PBBs were removed from the market in the early 1970s because of poisonings in Michigan, USA attributed to the inadvertent mixing of a bag of a commercial PBB mixture, into animal feed. The brominated flame retardants (BFRs) are currently the largest market group because of their low cost and high performance efficiency. More than 75 different BFRs are recognised commercially.

14.11.2 Bromoform and Vanadium Bromoperoxidase

Vanadium bromoperoxidase is an enzyme that brings about bromination of organic compounds and is found in seaweeds and marine algae. This enzyme produces the major part of naturally found organobromine compounds in the oceans.

The active site of vanadium bromoperoxidase features a vanadium oxide (VO_4) centre attached to the protein using a histidine side chain. The enzyme catalyses oxidation of bromide (which constitutes 0.0067% of sea water) by the formation of hydrogen peroxide. The formed electrophilic bromonium cation (Br^+) attacks the C–H bonds of hydrocarbons.[17]

$$R\text{–}H + Br+ + H_2O_2 \longrightarrow R\text{–}Br + H_2O + OH^- \quad (14.34)$$

The major organobromine compounds created by the action of vanadium bromoperoxidases are approximately 1–2 million tons of bromoform and 56,000 tons of bromomethane per year. [18]

14.12 Properties, Reactions and Applications of Iodine-Based Compounds

To combat iodine deficiency disorders, iodate and iodide are added to table salt for fortifying the iodine content. While iodide is the source of iodine used in North America, iodate is used in Europe and Asian countries. There are different methods for preparing iodates from iodine as shown below.[19]

$$I_2 + 6\,H_2O + 5\,Cl_2 \longrightarrow HIO_3 + 10\,HCl \quad (14.35)$$

$$HIO_3 + NaOH \longrightarrow NaIO_3 + H_2O \quad (14.36)$$

$$2\,HIO_3 \xrightarrow[\Delta]{200°C} I_2O_5 + H_2O \quad (14.37)$$

$$3\,I_2 + NaOH \longrightarrow NaIO_3 + 5\,NaI + H_2O \quad (14.38)$$

$$3I_2 + 3Na_2CO_3 \longrightarrow NaIO_3 + 5\,NaI + 3\,CO_2 \quad (14.39)$$

14.12.1 Iodine Pentoxide

I_2O_5 is the most stable among halogen oxides melting at 300°C. I_2O_5 is well known for its rapid, room temperature oxidation of carbon monoxide to carbon dioxide, while it gets reduced to I_2 in the process. This reaction which can be monitored by iodometric titrations is an elegant way for estimating the amount of CO present in a system.[20]

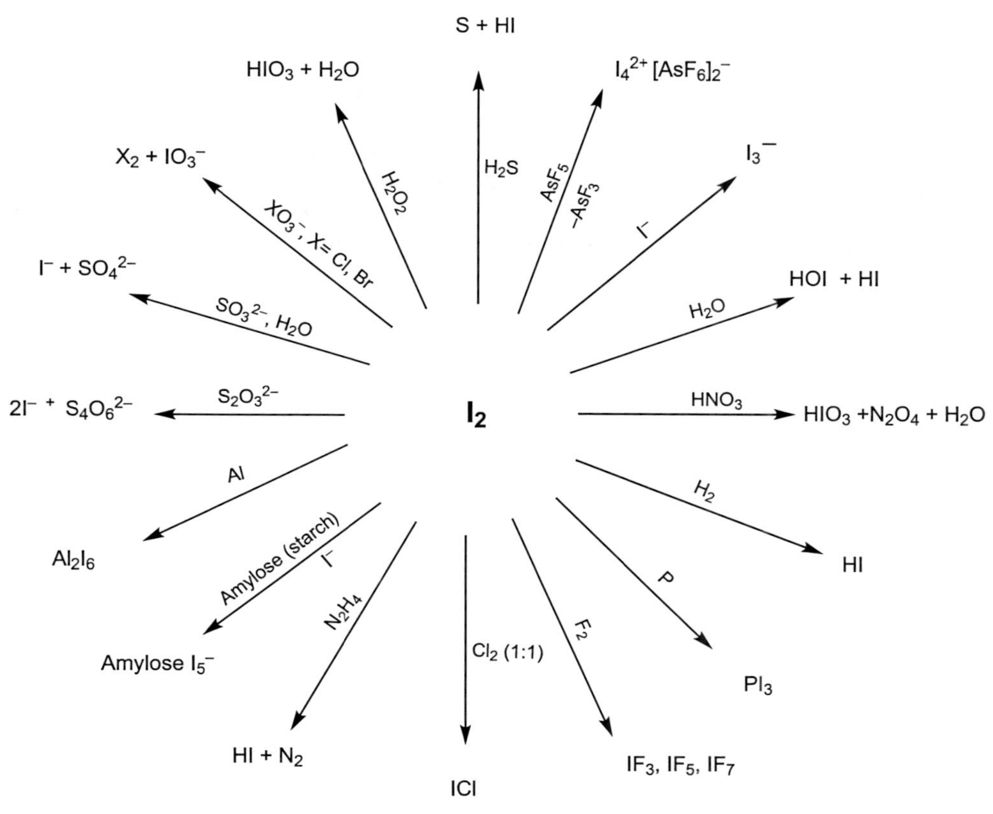

Scheme 14.5 General reactions of iodine

$$5CO + I_2O_5 \longrightarrow I_2 + 5\,CO_2 \qquad (14.40)$$

As sanitisers and antibacterials, iodine-based compounds are increasingly being used as they are less likely to form carcinogenic side products and are more environmentally friendly than chlorine. Iodophores (a preparation containing iodine complexed with a solubilising agent) also have a broader spectrum and are active against bacteria, fungi, protozoa, viruses, yeast and moulds. Disadvantages of iodophores include reduced efficacy at low temperatures, higher cost and staining effect on plastic and reaction with starch.

14.12.2 Hypervalent Iodine Compounds

Iodine-based compounds having iodine in higher oxidation states than 0, especially +3 and +5 are known as hypervalent iodine compounds. While chlorine, bromine and iodine are known to form hypervalent molecules such as ClF_3, BrF_5, IF_7, and so on, the ones which are easy to store and handle and which form easily isolable organohypervalent compounds belong to iodine. Hence, there is a lot of interest in preparing and utilising hypervalent iodine compounds, especially organic ones.

Hypervalent organo-iodine compounds have emerged as environmentally non-toxic and versatile reagents for many organic transformations. Chemical reactivity and structures of these compounds have been found to be similar to those of main group metal based reagents of Hg(II), Tl(III) and Pb(IV) and of some transition metal ions, but without the toxicity and environmental issues. These hypervalent iodine compounds undergo reactions such as ligand exchange, oxidative addition, reductive elimination and ligand coupling.

Inorganic hypervalent iodine compounds

While reaction of I_2 with NaOH gives hypoiodate, IO^-, it disproportionates readily to give I^- and IO_3^-. Iodite IO_2^-, has also been reported to form transiently on oxidation of IO^- with oxidants such as tert-butyl hydroperoxide (TBHP) but has not been isolated in the pure form since in the presence of a base, it get further oxidised to IO_3^-. Iodite formed from the reaction of I_2 and NaOH along with TBHP has been found to be an excellent catalyst for the conversion of alcohols and aldehydes to acids in water.[21]

$$R\text{-}CH_2OH \text{ OR } RCHO \xrightarrow[\text{aq.TBHP (4 equiv.), } H_2O,\ 70^\circ C,\ 10\text{–}16\text{ h}]{I_2 \text{ (10 mol\%), NaOH (20 mol\%)}} RCOOH \qquad (14.41)$$

The only stable and easily handled inorganic hypervalent iodine oxides are IO_3^- and its anhydride I_2O_5. Among hypervalent iodine halides, the relatively unstable, ICl_3, IF_3 and IF_5 are useful chlorinating and fluorinating agents.

Catalytic reactions, utilising iodide anion as the pre-catalyst are also known. Iodide in the presence of an oxidiser usually gets converted to active catalysts (IO^-) or the iodine (III) species IO_2^-. This catalytic procedure has recently been applied to the oxidative coupling of carbonyl compounds with carboxylic acids using Bu_4NI as catalyst and *tert*-butyl hydroperoxide (TBHP) as the oxidant and also oxidative cyclo-etherification reactions of oxo-substituted phenols using H_2O_2 as oxidant. Control experiments indicate that either tetrabutylammonium hypoiodite ($Bu_4N^+IO^-$) or iodite ($Bu_4N^+IO_2^-$), which are generated *in situ* from tetrabutylammonium iodide and hydrogen peroxide, are the active species in these oxidations.[22]

Bu_4NI (10 mol%), 30% H_2O_2, EtOAc, RT

R = aryl, alkyl, hetroaryl
n = 1, 2

(14.42)

$R_1CH_2CHO + R_2COOH$ — Bu_4NI (10 mol%), TBHP (1.1 equiv.), EtOAc, 50°C

R_1 = benzyl, alkyl, alkenyl
R_2 = aryl, Me, vinyl

(14.43)

Bu_4NI (10 mol%), 30% H_2O_2, THF, RT

(14.44)

Based on this strategy, highly enantioselective oxidative cyclo-etherification of substrates has also been achieved using H_2O_2 as the oxidant in the presence of the chiral quaternary ammonium iodide catalyst. This cycloetherification leads to the chiral 2-acyl-2,3-dihydrobenzofuran skeleton, which is a key structural unit in some biologically active compounds.

Catalyst (10 mol%), 30% H_2O_2, Et_2O/H_2O

Catalyst=

Ar = 3,5-[3,5-$(CF_3)_2C_6H_3$]C_6H_3

(14.45)

Organic hypervalent iodine compounds

In contrast to inorganic hypervalent iodine compounds, there are many examples of organo iodine compounds having iodine in the +3 and +5 oxidation states which are stable and can be isolated and used as reagents. The IUPAC nomenclature suggests that these compounds be termed as λ^3 and λ^5 iodanes. The λ^3 iodanes are found to have a T-shape while some λ^5 iodanes have a square pyramidal shape with the organic substituent in the apical position and the four electronegative groups in the basal positions. The first example of a λ^3 hypervalent organo iodine compound was dichloroiodobenzene, $PhICl_2$, which was prepared by C Willgerodt in 1886 by direct oxidation of iodobenzene using Cl_2. Figure 14.7 shows some of the well known examples of λ^3 and λ^5 hypervalent organoiodine compounds.[23]

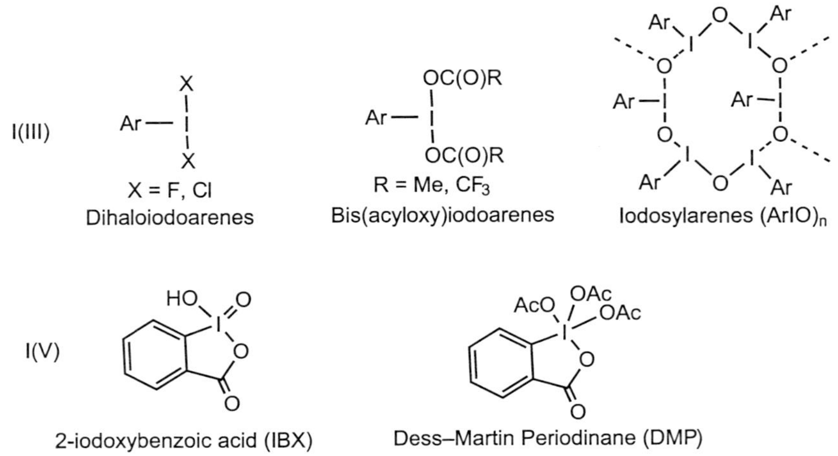

I(III)

X = F, Cl
Dihaloiodoarenes

R = Me, CF_3
Bis(acyloxy)iodoarenes

Iodosylarenes $(ArIO)_n$

I(V)

2-iodoxybenzoic acid (IBX)

Dess–Martin Periodinane (DMP)

Fig. 14.7 Different examples of hypervalent organoiodine compounds

Dichloroiodoarenes are heat and light-sensitive yellow solids due to which they cannot be stored for extended periods of time. They are mild chlorinating agents and difluoroiodoarenes are good fluorinating agents which are prepared from iodosylarenes by reaction with HF. Difluoroiodoarenes are difficult to handle in air as they are hygroscopic in nature. They are also prepared from aryliodides using XeF_2 and anhydrous HF. Scheme 14.6 shows the standard synthesis of dihaloiodoarenes.

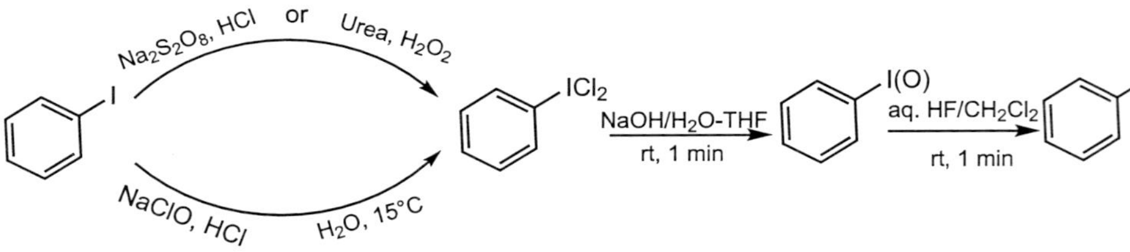

$Na_2S_2O_8$, HCl or Urea, H_2O_2

NaClO, HCl H_2O, 15°C

ICl_2 NaOH/H_2O-THF rt, 1 min

I(O) aq. HF/CH_2Cl_2 rt, 1 min

IF_2

Scheme 14.6 Synthesis of dichloro- and difluoroiodoarenes

Bis(acyloxy)iodo arenes are a well known class of hypervalent iodine(III) compounds and the most well known among this class are diacetoxyiodobenzene (DIB) [also known as phenyliodine diacetate (PIDA)] and bis(trifluoroacetoxy)iodo benzene (BTI or PIFA). This class of compounds can be prepared from benzene or aryl iodides by oxidation in the presence of the appropriate carboxylic acid or by exchange of the acid group using a stronger carboxylic acid. Iodosyl benzene PhI(O), another well-known iodine (III) reagent, is prepared by the alkaline hydrolysis of diacetoxyiodobenzene. This compound exists as a polymer containing bridging oxygen units but has poor solubility due to its polymeric nature. Reactions with this compound therefore requires initial depolymerisation by adding a hydroxylic solvent.[24]

All the λ^3 organoiodanes are mild to moderate oxidisers. While the dihaloiodoarenes are basically mild chlorinating and fluorinating agents for organic substrates, they as well

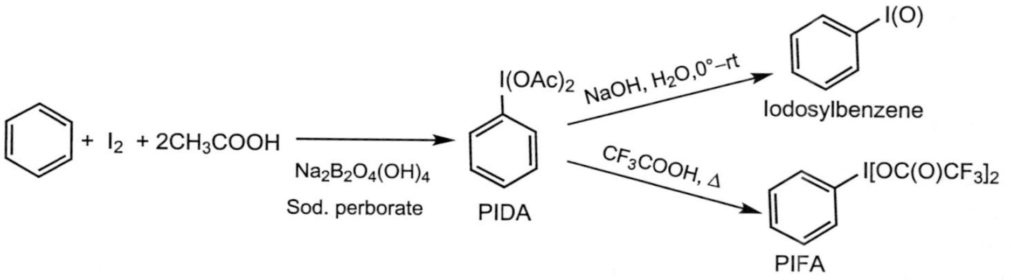

Scheme 14.7 Synthesis of bis(acyloxy)iodo arenes and iodosylbenzene

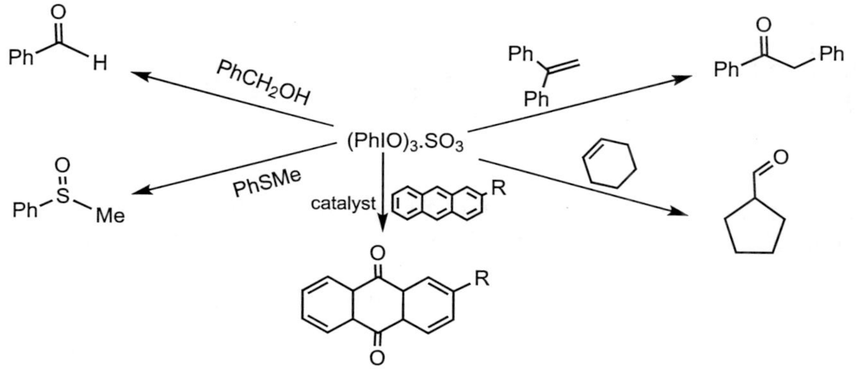

Scheme 14.8 Oxidation reactions of iodosylbenzene sulfate

as iodosyl benzene function as oxidisers for other substrates. A few such oxidations of iodosylbenzene sulfate are shown in Scheme 14.8.

An efficient procedure for the oxidation of alcohols with diacetoxyiodobenzene DIB in the presence of catalytic amounts of TEMPO (2,2,6,6-tetramethylpiperidin-1-oxyl) has been developed. This method shows a high degree of selectivity for the oxidation of primary alcohols to aldehydes, without any noticeable over-oxidation to carboxylic acids. There is also high chemoselectivity in the presence of either secondary alcohols or of other oxidisable units. The example shown (Eq. 14.46) converts a primary secondary 1,5-diol to the corresponding δ-lactone.

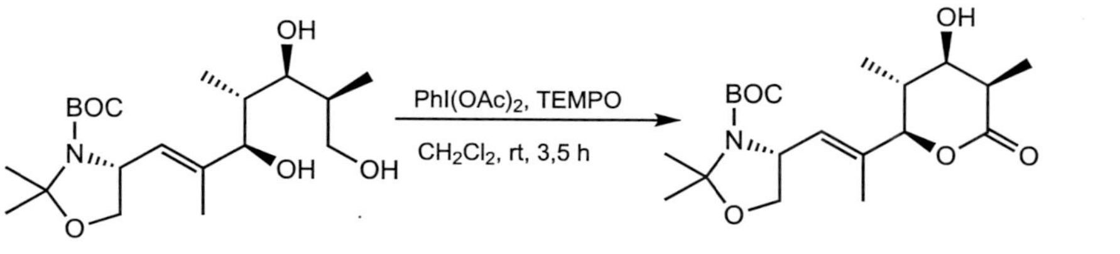

$PhI(OAc)_2$, TEMPO; CH_2Cl_2, rt, 3,5 h (14.46)

Non-phenolic electron-rich aromatic substrates can also be oxidatively coupled using $PhI(OC(O)CF_3)_2$ and $BF_3.OEt_2$.

$$\text{R}_3\text{C}_6\text{H}_3 \xrightarrow[\text{CH}_2\text{Cl}_2,\ -78^\circ\text{C}]{\text{PhI[OC(O)CF}_3]_2,\ \text{BF}_3\text{.OEt}_2} \text{biaryl} \quad (14.47)$$

R = Me, Et, I

2-Iodoxybenzoic acid [IBX, 1-hydroxy-1,2-benziodoxole-3(1H)-one-1-oxide] the first example of a λ^5 iodane was first synthesised in 1893 but is rarely used due to its poor solubility in most organic solvents and tendency to explode when heated above 200°C. IBX oxidations show a high degree of chemoselectivity for oxidation of primary alcohols to aldehydes in the presence of secondary alcohols.[25]

$$\xrightarrow[\text{rt, 2 h}]{\text{IBX, DMSO}} \quad (14.48)$$

Dess and Martin in 1983 modified IBX to a more soluble form which was named Dess–Martin periodinane (DMP).[26] While IBX is soluble only in DMSO and trifluoroacetic acid (TFA), DMP is soluble also in $CHCl_3$, CH_2Cl_2, CH_3CN and sparingly soluble in ether and hexane. DMP has thereafter been extensively used in the oxidation of alcohols and related reactions due to its high solubility, mild reactivity and chemoselectivity. The following scheme (Eq. 14.49) shows how IBX is synthesised and how it is converted to DMP.[26]

$$\text{2-IC}_6\text{H}_4\text{C(O)OH} \xrightarrow{\text{KBrO}_3,\ \text{H}_2\text{SO}_4 \text{ or Oxone, H}_2\text{O, 70}^\circ\text{C}} \text{IBX} \xrightarrow[\text{0.5\% TsOH, 80}^\circ\text{C, 2 h}]{\text{Ac}_2\text{O}} \text{DMP} \quad (14.49)$$

Oxone = $2KHSO_5\text{-}KHSO_4\text{-}K_2SO_4$

IBX in DMSO has been found to convert 1,2 diols to α-ketols or α-diketones without cleaving the glycolic C–C bond. In TFA, its behaviour is similar to that of DMP resulting in aldehydes or ketones where the glycolic C–C bond is cleaved.[27]

$$\text{PhCHO} \xleftarrow{\text{DMP}} \text{PhCH(OH)CH(OH)Ph} \xrightarrow[\text{DMSO}]{\text{IBX}} \text{PhCOCOPh}; \quad \xrightarrow[\text{TFA}]{\text{IBX}} \text{PhCHO} \quad (14.50)$$

14.12.3 Polyiodides

Iodine readily forms polyiodides. The most well known among these is I_3^-. Formation of brown I_3^- is very common in iodine chemistry and is a way of keeping iodine solubilised in aqueous medium for estimation. Another coloured complex is the amylose– I_5^- complex (dark purple) which is formed when starch is added to I_3^-.

Typically, genuine banknotes are printed on paper based on cotton fibres, and do not contain the starches that react with iodine. This property is utilised as one of the ways to identify counterfeit bank notes which are made on normal copier paper or other starch-based papers. A pen containing an iodine solution (that reacts with the starch in wood-based paper) creates a dark purple or black stain. When the solution is applied to real notes which are made with fibre-based paper, no discolouration occurs and ink shows only yellow colour.[28]

14.12.4 Iodine-Based Hormones

The thyroid gland produces two tyrosine based hormones namely triiodothyronine (T_3) and its prohormone thyroxine (T_4) having three and four iodine atoms in their structure. These hormones are primarily responsible for the regulation of metabolism. A deficiency of iodine affects the production of the hormones T_3 and T_4, resulting in the enlargement of thyroid glands, a disease known as goitre. The major thyroid hormone is thyroxine (T_4), which has a longer half-life than triiodothyronine. In human beings, the ratio between T_4 and T_3 released into the blood is between 14:1 and 20:1. The enzymes, deiodinases, convert T_4 to the active form T_3. The deiodinases are selenium containing enzymes, thus dietary selenium is essential for T_3 production.[29]

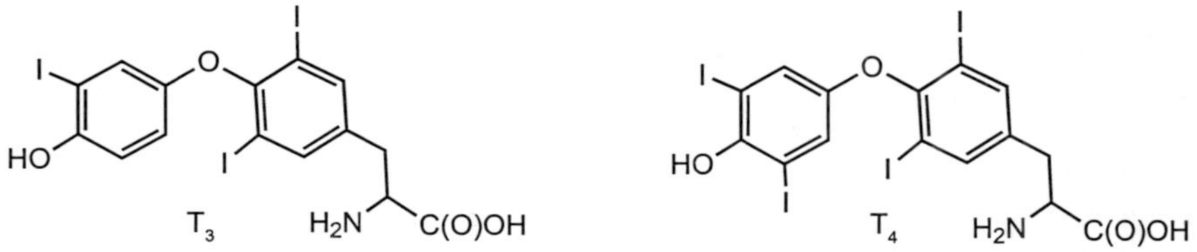

Fig. 14.8 Triiodothyronine (T_3) and thyroxine (T_4)

14.12.5 Radioactive Iodine Isotopes

Although there are 36 different radioactive isotopes of iodine, only four are used as tracers or as therapeutic agents in clinical practice. These are I-123, I-124, I-125 and I-131, with the I-131 being the most commonly used. Iodine-131, with a half-life of 8.02 days, is used in many applications. These include diagnostic and therapeutic thyroid applications and as industrial tracers. Iodine-131 is also used to label antibodies for therapeutic applications in cancer treatment.

Two main routes for iodine-131 production are fission of the uranium-235 isotope and the so-called (n, γ) reaction on TeO_2 or metallic tellurium. As the yield of iodine-131 is high and since the radioiodine isotopes with atomic mass higher than 131 are short-lived, iodine-131 is easily obtained in the pure form.

On the other hand, the (n, γ) reaction on tellurium-130 leads to the formation of tellurium-131m and tellurium-131g which convert to ^{131}I. The material used for irradiation is either tellurium oxide or the pure metal, depending on whether a wet chemical separation or a dry distillation method is used. On undergoing decay, ^{131}I (89% of the time) gives off 971 keV of energy by converting to the stable ^{131}Xe (xenon) in two steps, with β-decay followed by γ-decay.[30]

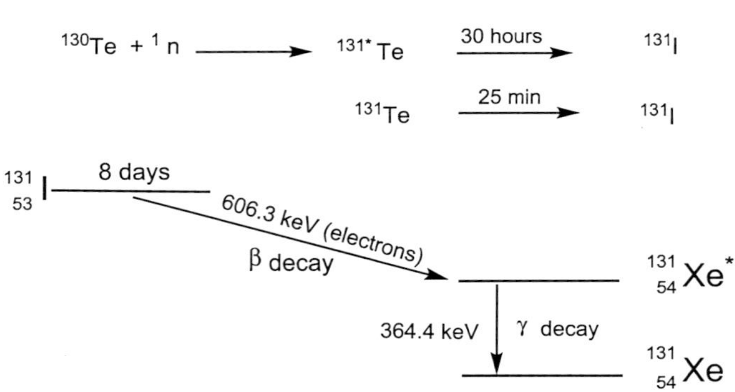

Scheme 14.9 Radioactive decay of iodine-131

Iodine-131 is the most feared fission product after an atomic bomb explosion or an accident like Chernobyl or Fukushima. The danger comes from its volatility thus being present in the air and high radioactivity with a short decay period. If breathed, it concentrates in the thyroid gland and is the cause of cancers of this sensitive gland that uses iodine to make thyroxine. During the initial days following an accidental release, inhalation of radioactive iodine is the primary route of assimilation in the body. Afterwards it is through food and especially milk that assimilation occurs. Inhalation risk is reduced by wearing masks, closing windows and staying indoors. Milk contamination is avoided by keeping the cows in the barn and feeding them with uncontaminated fodder. One method which can protect human beings and animals against the risk from radio-iodine is by consciously taking a dose of potassium iodide (KI) before exposure to radioactive iodine. The non-radioactive iodide from KI saturates the thyroid gland, causing less of the radioiodine to be absorbed and stored in the thyroid. This method has been found reduce the effects of radio-iodine by 99% and is an inexpensive readily administrable way to protect living beings from radioactive iodine poisoning.[31]

The halogen bond

Similar to the well-known hydrogen bond, there exists a weak bond involving halogens and other electron donor atoms.[32] Even though the term 'halogen bond' is of rather recent origin, the phenomenon was well known even in the 1960s. Odd Hassel, the Norwegian Nobel prize recipient of 1969, showed in 1954 that molecular complexes in the molar ratio 1:1 of dioxane and bromine associated with polarisation induced $Br^{\delta+}\cdots O^{\delta-}$ interactions. The electron density on a halogen atom is anisotropically distributed when it is involved in a covalent bond with same or other atoms. For example, in a carbon–halogen covalent bond, there is positive polarisation in the region of the halogen atom that is furthest away from the carbon atom. The equatorial region of the halogen atom is consequently negatively polarised as the overall charge on the atoms should be zero.[33] By this way, the positively polarised region of the halogen (electrophilic region) is able to initiate an electrostatic contact with a negatively polarised (nucleophilic region) atom or electronegative species. The region of lower electron density is also called a σ hole.

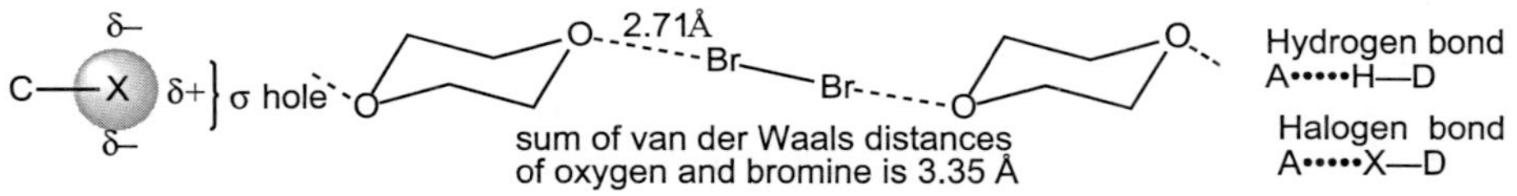

δ– C—X δ+ } σ hole δ– · 2.71Å · sum of van der Waals distances of oxygen and bromine is 3.35 Å · Hydrogen bond A•••••H—D · Halogen bond A•••••X—D

Halogens taking part in halogen bonding are mainly iodine, bromine, chlorine and rarely fluorine and follow the general trend: F < Cl < Br < I. An IUPAC recommendation states that a halogen bond exists when there is evidence of a net attractive interaction between an electrophilic region associated with a halogen atom in a molecular entity and a nucleophilic region in another, or the same, molecular entity. The core feature of the halogen bond is the electrophilic behaviour of halogen atoms. Structural studies have indicated that the halogen bonding has a specific role to play in the binding of the iodine-containing thyroid hormones triiodothyronine and thyroxine (Fig. 14.8) to the body's hormone receptor sites. Activity of small molecules such as the anesthetic, Halothane or the disinfectant Triclosan has been found to involve halogen bonding between C–Br or C–Cl bonds with oxygen atoms of the protein chains.[34] The short Br···O interaction between the bromine of a brominated uracil residue and the phosphate oxygens in a four-stranded DNA junction is also an example of non-covalent halogen bonding[35]

The strength of the halogen bond has been estimated roughly to be in the range of 0–10 kcal/mol comparable to medium hydrogen bonds (3–14 kcal/mol) or dipole–dipole interactions (1–12 kcal/mol). The typical distance of interaction has been found to be around 20% reduction of the van der Waals radii of the donor and acceptor atoms. When compared to hydrogen bonds, the unique features of halogen bonds are directionality (AXD angle ~180°), hydrophobicity, tunability and variable size of the donor atoms. Some examples of halogen bond involving iodine with energies varying in the 0–10 kcal/mol are given.

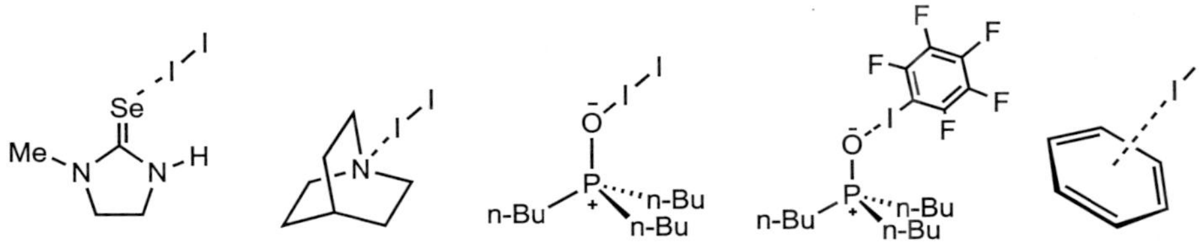

8.9 kcal/mol · 7.2 kcal/mol · 3.8 kcal/mol · 1.5 kcal/mol · 0.2 kcal/mol

Questions for self study (multiple choice)

14.1. Which one of the following is correct when the given acids are arranged in the increasing order of acid strengths?
(a) HOClO < HOCl < $HOClO_3$ < $HOClO_2$
(b) $HOClO_2$ < $HOClO_3$ < HOClO < HOCl
(c) $HOClO_3$ < $HOClO_2$ < HOClO < HOCl
(d) HOCl < HOClO < $HOClO_2$ < $HOClO_3$

14.2. Among radioactive isotopes of iodine, the one which is harmful and is generated in large quantities during an atomic bomb explosion is
(a) iodine-123 (b) iodine-124 (c) iodine-125 (d) iodine-131

14.3. The boiling points of hydrogen halides vary in the order, HF > HI > HBr > HCl. The higher boiling point of HF is due to
(a) Strong hydrogen bonds between molecules
(b) The bond energy of HF molecules is greater than the rest
(c) Nuclear shielding is reduced in fluorine which polarises the HF molecule
(d) The electronegativity of fluorine is higher than the rest of the halogens

14.4. Which of the following orders is not correct for the property stated against it?
(a) $F_2 > Cl_2 > Br_2 > I_2$: Bond dissociation energy
(b) $F_2 > Cl_2 > Br_2 > I_2$: Oxidising power
(c) F > Cl > Br > I: Electronegativity
(d) HI > HBr > HCl > HF: Acidic property in water

14.5. Which of the following has the greatest electron affinity?
(a) I (b) Br (c) Cl (d) F

14.6. Which of the following is used for the preparation of chlorine gas?
(a) Both MnO_2 and $KMnO_4$
(b) Only $KMnO_4$
(c) Only MnO_2
(d) Either MnO_2 or $KMnO_4$

14.7. Which of the following displaces Br_2 from an aqueous solution of bromide ions?
(a) I_2 (b) Cl_2 (c) I_3^- (d) SO_2

14.8. Bleaching powder is obtained by the action of chlorine gas and
(a) Dil. solution of $Ca(OH)_2$ (b) Conc. solution of $Ca(OH)_2$
(c) Dry CaO (d) Dry slaked lime

14.9. The bleaching action of chlorine is due to
(a) reduction (b) hydrogenation
(c) chlorination (d) oxidation

14.10. Which of the following elements is extracted commercially by the electrolysis of an aqueous solution of its compound?
(a) Cl (b) Br (c) Al (d) Na

Solved concept-based questions

14.11. In reactions involving $KMnO_4$ under acidic conditions, one does not use HCl as the acid. Why? Another haloacid can in principle be used for the same reactions, but is not used for two reasons. Explain.

$KMnO_4$ being a very strong oxidising agent can oxidise Cl^- to Cl_2. This is true for other halide ions as well except fluoride. But fluoride cannot be used as the HF is not a strong acid in aqueous solution and also it can react with glass.

14.12. Chlorine water on standing loses its yellow colour. Why?

Dissolved Cl_2 in water decomposes to HClO and HCl both of which are colourless. HClO also slowly converts to HCl and gives off nascent O which is responsible for the bleaching action.

14.13. ICl is more reactive than I_2. Why?

The bond dissociation enthalpy of ICl (211.3 kJ/mol) is less compared to that of Cl_2 (242.6 kJ/mol) and more compared to I_2 (151.1 kJ/mol). So strictly speaking, ICl should be less reactive than I_2. But due to the the electronegativity difference between I and Cl, and less effective overlap of orbitals with different sizes, ICl is highly polar and it has dipole moment. It is a source for reactive I^+ species and hence its higher reactivity.

14.14. Only one hydrogen halide is liquid at near room temperature. Which one and why?

HF. Due to strong hydrogen bonding between FHF^- and H^+ units.

14.15. Which oxide of halogen is used for the estimation of CO in automobile exhaust gases and why?

I_2O_5. Iodine pentoxide easily oxidises carbon monoxide to carbon dioxide at room temperature:

$$5CO + I_2O_5 \rightarrow I_2 + 5CO_2$$

This reaction can be used to analyse the concentration of CO in a gaseous sample by estimation of I_2 by iodometric titration.

14.16. While Cl_2 and Br_2 solutions in water can be used for bleaching purpose, F_2 cannot be used. Why?

F_2 oxidises water to O_2. Hence it cannot be used.

$$2\,F_2 + 2\,H_2O \rightarrow 4\,H^+ + 4\,F^- + O_2$$

14.17. The covalent radius of S and Cl are comparable (102 and 99 pm); so are the sizes of the HSO_4^- anion and ClO_4^- anion (221 and 225 pm). However, perchloric acid (pK_a −10) is a stronger acid than sulfuric acid (pK_a −1.9). Why?

The number of resonance structures possible for ClO_4^- ion is more than that possible for HSO_4^-. Greater resonance delocalisation makes the perchlorate anion more stable. The charge is dispersed over five atoms (the Cl and four O atoms) in the perchlorate anion, but only over four atoms in the hydrogen sulfate anion. Hence formation of the former is favoured over the latter resulting in easier cleavage of the H^+ from perchloric acid compared to sulfuric acid.

14.18. The pK_a values indicating acidity of chlorus acid, chloric acid and perchloric acid are +2, −1 and −10 respectively. Explain why the acidity increases in this order.

If you consider the anions of these acids one can see that the number of Cl=O bonds increase from 1–3 on going from chlorus acid, chloric acid to perchloric acid. The stability of the anion also increases in this order as the negative change can be shared by more resonance structures when there are more Cl=O bonds. Hence the increase in the acidity.

14.19. In titrations involving release of I_2, excess of I^- is used. Why?

I^- will react with I_2 formed to form I_3^- which is more soluble in water than I_2. In this way, the possibility of volatile I_2 being lost is reduced and the solubility of I_2 is enhanced.

14.20. Concentrated HNO_3 or conc. HCl do not dissolve gold on their own. Then, how does aqua regia (HNO_3:HCl 1:3) dissolve gold?

Although individually both HNO_3 and HCl cannot dissolve gold, together each acid plays a different role. Nitric acid, being a powerful oxidiser, will dissolve a very small, undetectable amount of gold, oxidising it to gold(III) ions. The hydrochloric acid on the other hand, provides a supply of chloride ions (Cl^-), which with the Au^{3+} ions produce chloroaurate anions ($AuCl_4^-$), which is soluble in the solution. The reaction with hydrochloric acid is an equilibrium reaction and it favours formation of chloroaurate anions ($AuCl_4^-$). Thus the forward reaction is favoured as more and more gold is made to go into solution and this allows further oxidation of gold to take place. Finally all of the gold dissolves and get converted to chloroauric acid.

Problems and exercises

14.21. Write three differences in the reactivity of Cl_2 and ClO_2.

14.22. Starting from Cl_2 how is $NaClO_2$ prepared? What is the use of $NaClO_2$ and why is it important?

14.23. Using the set of reagents provided, show the shortest synthetic route for the preparation of the given polyurethane.

$$\left[C(=O)-NH-C_6H_4-NH-C(=O)-OCH_2CH_2O \right]_n$$

Ethylene glycol, m-dinitrobenzene, carbon monoxide, hydrogen, chlorine.

14.24. How are impure silicon metal and impure TiO_2 purified? Show reaction schemes and explain in 1–2 lines.

14.25. Which is a better disinfecting agent for water in a swimming pool, Cl_2 or ClO_2? Justify your answer. Starting from chlorine gas how is pure Cl_2-free ClO_2 prepared?

14.26. For the following electrophilic chlorination, choose the most suitable and convenient chlorinating agent from the given list. Justify your selection of reagent.

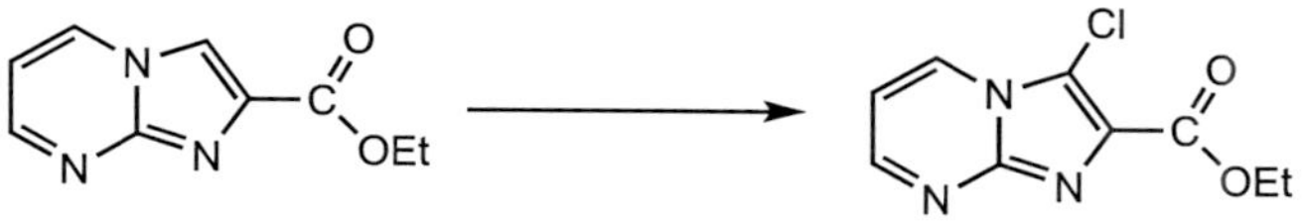

[Cl_2, HOCl, t-BuOCl, CBMG (Palau chlor), NCS (N-chlorosuccinimide)]

With reference to Cl, Br and I, fill in the blanks and rewrite the full statement.

14.27. Among the given compounds, the best electrophilic chlorinating agent for selective chlorination of unsaturated organic heterocyclic compounds is .... (TCICA, DCDMH, PBDE, CBMG, DAST).

(a) Among the given compounds, the best compound which can be used as a fire suppressing agent is .... (PBDE, CBMG, NFOBS, NaBr).

(b) The number of iodine atoms present in the prohormone thyroxine is ....

(c) The radioactive isotope of iodine I-131 decays by the release of...................... radiation /rays.

14.28. The Castner–Kellner chloralkali process has two separate reaction chambers. Why? Give equations for the additional reactions that are likely to happen if the electrolysis is carried out in a single chamber and without temperature control.

14.29. Using the set of reagents provided, show the shortest synthetic route for preparation of the given resin for making epoxy polymers.

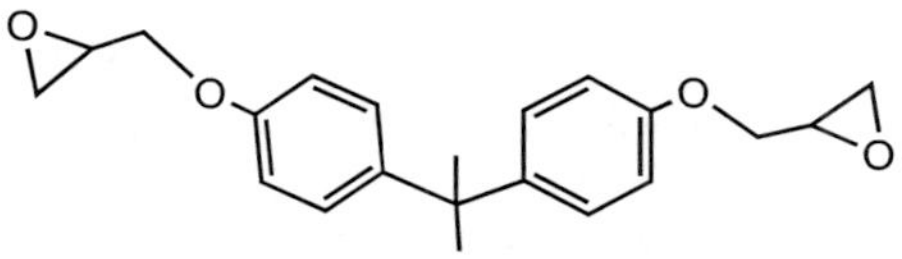

Bisphenol A diglycidyl ether

(Bisphenol-A, propylene, HOCl, NaOH, chlorine)

14.30. A hot aqueous solution of a sodium halide (A) on electrolysis was found to give compound B as a sodium salt. Compound B on anodic oxidation using a platinum electrode was found to give compound C, also as a sodium salt. Compound C on reaction with HCl gave sodium chloride and an acid D. Reaction of D with P_2O_5 resulted in E in which the halogen was in its highest oxidation state. Identify compounds A to E. Draw the possible structure of E.

14.31. Reaction of methane with water over a nickel catalyst at high temperature and pressure gave two gaseous products. One of these gaseous products A, on reaction with a dihalogen over activated carbon as catalyst gave compound B. Compound B on reaction with 1,4-diaminobenzene gave compound C. Compound B also reacts with methylamine (followed by reaction with a HCl scavenger) to give a highly toxic gaseous pesticide precursor D, whose leakage from a factory was responsible for the death of many people during the Bhopal gas disaster in India. Identify compounds A–D.

14.32. Halogen X_2 reacts with five moles of halogen Y_2 in water to give a halic acid A. Compound A on dehydration gave a halogen oxide B. Compound B has been found to react with CO to give CO_2, itself getting reduced to X_2. Identify X, Y, A and B.

14.33. Match the precursor with the correct product

(i) $CHCl_3$	(A) HFC 134a
(ii) $HC\equiv CH$	(B) CF_3COOH
(iii) $Cl_2C{=}CCl_2$	(C) PTFE (Teflon)
(iv) $Cl_2C{=}CHCl$	(D) PVDF (Polyvinylidene fluoride)
	(E) PVDC(Polyvinylidene chloride)

Answer key for self study questions

14.1. (d) 14.2. (d) 14.3. (a) 14.4. (a) 14.5. (c) 14.6. (a) 14.7. (b) 14.8. (d) 14.9. (d) 14.10. (a)

Supplementary reading

1. Schmittinger P, Flokiewicz T, Curlin L C, Luke B, Scannel R, Navin T, Zelfel E and Bartsch R, Chlorine, *Ullmann's encyclopedia of industrial chemistry*, 2011, Wiley-VCH, Weinheim.
2. Mills J F, Frim R, Ukeles S D and Yoffe D, Bromine, *Ullmann's encyclopedia of industrial chemistry*, 2015, Wiley-VCH, Weinheim.
3. McCall A S, Cummings C F, Bhave G, Vanacore R, McCallow A P and Hudson B G, Bromine is an essential trace element for assembly of collagen IV, scaffolds in tissue development and architecture, *Cell*, 2014, Vol. 157, 1380.
4. Mertz W, The essential trace elements, *Science,* 1981, Vol. 213, 1332.
5. Lyday P A and Kaiho T, Iodine and iodine compounds, *Ullmann's encyclopedia of industrial chemistry*, 2015, Wiley-VCH, Weinheim.
6. Hiegel G A, Trichloroisocyanuric acid, *Encyclopedia of reagents in organic synthesis*, 2001, John Wiley & Sons, NY.
7. Tilstam and Weinmann H, Trichloroisocyanuric Acid: A Safe and Efficient Oxidant, *Org. Proc. Res. Dev*, 2002, Vol. 6, 384.
8. Yoffe D, Frim R, Ukeles S D, Dagani M J, Barda H J, Benya T J and Sanders D C, Bromine compounds, *Ullmann's encyclopedia of industrial chemistry*, 2013, Wiley-VCH, Weinheim.
9. White C E, Laboratory preparation of iodine, *J. Chem. Educ.*, 1935, Vol. 12,
10. Vogt H, Balej J, Bennett J E, Wintzer P, Sheikh S A, Gallone P, Vasudevan S and Pelin K, Chlorine Oxides and chlorine oxygen acids, *Ullmann's encyclopedia of industrial chemistry*, 2010, Wiley-VCH, Weinheim.
11. Baum K and Beard C D, Reactions of dichlorine heptoxide with alcohols, *J. Am. Chem. Soc.*, 1974, Vol. 96, 3233.
12. Barros J C, Trichloroisocyanuric acid, *Syn. Lett*, 2005, Vol. 13, 2115.
13. Rodriguez R A, Pan C-M, Yabe Y, Kawamata U, Eastgate M D and Baran P S, Palau'chlor: A practical and reactive chlorinating agent, *J. Am. Chem. Soc.*, 2014, Vol. 136, 6908.

14. Chang S L, Destruction of Microorganisms, *Journal AWWA*, 1944, Vol. 36, 1192.
15. Prutz W A, Hypochlorous acid interactions with thiols, nucleotides, DNA and other biological substrates, *Archives of biochemistry and biophysics*, 1996, Vol. 332, 110.
16. *Chemistry in the cupboard: Dettol*, Royal Society of chemistry, 2010.
17. Winter J M and Moore B S, Exploring the chemistry of vanadium-dependent haloperoxidases, *J. Biol. Chem.*, 2009, Vol. 284, 18577.
18. Gribble G W, The diversity of naturally occurring organobromine compounds, *Chem. Soc. Rev.*, 1999, Vol. 28, 335.
19. Selte K and Kjekshus A, Iodine oxides Part III. The crystal structure of I_2O_5, *Acta Chem. Scand.*, 1970, Vol. 24, 1912.
20. Conradice J, Bary S and Ghosh A, Mechanism of oxygen transfer between main group elements, *Eur. J. Inorg. Chem.*, 2015, Vol. 24, 4138.
21. Hazra S, Deb M and Elias A J, Iodine catalysed oxidation of alcohols and aldehydes to carboxylic acids in water: A metal-free route to the synthesis of furandicarboxylic acid and terephthalic acid, *Green Chem.*, 2017, Vol. 19, 5548.
22. Uyanik M, Suzuki D, Yasui T and Ishihara K, In situ generated hypoiodite catalysts for the direct α-oxyacylation of carbonyl compounds with carboxylic acids, *Angew. Chem. Int. Ed.*, 2011, Vol. 50, 5331.
23. Yoshimura A and Zhdankin V V, Advances in synthesis of hypervalent iodine compounds, *Chem. Rev.*, 2016, Vol. 116, 3328.
24. Yusubov M S and Zhdankin V V, Iodine catalysis: A green alternative to transition metals in organic chemistry and technology, *Resource Efficient Technologies*, 2015, Vol. 1, 49.
25. Satam, V, Harad A, Rajule R and Pati H, 2-Iodoxybenzoic acid (IBX): an efficient hypervalent iodine reagent, *Tetrahedron*, 2010, Vol. 66, 7659.
26. Dess D B and Martin J C, Readily accessible 12-I-5 oxidant for the conversion of primary and secondary alcohols to aldehydes and ketones, *J. Org. Chem.*, 1983, Vol. 48, 4155.
27. Moorthy J, Singhal N and Senapati K, Oxidative Cleavage of Vicinal Diols: IBX can do what Dess–Martin Periodinane (DMP) can., *Org. Biomol. Chem.*, 2007, Vol. 5, 767.
28. Teitelbaum R C, Ruby S L and Marks T J, On the structure of starch iodine, *J. Am. Chem. Soc.*, 1978, Vol. 100, 3215.
29. Schweiser U and Steegborn C, Thyroid hormones: From crystal packing to activity to reactivity, *Angew Chem. Int. Ed.*, 2015, Vol. 54, 12856.
30. Robbins J and Schneider A B, Thyroid cancer following exposure to radioactive iodine, *Review in endocrine and metabolic disorders*, 2000, Vol. 1, 197.
31. Guiraand–Vitaux F, Elbart M, Colas–Linhart N and Hindie E, Thyroid cancer after Chernobyl: Is iodine-131 the only culprit? Impact on clinical practice, *Bulletin du cancer*, 2008, Vol. 95, 191.
32. Cavallo G, Metrangolo P, Milani R, Pilati T, Priimagi A, Resnati G and Terraneo G, The halogen bond, *Chem. Rev.*, 2016, Vol. 116, 2478.
33. Desiraju G R, Vittal J J and Ramanan A, *Crystal Engineering, A text book*, World Scientific, Singapore, 2011.
34. Parisini E, Metrangolo P, Pilati T, Resnati G and Terraneo G, Halogen bonding in halocarbon-protein complexes: a structural survey, *Chem. Soc. Rev.*, 2011, Vol. 40, 2267.
35. Metrangolo P and Resnati G, Halogen versus hydrogen, *Science*, 2008, Vol. 321, 918.

The Chemistry of Noble Gases

CHAPTER 15

Unique Properties of Noble Gases and Their Compounds

- *The relative abundance of noble gases in the atmosphere decreases in the order Ar > Ne > He > Kr/Xe > Rn.*
- *Except for helium and neon, all other noble gases have been made to undergo reactions leading to novel compounds.*
- *These compounds are neutral, anionic and cationic as well as those in which a noble gas atom or compound acts as a donor ligand.*
- *Most of the noble gas compounds are endothermic and often oxidising in nature.*
- *The stabilisation of noble gas compounds has been mostly achieved by highly electronegative substituents such as F, O, Cl or anions having high group electronegativities (example, teflate anion F_5TeO^-, EN 3.87).*
- *The very low bond enthalpy and high standard reduction potential of many of the noble gas compounds make them very strong oxidisers. KrF_2 is the strongest oxidising agent known so far with a higher E_0 of 3.5 V compared to F_2 (E_0 2.87 V). The bond enthalpy of KrF_2 is 49 kJ/mol compared to 154 kJ/mol of F_2.*
- *The high oxidising strength of KrF_2 has been utilised for making high oxidation state metal complexes such as Au_2F_{10} and AgF_3.*
- *One of the most unstable of noble gas compounds known so far is HArF, the only covalent compound of argon which is stable only below 17 K (–256°C). The shapes of most of the noble gas compounds comply with the VSEPR rules. A few exceptions are XeF_6, XeO_2 and XeF_8^{2-}.*
- *The useful applications of xenon include its use as an efficient inhalation anesthetic, as media for coloured illuminating displays and as fluorides especially XeF_2 as an oxidising and fluorinating reagent.*

- *Excimer lasers are ultraviolet lasers, commonly used in the production of microelectronic devices, semiconductor-based integrated circuits, Lasik eye surgery and micromachining. It uses a combination of a noble gas such as argon, krypton or xenon and a reactive gas such as fluorine or chlorine which under electrical stimulation and high pressure, generates a pseudo-molecule (excimer) having only a transient existence and gives rise to laser light of a specific frequency in the UV region.*
- *Liquid helium is produced commercially for use in superconducting magnets such as those used in nuclear magnetic resonance (NMR) and magnetic resonance imaging (MRI).*

15.1 About the Elements

Noble gases are colourless, odourless, tasteless and nonflammable gases under standard conditions. Noble gases are widely used in different fields, from incandescent lighting to excimer lasers. The inertness of noble gases makes them suitable in applications where chemical reactions should be avoided. Argon, for example is used in incandescent light bulbs to prevent the tungsten filament from getting oxidised. Xenon is used as a modern anesthetic due to its high lipid solubility, making it more active than N_2O. Being an unreactive gas, it is easily eliminated from the body, helping in quicker recovery from anesthesia. XeF_2 is a good fluorinating agent. After the explosion risks caused by the flammability of hydrogen became known, it was replaced with helium in airships and balloons. Argon being a dense inert gas is used often as a shielding gas in welding and other high-temperature industrial processes. It is also used in graphite electric furnaces to prevent the graphite from burning. As magnesium reacts with N_2, for effective blanketing on molten magnesium, argon is used instead of nitrogen. Many of the noble gases are also used in fluorescent lighting, and other gas discharge tubes.

The noble gases have weak interatomic forces, and consequently have very low melting and boiling points. Helium has several unique properties when compared with other elements. Its boiling and melting points are lower than those of any other known substance; it is the only element known to exhibit superfluidity and one that cannot be solidified by cooling under standard conditions. For solidification, helium requires a pressure of 25 atm. at a temperature of 0.95 K. Due to the very weak interatomic forces, helium will remain a liquid at atmospheric pressure all the way from its liquefaction point down to absolute zero. Unlike the noble gases up to xenon, which have multiple isotopes that are radioactively stable, radon does not have any stable isotopes. The longest-lived isotope of radon, ^{222}Rn, has a half-life of 3.8 days and it decays to form polonium and helium. The polonium decays ultimately to lead.

15.2 Isolation and Purification of the Elements

The discovery of noble gases begins from a note from Lord Raleigh in the journal *Nature* in 1892, in which he requested the readers from the chemistry community to help him in explaining an observation while trying to determine the density of nitrogen.[1] According to Raleigh he was puzzled by the fact that density of nitrogen gas which he collected and measured from two different sources were not agreeing. The relative difference was about 1/1000 part which is small but he felt that there were no errors in his experiments and the difference could only be attributed to a variation in the character of the gas.

In his first method he used atmospheric air. To remove O_2 from the air, he passed it over red hot copper which was freshly obtained by reducing copper oxide and removed CO_2 in the air by treating with K_2CO_3. The gas was then passed through a wash bottle containing sulfuric acid and then P_2O_5 to remove moisture and this process was repeated to get nitrogen gas in the expectedly pure form.

In his second method, which was suggested to him by Prof. Ramsay, he used liquid ammonia as the source of nitrogen through which air was bubbled and the rest of the experimental setup remained the same. The ammonia method was quite convenient, but the nitrogen obtained by means of this method was found to be 1/1000 part lighter than the nitrogen of the first method. The question was, how did this discrepancy arise?

Ramsey found a solution to this problem. He purified a sample of atmospheric nitrogen gas to remove carbon dioxide, moisture, and other contaminants. This gas was then passed over hot magnesium, which is well known to react with nitrogen to form magnesium nitride.

$$3\,Mg + N_2 \longrightarrow Mg_3N_2 \tag{15.1}$$

Ramsey was left with a small gaseous residue that occupied roughly $1/80^{th}$ of the original volume. He excited this gas in an electric discharge tube and found that the resulting emission spectrum contained lines that differed from those of all known gases. By 1894, after a series of experiments, Rayleigh and Ramsey together announced the discovery of a new element. As this gas did not react with any element or compound they tried, they named it *argon* from the Greek word meaning the 'lazy one'. Their work was published in 1895 in the *Philosophical Transactions* of the Royal Society. In 1904 Ramsey received the Nobel Prize 'in recognition of his services in the discovery of the inert gaseous elements in air' and Lord Raleigh received the Nobel Prize in the same year for physics 'for his investigations of the densities of the most important gases and for his discovery of argon in connection with these studies'.

The discovery of argon demanded the need to have a new group of elements in the periodic table to be inserted between halogens and alkali metals. Shortly after the discovery of argon in 1894, Ramsey detected another unreactive gas when a mineral of uranium was heated which was identified as helium. However, the first evidence of helium was observed on 18 August 1868, as a bright yellow line of wavelength 587.49 nm in the spectrum of the sun detected by the French astronomer Jules Janssen during his study of a total solar eclipse carried out in Guntur, Andhra Pradesh, India. This line was initially assumed to be that of sodium. On 20 October of the same year, the English astronomer Norman Lockyer observed a yellow line in the solar spectrum. Lockyer and the English chemist Edward Frankland named the element helium (from the Greek *helios*, 'sun').

Further careful experiments with liquid air led Ramsey to a third gas, which he named *krypton* meaning 'the hidden one'. Experiments with liquid air in the period 1894–1898 led Ramsey to a fourth gas, *neon* ('the new one'), and finally to a fifth gas, *xenon* ('the stranger'). Ramsey sent a sample of argon to Moissan, who in this period had isolated fluorine gas to see whether it would react with fluorine. Moissan's attempts to react argon with fluorine were not successful, and similar unsuccessful attempts by other chemists eventually led these gases to be labelled as *inert gases*. Only in 1962, after the path breaking work of Neil Bartlett, these gases were more aptly called noble gases.

In 1899, Rutherford detected a radioactive gas being released by thorium. The same year, Pierre and Marie Curie detected a radioactive gas emanating from radium. In 1900, Friedrich Dorn at Halle, Germany, noted that a gas was accumulating inside ampoules of radium. All these three observations were pointing in the direction of a radioactive noble gas which was named *radon*. The one from radium was the longer-lived isotope radon-222 which has a half-life of 3.8 days, and was the same isotope which the Curies observed. The radon gas that Rutherford detected from thorium was radon-220 which has a shorter half-life of 56 seconds. By 1908, enough radon was collected to determine its properties and to report that it was the heaviest gas known.

15.3 Chemistry of the Noble Gases

15.3.1 The Yost and Kaye Experiments

Linus Pauling had the intuition, based on an earlier note by von Antropoff that xenon should react with the most reactive of elements of that time, fluorine (F_2). In 1932, he managed to convince one of his highly experienced inorganic chemistry colleagues at CalTech, Prof. D M Yost to try out this reaction. The required amount of xenon was borrowed from Prof. Fred J Allen of Purdue University by the initiative of Pauling himself. Prof. Yost along with A L Kaye, a student trained from MIT, made a reactor out of quartz with copper electrodes and passed electric discharge though the mixture of Xe and F_2 (1:20 molar ratio) many times. However, their repeated attempts were in vain as they could not isolate any xenon–fluorine compounds. Disappointed, they sent a note to the *Journal of American Chemical Society* describing the reaction and their inability to isolate any xenon–fluorine compounds. However, they mentioned in the note that they did notice that the side tube of their reactor was appreciably attacked after some time and suggested that it might be due to some reactive xenon fluoride but without definite evidence.[2]

Twenty nine years later, in 1962, Neil Bartlett, who was then a fresh lecturer at the University of British Columbia, just then had success with a reaction of O_2 and PtF_6 resulting in $O_2^+PtF_6^-$ (dioxygenyl hexafluoroplatinate, a red solid). He got the gut feeling that PtF_6 should react with xenon as well because he noticed that the first ionisation energy for O_2 (1175 kJ mol^{-1}) is very close to the ionisation energy for Xe (1170 kJ mol^{-1}). However he was reluctant to suggest it to his students since Pauling had already indicated from the Yost and Kaye experiments that xenon is unlikely to react. He decided to carry out the reaction by himself and conducted it when his students had gone for dinner. When he allowed the dark red PtF_6 and the colourless xenon gases to mix in the vacuum line, he saw the immediate formation of a yellow coloured compound. His initial assumption of formation of $Xe^+ PtF_6^-$ turned out to be incorrect. However after more than a decade of detailed studies, it was found that the yellow compound he made was $[XeF]^+[Pt_2F_{11}]^-$. Studies carried out later by

Table 15.1 Physical properties of helium

Helium	*Atomic mass*	*Electronegativity (Pauling scale)*	*I[st] ionisation energy*	*Electronic configuration*	*van der Waals radius*
Symbol: He	4.002 g.mol^{-1}	No data	2372.3 kJ.mol^{-1}	$1s^2$	0.140 nm
At. no. 2	*Atomic radius*	*Melting point*	*Boiling point*	*Electron affinity*	*Density*
	0.31 Å	–272.20°C	–268.92°C	–50.00 kJ.mol^{-1}	0.145 g.cm^{-3}

Table 15.2 Physical properties of neon

Neon	*Atomic mass*	*Electronegativity (Pauling scale)*	*I[st] ionisation energy*	*Electronic configuration*	*van der Waals radius*
Symbol: Ne	20.179 g.mol^{-1}	No data	2080.7 kJ mol^{-1}	[He] $2s^2$ $2p^6$	0.154 nm
At. no. 10	*Atomic radius*	*Melting point*	*Boiling point*	*Electron affinity*	*Density*
	0.38 Å	–248.59°C	–246.04°C	–120.00 kJ.mol^{-1}	1.207 g.cm^{-3}

Table 15.3 Physical properties of argon

Argon	*Atomic mass*	*Electronegativity (Pauling scale)*	*I[st] ionisation energy*	*Electronic configuration*	*van der Waals radius*
Symbol: Ar	39.948 g.mol^{-1}	No data	1520.6 kJ.mol^{-1}	[Ne] $3s^2$ $3p^6$	0.188 nm
At. no. 18	*Atomic radius*	*Melting point*	*Boiling point*	*Electron affinity*	*Density*
	0.71 Å	–189.34°C	–185.84°C	–96.00 kJ.mol^{-1}	1.395 g.cm^{-3}

Table 15.4 Physical properties of krypton

Krypton	*Atomic mass*	*Electronegativity (Pauling scale)*	*I[st] ionisation energy*	*Electronic configuration*	*van der Waals radius*
Symbol: Kr	83.798 g.mol^{-1}	3.00	1350.8 kJ.mol^{-1}	[Ar]$3d^{10}$ $4s^2$ $4p^6$	0.202 nm
At. no. 36	*Atomic radius*	*Melting point*	*Boiling point*	*Electron affinity*	*Density*
	0.88 Å	–157.37°C	–153.41°C	–60.00 kJ.mol^{-1}	2.413 g.cm^{-3}

Table 15.5 Physical properties of xenon

Xenon	*Atomic mass*	*Electronegativity (Pauling scale)*	*I[st] ionisation energy*	*Electronic configuration*	*van der Waals radius*
Symbol: Xe	131.293 g.mol^{-1}	2.6	1170.4 kJ.mol^{-1}	[Kr]$4d^{10}$ $5s^2$ $5p^6$	0.216 nm
At. no. 54	*Atomic radius*	*Melting point*	*Boiling point*	*Electron affinity*	*Density*
	1.08 Å	–111.75°C	–108.09°C	–80.00 kJ.mol^{-1}	2.942 g.cm^{-3}

using SF_6 as a diluent resulted in the initially predicted compound $Xe^+PtF_6^-$ as a mustard yellow compound.[3–6]

$$O_2 + PtF_6 \longrightarrow [O_2]^+ [PtF_6]^- \tag{15.2}$$

$$Xe + 2\,PtF_6 \xrightarrow[25^\circ C]{} [XeF]^+ [PtF_6]^- + PtF_5 \xrightarrow[60^\circ C]{} [XeF]^+ [Pt_2F_{11}]^- \tag{15.3}$$

$$Xe + PtF_6 \xrightarrow[\text{excess } SF_6]{} [Xe]^+ [PtF_6]^- \tag{15.4}$$

Few months after Bartlett's discovery, Rudolf Hoppe from Munich, Germany, mixed xenon and fluorine in 1:1 molar ratio and after passing electrical discharge saw the formation of colourless crystals of XeF_2.[7] Shortly afterwards John Holloway from Scotland showed how easy it is to prepare XeF_2 by UV radiation, by mixing equimolar amounts of xenon and fluorine and keeping the flask on the terrace of their lab (European summer).[8] Another method of preparing xenon difluoride was using NiF_2 as catalyst at 120°C. In the same year, scientists working at the Argonne National Laboratory, USA found that xenon reacts with F_2 to form XeF_4.[9] Since then, many new compounds of xenon, some of krypton and one of argon have been isolated. Different methods of preparing XeF_2 from xenon are given in Scheme 15.1.[10]

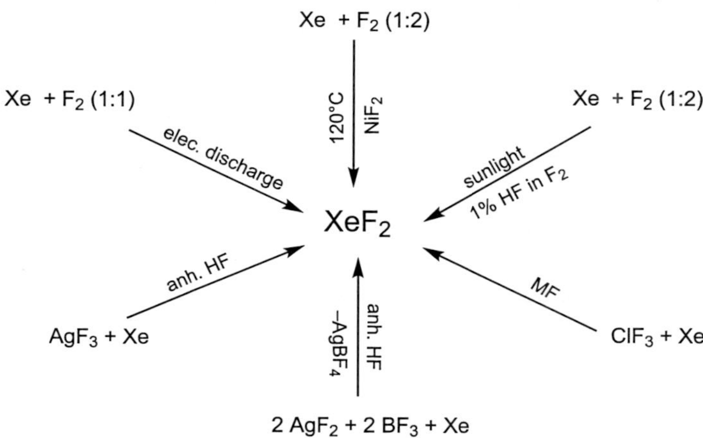

Scheme 15.1 Different methods of synthesis of XeF_2 from xenon

It was thereafter found that the stoichiometric ratio of xenon and fluorine has a significant role to play in deciding the nature of the xenon fluoride formed. Among the three possible products, the one with the lowest melting point and least stability is XeF_6. Xenon difluoride further reacts with F_2 at 400°C to give the square planar XeF_4 and with more F_2 at 300°C gives XeF_6 which has a distorted octahedral shape in the gas phase. ^{129}Xe- and ^{19}F-NMR spectroscopy indicates that in solution the compound assumes a tetrameric structure. Four equivalent xenon atoms are arranged in a tetrahedron surrounded by a fluctuating array of 24 fluorine atoms that interchange positions.[11,12]

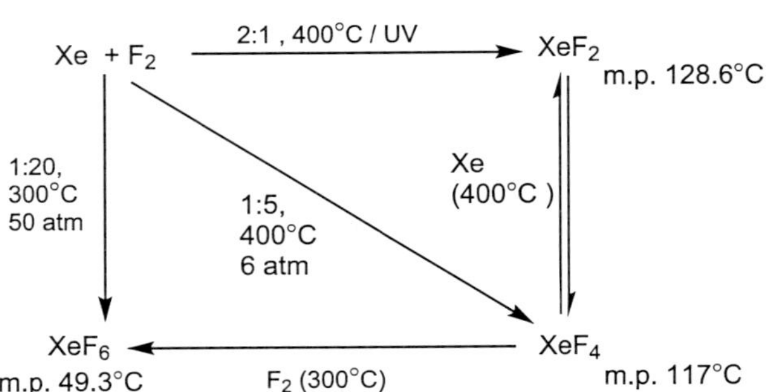

Scheme 15.2 Synthesis of XeF_2, XeF_4 and XeF_6 and their interconversions

Among the three fluorides, purifying XeF_4 is a relatively difficult task as one cannot have excess or less of fluorine; therefore traces of XeF_2 and XeF_6 are also formed. An easy method to purify XeF_4 involves making use of the lesser reactivity of AsF_5 with XeF_4. AsF_5 in BrF_5 converts XeF_2 and XeF_6 to ionic solids while XeF_4 does not react. The unreacted XeF_4 can be easily separated from the other two by application of vacuum.[12–14]

$$\left.\begin{matrix} XeF_2 \\ XeF_4 \\ XeF_6 \end{matrix}\right\} \xrightarrow[BrF_5,\ 273K]{AsF_5\ (excess)} \left.\begin{matrix} [Xe_2F_3]^+AsF_6^-;[XeF]^+AsF_6^- \\ XeF_4 \\ [XeF_5]^+\ AsF_6^- \end{matrix}\right\} \xrightarrow[293\ K]{Vacuum} XeF_4 \qquad (15.5)$$

It was also observed that from a mixture of XeF_2, XeF_4 and XeF_6, only XeF_6 reacts with NaF forming the ionic compound $Na_2[XeF_8]$. XeF_2 and XeF_4 can then be removed by applying vacuum. Na_2XeF_8 on warming, decomposes to give XeF_4, F_2 and NaF.

The oxides of xenon are quite unstable. Among these, XeO_3 has relatively better stability till 25°C and is best obtained by the reaction of XeF_6 with $HO(O)PF_2$. Previous methods include slow hydrolysis of XeF_6 and subsequent removal of HF as well as slow hydrolysis of XeF_4. Reaction of XeO_3 with NaOH gives sodium perxenate. Sodium perxenate is also obtained from XeF_6 and NaOH and it is the source for making the tetrahedral molecule XeO_4. XeO_4 is a yellow crystalline compound that is stable below –36°C. While XeO was predicted by theoretical studies to be unstable, XeO_2 which is stable only below –78°C, was prepared in 2011 by a careful reaction of XeF_4 and H_2O along with 2 M H_2SO_4. All the noble gas oxides are endothermic and decompose explosively into O_2 and Xe.[15–18]

$$XeF_6 + 3\,H_2O \longrightarrow XeO_3 + 6\,HF \tag{15.6}$$

$$4\,XeF_4 + 8\,H_2O \longrightarrow 2\,XeO_3 + 2\,Xe + O_2 + 16\,HF \tag{15.7}$$

$$n\,XeF_4 + 2n\,H_2O \xrightarrow{\text{dil } H_2SO_4} (XeO_2)_n + 4n\,HF \tag{15.8}$$

$$4\,XeF_6 \xrightarrow{HOP(O)F_2} 4\,XeO_3 \xrightarrow[64\%]{NaOH} \underset{+Xe}{Na_4XeO_6} \xleftarrow[54\%]{2M\ NaOH} 4\,XeF_6 \tag{15.9}$$

$$Na_4XeO_6 + 2\,H_2SO_4 \longrightarrow XeO_4 + 2\,H_2O + 2\,Na_2SO_4 \tag{15.10}$$

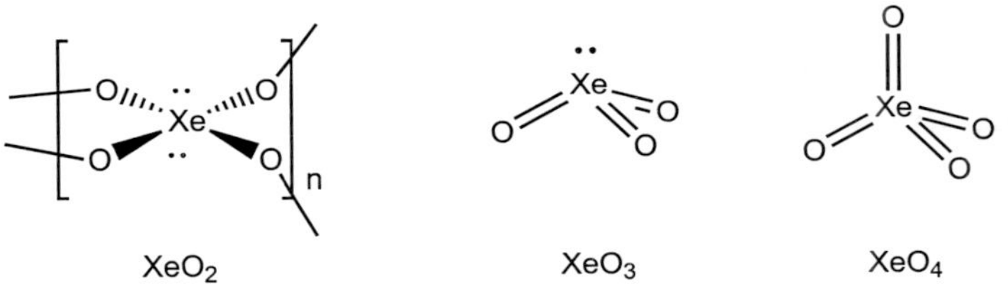

XeO_2 XeO_3 XeO_4

Fig. 15.1 Shapes of xenon oxides

Table 15.6 Comparison of oxides of xenon

Oxide	Xe = O	XeO_2 (2011)	XeO_3 (1982)	XeO_4 (1964)
Stability	Unstable; not known	Stable below –78°C	Stable below 25°C	Stable below –36°C
Shape	–	Predicted: bent Actual: extended square planar	Triangular pyramid	Tetrahedral
Heat of formation	–	487 kJ/mol endothermic	402 kJ/mol endothermic	643 kJ/mol endothermic
Preparation		$XeF_4 + H_2O$ + dil. H_2SO_4	$XeF_6 + HOP(O)F_2$	$Na_2XeO_6 + H_2SO_4$

An elegant way to stabilise xenon–oxygen single bonds was by the route of teflic acid (F_5TeOH). The reaction of hexahydroxytellurium $Te(OH)_6$ with fluorosulfonic acid results in teflic acid, F_5TeOH (m.p. 40°C, b.p. 60°C). The $F_5TeO–$ group is well suited for stabilisation of high valent, strong oxidant centres, and tends to provide stable compounds that have fluorine analogues. This group is predicted to have a very high group electronegativity and calculations predict it to be in the range of 3.87–3.88 (Allred–Rochow scale).[19]

$$Te + 3\,F_2 \longrightarrow TeF_6 \tag{15.11}$$

$$TeF_6 + 6\,H_2O \longrightarrow Te(OH)_6 + 6\,HF \tag{15.12}$$

$$Te(OH)_6 + 5\,HSO_3F \longrightarrow HOTeF_5 + 5\,H_2SO_4 \quad (15.13)$$

$$3\,HOTeF_5 + BCl_3 \xrightarrow{-80\,°C} B(OTeF_5)_3\ (\text{m.p.}\,37°C) + 3HCl \quad (15.14)$$

$$3\,XeF_4 + 4\,B(OTeF_5)_3 \xrightarrow{n\text{-}C_6F_{14}} 3\,Xe(OTeF_5)_4 + 4BF_3 \quad (15.15)$$

$$XeF_6 + 2\,B(OTeF_5)_3 \xrightarrow{n\text{-}C_6F_{14}} Xe(OTeF_5)_6 + BF_3 \quad (15.16)$$

$$KrF_2 + 2/3\,B(OTeF_5)_3 \xrightarrow[SO_2ClF]{-90\ \text{to}\ -112°C} Kr(OTeF_5)_2 + BF_3 \quad (15.17)$$

15.3.2 Cations and Anions of Noble Gases

The standard method to prepare xenon-based cations is by using very strong fluorinated Lewis acids such as AsF_5, SbF_5 or some transition metal fluorides such as PtF_5. On the other hand, the standard method to make xenon-based anions is by using fluoride ion salts such as NaF, KF, RbF, CsF, Me_4NF, NOF and NO_2F. However, reactivity varies from XeF_2 to XeF_6 and reagent to reagent. Also, compounds are formed where the cation or anion takes a molecule of the neutral reactant to be part of the product. The reaction of XeF_2 with AsF_5 results in both XeF^+ and $Xe_2F_3^+$. The anion XeF_3^- has so far not been prepared but has been identified in the mass spectral ionisation of XeF_2. The increase in Xe–F bond energy, for the cation formation (134 kJ/mol for XeF_2 versus 196 kJ/mol for $Xe_2F_3^+$) is a factor which favours the cation formation. The $Xe_2F_3^+$ molecule is V-shaped.[20]

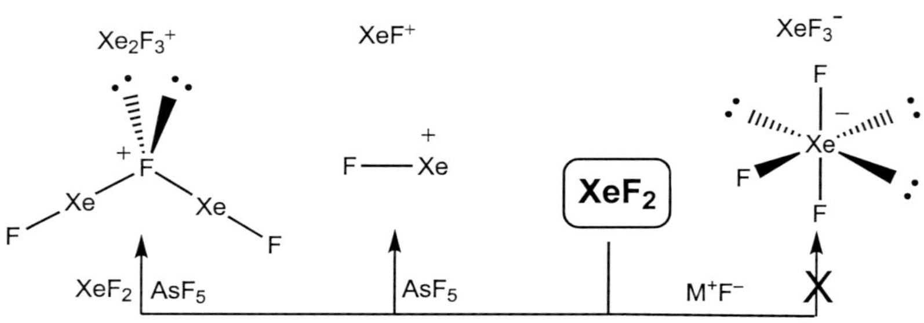

Scheme 15.3 Cations obtained from XeF_2

The fluorine donor ability of XeF_4 has been found to be less than that of XeF_2 and XeF_6. This observation is in agreement with the enthalpies of ionisation. As a result, the reaction of

AsF_5 does not proceed with XeF_4 at low temperatures and stronger Lewis acids such as SbF_5 at 50°C are required to convert XeF_4 to the T-shaped XeF_3^+.[21]

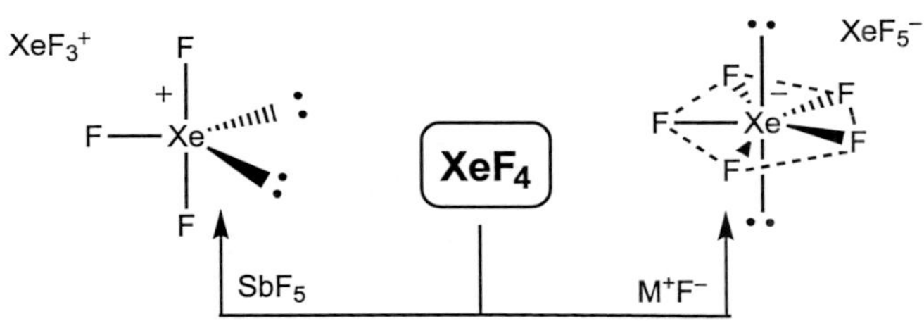

Scheme 15.4 Cations and anions obtained from XeF_4

XeF_6 forms two stable cations, namely, XeF_5^+ and $Xe_2F_{11}^+$. These are prepared by the reaction of strong binary fluorides such as AsF_5 and SbF_5. It also forms two anions namely XeF_7^- and $Xe_2F_{13}^-$.

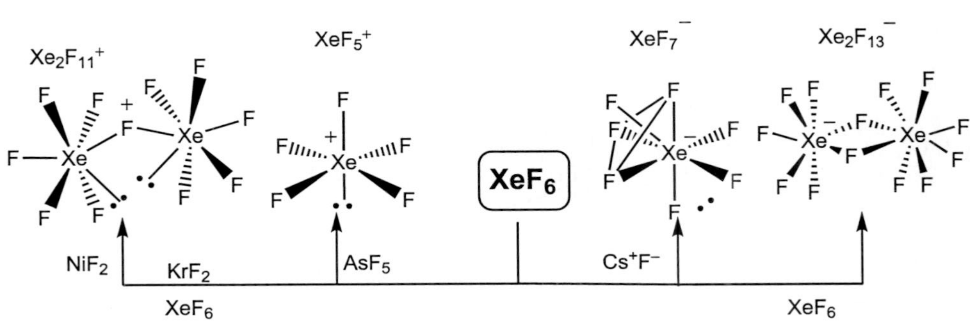

Scheme 15.5 Cations and anions obtained from XeF_6

15.3.3 Xenon Oxyfluorides

The three well known xenon oxyfluorides are $XeOF_4$, XeO_2F_2 and $XeOF_2$. Evidence for the formation of a few more neutral and ionic oxyfluorides exists. The most stable among the xenon oxyfluorides is the square pyramidal $XeOF_4$. A high yielding and safe method for its synthesis involves the use of F_3PO. $XeOF_4$ reacts with Me_4NF forming the explosive compound $XeOF_5^-$.[22,23]

The see-saw shaped and relatively unstable xenon dioxodifluoride is prepared from $XeOF_4$. A safe method for its synthesis involves the use of N_2O_5.[24]

$XeF_6 + F_3P{=}O \xrightarrow{-PF_5} XeOF_4$

$XeF_6 + O(SiF_3)_2 \xrightarrow{-SiF_4} XeOF_4$

$XeF_6 + CsNO_3 \xrightarrow{-CsF,\ -FNO_2} XeOF_4$

$XeF_6 + H_2O \text{ or } SiO_2 \xrightarrow{-HF} XeOF_4$

Scheme 15.6 Methods for preparation of $XeOF_4$

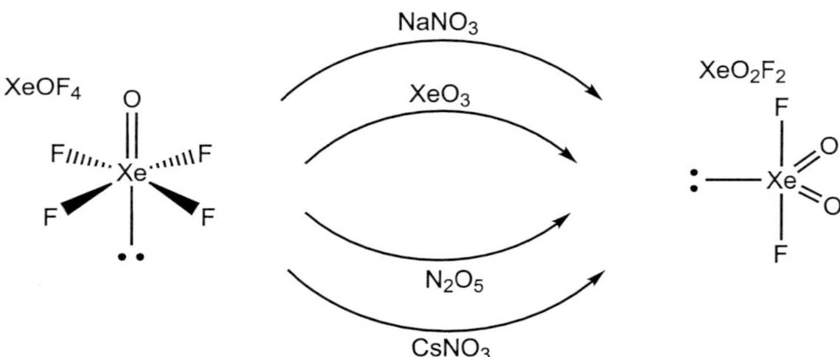

Scheme 15.7 Different ways of conversion of $XeOF_4$ to XeO_2F_2

In two studies, $XeOF_2$ was reported as the product of the co-condensation of H_2O and XeF_4 vapours at low temperatures. Co-deposited thin films were characterised by infrared spectroscopy. However there was a lack of complete evidence for its formation. Reproducible high yield and high-purity routes to $XeOF_2$ synthesis in synthetically useful quantities has been achieved by first making a CH_3CN adduct, separating and then removing the CH_3CN under vacuum.

$$XeF_4 + H_2O \xrightarrow[-45^\circ C]{CH_3CN} F_2OXe{:}NCCH_3 \xrightarrow[-CH_3CN]{\text{vacuum}} XeOF_2 \qquad (15.18)$$

15.3.4 The Chemistry of Xenon Difluoride, XeF_2

One of the most stable, easy to prepare, and easy to handle noble gas compounds is the linear XeF_2 molecule whose chemistry has been explored in detail. XeF_2 has been made from a mixture of xenon and fluorine using different sources of energy such as heat, UV light, direct sunlight, high density γ radiation, electric discharge, and so on. For thermal methods of the preparation, a large excess of xenon over fluorine is required to minimise the formation of higher fluorides. One of the most elegant and easy syntheses of XeF_2 yielding a pure product is a photochemical reaction of a gaseous mixture of xenon and fluorine with an excess of xenon. The mixture when exposed to direct sunlight (Scheme 15.1) or to UV light from a mercury lamp, forms pure XeF_2 and this reaction can be made to occur in gram to kilogram scale with addition of about 1 mol % of HF as a catalyst. Other reagents include ClF_3, AgF_2 and AgF_3.[25]

XeF_2 is a good fluoride ion donor. This finding is in agreement with the enthalpies of its ionisation [XeF_x $(g) \rightarrow XeF^+_{x-1}$ (g) + $F^-(g)$; ΔH_0 for XeF_2 = 9.45 eV]. Many fluoride ion acceptor Lewis acids such as AsF_5, SbF_5 and BiF_5 can withdraw a fluoride ion from XeF_2 forming the XeF^+ species. In the case of excess XeF_2 over Lewis acid, $Xe_2F_3^+$ species is also formed. The extra high melting point and stability of XeF_2 has recently been explained based on charge shift bonding which considers contributions from both ionic and covalent bonding.[26]

The use of XeF_2 as an oxidative fluorinating reagent has been explored in organic chemistry in detail. XeF_2 shows different types of reactions with organic substrates such as fluorodecarboxylation, fluorodeiodination, perfluoroalkylation, conversion of aryl aldehydes and ketones to aryl difluorinated ethers, and oxidofluorination of double and triple bonds.

The oxidative fluorinating ability of XeF_2 has also been widely utilised in inorganic chemistry, especially to make high oxidation state covalent fluorides of many compounds of *p*-block elements. The relatively high standard redox potential of XeF_2 makes it possible to oxidise water to oxygen. The relative redox potential of Xe–F bonds is compared with a few other xenon compounds in the following redox reactions.

$$XeO_2(OH)_4 + 2\,H^+ + 2\,e^- \rightleftharpoons XeO_3 + 3\,H_2O \qquad 2.38\text{ V} \qquad (15.19)$$

$$XeF_2 + 2\,H^+ + 2e^- \rightleftharpoons Xe\,(g) + 2\,HF \qquad 2.20\text{ V} \qquad (15.20)$$

$$XeO_3 + 6\,H^+ + 6e^- \rightleftharpoons Xe\,(g) + 3\,H_2O \qquad 2.10\text{ V} \qquad (15.21)$$

Schemes 15.8 and 15.9 consolidate some of the well known reactions of XeF_2 with inorganic and organic compounds.

A reaction occurs between organic iodides and XeF_2 where exchange of F for I occurs. 4-iodocubane carboxylic ester gets converted to fluorocubane carboxylate in 58% yield with XeF_2 at 45°C in dichloromethane (Scheme 15.9). Oxidative decarboxylation takes place when RCOOH is treated with XeF_2 which is essentially a Hunsdiecker reaction, and is useful in

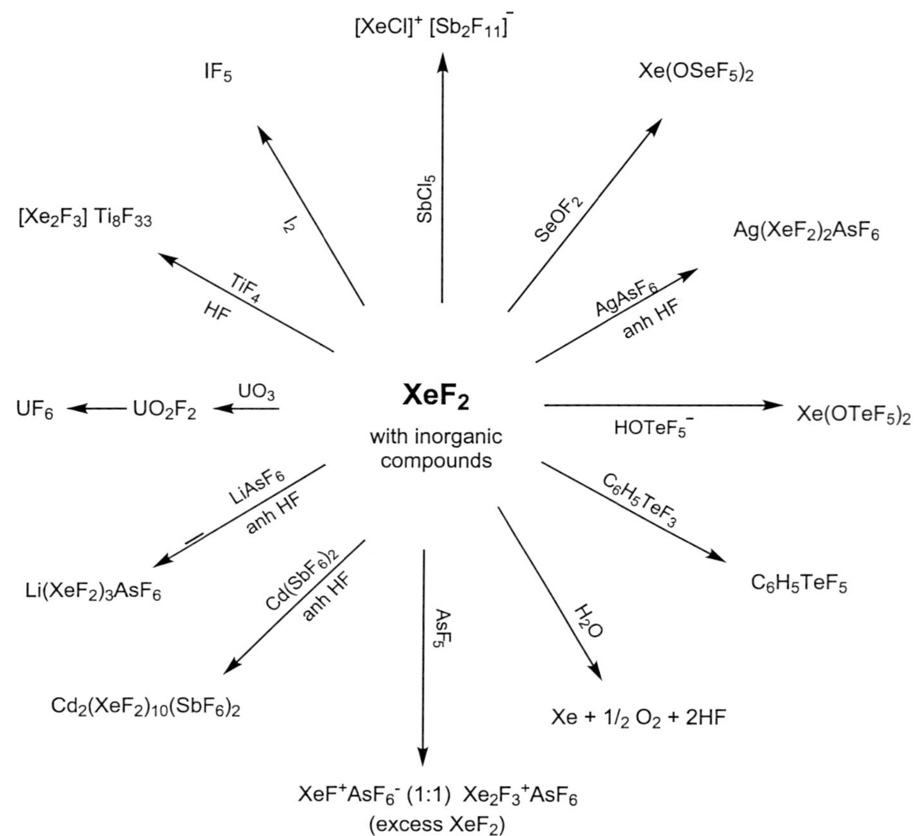

Scheme 15.8 Reactions of XeF_2 with inorganic compounds

the preparation of a wide range of fluorides. Perfluoroalkyl groups can also be introduced in aryl substrates by a fluoro-decarboxylative route. Para-dichlorobenzene undergoes efficient perfluoroalkylation with trifluoroacetic acid and analogous acids. The reaction possibly goes through the generation of a perfluoroalkyl radical species. Alkenes and alkynes can be converted to geminal difluorides with XeF_2 and 1,2-diketones undergo acid catalysed fluorination with XeF_2 (Scheme 15.9).

An interesting reaction is the one between XeF_2 and aryl aldehydes and ketones which is accompanied by a skeletal rearrangement. Aryl aldehydes and ketones get converted to difluoromethyl ethers when reacted with XeF_2 at room temperature in the presence of catalytic amounts of HF. Benzyl alcohol, in a similar skeletal rearrangement gets converted to fluoromethoxy benzene.[27]

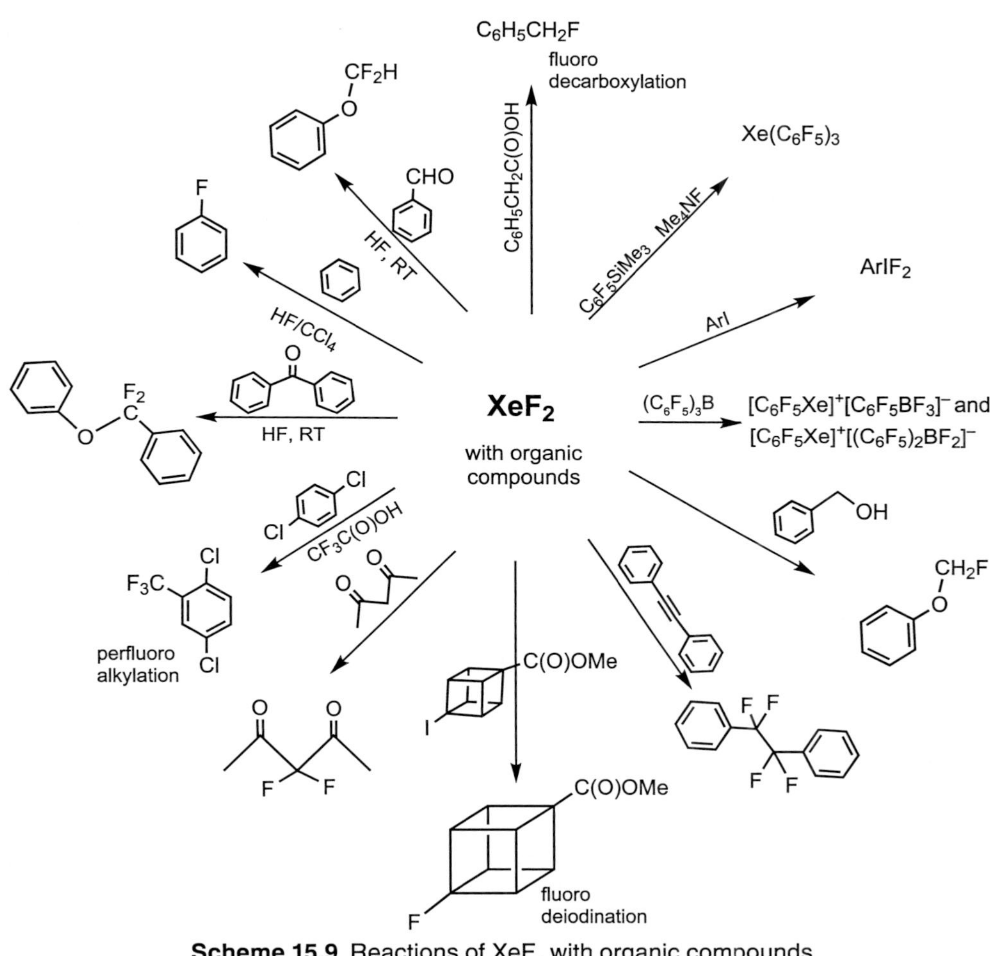

Scheme 15.9 Reactions of XeF_2 with organic compounds

15.3.5 Organoxenon Chemistry

Attempts to prepare organoxenon compounds were first made in 1979 when Lagow reported the synthesis of a highly unstable $Xe(CF_3)_2$. However, this compound had characterisation and reproducibility problems. After many failed attempts by various research groups, one method that showed promise was xenodeborylation by substitution of boron in organoboron compounds. To nucleophilically introduce organic groups on to XeF_2, the lack of permanent dipole of XeF_2 and the low electrophilicity of its Xe centre have to be overcome. The successful way to surmount these problems was the use of organic group transfer reagents with a Lewis acidic centre such as $B(C_6F_5)_3$, which polarises the hypervalent F–Xe–F bond and increases the electrophilicity of the xenon atom. $B(C_6F_5)_3$ was found to react with xenon difluoride with the formation of arylxenonium fluoroborates. The interaction of the Lewis acidic boron with one of the electron rich fluorine substituents of XeF_2 polarises the Xe–F bond, the hypervalent bond becomes asymmetric due to which the electrophilicity of the Xe increases.[28]

The first compounds containing a C–Xe(II) bond were reported in 1989.

$$(C_6F_5)_3B + XeF_2 \xrightarrow[n = 1 \text{ and } 2\ CH_2Cl_2;\ n = 3\ CH_3CN]{CH_3CN \text{ or } CH_2Cl_2} [C_6F_5Xe]^+ [(C_6F_5)_nBF_{4-n}]^- \quad (15.22)$$

Another access to such salts was based on the reaction of XeF_2 with in situ generated aryl difluoroboranes and this reaction was extended to perfluoroalkenes as well.

$$K^+[ArBF_3]^- + BF_3 \xrightarrow[-KBF_4]{CH_2Cl_2} ArBF_2 \xrightarrow[Ar\ =\ C_6F_5]{XeF_2\ /\ CH_2Cl_2} [ArXe]^+ [BF_4]^- \quad (15.23)$$

$$K^+[CF_2 = CFBF_3]^- + BF_3 \xrightarrow[-KBF_4]{CH_2Cl_2} CF_2 = CFBF_2 \xrightarrow[CH_2Cl_2]{XeF_2} [CF_2{=}CFXe]^+ [BF_4]^- \quad (15.24)$$

Alkyne derived xenon cations were prepared by using alkynyl lithium or trimethylsilyl alkyne derivatives.

$$t\text{-}BuC{\equiv}CLi \xrightarrow[-100^\circ C]{BF_3/CH_2Cl_2} Li^+ [t\text{-}BuC{\equiv}CBF_3]^- \xrightarrow[-40^\circ C]{XeF_2/CH_2Cl_2} [t\text{-}BuC{\equiv}CXe]^+ [BF_4]^- \quad (15.25)$$

$$RC{\equiv}CSiMe_3 + XeF_2 + BF_3.OEt_2 \xrightarrow[-70 \text{ to } -50^\circ C]{CH_2Cl_2} [RC{\equiv}CXe]^+ [BF_4]^- \quad (15.26)$$

R = Et, Pr, t-Bu, Me_3Si

While all of the above compounds indicated xenon–carbon cationic compounds, by careful choice of reagents, neutral xenon–carbon compounds were also made. The following reactions show the synthesis of the first neutral xenon–carbon compound which was prepared from pentafluorophenyl xenonium cation which itself was obtained from the reaction of XeF_2 and $B(C_6F_5)_3$.[29]

$$XeF_2 + B(C_6F_5)_3 \longrightarrow [C_6F_5Xe]^+ [C_6F_5BF_3]^- + [C_6F_5Xe]^+ [(C_6F_5)_2 BF_2]^- \quad (15.27)$$

The pentafluorophenyl xenonium cation is stable in CH_3CN and the crystal structure of its acetonitrile complex was determined.

$$[C_6F_5Xe]^+ [(C_6F_5)_2 BF_2]^- + M^+ AsF_6^- \longrightarrow \underset{\text{Stable}}{[C_6F_5Xe]^+ [AsF_6]^-} \quad (15.28)$$

$$[C_6F_5Xe]^+[AsF_6]^- + M^+OC(O)C_6F_5^- \longrightarrow C_6F_5\text{–Xe–O–C(=O)–}C_6F_5 \quad (15.29)$$

First stable Xe–C bond which is covalent m.p. 85°C

By using pentafluorophenyl trimethylsilane as the transfer reagent, a neutral xenon–carbon bonded compound having two xenon carbon bonds, $(C_6F_5)_2Xe$, was prepared as a colourless solid that decomposed spontaneously at 20°C. Attempts to prepare the analogous CF_3–Xe compound by the same methodology was unsuccessful resulting in C_2F_6. By anion exchange of $(C_6F_5Xe)AsF_6$ with $(Me_4N)F$, the linear compound $(C_6F_5)XeF$ was also prepared.

$$XeF_2 + 2\,C_6F_5SiMe_3 \xrightarrow[-60\text{ to }-40°C]{[NMe_4]F} Xe(C_6F_5)_2 + 2\,Me_3SiF \quad (15.30)$$

$$XeF_2 + C_6F_5SiMe_3 \xrightarrow[-60\text{ to }-40°C]{[NMe_4]F} C_6F_5XeF + Xe(C_6F_5)_2 + Me_3SiF \quad (15.31)$$

$$[C_6F_5Xe][AsF_6]_{(s)} + [NMe_4]F \xrightarrow[{[NMe_4]^+[AsF_6]^-}]{CH_2Cl_2\,/\,-78°C} C_6F_5XeF \quad (15.32)$$

The first example of a xenon–carbon compound having xenon in the tetravalent state was prepared starting from XeF_4.

$$XeF_4 + C_6F_5BF_2 \xrightarrow[-55\text{ to }-40°C]{CH_2Cl_2} [C_6F_5XeF_2]^+[BF_4]^- \quad (15.33)$$

15.3.6 Xenon-Nitrogen Compounds

The first example of a compound having a xenon–nitrogen bond was prepared by the reaction of xenon difluoride with bis-fluorosulfonylamine at 0°C. The white solid undergoes thermal decomposition at 70°C.

$$XeF_2 + HN(SO_2F)_2 \xrightarrow[273\text{ K}]{CF_2Cl_2} FXeN(SO_2F)_2 + HF \quad (15.34)$$

This study was followed by the synthesis of a similar compound $Xe[N(SO_2CF_3)_2]$ as a white solid that is stable up to 72°C. The reactions of $XeF^+AsF_6^-$ and $Xe_2F_3^+AsF_6^-$ with HCN, CH_3CN, CH_2FCN, C_2H_5CN, C_2F_5CN, C_3F_7CN and C_6F_5CN were carried out by combining stoichiometric amounts of the reactants in anhydrous HF and warming to –20 to –10°C and dissolution in the solvent.

$$XeF^+AsF_6^- + RCN \longrightarrow R\text{–C}{\equiv}\text{N–}\overset{+}{Xe}\text{–F}\; AsF_6^- \quad (15.35)$$

The fluoro(perfluoropyridine)xenon(II) cations, 4-$RC_5F_4N^-XeF^+$ (R = F or CF_3), have also been observed in HF and BrF_5 solutions (stable up to –30°C) and their AsF_6^- salts have been isolated from BrF_5 solutions.

15.3.7 Xenon-Chlorine Compounds

$XeCl_2$ has only been detected in a matrix at extremely low temperatures. However, $(C_6F_5Xe)_2Cl^+$ was prepared from $C_6F_5Xe^+$ and $(CH_3)_3SiCl$. In the solid state it shows contact distances between Xe and Cl of 284.7(1) and 278.4(2) pm and a Xe–Cl–Xe bond angle of 116.96(8)°. $C_6F_5Xe–Cl$ has also been isolated by a reaction carried out under strict conditions starting with $C_6F_5Xe^+$ and the HCl–pyridine adduct 4-$ClC_5H_4N{\cdot}HCl$. However due to instability in solution, no detailed structural information has been provided. As the bond strength in XeF^+ is stronger than in XeF_2, salts with the $XeCl^+$ ion were tried. Seppelt and coworkers reacted $[XeF]^+[SbF_6]^-$ in HF/SbF_5 solution with small amounts of $SbCl_5$. The compound $XeCl^+\ Sb_2F_{11}^-$ obtained was structurally characterised and shows a Xe–Cl distance of 230.9(2) pm.[30]

$$\underset{\text{yellow}}{XeF^+\ SbF_6^-} \xrightarrow[HF/SbF_5]{SbCl_5} \underset{\text{orange}}{XeCl^+\ Sb_2F_{11}^-} \tag{15.36}$$

15.4 Chemistry of Krypton

Along with xenon, krypton is the other noble gas which forms chemical compounds in macroscopic amounts. The first krypton compound to be prepared was KrF_2. All krypton compounds currently known are prepared from KrF_2.

Krypton difluoride is a powerful oxidising and fluorinating agent. It is more powerful than even elemental fluorine or xenon fluorides with a redox potential of 3.27 V, making it the most powerful oxidising agent known. It can oxidise gold to its highest-known oxidation state, +5. It has been observed that KrF_2 can bring about a rapid room temperature oxidative fluorination of xenon gas to XeF_6 and I_2 to IF_7.[31–33]

$$\underset{\text{liquified mixture}}{Kr + F_2\ (1{:}4)} \xrightarrow[\text{near UV, 77K}]{} KrF_2 \tag{15.37}$$

$$7\ KrF_2 + 2\ Au \longrightarrow 2\ KrF^+\ AuF_6^- + 5\ Kr \tag{15.38}$$

$$2\ KrF^+\ AuF_6^- \xrightarrow[60°C]{} \underset{\text{Gold(V) fluoride}}{Au_2F_{10}} + 2\ Kr + F_2 \tag{15.39}$$

$$3\ KrF_2 + 2\ Ag \longrightarrow \underset{\text{Silver(III) fluoride}}{2\ AgF_3} + 3\ Kr \tag{15.40}$$

$$KrF_2 + HC{\equiv}NH^+AsF_6^- \xrightarrow[BrF_5]{} [HC{\equiv}N{-}Kr{-}F]^+ AsF_6^- \quad \text{stable upto } -55°C \tag{15.41}$$

$$Xe + 3\,KrF_2 \longrightarrow XeF_6 + 3\,Kr \tag{15.42}$$

$$I_2 + 7\,KrF_2 \longrightarrow 2\,IF_7 + 7\,Kr \tag{15.43}$$

The reason for KrF_2 being the strongest fluorinating and oxidising agent is evident from the Kr–F bond enthalpy which is much lower than that of F–F or Xe–F and its redox potential. The standard redox potential of KrF_2 is estimated to be 3.27 V which is higher than XeF_2 (2.32 V), F_2 (2.87 V), H_4XeO_6 (2.42 V) and XeO_3 (2.12V).

Table 15.7 Comparison of bond enthalpies of noble gas compounds with halogens

Bond	Bond enthalpy
F–F	154 kJ/mol
I–I	149 kJ/mol
Xe–F	131 kJ/mol
Xe–O	84 kJ/mol
Kr–F	49 kJ/mol

15.5 Chemistry of Argon

The discovery of the first compound of argon HArF, argon fluorohydride, is credited to a group of Finnish scientists, led by Markku Räsänen. On 24 August 2000, they announced their discovery of argon fluorohydride in the journal *Nature*. This compound was synthesised by mixing argon and hydrogen fluoride on a surface of cesium iodide at 8 K (–265°C), and exposing the mixture to ultraviolet radiation causing the gases to combine to form HArF. The infrared spectrum of the resulting gas mixture indicated that it contains chemical bonds, though very weak ones. Chemical bonds of argon fluorohydride are stable only if the substance is kept at temperatures below 17 K (–256°C). On increasing the temperature, it decomposes into argon and hydrogen fluoride.[34]

15.6 Noble Gas Elements and Compounds as Ligands

Stefan Seidel and Konrad Seppelt of the Free University of Berlin made the first example of a compound where a noble gas molecule was acting as a ligand in a serendipitous manner. They had originally set out to prepare the elusive gold(I) fluoride (AuF) by reducing AuF_3 with a weakly coordinating agent, arsenic trifluoride (AsF_3) in a solution of HF/SbF_5. This reaction resulted in a novel derivative of AuF. To obtain AuF itself, the researchers decided to replace AsF_3 in their reaction with xenon, another mild reducing agent. To their surprise, the reduction reaction stopped at the Au^{2+} stage and produced an unexpected cation, $AuXe_4^{2+}$.

They could grow dark red crystals of this cation in the reaction solution at –78°C. The cation decomposed when the xenon was removed from the reaction mixture by vacuum, but was stable at room temperature if a positive xenon pressure was maintained. The $AuXe_4^{2+}$ ion is stabilised by interactions with the fluoride atoms of the counter ions. The Au–Xe bond length varies between 272.8 to 275.0 pm. It is also unusual in that it uses xenon as a ligand, and in that it contains gold in the +2 oxidation state. It is produced by reduction of AuF_3 in the presence of fluoroantimonic acid and xenon, and crystallised at low temperature.[35]

$$AuF_3 + 6\,Xe + 3H^+ \xrightarrow[HF/SbF_5]{} [AuXe_4]^{2+}(Sb_2F_{11})_2^- + Xe_2^+ + 3\,HF \qquad (15.44)$$

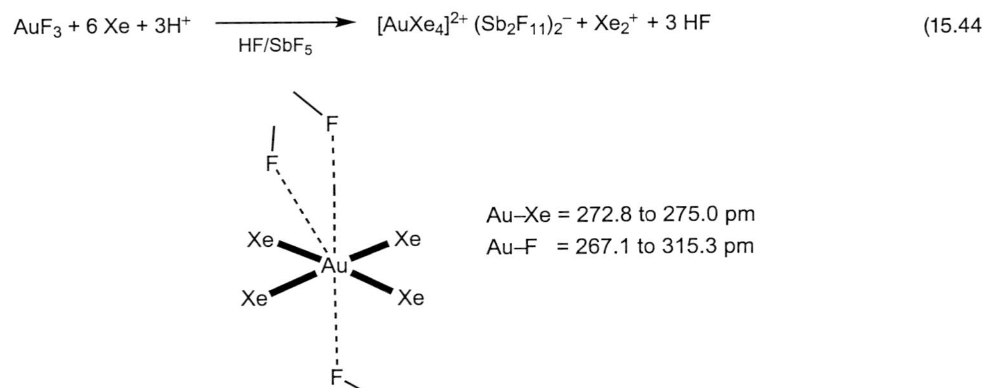

Fig. 15.2 Structure of $(AuXe_4)^{2+}(Sb_2F_{11})_2^-$

This path breaking discovery was followed by more examples of gold complexes of xenon with xenon binding as one or two (*cis* and *trans*) ligands on gold and also gold in +1 and +3 oxidation states. Seppelt and coworkers tried to extend this chemistry to other transition metals and found that they can make mercury–xenon complexes as well. The first example of a mercury–xenon compound was prepared in 2003. The crystal structure of the compound indicated the presence of a Hg–Xe bond of length 276.93 pm. The coordination sphere of Hg^{2+} is completed with six fluorine atoms belonging to the counter ions at distances between 227 and 259 pm.[36]

$$HgF_2 + \text{excess } SbF_5 \xrightarrow[60°C]{Xenon} [HgXe]^{2+}[SbF_6]^-[Sb_2F_{11}]^- \qquad (15.45)$$

XeF_2, XeF_4 and KrF_2 can act as ligands in coordination complexes of metals. For example, in HF solution, $Mg(AsF_6)_2$ reacts readily giving complexes, most of which have been structurally characterised. The xenon and krypton fluorides retain their original geometry in the complexes formed.[37–39]

$$Mg(AsF_6)_2 + 4\,XeF_2 \xrightarrow[HF]{} [Mg(XeF_2)_4](AsF_6)_2 \qquad (15.46)$$

$$Mg(AsF_6)_2 + XeF_2 + XeF_4 \xrightarrow[HF]{} [Mg(XeF_2)(XeF_4](AsF_6)_2 \qquad (15.47)$$

$$Mg(AsF_6)_2 + 4\,KrF_2 \xrightarrow[HF,\ BrF_5]{} [Mg(KrF_2)_4](AsF_6)_2 \qquad (15.48)$$

Bartlett et al. in 1991 reported the first metal complex salt, $[Ag(XeF_2)_2]AsF_6$, in which XeF_2 functions as a transition metal ligand. The Ag(I)–XeF_2 complex salt, $[Ag(XeF_2)_2]PF_6$, was obtained by the reaction of XeF_2 and $AgPF_6$ in anhydrous hydrogen fluoride (aHF). This was followed by a large number of similar complexes.

15.7 Noble Gases for Making Excimer Lasers

Noble gas compounds along with halogens are used to make a unique type of laser known as the *excimer laser* or sometimes as the *exciplex laser*. The excimer laser uses gases such as F_2 and Cl_2, mixed with inert gases such as argon, krypton or xenon. When this mixture is electrically excited, the mixture of gases emits ultraviolet light of a specific frequency, which can make very precise, minute changes to irradiated material, such as polymers. The term excimer is short for *excited dimer*, while exciplex is short for *excited complex*. The excited dimer term comes from the fact that the first excimer was made with a xenon dimer (Xe_2) excited by an electron beam to give stimulated emission at 172 nm wavelength. This invention was made by three Russian scientists of the Lebedev Physical Institute in Moscow in 1970 (Nikolai Basov, V A Danilychev and Yu M Popov). A significant improvement, developed by many groups in 1975 was the use of noble gases along with halogens, thereby getting access to a range of UV frequencies (Table 15.3).[40]

Two important and path-breaking applications of excimer lasers were independently discovered by two research groups in America in 1981–82 and in both, the leading scientists were persons of Indian origin. In 1981, Rangaswamy Srinivasan (originally from the University of Madras, India and Ph.D. from University of Southern California) while working at the IBM TJ Watson Research Centre, discovered that an ultraviolet excimer laser could be used to etch a living tissue without causing any thermal damage to the surrounding areas. He named the phenomenon Ablative Photo-Decomposition (APD). Three years later, Rangaswamy collaborated with an ophthalmic surgeon to use a ArF (193 nm) excimer laser to reshape the cornea, to arrive at a novel method of eye surgery called Lasik surgery (*laser-assisted in situ keratomileusis*) which has since then helped millions of people as a vision correction surgery and permanent alternative to eye glasses or contact lenses. An ophthalmologist uses the excimer laser to reshape the eye's cornea in order to improve visual acuity.[41]

Table 15.8 Wavelength and relative power of excimer lasers

Excimer noble gas/noble gas–halogen combination	Wavelength of laser produced	Relative power mW
Xe_2^*	172 and 175 nm	NA
ArF	193 nm	60
KrCl	222 nm	25
KrF	248 nm	100
XeBr	282 nm	NA
XeCl	308 nm	50
XeF	351 nm	45

The UV radiation of a specific frequency from an excimer laser is absorbed efficiently by biological matter and organic compounds. The excimer laser does not burn or cut but adds sufficient energy to break the molecular bonds of the surface tissue, which effectively causes the disintegration of the molecules into the air in a tightly controlled manner by the process of ablation rather than burning. Thus, excimer lasers can remove exceptionally fine layers of surface material with almost no heating and the surrounding material is left intact.

It is recorded that Srinivasan brought his Thanksgiving turkey leftovers to the lab the day after Thanksgiving in 1981. He and his colleagues used the ArF excimer laser at 193 nm to etch patterns in whatever bone, cartilage or meat was on the turkey sample. By using the excimer laser, they were getting an extremely clean cut, with no evidence of damage to the surrounding tissue. To visually demonstrate this effect to the medical technology field in a dramatic way, Srinivasan and coworkers produced a highly magnified electron micrographic image of a single human hair, which has been etched by the excimer laser.[41]

The other path-breaking discovery was made by Dr Kanti Jain (an engineering degree holder from IIT Kharagpur, India with a Ph.D. from the University of Illinois, USA). The most widespread industrial application of excimer lasers has been in deep ultraviolet photolithography, a critical technology used in the manufacturing of microelectronic devices (integrated circuits). In a pioneering development in 1982, deep-UV excimer laser lithography was proposed and demonstrated at IBM by Kanti Jain. He published the first paper and gave the first talk on excimer laser lithography. Laser lithography is used extensively today to make microchips for the computer and electronics industry. Jain developed the technology while working at the IBM San Jose Research Laboratory. It is one of the critical technologies required for microelectronic chip manufacturing and present day state-of-the-art lithography tools use deep ultraviolet light from the KrF and ArF excimer lasers with wavelengths of 248 and 193 nanometers which has enabled transistor sizes to shrink below 45 nm. Semiconductor electronic devices fabricated using laser lithography total $400 billion in annual production in 2017.[42]

Questions for self study (multiple choice)

15.1. Among noble gases, those gases which have been found to form isolable and non-radioactive compounds by chemical reactions are

(a) Xe, Kr, Ar (b) Xe, Kr, Ne
(c) Xe, Kr, He (d) Xe, Ar, Ne

15.2. Among the given reactions, which one is not a method for making XeF_2?

(a) $Xe + AgF_2 + BF_3$ (b) $Xe + ClF_3$
(c) $Xe + AuF_3 + H^+$ (d) $Xe + F_2 + NiF_2$ (catalyst)

15.3. Xenon-based product(s) obtained from the reaction of XeF_2 with AsF_5 is (are)

(a) XeF_4 (b) $XeF^+ + Xe_2F_3^+$
(c) $XeF^+ + XeF_3^+$ (d) $XeF_3^+ + Xe_2F_3^+$

15.4. The bond enthalpy of noble gas compounds varies in the order

(a) Kr–F < Xe–O < Xe–F (b) Kr–F > Xe–O > Xe–F
(c) Kr–F < Xe–F < Xe–O (d) Xe–O < Kr–F < Xe–F

15.5. The melting points of xenon fluorides varies in the order
(a) $XeF_2 < XeF_4 < XeF_6$ (b) $XeF_6 < XeF_4 < XeF_2$
(c) $XeF_4 < XeF_6 < XeF_2$ (d) $XeF_6 < XeF_2 < XeF_2$

15.6. Which among the following statements is not true for XeO_3?
(a) It is formed by the hydrolysis of XeF_6
(b) It is formed by the hydrolysis of XeF_4
(c) It reacts with NaOH giving perxenates (Na_4XeO_6)
(d) It reacts with H_2SO_4 giving XeO_4

15.7. Which among the following statements is not true for XeO_2?
(a) It is prepared by the reaction of XeF_4 and dil. H_2SO_4
(b) It is the least stable among XeO_2, XeO_3 and XeO_4
(c) It has a bent structure
(d) It decomposes to give Xe and O_2

15.8. LASIK eye surgery makes use of an excimer laser made of
(a) Kr and F (b) Xe and F
(c) Ar and F (d) Kr and Cl

15.9. Metals on which xenon gas has been found to act as a ligand are
(a) Au and Ag (b) Au and Pt
(c) Au and Hg (d) Au and Cu

15.10. The noble gas compound which is known to convert I_2 to IF_5 is
(a) KrF_2 (b) XeF_2
(c) $XeOF_5^-$ (d) $XeOF_4$

15.11. The very first noble gas compound to be prepared was
(a) XeF_2 (b) $[XeF]^+[Pt_2F_{11}]^-$
(c) $[Xe]^+[PtF_6]^-$ (d) $[Xe]^+[Pt_2F_{11}]^-$

Solved concept-based questions

15.12. Xenon is the only element among noble gases which form compounds readily. Why?

The first ionisation potential of xenon is the lowest among all non-radioactive noble gases other than radon. Radon, which has a lower first ionisation potential has a very short half life and hence cannot be handled easily. The longest-lived isotope, radon-222, has a half-life of only 3.82 days.

15.13. Noble gases have very low boiling points. Why?

Noble gases are monoatomic and except weak dispersion forces, no interatomic forces exist between the atoms for the smaller elements such as He and Ne. The electron clouds of the larger noble gas atoms become polarised which leads to weak van der Waals forces among the atoms. Overall therefore they have low boiling points.

15.14. XeF_2 is a very convenient fluorinating agent. Why?

It is a crystalline solid which is easily prepared (by mixing Xe And F_2 in the proper molar ratio and keeping the flask in sunlight). The side product of the fluorination is xenon gas which is unreactive and easily separated from the products. In addition, this evolved xenon can be mixed again with F_2 and XeF_2 regenerated.

15.15. Structures of xenon fluorides cannot be explained by the conventional valence bond theory. Why?

The conventional valence bond theory is based on elements acquiring the octet. The noble gases already have the octet configuration and the compounds they form are hypervalent compounds in which the valence electron count is an even number above 8.

15.16. Oxygen is less electronegative than fluorine. But unlike fluorine, it is able to form compounds with xenon in its highest formal oxidation state of +8 (XeF_6 versus XeO_4). Why?

Oxygen is only second to fluorine in the electronegativity scale. However, it can form double bonds which reduces crowding around the central atom. Therefore oxygen is able to form compounds with xenon with a formal oxidation state of 8.

15.17. Other than the presence of fluorine, what is common when one considers the type of chemical bonding involved in XeF_2, ClF_3, SF_4 and PF_5?

They all have 10 valence electrons (they are hypervalent compounds).

15.18. Xenon oxides are acidic. Why?

Xenon oxides on hydrolysis give hydroxenous and hydroxenic acids. Hence they are acidic.

15.19. Why is the teflate group ($OTeF_5$) able to stabilise Xe–O single bonds?

Teflate has a group electronegativity which is similar to that of fluorine. Calculations predict the group electronegativity around 3.87, close to that of fluorine.

Problems and exercises

15.20. What is Teflic acid? How is it prepared from tellurium and how is it used for the synthesis of the only known example of a krypton–oxygen compound?

15.21. Suggest the best fluorine-based reagents from the following reactions.

(a) Ag ⟶ AgF_3

(b) [2-fluoro-1-indanone] ⟶ [3,3,4-trifluoro-chromane structure]

(c) XeF_6 ⟶ XeO_3

(d) XeF_4 ⟶ XeF_3^+

15.22. Which the strongest known oxidising agent? Give a reason justifying your answer.

15.23. Given a mixture of XeF_2 and XeF_6, suggest a method to separate pure XeF_2 by means of chemical methods.

15.24. Write the structure of the product in the following reactions of XeF_2.

$$C_6H_5COCH_3 \xrightarrow[\text{HF, RT}]{XeF_2}$$

$$H_3C\text{-}C_6H_4\text{-}CH_2OH \xrightarrow[\text{HF, RT}]{XeF_2}$$

15.25. Arrange the following compounds in the decreasing order of M–F bond enthalpy.

Kr–F, Si–F, C–F, Xe–F, F–F

15.26. How is XeO_2F_2 prepared starting from xenon gas?

15.27. Give four elements or compounds with which xenon gas has been found to directly react forming isolable compounds. Show the formula of the products formed in each case.

15.28. How are XeF_3^+, $Xe_2F_3^+$ and XeO_4 prepared from xenon?

Answer key for self study questions

15.1. (a) 15.2. (c) 15.3. (b) 15.4. (a) 15.5. (b) 15.6. (d) 15.7. (c) 15.8. (c) 15.9. (c) 15.10. (b)

Supplementary reading

1. Rayleigh, Density of nitrogen, *Nature*, 1892, Vol. 46, 512.
2. Yost D M and Kaye A L, An attempt to prepare a chloride or fluoride of xenon, *J. Am. Chem. Soc.*, 1933, Vol. 55, 3890.
3. Christe K O, Bartlett's discovery of noble gas fluorides, a milestone in chemical history, *Chem. Commun.*, 2013, Vol. 49, 4588.
4. Bartlett N and Lohmann D H, Dioxygenylhexafluoroplatinate (V), $O_2^+[PtF_6]^-$, *Proceedings of the Chemical Society*, 1962, Vol. 3, 115.
5. Bartlett N, Xenon hexafluoroplatinate$Xe^+[PtF_6]^-$, *Proceedings of the Chemical Society,* 1962, Vol. 6, 218.
6. Graham L, Graudejus O, Jha N K and Bartlett N, Concerning the nature of $XePtF_6$, *Coord. Chem. Rev.*, 2000, Vol. 197, 321.
7. Hoppe R, Daehne W, Mattauch H and Roedder K, Fluorination of xenon, *Angew. Chem. Int. Ed.*, 1962, Vol. 11, 599.
8. Holloway J H, Synthesis of xenon difluoride by exposure of xenon–fluorine mixture to daylight at room temperature, *Chem. Commun.*, 1966, 22.
9. Classen H H, Selig H and Malm J G, Xenon tetrafluoride, *J. Am. Chem. Soc.*, 1962, Vol. 84, 3593.
10. Betz R and Schrobilgen G J, The synthesis of XeF_2. A predictable outcome with a significant chemistry, *Z. Anorg. Allg. Chem.*, 2012, 1385.

11. Ibers J A and Hamilton W C, Xenon tetrafluoride: Crystal structure, *Science*, 1963, Vol. 139, 3550.
12. Hoyer S, Emmler K and Seppelt T, The structure of xenon hexafluoride in the solid state, *J. Fluor. Chem.*, 2006, Vol. 127, 1415.
13. Moody G J, A decade of xenon chemistry, *J. Chem. Educ.*, 1974, Vol. 51, 628.
14. Grochala W, Atypical compounds of gases, which have been called noble, *Chem. Soc. Rev.*, 2007, Vol. 36, 1632.
15. Selig H, Malm J G, Claassen H H, Chernick C L and Huston J L, Xenon tetroxide—preparation and some properties, *Science*, 1964, Vol. 143, 1322.
16. Huston J L, Studier M H and Sloth E N, Xenon tetroxide—mass spectrum, *Science*, 1964, Vol. 143, 1162.
17. Templeton D H, Zalkin A, Forrester J D and Williamson S M, Crystal and molecular structure of xenon trioxide, *J. Am. Chem. Soc.*, 1963, Vol. 85, 817.
18. Brock D S and Schrobilgen G J, Synthesis of the missing oxide of xenon, XeO_2, and its implications for earth's missing xenon, *J. Am. Chem. Soc.*, 2011, Vol. 133, 6265.
19. Birchall T, Myers R D, Waard H D and Schrobilgen G J, Multinuclear NMR and Mössbauer study of pentafluorooxotellurate ($OTeF_5$) derivatives of tellurium, iodine, and xenon: spectroscopic determination of the relative electronegativities of fluoride and pentafluorooxotellurate, *Inorg. Chem.*, 1982, Vol. 21, 1068.
20. Vasdev N, Moran, M D, Tuononen H M, Chirakal R, Suontamo R J, Bain A D and Schrobilgen G J, NMR spectroscopic evidence for the intermediacy of XeF_3^- in XeF_2/F^- exchange, attempted syntheses and thermochemistry of XeF_3^-salts, and theoretical studies of the XeF_3^-anion, *Inorg. Chem.*, 2010, Vol. 49, 8997.
21. Mckee D E, Zalkin A and Bartlett N, Crystal structure of $[XeF_3^+]$ $[Sb_2F_{11}^-]$, *Inorg. Chem.*, 1973, Vol. 12, 1713.
22. Gillespie R J and Schrobilgen G J, Trifluoroxenon(1+), trifluorooxoxenon(1+0), and fluorodioxoxenon(1+) cations. Preparation and characterisation by fluorine-19 nuclear magnetic resonance spectroscopy, *Inorg. Chem.*, 1974, Vol. 13, 2370.
23. Christe K O, Dixon D A, Sanders J C P, Schrobilgen G J, Tsai S S and Wilson W W, On the structure of the $[XeOF_5]^-$anion and of heptacoordinated complex fluorides containing one or two highly repulsive ligands or sterically active free valence electron pairs, *Inorg. Chem.*, 1995, Vol. 34, 1868.
24. Christe K O and Wilson W W, New syntheses and properties of XeO_2F_2, $Cs^+XeO_2F_3^-$, and $NO_2^+[XeO_2F_3.nXeO_2F_2]^-$, *Inorg. Chem.*, 1988, Vol. 27, 3763.
25. Williamson S M, Sladky F O and Bartlett N, Xenon difluoride, *Inorg. Synth.*, 1968, Vol. 11, 147.
26. Braida B and Hilberty P C, The essential role of charge shift bonding in hypervalent prototype XeF_2, *Nature Chem.*, 2009, Vol. 5, 417.
27. Tius M A, Xenon difluoride in synthesis, *Tetrahedron*, 1995, Vol. 51, 6605.
28. Koppe K, Haner J, Mercier H P A, Frohn H J and Schrobilgen G J, Xenon(IV)–Carbon bond of $[C_6F_5XeF_2]^+$; Structural characterisation and bonding of $[C_6F_5XeF_2][BF_4]$, $[C_6F_5XeF_2][BF_4]{\bullet}2HF$, and $[C_6F_5XeF_2][BF_4]{\bullet}nNCCH_3$ (n = 1, 2); and the fluorinating properties of $[C_6F_5XeF_2][BF_4]$, *Inorg. Chem.*, 2014, Vol. 53, 11640.
29. Frohn S-J and Bardin V V, Preparation and reactivity of compounds containing a carbon–xenon bond, *Organometallics*, 2001, Vol. 20, 4750.
30. Seidel S and Seppelt S, The $XeCl^+$ Ion: $[XeCl]^+[Sb_2F_{11}]^-$, *Angew. Chem. Int. Ed.*, 2001, Vol. 40, 4225.
31. MacKenzie D R, Krypton Difluoride: Preparation and Handling, *Science*, 1963, Vol. 141, 1171.

32. Lehmann J, Mercier H P A and Schrobilgen G J, The chemistry of krypton, *Coord. Chem. Rev.*, 2002, Vol. 233–234, 1.
33. Lehmann J F, Dixon D A and Schrobilgen G J, X-ray crystal structures of α-KrF_2, [KrF][MF_6] (M = As, Sb, Bi), [Kr_2F_3][SbF_6]·KrF_2, [Kr_2F_3]$_2$[SbF_6]$_2$·KrF_2, and [Kr_2F_3][AsF_6]·[KrF][AsF_6]; Synthesis and characterisation of [Kr_2F_3][PF_6]·nKrF_2; and theoretical studies of KrF_2, KrF^+, $Kr_2F_3^+$, and the [KrF][MF_6] (M = P, As, Sb, Bi) ion pairs, *Inorg. Chem.*, 2001, Vol. 40, 3002.
34. Khriachtchev L, Petterson M, Runeberg N, Lundell J and Räsänen M, A stable argon compound, *Nature*, 2000, Vol. 406, 874.
35. Seidel S and Seppelt K, Xenon as a complex ligand:The tetra xenono gold(II) cation in $AuXe_4$+($Sb_2F_{11}^-$)$_2$, *Science*, 2000, Vol. 290, 117.
36. Hwang I C, Seidel S and Seppelt K, Gold(I) and Mercury(II) Xenon Complexes, *Angew. Chem. Int. Ed.*, 2003, Vol. 42, 4392.
37. Tramšek M, Benkič P and Žemva B, The first compound containing a metal center in a homoleptic environment of XeF_2 molecules, *Angew. Chem. Int. Ed.*, 2004, Vol. 43, 3456.
38. Tavcar G and Tramsek M, XeF_2 as a ligand to a metal center, an interesting filed of noble gas chemistry, *J. Fluor. Chem.*, 2015, Vol. 174, 14.
39. Lozinsek M, Mercier H P A, Brock D S, Zemva B and Schrobilgen G A, Coordination of KrF_2 to the naked metal cation Mg^{2+}, *Angew. Chem. Int. Ed.*, 2017, Vol. 56, 6251.
40. Basting D, Djeu N and Jain K, Historical review of excimer laser development, *Excimer Laser Technology*, Basting D and Marowsky G, Eds., Springer, 2005, 8.
41. Srinivasan R and Bodil B, Ultraviolet laser ablation of organic polymers, *Chem. Rev.*, 1989, Vol. 89, 1303.
42. Jain K, Wilson C G and Lin B J, Ultrafast deep UV Lithography with excimer lasers, *IEEE Electron Device Letters*, 1982, Vol. 3, 53.

APPENDIX

SOLUTIONS TO PROBLEMS AND EXERCISES

CHAPTER 2

2.12. $PhIF_2 < BrF_3 < ClF_3$

2.13. $SOF_2 < SOCl_2 < SOBr_2$

2.14. NSF_3 = Tetrahedral; IOF_4^- = Square pyramidal; $IO_2F_5^{2-}$ = Pentagonal bypyramid; $Kr_2F_3^+$ = Bent

2.15.

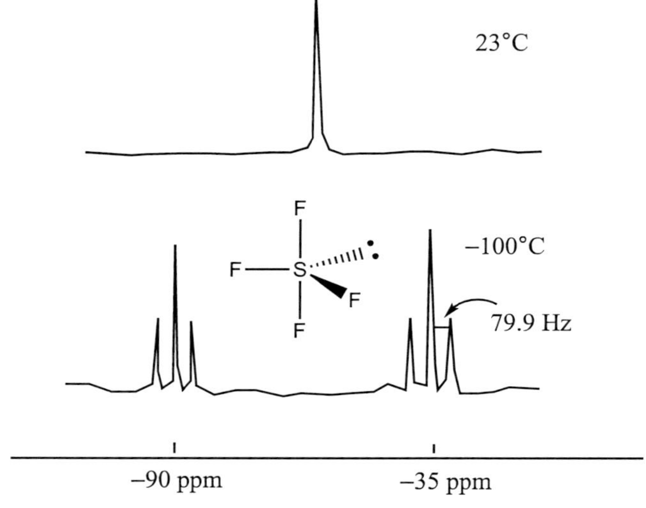

2.16. A lone pair is a pair of localised electrons that do not take part in bonding. An electron pair domain is a region in space where there is an enhanced probability of finding an electron pair. The domain extends around the most probable position of finding an electron pair as determined by the Pauli principle. A lone pair domain is more spread out and larger than a bonding pair domain as the electrons of a lone pair are subject to the attraction of only one atomic nucleus while those of a bonding pair are subject to the attraction of two nuclei.

2.17. NH_4^+ Not hypervalent

ICl Not hypervalent

IF_7 Hypervalent (14-I-7)

TeF_6 Hypervalent (12-Te-6)

2.18. Charge shift bonding is the bonding situation wherein the large covalent–ionic interaction energy is also a major stabilising factor for the bond. XeF_2 is a molecule having relatively high stability which is indicated by its large atomisation energy and melting point which is explained from an exceptionally large resonance energy arising from the mixing of several VB structures which are covalent and ionic.

2.19. Pauling explained PF_5 by using resonance structures drawn with no more than four covalent electron pair bonds and completed with ionic bonds to obey the octet rule. By MO theory one can use the 3c–4e bonds to account for the additional electrons which are accommodated in non-bonding orbitals.

2.20. In silyl ethers, the electronegativity of the central atom O is higher than that of the substituent Si atoms. Localisation of the lone pairs, and their ability to repel other electron pairs, is greatest when the substituents have an electronegativity similar to or greater than, that of the central atom. Large bond angles are therefore due to weak pairing of electrons on oxygen as a result of weakly electronegative SiH_3 group as well as electrostatic repulsions between bulky SiH_3 groups.

CHAPTER 3

3.19. Increasing order of Lewis acidity = $BF_3 < BCl_3 < BBr_3$

Resistance to hydrolysis of the B–X bond = $BF_3 > BCl_3 > BBr_3$

3.20. (a) Reaction with Br_2 (b) Heating of borazine gives polyborazinylene

3.21. For each of the following applications/ chemical reactions, write/draw against each the most appropriate boron containing compound as reagent/catalyst from the list provided.

(a) Preparing $C_6H_7^+$	$H(CHB_{11}Cl_{11})$
(b) Olefin polymerisation (using metallocenes)	$B(C_6F_5)_3$
(c) Alkyl ether cleavage	BBr_3
(d) Control rods in nuclear reactors	B_4C

3.22. (a) 13 pairs – hypho (b) 8 pairs – closo (c) 11 pairs – arachno

(d) 13 pairs – nido (e) 12 pairs – arachno (f) 13 pairs – closo

(g) 13 pairs – arachno

3.23.

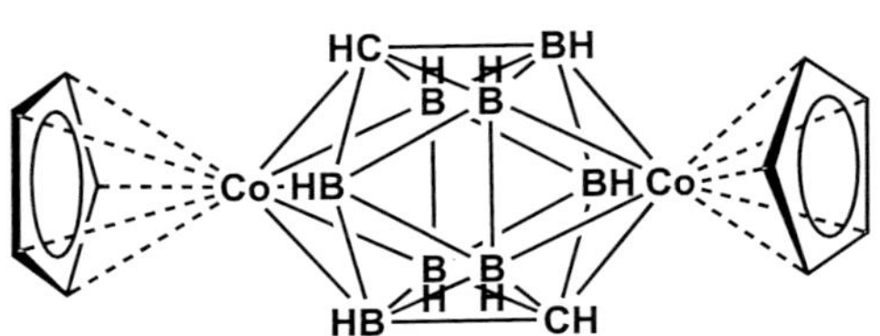

As per Jemmis rule: $m + n + o + p = 3 + 22 + 2 + 2 = 29$

12CH =36 *e*, 8 BH = 16 *e*: total 52 *e* = 26 pairs; 6 electrons required to get 29: Therefore each cobalt is in the +3 oxidation state.

3.24. Orthoboric acid is best described as a Lewis acid as it does not dissociate releasing H^+, but donates a proton to another molecule of water while itself acts as a Lewis acid.

3.25. (a) A (b) B (c) B (d) B

Reason: 2-pyridyl boronic acid is unstable—it undergoes protodeborylation. Also alkenyl and alkynyl boronic acids are unstable.

3.26.

$$BF_3 + LiAlH_4 \longrightarrow B_2H_6 \xrightarrow{\text{pyrolysis}} B_{10}H_{14}$$

$$B_{10}H_{14} + 2Et_2S \xrightarrow[-H_2]{n\text{-}Pr_2O} B_{10}H_{12}(Et_2S)_2 \xrightarrow[-H_2,\ -2Et_2S]{HC\equiv CH} C_2B_{10}H_{12}$$

Carboranes are stable to air and moisture, unlike boranes. Also the C–H bonds of carboranes provide scope for derivatising carboranes with organic substituents and functional groups by conventional reaction methods.

3.27. $H(CHB_{11}F_{11})$. This superacid was found to protonate CO_2.

$$CO_2 + H(CHB_{11}F_{11}) \longrightarrow [H(CO_2)_2]^+ (CHB_{11}F_{11})^-$$

3.28.

3.29. (a) Boron-10 is a naturally occurring isotope of boron.

(b) The particles emitted after it is hit by thermal neutrons are harmless and non-radioactive (He and ^7Li).

(c) The combined path-length of the daughter nuclei is only 12 microns which is the diameter of a single living animal cell.

(d) Boron has no known toxicity and its water soluble derivatives with high boron content can be easily prepared.

3.30. Hexagonal boron nitride is known as white graphite as it is white (unlike graphite), with most of the other properties being comparable. Unlike graphite which conducts electricity, hexagonal boron nitride does not conduct electricity.

CHAPTER 4

4.17. Al_2O_3 is used as the catalyst for the Claus process which converts H_2S to elemental sulfur.

$$2H_2S + SO_2 \xrightarrow[200\text{–}300^\circ C]{Al_2O_3} 3\,S + 2H_2O$$

4.18.

$$C_6H_6 \xrightarrow[AlCl_3]{\text{ethylene oxide}} C_6H_5CH_2CH_2OH$$

4.19.

$$C_6H_6 \xrightarrow[AlCl_3]{H_2C{=}CH_2} C_6H_5CH_2CH_3$$

Ethyl benzene is used for making styrene and thereby polystyrene.

4.20.

$$NaAlH_4 + 2\,CH_3OCH_2CH_2OH \xrightarrow[-\,H_2]{} \text{Red Al}$$

It is not pyrophoric, has better solubility in organic solvents than LAH and has a better shelf life. Red-Al can be used to reduce aliphatic *p*-toluenesulfonamides ($TsNR_2$) to the corresponding free amines, which is not done by LAH.

4.21. 7068 aluminium alloy having main additional elements Zn, Mg, Cu and Zr is the strongest commercially available aluminium alloy, which is comparable to the strength of some steels.

4.22.

$$[(i\text{-}Bu)_3Al]_2 \xrightarrow[-3\,H_2C{=}CMe_2]{3\,H_2C{=}CHMe} [(n\text{-}Pr)_3Al]_2 \xrightarrow[-3H_2C{=}CHMe]{3\,H_2C{=}CH_2} [Et_3Al]_2$$

$$Al_2(CH_2CH_3)_6 \xrightarrow[110^\circ C,\ 100\ bar]{6n\ H_2C{=}CH_2} Al_2[(CH_2CH_2)n(CH_2CH_3)]_6$$

$$Al_2[(CH_2CH_2)n(CH_2CH_3)]_6 \xrightarrow{\text{H elimination}} 6\ CH_3CH_2(CH_2CH_2)_{n-1}CH{=}CH_2$$

$$Al_2[(CH_2CH_2)n(CH_2CH_3)]_6 \xrightarrow{O_2} Al_2[(OCH_2CH_2)n(CH_2CH_3)]_6$$

$$Al_2[(OCH_2CH_2)n(CH_2CH_3)]_6 \xrightarrow{H_2O} 6\ CH_3CH_2(CH_2CH_2)_nOH + 2\ Al(OH)_3$$

4.23. $In_xGa_{1-x}N$ has a bandgap span from the infrared (0.70 eV) for InN to the ultraviolet (3.4 eV) of GaN. Therefore by varying the amount of indium in GaN one can vary the band gap and thereby produce all the colours of the rainbow.

4.24. Et_3Ga has an additional decomposition pathway to elemental gallium by β-hydride elimination which is not possible for Me_3Ga.

4.25. (i) GaAs has a direct bandgap which helps to absorb solar radiation efficiently.

(ii) It has no performance loss, unlike Si, on increase of day temperature.

(iii) The band gap (1.4 eV) which is closer to the ideal value helps to have a much better conversion efficiency for GaAs than silicon.

4.26.

	Si	GaAs
Bandgap (eV)	1.1	1.4
Temp. coeff.	–0.41%/°C	0.0
Electron mobility	1400 $cm^2/V.s$	8500 $cm^2/V.s$

4.27. A = Al, B = $AlCl_3$; C = $LiAlH_4$; D = $CH_3CH_2NH_2$

4.28. The best compounds for making GaN are Me_3Ga or Et_3Ga and NH_3. Ammonia is made from N_2 by Haber–Bosch Process. Gallium is first converted to $GaCl_3$ by reaction with Cl_2 or $SOCl_2$. Reaction of $GaCl_3$ with Et_3Al gives Et_3Ga. Reaction of $GaCl_3$ with Me_2Zn gives $GaMe_3$ along with many other methods (see chapter).

4.29. (i) High electrical conductivity and high optical transparency among similar TCOs.

(ii) Resistant to moisture penetration.

(iii) Better chemical stability among similar TCOs.

(iv) Ease of deposition as a thin film.

4.30. A band gap between 1.0 and 1.7 eV is required for a material to work as a solar cell. GaN has a band gap of 3.4. By doping with indium, its band gap can be brought down to the desired range.

4.31. $3\ Fe_3O_4 + 8\ Al \longrightarrow 4\ Al_2O_3 + 9Fe + 3341\ kJ$

(i) Aluminium is less expensive compared to other reactive metals.

(ii) The passivation layer which forms on aluminium makes it easy to store and handle.

(iii) Low melting point (660°C) of aluminium makes it easy to melt.

(iv) The high boiling point (2519°C) of aluminium enables the reaction to reach temperatures above 2000°C.

(v) The low density of Al_2O_3 makes it float above the molten iron preventing oxidation of Fe.

4.32. Duralumin is basically an Al–Cu alloy. Magnesium is added to aluminium to make it sea water-corrosion resistant.

4.33.

M + Na + H_2	⟶	A = $NaAH_4$
A + LiCl	⟶	B = $LiAlH_4$ + NaCl
B + CH_2=CH-CH_3	⟶	C = LiAl(n-Pr)$_4$
3 C + $AlCl_3$	⟶	D = 4 Al(n-Pr)$_3$ + 3LiCl
D + 3 CH_2=CH_2	⟶	E = Al(Et)$_3$
E+ O_2 + H_2O	⟶	F = Al(OH)$_3$ + EtOH

4.34. Mercury will not react with water and will remain liquid in ice cold water (m.p. −38.8°C). Cesium will react vigorously with water converting to hydroxide and liberating hydrogen gas. Gallium will simply solidify in ice cold water (m.p. 29.8°C).

Chapter 5

5.16. By using ^{13}C-NMR spectroscopy, one can distinguish between C_{60} and C_{70}. C_{60} gives one signal at 142.8 ppm indicating that all the 60 carbon atoms are spectroscopically identical. However C_{70} gives five lines in the range 150.7 to 130.3 ppm as expected for the D_5h molecule.

5.17. OsO_4 is a powerful as well as a selective oxidant and the osmylation of polycyclic aromatic hydrocarbons have been known for many years. For example, anthracene adds 2 equivalents of OsO_4 in the presence of pyridine.

5.18. Bingel reaction is a cyclopropanation reaction of the double bond of a fullerene carried out with bromodiethyl malonate. This general reaction allows the introduction of many useful extensions on the fullerene.

5.19. Pentamethyl derived C_{60} was prepared using the methyl Grignard reagent along with CuBr and SMe_2. This method of pentaalkylation increased the electron cloud present in the five membered ring considerably. Thus, they are at present converting it almost to a cyclopentadienyl system. The first ferrocene–fullerene hybrid molecule (called buckyferrocene) was prepared by the reaction of iron cyclopentadienyl dicarbonyl dimer and Me_6HC_{60}.

5.20. The high electrical conductivity of CNT is utilised in CFRP polymer composite as an additive to make it conducting. For the aeronautical industry, where CFRP is increasingly being used, this helps to replace aluminium. The problem of non-conductivity of CFRP is thus taken care of.

5.21. Hopcalite is a mixture of the oxides of manganese, copper, cobalt and silver [MnO_2 (50%) CuO (30%) Co_2O_3 15%) and Ag_2O (5%)]. This mixture has been used in gas masks to oxidise CO to CO_2 present in air and this occurs at room temperature.

5.22. Cu(I) ammonia solutions are able to absorb CO from air. The copper–ammonium salt wash developed by BASF uses aqueous ammonia solutions of copper carbonate, formate, acetate, lactate, or a mixture of some of these. The absorption–desorption reaction is highly dependent on the temperature, pressure and acidity of the medium.

$$[Cu(NH_3)_2]^+ + CO + NH_3 \underset{350\ K}{\overset{283\text{–}303\ K,\ 20\ MPa}{\rightleftharpoons}} [Cu(NH_3)_3CO]^+$$

5.23. Carbon monoxide reacts with I_2O_5 getting oxidised to CO_2 and the reduction gives I_2, the formation of which can be monitored by reacting with starch.

5.24. K_2CO_3 and monoethanolamine, $HOCH_2CH_2NH_2$

5.25. Grubbs 1st generation catalyst has two phosphines *trans* to each other, while in Grubbs 2nd generation catalyst an NHC and phosphine are *trans* to each other. Being a better σ donor, the NHC is able to labilise the phosphine *trans* to it by the *trans effect*, which does not happen for Grubbs 1st generation catalyst. Also the steric bulkiness of NHC provides extra air- and moisture-stability to the catalyst.

5.26. cAACs react with H_2, NH_3 and CO which is not shown by NHCs.

Chapter 6

6.18. Since there is only one product and there are no possible isomers, the Badger–Mobil process for ethylbenzene is said to be a simple acid catalysis where there is no shape selectivity. The small amount of diethyl benzene formed is most likely to be 1,4-diethylbenzene due to the shape selectivity of ZSM-5.

6.19. Kaolinite has a two-layer structure with the formula $Al_2(OH)_4Si_2O_5$ where the aluminum ions are in the octahedral sites and the OH groups are on the periphery. The peripheral OH groups make kaolinite sticky as they can get involved in hydrogen bonding. In phyllosilicate mica, $KAl_2(OH)_2[Si_3AlO_{10}]$, one of the tetrahedral silicon atoms is replaced by a tetrahedral aluminium and among counterions, we have octahedral aluminium as well. Molecular sieves 4A has alternating tetrahedral silicon and aluminium in its open framework structure with counterions for each aluminium present.

6.20. Me_2SiCl_2 along with small amounts of Me_3SiCl is prepared by the Rochow–Mueller process involving Si/Cu alloy and MeCl. Hydrolysis of Me_2SiCl_2 gives D_4 $[Me_2SiO]_4$ and hydrolysis of Me_3SiCl gives $Me_3SiOSiMe_3$. Acid or base catalysed polymerisation of a mixture of $[Me_2SiO]_4$ and $Me_3SiOSiMe_3$ in the ratio 50:1 will eventually give the linear polymer $Me_3SiO(SiMe_2O)_{200}SiMe_3$.

6.21.

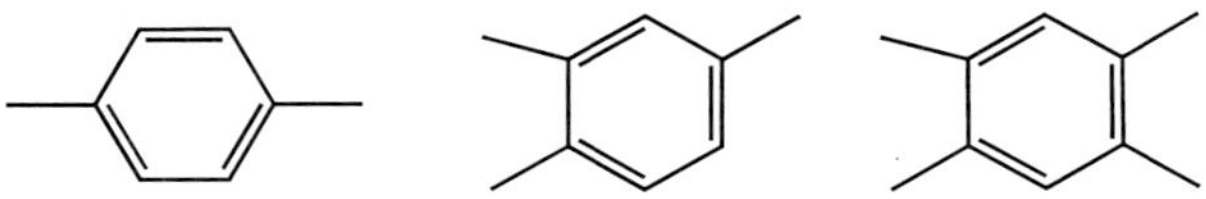

6.22.

$$SiO_2 + 4C \longrightarrow Si(Fe) \xrightarrow{Cl_2} SiCl_4 \xrightarrow{redn} \text{Pure silicon}$$

$$\text{Pure silicon} \xrightarrow[Cu]{MeCl} Me_2SiCl_2 + Me_3SiCl$$

$$Me_3SiCl + NH_3 \longrightarrow (Me_3Si)_2NH$$

6.23. A = SiH_4 B = Cl_3SiH C = $SiCl_4$

6.24. By reacting with coke (carbon) at around 2000°C, ferrosilicon is prepared. It is used in the steel industry to remove dissolved oxygen in steel, as oxygen will react with carbon, forming carbon monoxide and dioxide.

6.25. (i) c (ii) d (iii) b (iv) e (v) a

6.26. $Si + Cl_2 \longrightarrow SiCl_4 \xrightarrow{Si} Si_3Cl_8 \xrightarrow{SiCl_4} 2\ Si_2Cl_6$

The most important use of Si_2Cl_6 is for the reduction of phosphine oxides $R_3P{=}O$ to R_3P due to its high oxophilicity.

6.27.

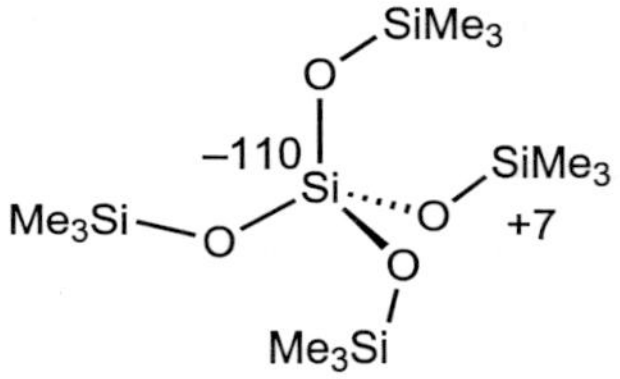

SiMe3
O
–110
Si
SiMe3
+7
Me3Si—O
O
O
Me3Si

6.28. A = Me_2SiCl_2 B = Me_3SiCl C = $[Me_2SiO]_4$ (cyclic) D = $Me_3SiOSiMe_3$
E = $Me_3SiO(SiMe_2O)_4SiMe_3$ (linear)

6.29.

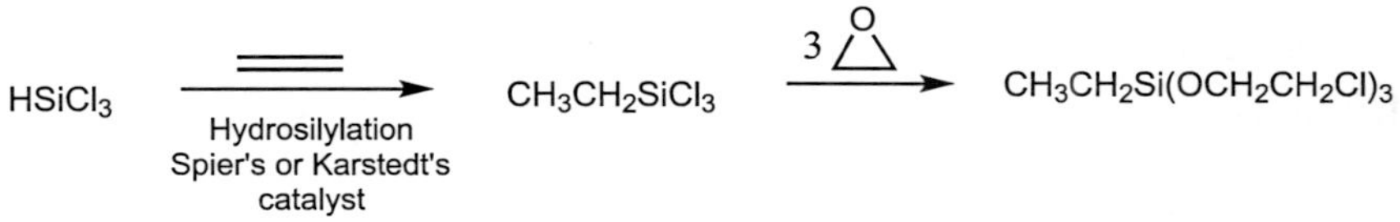

$HSiCl_3 \xrightarrow[\text{Hydrosilylation Spier's or Karstedt's catalyst}]{=} CH_3CH_2SiCl_3 \xrightarrow{3\ \text{(ethylene oxide)}} CH_3CH_2Si(OCH_2CH_2Cl)_3$

CHAPTER 7

7.10. A = PbS B = PbO C = Pb_3O_4 D = PbO_2

7.11. (a) Intrinsic semi-conductivity (b) Infra red transparency
(c) Glass forming ability (d) Highest refractive index (4.0026)
(e) Low chromatic dispersion (f) Ability of GeO_2 to act as catalyst for PET manufacture.

7.12.

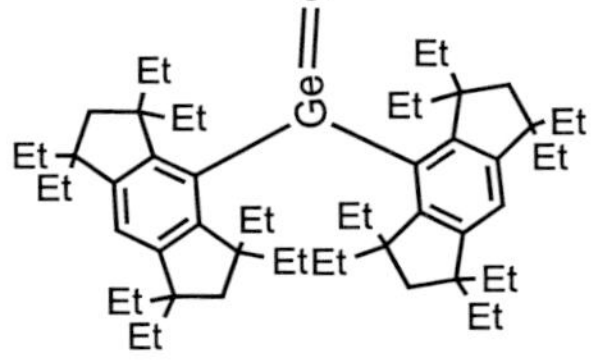

O
Ge
Et Et Et Et Et Et Et Et Et Et Et Et Et Et Et Et

Ge=O Bond order = 1.252

7.13. $(n\text{-}Bu)_4Sn + SnCl_4 \xrightarrow{210°C} n\text{-}Bu_3SnCl \xrightarrow{aq.\ NaOH} (n\text{-}Bu_3)SnOSn(n\text{-}Bu)_3 + NaCl$

$$x\,(n\text{-}Bu_3Sn)_2O + 2\,(MeSiHO)_x \longrightarrow 2\,(n\text{-}Bu)_3SnH + 2\,(MeSiO_{1.5})_x$$

The Sn–H bond of Bu_3SnH is relatively weak and nonionic and can cleave homolytically.

7.14. Me_2GeCl_2 can be prepared by a method similar to the Rochow–Mueller process for Me_2SiCl_2.

$$\underset{10:4\ \text{wt ratio}}{Ge + Cu} \xrightarrow[400°C]{CH_3Cl} Me_3GeCl\ (20\%) + Me_2GeCl_2\ (61\%)$$

Hydrolysis of Me_2GeCl_2 in aqueous alkaline medium gives $[Me_2GeO]_4$. However, unlike the corresponding silicon compound, $[Me_2GeO]_4$ is highly soluble in water. A polymer formed from the same was also found to be water soluble.

7.15. High density and low cost : Nuclear shield; Low melting point : Solders

7.16. PET synthesis using antimony-based catalysts results in yellow colouring of PET bottles which does not happen when GeO_2 is used as the catalyst.

7.17. Germanium has a low band gap (0.7 eV) which restricts its use as a semiconductor in high temperature applications. However germanium grown on silicon (silicon–germanium) transistors are faster than silicon transistors, they use much less power than standard transistors which holds promise in hand-held devices which needs charged power.

7.18. It is easily regenerated by halogen exchange reaction of $(CH_3)_3SnCl$ with KF or NaF in a water ethanol mixture.

7.19.

H—E≡E—H   Alkyne type
H—E≡E—H (trans)   Trans bent
E=EH₂   Vinylidene
E(μ-H)E—H   Mono-hydride bridged
E(μ-H)₂E   Bis-hydride bridged

Out of these *trans* bent and vinylidene type of isomers have been made so far with germanium.

7.20.

Germylene
Ge (II)

Germylone
Ge (0)

Eind₂Ge=O
Germanone
Ge (IV)

CHAPTER 8

8.24. Activated carbon has micro and transitional pores of 10 to 500 Å pore diameter while molecular sieving carbon has only uniform super-micro pores of less than 10 Å (1 nm) pore diameter. Carbon molecular sieves bring about molecular separations based on rate of adsorption rather than the differences in adsorption capacity.

8.25. Although C≡O bond enthalpy is higher than N≡N bond enthalpy, the formation of C≡O bond in a reaction are not explosive and does not involve the release of large quantities of energy when compared to the formation of N_2 (3 × C–O single bond enthalpy is slightly less than that of C≡O). But a comparison to the corresponding enthalpy of 3 × N–N single bond indicates that the latter, 477 kJ/mol, is almost ½ of the N≡N bond enthalpy of 945 kJ/mol).

8.26. Ammonia is first converted to urea by reaction with CO_2. Urea on heating forms cyanic acid (HCNO) and ammonia. HCNO trimerises to give cyanuric acid $C_3N_3(OH)_3$ which condenses with ammonia forming melamine, $C_3N_3(NH_2)_3$.

Ammonia is oxidised to NO which reacts further with O_2 forming NO_2. NO_2 dissolves in water giving HNO_3. Nitration of glycerine using HNO_3/H_2SO_4 mixture gives nitroglycerine.

8.27. $(NI_3.NH_3)_n$ is called touch powder as it explodes with violet fumes of I_2 when the dry compound is touched with a stick or disturbed. It is prepared by the reaction of aq. NH_3 and iodine in water.

8.28. It can be concentrated by dehydration with conc. H_2SO_4 or better by treatment with anhydrous $Mg(NO_3)_2$. Hydrated magnesium nitrate on heating gives MgO which gets reconverted to $Mg(NO_3)_2$ on treatment with HNO_3.

8.29. The iron group of metals, and among them ruthenium, shows the best catalytic activity for the Haber–Bosch process.

8.30. N_2O reacts with $NaNH_2$ giving sodium azide. It is a well known inhalation anesthetic. It is used as a source of atomic oxygen for selected reactions

8.31. Since it is a gas it has to be NO, N_2O, NO_2 or N_2O_4. If it is a gas which is brown at high temperature and colourless at low temperature, it is NO_2 or N_2O_4. If it is a colourless gas which on exposure to oxygen forms brown fumes, it is NO. If the gas reacts with $NaNH_2$ forming a white solid which explodes when grounded or subjected to shock, the gas is N_2O.

8.32. KAAP is Kellogg advanced ammonia process which makes use of $Ru_3(CO)_{12}$ along with rubidium nitrate as catalyst and promoter respectively. This catalyst has a high surface area and has an activity 10–20 times more than Fe_3O_4. The KAAP process operates at a lower pressure (90 bar pressure versus 140–170 bar pressure) thus reducing running expenses.

8.33. NaCN will give Prussian blue colour/precipitate with a mixture of Fe^{2+} and Fe^{3+} solutions. NaNCS will give intense blood red colour with ferric ion solution. NaN_3 on reaction with brown I_3^- will give brisk effervescence of N_2 gas and decolourisation of I_3^- to colourless.

8.34. In principle a reaction of azide N_3^- with pentazenium N_5^+ should result in N_8. Considering the high instability associated with N_5^+, the compound if formed will be highly unstable decomposing to 4 molecules of N_2.

8.35. The copper(I)-catalysed azide-alkyne cycloaddition (CuAAC) preferentially gives the 1,4-regio isomer of 1,2,3-triazole while a ruthenium catalysed reaction (RuAAC) preferentially gives the 1,5-regio isomer of the triazole as the product.

8.36.

$$ClCH_2CH_2Cl \xrightarrow{NH_3} H_2NCH_2CH_2NH_2$$

$$CH_4 + NH_3 \xrightarrow[1400°C]{Pt} HCN + H_2 \quad \text{(Degussa process)}$$

$$H_2NCH_2CH_2NH_2 + HCN + HCHO \longrightarrow (NCCH_2)_2NCH_2CH_2N(CH_2CN)_2$$

$$(NCCH_2)_2NCH_2CH_2N(CH_2CN)_2 \xrightarrow{\text{Aq. NaOH}} (NaOOCCH_2)_2NCH_2CH_2N(CH_2COONa)_2$$

8.37. A = Cl_2 B = $ClCH_2CH_2Cl$ C = $H_2NCH_2CH_2NH_2$ D = $[Ni(en)_3]^{2+}$

8.38.

A B C

(A: hexamethylenetetramine cage; B: 1,3,5-trinitro-1,3,5-triazinane with three NO_2 groups; C: 1,3,5,7-tetranitro-1,3,5,7-tetrazocane with four NO_2 groups)

8.39.

Picric Acid	EGDN	FOX–7
C6H3N3O7	C2H4N2O6	C2H4N4O4
$\frac{-1600[12+1.5-7]}{229.1} = -45.39$	$\frac{-1600[4+2-6]}{152.06} = 0$	$\frac{-1600[4+2-4]}{148.08} = -21.6$

Increasing order of strength based on %OB = Picric acid < Fox-7 < EGDN

Chapter 9

9.20. In all the four allotropes, the phosphorus is tricoordinate in nature. Also going from white to red to violet to black, the strain on the individual phosphorus atoms decreases

9.21. A = PCl_3 B = PPh_3 C = $PPh_3Me^+I^-$ D = $Ph_3P{=}CH_2$

9.22. SO_2Cl_2. As the side product formed after oxidative chlorianation is SO_2 which is easily expelled from the reaction medium.

9.23. P_4 reacts with S_8 giving P_4S_{10}. P_4S_{10} on reaction with ethanol forms dithiophosphoric acid.

9.24. (a) H_3PO_4 and NH_4OH (b) chlorophosphazene polymer $[Cl_2PN]_n$

9.25. Pesticides have P=S units and these should get oxidised to P=O units before the compounds can affect the active site of acetyl cholinesterase enzyme. These oxidations occur in the system by oxidising agents such as cytochrome P-450. Therefore the level of toxicity is less for pesticides than nerve gases where the P=O bond is already in place.

9.26. The halocyclophosphazene ring skeleton has substituents only at alternating atoms and these substituents are also oriented away from the mean plane of the ring. Hence the growth of the dendrimer chain will have less steric hindrance when compared to analogous cyclic carbon based rings.

9.27. The reactive P–Cl bonds of chlorinated cyclophosphazenes help in substituting the ring with different types of substituents, some hydrophilic and some hydrophobic. The substituents on cyclosiloxanes are mostly alkyl and aryl units which are prefixed. Also degradation the phosphazene core leads only to harmless products.

9.28.

P_4S_{10} + EtOH → (HS)P(=S)(OEt)$_2$

Chlorpyriphos

9.29. $\%V_{bur}$ is defined as the per cent of the total volume of a sphere occupied by a ligand when bound to a metal. The volume of this sphere represents the potential coordination sphere around the metal occupied by a ligand under consideration. The sphere has a defined radius (fixed at 3.50 Å) and the metal to phosphine distance is fixed at 2.00 Å and 2.28 Å. Since cone angles are defined with the M–P distance of 2.28 Å it is the one best suited for comparison with cone angles.

9.30.

H_3PO_2	Electroless nickel plating
H_3PO_3	+3 formal oxidation state of P
H_3PO_4	Sodium tripolyphosphate synthesis

$H_4P_2O_6$	Presence of P–P bond but not reducing
$H_4P_2O_7$	Pyrophosphoric acid
$H_4P_2O_4$	+2 formal oxidation state of P

9.31.

A: $Mes_2P^{\oplus}$ … $B(C_6F_5)_2$ (with O, O$^{\ominus}$)

B: $Mes_2P^{\oplus}$ … $B(C_6F_5)_2^{\ominus}$

C: $Mes_2P^{\oplus}$ … $B(C_6F_5)_2^{\ominus}$

9.32. $CH_2(COOH)_2 + P_4O_{10} \longrightarrow O{=}C{=}C{=}C{=}O$

Carbon suboxide

9.33. $NH_3 + ClCH_2C(O)OH \longrightarrow H_2N\text{-}CH_2\text{-}COOH$

$PCl_3 + HCHO \longrightarrow Cl_2P(O)CH_2Cl \xrightarrow[-HCl]{H_2O} (HO)_2P(O)CH_2Cl$

$(HO)_2P(O)CH_2Cl + H_2N\text{-}CH_2\text{-}COOH \longrightarrow (HO)_2P(O)\text{-}CH_2\text{-}NH\text{-}CH_2\text{-}C(O)OH$

9.34.

A: P_4S_{10}

B: $S{=}P(OEt)_2(SH)$

C: $S{=}P(OEt)_2Cl$

D: $O_2N\text{-}C_6H_4\text{-}O\text{-}P(S)(OR)_2$

9.35. A=PCl_5; B=$N_3P_3Cl_6$; C =$N_4P_4Cl_8$; D= $[N{=}PCl_2]_n$

9.36.

A B C D E

E D C B A

7-8 24.3 26.9 28.4 33.4

9.37.

(a) $HClO_4 \xrightarrow{P_2O_5} Cl_2O_7$

(b) $Sn \xrightarrow{PCl_5} SnCl_4$

(c) $PhCH_2N_3 \xrightarrow{PPh_3,\ H_2O} PhCH_2NH_2$

(d) $PhC(=O)OEt \xrightarrow{P_4S_{10}} PhC(=S)OEt$

(e) $HCHO \xrightarrow{PCl_3} Cl_2P(=O)CH_2Cl$

9.38.

$C_6H_6 \xrightarrow[\text{cold, dark}]{Cl_2/AlCl_3} C_6Cl_6 \xrightarrow{KF/HF} C_6F_6 \xrightarrow[\text{Mg, PMe}_3]{Cp_2ZrCl_2} C_6F_5H \xrightarrow{Br_2/AlCl_3} C_6F_5Br \xrightarrow{Mg} C_6F_5MgBr \xrightarrow{AlCl_3} (C_6F_5)_3Al$

9.39. A = PCl_3 B = $PhPCl_2$ C = $PhPCl_4$ D = $PhPF_4$

9.40. $Ca_5(PO_4)_3OH + 5\ H_2SO_4 + 10\ H_2O \longrightarrow 3\ H_3PO_4 + 5\ CaSO_4.2H_2O + H_2O$

$H_3PO_4 + Na_2CO_3 \longrightarrow Na_2HPO_4 + CO_2 + H_2O$

$Na_2HPO_4 + H_3PO_4 \longrightarrow 2\ NaH_2PO_4$

$2\ Na_2HPO_4 + NaH_2PO_4 \longrightarrow Na_5P_3O_{10} + 2\ H_2O$

Chapter 10

10.13. A = Bi B = BiF_3 C = BiF_5

10.14. A = Sb B = $SbCl_3$ C = $SbCl_5$ D = SbF_5

10.15. (a) BiH_3 is most unstable and decomposes below 0°C.

(b) Heat of formation is highest for BiH_3.

10.16. Antimony compounds are fire retardant additives, used in the making of electrodes for lead acid batteries and fluorinating agents (Swarts reagent). Bismuth and its compounds are used as non-toxic replacements for lead. Its subcitrate and subsalicylate are used in the treatment of gastrointenstinal diseases.

10.17. The dissociation constants of arsenic acid are comparable to that of phosphoric acid. Arsenate units can replace phosphate units in many biochemical reactions. But polyarsenates hydrolyse orders of magnitude faster than do polyphosphates which affects the synthesis and chemistry of ATP and ADP if the phosphate units get replaced by arsenate units.

10.18. P_4O_6 and As_4O_6 have comparable structures. However the structure of As_4O_{10} differs from that of P_4O_{10}. The structure of As_4O_{10} consists of tetrahedral AsO_4 and octahedral

AsO_6 units linked by sharing of corners and therefore, it is completely different from the structure of P_4O_{10}.

10.19. The treatment of arsenic solution with HCl and H_2S will result in a yellow precipitate of As_2S_3. The reaction of an arsenic compound with H_2SO_4 and Zn gives AsH_3, which can be burned and decomposed on a cold glass surface deposited as an arsenic mirror. This mirror is soluble in sodium hypochlorite solution unlike an antimony mirror.

10.20. Peltier effect is a heating or cooling effect, different from the ordinary resistance heating, and occurs whenever an electric current flows through a junction between two different metals. A substantial lowering of temperature due to the Peltier effect was observed in a thermocouple incorporating bismuth telluride, Bi_2Te_3.

Chapter 11

11.21. As an oxidising agent, it gets reduced to water while as a reducing agent it gets oxidised to O_2.

11.22.

$$2\,NH_3 + H_2S + 1/2\,S_8 \longrightarrow (NH_4)_2S_5$$

$$(NH_4)_2S_5 + Cp_2TiCl_2 \xrightarrow{CHCl_3} Cp_2TiS_5 + NH_4Cl$$

$$Cp_2TiS_5 + SCl_2 \longrightarrow Cp_2TiCl_2 + S_6$$

$$S_8 + O_2 \longrightarrow SO_2 \xrightarrow[V_2O_5]{O_2} SO_3 \xrightarrow{HCl} ClSO_3H \xrightarrow{C_6H_6} C_6H_5SO_3H$$

11.23. A = Cl_2 B = SCl_2 C = NaF D = SF_4

11.24. $(NH_4)_2S_5$ is converted to Cp_2TiS_5. The reaction of this with SCl_2 gives S_6 and reaction with SO_2Cl_2 gives S_{10}.

11.25. A = S_4N_4 B = $S_4N_4F_4$ C = $S_4N_4H_4$

11.26.

Cyclohexadiene ⟶ dichlorocyclohexene (Cl, Cl)

$$Ph_2PCl \longrightarrow Ph_2PCl_3$$

SO_2Cl_2 is the best reagent suitable for both the reactions as stoichiometry can be controlled preventing saturation of both the double bonds of cyclohexadiene and SO_2 gas (being the side product) is easily removed.

11.27. Because there are six fluorine atoms forming strong S–F covalent bonds surrounding the relatively small sulfur atom, it is well shielded from attacking molecules to a large extent. In contrast the size of tellurium is larger and so gets attacked by water molecules leading to hydrolysis.

11.28. A = $SOCl_2$ B = SO_2 C = PCl_3 D = H_3PO_3 E = PH_3

11.29. P = Cl_2 Q = SCl_2 R = S_4N_4 S = $Ph_3P{=}N{-}S_3N_3$

11.30. A = SF_4 or deoxofluor or DAST B = $SOCl_2$ C = Oxidation (for example, $KMnO_4$) followed by reaction with SF_4

CHAPTER 12

12.12. SO_2 and SeO_2 in the gaseous phase have comparable bent structures. The solid state structure of SeO_2 is polymeric as shown and TeO_2 structure has no double bonds. SO_2 is reducing in nature while SeO_2 and TeO_2 are oxidising in nature.

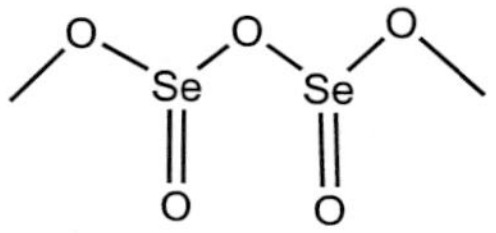

Solid phase

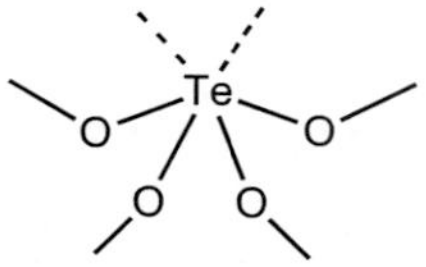

α TeO_2

12.13.

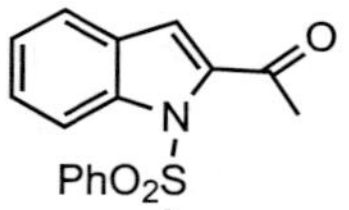

A

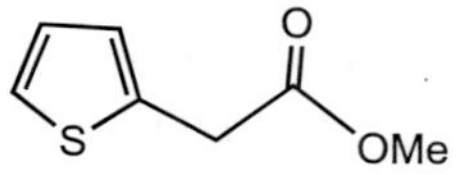

B

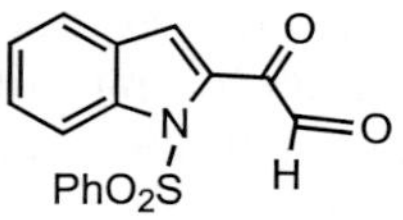

A

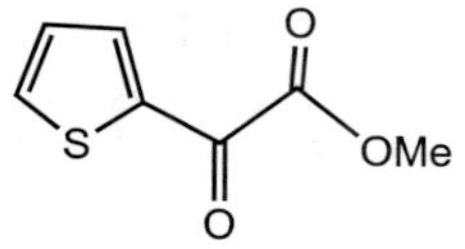

B

12.14. Stability decreases on going from thiols to selenols and hydrogen bonding also decreases on going from thiols to tellurols.

12.15.

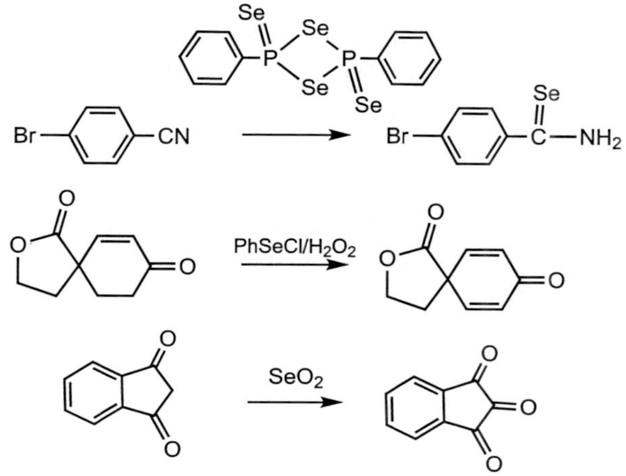

PhSeCl/H_2O_2

SeO_2

12.16. It satisfies the criteria met by the other 20 common amino acids. It is encoded by DNA, has its own unique codon (UGA), it has an unique *t*-RNA which brings the residue to the ribosome and it is cotranslationally inserted into the polypeptide chains of selenoproteins.

12.17. Glutathione peroxidase has been found to be quite effective in protecting cells from the oxidative breakdown brought about by low concentrations of ascorbic acid. It scavenges H_2O_2 and hydroperoxides preventing oxidative stress and damage. Selenium present in glutathione peroxidase is responsible for destroying lipid-damaging peroxides.

12.18.

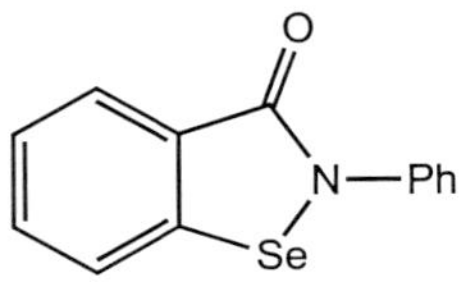

O
N—Ph
Se

It is a synthetic mimic of glutathione peroxidise thereby helping to remove peroxides. This compound also has additional properties such as anti-inflammatory, antioxidant and cytoprotective activity.

12.19. The unusual ability of selenium to rapidly become oxidised and then rapidly get reduced back to the original oxidation state is termed as selenium paradox. This chemistry is in contrast to that of sulfur and tellurium.

12.20. Chemical reactions with selenium are faster than with sulfur and enhanced chemical activity is observed for selenoenzymes with accelerated rates compared to sulfur-based enzymes. Selenium has the unusual ability to rapidly become oxidised and then rapidly get reduced back to the original oxidation state, unlike sulfur and tellurium. Unlike sulfur, selenium provides resistance to inactivation of its enzymes by oxidation.

12.21.

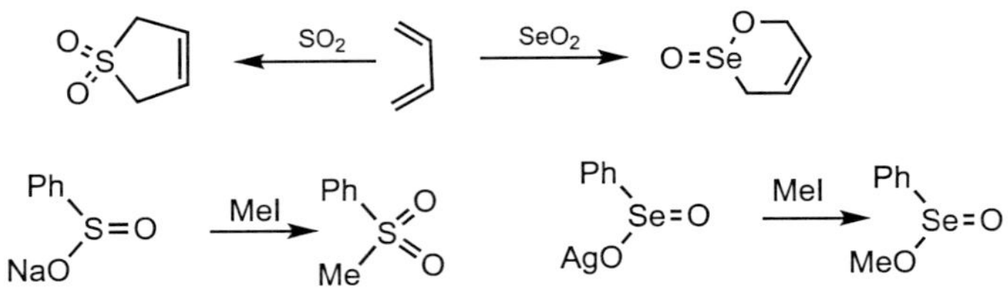

O, S, O ← SO_2 — (butadiene) — SeO_2 → O=Se, O

Ph, NaO, S=O —MeI→ Ph, Me, S, O, O

Ph, AgO, Se=O —MeI→ Ph, MeO, Se=O

12.22. Chlorination of S results in S_2Cl_2 which on further chlorination in the presence of a catalyst gives SCl_2. Attempts to make sulfur tetrachloride indicated possible formation of an ionic species $SCl_3^+Cl^-$. Chlorination of Te preferentially forms $TeCl_4$ as a tetrameric solid with a cubic unit. Te_2Cl_2 although predicted, is not characterised and $TeCl_2$ like $SeCl_2$ is known only in the gas phase from thermal decomposition of the tetrachloride. Selenium on the other hand, has been found to form stable Se_2Cl_2 and $SeCl_4$.

12.23. CdTe has a band gap of 1.5 eV which is ideal for solar cells. Also its thin films can be manufactured easily and inexpensively making it the best competitor for silicon based solar cells.

12.24. Silver–indium–antimony–tellurium (AgInSbTe) alloy is the material used in rewritable CDs. In the original state it has a polycrystalline structure and reflective properties. When writing a CD, it is heated to 500–700°C making the alloy change its polycrystalline structure and become amorphous. To erase the disc, the write beam heats the amorphous regions with low power of about 200°C and the alloy returns to the polycrystalline state becoming reflective again.

12.25. The products formed are both cyclic, the first being a CT complex and the tellurium compound having a 4-membered ring.

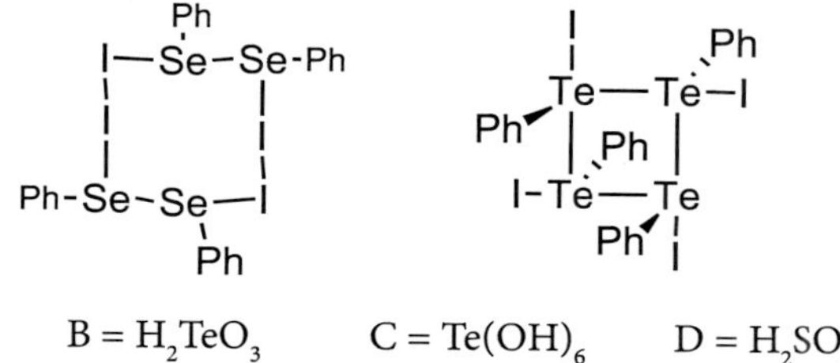

12.26. A = TeO_2 B = H_2TeO_3 C = $Te(OH)_6$ D = H_2SO_4

Chapter 13

13.21. (i) Exchanging of H by F exerts a large electronic effect on neighbouring carbon centres altering both dipole moment and pK_a.

(ii) The stability and reactivity of functional groups on the fluorinated molecule are affected. The presence of C–F bonds aids hydrophobic interactions.

(iii) Covalently bound fluorine also increases the lipophilicity, thus increasing its bioavailability.

(iv) C–F bonds have high oxidation and thermal stability. In the body it has been found to be not susceptible to oxidation by cytochrome P-450.

(v) The NMR activity of ^{19}F is convenient for characterisation and also to obtain valuable information about active site interactions of a drug.

13.22.

$$UO_2 \xrightarrow{\text{anhydrous HF}} UF_4 \xrightarrow{F_2 \text{ gas}} UF_6 \text{ (mixture of } ^{235}U \text{ and } ^{238}U)$$

Separation carried out by centrifugation or laser separation process to get $^{235}UF_6$

$$^{235}UF_6 + H_2 \longrightarrow {}^{235}UF_4 + HF$$

$$^{235}UF_4 \xrightarrow{Mg} {}^{235}U$$

13.23.

$$2\,KMnO_4 + 2\,KF + 10\,HF + 3\,H_2O_2 \longrightarrow 2\,K_2MnF_6 + 8\,H_2O + 3\,O_2$$

$$2\,K_2MnF_6 + 4\,SbF_5 \longrightarrow 2\,KSbF_6 + 2\,MnF_3 + F_2$$

A weaker Lewis acid such as MnF_4 can be displaced from its salts by stronger Lewis acids such as SbF_5. Once displaced, MnF_4 being thermodynamically unstable, decomposes to the stable lower fluoride MnF_3 with the elimination of fluorine gas.

13.24.

$$HC\equiv CH \xrightarrow{HF} CH_3CHF_2 \xrightarrow{Cl_2} CH_3CF_2Cl \xrightarrow[-HCl]{} F_2C{=}CH_2 \xrightarrow[\text{polymerisation initiator}]{} \left[\!-CF_2-CH_2-\right]_n$$

13.25. (a) PTFE (b) PVDF (c) FEP and PVDF (d) PVDF

13.26

A: $C_6H_5COOH \xrightarrow{SF_4} C_6H_5CF_3$

C: cyclohexanone $\xrightarrow{CF_3SiMe_3}$ 1-(trifluoromethyl)cyclohexanol (F_3C, OH)

B: $C_6H_5CCl_3 \xrightarrow{SbCl_5/SbF_5} C_6H_5CF_3$

D: $C_6H_5CH_3 \xrightarrow{\text{(a) } KMnO_4 \text{ (b) } SF_4} C_6H_5CF_3$

13.27.

B

Compound B having the aryl fluorine bond will be more potent. The reason for the same is that it reaches the blood circulation without getting oxidised by cytochrome P-450.

13.28.

$$CH_4 + Cl_2 \xrightarrow{400\text{–}500°C} CH_3Cl,\ CH_2Cl_2,\ CHCl_3,\ CCl_4$$

Chloroform separated by distillation

$$CHCl_3 + HF \longrightarrow HCF_2Cl \xrightarrow[\text{pyrolysis}]{} F_2C{=}CF_2$$

$$F_2C{=}CF_2 + CH_3F \xrightarrow[SbF_5]{} H_2C{=}CFCF_3$$

CHAPTER 14

14.21. Cl_2 reacts with water forming HCl and HClO while ClO_2 does not react with water.

Cl_2 reacts with ammonia and amines forming chloramine compounds while ClO_2 does not react.

Cl_2 chlorinates acetaldehyde to trichloroacetaldehyde while ClO_2 oxidises it to acetic acid.

14.22.

$$Cl_2 + NaOH \longrightarrow NaClO + NaCl + H_2O$$

$$3\,NaClO \xrightarrow[-NaCl]{\text{disproportionation}} NaClO_3 \xrightarrow[\text{strong acid soln}]{\text{reduction}} ClO_2 \xrightarrow[\text{reduction}]{NaOH} NaClO_2$$

Since ClO_2 is difficult to handle and is also an explosive, it is generated from sodium chlorite, $NaClO_2$. ClO_2 is prepared by the oxidation of $NaClO_2$.

14.23.

$$\left[-\underset{\substack{\|\\O}}{C}-NH-C_6H_4-NH-\underset{\substack{\|\\O}}{C}-OCH_2CH_2O-\right]_n$$

$$O_2N-C_6H_4-NO_2 \xrightarrow[\text{catalyst}]{H_2} H_2N-C_6H_4-NH_2 \qquad CO + Cl_2 \longrightarrow COCl_2$$

$$H_2N-C_6H_4-NH_2 \xrightarrow{COCl_2} OCN-C_6H_4-NCO$$

$$OCN-C_6H_4-NCO + HOCH_2CH_2OH \longrightarrow \left[-\underset{\substack{\|\\O}}{C}-NH-C_6H_4-NH-\underset{\substack{\|\\O}}{C}-OCH_2CH_2O-\right]_n$$

14.24. Impure silicon is reacted with chlorine to convert it to $SiCl_4$. Impure TiO_2 also reacts with Cl_2 forming $TiCl_4$. Both $SiCl_4$ and $TiCl_4$ are liquids which can be repeatedly distilled to purify them. $SiCl_4$ on reduction with H_2 gives pure silicon and $TiCl_4$ on hydrolysis gives pure TiO_2.

14.25. ClO_2 is a better disinfecting agent for water as it does not react with water or react with amines and ammonia converting them to chloramines which can be skin irritants. Also solubility of ClO_2 in water is much more than that of Cl_2.

14.26. CBMG is the most suitable chlorinating agent for aromatic heterocyclic compounds.

14.27. (a) CBMG (b) PBDE (c) 4 (d) β and γ rays

14.28. The sodium formed if not removed will react with water to form NaOH and H_2 and separation of H_2 and Cl_2 will not be possible which can also explosively react under UV radiation. NaOH will also react with Cl_2 forming NaClO at low temperatures and $NaClO_3$ at higher temperatures.

14.29. Cl_2 + propylene ⟶ $H_3C-CH(Cl)-CH_2Cl$ $\xrightarrow{\Delta}$ $H_2C=CH-CH_2Cl$ $\xrightarrow{HOCl}$ (epoxide)$-CH_2Cl$

(epoxide)$-CH_2Cl$ + HO–(bisphenol A)–OH ⟶ Bisphenol A diglycidyl ether

14.30. A = NaCl B = $NaClO_3$ C = $NaClO_4$ D = $HClO_4$ E = Cl_2O_7

14.31. A = CO B = $COCl_2$

C = O=C=N–C_6H_4–N=C=O D = MIC (Methyl isocyanate)

14.32. X = I_2 Y = Cl_2 A = HIO_3 B = I_2O_5

14.33. (i) $CHCl_3$ (C) PTFE (Teflon)

(ii) $HC\equiv CH$ (D) PVDF (Polyvinylidene fluoride)

(iii) $Cl_2C=CCl_2$ (B) CF_3COOH

(iv) $Cl_2C=CHCl$ (A) HFC 134a

Chapter 15

15.20.

$$Te + 3\,F_2 \longrightarrow TeF_6 \xrightarrow{H_2O} Te(OH)_6 + 6\,HF$$

$$Te(OH)_6 + 5\,HSO_3F \longrightarrow HOTeF_5 + 5\,H_2SO_4$$

$$3\,HOTeF_5 + BCl_3 \xrightarrow[-80°C]{} B(OTeF_5)_3 + 3\,HCl$$

$$KrF_2 + 2/3\,B(OTeF_5)_3 \xrightarrow[-90\ \text{to}\ -112°C]{} Kr(OTeF_5)_2 + BF_3$$

15.21. (a) KrF_2 (b) XeF_2 (c) $PF_2(O)OH$ (d) SbF_5

15.22. KrF_2 is the strongest oxidising agent. KrF_2 can oxidise xenon gas to XeF_6 and I_2 to IF_7. It has a standard redox potential of 3.5 V highest among all known oxidisers.

15.23. Reaction of the mixture with NaF makes XeF_6 to form the ionic compound $Na_2(XeF_8)$. XeF_2 can be removed by applying vacuum from this mixture.

15.24.

A $C_6H_5COCH_3$ $\xrightarrow[HF,\ RT]{XeF_2}$ $C_6H_5-O-CF_2CH_3$

B $H_3C-C_6H_4-CH_2OH$ $\xrightarrow[HF,\ RT]{XeF_2}$ $H_3C-C_6H_4-O-CH_2F$

15.25. Si–F > C–F > F–F > Xe–F > Kr–F

15.26.

$$Xe \xrightarrow[1:20]{F_2} XeF_6 \xrightarrow{F_3P=O} XeOF_4 \xrightarrow{N_2O_5} XeO_2F_2$$

15.27. (a) F_2: Products XeF_2, XeF_4, XeF_6

(b) PtF_6: Products $XeF^+Pt_2F_{11}^-$ or $XeF^+PtF_6^-$ (with SF_6 diluent)

(c) KrF_2: Product XeF_6

(d) AuF_3: Product $[AuXe_4]^{2+}$

(e) HgF_2

15.28. XeF_3^+ from XeF_4 and SbF_5

$Xe_2F_3^+$ from 2 XeF_2 and AsF_5

XeO_4 from XeF_6 and NaOH giving Na_4XeO_6 followed by reaction of Na_4XeO_6 with H_2SO_4

Index

A

B

C

D

I

J

K

L

M

N

O

P

Q

R

S